Prealgebra

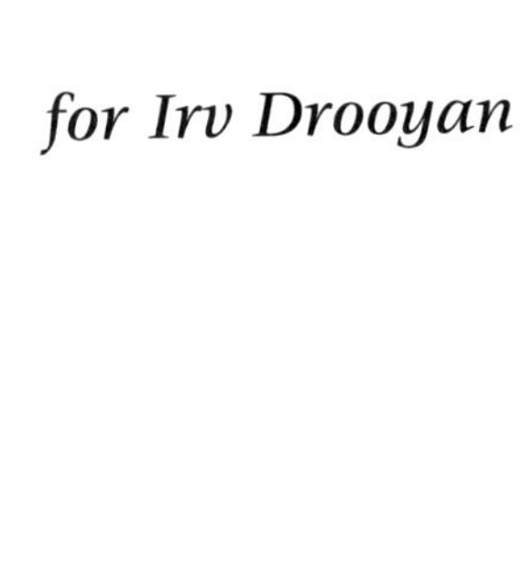

for Irv Drooyan

2ND EDITION

Prealgebra

Katherine Yoshiwara
Los Angeles Pierce College

Bruce Yoshiwara
Los Angeles Pierce College

Australia • Canada • Mexico • Singapore • Spain • United Kingdom • United States

Publisher: *Bob Pirtle*
Senior Editor: *Jennifer Huber*
Sr. Assistant Editor: *Rachael Sturgeon*
Editorial Assistant: *Carrie Dodson*
Technology Project Manager: *Star MacKenzie*
Sr. Marketing Manager: *Leah Thomson*
Marketing Assistant: *Jessica Perry*
Advertising Project Manager: *Bryan Vann*
Project Manager, Editorial Production: *Mary Vezilich*
Print/Media Buyer: *Barbara Britton*
Production Service: *Matrix Productions*
Text Designer: *Carolyn Deacy*
Illustrator: *Asterisk Inc.*
Cover Designer: *Roy Neuhaus*
Cover Photo: *Shigeru Tanaka*
Cover Printer: *Phoenix Color Corp.*
Compositor: *Better Graphics*
Printer: *Quebecor World–Dubuque*

Printed in the United States of America
4 5 6 7 06

For more information about our products, contact us at:
Thomson Learning Academic Resource Center
1-800-423-0563
For permission to use material from this text, contact us by:
Phone: 1-800-730-2214 **Fax:** 1-800-730-2215
Web: http://www.thomsonrights.com

Library of Congress Control Number: 2002106710

Student Edition: ISBN 0-534-36831-x
Instructor's Edition: ISBN 0-534-39670-4

Brooks/Cole–Thomson Learning
511 Forest Lodge Road
Pacific Grove, CA 93950
USA

Asia
Thomson Learning
5 Shenton Way #01-01
UIC Building
Singapore 068808

Australia/New Zealand
Thomson Learning
102 Dodds Street
Southbank, Victoria 3006
Australia

Canada
Nelson
1120 Birchmount Road
Toronto, Ontario M1K 5G4
Canada

Europe/Middle East/Africa
Thomson Learning
High Holborn House
50/51 Bedford Row
London WC1R 4LR
United Kingdom

Latin America
Thomson Learning
Seneca, 53
Colonia Polanco
11560 Mexico D.F.
Mexico

Spain/Portugal
Paraninfo
Calle/Magallanes, 25
28015 Madrid, Spain

Contents

CHAPTER 7

Fractions 463

CHAPTER 8

Proportion and Percent 529

CHAPTER 9

Applications to Problem Solving 609

APPENDIX

Review of Arithmetic Skills 675

Answers to Odd-Numbered Problems 711

Index 744

Preface

Many schools and colleges are struggling with low success rates in their elementary algebra classes, even when students have completed a preliminary course in arithmetic. Some students find the transition to the more abstract concepts of algebra too abrupt, or the accelerated pace of a one-semester algebra course too rapid, for them to come to grips with the many new ideas presented.

Prealgebra was written to help bridge the gap between arithmetic and algebra. A good grasp of arithmetic is necessary but, for some students, not sufficient for success in algebra. The next reasonable step is to identify the concepts and skills that give students the most trouble in elementary algebra. Each chapter of *Prealgebra* presents one of these fundamental ideas gradually, from many viewpoints and in as many realistic contexts as possible. This approach is in keeping with the NCTM and AMATYC standards, which call for an increased attention to conceptual understanding in place of rote application of algorithms, and recommend that, as a transition between arithmetic and algebra, "students explore algebraic concepts in an informal way to build a foundation for the subsequent formal study of algebra."

Content

Chapters 1 and 2 concentrate on the notion of variable and the meaning of equation. After using short phrases and then single letters to represent variable quantities, students are immediately engaged in writing and evaluating simple algebraic expressions. Choosing an appropriate equation to model a situation and then writing simple models follow naturally. Unless students can understand that algebra is a language and begin to use algebraic equations to describe relationships between variables, solving equations will remain for them a mysterious and meaningless task.

Chapters 3 and 4 tackle the order of operations and computations with signed numbers. These skills are considered in concrete situations, not merely as rules for simplifying textbook expressions. Students practice working with signed numbers by solving equations and word problems, thus reviewing and extending their work from earlier chapters. The order of operations is used to explain the solution of multi-step equations and leads to a greater variety of modeling problems.

Exponents and roots are introduced in Chapter 5 through formulas for area and volume. Scientific notation and the Pythagorean Theorem are presented as applications of these ideas. Chapter 6 studies graphing. Bar graphs were introduced in Chapter 1 as a way to visualize variables, and line graphs follow easily from there. Display of data, including histograms and boxplots, and graphs of equations occupy the rest of the chapter. Here the emphasis is on reading and interpreting graphs, and using graphs to describe variable relationships.

Algebraic fractions are another troublesome topic, probably because many students have only a shaky grasp of arithmetic fractions. Chapter 7 reviews arithmetic fractions as an introduction to simple algebraic fractions. The simpler laws of exponents are discovered while studying fractions.

Chapters 8 and 9 use the problem-solving skills developed throughout the course to investigate simple examples of topics encountered in elementary algebra: proportions and percent, interest, mixtures, and motion problems. Equations with variables on both sides and the distributive law are needed for the more sophisticated models in Chapter 9. All of these topics are approached intuitively, with many numerical (as opposed to algebraic) examples.

Organization

Each chapter of the book explores the material from many angles, returning to the same ideas in different settings. Modeling and problem solving are the unifying thread that runs through the text and motivates the introduction of new mathematical techniques. An effort has been made to increase gradually both the amount of reading and the formality of the presentation (if not the mathematics!) as the student proceeds through the text. Early lessons require very little reading, and the problems are mostly short-answer, requiring little computation. Soon, however, students are asked to read short passages and to answer questions about the reading. Some lessons (particularly the lessons on geometric formulas) are designed to be completed by the student in the classroom, either working in small groups or as guided discovery lessons. Beginning in Chapter 5 the lessons are written and formatted more like sections in a typical algebra text, although there is less material covered.

Students will need a four-function or scientific calculator for routine computation. It is expected that students will have some competence in arithmetic but will probably need to review fractions and percents. The text does not begin with a formal review of arithmetic; instead, arithmetic skills are reviewed as they are encountered in context or as needed for the development. Each exercise set includes a Building Number Skills section that incorporates mental arithmetic or calculator skills into the current material. In addition, an Appendix containing ten lessons that review operations on fractions, decimals, and percent has been included for student reference.

Using the Text

There is more than enough material for most prealgebra courses, allowing instructors to choose the topics that best fit their needs. A three-unit course might include Chapters 1 through 5, or by omitting or skimming some lessons, might also cover parts of Chapters 6 and 7. A five-unit course can include nearly all of the first seven chapters and parts of Chapters 8 and 9.

New to the Second Edition

- The four Units of the first edition have been reorganized into nine Chapters to afford instructors more flexibility.
- New Chapter Summaries highlight the key concepts in each lesson to help students identify important topics.
- Building Number Skills sections now appear in every Homework Set. These exercises have been reorganized and supplemented to provide a complete arithmetic review strand.

- Each Example is now followed by an exercise to allow students to practice each skill as it is presented.
- New material on graphing has been added, starting with bar graphs in Chapters 1–3, and building to a new Chapter 6, which is devoted to graphs of equations and display of data.
- Treatment of area and perimeter has been expanded and is now set against applications familiar to students, including quilt patterns.
- Tables of variables are used more frequently to give students greater practice and to provide an early familiarity with the idea of function.
- Grids for graphing are now provided in the Homework problems to encourage students to spend their time analyzing graphs rather than drawing and scaling the axes.
- To reinforce skills as they progress, more review exercises have been included in each Homework Set.
- In response to reviewer suggestions, a new section on slope has been added to Chapter 8.

Student Resource Manual

Many prealgebra instructors find that concrete manipulatives are a useful tool for helping students visualize abstract ideas, and some schools now incorporate a lab component in their prealgebra courses. We have prepared a selection of lessons using manipulatives as a lab manual to accompany *Prealgebra*. The text itself does not depend on these manipulatives labs, so their use is entirely optional. The *Complete Solutions Manual* includes notes on each lab for teachers who may be new to manipulatives.

The labs use algeblocks, fraction blocks, and other manipulatives to model most of the topics in the text, including area and perimeter, solving equations, operations on signed numbers and fractions, and mixtures. The manipulatives used in the labs are readily available commercially, but the manual also includes a cardboard insert with cutout patterns for the pieces needed.

For the Instructor

- **Annotated Instructor's Edition** (0-534-39670-4) This special version of the complete student text contains a *Resource Integration Guide* as well as answers printed next to the respective exercises.
- **Test Bank** (0-534-38604-0) The *Test Bank* includes 8 tests per chapter as well as 3 final exams. The tests are made up of a combination of multiple-choice, free-response, true/false, and fill-in-the-blank questions.
- **Complete Solutions Manual** (0-534-38601-6) The *Complete Solutions Manual* contains teaching notes, reading questions, and complete worked-out solutions to all of the problems in the text.
- **BCA Instructor Version** (0-534-38605-9) With a balance of efficiency and high performance, simplicity, and versatility, *Brooks/Cole Assessment* (BCA) gives you the power to transform the learning and teaching experience. BCA is a totally integrated testing and course-management system accessible by instructors and students anytime, anywhere. Delivered in a browser-based format without the need for

any proprietary software or plug-ins, BCA uses correct mathematics notation to provide the drill of basic skills that students need, enabling the instructor to focus more time in higher-level learning activities (i.e., concepts and applications). Students can have unlimited practice in questions and problems, building their own confidence and skills. Results flow automatically to a grade book for tracking so that instructors will be better able to assess student understanding of the material, even prior to class or an actual test.

- **BCA Tutorial** (included with BCA Instructor Version) This interactive tutorial software is text-specific and delivered through a browser via the Web. Like *BCA Testing*, *BCA Tutorial* is an intuitive mathematical guide, even for students with little technological proficiency. The *BCA Tutorial* tracking program enables instructors to monitor student progress carefully.
- **Text-Specific Videotapes** (0-534-38599-0) This set of videotapes is available to adopters of the text. Each tape offers one chapter of the text and is broken down into 10- to 20-minute problem-solving lessons that cover each section of the chapter.
- **Make the Grade w/InfoTrac** Every new copy of this text is packaged with *Make the Grade*, our suite of online tutorial services. With *Make the Grade*, your students have dynamic, flexible online tutorial resources at their fingertips. By entering a PIN code packaged with their textbook, students gain access to *BCA Tutorial*, a text-specific tutorial with step-by-step explanations, exercises and quizzes, and *vMentor*, live, one-on-one help from an experienced mathematics tutor.

  In addition to robust tutorial services, your students also receive anytime, anywhere access to **InfoTrac College Edition**. This online library offers the full text of articles from almost 4000 scholarly and popular publications, updated daily and going back as much as 22 years. Both adopters and their students receive unlimited access for four months.
- **MyCourse 2.0** (0-534-16641-5) Our new FREE online course builder! Whether you want only the easy-to-use tools to build it or the content to furnish it, we offer you a simple solution for a custom course Web site that allows you to assign, track, and report on student progress; load your syllabus, and more.

The above items are available to qualified adopters. Please consult your local sales representative for details.

For the Student

- **Student Solutions Manual** (0-534-38603-2) The *Student Solutions Manual* provides worked-out solutions to the odd-numbered problems in the text.
- **Student Resource Manual** (0-534-40109-0) The *Student Resource Manual* offers manipulative labs. Many instructors find that concrete manipulatives are useful tools for helping students visualize abstract ideas. These labs use algeblocks, fraction blocks, and other manipulatives to model most of the topics in the text.
- Website **www.brookscole.com/mathematics** The Brooks/Cole Mathematics Resource Center offers book-specific student and instruc-

tor resources, discipline-specific links, and a complete catalog of Brooks/Cole mathematics products.

- **BCA Tutorial** (Student Version 0-534-38600-8) This is interactive tutorial software that's so sophisticated, it's simple! This text-specific software is delivered via the Web and is offered in both student and instructor versions. Like *BCA Testing*, *BCA Tutorial* is browser-based, making it an intuitive mathematical guide even for students with little technological proficiency. *BCA Tutorial* allows students to work with real math notation in real time, providing instant analysis and feedback.
- **Interactive Video Skillbuilder CD** (0-534-38602-4) Think of it as portable office hours! The *Interactive Video Skillbuilder CD-ROM* contains more than eight hours of video instruction. The problems worked during each video lesson are shown next to the viewing screen so that students can try working them before watching the solution. To help students evaluate their progress, each section contains a 10-question Web quiz (the results of which can be emailed to the instructor) and each chapter contains a chapter test, with answers to each problem on each test. Also includes *MathCue Tutorial* software. This dual-platform software presents and scores problems and tutors students by displaying annotated, step-by-step solutions. Problem sets may be customized as desired.
- **vMentor** Offer your students customized, interactive homework help and tutorial services online with *vMentor.* Whether it's one-to-one tutoring help with daily homework, scheduled group tutoring sessions or exam-review tutorials, vMentor lets students interact with experienced instructors and other students right from their computer, at school or at home. *vMentor* is available through *Make the Grade*, which is packaged with each text.

Acknowledgments

We would like to thank the following reviewers for their helpful comments and suggestions: Kelly D. Brooks, Pierce College; Michael Coffey, Pierce College at Puyallup; Jennifer M. Dollar, Grand Rapids Community College; Sharon J. Edgmon, Bakersfield College; Joanne Kendall, Blinn College; Theodore Panitz, Cape Cod Community College; Nicole Pfeifer, Spokane Community College; Mary Jane Sterling, Bradley University.

Katherine Yoshiwara
Bruce Yoshiwara

CHAPTER

1

Variables

1.1 Variables and Bar Graphs

A. What Is a Variable?

A **variable** is a numerical quantity that can take on different values at different times or in different situations. For example, a person's age is a variable.

EXAMPLE 1  

a. Consider the table below. Do you see a pattern that relates Delbert's age to Francine's age?

Francine's age	*Delbert's age*
10	16
13	19
17	23

Yes: Delbert's age is always 6 more than Francine's age.

b. Write a mathematical sentence that expresses Delbert's age in terms of Francine's age.

Delbert's age = **Francine's age + 6**

In Example 1, *Francine's age* and *Delbert's age* are variables. Quantities that do not change, such as the 6-year age difference between Delbert and Francine, are called **constants** in algebra, in order to distinguish them from variables. Try Exercise 1, which shows the discount price and the regular price at a sporting goods store. (Exercise 1 appears in the margin on page 2.)

EXERCISE 1

The table shows the regular price, in dollars, of several items at a sporting goods store and the discount price with a coupon.

Regular price ($)	*Discount price ($)*
9.99	7.99
10.99	8.99
11.49	9.49

a. Is there a pattern that relates the discount price to the regular price?

b. Write a mathematical sentence that expresses the discount price in terms of the regular price:

Discount price =

EXERCISE 2

The table shows the weight of a salad and the price charged by the cafeteria.

Weight (oz)	*Price ($)*
4	0.40
6	0.60
8	0.80

a. Is there a pattern that relates the price of the salad to its weight?

b. Write a mathematical sentence that expresses the price of the salad in terms of its weight:

Price =

EXAMPLE 2

a. Look at the table below. Do you see a pattern that relates the total utility bill to your share?

Utility bill ($)	*Your share ($)*
28	7
36	9
42	10.50

Yes: Your share is one-fourth of the total bill, so divide the total bill by 4 to get your share.

b. Write a mathematical sentence that expresses your share of the bill in terms of the total bill:

Your share = **utility bill ÷ 4**

In Example 2, *Your share* and *utility bill* are variables, because they can change from month to month. Now try Exercise 2 in the margin.

ANSWERS TO 1.1A EXERCISES

1a. The discount price is $2 less than the regular price.

1b. Discount price = regular price − 2

2a. The price of the salad is $0.10 times its weight.

2b. Price of salad = weight × 0.10

HOMEWORK 1.1A

1. Write a mathematical sentence that expresses the number of laps Shirley has swum in terms of the number of laps Janet has swum.

Janet's laps	*Shirley's laps*
18	10
22	14
30	22

Shirley's laps =

2. Write a mathematical sentence that expresses the number of miles Cara has bicycled in terms of the number of miles Jamil has bicycled.

Jamil's miles	*Cara's miles*
2	5
6	9
12	15

Cara's miles =

3. Write a mathematical sentence that gives the number of batteries in terms of the number of packages.

Packages	*Batteries*
5	30
8	48
12	72

Number of batteries =

4. Write a mathematical sentence that gives the number of study groups in terms of the number of students:

Students	*Study groups*
20	5
28	7
40	10

Number of study groups =

5. Write a mathematical sentence that tells how many pages you have left to read out of a 200-page assignment:

Pages read	*Pages left*
30	170
85	115
160	40

Number of pages left =

6. Write a mathematical sentence that tells how many minutes are left in a 2-hour (120-minute) game:

Minutes played	*Minutes left*
15	105
40	80
97	23

Number of minutes left =

7. Write a mathematical sentence that expresses the amount of your wages in terms of the number of hours you worked:

Hours worked	*Wages*
5	35
12	84
20	140

Wages =

8. Write a mathematical sentence that expresses your weekly study hours in terms of the number of units you are taking:

Units	*Study hours*
6	18
12	36
18	54

Study hours =

Building Number Skills

We will use a calculator to help us with many computations involving decimals and fractions. First, you should become familiar with your calculator keypad. Make sure you can locate the operation keys, the decimal point, and the equal sign.

In algebra, it is best to use a scientific calculator, rather than a four-function calculator. To find out whether your calculator is scientific, try the following test. Key in

20 [+] 30 [÷] 5 [=]

If your calculator returns 26 as the result, you have a scientific calculator. If you get 10 as the answer, you have a four-function calculator.

Use your calculator to compute.

9. 16.9 + 1.4

10. 8.1 + 0.5

11. 9.6 − 0.9

12. 6.1 − 5.8

13. 2.5 × 18

14. 3.6 × 3.6

15. 18 ÷ 24

16. 35 ÷ 50

17. 11.2 ÷ 1.4

18. 39 ÷ 2.6

B. Bar Graphs

We can use a **bar graph** to display the values of a variable. The height of the bar illustrates the value at each time or in each situation.

EXAMPLE 3  

Figure 1.1 shows the number of new homes sold each month in Columbus County last year.

Figure 1.1 New homes sold each month

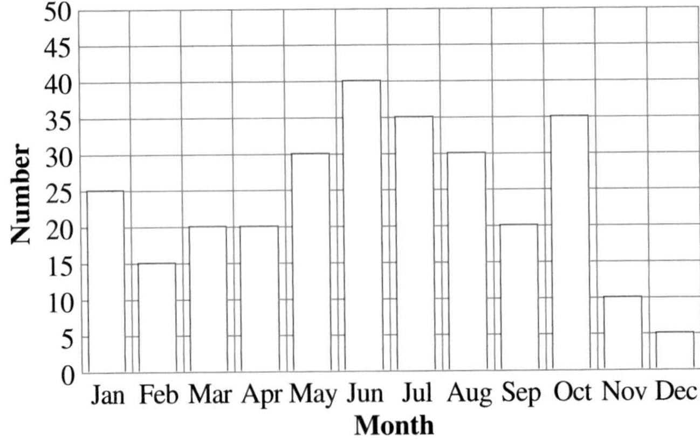

a. What variable is shown in this graph?

Number of new homes

b. How many houses were sold in March?

Twenty houses were sold in March.

c. In which month(s) were 35 houses sold?

Thirty-five houses were sold in July and in October.

EXERCISE 3

Refer to Figure 1.1 showing new homes sales in Columbus County.

a. How many houses were sold in February?

b. In which month(s) were 20 houses sold?

c. In which month were the fewest houses sold, and how many houses were sold that month?

d. In which month were the most houses sold, and how many houses were sold that month?

The most houses were sold in June; 40 houses were sold that month.

e. How much did housing sales decline from October to November?

Housing sales declined by 25 (from 35 to 10) from October to November.

f. Make a table of values showing the number of new homes sold each month.

Month	*Houses sold*	*Month*	*Houses sold*
January	**25**	July	**35**
February	**15**	August	**30**
March	**20**	September	**20**
April	**20**	October	**35**
May	**30**	November	**10**
June	**40**	December	**5**

We can compare the values of two variables with a double-bar graph.

EXAMPLE 4

Motorola and Nokia are competing wireless phone companies. Figure 1.2 shows the portion of the market captured by each company from 1994 to 2000.

Figure 1.2 Market share of wireless phones

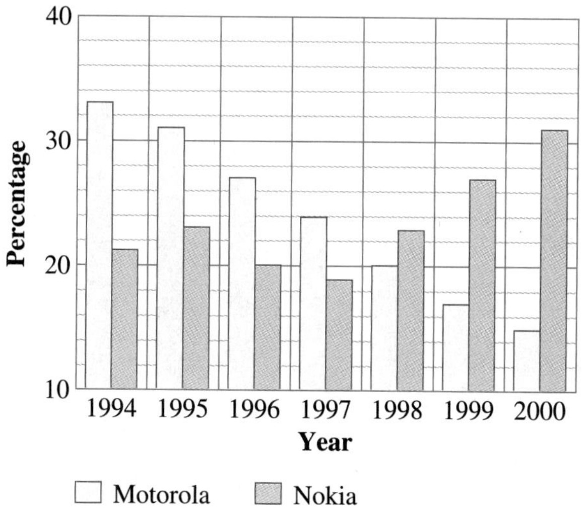

(Source: Gartner Dataquest)

a. What was Motorola's share of the market in 2000?

15%

b. In which year shown did Motorola have a 24% market share?

1997

EXERCISE 4

Refer to Figure 1.2.

a. What was Nokia's share of the market in 2000?

b. In which of the years shown did Nokia have its lowest market share? What was that share?

c. In which of the years shown did Nokia have its largest lead over Motorola in market share? What was that lead?

d. In which year did Nokia overtake Motorola in market share?

c. In which of the years shown did Motorola have its largest market share? What was that share?

In 1994 Motorola had 33% of the market.

d. By how much did Motorola's market share exceed Nokia's in 1995?

8 percentage points

e. In which of the years shown did Motorola have its largest lead over Nokia in market share? What was that lead?

In 1994 Motorola had a 12% lead in market share.

ANSWERS TO 1.1B EXERCISES

3a. 15 houses

3b. March, April, September

3c. December, 5 houses

4a. 31%

4b. In 1997 Nokia had 19% of the market.

4c. In 2000 Nokia led Motorola by 16% of the market.

4d. 1998

HOMEWORK 1.1B 

1. Figure 1.3 shows the percent of children born outside of marriage in various European countries.

Figure 1.3 Children born outside of marriage

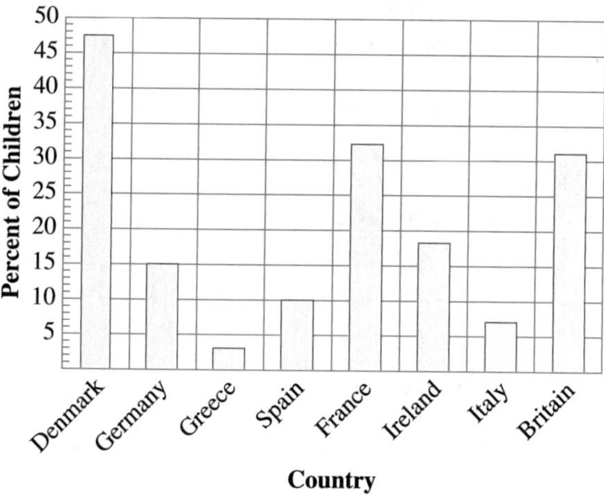

a. What variable is displayed in this graph?

b. What percent of children are born outside of marriage in Germany?

c. In which country are 18% of children born outside of marriage?

d. In which country is the largest percent of children born outside of marriage?

e. In which country is the smallest percent of children born outside of marriage?

2. Figure 1.4 shows the percent of the original rain forest that remains in several areas that are undergoing heavy deforestation.

Figure 1.4 Remaining rain forest

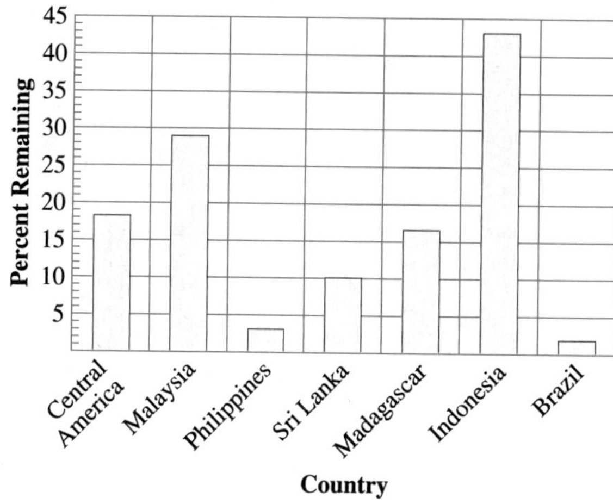

a. What variable is displayed in this graph?

b. What percent of the Central American rain forest remains?

c. In what region does only 3% of the rain forest remain?

d. In what region is the largest percent of the original rain forest left?

e. In what region is the smallest percent of the original rain forest left?

3. Figure 1.5 shows the average temperature in Chicago during each month of the year.

Figure 1.5 Average temperature in Chicago

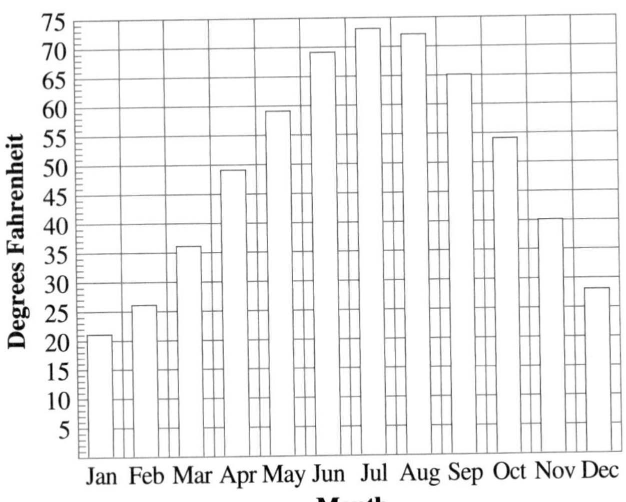

a. What variable is displayed in this graph?

b. What is the average temperature during October?

c. In what month is the average temperature 59°?

d. Which month experiences the greatest temperature increase over the previous month?

e. Which month experiences the greatest temperature decrease over the previous month?

4. Figure 1.6 shows the average precipitation (in inches) each month in Tampa, Florida.

Figure 1.6 Average precipitation in Tampa

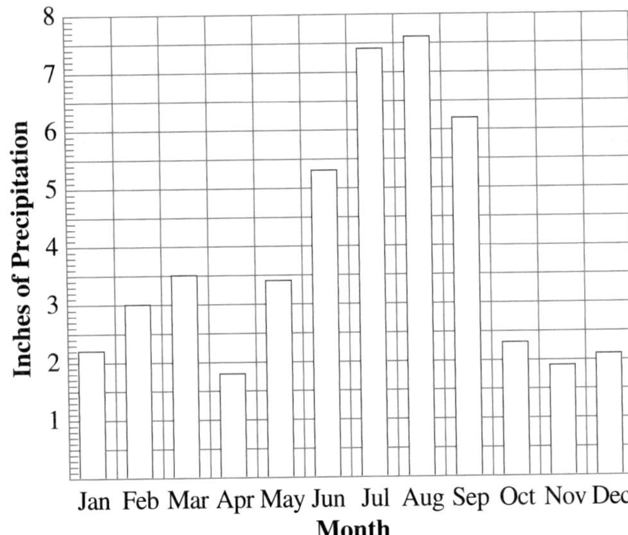

a. What variable is displayed in this graph?

b. In which month does Tampa receive the most precipitation?

c. In which month does Tampa receive the least precipitation?

d. What is the increase in precipitation from April to May?

e. What is the decrease in precipitation from September to October?

5. Figure 1.7 shows the leading causes of accidental deaths in 1993 for children under 5 years old.

Figure 1.7 Accidental deaths of children under 5 years of age

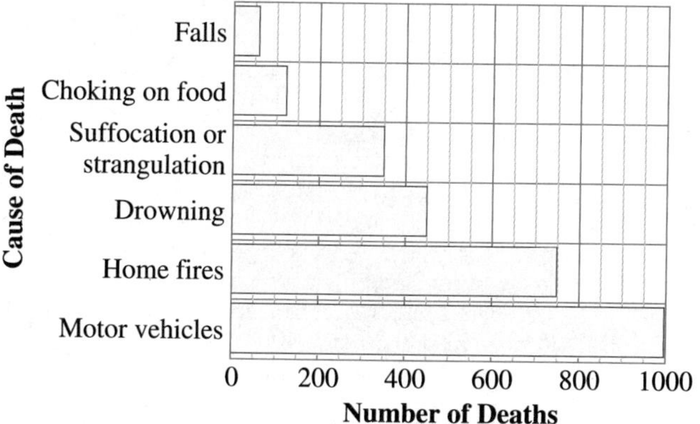

(*Source:* National Safety Council)

a. Make a table showing the number of deaths for each cause.

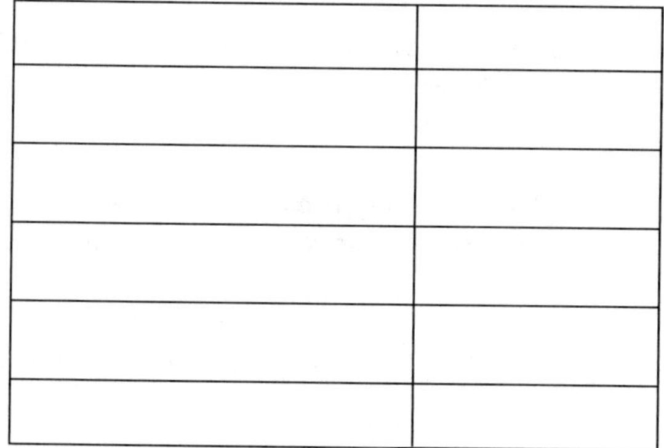

b. What caused the most deaths for children under age 5?

c. About 3400 accidental deaths of children under age 5 were recorded in 1993. What portion of those deaths were caused by drowning? (Give your answer as a decimal fraction rounded to the nearest thousandth.)

d. How many accidental deaths were due to causes other than those shown in the bar graph?

6. Figure 1.8 shows the number of annual paid vacation days received by workers in ten industrial nations.

Figure 1.8 Paid vaction days per year

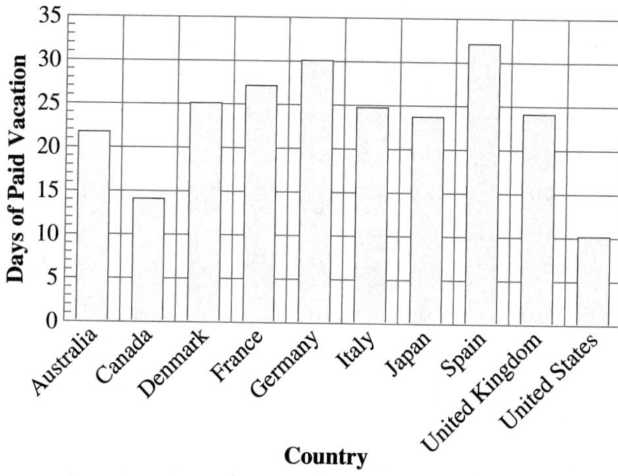

(*Source:* Union Bank of Switzerland, *Prices and Earnings around the Globe, 1991*)

a. In which nation do workers receive the most vacation days? In which nation do they receive the fewest?

b. Make a table showing the number of vacation days in each country. List the countries in order from most vacation days to fewest.

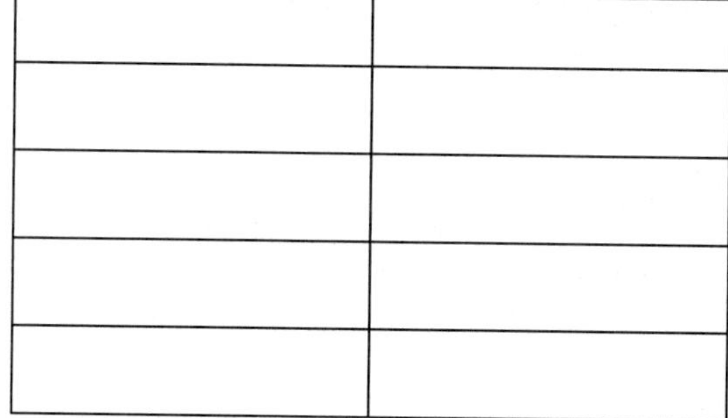

c. How many more vacation days would you get in France than in Australia?

d. Which countries get at least double the amount of vacation we get in the United States? Do any countries get at least double the amount of vacation Canadians get?

7. Figure 1.9 is a double-bar graph showing the average life expectancy in 1900 and now, for four countries.

Figure 1.9 Life expectancy

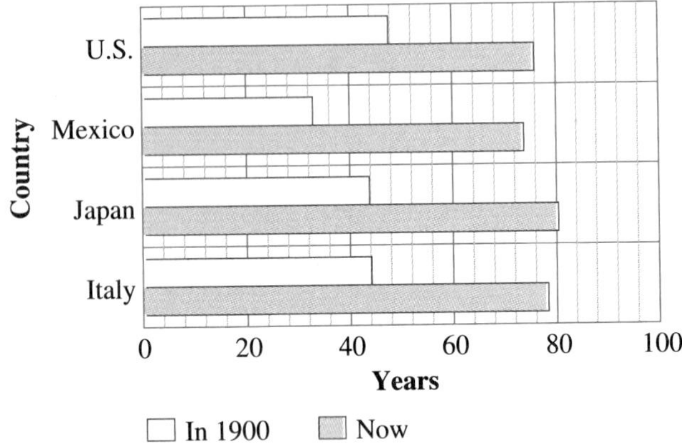

a. Which country had the highest life expectancy in 1900? Which had the lowest?

b. Which country has the highest life expectancy now? Which has the lowest?

c. Which country experienced the greatest increase in life expectancy over the last century? By how many years?

d. How much has the gap in life expectancy among the countries listed decreased over the last century?

8. Figure 1.10 shows the average weekly working hours in 1900 and now, for four countries.

Figure 1.10 Average weekly working hours

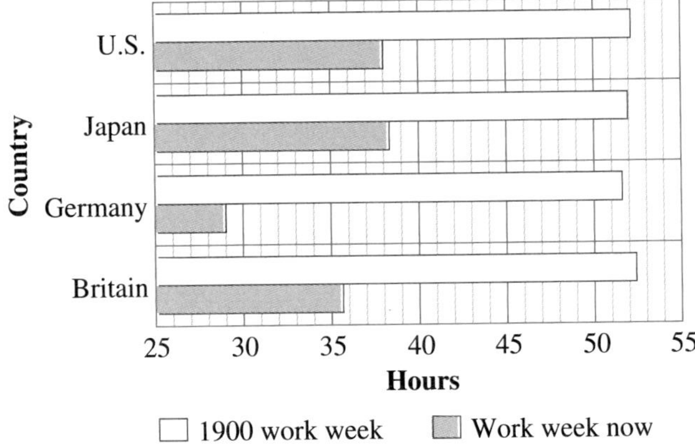

a. Which country had the longest average work week in 1900? Which had the shortest?

b. Which country has the longest average work week now? Which has the shortest?

c. In which country did the number of working hours decrease the most over the century? By how many hours?

d. How much has the gap in working hours among the countries listed increased over the last century?

9. Find three new mathematical terms introduced in this lesson. List these terms in your glossary and give an explanation of each term in your own words.

10. Write a short summary of the topics discussed in this lesson.

Building Number Skills

A percent is just a fraction whose denominator is 100. (Percent means "divided by 100.") Thus, 15% means $\frac{15}{100}$, or $15 \div 100$. (See Appendix A.9 to review percents.) Use your calculator to verify that $15 \div 100 = 0.15$.

Use your calculator to convert each percent to a decimal fraction.

11. 25% **12.** 80% **13.** 62% **14.** 17% **15.** 6%

16. 1% **17.** 150% **18.** 118% **19.** 4.6% **20.** 32.5%

Do you see a shortcut? How can you convert a percent to a decimal fraction without using your calculator?

Convert each percent to a decimal fraction without using a calculator.

21. 60% **22.** 75% **23.** 9% **24.** 2% **25.** 8.5%

26. 9.1% **27.** 250% **28.** 110% **29.** 0.4% **30.** 0.68%

1.2 Algebraic Expressions

A. Using Letters for Variables

We often use a single letter to represent a variable quantity.

EXAMPLE 1  

Francine was 22 years old when her son Reginald was born.

a. Fill in the table below (the variables are *Reginald's age* and *Francine's age*:

Reginald's age	1	4	7	12	18	25
(Calculation)	1 + 22	4 + 22	7 + 22	12 + 22	18 + 22	25 + 22
Francine's age	**23**	**26**	**29**	**34**	**40**	**47**

b. Explain how to find Francine's age if you know Reginald's age.

Add 22 to Reginald's age.

c. Write a mathematical sentence for Francine's age:

Francine's age = **Reginald's age + 22**

d. Let a stand for Reginald's age in years and write an expression for Francine's age in terms of a:

Francine's age = $\boldsymbol{a + 22}$

e. Check that your expression for Francine's age gives the correct values when we replace Reginald's age by a in the table in part (a):

a	1	4	7	12	18	25
$a + 22$	23	26	29	34	40	47

A letter used as a variable must always stand for a number. In Example 1, the variable a stands for *Reginald's age*, which is a number. We cannot let a variable stand for, say, Reginald himself or for Reginald's occupation.

EXERCISE 1

A belt should be 3 inches longer than the maximum waist size.

a. Fill in the table below. (The variables are *Waist size* and *Belt length*):

Waist size (in.)	24	28	32	36	40
(Calculation)					
Belt length (in.)					

b. Explain how to find the belt length if you know the waist size.

c. Write a mathematical sentence for the belt length:

Belt length =

d. Let w stand for the waist size in inches and write an expression for the belt length in terms of w.

e. Check that your expression for the belt length gives the correct values when we replace the waist size by w in the table in part (a):

w	24	28	32	36	40

ANSWERS TO 1.2A EXERCISES

1a.

Waist size (in.)	24	28	32	36	40
(Calculation)	24 + 3	28 + 3	32 + 3	36 + 3	40 + 3
Belt length (in.)	27	31	35	39	43

1b. Add 3 to the waist size.

1c. Belt length = waist size + 3

1d. Belt length = $w + 3$

1e. When $w = 24$, $w + 3 = 24 + 3 = 27$, and so on.

HOMEWORK 1.2A

Use your calculator as needed for the problems below.

1. Delbert has a coupon for $12 off at Video World.
   a. Fill in the table below (the variables are *Regular price* and *Delbert's price*):

Regular price	18	25	54	76	115	130
(Calculation)						
Delbert's price						

   b. Explain how to find Delbert's price if you know the regular price.

   c. Write a mathematical sentence for what Delbert pays:

   Delbert's price =

   d. Let p stand for the regular price and write an expression for Delbert's price in terms of p:

   Delbert's price =

   e. Check that your expression gives the correct values for Delbert's price when we replace the regular price by p in the table in part (a):

p	18	25	54	76	115	130

2. Mariel is driving 400 miles to her sister's house.
   a. Fill in the table below (the variables are *Miles driven* and *Miles left*):

Miles driven	30	80	150	225	260	310
(Calculation)						
Miles left						

   b. Explain how to find the number of miles Mariel has left to drive if you know how many miles she has driven.

   c. Write a mathematical sentence for the number of miles left to drive:

   Miles left =

   d. Let d stand for the number of miles Mariel has driven and write an expression for the number of miles she has left in terms of d:

   Miles left =

   e. Check that your expression gives the correct values for the miles left when we replace the miles driven by d in the table in part (a):

d	30	80	150	225	260	310

3. Adria's diet allows one-fifth of her total daily calories to come from fat.

a. Fill in the table below (the variables are *Total calories* and *Calories from fat*). (*Hint*: To find one-fifth of a number, you can divide the number by 5.)

Total calories	400	550	700	925	1200	1500
(Calculation)						
Calories from fat						

b. Explain how to find the number of calories from fat allowed if you know how many calories Adria plans to eat.

c. Write a mathematical sentence for the number of calories from fat:

Calories from fat =

d. Let c stand for the Total number of Calories Adria plans to eat and write an expression for the number of calories from fat allowed in terms of c:

Calories from fat =

e. Check that your expression gives the correct values for Calories from fat when we replace Total calories by c in the table in part (a):

c	400	550	700	925	1200	1500

4. Marvin goes to a restaurant with two friends and agrees to split the bill equally.

a. Fill in the table below (the variables are *Total bill* and *Marvin's share*):

Total bill	21	27	30	42	57	72
(Calculation)						
Marvin's share						

b. Explain how to find Marvin's share of the bill.

c. Write a mathematical sentence for Marvin's share of the bill:

Marvin's share =

d. Let B stand for the Total bill and write an expression for Marvin's share in terms of B:

Marvin's share =

e. Check that your expression gives the correct values for Marvin's share when we replace Total bill by B in the table in part (a):

B	21	27	30	42	57	72

5. Marvin and his friends agree to tip 15% of the total bill.

a. Fill in the table below (the variables are *Total bill* and *Tip*). (*Hint*: "15% of the bill" means 0.15 × the bill. See Appendix A.9 to review finding a percent of a number.)

Total bill	20	26	32	48	52	60
(Calculation)						
Tip						

b. Explain in words how to find the tip.

c. Write a mathematical sentence for the amount of the tip:

Tip =

d. Let B stand for the total bill and write an expression for the tip in terms of B:

Tip =

e. Check that your expression gives the correct values for the tip when we replace the total bill by B in the table in part (a):

B	20	26	32	48	52	60

6. Rae has been collecting quarters to take to the bank.

a. Fill in the table below (the variables are *Number of quarters* and *Value of quarters*):

Number of quarters	6	9	15	32	68	100
(Calculation)						
Value of quarters						

b. Explain how to find the value of the quarters if you know how many quarters Rae has.

c. Write a mathematical sentence for the value of the quarters:

Value of quarters =

d. Let q stand for the number of quarters Rae has and write an expression for the value of the quarters in terms of q:

Value of quarters =

e. Check that your expression gives the correct values for the value of the quarters when we replace the number of quarters by q in the table in part (a):

q	6	9	15	32	68	100

Read the bar graphs to answer the problems. (See Lesson 1.1B to review bar graphs.)

7. Figure 1.11 shows the graduation rates for all students and for football players at several universities and conferences.

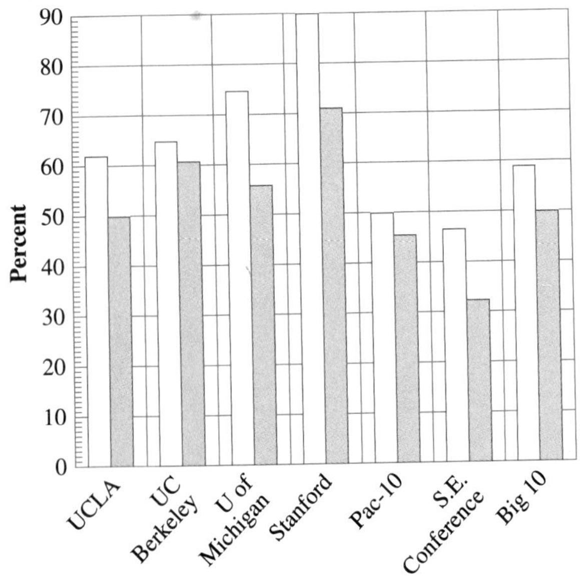

Figure 1.11 Graduation rates

a. What percent of all students graduate at UCLA?

b. What percent of football players graduate in the Big 10 Conference?

c. Which institution has the lowest graduation rate for all its students?

d. Which institution has the lowest graduation rate for its football players?

e. At which institution is there the greatest difference between the graduation rate for football players and the graduation rate for all students?

8. Figure 1.12 shows the number of homicides committed in the United States each year with guns versus all other weapons.

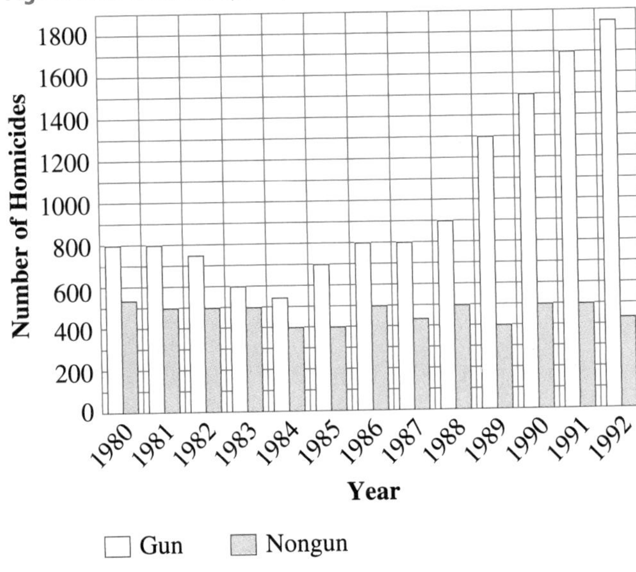

Figure 1.12 Homicide by weapon for offenders aged 10–17

a. How many homicides were committed with guns in 1988?

b. How many homicides were committed with other weapons in 1980?

c. In what year were the fewest homicides committed with guns?

d. In what year were the most homicides committed with other weapons?

e. What was the increase in the number of homicides committed with guns from 1984 to 1992?

9. The table shows the age distribution of the U.S. population in 1900 and in 1999.

Age group	*% of population in 1900*	*% of population in 1999*
Under 5	12	7
5–14	22	13
15–24	20	15
25–34	16	14
35–44	12	16
45–54	9	13
55–64	5	9
65 and over	4	13

a. Make a double-bar graph displaying this information. Use the grid in Figure 1.13.

Figure 1.13 U.S. age distribution

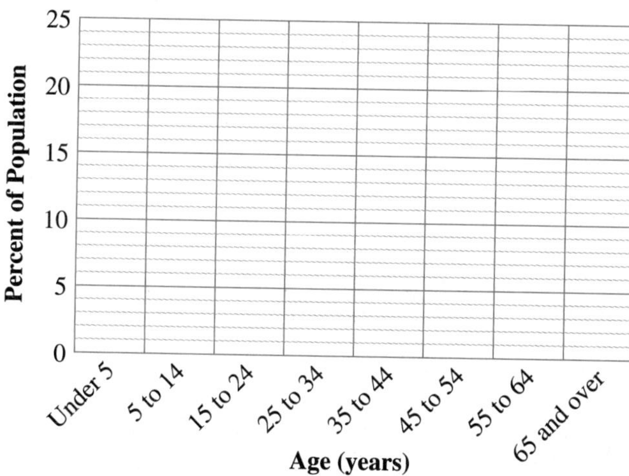

b. Which was the largest age group in 1900? In 1999?

c. What percent of the population was under 25 in 1900? In 1999?

d. What percent of the population was over 44 in 1900? In 1999?

10. The table shows the percent of women of various ages who were never married. The figures are given for three different decades.

Age in years	*1970*	*1980*	*1990*
20–24	35.8	50.2	62.8
25–29	10.5	20.9	31.1
30–34	6.2	9.5	16.4

(*Source:* U.S. Census)

a. Make a triple-bar graph displaying this information. Use the grid in Figure 1.14.

Figure 1.14 Women who have never married

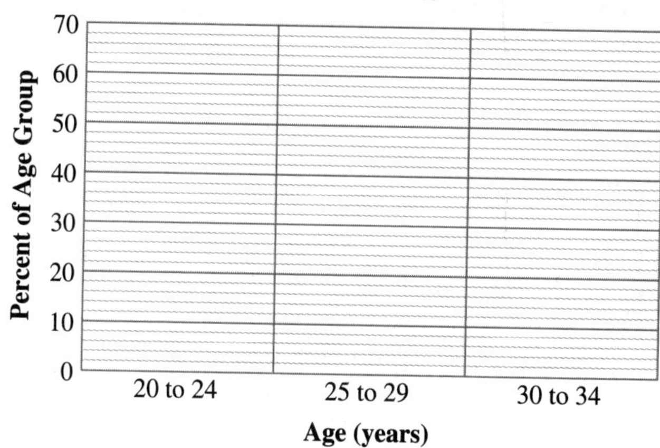

b. In any one of the 3 years listed, what happens to the percent as the women's age increases? Why does this make sense?

c. For any one of the three age groups listed, what happens to the percent over time? What conclusion can you draw from this observation?

Building Number Skills

To take $\frac{1}{3}$ of something means to divide it into three equal parts and to take one of them. In other words, to *multiply* by $\frac{1}{3}$ is the same as to *divide* by 3. For example, $\frac{1}{3} \times 12 = 12 \div 3 = 4$. (See Appendix A.2 to review fractions.)

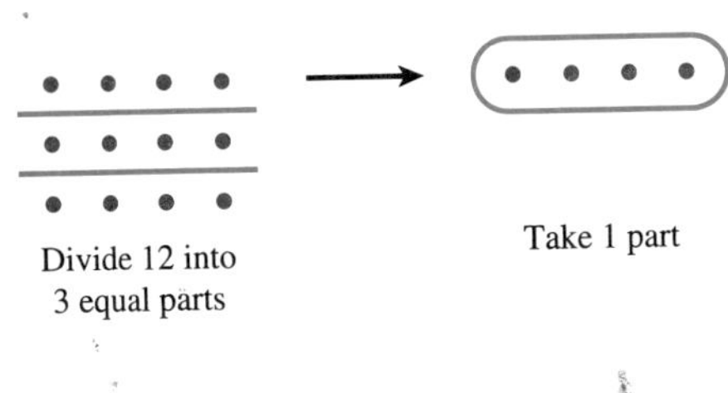
Divide 12 into
3 equal parts
Take 1 part

Find each fraction by writing it as a division.

Example: $\frac{1}{5}$ of $20 = \frac{1}{5} \times 20 = 20 \div 5 = 4$

11. $\frac{1}{4}$ of 48 **12.** $\frac{1}{3}$ of 21 **13.** $\frac{1}{2}$ of 18 **14.** $\frac{1}{5}$ of 60 **15.** $\frac{1}{8}$ of 40

16. $\frac{1}{9}$ of 36 **17.** $\frac{1}{6}$ of 42 **18.** $\frac{1}{7}$ of 35 **19.** $\frac{1}{12}$ of 72 **20.** $\frac{1}{10}$ of 80

Use your calculator to find each fraction.

21. $\frac{1}{3}$ of 234 **22.** $\frac{1}{4}$ of 372 **23.** $\frac{1}{5}$ of 685 **24.** $\frac{1}{8}$ of 448 **25.** $\frac{1}{9}$ of 243

26. $\frac{1}{6}$ of 174 **27.** $\frac{1}{15}$ of 240 **28.** $\frac{1}{12}$ of 300 **29.** $\frac{1}{16}$ of 208 **30.** $\frac{1}{18}$ of 594

B. What Is an Algebraic Expression?

When we use operation symbols to combine numbers and variables, we create **algebraic expressions.** Some examples of algebraic expressions are

$$8 + b, \qquad r \times t, \qquad 12 \div d$$

We can use variables in mathematical expressions the same way we use numbers. We can do this because a variable really *is* a number, but one whose value is unknown or is not specified.

Sums and Products

When we add two numbers or variables together, the result is called the **sum**, and the things added together are called **terms**. When we multiply two numbers or variables together, the result is called the **product**, and the things multiplied together are called **factors.**

Terms → $\underbrace{a + b}_{\text{Sum}}$ **Factors** → $\underbrace{a \times b}_{\text{Product}}$

In algebra, we use a raised dot or parentheses instead of a cross to indicate multiplication. For example, we can write

$$2 \cdot 3 \quad \text{or} \quad 2(3) \quad \text{to mean} \quad 2 \times 3$$

We can show the product of two variables, or a number and a variable, simply by writing them next to each other, without any symbol between them. For example,

$$5g \quad \text{means} \quad 5 \times g$$

and

$$xy \quad \text{means} \quad x \times y$$

CAUTION!

Although $2b$ means $2 \times b$, 25 does not mean 2×5; it means twenty-five. We cannot write the product of two numbers without a dot or other multiplication symbol.

EXAMPLE 2  

Write each phrase as an algebraic expression.

a. The product of 0.05 and w **b.** The sum of D and $\frac{3}{4}$

c. The product of P and r **d.** The sum of l and w

Solution

a. $0.05w$ **b.** $D + \frac{3}{4}$

c. Pr **d.** $l + w$

In Example 2a, we do not write a dot or parentheses between 0.05 and w. The simplest way of writing an algebraic expression is usually best. We could also write $w(0.05)$, because we get the same result when we multiply the factors in either order. However, in algebra we prefer to write the number before the variable in a product. In the case of $1 \times x$, we write x instead of $1x$, because x is simpler. Thus, x always means $1x$.

EXERCISE 2

Write each phrase as an algebraic expression.

a. The sum of Q and 1.3

b. The product of 1.07 and P

In Example 2b, we could also write $\frac{3}{4} + D$. In a sum, it does not matter which term is written first. (You may want to review Appendix A.10 on the laws of arithmetic before continuing.)

Differences and Quotients

When we subtract one number or variable from another, the result is called the **difference.** In subtraction, it *does* matter which term appears first: $7 - 3$ does *not* mean the same thing as $3 - 7$.

Here are two common ways to describe subtraction:

p subtracted from 12	means	$12 - p$
The difference of 12 and p	means	$12 - p$

When we divide one number or variable by another, the result is called the **quotient.** In algebra we often use a fraction bar instead of the symbol ÷ to denote division. For example,

$$\frac{20}{5} \quad \text{means} \quad 20 \div 5.$$

The fraction bar and the symbol ÷ both mean "divided by." Just as in subtraction, the order of the numbers matters:

$$20 \div 5 = \frac{20}{5} = 4, \quad \text{but} \quad 5 \div 20 = \frac{5}{20} = \frac{1}{4}$$

EXERCISE 3

Write each phrase as an algebraic expression.

a. The ratio of 100 to p

b. d subtracted from 19.95

EXAMPLE 3  

Write each phrase as an algebraic expression.

a. 16 subtracted from z **b.** m divided by 6

c. The difference of R and C **d.** The quotient of P and n

Solution

a. $z - 16$ **b.** $\frac{m}{6}$, or $m \div 6$

c. $R - C$ **d.** $\frac{P}{n}$, or $P \div n$

Sometimes English phrases for arithmetic operations can be tricky. We must be especially careful with subtraction and division. In Example 3a, the expression $16 - z$ is not a correct answer. In Example 3b, m divided by 6 means $\frac{m}{6}$, but m divided *into* 6 means $\frac{6}{m}$.

Another common way of referring to division uses the word **ratio.** The ratio of T to 100 means the quotient $\frac{T}{100}$.

This lesson contains several new mathematical words and their definitions. List these new words in your glossary and give a brief definition of each word. You should make a habit of adding new words to your glossary after you read each new lesson.

ANSWERS TO 1.2B EXERCISES

2a. $Q + 1.3$

2b. $1.07P$

3a. $\frac{100}{p}$

3b. $19.95 - d$

HOMEWORK 1.2B

1. Choose the correct operation (addition, subtraction, multiplication, or division) associated with each word:

a. product ____________________

b. difference ____________________

c. terms ____________________

d. ratio ____________________

e. factors ____________________

f. sum ____________________

g. quotient ____________________

2. Write each phrase as an algebraic expression:

a. The product of b and c

b. The ratio of b to c

c. b subtracted from c

d. The sum of b and c

e. b divided into c

f. The difference of b and c

g. b divided by c

h. The quotient of b and c

3. Name two ways besides "×" to indicate multiplication. Give an example of each.

4. Name two ways to show division. Give an example of each.

5. When two numbers or variables are added together, the two numbers or variables are called ____________. The result of the addition is called the ____________.

6. When two numbers or variables are multiplied together, the two numbers or variables are called terms. The result of the multiplication is called the ____________.

Write each phrase as an algebraic expression.

7. The sum of 8 and z

8. The quotient of w and 12

9. 5 subtracted from h

h-5

10. The product of q and 6

11. The ratio of t to 20

t/20

12. The difference of m and 15

13. 16 divided by v

14. 4 divided into d.

Write English phrases for each algebraic expression. (Many answers are possible.)

15. $7.1 - a$

16. $s + 0.3$

17. $15j$

18. $\frac{36}{u}$

19. $\frac{A}{c}$

20. $M - D$

21. $y + h$

22. ng

23. $H - 2.5$

24. $\frac{x}{30}$

25. $\frac{3}{5}b$

26. $z + \frac{5}{2}$

For problems 27–30, you may want to review Appendix A.1 on fractions.

27. Francine is saving up to buy a car, and deposits $\frac{1}{5}$ of her spending money each month into a savings account. (*Hint*: To find $\frac{1}{5}$ of a number, you can divide the number by 5.)

a. If m stands for Francine's spending money, write an expression for the amount she saves:

Amount Francine saves =

b. Fill in the table:

Spending money ($)	m	20	25	60	80
Amount saved ($)					

28. The Pep Club donates $\frac{1}{6}$ of the money it raises to charity.

a. If R stands for the amount of money the Pep Club raised, write an expression for the amount they donated to charity:

Amount donated =

b. Fill in the table:

Money raised ($)	R	30	120	240	300
Amount donated ($)					

29. Delbert wants to enlarge his class photograph. Its height is $\frac{1}{4}$ of its width. The enlarged photo should have the same shape as the original.

a. If W stands for the width of the enlarged photo, write an expression for its height:

Height =

b. Fill in the table:

Width of photo (in.)	W	6	10	24	30
Height of photo (in.)					

30. The amount of soy flour in Lucy's protein bread is $\frac{1}{8}$ of the amount of wheat flour.

a. If w stands for the amount of wheat flour, write an expression for the amount of soy flour:

Amount of soy flour =

b. Fill in the table:

Wheat flour (cups)	w	2	4	6	10
Soy flour (cups)					

Building Number Skills

To take $\frac{2}{3}$ of something means to divide it into 3 equal parts and take 2 of them. Here is $\frac{2}{3}$ of 12:

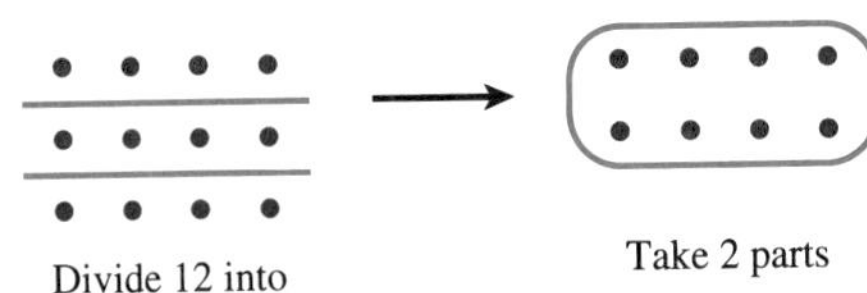
Divide 12 into
3 equal parts
Take 2 parts

To take $\frac{2}{3}$ of 12, divide 12 by 3, then multiply the result by 2:

$$\frac{2}{3}(12) = 2\left(\frac{12}{3}\right) = 2(4) = 8$$

Find each fraction by dividing by the denominator, then multiplying by the numerator. (You may be able to do this mentally.)

Example: $\frac{4}{5}$ of $20 = \frac{4}{5}(20) = 4\left(\frac{20}{5}\right) = 4(4) = 16$

31. $\frac{2}{3}(18)$ **32.** $\frac{3}{5}(35)$ **33.** $\frac{5}{2}(28)$ **34.** $\frac{7}{4}(32)$ **35.** $\frac{3}{8}(24)$

36. $\frac{3}{4}(48)$ **37.** $\frac{7}{3}(21)$ **38.** $\frac{5}{8}(40)$ **39.** $\frac{7}{9}(36)$ **40.** $\frac{3}{10}(80)$

You can also use your calculator to find a fraction of a whole number. For example,

$$\frac{3}{4}(42) = 3\left(\frac{42}{4}\right) = 3(10.5) = 31.5$$

On your calculator, enter

3 [×] 42 [÷] 4 [=]

Use your calculator to find each product.

41. $\frac{5}{7}(28)$ **42.** $\frac{7}{8}(32)$ **43.** $\frac{5}{3}(18)$ **44.** $\frac{6}{5}(35)$ **45.** $\frac{4}{5}(16)$

46. $\frac{3}{4}(25)$ **47.** $\frac{3}{8}(36)$ **48.** $\frac{3}{2}(11)$ **49.** $\frac{9}{4}(13)$ **50.** $\frac{9}{8}(14)$

1.3 Using Algebraic Expressions

A. Other Ways to Express Operations

There are usually several different ways to write an algebraic expression in English. Here are some English phrases that are used for each of the four arithmetic operations.

Addition

English Phrase	*Algebraic Expression*
The sum of 5 and x	$5 + x$
3 more than n	$n + 3$
q increased by 5	$q + 5$
15 plus w	$15 + w$
The total of 8 and D	$8 + D$
M added to 12	$12 + M$
1.7 greater than B	$B + 1.7$
Exceeds g by 3	$g + 3$

Subtraction

English Phrase	*Algebraic Expression*
The difference of b and 2	$b - 2$
5 less than z	$z - 5$
n minus 2.5	$n - 2.5$
9 reduced by s	$9 - s$
H subtracted from K	$K - H$
Q decreased by 16	$Q - 16$

Multiplication

English Phrase	*Algebraic Expression*
The product of 6 and w	$6w$
5 times v	$5v$
$\frac{3}{4}$ of c	$\frac{3}{4}c$
Twice p	$2p$

Division

English Phrase	*Algebraic Expression*
The quotient of a and 7	$\frac{a}{7}$
G divided by 10	$\frac{G}{10}$
m divided into 15	$\frac{15}{m}$
The ratio of V to 5	$\frac{V}{5}$
S split 7 ways	$\frac{S}{7}$
Dollars, d, per hour, h	$\frac{d}{h}$

EXERCISE 1
Write algebraic expressions for each phrase.
a. 6 greater than h

b. The product of 9 and c

ANSWERS TO 1.3A EXERCISES
1a. $h + 6$ or $6 + h$

1b. $9c$

EXAMPLE 1 

Write algebraic expressions for each phrase.

a. $\frac{5}{8}$ of M

b. 4 less than w

Solution

a. $\frac{5}{8}M$

b. $w - 4$

In Example 1a, notice that "of" means multiplication when used with fractions. The same is true for percents. For example, "35% of P" means $0.35P$.

HOMEWORK 1.3A 

For Problems 1–20, write an algebraic expression for each English phrase.

1. 8 less than v

2. 4 divided into c

3. H increased by 14

4. 10 times g

5. t divided by 5

6. 20 more than J

7. $\frac{2}{3}$ of w

8. 16 reduced by h

9. The ratio of 15 to b

10. The total of 32 and f

11. The difference of R and 2.6

12. 17.5 divided by q

13. P split 3 ways

14. Z reduced by $\frac{3}{2}$

15. 18% of N

16. 4% of G

17. d subtracted from r

18. The product of y and z

19. The total of P and I

20. m less than k

21. Write four different English phrases that mean $x - 3$.

22. Write four different English phrases that mean $v \div 5$.

23. Write four different English phrases that mean $9h$.

24. Write four different English phrases that mean $t + 2$.

25. Write an English phrase for each of the following, then simplify. Which expressions are the same?

a. $24 \div 6$

b. $6 \div 24$

c. $\frac{6}{24}$

d. $\frac{24}{6}$

26. Write an English phrase for each of the following, then simplify. Which expressions are the same?

a. $\frac{36}{4}$

b. $\frac{1}{4}(36)$

c. $36 \div 4$

d. $0.25(36)$

For problems 27–30, you may want to review Appendix A.2 on fractions.

27. If the governor vetoes a bill passed by the State Assembly, $\frac{2}{3}$ of the members present must vote for the bill in order to overturn the veto.

a. If p stand for the number of Assembly members present, write an expression for the number of votes needed to overturn a veto:

Votes needed =

b. Fill in the table:

Members present	p	90	96	120	129
Votes needed					

28. The case for a compact disc (CD) is $\frac{3}{8}$ inch thick. You keep your CDs lined up side-by-side on a bookshelf.

a. If you own n CDs, write an expression for the length of shelf space you will need:

Length of shelf space =

b. Fill in the table:

Number of CDs	n	24	40	48	72
Length of shelf space					

29. Write an expression for the sale price in terms of the regular price, p. (*Hint*: The sale price is a fraction of the regular price.)

Regular price	*Sale price*
4	3
8	6
20	15
100	75

Sale price =

30. Write an expression for the number of cups of flour in terms of the number of cups of milk, M. (*Hint*: The amount of flour is an improper fraction of the amount of milk. See Appendix A.4 to review improper fractions.)

Cups of milk	*Cups of flour*
2	3
5	7.5
6	9
8	12

Cups of flour =

Building Number Skills

Calculate each percent. (See Appendix A.9 to review finding a percent of a number.)

Example: 65% of 80 = 0.65 × 80 = 52

31. 70% of 30

32. 40% of 90

33. 15% of 60

34. 85% of 70

35. 32% of 58

36. 64% of 26

37. 7% of 83

38. 9% of 64

39. 2% of 18

40. 5% of 36

B. Evaluating Algebraic Expressions

If we know the value of the variable for a particular situation, we can substitute this value into the expression and simplify the result. This process is called **evaluating** the algebraic expression.

EXAMPLE 2 

If Martha puts g gallons of gas in her car, then she can drive $22g$ miles.

a. How many miles can Martha drive if she buys 8 gallons of gas?

$22 \cdot 8$, or 176 miles

b. Explain how to find the number of miles Martha can drive if she buys g gallons of gas:

Multiply the number of gallons, g, by 22.

c. Evaluate the expression $22g$ for the given values of g:

g	4	7.5	9	10	12.5	14
$22g$	**88**	**165**	**198**	**220**	**275**	**308**

EXERCISE 2

There is a room tax on all hotel bills. When the bill is B dollars, the tax is $0.08B$ dollars.

a. How much is the tax on $12?

b. Explain how to find the tax when the bill is B dollars.

c. Evaluate the expression $0.08B$ for the given values of B:

B	10.00	20.00	50.00	120.00
$0.08B$				

EXAMPLE 3  

a. Evaluate $y + 14$ for $y = 7$.

b. Evaluate $3xz$ for $x = 5$ and $z = 4$.

Solution

a. Replace y by **7** to get

$$y + 14 = \mathbf{7} + 14 = 21$$

b. Replace x by **5** and z by **4** to get

$$3xz = 3(\mathbf{5})(\mathbf{4}) = 60$$

EXERCISE 3

Evaluate each expression for $a = 0.3$, $b = 1.2$, and $c = 6$.

a. $8.4 - b$ **b.** $\frac{9}{a}$ **c.** $5bc$

ANSWERS TO 1.3B EXERCISES

2a. $0.96

2b. Multiply the bill, B, by 0.08.

2c.

B	10.00	20.00	50.00	120.00
$0.08B$	0.80	1.60	4.00	9.60

3a. 7.2 **b.** 30 **c.** 36

HOMEWORK 1.3B 

1. Uncle Herb is on a diet. If he consumes c calories for lunch, then he can have $1200 - c$ calories for dinner.
   a. If Uncle Herb has 700 calories for lunch, how many calories can he have for dinner?

   b. Explain how to find the number of calories Uncle Herb can have for dinner if he consumes c calories for lunch.

   c. Evaluate the expression $1200 - c$ for the given values of c:

c	200	350	425	515	640	870
$1200 - c$						

2. Yasmin needs $9\frac{1}{2}$ hours per week to study for her math class. If she has t hours of study time available per week, that leaves her $t - 9\frac{1}{2}$ hours for her other classes.
   a. If Yasmin has 15 study hours per week, how many hours does she have for her other classes?

   b. Explain how to find the number of study hours Yasmin has for her other classes if she has t hours of weekly study time.

   c. Evaluate the expression $t - 9\frac{1}{2}$ for the given values of t. (*Hint*: How do you express $9\frac{1}{2}$ as a decimal number? Now use your calculator.)

t	12	16	20	22	25	31
$t - 9\frac{1}{2}$						

3. Quentin has to drive 600 miles. If he drives at r miles per hour, it will take $\frac{600}{r}$ hours to reach his destination.
   a. How long will it take if he drives at 60 miles per hour?

   b. Explain how to find how long the journey will take at r miles per hour.

   c. Evaluate the expression $\frac{600}{r}$ for the given values of r:

r	30	40	45	50	80
$\frac{600}{r}$					

4. Rose and her two sisters, Iris and Violet, plan to sell the family farm and move to Florida. If they make s dollars on the sale, each sister's share will be $\frac{s}{3}$.
   a. How much will Rose get if they make $30,000?

   b. Explain how to find Rose's share of s dollars.

   c. Evaluate the expression $\frac{s}{3}$ for the given values of s:

s	36,000	45,000	51,000	62,700	112,200
$\frac{s}{3}$					

5. Kranberry Kooler is 10% cranberry juice, so the amount of cranberry juice in q quarts of Kranberry Kooler is $0.10q$.

 a. How much cranberry juice is in 15 quarts of Kranberry Kooler?

 b. Explain how to find the number of quarts of cranberry juice needed for q quarts of Kranberry Kooler.

 c. Evaluate the expression $0.10q$ for the given values of q:

q	10	20	45	58	260	1250
$0.10q$						

6. The registrar at City College predicts a 4% enrollment increase for all departments next year. A department that has p students this year can expect $1.04p$ students next year.

 a. If the English department has 5200 students enrolled this year, how many should they expect next year?

 b. Explain how a department with p students this year can predict next year's enrollment.

 c. Evaluate the expression $1.04p$ for the given values of p:

p	600	1300	1700	2200	2600	3500
$1.04p$						

7. Camilla wants to deposit her savings of $2000 into a 1-year treasury bill (T-bill). If the T-bill offers an interest rate of r, then Camilla's savings will earn $2000r$ dollars for the year.

 a. How much will Camilla's savings earn if the T-bill earns 7% interest? (Don't forget to write the percent in decimal form, 0.07.)

 b. Explain how to compute Camilla's interest if the T-bill pays an interest rate of r.

 c. Evaluate the expression $2000r$ for the given values of r:

r	6%	6.5%	8%	9.25%	10%
$2000r$					

8. Human hair grows approximately 0.085 centimeters per day. If Sinead shaves her head, then after t days her hair will be $0.085t$ centimeters long.

 a. How long will Sinead's hair be after 30 days?

 b. Explain how to find the length of Sinead's hair after t days.

 c. Evaluate the expression $0.085t$ for the given values of t:

t	7	60	180	250	360
$0.085t$					

For Problems 9–26, evaluate each expression for $x = 6$, $y = 3$, and $z = 8$.

9. $z - y$

10. $x + x$

11. $9.4 + x$

12. $12.7 - z$

13. $5x$

14. $\frac{1}{2}z$

15. xz

16. $2xyz$

17. $0.4y$

18. xxx

19. $\frac{12}{x}$

20. $\frac{z}{4}$

21. $\frac{y}{9}$

22. $\frac{x}{z}$

23. $\frac{2.4}{x}$

24. $0.8z$

25. $z - 0.8$

26. $\frac{z}{0.8}$

27. Petite-size skirts are hemmed shorter than regular-size skirts in the same style. Write an expression for the length of a petite skirt if the regular skirt has length L inches:

Length of regular skirt	24	26.5	31	34	L
Length of petite skirt	21.5	24	28.5	31.5	

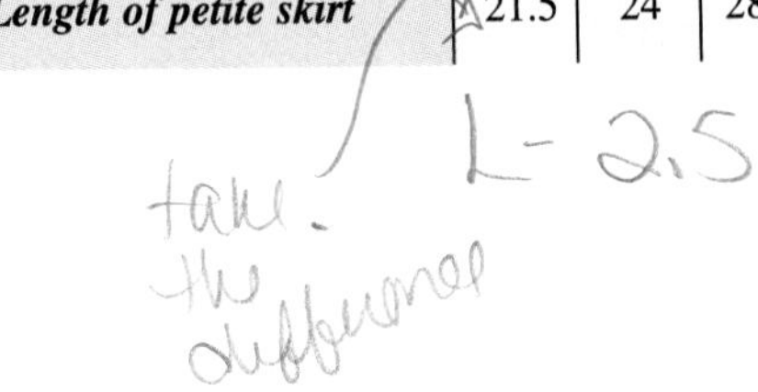

28. If Evyn writes a check for d dollars, write an expression for the amount of money left in her checking account:

Amount of check	20	40	50	70	d
Amount left	80	60	50	30	

29. The decorators for a new hotel are buying carpet. Write an expression for the cost of carpeting a room of area A square feet:

Area of room	100	150	200	360	A
Cost of carpet	800	1200	1600	2880	

30. The engineers in Dana's office are splitting the cost of a wedding gift for their secretary. If m people contribute, write an expression for each person's share:

Number of people	2	3	4	8	m
Each person's share	24	16	12	6	

Building Number Skills

Find each product, using your calculator only when necessary. Remember to convert each percent to a decimal fraction.

31. $\frac{1}{5}(75)$

32. 20% of 35

33. 0.3(20)

34. $\frac{1}{3}(45)$

35. $\frac{3}{4}(24)$

36. 0.58(1000)

37. 80% of 60

38. $\frac{5}{8}(40)$

39. 0.8(74)

40. $\frac{4}{5}(36)$

41. $\frac{1}{4}(39)$

42. 18% of 67

43. 43% of 17

44. 0.7(52)

45. $\frac{3}{8}(95)$

46. $\frac{1}{6}(27)$

1.4 Writing Algebraic Expressions

A. "Translating" English to Algebra

Algebraic expressions can be used to describe many situations involving variables.

EXAMPLE 1  

Choose the correct algebraic expression from the box for each situation described below.

$n + 6$	$n - 6$	$6 - n$
$6n$	$\frac{n}{6}$	$\frac{6}{n}$

a. Rashad is 6 years younger than Shelley. If Shelley is n years old, how old is Rashad?

We should subtract 6 from Shelley's age to get $\quad n - 6$**.**

b. Each package of sodas contains 6 cans. If Antoine bought n packages of sodas, how many cans did he buy?

We should multiply the number of packages by 6 to get $\quad 6n$**.**

c. Lizette and Patrick together own 6 cats. If Lizette owns n cats, how many cats does Patrick own?

We should subtract Lizette's cats from the total to get $\quad 6 - n$**.**

EXERCISE 1

Choose the correct algebraic expression from the box for each situation described below.

$v + 5$	$v - 5$	$5 - v$
$5v$	$\frac{v}{5}$	$\frac{5}{v}$

a. The volume of a standard vat is v liters. What is the total capacity of 5 standard vats?

b. The maximum velocity of a jet ski is v miles per hour. How fast can the jet ski travel against a current, if the speed of the current is 5 miles per hour?

c. Each batch of cookies requires v teaspoons of vanilla. How many batches can be made with 5 teaspoons of vanilla?

The first step in solving problems with algebra is to describe the problem in mathematical terms. This usually involves writing algebraic expressions. We can think of this process in three steps.

Steps for Writing Algebraic Expressions

Step 1 Identify the unknown quantity. Remember that it must have *numerical* values.

Step 2 Choose a variable to represent the unknown quantity.

Step 3 Translate the English phrase into an algebraic expression, using the variable and appropriate operation symbols.

EXAMPLE 2 

Write an algebraic expression for "6 feet taller than the height of the roof."

Solution

Step 1 **The height of the roof** is unknown.
Step 2 Height of the roof: h
Step 3 "Taller than" suggests addition: $h + 6$

EXERCISE 2
Write an algebraic expression for "two-thirds of the weight of the best-selling laptop."

ANSWERS TO 1.4A EXERCISES
1a. $5v$ liters
1b. $v - 5$ miles per hour
1c. $\frac{5}{v}$ batches
2. $\frac{2}{3}w$, where w is the weight of the best-selling laptop

HOMEWORK 1.4A

Choose the correct algebraic expression from the options given.

$m + 15 \qquad m - 15 \qquad 15 - m$

1. Carol weighs 15 pounds less than Garth. If Garth weighs m pounds, how much does Carol weigh?

2. Amber and Beryl together planted 15 trees. If Amber planted m trees, how many trees did Beryl plant?

3. Fred earned $15 more this week than last week. If he earned m dollars last week, how much did he earn this week?

4. Meg bicycled 15 miles farther than Kwan. If Kwan rode m miles, how far did Meg ride?

5. There are 15 children in Amy's swim class. If there are m girls, how many are boys?

6. The sale price of a sweater is $15 less than the regular price. If the regular price is m dollars, what is the sale price?

Choose the correct algebraic expression from the options given.

$12p \quad \frac{12}{p} \quad \frac{p}{12}$

7. Julian earns $12 an hour. If he works for p hours, how much will he make?

8. Farmer Brown collected p eggs this morning. How many dozen is that?

9. Melissa bought 12 colored markers. If their total cost was p dollars, how much did each marker cost?

10. Rosalind is baby-sitting for p children. If she brings 12 puzzles, how many will each child get?

11. Hector has to read 12 chapters in his history text. If he has p days to complete the assignment, how many chapters should he read per day?

12. Roma swims 12 laps per day. After p days, how many laps will she have swum?

Choose the correct algebraic expression from the options given.

$c + 36 \quad c - 36 \quad 36 - c$

$36c \quad \frac{36}{c} \quad \frac{c}{36}$

13. Renee made $36 for c hours of work. How much does she earn per hour?

14. A pair of shorts and a shirt cost $36. If the shorts cost c dollars, how much did the shirt cost?

15. Elba's Scrabble total was 36 points higher than Carla's. If Carla's total was c points, what was Elba's total?

16. Marc's business will use c padded mailing envelopes this month. If there are 36 envelopes in a package, how many packages should he order?

17. Heather had 36 math problems to work. She finished c of them. How many more does she have to work?

18. Takiji is 36 years older than Seiki. If Takiji is c years old, how old is Seiki?

19. If the 36 members of the Ski Club each pay c dollars in dues, how much money will the club raise?

20. If c roommates share the $36 electricity bill, how much should each pay?

Write an algebraic expression for each word phrase. Follow the three steps on page 34.

21. 3 times the cost of a water filter

22. 8 more than the number of students

23. The radius of the circle decreased by 6 centimeters

24. 10 times the height of the triangle

25. $\frac{2}{5}$ of the checking account balance

26. $3\frac{1}{4}$ inches taller than last year's height

27. The price of the pizza divided by 4

28. \$30 more than a bus ticket

29. The perimeter of the triangle diminished by 15 feet

30. The quotient of the volume of the sphere and 8

Building Number Skills

A fraction bar is actually a division symbol. Thus, $\frac{3}{4}$ means $3 \div 4$. Use your calculator to verify that $\frac{3}{4} = 0.75$.

Use your calculator to convert each fraction to decimal form.

31. $\frac{1}{4}$ **32.** $\frac{1}{2}$ **33.** $\frac{5}{8}$ **34.** $\frac{3}{8}$ **35.** $\frac{2}{5}$

36. $\frac{4}{5}$ **37.** $\frac{7}{20}$ **38.** $\frac{13}{20}$ **39.** $\frac{5}{16}$ **40.** $\frac{9}{16}$

A mixed number like $5\frac{1}{4}$ means $5 + \frac{1}{4}$. To write $5\frac{1}{4}$ as a decimal, first change $\frac{1}{4}$ to a decimal, then add 5. On your calculator, key in

1 [÷] 4 [+] 5 [=]

to get 5.25.

Change each mixed number to decimal form. (See Appendix A.4 to review mixed numbers.)

41. $6\frac{1}{4}$ **42.** $12\frac{1}{2}$ **43.** $4\frac{1}{8}$ **44.** $2\frac{3}{8}$ **45.** $7\frac{2}{5}$

46. $9\frac{4}{5}$ **47.** $15\frac{7}{16}$ **48.** $1\frac{3}{4}$ **49.** $2\frac{9}{20}$ **50.** $5\frac{3}{5}$

B. Some Common Algebraic Expressions

There are several useful variable relationships that deserve special attention.

1. The **distance** traveled by an object moving at a constant speed r for a time t is given by $\boldsymbol{rt}$.

EXAMPLE 3

A train travels at 70 miles per hour for 3 hours. How far did the train travel?

Solution

We are looking for a *distance* (how far). We have $r = 70$ and $t = 3$. The distance traveled is

$$\boldsymbol{rt} = (70)(3) = 210 \text{ miles}$$

2. The **time** it takes an object traveling at a constant speed r to travel a distance d is given by $\boldsymbol{\frac{d}{r}}$.

EXAMPLE 4 

Suppose you drive at an average speed of 50 miles per hour on a trip of 250 miles. How long will your trip take?

Solution

We are looking for a *time* (how long). We have $r = 50$ and $d = 250$. The time your trip takes is

$$\frac{\boldsymbol{d}}{\boldsymbol{r}} = \frac{250}{50} = 5 \text{ hours}$$

3. If you invest an amount of money (called the principal) P in an account that pays interest at annual rate r, then the **interest** earned after t years is given by $\boldsymbol{Prt}$.

EXAMPLE 5

Three years ago, you invested \$400 in an account that pays 5% interest. How much interest has your account earned?

Solution

We are looking for *interest.* We have $P = 400$, $r = 0.05$, and $t = 3$. The interest earned on your account is

$$\boldsymbol{Prt} = (400)(0.05)(3) = \$60$$

4. To find a **percentage** of a whole amount, multiply the percentage rate r times the whole amount W: $\boldsymbol{rW}$.

EXAMPLE 6

Your insurance pays for 80% of your dentist bill. The bill is \$120. How much will your insurance pay?

Solution

We are looking for a *percentage.* We have $r = 0.80$ and $W = 120$. The portion your insurance pays is

$$\boldsymbol{rW} = (0.80)(120) = \$96$$

5. To find the **average** of several values, divide the sum S of the values by the number n of values: $\dfrac{\boldsymbol{S}}{\boldsymbol{n}}$.

EXAMPLE 7

Last week you worked a total of 27 hours on 6 days. What was the average number of hours you worked per day?

Solution

We are looking for an *average*. We have $S = 27$ and $n = 6$. Your average work day is

$$\frac{\boldsymbol{S}}{\boldsymbol{n}} = \frac{27}{6} = 4.5 \text{ hours long}$$

Use the algebraic expressions discussed above to answer the questions in exercises 3–7 on the next page. First decide which expression is appropriate for the situation described in the problem.

EXERCISE 3
How far can a hiker travel in 12 hours if she moves at 6 kilometers per hour?

EXERCISE 4
How long will it take a hiker to travel 57 kilometers if she moves at 6 kilometers per hour?

EXERCISE 5
You loan your cousin \$1200, and he agrees to pay you back the full amount plus 9% annual interest after 5 years. How much interest will he pay?

EXERCISE 6
Delbert must make a down payment of 20% of the house price in order to qualify for a mortgage. If the price of the house is \$257,000, what is Delbert's down payment?

EXERCISE 7
Percy squeezes 9 ounces of lime juice from five limes. What is the average amount of juice per lime?

Writing Algebraic Expressions

The five common algebraic expressions discussed above can be helpful when we are working with variables. For the situations in Example 8 and Exercise 8, first decide which of the five expressions is appropriate.

EXAMPLE 8 

a. Write an expression for the interest earned in 5 years by an amount of money invested in an account that pays 7% annual interest:

Use the expression for interest, with $r = 0.07$ and $t = 5$:

$$P(0.07)(5) \quad \text{or} \quad 0.35P$$

b. How much interest will be earned if \$500 is invested? If \$3000 is invested?

Evaluate the expression for the given values of P:

If $P = 500$, the interest is $0.35(500) = 175$.
If $P = 3000$, the interest is $0.35(3000) = 1050$.

EXERCISE 8
a. Delbert's French class has taken 12 quizzes this semester. Write an expression for Delbert's quiz average.

b. If Delbert has earned 996 quiz points, what is his quiz average?

In Example 9, think about which operation (addition, subtraction, multiplication, or division) is useful for the problem.

EXAMPLE 9

a. The cost of the conference was $2000 over budget. Choose a variable for the budget, and then write an expression for the cost of the conference:

Let b stand for the amount of the budget.

$$\textbf{Cost of conference} = b + 2000$$

b. If the budget was $9000, what was the cost of the conference?

Evaluate $b + 2000$ for $b = 9000$:

$$b + 2000 = 9000 + 2000 = 11{,}000$$

The cost of the conference was $11,000.

EXERCISE 9

a. A bag of chocolate chip cookies weighs 1 pound. Choose a variable for the price of the bag and then write an expression for the cost per ounce of the cookies. (*Hint*: How many ounces are in a pound?)

b. If Hector pays $5.60 for one bag of cookies, how much is he paying per ounce?

ANSWERS TO 1.4B EXERCISES

3. 72 kilometers

4. $9\frac{1}{2}$ hours

5. $540

6. $51,400

7. $\frac{9}{5}$ ounces = 1.8 ounces

8. **a.** $\frac{S}{12}$ **b.** 83

9. **a.** $\frac{p}{16}$ **b.** $0.35

HOMEWORK 1.4B

For Problems 1–16, use the common algebraic expressions to (a) write a variable expression and (b) evaluate your expression.

1. a. Write an expression for the distance traveled in t hours by a small plane flying at 150 miles per hour.

b. How far will the plane fly in 3 hours? In 8 hours? In $4\frac{1}{2}$ hours?
(*Hint*: Write $4\frac{1}{2}$ as a decimal number.)

2. a. Write an expression for the distance traveled in t seconds by a car moving at 88 feet per second.

b. How far will the car travel in 5 seconds? In $\frac{1}{2}$ second? In $\frac{1}{2}$ minute?

(*Hint*: How many seconds are there in a minute? In half a minute?)

3. a. Write an expression for the time it takes to travel from Salt Lake City to Kansas City (approximately 1000 miles) at different speeds.

b. How long would it take to travel from Salt Lake City to Kansas City by car at 50 miles per hour? By plane at 400 miles per hour? By wagon train at 8 miles per hour?

4. a. Write an expression for the time it takes for the different members of the track team to run 800 meters at different speeds.

b. How long would it take to run 800 meters at a speed of 5 meters per second? At 6.4 meters per second? At 8 meters per second?

5. a. Write an expression for the amount of interest earned by $1600 deposited in an account that pays 4% annual interest rate. (Don't forget to change 4% to a decimal.)

b. How much interest will the account earn after 1 year? After 2 years? After 5 years?

6. a. Suppose your credit union loans you $3000 to be repaid with 4% annual interest when you finish school. Write an expression for the amount of interest you will owe.

b. How much interest will you owe if you finish school in 2 years? In 4 years? In 7 years?

7. a. Milton's great-aunt plans to put $6000 in a trust fund for Milton until he turns 21 years old, 3 years from now. She has a choice of several different accounts. Write an expression in terms of r for the amount of interest the money will earn in 3 years.

b. How much interest will the $6000 earn in an account that pays 5% interest? $8\frac{1}{4}$% interest? 12% interest? (*Hint*: For help changing these percents to decimals, see Appendix A.9.)

8. a. Write an expression in terms of r for the amount of interest earned by $1600 deposited in an interest-bearing account for 1 year.

b. How much interest will $1600 earn after 1 year in an account that pays 3% interest? $5\frac{1}{2}$% interest? 6.3% interest?

9. a. Write an expression for the amount of grape juice in a fruit punch that is 20% grape juice.

b. How much grape juice is in 3 quarts of fruit punch? In 14 quarts? In 60 quarts?

10. a. Eggnog is 70% milk. Write an expression for the amount of milk in a container of eggnog.

b. How much milk is in a 6-ounce glass of eggnog? In a 1-quart carton? In a 3-gallon bowl?

11. a. Laureen took 12 quizzes in her math class this semester and earned a total of S points. Write an expression for Laureen's quiz average.

b. What is Laureen's average if she earned 96 points? If she earned 78 points? If she earned 111 points?

12. a. Errol has saved $1200 for his vacation this year. If he goes on vacation for n days, write an expression for the average amount he can spend each day.

b. How much can Errol spend each day on average if he takes a 10-day vacation? A 15-day vacation? A 20-day vacation?

13. a. Amanda's bread recipe calls for 3 cups of flour more than she has. Choose a variable for the amount of flour Amanda has, then write an expression for the amount of flour required.

$f + 3$

b. If Amanda has 8 cups of flour, how much flour does the recipe call for?

14. a. Garth received 432 fewer votes than his opponent in the election. Choose a variable for the number of votes that Garth's opponent received and then write an expression for the number of votes Garth received.

b. If Garth's opponent received 3297 votes, how many votes did Garth receive?

15. a. To find his homework grade, Jamal should divide his total points by 600. Choose a variable for Jamal's total points and then write an expression for Jamal's homework grade.

b. If Jamal earned 480 homework points, what is his homework grade?

16. a. Each member of the Forensics Club pays $5 in dues. Choose a variable for the number of members, then write an expression for the amount the club earns in dues.

b. If the Forensics Club has 38 members, how much does it earn in dues?

17. **a.** The bar graph in Figure 1.15 shows the speed of rush-hour traffic in each city. Fill in the first column of the table.

Figure 1.15 Rush-hour traffic speed

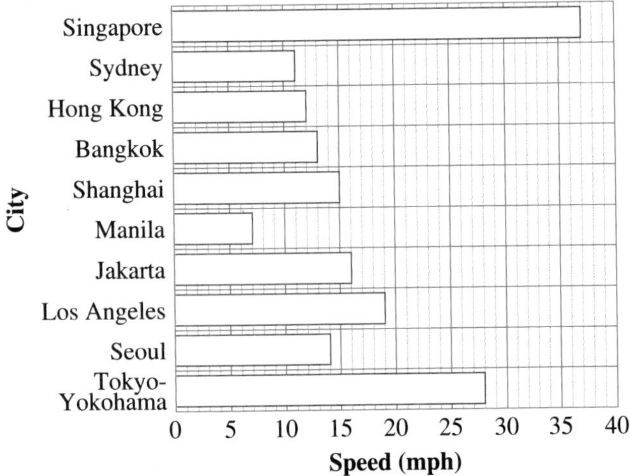

b. If you commuted 10 miles to work each day, how long would it take you to get home at rush hour in each city? Complete the table.

City	*Rush hour traffic flow (mph)*	*Commute time (h)*
Singapore		
Sydney		
Hong Kong		
Bangkok		
Shanghai		
Manila		
Jakarta		
Los Angeles		
Seoul		
Tokyo-Yokohama		

c. Which city has the slowest traffic flow at rush hour? Which city has double the slowest speed?

d. In which city is the rush hour speed $\frac{4}{5}$ of the speed in Shanghai? In which city is the speed 175% of the speed in Jakarta?

18. **a.** The table shows the annual rainfall in Los Angeles for the decade from 1987 to 1996. Use the grid in Figure 1.16 to make a bar graph for the annual rainfall in Los Angeles.

Year	*Rainfall (in.)*
1987	9.11
1988	11.57
1989	4.56
1990	6.49
1991	15.07
1992	22.65
1993	23.44
1994	8.69
1995	24.06
1996	17.75

(*Source*: National Weather Service, WeatherData Inc.)

Figure 1.16 Los Angeles annual rainfall

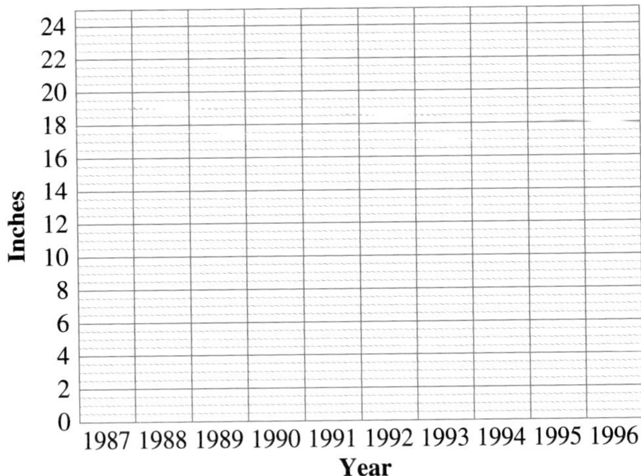

b. In which year did Los Angeles receive the most rain? Which year was the driest?

c. In which three consecutive years was the total rainfall the smallest? In which three-year interval was rainfall the greatest?

d. In which year did the rainfall decrease the most from the previous year?

e. Calculate the average rainfall per year for the decade shown. How does that average compare to 15 inches, the average annual rainfall for all years since 1878?

Building Number Skills

Some fractions do not come out evenly in decimal form. For example,

$$\frac{1}{3} = 1 \div 3 = 0.333\ldots$$

where the 3s continue forever. We use a bar over the repeated digit (or digits) to denote this, so we write $\frac{1}{3} = 0.\overline{3}$.

Use your calculator to convert each fraction to decimal form and use the repeater bar to write your answer. (See Appendix A.5 to review the decimal form of a fraction.)

Example: $\frac{5}{12} = 0.41666\ldots = 0.41\overline{6}$

19. $\frac{2}{3}$ **20.** $\frac{1}{6}$ **21.** $\frac{5}{6}$ **22.** $\frac{5}{9}$ **23.** $\frac{7}{9}$

24. $\frac{3}{11}$ **25.** $\frac{8}{11}$ **26.** $\frac{7}{12}$ **27.** $\frac{1}{12}$ **28.** $\frac{7}{15}$

1.5 Equations

In arithmetic an equal sign usually tells you to perform a calculation. If you were given a problem like

$$5 \times 12 = \_\_\_\_$$

you would respond by filling in the blank: $5 \times 12 = 60$. In this problem the equal sign is like a command to perform some mathematical task; it prompts you to give an answer. An equal sign can also be used to show the result of a simplification. It says: Here is a simpler form of the same expression. An example of this use of the equal sign is

$$8 + 15 + 3 + 24 = 50$$

In algebra, we often use the equal sign in a completely different way, to show a relationship between two algebraic expressions. When we write

Delbert's age = Francine's age + 6

we are describing the relationship between Delbert's age and Francine's age. This description is a mathematical statement that two algebraic expressions are equal. Such a statement is called an **equation.** We can also use variables to write equations. For example, we might write

$$D = F + 6$$

where D and F stand for Delbert's age and Francine's age. Finding relationships between variables and describing them as equations is an important algebraic skill.

EXERCISE 1

The value-added tax (VAT) in England is 17.5% of the price. Fill in the following table and answer the questions.

Price (£)	6	10	14	
VAT (£)	1.05			5.60

Price (£)	92		p
VAT (£)		24.85	$0.175p$

a. Explain how to find the VAT when you know the price before tax.

b. Explain how to find the price before tax when you know the VAT.

c. Let V stand for the VAT. Write an equation for V in terms of p.

d. How can you find V if you know p?

e. How can you find p if you know V?

EXERCISE 2

a. Describe the relationship between the two variables shown in the table.

t	0.5	1	2	3
d	15	30	60	90

b. Write an equation that expresses the second variable, d, in terms of the first variable, t.

EXAMPLE 1

Tamir and his three roommates split their weekly grocery bill equally. Fill in the following table and answer the questions.

Grocery bill	120	140	164	**192**	220	**236**	g
Tamir's share	30	**35**	**41**	48	**55**	59	$\frac{g}{4}$

a. Explain how to find Tamir's share when you know the grocery bill.

Divide the grocery bill by 4.

b. Explain how to find the grocery bill when you know Tamir's share.

Multiply Tamir's share by 4.

c. Let T stand for Tamir's share. Write an equation for T in terms of g:

$$T = \frac{g}{4}$$

d. How can you find T if you know g?

Divide g by 4.

e. How can you find g if you know T?

Multiply T by 4.

EXAMPLE 2

a. Describe the relationship between the two variables shown in the table.

w	4	7	10	12
g	9	12	15	17

g is 5 more than w.

b. Write an equation that expresses the second variable g, in terms of the first variable, w. (In other words, write an equation starting with $g = \ldots$.)

$$g = w + 5$$

ANSWERS TO 1.5 EXERCISES

1.

Price (£)	6	10	14	**32**	92	**142**	p
VAT (£)	1.05	**1.75**	**2.45**	5.60	**16.10**	24.85	$0.175p$

1a. Multiply the price by 0.175. **1b.** Divide the VAT by 0.175.

1c. $V = 0.175p$ **1d.** Multiply p by 0.175. **1e.** Divide V by 0.175.

2a. d is 30 times t. **2b.** $d = 30t$

HOMEWORK 1.5 

Fill in each table and answer the questions.

1. Rachel is 8 years older than Simone.

Simone's age	6	10	15	18	30		s
Rachel's age	14	18	23	26	38	32	

a. Explain how to find Rachel's age when you know Simone's age.

Subtract

b. Explain how to find Simone's age when you know Rachel's age.

c. Let r stand for Rachel's age. Write an equation for r in terms of Simone's age, s:

$r =$

R = S+8

d. How can you find r if you know s?

add

e. How can you find s if you know r?

2. Farnaz deposits one-third of her paycheck into her savings account.

Paycheck	30	60	120		240		p
Deposit	10			70		100	

a. Explain how to find Farnaz's deposit when you know her paycheck.

b. Explain how to find Farnaz's paycheck when you know her deposit.

c. Let d stand for Farnaz's deposit. Write an equation for d in terms of her paycheck, p:

$d =$

d. How can you find d if you know p?

e. How can you find p if you know d?

3. Delbert's checking account has an outstanding check (not yet received by the bank) for \$76. Consequently, his actual balance is \$76 less than the balance shown on his bank statement.

Bank statement	100	138	188		332		b
Actual balance	24			200		276	

a. Explain how to find the actual balance from the bank statement.

b. Explain how to find the bank's figure for the balance if you know the actual balance.

c. Let a stand for the actual balance. Write an equation for a in terms of the bank statement, b:

$a =$

d. How can you find a if you know b?

e. How can you find b if you know a?

4. Francine is traveling on an express train at 65 miles per hour.

Hours traveled	1	2	3.5		6		t
Miles traveled	65			325		650	

a. Explain how to find the distance Francine traveled if you know how many hours she traveled.

b. Explain how to find the number of hours Francine traveled if you know how many miles she traveled.

c. Let d stand for the distance Francine traveled. Write an equation for d in terms of the number of hours she traveled, t:

$d =$

d. How can you find d if you know t?

e. How can you find t if you know d?

For each table (a) describe the relationship between the variables and (b) write an equation that expresses the second variable in terms of the first variable.

5.

r	q
25	17
21	13
16	8
10	2

6.

v	b
4	36
6	54
8	72
10	90

7.

h	m
6	2
12	4
21	7
33	11

8.

s	t
0	4
8	12
12	16
16	20

9.

c	k
1.2	1.8
1.4	2.0
2.3	2.9
2.6	3.2

10.

x	z
17.5	15
16	13.5
12.7	10.2
9.3	6.8

11.

a	d
20	15
12	9
8	6
4	3

12.

u	s
6	4
15	10
27	18
36	24

Choose the equation that best describes each situation. In each case *p* represents the unknown quantity.

$p + 8 = 40 \qquad p - 8 = 40 \qquad 8p = 40$

13. 8 times a number is 40. What is the number?

14. A number decreased by 8 is 40. What is the number?

15. 8 less than a number is 40. What is the number?

16. 40 exceeds a number by 8. What is the number?

17. Al scored 8 points more than Katy scored. If Al scored 40 points, how many points did Katy score?

18. Raul got a discount of $8 on his calculator and paid $40. How much is the regular price?

19. Eight tickets to the rodeo cost $40. How much does one ticket cost?

20. If Jaye's retriever puppy gains 8 pounds, it will weigh 40 pounds. How much does the puppy weigh now?

Choose the equation that best describes each situation. In each case *b* represents the unknown quantity.

$$b - 18 = 10 \qquad 18 - b = 10$$

21. 18 less than a number is 10. What is the number?

22. 18 reduced by a number is 10. What is the number?

23. The difference of 18 and a number is 10. What is the number?

24. The temperature dropped 18° over night, and now it is 10°. What was the temperature yesterday?

25. Negin was enrolled in 18 units, but she dropped some classes and now she has 10 units. How many units did she drop?

26. Dorrie bought an $18 video on sale for $10. What was the discount?

Choose the equation that best describes each situation. In each case *t* represents the unknown quantity.

$$16t = 80 \qquad \frac{t}{16} = 80$$

27. How many pounds does an 80-ounce box of laundry soap weigh?

28. How many ounces does an 80-pound child weigh?

29. If Sarmila works 16 hours a week, how many weeks will it take her to complete 80 hours of cooperative education?

30. Each member of the Friends of the Youth Symphony must raise $80 to keep the symphony out of debt. If there are 16 members, how much money do they need?

Choose the equation that best describes each situation. In each case *n* represents the unknown quantity.

$n + 5 = 30$	$n - 5 = 30$
$5n = 30$	$\frac{n}{5} = 30$

31. 5 less than a number is 30.

32. The quotient of a number and 5 is 30.

33. The product of a number and 5 is 30.

34. 5 more than a number is 30.

35. The price of a concert ticket increased $5 this year and is now $30. How much did a ticket cost last year?

36. Amir spent $5 and now has $30. How much did he have before he spent $5?

37. Marty jogged the same course 5 days this week for a total of 30 miles. How far did he jog each day?

38. Five brothers split the cost of a new TV, each paying $30. How much was the TV?

Building Number Skills

Fill in the table with equivalent values. You should memorize these conversions!

Fraction	*Decimal*	*Percent (%)*	*Fraction*	*Decimal*	*Percent (%)*
$\frac{1}{2}$			$\frac{3}{5}$		
$\frac{1}{3}$			$\frac{4}{5}$		
$\frac{2}{3}$			$\frac{1}{8}$		
$\frac{1}{4}$			$\frac{3}{8}$		
$\frac{3}{4}$			$\frac{5}{8}$		
$\frac{1}{5}$			$\frac{7}{8}$		
$\frac{2}{5}$			1		

Without using your calculator or the table above, give the decimal form of each common fraction and the common fraction for each decimal.

39. 0.125

40. $\frac{3}{5}$

41. $\frac{3}{8}$

42. 0.75

43. $0.\overline{6}$

44. $\frac{1}{4}$

45. $\frac{2}{5}$

46. 0.8

47. $\frac{1}{3}$

48. 0.375

49. 0.625

50. $\frac{7}{8}$

1 Summary

Lesson 1.1

A **variable** is a numerical quantity that can take on different values at different times or in different situations.

Quantities that do not change are called **constants.**

The values of two variables may be related by a mathematical sentence.

A **bar graph** displays the values of a variable. The height of the bar illustrates the value at each time or in each situation.

Lesson 1.2

We often use a single letter to represent a variable quantity.

When we use operation symbols to combine numbers and variables, we create **algebraic expressions.**

When we add two numbers or variables together, the result is called the **sum,** and the things added together are called **terms.**

When we multiply two numbers or variables together, the result is called the **product,** and the items multiplied together are called **factors.**

In algebra we use a raised dot or parentheses instead of a cross to indicate multiplication.

When we subtract one number or variable from another, the result is called the **difference.**

When we divide one number or variable by another, the result is called the **quotient.**

The fraction bar and the symbol ÷ both mean "divided by."

A **ratio** also denotes division: $\frac{a}{b}$ means $a \div b$.

In subtraction and division, the order of the numbers makes a difference.

Lesson 1.3

There are usually several different ways to write an algebraic expression in English.

To **evaluate** an algebraic expression, substitute the value of the variable and simplify.

Lesson 1.4

Steps for Writing an Algebraic Expression

Step 1 Identify the unknown quantity. Remember that it must have *numerical* values.

Step 2 Choose a variable to represent the unknown quantity.

Step 3 Translate the English phrase into an algebraic expression, using the variable and appropriate operation symbols.

Some Common Algebraic Expressions

The **distance** traveled by an object moving at a constant speed r for a time t is given by $\boldsymbol{rt}$.

The **time** it takes an object traveling at a constant speed r to travel a distance d is given by $\frac{d}{r}$.

If you invest an amount of money (called the principal) P in an account that pays interest at annual rate r, then the **interest** earned after t years is given by $\boldsymbol{Prt}$.

To find a **percentage** of a whole amount, multiply the percentage rate r times the whole amount W: $\boldsymbol{rW}$.

To find the **average** of several values, divide the sum S of the values by the number n of values: $\frac{S}{n}$.

Lesson 1.5

An **equation** is a statement that two algebraic expressions are equal.

An equation can be used to describe the relationship between two variables.

CHAPTER 1 REVIEW 

For Problems 1–12, match each word with its mathematical meaning.

variable	sum	difference	equation
product	factors	quotient	terms
algebraic expression	constant	evaluate	ratio

1. The result of multiplying two expressions together

2. A numerical quantity that does not change

3. Expressions that are multiplied together

4. A meaningful combination of variables, numbers, and operation symbols

5. Another word for quotient

6. The result of subtracting one expression from another

7. A numerical quantity that can take on different values

   variable

8. The result of adding two expressions together

9. To substitute specific values for the variables

10. The result of dividing one expression by another

11. Expressions that are added together

    terms

12. A statement that two algebraic expressions are equal

For Problems 13–16, write algebraic expressions for each situation.

13. Chelsea has studied dance for 3 years longer than Brianna.
    a. Explain in words how to find the number of years Chelsea has studied dance if you know how long Brianna has studied dance.

    b. Choose a variable and write an expression for the number of years Chelsea has studied dance.

14. The tuition for in-state students is $\frac{2}{5}$ of the tuition for out-of-state students.
    a. Explain in words how to find the tuition for in-state students if you know the tuition for out-of-state students.

    b. Choose a variable and write an expression for in-state tuition.

15. Vanda deposits \$500 from her paycheck into her savings account every month.
    a. Explain in words how to find the amount Vanda has left for living expenses every month.

    b. Choose a variable and write an expression for the amount Vanda has for monthly living expenses.

16. Zsa-Zsa has started 200 hours of community service work.
    a. Explain in words how to find the number of hours Zsa-Zsa has left to do.

    b. Choose a variable and write an expression for the number of hours of community service Zsa-Zsa has left.

17. The bar graph in Figure 1.17 shows the high and low temperatures at Fresno Air Terminal for the week of July 4, 1994.

    a. On which day was the lowest high temperature recorded? What was that temperature?

    b. On which day was the highest low temperature recorded? What was that temperature?

    c. On which day did the high temperature increase the most from the previous day? How much did it increase?

    d. On which day was the difference between the high and the low temperatures the greatest? What was that difference?

Figure 1.17 High and low temperatures in Fresno

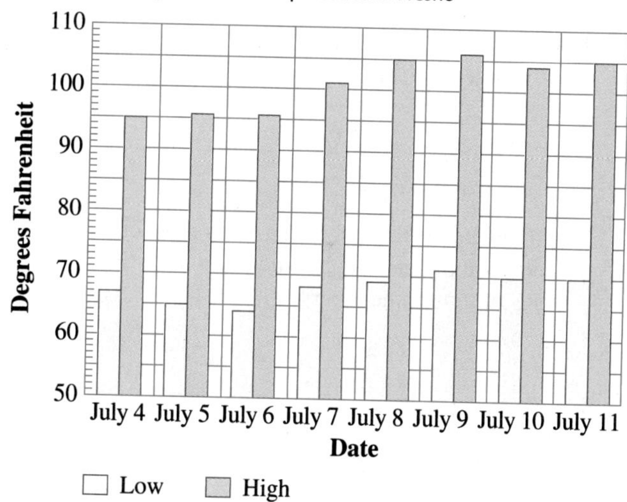

18. Codes have been used throughout history to guard secrets. Today they are also used to secure financial transactions on the Internet. The simplest type of code, or cipher, substitutes a different symbol for each letter of the alphabet. Here is a simple substitution cipher:

Original letter: A B C D E F G H I J K L M N O P Q R S T U V W X Y Z
Code letter: k l m n o p q r s t u v w x y z a b c d e f g h i j

a. Do you see the pattern for this substitution cipher? Decode the following message written in the cipher:

`wkdr sc zygob`

Some letters of the alphabet occur more frequently than others in written English. By searching for repeated letters, cryptanalysts (code breakers) use *frequency analysis* to break codes based on substitution ciphers. The bar graph in Figure 1.18 shows the frequencies, in percent, of the nine most common letters in English.

Figure 1.18 The most common English letters

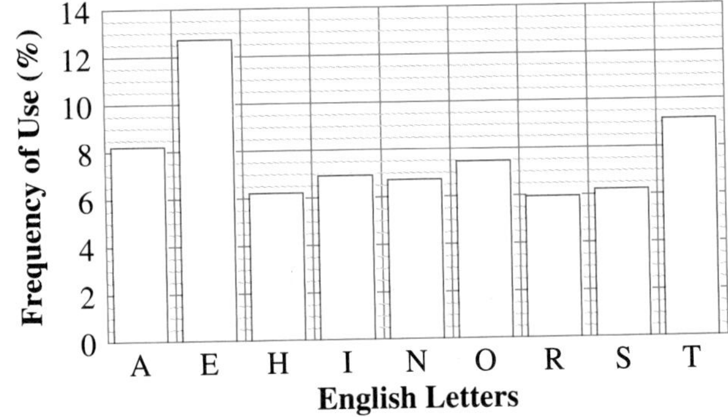

b. Which letter occurs most frequently in English? Which letter is next?

c. Complete the table showing the frequencies of the letters in the bar graph. Find the sum of the frequencies of those nine letters.

d. What is the average frequency for each of the remaining 17 letters in the alphabet?

Letter	*Frequency (%)*
A	
E	
H	
I	
N	
O	
R	
S	
T	

Evaluate each expression for the given values of the variables.

19. $1.5a$, for $a = 2.8$

20. $\frac{b}{5}$, for $b = 8$

21. $\frac{6.4}{c}$, for $c = 1.6$

22. $0.2 - d$, for $d = 0.05$

23. $2pq$, for $p = 3$, $q = 14$

24. $\frac{K}{M}$, for $K = 60$, $M = 45$

State the variable expression we studied for each quantity. State what each variable stands for.

25. The distance traveled by a moving object:
distance =

26. The time needed to travel a given distance:
time =

27. The interest earned on an account:
interest =

28. The percentage of a whole amount:
percentage =

Use the variable expressions in Problems 25–28 to answer each question.

29. About 22% of the land area of North America is considered arable (suitable for agriculture). If North America has an area of 9,400,000 square miles, how much land is arable?

30. Hamilton deposited \$800 in an account that pays 5% annual interest. How much interest will his account earn in two years?

31. How long will it take Luigi to roller-blade once around Balboa Park, a distance of 12 miles, if he skates at 8 miles per hour?

32. The fastest moving insects are large tropical cockroaches that have been clocked at 48 inches per second. How many inches per hour is that? How many miles per hour is that? (There are 5280 feet in a mile.)

For problems 33–36, (a) Write an English phrase describing the unknown quantity and (b) choose the correct equation to describe the situation.

$50 - n = 10$	$\frac{n}{50} = 10$
$n - 50 = 10$	$50n = 10$

33. If 50 glasses of wine use up 10 bottles, what fraction of a bottle is in each glass?

34. There are 50 seats on the tour bus; 10 are empty. How many people are on the bus?

35. Dabney raised the rent on each apartment in his building by \$10 to pay for his property tax increase. If there are 50 units in his building, what was the tax increase?

36. Carl spent \$50 at the hardware store and has \$10 left. How much did Carl have before he went to the hardware store?

For each table, write an equation that expresses the second variable in terms of the first variable.

37.

d	v
0.2	0.6
0.7	1.1
1.1	1.5
2.0	2.4

38.

t	s
4	6
10	15
16	24
25	37.5

39.

x	y
3	5.5
5.5	8
8	10.5
10	12.5

40.

H	K
4	16
7	13
12	8
15	5

Building Number Skills

Convert each percent to a decimal fraction.

41. 130% **42.** 7% **43.** 29% **44.** 11%

Compute each fraction by first writing it as a division.

45. $\frac{1}{3}$ of 36 **46.** $\frac{1}{5}$ of 45 **47.** $\frac{1}{8}$ of 48 **48.** $\frac{1}{9}$ of 27

Find each product.

49. $\frac{2}{5}(25)$ **50.** $\frac{5}{6}(42)$ **51.** $\frac{7}{4}(20)$ **52.** $\frac{5}{3}(24)$

Find each percent.

53. 125% of 60 **54.** 83% of 27 **55.** 6% of 39 **56.** 13% of 13

Convert each mixed number to decimal form.

57. $5\frac{3}{4}$ **58.** $8\frac{1}{16}$ **59.** $7\frac{3}{5}$ **60.** $1\frac{5}{8}$

Convert each fraction or mixed number to decimal form.

61. $\frac{2}{11}$ **62.** $\frac{4}{9}$ **63.** $4\frac{1}{6}$ **64.** $2\frac{5}{12}$

CHAPTER

2

Equations

2.1 Solving Equations

A. What Is a Solution of an Equation?

An **equation** is a statement that two expressions are equal. It can be true or false. For example, the equation

$$3 + 4 = 7$$

is true, but the equation

$$16 - 10 = 2$$

is false. If an equation contains a variable it can be true *or* false, depending on the value of the variable. The equation

$$x + 5 = 9$$

is true if the value of x is 4, but false if x has any other value.

A **solution** of an equation is a value of the variable that makes the equation true. For example, the solution of the equation $x + 5 = 9$ is 4 because $4 + 5 = 9$.

EXAMPLE 1 

Decide whether 28 is a solution of the equation $6z = 178$

Solution

Substitute 28 for z in the equation.

$$\mathbf{6(28) = 178\ ?}$$

Simplify both sides of the equation to see if a true statement results.

$$\mathbf{168 = 178} \qquad \textbf{(False!)}$$

The statement is *false*, so 28 is *not* a solution of the equation $6z = 128$.

EXERCISE 1
Decide whether 221 is a solution of the equation $\frac{w}{17} = 13$

EXERCISE 2
Decide whether 30.1 is a solution of the equation $r + 12.5 = 17.6$

EXAMPLE 2 

Decide whether 18.2 is a solution of the equation $b - 13.5 = 4.7$

Solution
Substitute 18.2 for b in the equation.

$$\mathbf{18.2 - 13.5 = 4.7 \ ?}$$

Simplify both sides of the equation to see if a true statement results.

$$\mathbf{4.7 = 4.7} \quad \textbf{(True)}$$

The statement is *true,* so 18.2 *is* a solution of the equation $b - 13.5 = 4.7$

ANSWERS TO 2.1A EXERCISES

1. Yes
2. No

HOMEWORK 2.1A 

1. What is an equation? (Add this definition to your glossary.)

2. What is a solution of an equation? (Add this definition to your glossary.)

Which of the following are equations? Explain why or why not.

3. $8 + 6 = 14$
4. $9 - 5 = 4$
5. $13 - 7 = 2$
6. $2 + 10 = 8$
7. $25 = 12 + 4 + 9$
8. $18 = 4 + 5 + 9$
9. $9 + 7 = 12 + 4$
10. $20 - 2 = 12 + 6$

11. $9 + 7 + 6$

12. $12 + 6 - 4$

13. $x - 5 = 9$

14. $11 + x = 17$

15. $24 = 7x$

16. $15 = 6x$

Decide whether the given value is a solution of the equation.

17. $x + 19 = 36$, for $x = 17$

18. $x - 14 = 8$, for $x = 22$

19. $t - 7 = 25$, for $t = 18$

20. $t + 15 = 21$, for $t = 36$

21. $12v = 60$, for $v = 5$

22. $8v = 96$, for $v = 14$

23. $\frac{z}{15} = 3$, for $z = 5$

24. $\frac{z}{18} = 4$, for $z = 72$

25. $3.4 - q = 0.8$, for $q = 1.6$

26. $11.3 - q = 6.9$, for $q = 4.4$

27. $5n = 28.2$, for $n = 23.2$

28. $7n = 61.6$, for $n = 54.6$

(a) Evaluate the expression to fill in the table and, (b) use the table to solve the equation.

29. a.

x	0.6	0.8	1.0	1.2	1.4	1.6
$1.6x$						

b. $1.6x = 1.92$

30. a

y	18	30	42	54	66	78
$\frac{y}{24}$						

b. $\frac{y}{24} = 2.25$

31. a.

z	14	21	28	35	42	49
$87 - z$						

b. $87 - z = 38$

32. a.

w	3.4	3.8	4.2	4.6	5.0	5.4
$w - 0.58$						

b. $w - 0.58 = 3.62$

For part (a) write an algebraic expression. For part (b) write an equation.

33. a. Megan's cat Silky lost 4.5 pounds. If Silky used to weigh w pounds, write an expression for her new weight.

b. Silky's current weight is n pounds. Write an equation for n in terms of w.

c. Fill in the table:

w	20		15
n		13.5	

34. a. Rocio has $20,000 for a down payment on a house. If the house costs p dollars, write an expression for the amount Rocio will need to borrow.

b. Rocio borrowed b dollars to buy the house. Write an equation for b in terms of p.

c. Fill in the table:

p	100,000		180,000
b		130,000	

35. a. In 2001, one British pound was worth $1.60. Write an expression for the value in dollars of p British pounds.

b. Marion received d American dollars in exchange for p British pounds. Write an equation for d in terms of p.

c. Fill in the table:

p	5		20
d		16	

36. a. The gas company charges $0.725 per therm, including taxes. Write an expression for the cost of t therms.

b. Ryan's gas bill was b dollars last month. Write an equation for b in terms of t.

c. Fill in the table:

t	8		20
b		7.25	

Building Number Skills

Convert each percent to a decimal fraction. (See Appendix A.9 to review percents.)

Example: $12\frac{1}{2}\% = 12.5\% = 0.125$ First change $12\frac{1}{2}$ to 12.5.

37. $8\frac{1}{4}\%$ **38.** $6\frac{1}{2}\%$ **39.** $108\frac{3}{5}\%$ **40.** $227\frac{1}{5}\%$ **41.** $\frac{5}{8}\%$

42. $\frac{3}{4}\%$ **43.** $33\frac{1}{3}\%$ **44.** $66\frac{2}{3}\%$ **45.** $10\frac{3}{8}\%$ **46.** $1\frac{7}{8}\%$

B. Finding Solutions

To **solve** an equation means to find its solution (or solutions). How can we find the solution of an equation? Some equations are easy to solve; you can *see* the solution just by thinking about it. For example, the equation $x = 3$ is very easy to solve; its solution is 3.

We can solve more complicated equations by transforming them into simple ones. Of course, we must do this in a way that the new equation has the same solution as the original equation.

Opposite Operations

We will use the fact that each of the four arithmetic operations has an opposite operation. Addition and subtraction are opposite operations. To see this, choose a value for x and fill in the Example column in Activity 1.

Activity 1

		Example
1. Start with any number.	x	________
2. Add 12 to get a new number.	$x + 12$	________
3. Now subtract 12 from the new number.	$x + 12 - 12$	________
4. You should get your original number back again.	x	________

You don't have to use 12 to add and subtract; any other number will work. ________

Multiplication and division are opposite operations. To see this, choose a value for x in Activity 2.

Activity 2

		Example
1. Start with any number.	x	________
2. Multiply by 3 to get a new number.	$3x$	________
3. Now divide the new number by 3.	$\frac{3x}{3}$	________
4. You should get your original number back again.	x	________

Try multiplying and dividing by some other number besides 3 (but not 0). ________

Isolating the Variable

Now let's use opposite operations to solve an equation. Consider the equation

$$x - 6 = 4$$

You can check that the solution to this equation is 10. We would like to transform the equation $x - 6 = 4$ into the simpler equation $x = 10$. We accomplish this by *undoing* any operations performed on x. Undoing the operation will *isolate* the variable on one side of the equal sign, with the solution on the other side.

EXAMPLE 3  

Solve the equation $x - 6 = 4$

Solution

Use the following steps.

Step 1 Ask yourself: What operation has been performed on x?

6 has been subtracted from x.

Step 2 Perform the opposite operation on both sides of the equation. The opposite operation for subtraction is addition.

$$\begin{array}{rcr} x - 6 & = & 4 \\ \mathbf{+6} & & \mathbf{+\ 6} \\ \hline \mathbf{x} & = & \mathbf{10} \end{array}$$

Add 6 to both sides of the equation.

The solution is $x = 10$.

To check that the solution is correct, we substitute our value for x—namely, 10—into the original equation to see whether a true statement results.

Check: Does $\mathbf{10} - 6 = 4$?
Yes. The solution is correct.

You should perform the check as part of the solution process.

To Solve an Equation

Step 1 Ask yourself: Which operation has been performed on the variable?

Step 2 Perform the opposite operation on both sides of the equation.

Step 3 Check your solution.

Here are examples showing how to undo each of the four arithmetic operations. Note carefully how we write down the solution step in each case. Then work Exercise 4 following Example 4.

EXAMPLE 4  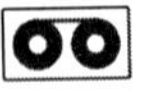

Solve the equation $4x = 32$

Solution

Step 1 What operation has been performed on x?

x has been multiplied by 4.

EXERCISE 3

Solve the equation $a - 3.2 = 7.2$

EXERCISE 4

Solve the equation $12p = 60$

Step 2 Perform the opposite operation on both sides of the equation.

$$\frac{4x}{\mathbf{4}} = \frac{32}{\mathbf{4}} \quad \text{Divide both sides by 4.}$$
$$\mathbf{x = 8}$$

The solution is $x = 8$.

Check: Does $4(\mathbf{8}) = 32$?
Yes. The solution is correct.

EXAMPLE 5 

Solve the equation $x + 8.3 = 24$

Solution

Step 1 What operation has been performed on x?

8.3 has been added to *x*.

Step 2 Perform the opposite operation on both sides of the equation.

$$\begin{array}{rcl} x + 8.3 & = & 24 \\ \mathbf{-\ 8.3} & & \mathbf{-8.3} \\ \hline \mathbf{x} & = & \mathbf{15.7} \end{array} \quad \text{Subtract 8.3 from both sides.}$$

The solution is $x = 15.7$.

Check: Does $\mathbf{15.7} + 8.3 = 24$?
Yes. The solution is correct.

EXERCISE 5
Solve the equation $m + 1.6 = 5.2$

EXAMPLE 6 

Solve the equation $\frac{x}{6} = 12$

Solution

Step 1 What operation has been performed on x?

***x* has been divided by 6.**

Step 2 Perform the opposite operation on both sides of the equation.

$$\mathbf{6}\left(\frac{x}{6}\right) = (12)\mathbf{6} \quad \text{Multiply both sides by 6.}$$
$$\mathbf{x = 72}$$

The solution is 72.

Check: Does $\frac{\mathbf{72}}{6} = 12$?
Yes. The solution is correct.

EXERCISE 6
Solve the equation $\frac{S}{7} = 12$

ANSWERS TO 2.1B EXERCISES
3. $a = 10.4$
4. $p = 5$
5. $m = 3.6$
6. $S = 84$

Even if you can solve these equations in your head, you should still learn to write down the solution steps. You will need to know a solution method to solve harder equations!

HOMEWORK 2.1B 

1. What does it mean to solve an equation?

2. How can you check whether your solution is correct?

3. Describe a three-step strategy for solving an equation.

4. Name the opposite operation for each of the four arithmetic operations.

Solve each equation. Check your solutions.

5. $b + 8 = 15$

6. $g + 4 = 9$

7. $3 + t = 5$

8. $6 + v = 23$

9. $4 = x + 1$

10. $19 = z + 16$

11. $h + 3.9 = 6.8$

12. $k + 1.7 = 1.9$

13. $w - 7 = 5$

14. $u - 1 = 2$

15. $16 = d - 8$

16. $20 = j - 11$

17. $p - 15 = 15$

18. $q - 9 = 9$

19. $x - 0.05 = 0.8$

20. $z - 0.3 = 0.04$

21. $5g = 30$

22. $8c = 96$

23. $78 = 6h$

24. $48 = 4b$

25. $15p = 75$

26. $12q = 108$

27. $2z = 9.6$

28. $3w = 14.4$

29. $\frac{m}{4} = 7$

30. $\frac{k}{5} = 8$

31. $\frac{x}{3} = 14$

32. $\frac{b}{2} = 19$

33. $13 = \frac{s}{16}$

34. $24 = \frac{c}{9}$

35. $\frac{v}{6} = 1.3$

36. $\frac{a}{7} = 2.5$

37. $d - 16 = 16$

38. $s + 23 = 23$

39. $0 = f - 37$

40. $6z = 0$

41. $\frac{x}{5} = 0$

42. $1 = \frac{c}{25}$

43. $47.3 = 24.8 + v$

44. $100b = 2.74$

45. $68.89 = 8.3n$

46. $0.04 + q = 0.3$

47. $\frac{w}{38} = 1.1$

48. $0.7 = y - 2.7$

Building Number Skills

In each pair of numbers, decide which is larger. (You may want to review place value in Appendix A.5.)

49. 8.28 or 8.3

50. 2.2 or 2.15

51. 4.0090 or 4.0800

52. 3.101 or 3.0110

53. 0.20 or 0.021

54. 0.105 or 0.0150

55. 606.060 or 600.066

56. 110.019 or 110.109

57. 0.0110 or 0.0101

58. 9.0909 or 9.0099

2.2 Area and Perimeter

A. What Are Area and Perimeter?

Perimeter

When we use the word *perimeter* in English, we mean the outer boundaries of a region or space. In mathematics, the **perimeter** of a two-dimensional figure is the *distance* around the border of the figure. The perimeter is actually a length. For example, if you want to put a fence around your front lawn, you need to know its perimeter to figure out how much fence to buy.

Imagine stretching a string tight around the outside of the figure, and then measuring the length of the string. That length is the perimeter of the figure.

If the figure is made up of straight-line segments, we can find the length of each side and then add them up. For example, Figure 2.1 has five sides, and the length of each side is shown on the figure.

Figure 2.1

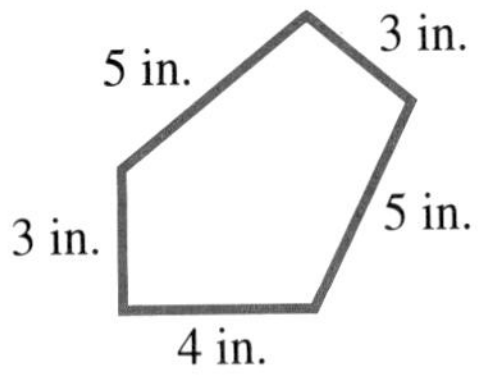

5 in.
3 in.
3 in.
5 in.
4 in.

The perimeter of the figure is

$$3 + 4 + 5 + 3 + 5 = 20 \text{ inches}$$

Area

In English, the word *area* means a region or locality. But in mathematics the **area** of a two-dimensional figure is a measure of the amount of space enclosed by the figure. For example, if you want to plant a lawn in your front yard, you need to know its area to figure out how much grass seed to buy.

We measure area in **square units** such as square inches or square miles. A square inch, for example, is a small square, all of whose sides are 1 inch long. A square inch and a square centimeter are shown in Figure 2.2. The more square units that fit inside a figure, the larger is its area.

Figure 2.2

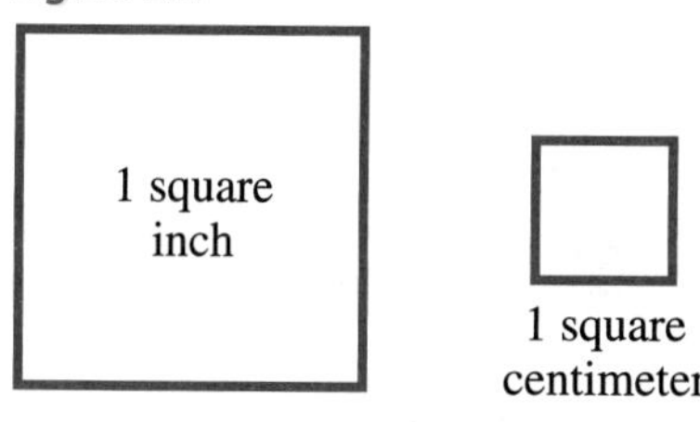

EXAMPLE 1

Find the perimeter and area of the rectangle in Figure 2.3.

Figure 2.3

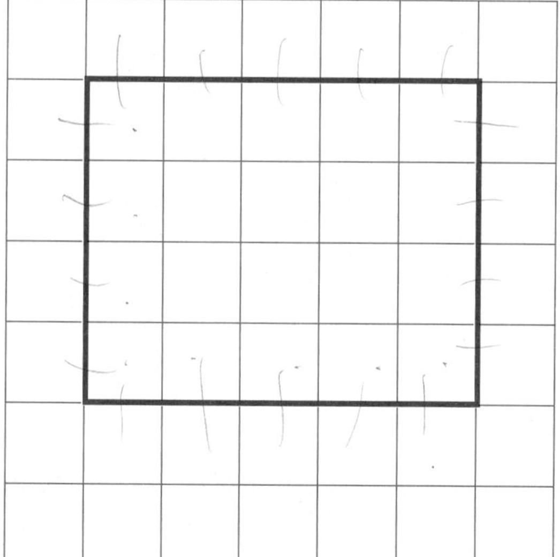

Solution

a. To find the perimeter, we measure the length of each side and add them together. The grid is marked off in centimeters, so the perimeter is

$$4 + 5 + 4 + 5 = 18 \text{ centimeters}$$

b. To find the area, we count the number of squares inside the rectangle. Each square is 1 square centimeter. There are four rows of five squares each, so the area is 20 square centimeters.

EXERCISE 1

Find the perimeter and area of the rectangle in Figure 2.4.

Figure 2.4

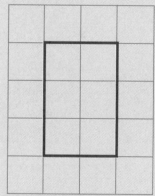

EXERCISE 2

Find the perimeter and area of Figure 2.6.

Figure 2.6

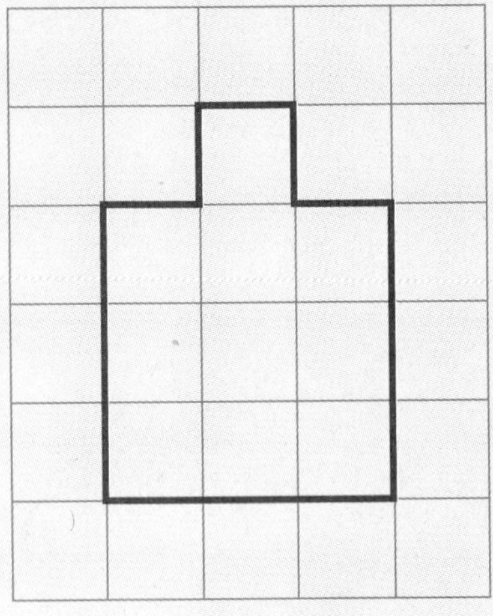

EXAMPLE 2

Find the perimeter and area of Figure 2.5.

Figure 2.5

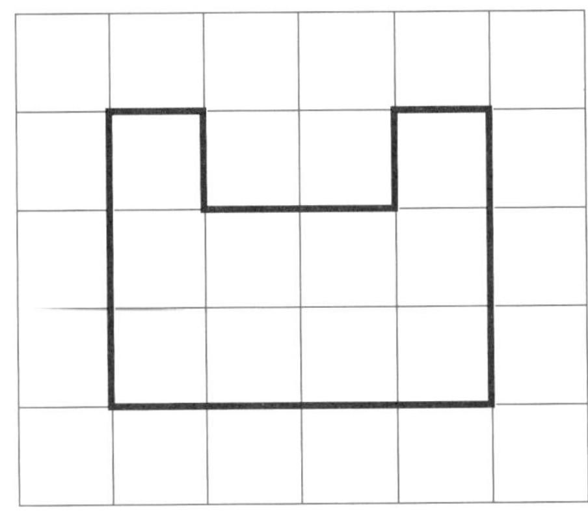

Solution

a. We measure the length of each side and add them to find the perimeter. Starting with the upper left corner and going around the figure clockwise, the sides give

$$1 + 1 + 2 + 1 + 1 + 3 + 4 + 3 = 16 \text{ centimeters}$$

b. There are ten squares inside the figure, so the area is 10 square centimeters.

ANSWERS TO 2.2A EXERCISES

1. Perimeter: 10 centimeters; area: 6 square centimeters
2. Perimeter: 14 centimeters; area: 10 square centimeters

HOMEWORK 2.2A

1. What does the perimeter of a figure measure? Name two typical units for perimeter.

2. What does the area of a figure measure? Name two typical units for area.

Find the perimeter and area of each figure below. The grids are marked off in centimeters.

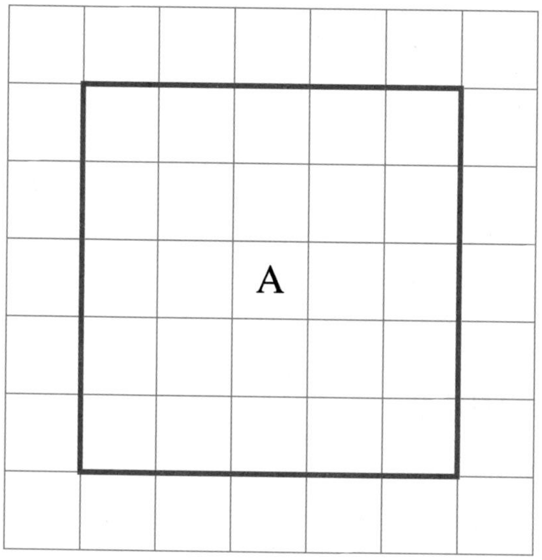

A

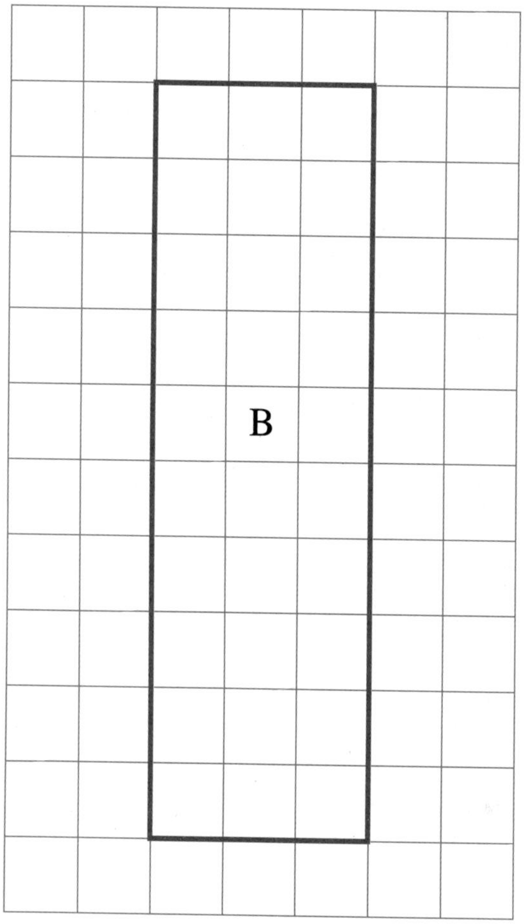

B

3. Figure A:

Perimeter: __________

Area: __________

4. Figure B:

Perimeter: __________

Area: __________

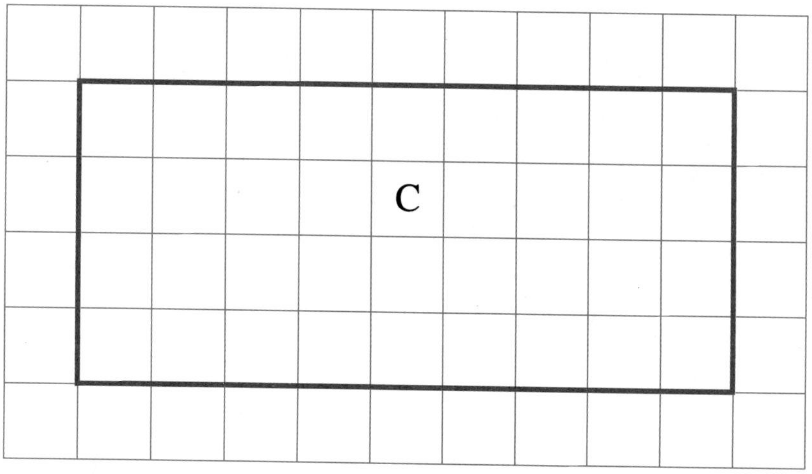

C

5. Figure C:

Perimeter: __________

Area: __________

6. Figure D:

Perimeter: __________

Area: __________

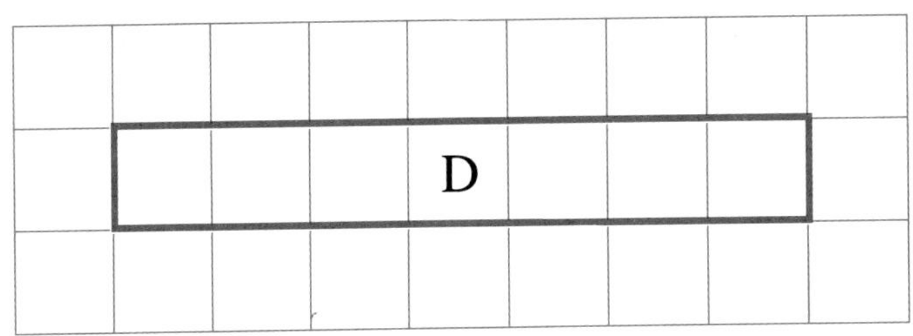

7. Figure E:

Perimeter: __________

Area: __________

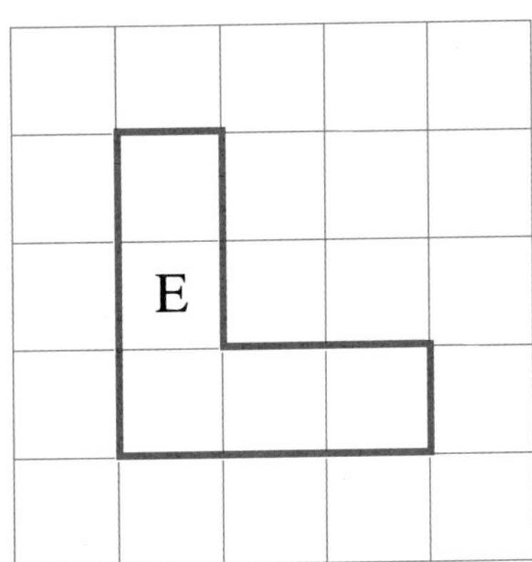

8. Figure F:

Perimeter: __________

Area: __________

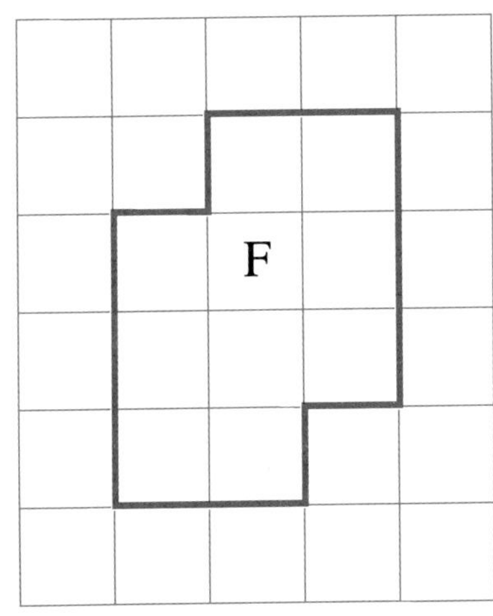

9. Figure G:

Perimeter: __________

Area: __________

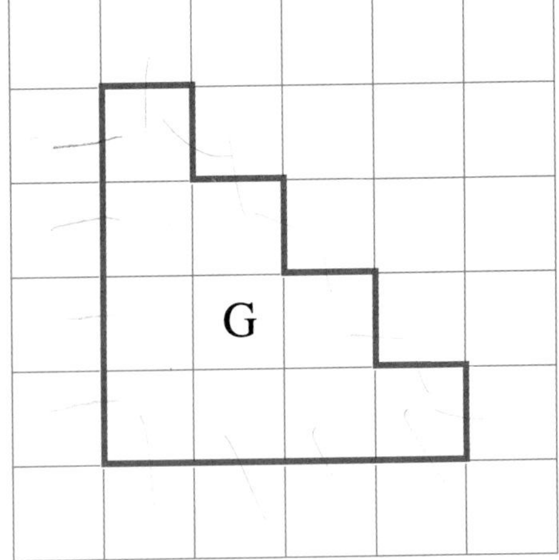

10. Figure H:

Perimeter: __________

Area: __________

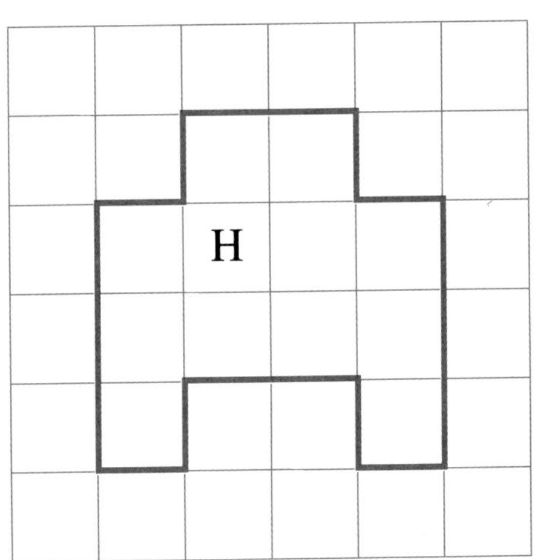

Solve each equation. Check your solutions. (See Lesson 2.1 to review solving equations.)

11. $59 = 32 + m$

12. $91 = 77 + n$

13. $12.3 = a - 9.1$

14. $3.6 = r - 7.2$

15. $4.2d = 33.6$

16. $2.3v = 16.1$

17. $\frac{t}{0.9} = 0.2$

18. $\frac{z}{2.4} = 0.3$

Write an algebraic expression and then evaluate.

19. a. Riki likes to read $\frac{2}{3}$ of her history assignment before dinner. Choose a variable for the length of the assignment and then write an expression for the number of pages Riki should read before dinner.

b. If Riki's history assignment is 78 pages long tonight, how many pages should she read before dinner? (*Hint:* See Lesson 1.2B if you have forgotten how to multiply by a fraction.)

20. a. Amit is knitting a sweater with 32-inch sleeves. Choose a variable for the length Amit has already knitted and then write an expression for the number of inches he has left to knit.

b. If Amit has knitted $15\frac{1}{4}$ inches on a sleeve, how many inches does he have left to knit on that sleeve? (*Hint:* Write the mixed number as a decimal.)

Building Number Skills

In each pair of numbers, decide which is larger.

21. $\frac{1}{8}$ or 0.12

22. $\frac{5}{8}$ or 0.65

23. $\frac{3}{8}$ or 0.38

24. $\frac{7}{8}$ or 0.87

25. $\frac{1}{3}$ or 0.333

26. $\frac{2}{3}$ or 0.666

27. $\frac{2}{9}$ or 0.23

28. $\frac{4}{9}$ or 0.45

29. $\frac{3}{11}$ or 0.27

30. $\frac{5}{11}$ or 0.455

B. Using a Ruler

We can use a ruler or tape measure to find lengths. Most rulers have scales in inches and in centimeters, as shown in Figure 2.7. Inches are divided into halves, then fourths, eighths, and sixteenths, so that the distance between two adjacent marks is $\frac{1}{16}$ of an inch. Centimeters are divided into tenths, and $\frac{1}{10}$ of a centimeter is called a **millimeter.**

Figure 2.7

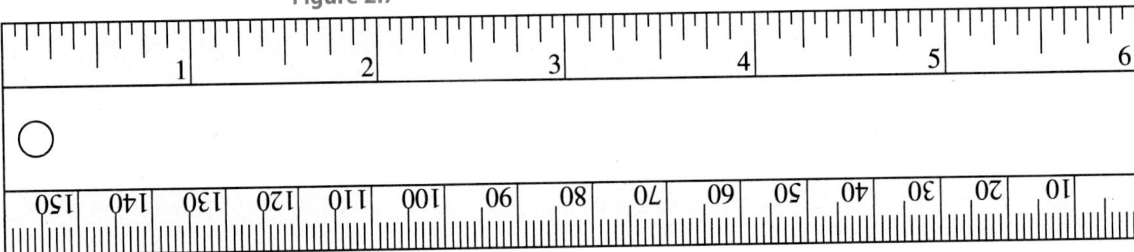

To measure a line segment with a ruler, line up the left end of the ruler with one end of the segment and read the number on the ruler at the other end of the segment.

EXAMPLE 3

a. The pencil in Figure 2.8 is approximately $3\frac{1}{4}$ inches long. We can never find the *exact* length of an object by measuring. We approximate the length by choosing the closest mark on the ruler to the end of the segment.

Figure 2.8

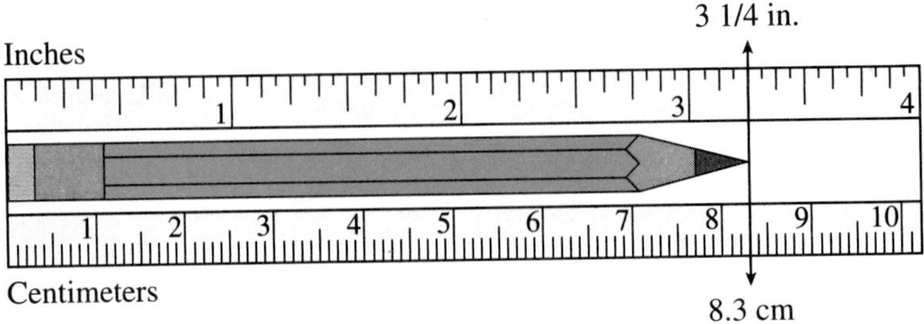

b. The same pencil is approximately 8 centimeters, 3 millimeters long. Or, because a millimeter is $\frac{1}{10}$ of a centimeter, we can say that the length is 8.3 centimeters.

EXERCISE 3

Use a ruler to find the length and width in inches of the credit card shown in Figure 2.9. Then measure the length and width in centimeters. (Your answers will be approximations.)

Figure 2.9

EXAMPLE 4

Find the perimeter in centimeters of Figure 2.10a. (The grid is marked off in centimeters.)

Figure 2.10a

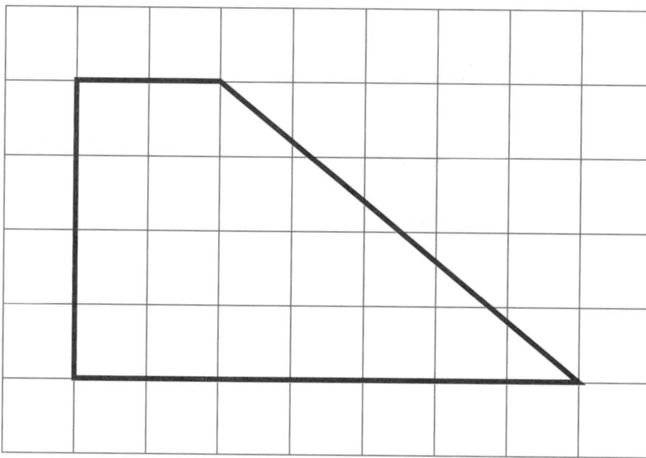

EXERCISE 4

Find the perimeter and area of Figure 2.11.

Figure 2.11

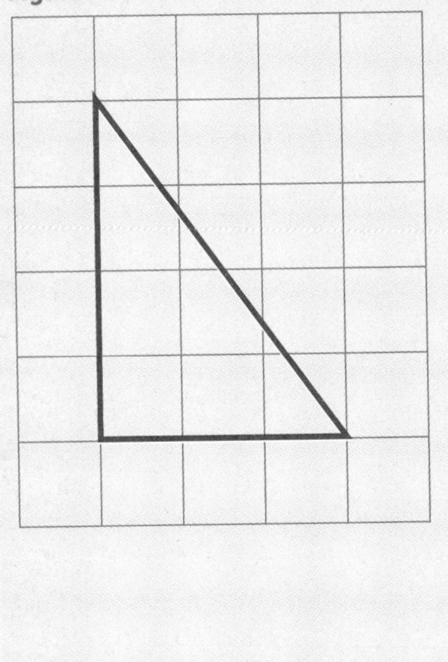

ANSWERS TO 2.2B EXERCISES

3. Length: $3\frac{3}{8}$ inches or 8.5 centimeters; width: $2\frac{1}{8}$ inches or 5.4 centimeters

4. Perimeter: 12 centimeters; area: 6 square centimeters

Solution

First measure the length of each side. The lengths of the three horizontal and vertical sides are easy to find by using the grid; their lengths are 2 centimeters, 4 centimeters, and 7 centimeters. We must measure the length of the fourth, slanting side with a centimeter ruler. The length is not a whole number of centimeters, so we estimate: It is approximately $6\frac{1}{2}$ centimeters. Finally, add the lengths of the sides to find the perimeter. The perimeter is approximately

$$2 + 4 + 7 + 6\tfrac{1}{2} = 19\tfrac{1}{2} \text{ centimeters}$$

Estimating Areas

Area measures the number of square units that fit inside a region or geometric figure. Sometimes the squares do not fit evenly inside the figure. In that case we have to estimate the area by piecing together parts of squares to make up whole squares.

EXAMPLE 5  

Find the area of the region shown in Figure 2.10a.

Solution

To find the area, we begin by adding up the whole squares inside the figure. There are 14 whole squares inside. Next, we try to put together pieces of squares inside the figure to make whole squares. The 8 partial squares make up 4 whole squares as shown in Figure 2.10b. The area is thus

$$14 + 4 = 18 \text{ square centimeters}$$

Figure 2.10b

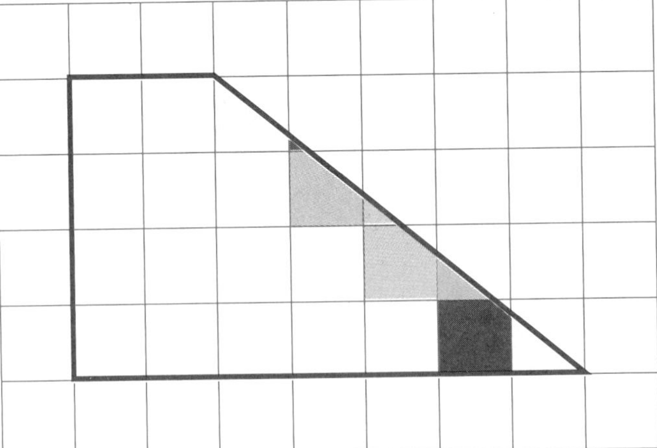

HOMEWORK 2.2B

Use a ruler to measure the length of each line segment. (a) Measure the length in inches and (b) measure the length in centimeters. Give your answers to the nearest millimeter.

1. ____

2. ______________________

3. ________________________________

4. ________________

Use a ruler for Problems 5 and 6. Your answers will be approximations.

5. a. Measure the length and width of your textbook in centimeters.

   b. Calculate the perimeter of your textbook.

6. a. Measure the length and width of a sheet of notebook paper in inches.

   b. Calculate the perimeter of a sheet of notebook paper.

Refer to a ruler to help you convert units.

7. a. 3 cm = _____ millimeters

   b. 1.8 cm = _____ millimeters

   c. 0.6 cm = _____ millimeters

   d. 0.25 cm = _____ millimeters

8. a. 25 mm = _____ centimeters

   b. 14 mm = _____ centimeters

   c. 7 mm = _____ centimeter

   d. 36 mm = _____ centimeters

9. a. $\frac{1}{2}$ in. $= \frac{?}{8}$ inch

   b. $\frac{3}{4}$ in. $= \frac{?}{16}$ inch

   c. $\frac{5}{8}$ in. $= \frac{?}{16}$ inch

   d. $\frac{5}{4}$ in. $= \frac{?}{8}$ inches

10. **a.** $\frac{6}{8}$ in. $= \frac{?}{4}$ inch

b. $\frac{8}{16}$ in. $= \frac{?}{2}$ inch

c. $\frac{6}{4}$ in. $= \frac{?}{2}$ inches

d. $\frac{6}{16}$ in. $= \frac{?}{8}$ inch

Estimate the area and perimeter of each figure. (The grids are in centimeters.)

11. Figure A:

Area: ___________

Perimeter: ___________

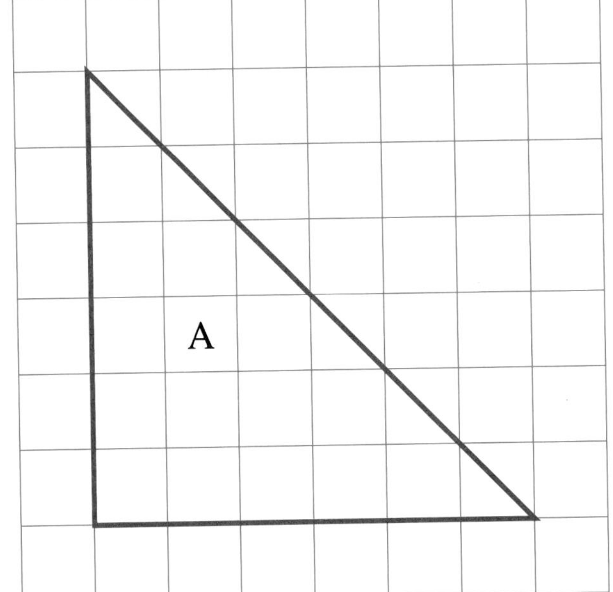

12. Figure B:

Area: ___________

Perimeter: ___________

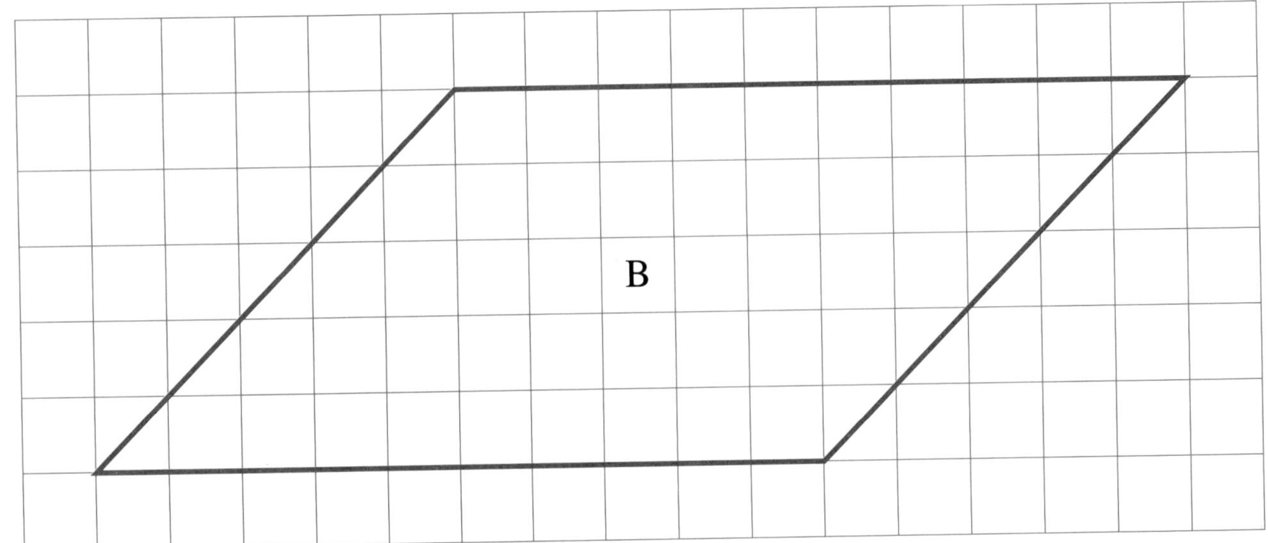

13. Figure C:

Area: __________

Perimeter: __________

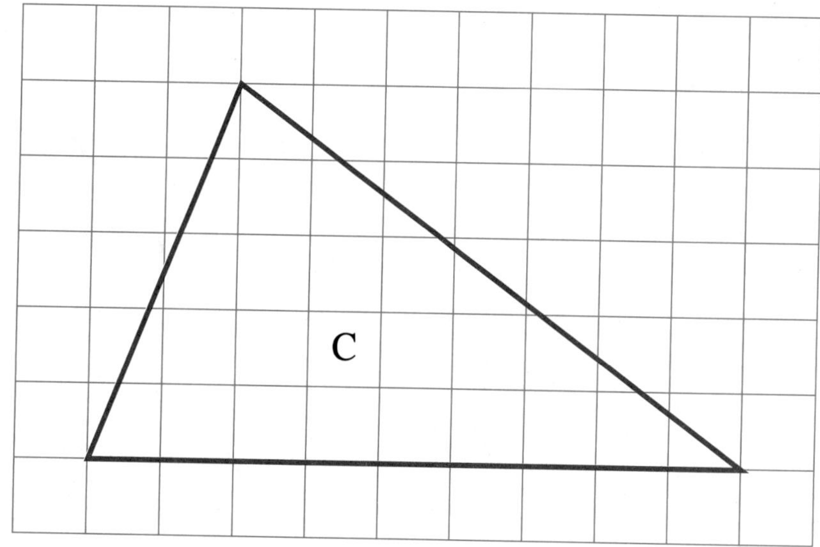

14. Figure D:

Area: __________

Perimeter: __________

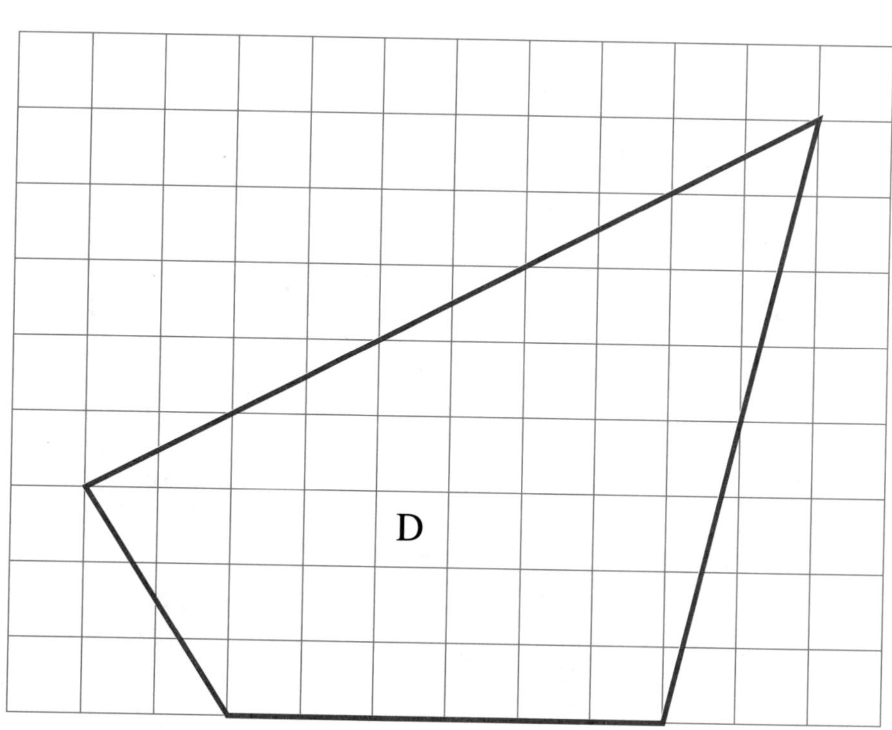

15. Figure E:

Area: ________

Perimeter: ________

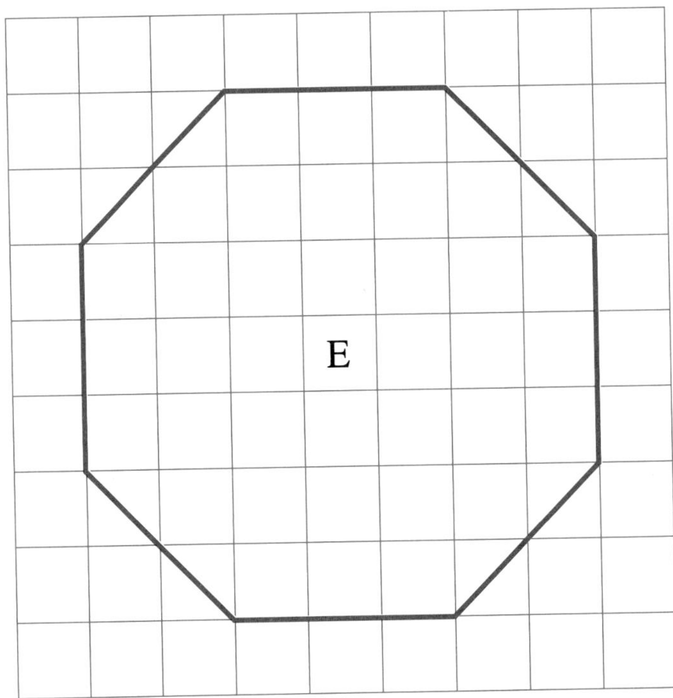

16. Figure F:

Area: ________

Perimeter: ________

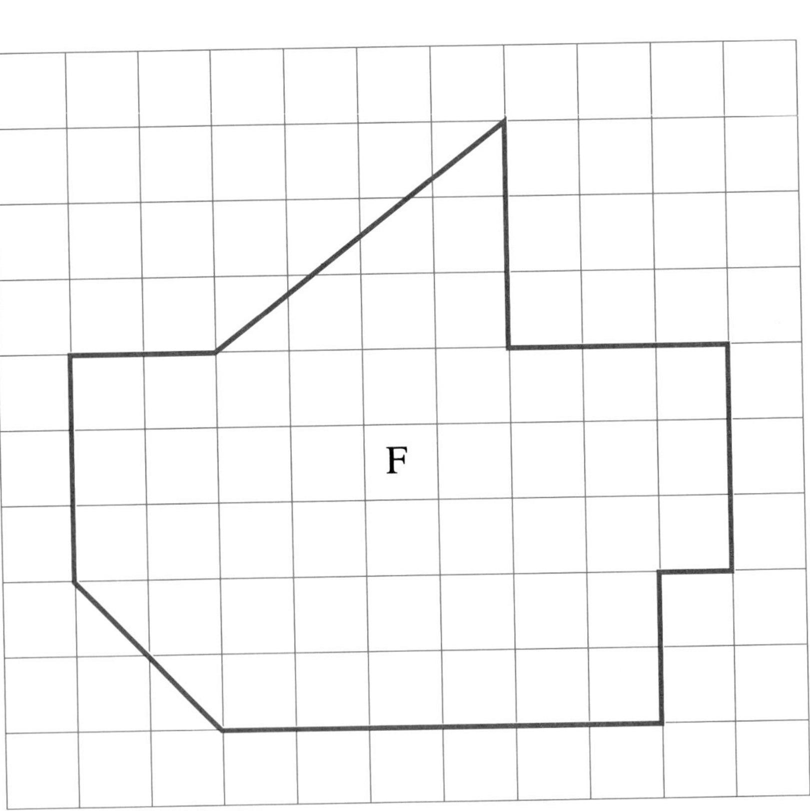

Choose the correct equation for each problem. (See Lesson 1.5 to review equations.)

17. The list price for a laptop computer is \$28.35 more than the price at Bullseye. If the list price is \$2652, what is the price at Bullseye?
a. $p + 28.35 = 2652$
b. $p - 28.35 = 2652$
c. $p - 2652 = 28.35$

18. By using low-fat ingredients, Marci cut 126 calories from her cheesecake recipe. If the low-fat recipe has 854 calories, how many calories did the original recipe have?
a. $c + 126 = 854$
b. $854 - c = 126$
c. $c - 126 = 854$

19. This year Fairfax County got 3.8 fewer inches of rain than last year. If the rainfall this year was 12.7 inches, what was it last year?
a. $3.8 - r = 12.7$
b. $r - 3.8 = 12.7$
c. $r + 3.8 = 12.7$

20. The price of a movie ticket is \$6.25. If the theater took in \$893.75 at the 7 P.M. showing, how many people bought tickets?
a. $6.25 + t = 893.75$
b. $t - 6.25 = 893.75$
c. $6.25t = 893.75$

Building Number Skills

When we round a number, we are really choosing the closest number with fewer digits. For example, 6.74 is between 6.7 and 6.8, but it is closer to 6.7. (You can review rounding in Appendix A.6.)

21. Is 2.43 closer to 2.4 or 2.5?

22. Is 7.56 closer to 7.5 or 7.6?

23. Is 21.08 closer to 21.0 or 21.1?

24. Is 59.04 closer to 59.0 or 59.1?

25. Is 8.237 closer to 8.2 or 8.3?

26. Is 4.862 closer to 4.8 or 4.9?

27. Is 48.8455 closer to 48.8 or 48.9?

28. Is 71.554 closer to 71.5 or 71.6?

29. Is 0.9098 closer to 0.91 or 0.90?

30. Is 0.3939 closer to 0.39 or 0.40?

2.3 Formulas

A. Evaluating Formulas

A **formula** is an equation that relates two or more variables. One frequently used formula is

$$\textbf{distance} = \textbf{rate} \cdot \textbf{time}$$

or

$$d = rt$$

(Before continuing, you may want to review the common algebraic expressions discussed in Lesson 1.4B.) This formula can be used to compute the distance you travel if you move at a constant speed, or *rate,* for a given length of time. For example, if you drive in your car for 2 hours at 50 miles per hour, you will travel 2(50), or 100 miles. Substituting values for the variables on the right side of the formula is called **evaluating** the formula.

EXAMPLE 1  

Find the distance traveled in 3.5 hours by a passenger train moving at 74 miles per hour.

Solution

Evaluate the formula for distance by substituting 74 for r and 3.5 for t and then simplify.

$$\begin{aligned} d &= rt \\ &= (74)(3.5) = 259 \text{ miles} \end{aligned}$$

EXERCISE 1

If Ann and Bill hike 16 kilometers per day, how far will they hike in a week?

When we use the distance formula, the rate, time, and distance must be expressed in compatible units. In Example 1, the speed is given in *miles per hour,* so the time must be expressed in *hours,* and the distance found will be in *miles.*

Here are some of the formulas we will investigate in the exercises:

1. $d = rt$ (*distance* = *rate* · *time*)

   Find the **distance** traveled: Multiply speed times travel time. (*Rate* means the same as *speed.*)

2. $P = R - C$ (*profit* = *revenue* − *costs*)

   Find the **profit** earned or lost: Subtract costs from revenue. (*Revenue* is the amount of money received.)

3. $s = p - d$ (*sale price* = *regular price* − *discount*)

   Find the **sale price:** Subtract the discount from the regular price.

4. $e = \dfrac{m}{g}$ $\left(\textit{efficiency} = \dfrac{\textit{miles}}{\textit{gallons}}\right)$

Find the **fuel efficiency** of a vehicle: Divide the number of miles traveled by the number of gallons of gasoline used.

5. $u = \dfrac{p}{n}$ $\left(\textit{unit cost} = \dfrac{\textit{total price}}{\textit{number of items}}\right)$

Find the **unit cost** of a commodity: Divide the total price by the number of items bought.

6. $P = rW$ (*part = percentage rate · whole*)

Find the **part** or **percentage** of a whole amount: Multiply the percentage rate (in decimal form) times the whole amount.

7. $I = Prt$ (*interest = principal · rate · time*)

Find the **interest** on a debt or an investment: Multiply the principal times the interest rate (in decimal form) times the number of years elapsed. (The *principal* is the amount of money invested.)

8. $A = \dfrac{S}{n}$ $\left(\textit{average value} = \dfrac{\textit{sum of values}}{\textit{number of values}}\right)$

Find the **average value:** Divide the sum of all the values by the number of values.

Formulas 6 and 7, $P = rW$ and $I = Prt$, involve percents. To calculate a percent, remember that a percent is a fraction with denominator 100. Thus, in Example 2,

$$3\% = \frac{3}{100} = 0.03$$

EXAMPLE 2 

Francisco deposited \$400 in a savings account that pays 3% interest. How much interest will his savings earn in 2 years?

a. Which formula is appropriate for this problem?

b. Use the formula to calculate the interest.

Solution

a. The problem is about interest, so we use the formula $I = Prt$.

b. The interest rate is 3%, or $\frac{3}{100}$, or 0.03. Substitute 400 for P, 0.03 for r, and 2 for t:

$$I = Prt$$
$$= (400)(0.03)(2) = 24$$

Francisco's savings will earn \$24 interest.

EXERCISE 2

Delbert traveled 250 miles in his new SUV on one tank of gas. If the tank holds 20 gallons of gas, what is the fuel efficiency of the vehicle?

a. Which formula is appropriate for this problem?

b. Use the formula to calculate the fuel efficiency.

ANSWERS TO 2.3A EXERCISES

1a. 112 kilometers

2a. $e = \dfrac{m}{g}$

2b. 12.5 miles per gallon

HOMEWORK 2.3A

1. What is a formula?

2. What does it mean to evaluate a formula?

State a formula for each of the following quantities and explain what the variables in each formula represent.

3. average value

4. distance

5. fuel efficiency

6. interest

7. part (of a whole)

8. profit

9. sale price

10. unit cost

Choose the appropriate formula and evaluate it to answer the question.

11. Ridwan bicycled for 6 hours at an average speed of 14 miles per hour. How far did he ride?

12. Elsie is on a cruise to Alaska. The cruise ship averaged 32 miles per hour for the first 2 days of the trip. How far did it travel?

13. Elias took in $8200 last week at his auto repair shop. His total costs for the week, including salaries and overhead, were $6835. What was his profit for the week?

14. Heather spent $475 to rent a booth at the Craft Fair and to buy materials for her product. Her receipts at the end of the fair came to $950. What was her profit from the fair?

15. Asa bought a $360 wool suit at a discount of $43.20. What was the sale price of the suit?

16. Mike bought a wind surfer on sale and got a $75 discount. If the wind surfer originally cost $430, what was the sale price?

17. Leon can fly his four-person airplane from Los Angeles to Santa Barbara and back, a distance of 180 miles, on 24 gallons of gas. What is the fuel efficiency, in miles per gallon, of Leon's airplane?

18. Marlene filled her car's 12-gallon gas tank on Monday and drove 348 miles before running out of gas. What is her car's fuel efficiency, in miles per gallon?

#4

19. Harley bought 18 chairs for his new restaurant at a total cost of $810. What did he pay per chair?

20. Doug bought a case of cat food at PetMart for $25.20. If a case contains 36 cans, how much did he pay per can?

#5

21. Isabelle spends 35% of her take-home pay on rent. If she takes home $1500 per month, what is her rent?

22. Naturelle shampoo is 4% almond extract. How much almond extract is in a 16-ounce bottle?

#6

Decide two things about each of the following: (a) Is it an equation? (b) If it is an equation, is it true?

23. $12 = 2 + 4 + 6$

24. $20 = 5(4)$

25. $9 + 6 + 8 - 4$

26. $1 + 6 + 7 - 2$

27. $7(3) = 10$

28. $15 = 8 + 9$

29. $24 - 9 = 13 + 2$

30. $6(5) = \frac{120}{4}$

31. $51 - 37 =$

32. $148 \div 6 =$

(a) Look for a pattern relating the numbers in the second row of the table to the numbers in the first row. (b) Write an equation relating the variables and fill in the rest of the table.

33.

t	w
40	25
55	40
60	45
80	65
90	
	95

34.

x	z
6	2
12	4
27	9
33	11
36	
	14

35.

D	H
3	12
5	20
6	24
9	36
11	
	64

36.

d	h
2	8
3	9
5	11
8	14
12	
	25

Building Number Skills

Before trying these problems, you may want to review Appendix A.6 on rounding.

(a) Round each number to thousandths and (b) round each number to tenths.

37. 28.4672

38. 13.1819

39. 5.66666

40. 7.33333

41. 0.0055

42. 0.0489

43. 4.99999

44. 3.88888

45. 1.0001

46. 6.2995

B. Solving Formulas

Sometimes we have to solve an equation when we use a formula.

EXAMPLE 3  

How long will it take Rudi to swim 1 mile if she can swim at a rate of 50 yards per minute?

Solution

We will use the formula $d = rt$. We are asked to find the *time* it will take Rudi to swim 1 mile, so t will be the variable in our equation. The rate at which Rudi swims is given in yards per minute, so we must first convert the distance, 1 mile, into yards: 1 mile = 1760 yards. Now substitute **50** for r and **1760** for d into the formula:

$$d = rt$$
$$\mathbf{1760} = \mathbf{50}t$$

Finally, solve the equation above for t. Since t has been multiplied by 50, we will divide both sides of the equation by 50:

$$\frac{1760}{\mathbf{50}} = \frac{50t}{\mathbf{50}} \quad \text{Divide both sides by 50.}$$
$$35.2 = t$$

It takes Rudi 35.2 minutes to swim 1 mile.

EXERCISE 3

What revenue is needed to make a profit of \$1100 if the costs total \$9100?

ANSWER TO 2.3B EXERCISE

3. \$10,200

HOMEWORK 2.3B 

Solve. (See Lesson 2.1 to review solving equations.)

1. $0.38 + f = 0.72$

2. $0.07 + c = 0.21$

3. $y - 43 = 13$

4. $s - 28 = 27$

5. $10.8 = 0.6u$

6. $4.8 = 0.2f$

7. $36 = \frac{p}{1.5}$

8. $65 = \frac{n}{0.8}$

(a) Choose the appropriate formula on pages 82 and 83 and write an equation. (b) Solve the equation and answer the question.

9. How far must a compact car travel on a tank of gas (14 gallons) to meet the government standards for fuel efficiency, which require 26 miles per gallon?

10. A jet airliner can fly approximately 0.16 mile on 1 gallon of gas. If the plane holds 25,000 gallons of gas, how far can it fly without refueling?

11. The Camp for Kids Foundation hopes to make $1500 profit at its next rummage sale. If the cost of putting on the sale is $215, how much revenue will they need?

12. Alisha bought a computer on sale and saved $68. If she paid $347, what was the regular price of the computer?

13. Of the freshman class at Hollins College, 36% is from out of state. If there are 162 freshmen from out of state, how large is the freshman class?

14. A telephone poll of 800 voters showed 496 in favor of a school bond. What percent of those polled favored the bond?

15. Millie invested $1300 in a certificate of deposit for 1 year and earned $98.80 interest. What interest rate did the certificate earn?

16. Scott earned $14.70 in interest from his credit union last year. He made no deposits or withdrawals from the account. If the interest rate is 4.9%, how much was in the account?

17. Of the citizens who voted, 45% voted for Senator Fogbank. If the Senator received 540 votes, how many people voted?

18. A fruit punch contains 36% ginger ale. If you have 9 quarts of ginger ale, how much fruit punch can you make?

19. Staci invested some money in a T-bill (treasury bill) account that pays $9\frac{1}{2}$% interest, and 1 year later the account had earned $171 interest. How much did Staci deposit in the T-bill?

20. Clive loaned his brother some money to buy a new truck, and his brother agreed to repay the loan in 1 year with 3% interest. Clive earned $75 interest on the loan. How much did Clive loan his brother?

Estimate the area and perimeter of each figure. The grids are in centimeters. (See Lesson 2.2 to review area and perimeter.)

21. Figure A:

Perimeter: __________

Area: __________

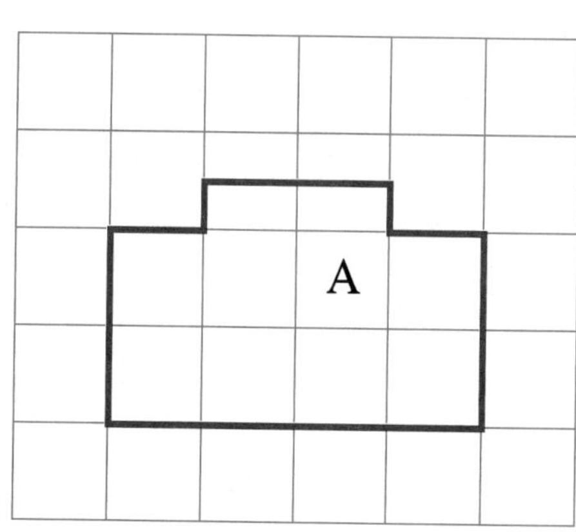

22. Figure B:

Perimeter: __________

Area: __________

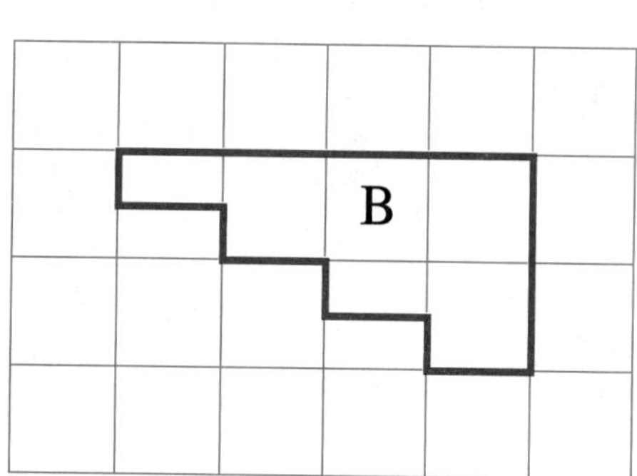

23. Figure C:

Perimeter: ________

Area: ________

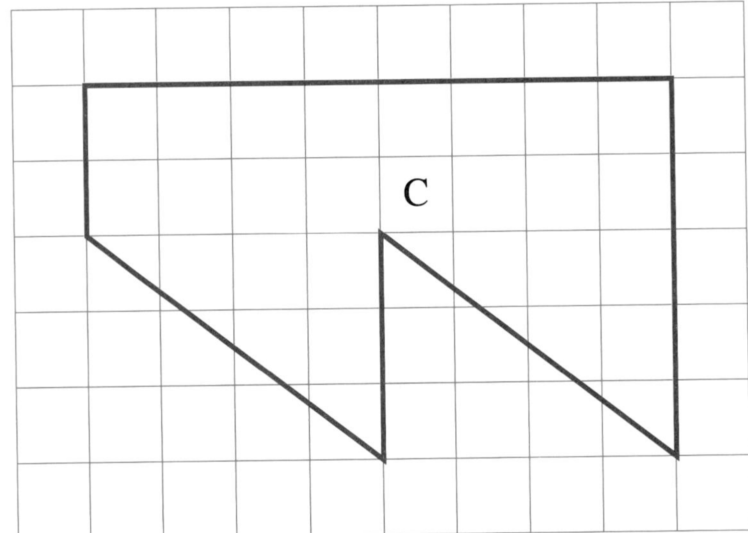

24. Figure D:

Perimeter: ________

Area: ________

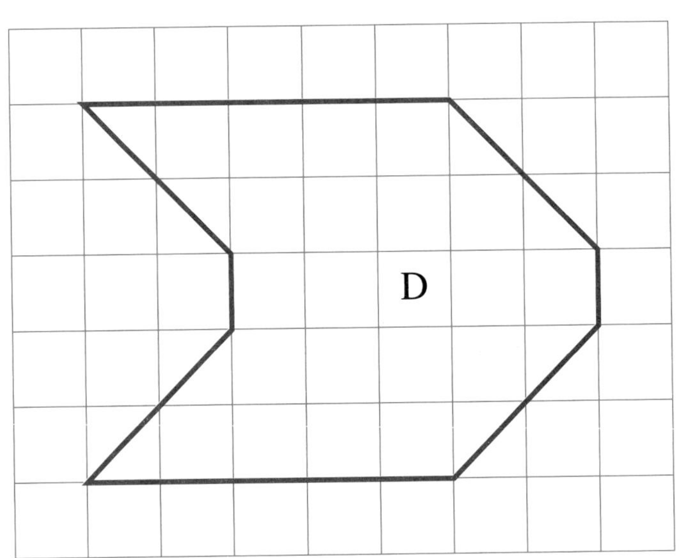

Building Number Skills

Write each fraction as a decimal. Round to hundredths. (See Appendix A.6 to review rounding.)

25. $\frac{5}{6}$ **26.** $\frac{7}{9}$ **27.** $\frac{2}{7}$ **28.** $\frac{4}{7}$ **29.** $\frac{5}{13}$

30. $\frac{9}{14}$ **31.** $\frac{13}{15}$ **32.** $\frac{13}{18}$ **33.** $\frac{7}{37}$ **34.** $\frac{9}{29}$

2.4 Some Geometric Formulas

A. Simple Figures

In Lesson 2.2 we found the area of a geometric figure by counting the square units enclosed by the figure. For many common figures, like rectangles, triangles, and parallelograms, we can use formulas to find their areas. In the activities for this lesson, you will discover the formulas for these figures.

Activity 1: Rectangles

A **rectangle** is a four-sided figure in which all the angles are right angles, or 90°. The opposite sides of a rectangle are parallel and equal in length.

For each of the rectangles shown, record the **length** and the **width** in the spaces provided. Then find the area of the rectangle by counting the square units enclosed by the figure.

1. 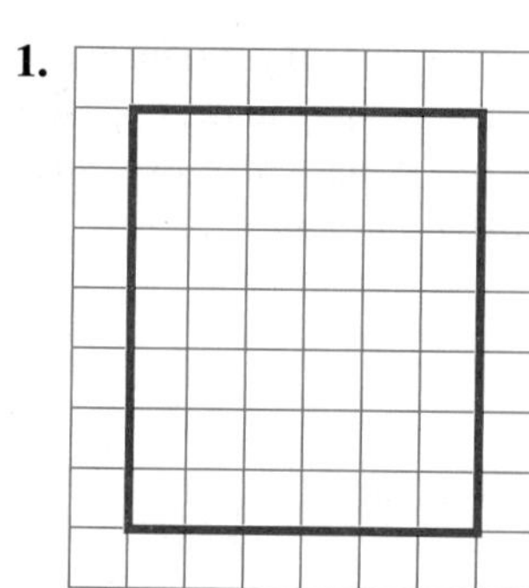

Length ________

Width ________

Area ________

2. 

Length ________

Width ________

Area ________

3. 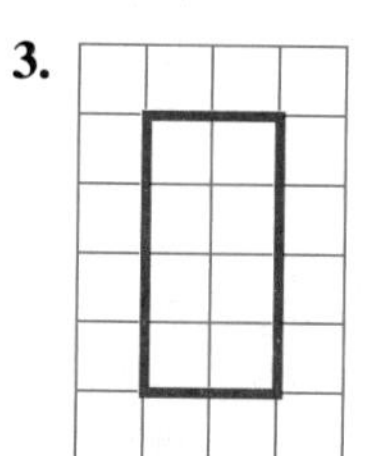

Length ________

Width ________

Area ________

4. 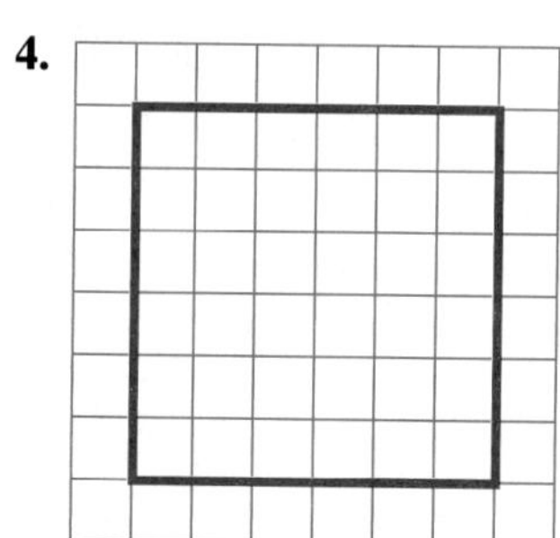

Length ________

Width ________

Area ________

5. a. Do you see a pattern relating the length and width of the rectangle to its area? Write a formula for the area of a rectangle in terms of its length and width.

b. Explain why your formula makes sense by describing a faster way of counting the square units. (*Hint:* Think about what multiplication means.)

Activity 2: Triangles

A **triangle** is a three-sided figure. Any one of the three sides can be designated as the **base** of the triangle. The perpendicular distance from the base to the opposite vertex is called the **height** of the triangle. (A **vertex** is a point where two sides meet.) The base and height of two triangles are illustrated in Figure 2.12.

Figure 2.12

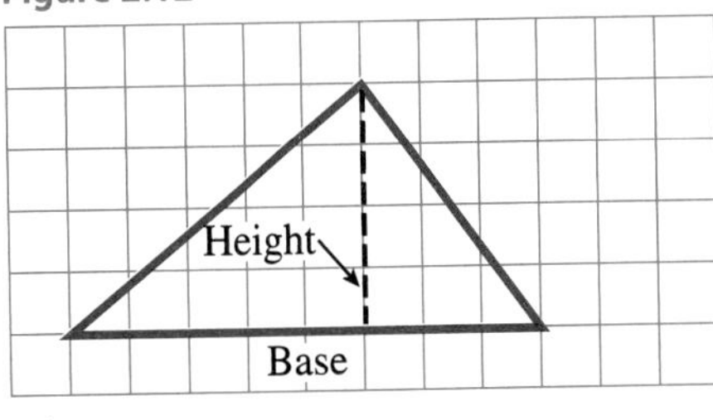

(a)

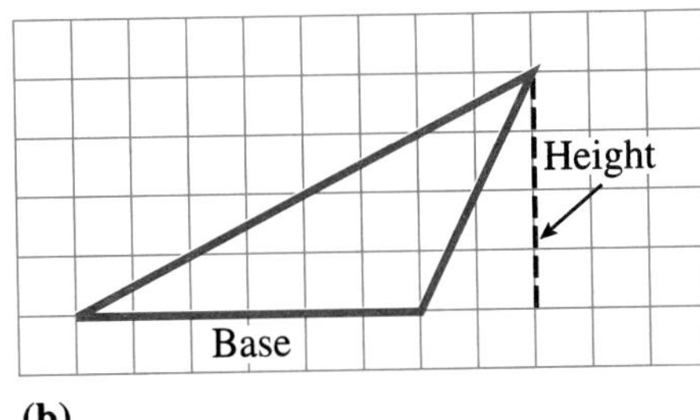

(b)

For each of the triangles shown, record the base and height in the spaces provided. Then find the area of the triangle by counting the square units enclosed by the figure. You will have to estimate by combining pieces of squares to make whole squares.

6. 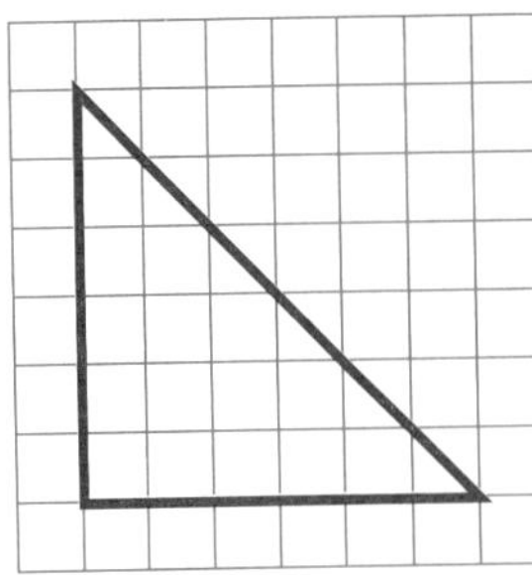

Base ________

Height ________

Area ________

7. 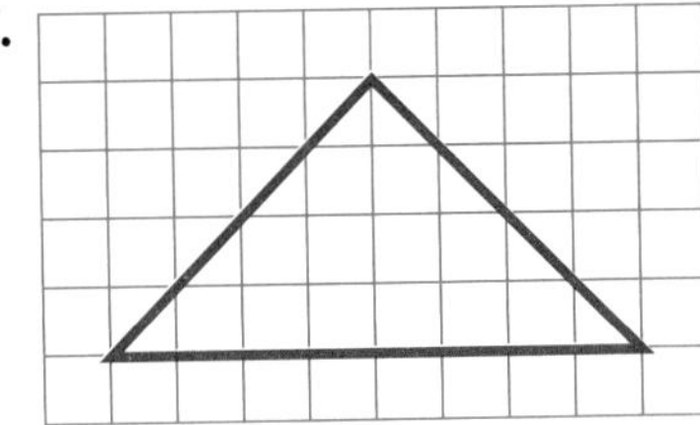

Base ________

Height ________

Area ________

8.

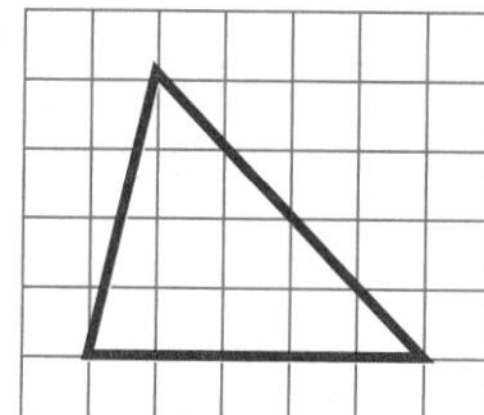

Base ________

Height ________

Area ________

9.

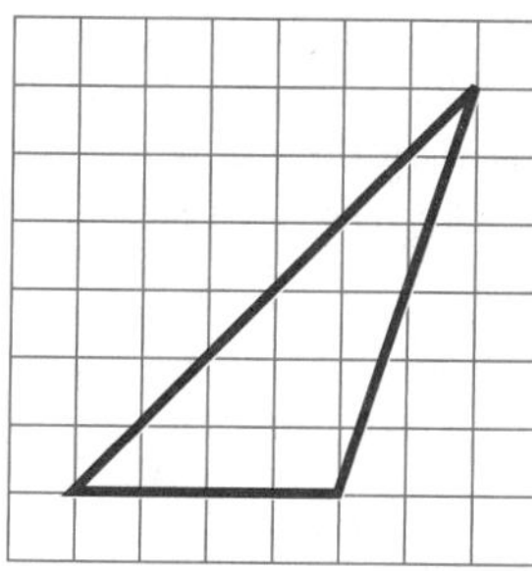

Base ________

Height ________

Area ________

10. a. Do you see a pattern relating the base and height of the triangle to its area? Write a formula for the area of a triangle in terms of its base and height.

b. Use the figures below to explain why your formula makes sense.

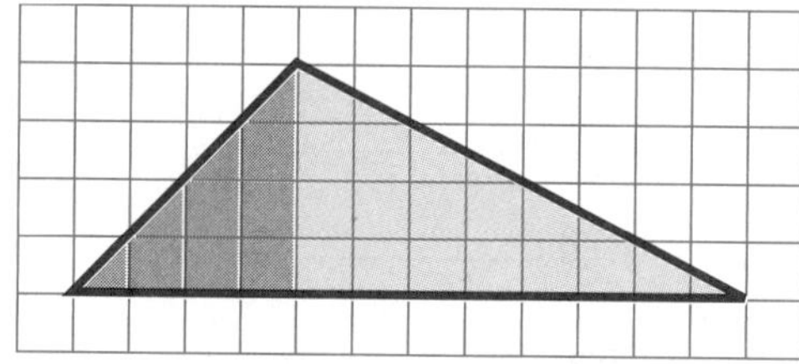

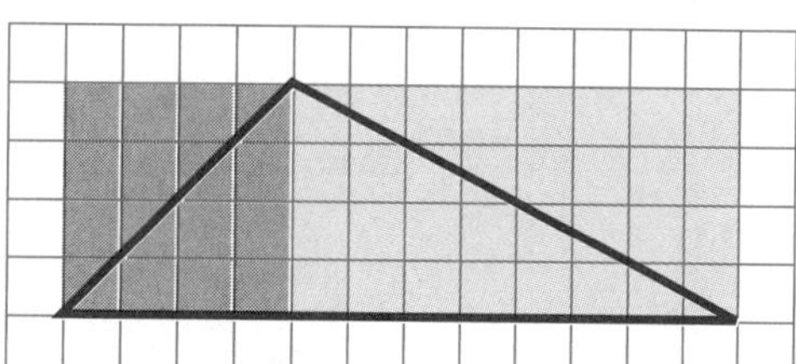

Activity 3: Parallelograms

A **parallelogram** is a four-sided figure whose opposite sides are parallel. The **base** and **height** of a parallelogram are illustrated in Figure 2.13.

Figure 2.13

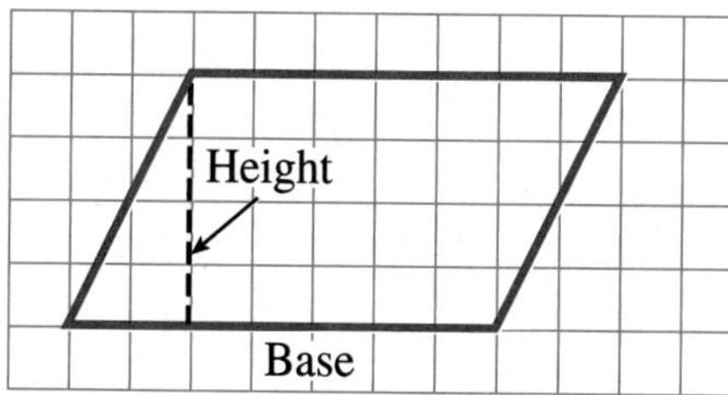

(a)

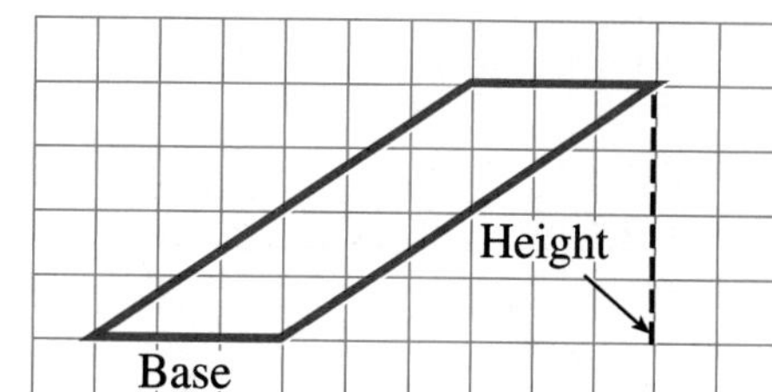

(b)

For each of the parallelograms shown, record the base and height in the spaces provided. Then find the area of the parallelogram by counting the square units enclosed by the figure. You will have to estimate by combining parts of squares to make whole squares.

11. 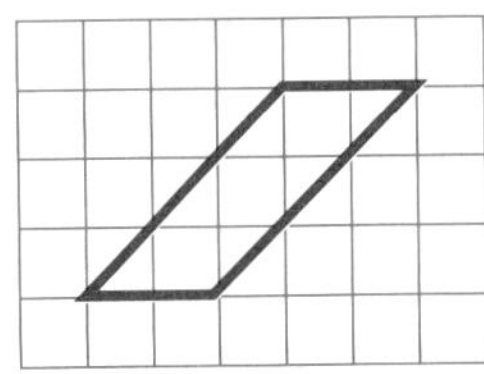

Base ________

Height ________

Area ________

12. 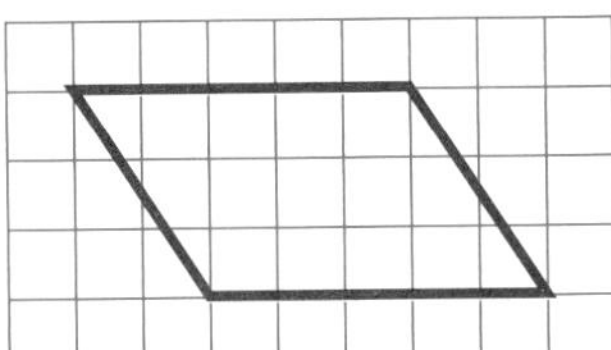

Base ________

Height ________

Area ________

13. 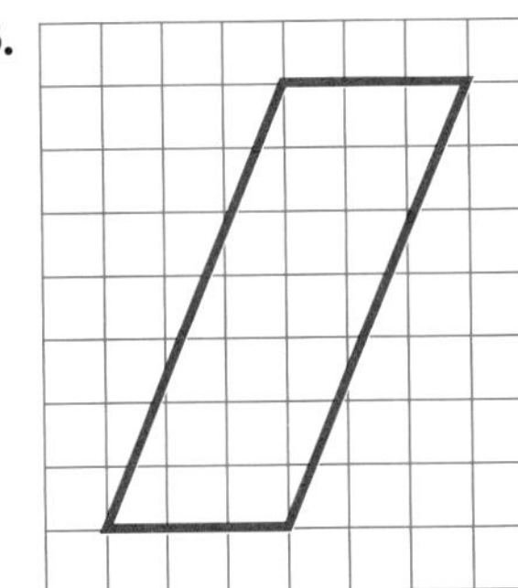

Base ________

Height ________

Area ________

14. 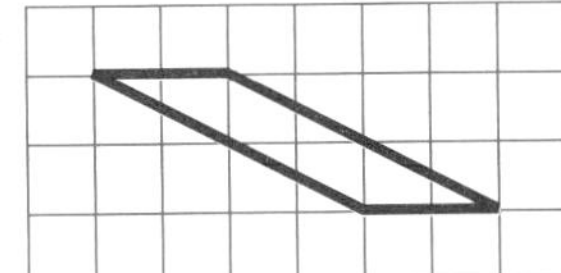

Base ________

Height ________

Area ________

15. a. Do you see a pattern relating the base and height of the parallelogram to its area? Write a formula for the area of a parallelogram in terms of its base and height.

b. Use the figures below to explain why your formula makes sense.

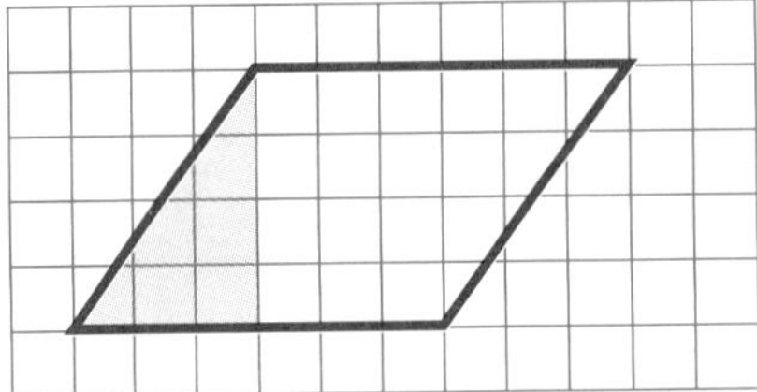

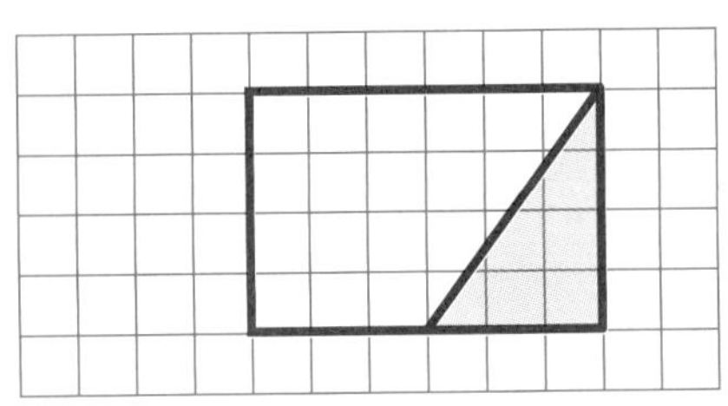

Now that we have formulas for the areas of rectangles, triangles, and parallelograms, we can use them to compute areas without having to count squares.

EXAMPLE 1  

Find the area of each shape in Figure 2.14.

Figure 2.14

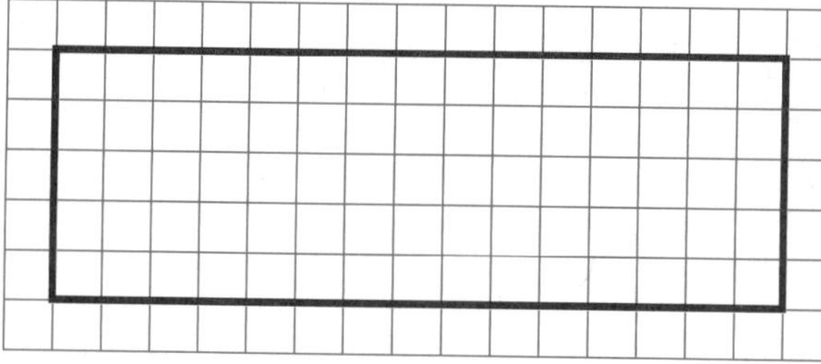

(a)

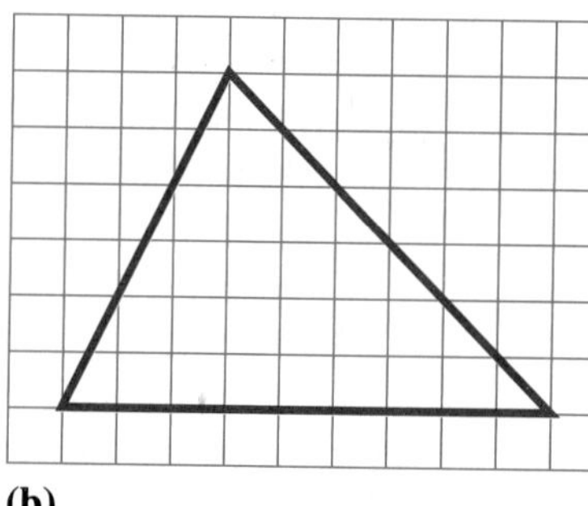

(b)

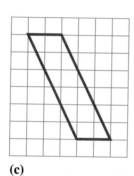

(c)

Solution

a. The rectangle is 15 units long and 5 units tall, so its area is

$$(\text{length})(\text{width}) = 15 \cdot 5 = 75 \text{ square units}$$

b. The triangle has a base 9 units long and a height 6 units tall, so its area is

$$\frac{1}{2}(\text{base})(\text{height}) = \frac{1}{2} \cdot 9 \cdot 6 = 27 \text{ square units}$$

c. The parallelogram has a base 2 units long and height 6 units tall, so its area is

$$(\text{base})(\text{height}) = 2 \cdot 6 = 12 \text{ square units}$$

EXERCISE 1

Find the area of each shape in Figure 2.15.

Figure 2.15

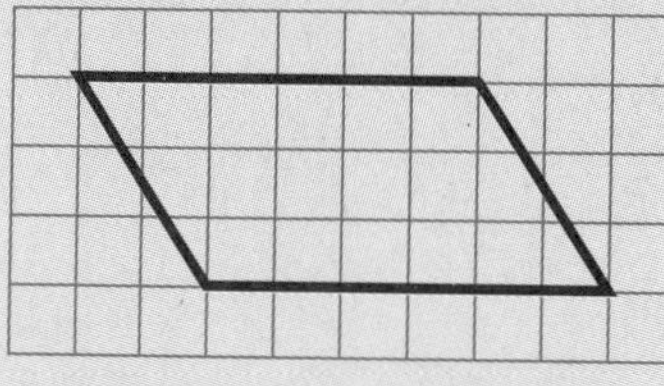

(a)

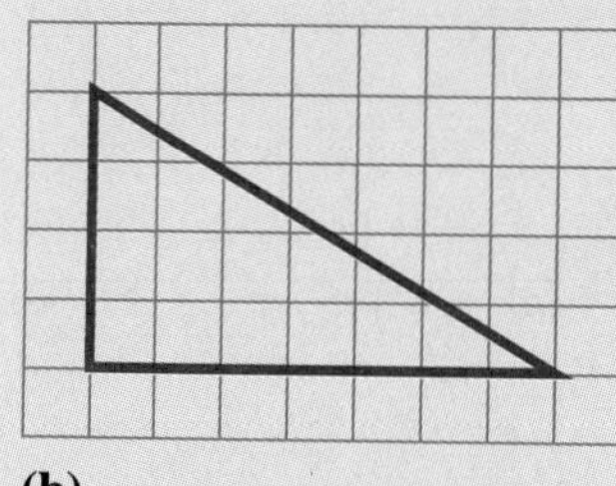

(b)

EXERCISE 2

The triangle in Figure 2.17a is called a **right triangle** because one of its angles is a right angle, or 90°. Figure 2.17b is a parallelogram. The opposite sides of a parallelogram are not only parallel, they are also equal in length. Find the area of each figure, using the given dimensions.

Figure 2.17

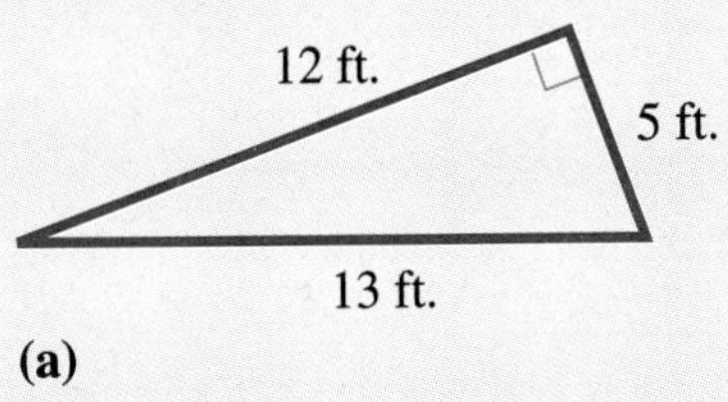

(a)

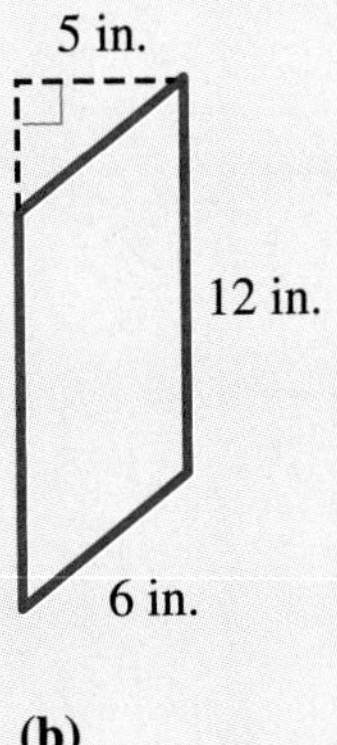

(b)

ANSWERS TO 2.4A EXERCISES

1a. 18 square units

1b. 14 square units

2a. 30 square feet

2b. 60 square inches

EXAMPLE 2

The triangle in Figure 2.16a is called an **isosceles triangle;** two of its sides are equal in length. The parallelogram in Figure 2.16b is called a **diamond** or **rhombus;** all its sides are equal in length. Find the area of each figure based on the given dimensions. The figures are not drawn to scale. (The small squares in the figures indicate right angles.)

Figure 2.16

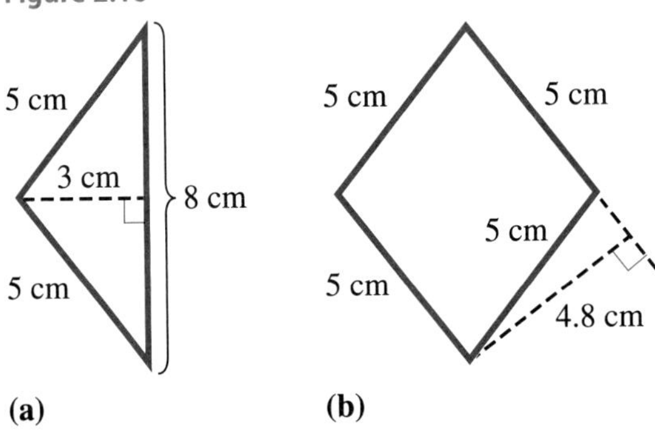

Solution

a. The triangle is not resting on its base. We can imagine rotating the triangle so that the 8-centimeters side is on the bottom, and the height is 3 centimeters. The area of the triangle is

$$\frac{1}{2}\text{(base)(height)} = \frac{1}{2}(8\text{ cm})(3\text{ cm}) = 12\text{ square cm}$$

(The two sides marked 5 centimeters were not used to compute the area.)

b. The upper right side can be considered the base of the parallelogram, and the altitude is 4.8 centimeters. The area is

$$\text{(base)(height)} = (5\text{ cm})(4.8\text{ cm}) = 24\text{ square cm}$$

HOMEWORK 2.4A 

Use the formulas you discovered in the lesson to find the areas of the figures.

1. 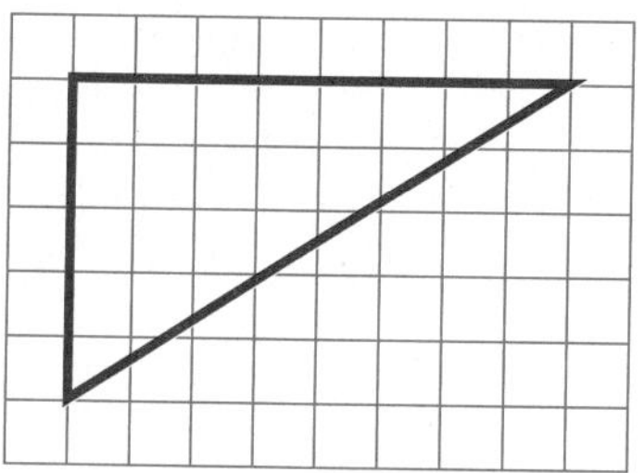

2. 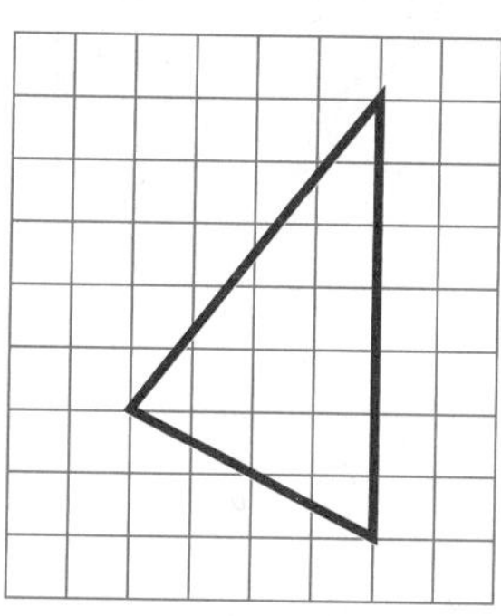

3. 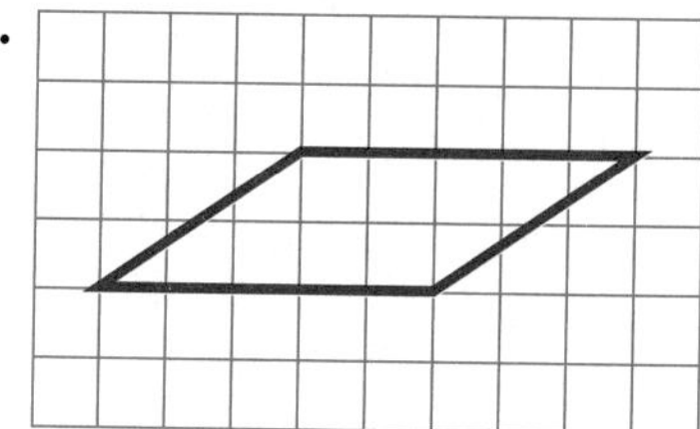

4. 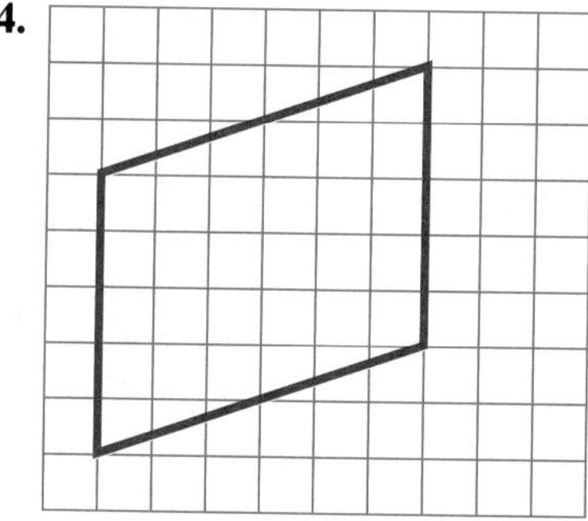

5. 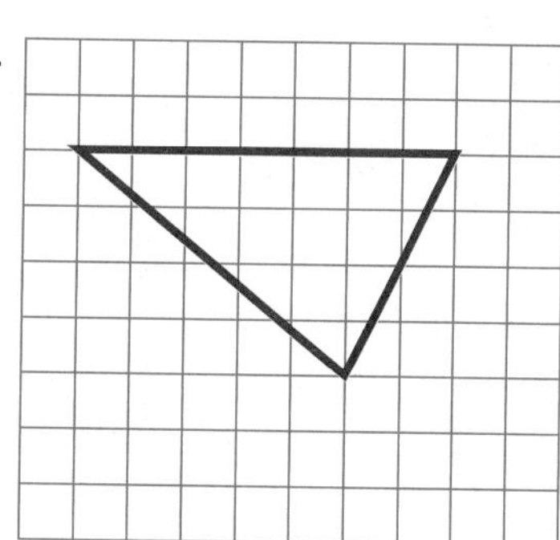

6. 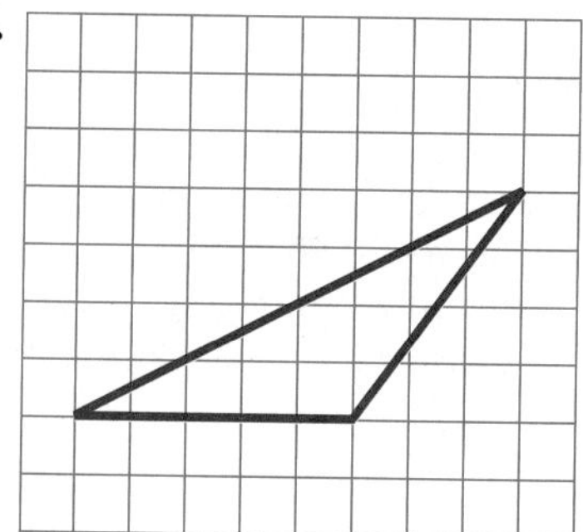

7. 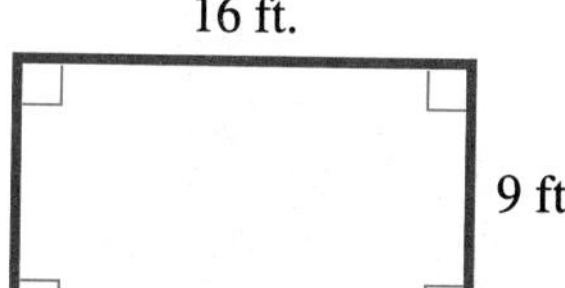

8. 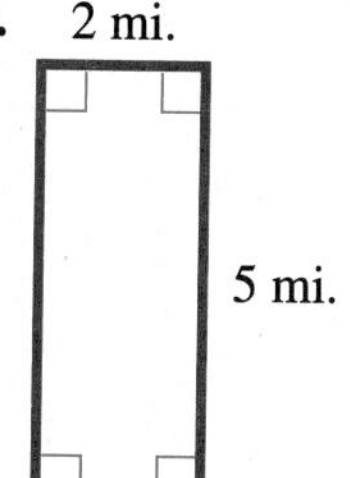

9. 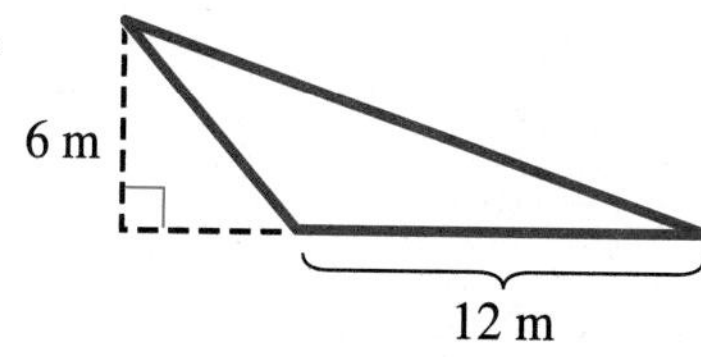

10.

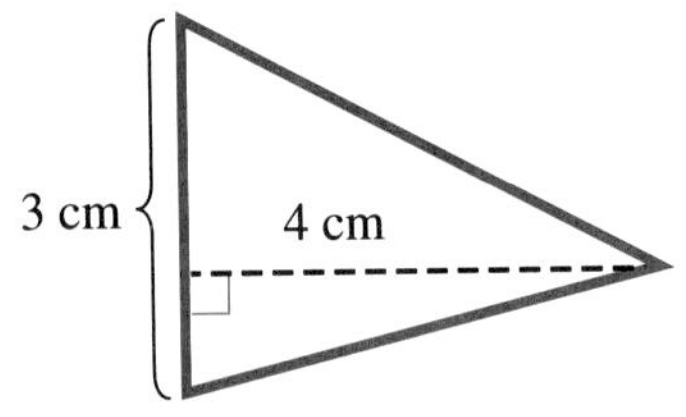

11.

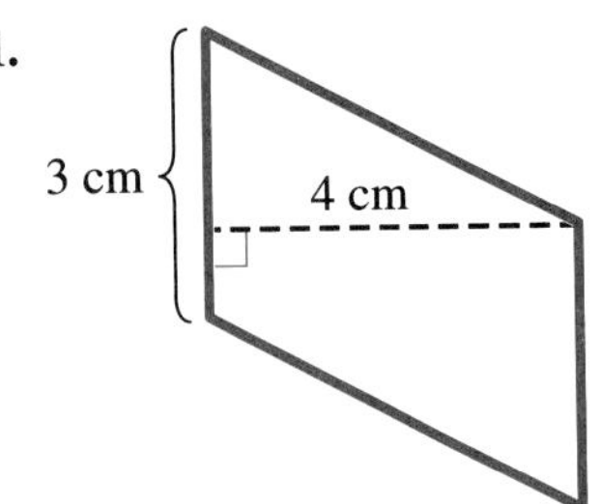

12.

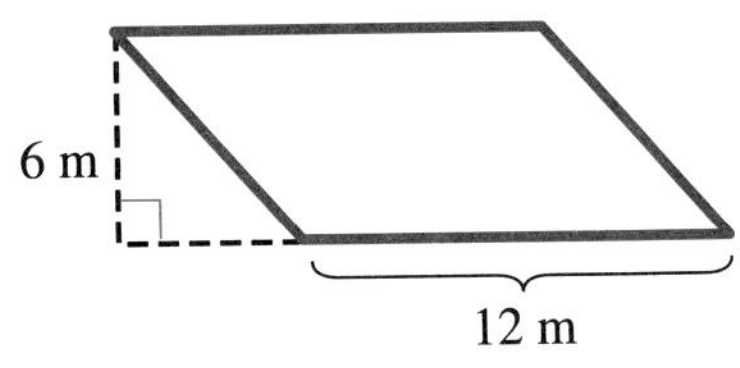

Choose the appropriate formula on pages 82 and 83 and evaluate it to answer the question.

13. Jay borrowed $20,000 from his father for the down payment on a house. He will pay it back in 5 years with 3% annual interest. How much interest will he owe his father?

14. Nurit won $2500 in the lottery 3 years ago and put it in an account that pays 6.5% annual interest. Now she plans to travel and wants to withdraw the money. How much interest will the account have earned?

15. Darren typed three essays in $2\frac{1}{2}$ hours. The essays were 500 words, 1500 words, and 4000 words long. What is Darren's average typing speed, in words per minute? (*Hint:* How many minutes are there in $2\frac{1}{2}$ hours?)

16. In Roza's graphics design class, there are five students who are 19 years old, six students who are 20, four students who are 23, one who is 24, and two who are 28. What is the average age of the students in Roza's class?

Look for a pattern relating the numbers in the second column of the table to the numbers in the first column. (*Hint:* Try multiplying the first column by a fraction.) Then write an equation for the second variable in terms of the first variable.

17.

x	y
6	4
12	8
15	10
24	16

18.

s	t
16	12
28	21
36	27
40	30

19.

N	R
8	12
12	18
20	30
24	36

20.

J	B
3	4
6	8
15	20
30	40

Building Number Skills

Use your calculator to find each fraction. Round your answer to hundredths if necessary. (See Appendix A.6 to review rounding.)

Example: $\frac{1}{3}$ of 17 $= \frac{1}{3} \times 17 = 17 \div 3 = 5.666\ldots \approx 5.67$

21. $\frac{1}{6}$ of 11
22. $\frac{1}{3}$ of 20
23. $\frac{1}{5}$ of 32
24. $\frac{1}{4}$ of 17
25. $\frac{1}{7}$ of 25

26. $\frac{1}{6}$ of 50
27. $\frac{1}{12}$ of 35
28. $\frac{1}{9}$ of 70
29. $\frac{1}{8}$ of 17
30. $\frac{1}{16}$ of 53

B. Compound Figures

A figure that can be divided into rectangles and triangles is called a **compound** figure. We can find the area of a compound figure by finding the areas of the triangles and rectangles and then adding those areas. For reference, we list here the area formulas we found in Part A:

Rectangle: $A = lw$

Triangle: $A = \frac{1}{2}bh$

Parallelogram: $A = bh$

EXAMPLE 3

Find the area of Figure 2.18a. Each square represents 1 square inch.

Figure 2.18

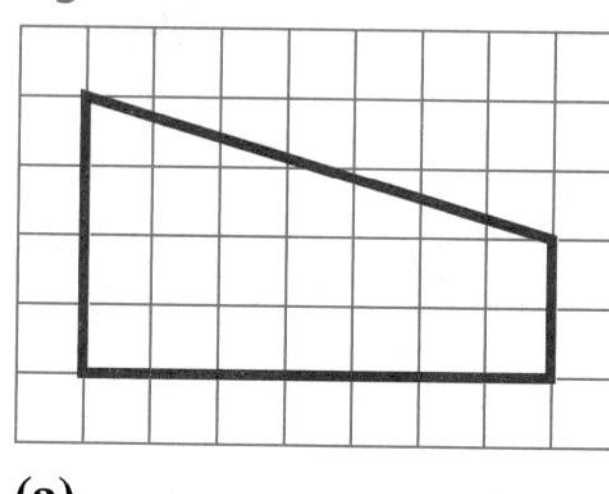

(a)

(b)

EXERCISE 3

Find the area of Figure 2.19.

Figure 2.19

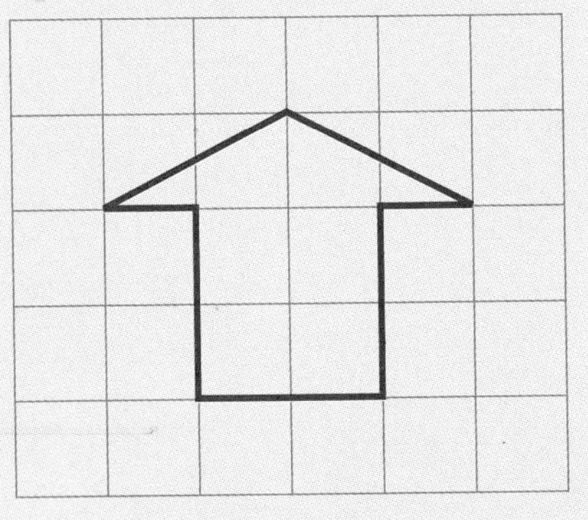

EXERCISE 4

Find the area of Figure 2.21.

Figure 2.21

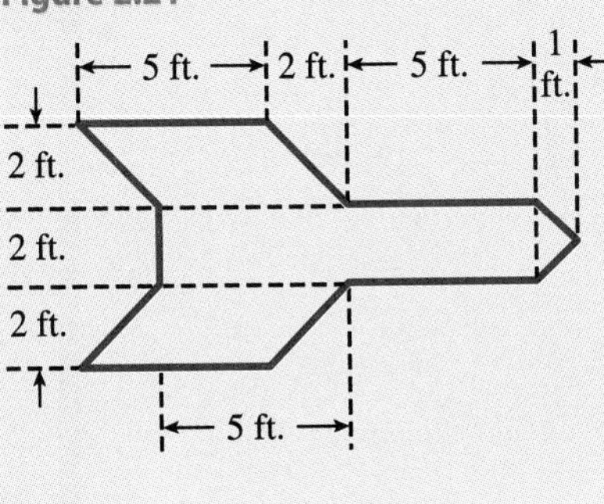

Solution

Divide the figure into a triangle and a rectangle as shown in Figure 2.18b. Then use the formulas to find the area of each piece:

$$\text{Area of triangle: } A = \frac{1}{2}bh$$
$$= \frac{1}{2}(7)(2) = 7 \text{ square inches}$$
$$\text{Area of rectangle: } A = lw$$
$$= (7)(2) = 14 \text{ square inches}$$

The total area of Figure 2.18a is the sum of the parts:

$$\text{Total area} = 7 + 14$$
$$= 21 \text{ square inches}$$

EXAMPLE 4  

Find the area of Figure 2.20.

Figure 2.20

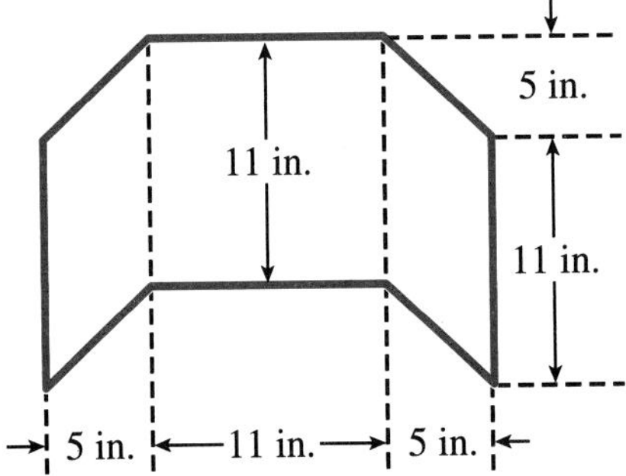

Solution

The figure has been divided into a square and two parallelograms. The square has length and width 11 inches, and each parallelogram has a base of 11 inches and a height of 5 inches. The total area of the figure is the sum of three parts.

$$\text{Total area} = (\text{area of square}) + (\text{area of parallelogram}) + (\text{area of parallelogram})$$
$$= (lw) + (bh) + (bh)$$
$$= (11 \cdot 11) + (11 \cdot 5) + (11 \cdot 5)$$
$$= 121 + 55 + 55$$
$$= 231 \text{ square inches}$$

Sometimes we can find an area by removing a piece from a larger figure.

EXAMPLE 5 

Find the area and the perimeter of Figure 2.22a.

Figure 2.22

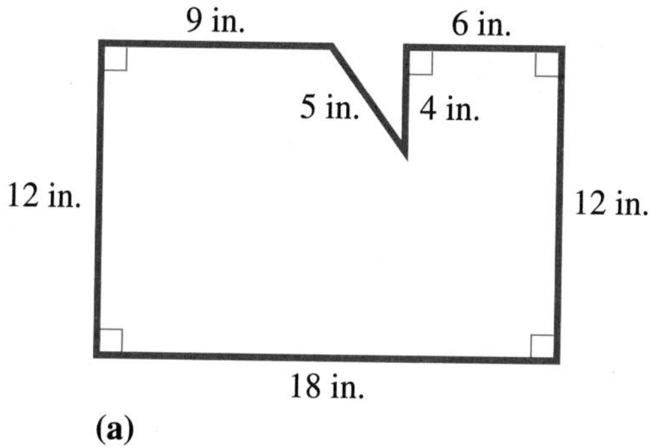

(a)

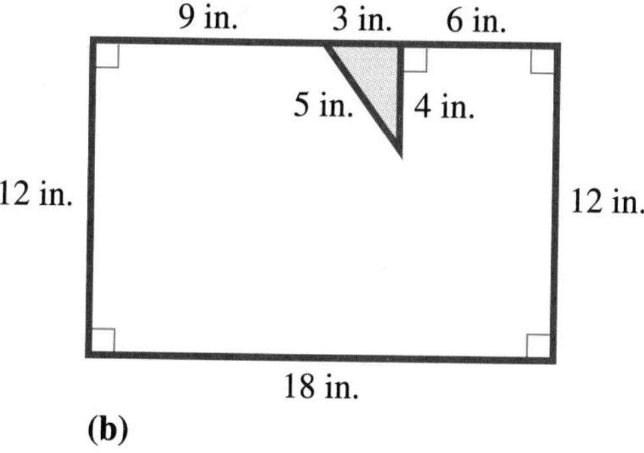

(b)

Solution

The figure is a rectangle with a triangle removed, as shown in Figure 2.22b. The top side of the rectangle is 18 inches long, the same as the base, so the top side of the triangle must be 3 inches. We subtract the area of the triangle from the area of the rectangle:

$$\begin{aligned} A &= (\text{area of rectangle}) - (\text{area of triangle}) \\ &= (lw) - \left(\frac{1}{2}bh\right) \\ &= (18 \cdot 12) - \left(\frac{1}{2} \cdot 4 \cdot 3\right) \\ &= 216 - 6 \\ &= 210 \text{ square inches} \end{aligned}$$

The perimeter of the original figure is the sum of its sides. Remember that the top of the triangle is *not* part of the original figure. The perimeter is

$$P = 9 + 5 + 4 + 6 + 12 + 18 + 12 = 66 \text{ inches}$$

EXERCISE 5

Find the area and perimeter of Figure 2.23.

Figure 2.23

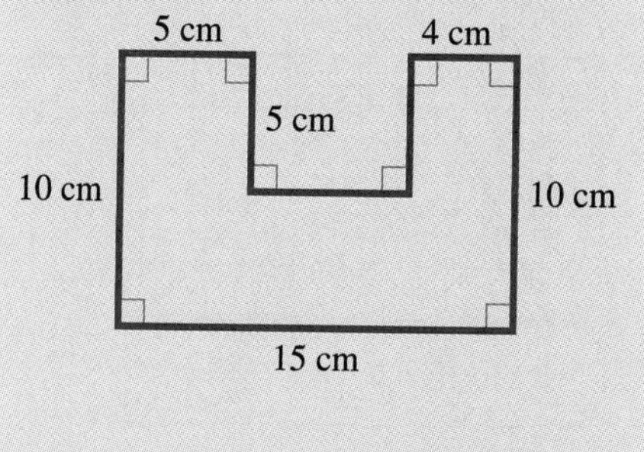

ANSWERS TO 2.4B EXERCISES

3. 6

4. 41 square feet

5. $A = 120$ square centimeters; $P = 60$ centimeters

HOMEWORK 2.4B 

Find the areas of the compound figures. The grids are in centimeters.

1. 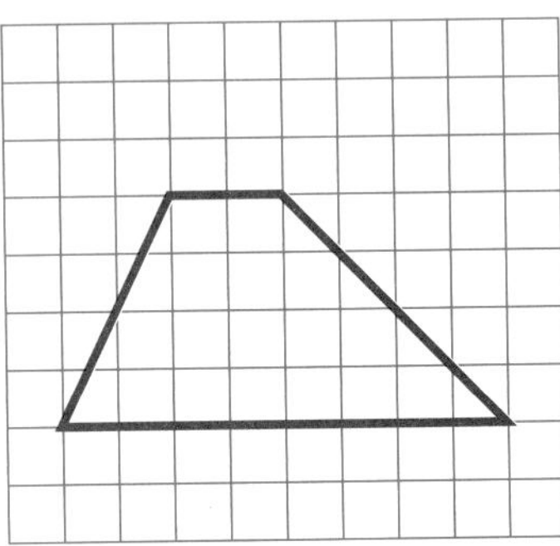

2. 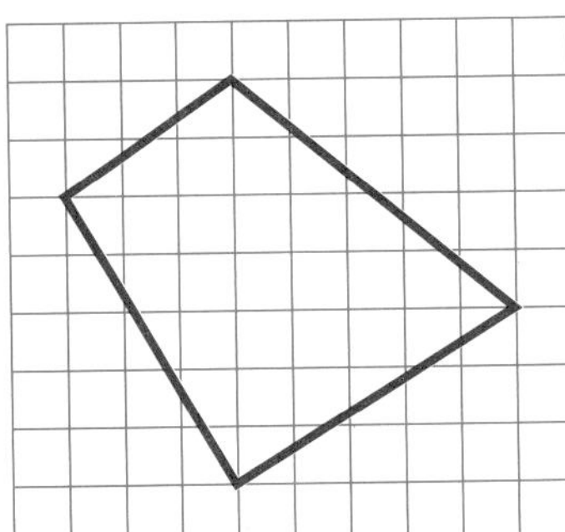

3. 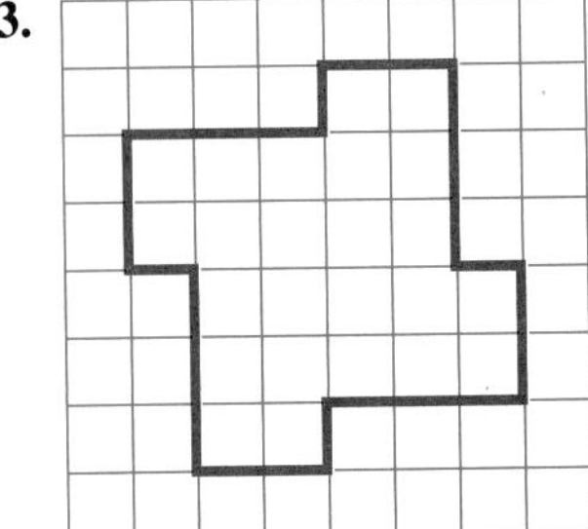

4. 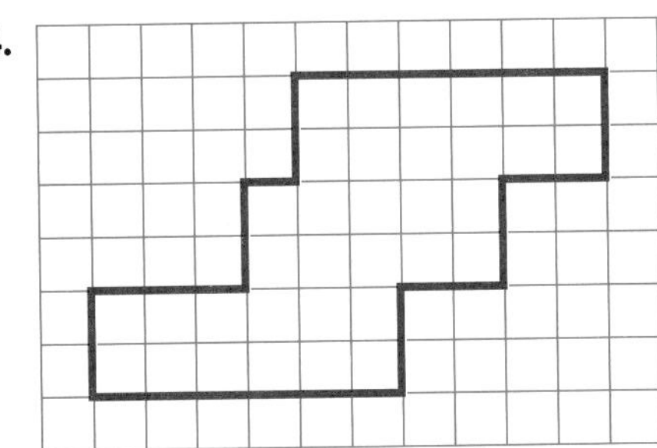

5. 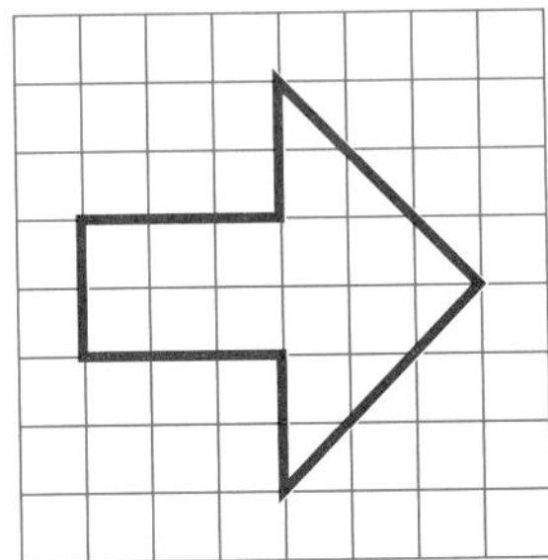

6. 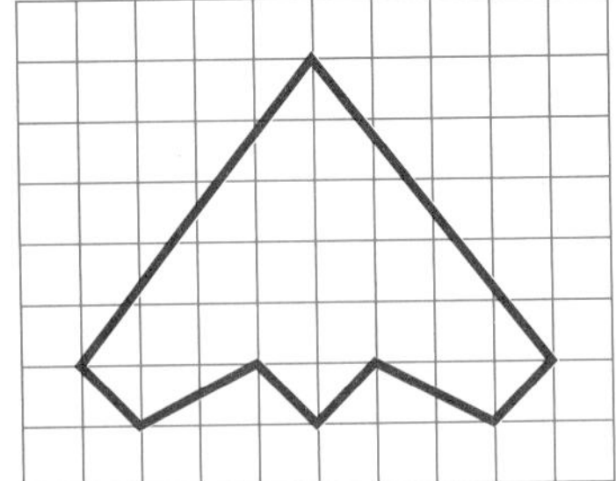

Find the area of each figure.

7.

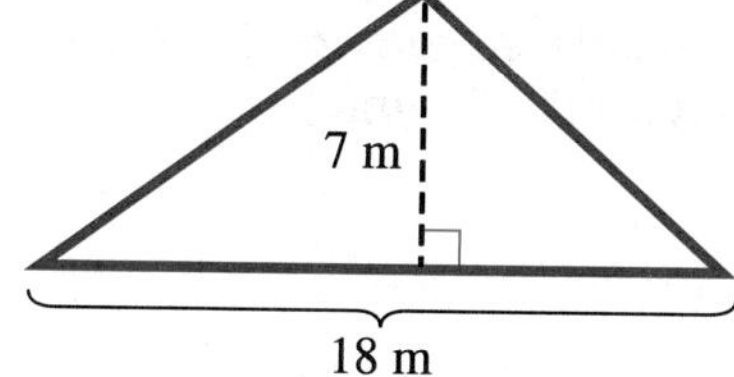

8.

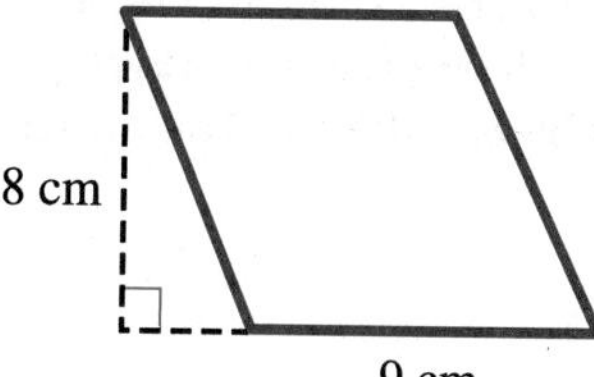

9.

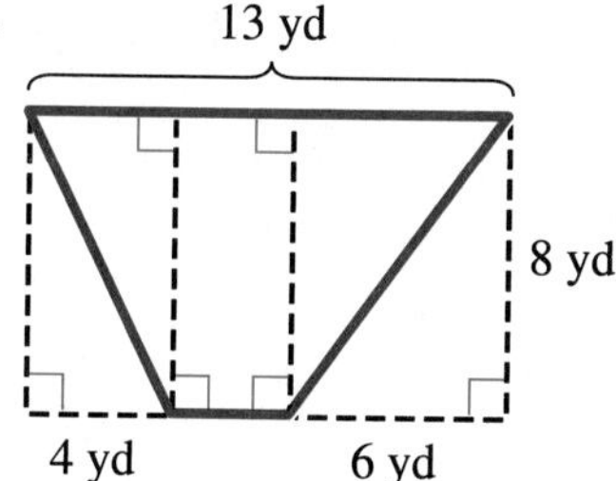

10.

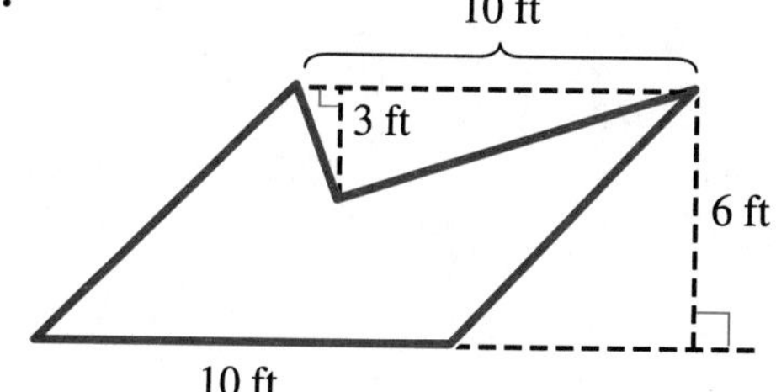

Find the area and perimeter of each figure.

11.

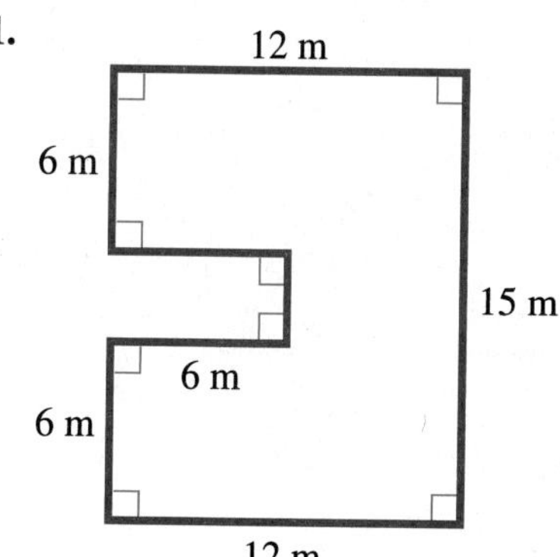

12.

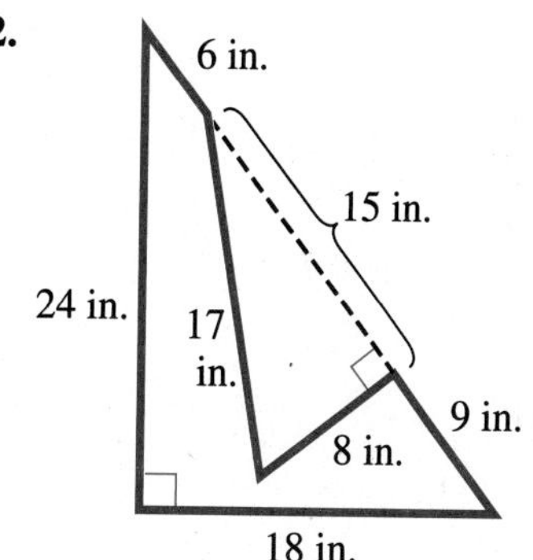

Patchwork quilts are made by sewing individual pieces of fabric together to make a design. The basic patchwork shapes are shown in Figure 2.24. All these shapes—except the last four (trapezoid, kite, hexagon, and octagon)—are examples of rectangles, triangles, and parallelograms. The exceptions can be considered compound figures. Find the area of each patchwork shape. (The dimensions are rounded to the nearest tenth of a unit.)

Figure 2.24

Patchwork Shapes

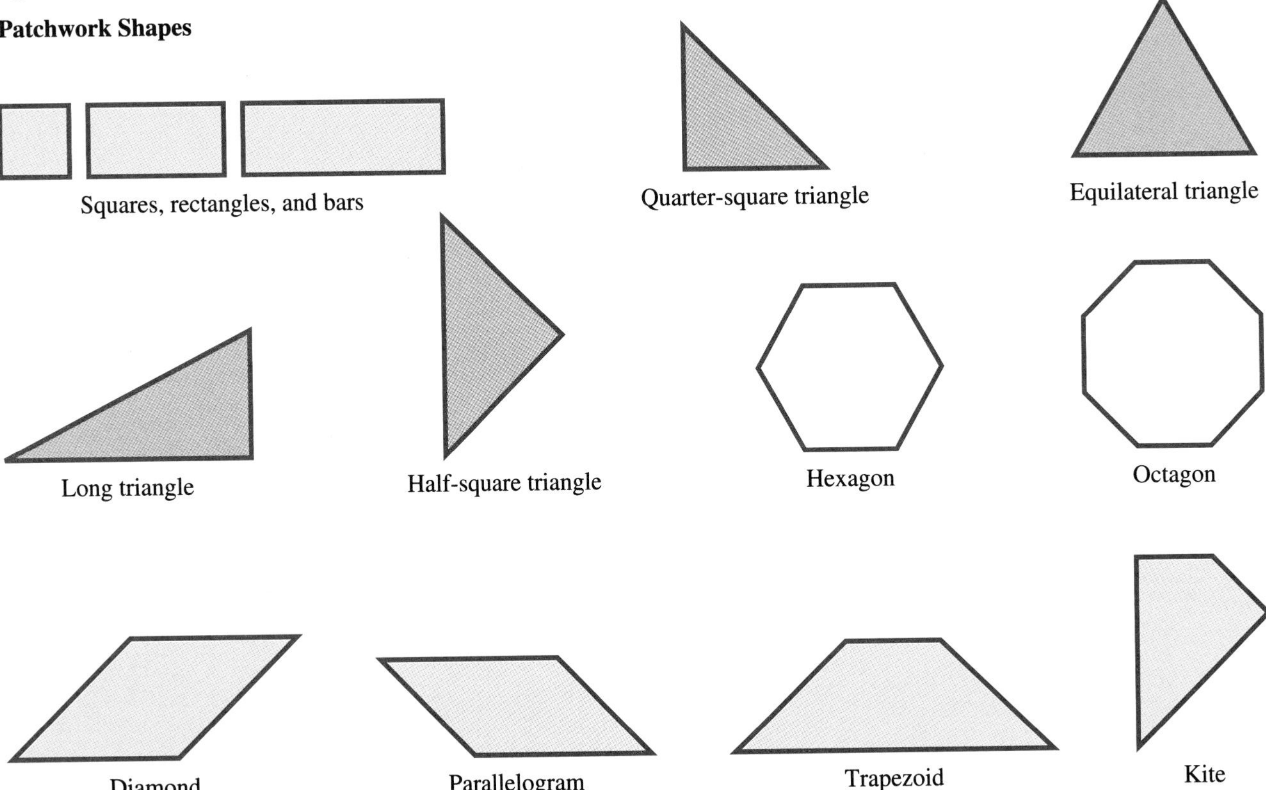

13. To make a kite, first cut out an isosceles right triangle and then trim a smaller isosceles right triangle from one corner, as shown in Figure 2.25.

Figure 2.25

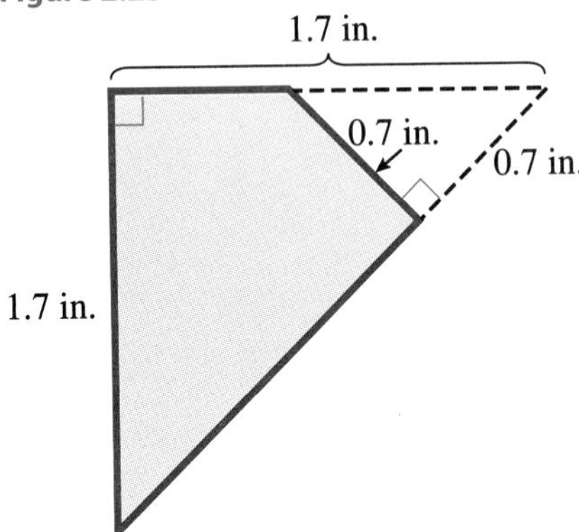

14. To make a trapezoid, start with a rectangle and cut off an isosceles right triangle from each upper corner, as shown in Figure 2.26.

Figure 2.26

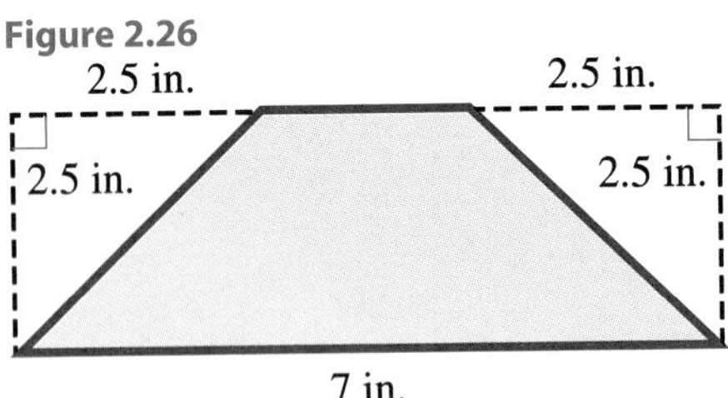

15. To make a hexagon, start with a diamond and trim an equilateral triangle from two opposite corners, as shown in Figure 2.27. (All three sides of an equilateral triangle are equal in length.)

Figure 2.27

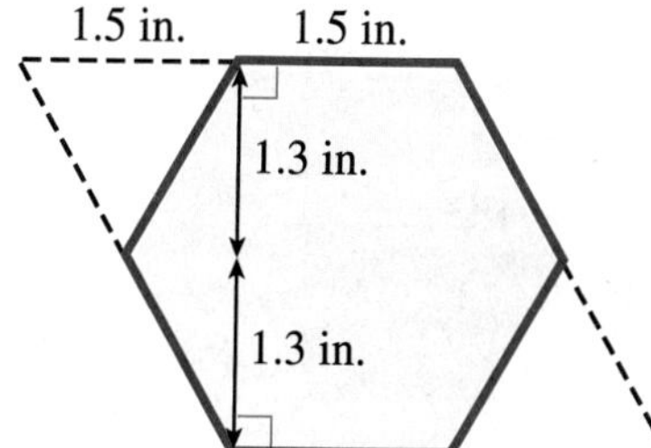

16. To make an octagon, start with a square and trim an isosceles right triangle from each corner, as shown in Figure 2.28.

Figure 2.28

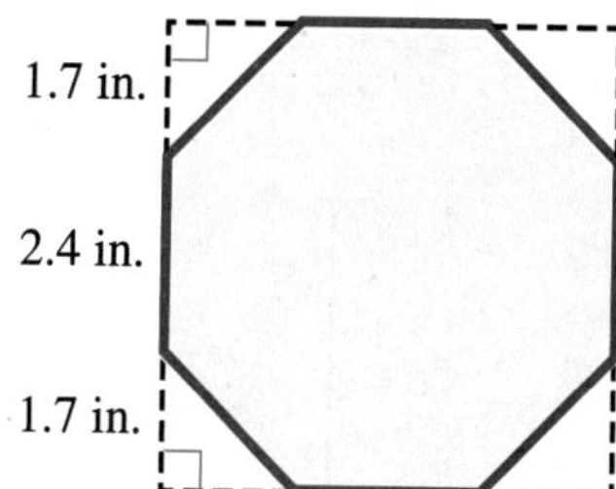

17. The trapezoid in Figure 2.29 is made up of a rectangle and two identical right triangles.

Figure 2.29

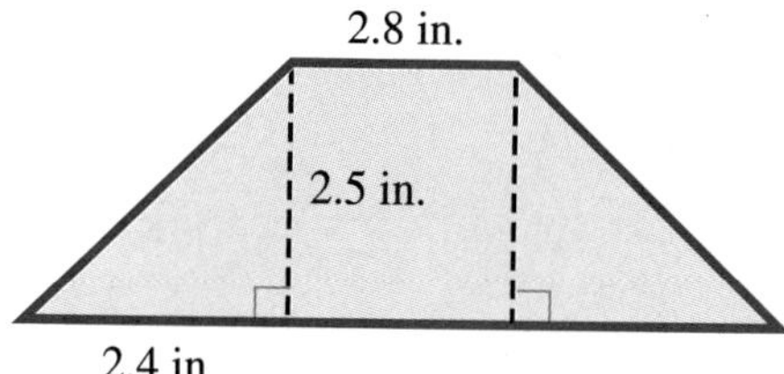

18. The kite in Figure 2.30 is made up of two identical right triangles.

Figure 2.30

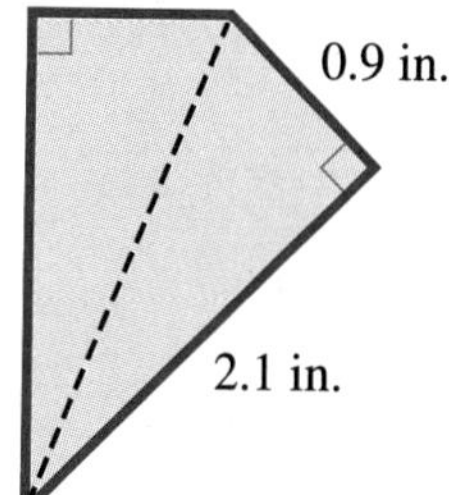

19. The octagon in Figure 2.31 is made up of a square and four identical triangles.

Figure 2.31

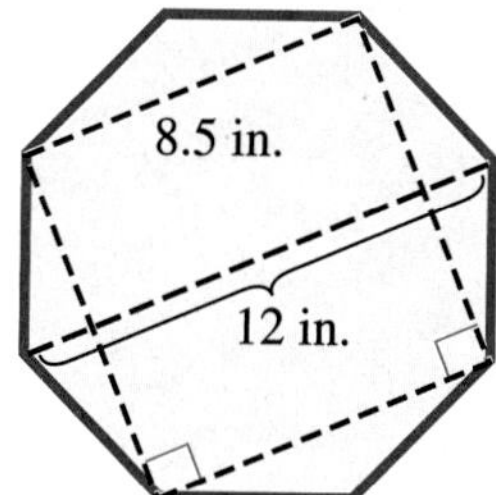

20. The hexagon in Figure 2.32 is made up of a rectangle and two identical triangles.

Figure 2.32

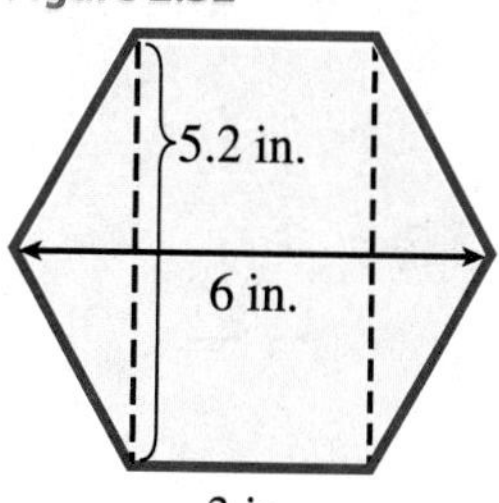

The patchwork pieces are assembled into blocks that make up the pattern of the quilt. Blocks are usually designed on a square grid. Find the total area of each color in the blocks shown. Each block is built on a grid of 1-centimeter squares.

21.

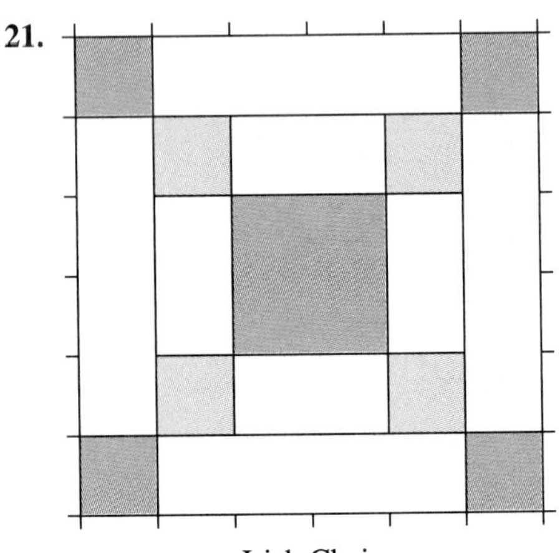

Irish Chain

22.

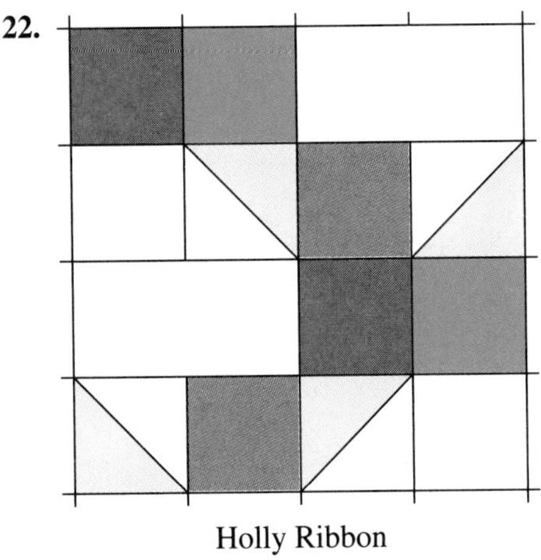

Holly Ribbon

23.

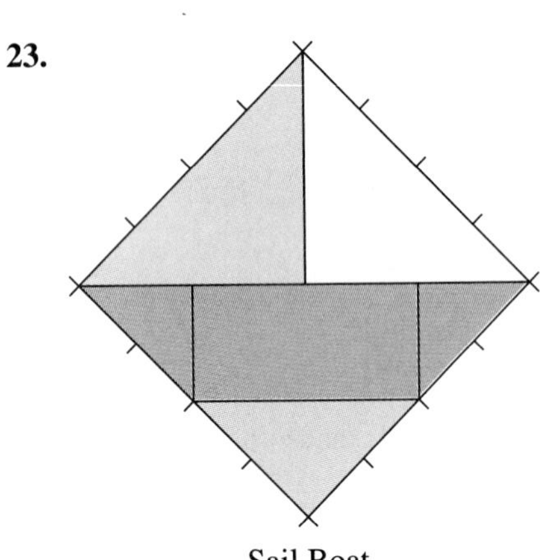

Sail Boat

24.

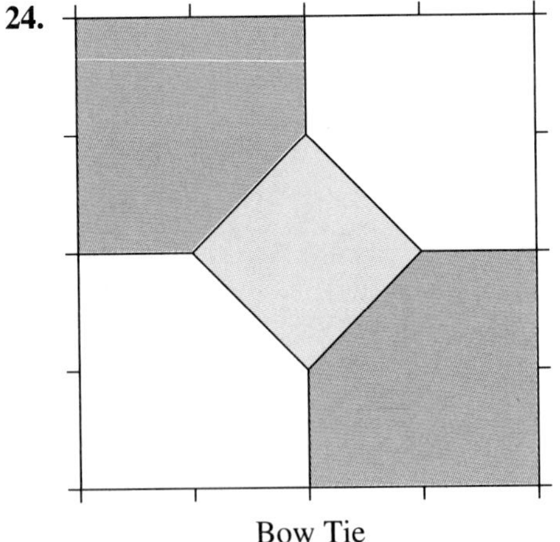

Bow Tie

(a) Choose the appropriate formula on pages 82 and 83 and write an equation. (b) Solve the equation and answer the question.

25. Agnes drove 364 miles in 7 hours. How fast did she drive?

26. A sprinter runs 200 meters in 20 seconds. What is the sprinter's speed?

27. A jet flies 2800 miles at a speed of 560 miles per hour. How long does the trip take?

28. How long will it take a cyclist traveling 13 miles an hour to cover 234 miles?

29. Dinora got 16 people to donate money to her favorite charity, and the average amount donated was $13.25. What was the total amount Dinora collected?

30. The average monthly rainfall in New Orleans, Louisiana, is $4\frac{7}{8}$ inches. How much rainfall does New Orleans receive in an entire year?

Building Number Skills

Use your calculator to find each product. Round your answers to thousandths if necessary. (See Appendix A.6 to review rounding.)

31. $\frac{2}{7}(20)$

32. $\frac{5}{7}(12)$

33. $\frac{7}{6}(25)$

34. $\frac{11}{6}(14)$

35. $\frac{4}{9}(38)$

36. $\frac{2}{9}(25)$

37. $\frac{5}{3}(8)$

38. $\frac{4}{3}(10)$

39. $\frac{11}{12}(37)$

40. $\frac{7}{12}(23)$

2.5 Problem Solving

A. Using Equations to Solve Problems

In this lesson we use equations to help us solve problems. Some problems are fairly easy to translate into mathematical language.

EXAMPLE 1  

The sum of 17 and a number is 45. What is the number?

Solution

Choose a variable—say n—to represent the unknown number. Then write an equation about the number n. Look for mathematical words to help you: "sum" indicates addition and "is" means equal.

The sum of 17 and a number is 45.

$$17 + n = 45$$

Finally, we solve the equation. Since 17 is added to n, we subtract 17 from both sides of the equation.

$$\begin{aligned} 17 + n &= 45 \\ -17 \quad &\quad -17 \\ n &= 28 \end{aligned}$$

The number is 28.

EXERCISE 1

The product of 13 and a number is 598.

a. Write an equation that describes this problem.

b. Solve your equation. What is the number?

ANSWERS TO 2.5A EXERCISE

1a. $13n = 598$

1b. 46

HOMEWORK 2.5A

(a) Translate each problem into an equation. Let n stand for the unknown number. (b) Solve your equation for the unknown number.

1. The product of a number and 6 is 162. What is the number?

2. The quotient of a number and 8 is 13. What is the number?

3. A number decreased by 24 is 38. What is the number?

4. 11 more than a number is 30. What is the number?

5. A number divided by 2.5 is equal to 6.6. What is the number?

6. 58 times a number is 34.8. What is the number?

7. 20 is 15.3 more than a number. What is the number?

8. 18 is the difference of a number and 7.1. What is the number?

See Lesson 1.1B to review bar graphs.

9. The bar graph in Figure 2.33 shows the average family size in various countries.

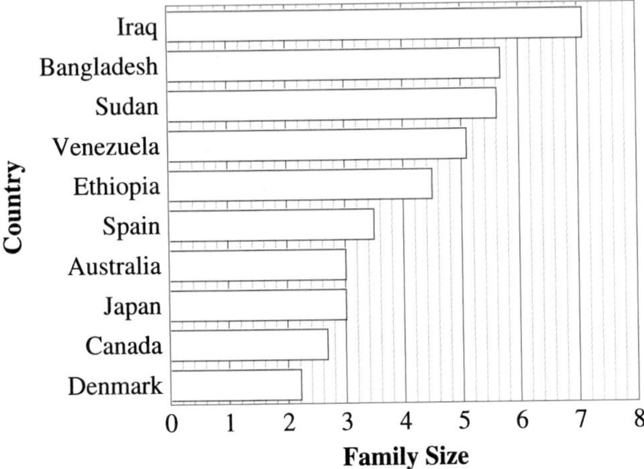

Figure 2.33 Average family size

a. Complete the table with the average family size in the countries listed.

Country	*Family size*	*Women's average age at marriage*
Iraq		22.3
Bangladesh		16.7
Sudan		20.7
Venezuela		21.3
Ethiopia		18.9
Spain		23.1
Australia		22.0
Japan		25.4
Canada		24.3
Denmark		25.6

b. Use the data from the table to make a bar graph showing the average age at marriage for women in the countries listed. Use the grid in Figure 2.34.

Figure 2.34 Women's age at marriage

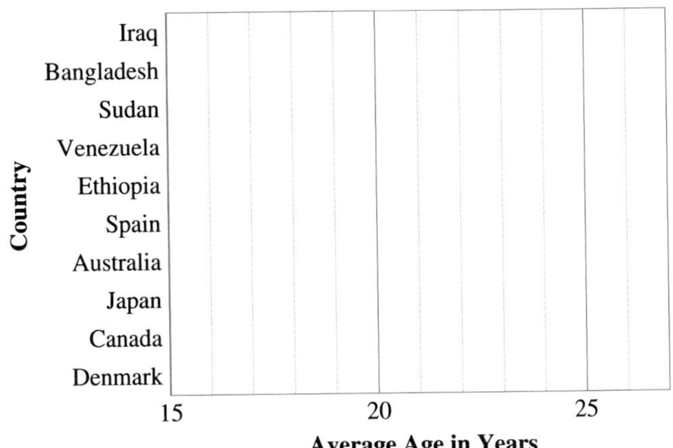

c. In which country do women marry at the youngest age? In which country are they oldest at marriage?

d. Do you think there is a connection between the age at which women marry and their family size? Is it a strong connection, or are there other factors involved? Support your answer with the statistics given here.

10. The bar graph in Figure 2.35 shows the unemployment rate in 1997 for full-time workers age 25 and over, broken down by educational attainment.

Figure 2.35 Unemployment rate by education

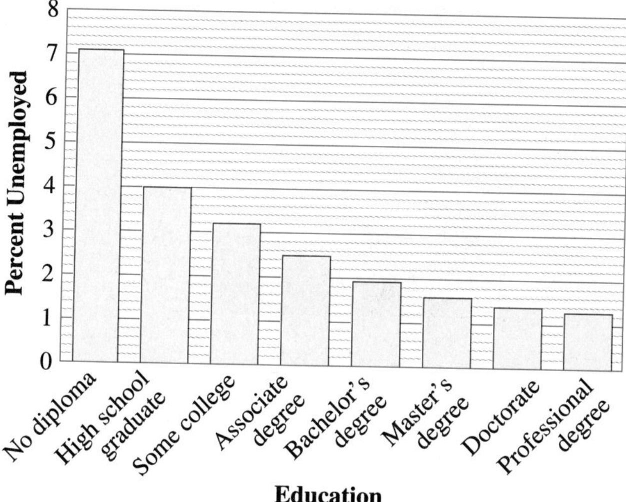

a. Complete the table with the unemployment rates for each group.

Education	*Annual earnings (thousand $)*	*Unemployment rate (%)*
No diploma	19.7	
High school graduate	26.0	
Some college	30.4	
Associate degree	31.7	
Bachelor's degree	40.1	
Master's degree	50.0	
Doctorate	62.4	
Professional degree	71.7	

b. Use the data from the table to make a bar graph showing median annual earnings by educational attainment. Use the grid in Figure 2.36.

Figure 2.36 Annual earnings by education

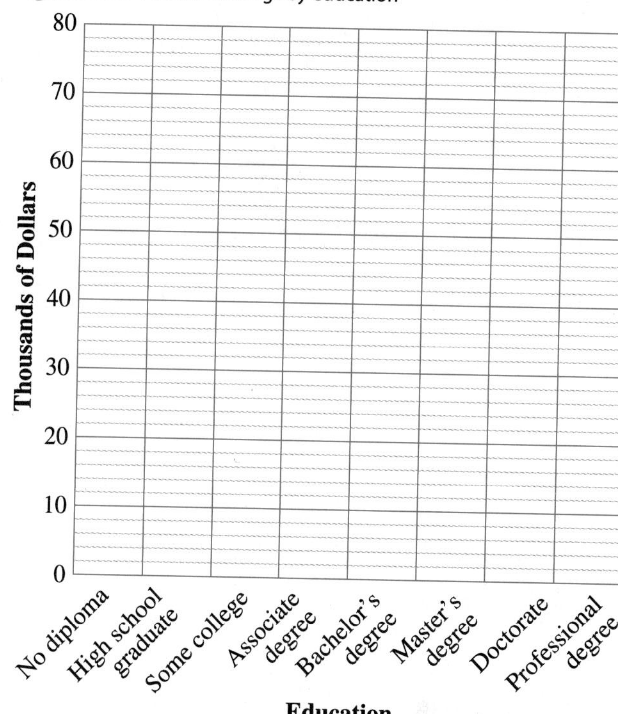

c. By what factor do the salaries of people with bachelor's degrees exceed those of high school graduates? (Divide the earnings of college graduates by the earnings of high school graduates.)

d. Describe the trend in annual earnings as educational attainment increases. Describe the trend in unemployment. Summarize the effect of higher educational attainment on earnings and unemployment.

Find the total area of each color in the quilt blocks. Each block is built on a 1-inch grid. (See Problems 21–24 of Homework 2.4B to read about quilt blocks.)

11.

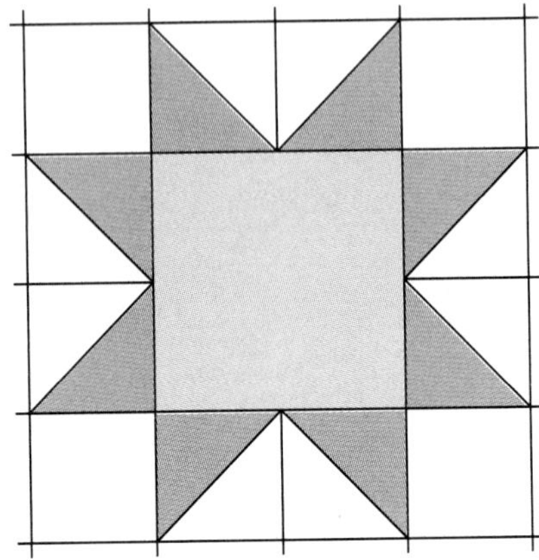

12.

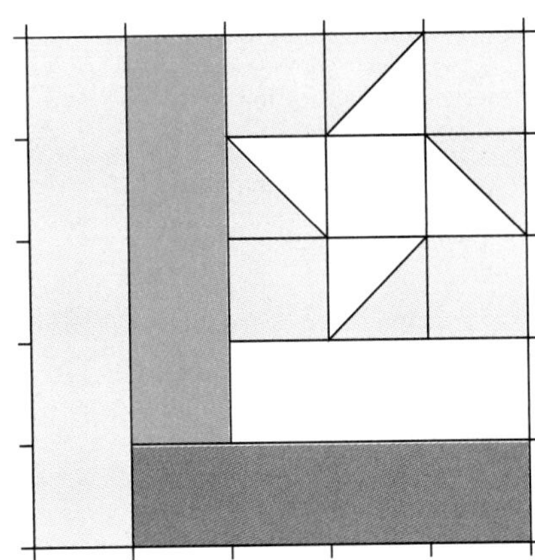

13.

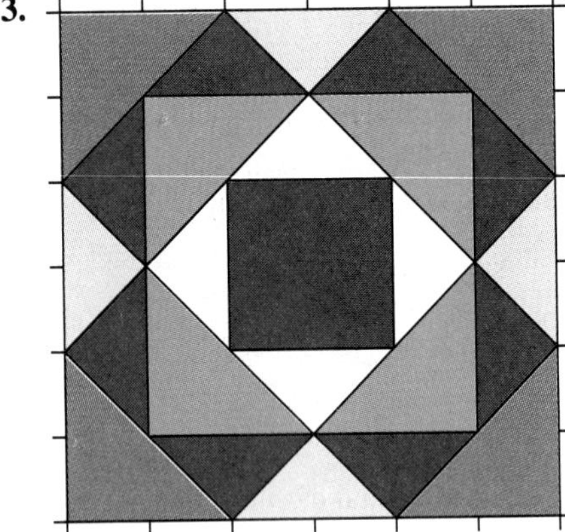

14.

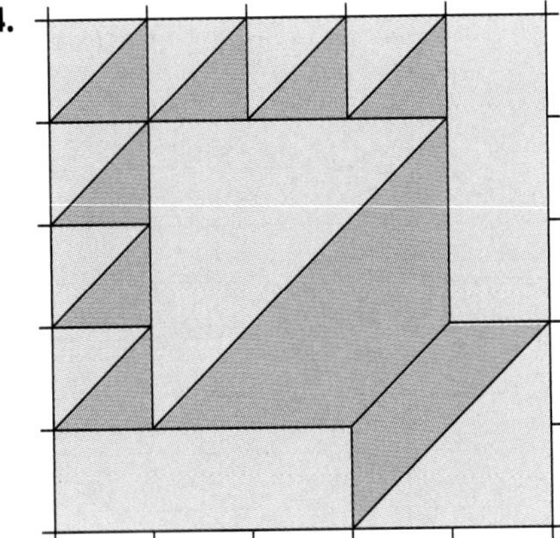

Building Number Skills

You can use your calculator to perform many calculations involving fractions. Just change each fraction into decimal form first.

Example: $3\frac{1}{4} - 2\frac{3}{4} = 3.25 - 2.75 = 0.5$ (Recall that to change $3\frac{1}{4}$ to a decimal number, you enter: 3 [+] 1 [÷] 4 [=].)

Use your calculator to convert each fraction to decimal form and then compute.

15. $\frac{3}{8} + \frac{4}{5}$

16. $\frac{4}{3} + \frac{1}{6}$

17. $\frac{11}{12} - \frac{1}{6}$

18. $\frac{3}{4} - \frac{5}{8}$

19. $8 - 2\frac{1}{3}$

20. $5 - 3\frac{7}{8}$

21. $4\frac{1}{6} - 2\frac{2}{3}$

22. $5\frac{3}{8} - 4\frac{3}{4}$

23. $6\frac{5}{8} + 3\frac{2}{5}$

24. $8\frac{4}{5} + 6\frac{1}{2}$

B. Modeling a Problem

For most problems we must write an equation using information given in the problem. Pay close attention to the method outlined in Example 2.

EXAMPLE 2

Robbie is 28 years younger than his Uncle Neil. If Robbie is 9 years old, how old is Neil?

Solution

Step 1a What are we asked to find in the problem?

Neil's age

Step 1b Choose a variable to represent the unknown quantity:

Neil's age: N

Step 2a Find something in the problem that we can express in *two different* ways:

Robbie's age: $N - 28$ (Robbie is 28 years younger than Neil.)

9 (Robbie is 9 years old.)

Step 2b Write an equation using the expressions from Step 2a:

$$N - 28 = 9$$

Step 3a Solve the equation.

$$\begin{aligned} N - 28 &= 9 \\ +28 \quad & \quad +28 \\ \hline N &= 37 \end{aligned}$$

Add 28 to both sides.

Step 3b Answer the question in the problem.

Neil is 37 years old.

EXERCISE 2

Megan and her cat Silky together weigh 83 pounds. Megan alone weighs 65 pounds.

a. Let W stand for Silky's weight in pounds. Find two different expressions for the sum of Megan's and Silky's weights and write an equation.

b. Solve your equation. How much does Silky weigh?

We summarize our method in three steps.

Steps for Modeling a Problem

Step 1 Identify the unknown quantity and choose a variable to represent it.

Step 2 Find some quantity that can be expressed in two different ways and write an equation.

Step 3 Solve the equation and answer the question in the problem.

There are two things to remember about writing equations to solve problems.

1. It is very important to specify *precisely* what the variable represents. You will have to write an equation about this variable, so you must have a clear idea of what it stands for.
2. Although the equation *includes* the variable, the two sides of the equation may actually be expressions for some other quantity. In Example 2 the variable N stands for Neil's age, but the equation gives two ways to express Robbie's age.

EXAMPLE 3  

The ratio of calories from saturated fat to calories from polyunsaturated fat in most margarine is 0.4. If 1 tablespoon of margarine contains 45 calories from polyunsaturated fat, how many calories from saturated fat does 1 tablespoon of margarine contain?

Solution

Step 1 Calories from saturated fat: s

Step 2 Write the ratio of calories from saturated fat to calories from polyunsaturated fat in two different ways:

$$\frac{s}{45} = 0.4$$

Step 3 Solve the equation (multiply both sides by 45):

$$\mathbf{(45)}\frac{s}{45} = 0.4\mathbf{(45)}$$

$$s = 18$$

There are 18 calories from saturated fat in 1 tablespoon of margarine.

EXERCISE 3

Frank must reduce his daily calorie intake by 260 calories. If his goal is 1350 calories per day, what is his current calorie intake?

a. What are we asked to find? Choose a variable for that quantity.

b. Find two different ways to express Frank's dieting goal and write an equation.

c. Solve your equation and answer the question.

ANSWERS TO 2.5B EXERCISES

2a. $w + 65 = 83$

2b. 18 pounds

3a. Frank's current calorie intake: c

3b. $c - 260 = 1350$

3c. 1610 calories per day

HOMEWORK 2.5B 

Follow the steps to solve each problem.

1. Lupé spent $24 at the Craft Fair. She now has $39 left. How much did she have before the Craft Fair?
   a. What are we asked to find? Choose a variable to represent it.
   b. Find two ways to express the amount of money Lupé had after the Craft Fair and write an equation.
   c. Solve the equation and answer the question in the problem.

2. The sale price on a washing machine is $29 less than the regular price. The sale price is $258. What is the regular price?
   a. What are we asked to find? Choose a variable to represent it.
   b. Find two ways to express the sale price of the washing machine and write an equation.
   c. Solve the equation and answer the question in the problem.

3. Danny weighs 32 pounds more than Brenda. If Danny weighs 157 pounds, how much does Brenda weigh?
   a. What are we asked to find? Choose a variable to represent it.
   b. Find two ways to express Danny's weight and write an equation.
   c. Solve the equation and answer the question in the problem.

4. Trinh is 14 years older than his brother Loc. If Trinh is 23, how old is Loc?
   a. What are we asked to find? Choose a variable to represent it.
   b. Find two ways to express Trinh's age and write an equation.
   c. Solve the equation and answer the question in the problem.

5. Miranda worked 20 hours this week and made $136. What is Miranda's hourly wage?

a. What are we asked to find? Choose a variable to represent it.

b. Find two ways to express Miranda's total earnings and write an equation.

c. Solve the equation and answer the question in the problem.

6. Bruce goes jogging on the same course every morning except Sundays, when he rests. His weekly mileage is 57 miles. How long is the course?

a. What are we asked to find? Choose a variable to represent it.

b. Find two ways to express Bruce's weekly mileage and write an equation.

c. Solve the equation and answer the question in the problem.

7. Struggling Students Gardening Service splits their profit equally among their eight members. If each member made $64 last week, what was the total profit?

a. What are we asked to find? Choose a variable to represent it.

b. Find two ways to express each member's share and write an equation.

c. Solve the equation and answer the question in the problem.

8. Judy's average score on 12 homework assignments was 17.5 points. What was the total number of points she earned?

a. What are we asked to find? Choose a variable to represent it.

b. Find two ways to express Judy's average score and write an equation.

c. Solve the equation and answer the question in the problem.

Solve each problem by writing and solving an equation. Follow the three steps on page 114.

9. Corey worked for 5 hours longer than Shant. If Corey worked for 18 hours, how long did Shant work?

10. Mac's bowling score is 28 points higher than Tyrone's. If Mac's score is 204, what is Tyrone's score?

11. Abby is 3 times as old as Rina. If Abby is 21 years old, how old is Rina?

12. For each camp counselor, there are 8 children. If there are 96 children, how many camp counselors are there?

13. Wiley has 17 fewer compact discs (CDs) than his brother, Will. If Wiley has 29 CDs, how many CDs does Will have?

14. Mishell weighs 28 pounds less than her dog, Bear. If Mishell weighs 96 pounds, how much does Bear weigh?

15. Caroline Gottrich divided her fortune equally among her three no-good sons and her cat. Each beneficiary received $350,000. How much was Caroline's fortune?

16. The ratio of your quiz score to 80 points is 0.85. What is your quiz score?

17. After writing a check for $2378, Averil's bank account shows a balance of $1978. How much was in Averil's account before the check cleared?

18. Akiko's great-grandmother says she was 14 years old when the hotel burned down. If Akiko's great-grandmother is 92 years old now, how long ago did the hotel burn down?

19. The bar graphs in Figure 2.37 give data about living conditions in several Pacific Rim cities. The number of people per room indicates how crowded the living conditions are (Figure 2.37a). Infant mortality gives the number of babies who die for every 1000 babies born (Figure 2.37b).

Figure 2.37 Living conditions in selected Pacific Rim cities

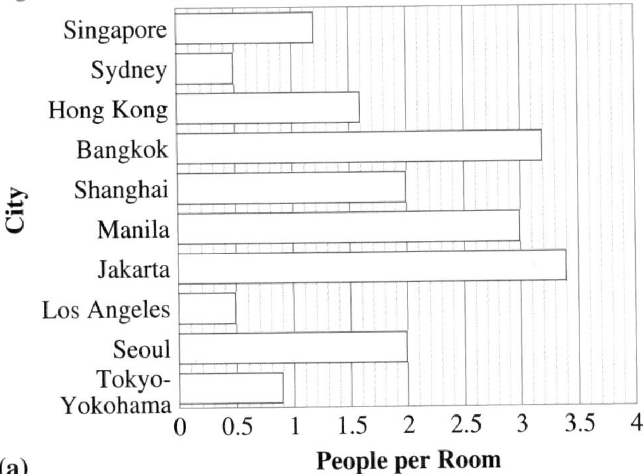

(a)

Figure 2.37 Infant mortality in selected Pacific Rim cities

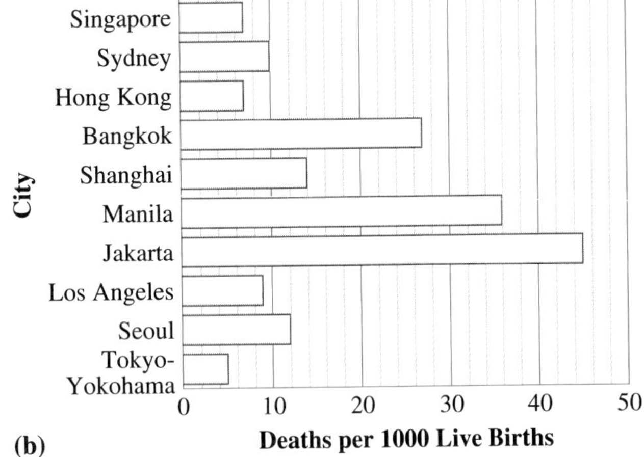

(b)

a. Use the bar graphs to complete the table. For the number of people per room, give your answers to the nearest tenth.

City	*People per room*	*Deaths per 1000 births*
Singapore		
Sydney		
Hong Kong		
Bangkok		
Shanghai		
Manila		
Jakarta		
Los Angeles		
Seoul		
Tokyo-Yokohama		

b. Which city has the highest infant mortality rate? Which city has the most crowded living conditions?

c. Which city has the lowest infant mortality rate? Which city has the least crowded living conditions?

d. Describe any relationship you see between the two variables, people per room and infant mortality. Name several other variables that could help explain the relationship.

20. The bar graph in Figure 2.38 shows the number of murders annually per 100,000 people in several Pacific Rim cities.

Figure 2.38 Annual murders in selected Pacific Rim cities

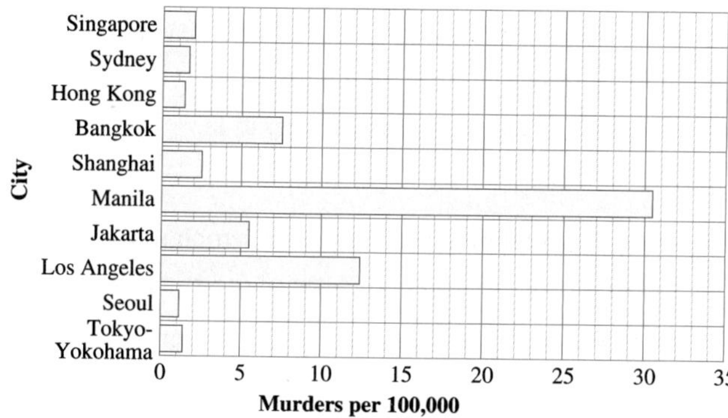

a. Use the bar graph to fill in the second column of the table. Estimate your answers to the nearest tenth.

b. Which city has the highest annual murder rate? Which city has the lowest rate?

c. Calculate the number of murders that occur in each city annually, and complete the table. (You will need to use the fact that 1 million = 10 × 100,000.)

City	*Population (millions)*	*Muders per 100,000 people*	*Murders*
Singapore	3.1		
Sydney	3.6		
Hong Kong	5.2		
Bangkok	7.0		
Shanghai	9.2		
Manila	9.2		
Jakarta	9.9		
Los Angeles	11.5		
Seoul	15.8		
Tokyo-Yokohama	28.7		

d. Which city has the most murders annually? Which city is second?

Building Number Skills

Use your calculator to convert each fraction to decimal form and then compute.

Example: $4\left(3\frac{1}{8}\right) = 4(3.125) = 12.5$

21. $\frac{7}{8}\left(\frac{4}{5}\right)$

22. $\frac{5}{12}\left(\frac{3}{10}\right)$

23. $15\left(2\frac{1}{3}\right)$

24. $9\left(5\frac{5}{6}\right)$

25. $\frac{3}{8} \div \frac{5}{16}$

26. $\frac{7}{16} \div \frac{1}{4}$

27. $1 \div \frac{5}{8}$

28. $1 \div \frac{4}{5}$

29. $6 \div 2\frac{2}{5}$

30. $12 \div 4\frac{1}{2}$

2 Summary

Lesson 2.1

An **equation** is a statement that two expressions are equal.

A **solution** of an equation is a value of the variable that makes the equation true.

To **solve** an equation means to find its solution (or solutions).

Each of the four arithmetic operations has an opposite operation.

To Solve an Equation

Step 1 Ask yourself: Which operation has been performed on the variable?

Step 2 Perform the opposite operation on both sides of the equation.

Step 3 Check your solution.

Lesson 2.2

The **perimeter** of a two-dimensional figure is the *distance* around the border of the figure.

To find the perimeter, we measure the length of each side and add the lengths.

The **area** of a two-dimensional figure is a measure of the amount of space enclosed by the figure.

To find the area of a figure, we count the number of square units enclosed by the figure.

To measure the length of a line segment with a ruler, line up the left end of the ruler with one end of the segment and read the number on the ruler at the other end of the segment.

To estimate the area of a figure, piece together parts of square units enclosed by the figure to make up whole squares.

Lesson 2.3

A **formula** is an equation that relates two or more variables. Some useful formulas:

$d = rt$	distance = rate · time
$P = R - C$	profit = revenue − costs
$s = p - d$	sale price = regular price − discount
$e = \dfrac{m}{g}$	efficiency = $\dfrac{\text{miles}}{\text{gallons}}$
$u = \dfrac{p}{n}$	unit cost = $\dfrac{\text{total price}}{\text{number of items}}$
$P = rW$	part = percentage rate · whole
$I = Prt$	interest = principal · rate · time
$A = \dfrac{S}{n}$	average value = $\dfrac{\text{sum of values}}{\text{number of values}}$

Depending on which variable we want to find, we may use a formula by evaluating or by solving.

Lesson 2.4

A **rectangle** is a four-sided figure in which all of the angles are right angles, or 90°.

A **triangle** is a three-sided figure. Any one of the three sides can be designated as the **base** of the triangle. The perpendicular distance from the base to the opposite vertex is called the **height** of the triangle.

An **isosceles triangle** has two sides of equal length.

In a **right triangle,** one of the angles is a right angle, or 90°.

A **parallelogram** is a four-sided figure whose opposite sides are parallel.

The opposite sides of a parallelogram are equal in length. A **rhombus (diamond)** is a parallelogram with all four sides of equal length. (See Figure 2.39.)

Figure 2.39

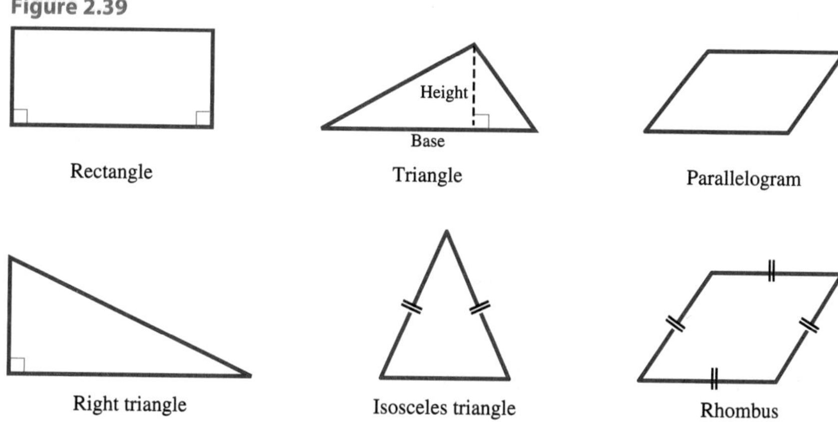

The areas of these figures can be computed using the following formulas:

Rectangle: $a = lw$

Triangle: $A = \frac{1}{2}bh$

Parallelogram: $A = bh$

A figure that can be divided into rectangles and triangles is called a **compound** figure.

We can find the area of a compound figure by finding the areas of the triangles and rectangles and then adding those areas.

Lesson 2.5

We can use equations to help us solve problems. We first translate the problem into mathematical language.

Steps for Modeling a Problem

Step 1 Identify the unknown quantity and choose a variable to represent it.

Step 2 Find some quantity that can be expressed in two different ways, and write an equation.

Step 3 Solve the equation, and answer the question in the problem.

CHAPTER 2 REVIEW 

Match each word to its definition.

equation	area	formula	revenue
solution	perimeter	parallelogram	principal

1. A value of the variable that makes an equation true

2. The distance around a geometric figure

3. A four-sided figure whose opposite sides are parallel

4. The amount of money invested in an interest-bearing account

5. An equation relating two or more variables

6. A statement that two algebraic expressions are equal

7. A measure of the amount of space enclosed by a geometric figure

8. The amount of money received from a sale or a business venture

Explain the difference between each pair of mathematical words. (Check that all these words are included in your glossary.)

9. variable, constant

10. expression, equation

11. terms, factors

12. evaluate, solve

13. area, perimeter

14. principal, interest

15. profit, revenue

16. total cost, unit cost

17. base, height

18. rectangle, parallelogram

Evaluate each expression for the given value(s) of the variable(s).

19. $45 - t$, for $t = 17$

20. $P - 87$, for $P = 111$

21. lw, for $l = 13$ and $w = 6$

22. bh, for $b = 12$ and $h = 15$

23. $\frac{560}{r}$, for $r = 35$

24. $\frac{d}{16}$, for $d = 288$

25. $x + 3.7$, for $x = 4.4$

26. $0.7 + z$, for $z = 9.5$

27. $4ac$, for $a = 2$ and $c = 3$

28. $3.14rr$, for $r = 10$

Write an algebraic expression for each English phrase. State what your variable represents.

29. 5 times the height of a normal doorway

30. The weight of a coffin divided by 6

31. \$8 dollars subtracted from the price of the book

32. 12 more than the number of chairs

33. The ratio of 156 to the number of sheep

34. The weight of the bicycle increased by 5 pounds

35. The regular fare reduced by \$250

36. Half the weight of standard laptop computer

37. 20 minutes less than the length of the play

38. The product of her annual income and 0.27

State what the variables represent in the following formulas, we have studied. (Include a sketch for Problems 47 and 48.)

39. $P = rW$

40. $s = p - d$

41. $d = rt$

42. $P = R - C$

43. $e = \frac{m}{g}$

44. $u = \frac{p}{n}$

45. $I = Prt$

46. $a = \frac{S}{n}$

47. $A = \frac{1}{2}bh$

48. $A = lw$

Write an equation that expresses the second variable in terms of the first variable.

49.

x	y
1	2
2	4
3	6
4	8

50.

a	b
1	3
2	4
3	5
4	6

51.

i	j
20	11
40	31
60	51
80	71

52.

c	d
20	5
40	10
60	15
80	20

53.

p	q
2	3
4	6
6	9
8	12

54.

r	s
3	4
6	8
9	12
12	16

55.

u	v
1	60
2	30
3	20
4	15

56.

m	n
1	9
2	8
3	7
4	6

Solve each equation and check your solutions.

57. $19 + a = 32$

58. $37 = b + 13$

59. $22 = c - 11$

60. $d - 26 = 36$

61. $13h = 52$

62. $132 = 11k$

63. $\frac{m}{4} = 8$

64. $18 = \frac{n}{6}$

65. $9.6 = 1.6p$

66. $4.8 = \frac{w}{1.2}$

Answer each question by applying a formula or definition.

67. A rectangle is 5 feet wide by 8 feet long. What is its perimeter?

68. What is the area of a rectangle whose length is 8 centimeters and whose width is 6 centimeters?

69. A gym bag costs $1.73 to produce and sells for $7. What is the profit from selling one bag?

70. Bart invested $5000 at a 5% annual interest rate for 1 year. How much interest did he earn?

71. The Co-op Shop donates 5% of its profits to charity. If the Co-op Shop made $12,900 in May, how much did it donate to charity that month?

72. A car travels 444 miles on 15 gallons of gasoline. What is its fuel efficiency (in miles per gallon)?

73. The area of a rectangle is 32 square meters and its width is 4 meters. What is the length of the rectangle?

74. Milk costs 3¢ per ounce. How much does a 32-ounce carton of milk cost?

75. How long will it take to travel 121 miles at a constant speed of 55 miles per hour?

76. Jurgen earned $76 interest in 1 year on an account that paid 3.8% annual interest. What was Jurgen's initial investment?

(a) Choose a variable to represent the unknown quantity, (b) write an equation for the problem, and (c) solve your equation and answer the question in the problem.

77. The difference in age between two churches is 173 years. How old is the older church when the younger church is 238 years old?

78. The sale price of a bracelet was $24.99, which was discounted $17 from the original price. What was the original price of the bracelet?

79. The Flying Saucer ride at the amusement park allows a total of 650 pounds on each saucer. If Mrs. Grape weighs 483 pounds, how much can her son weigh and still ride with his mother?

80. Harris Aircraft Company allows any of its employees to retire with full benefits when the sum of the employee's age plus her number of years with the company totals 65. Charlotte retired as soon as the company allowed, when she was 47 years old. How long was she with the company?

81. Six actors split their lottery winnings equally when a ticket they bought won the jackpot. If each actor received $2,535,000, how large was the jackpot?

82. A piece of string is cut into five equal pieces, each of length 40 centimeters. How long was the original piece of string?

83. A pendulum needs 4 seconds to swing back and forth 5 times. How long does the pendulum need to swing back and forth once?

84. Twelve calculators laid end to end form a line 81 inches long. How long is each calculator?

85. A poll reports that 65% of voters favor a new gun control bill. If there were 663 voters who said they favored the bill, how many voters in all were polled?

86. The average weight of Laurence's six textbooks is 12.6 pounds. What is the total weight of his texts?

Find the area and perimeter of each figure. Each square represents 1 square centimeter.

87. Figure A:

Perimeter: __________

Area: __________

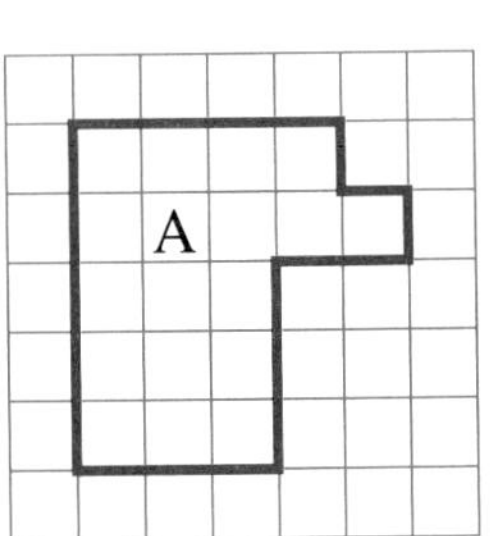

88. Figure B:

Perimeter: __________

Area: __________

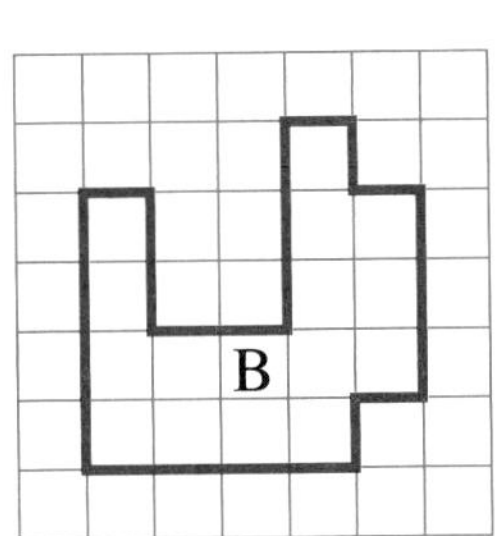

Find the area and perimeter of each figure. The grids are marked off in centimeters, so use a centimeter ruler to measure the lengths of the slanted sides.

89. Figure C:

Perimeter: __________

Area: __________

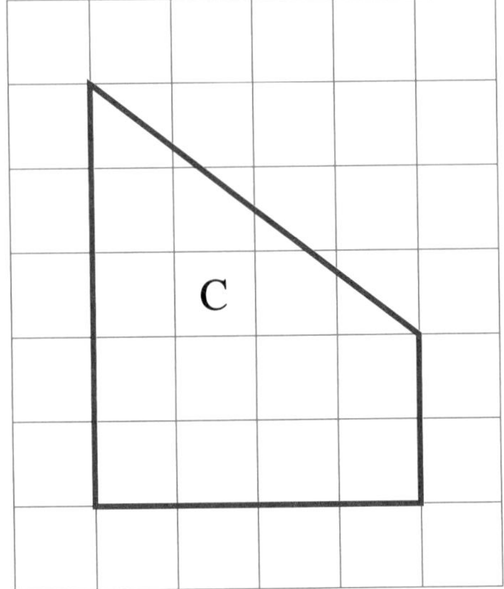

90. Figure D:

Perimeter: __________

Area: __________

D

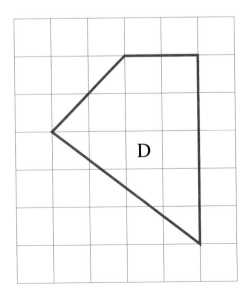

Find the area of each figure using formulas from Lesson 2.4. Each square represents 1 square centimeter.

91.

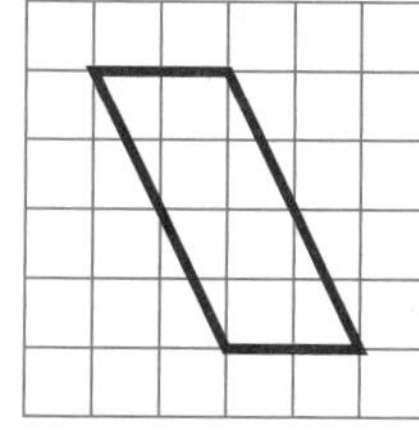

92.

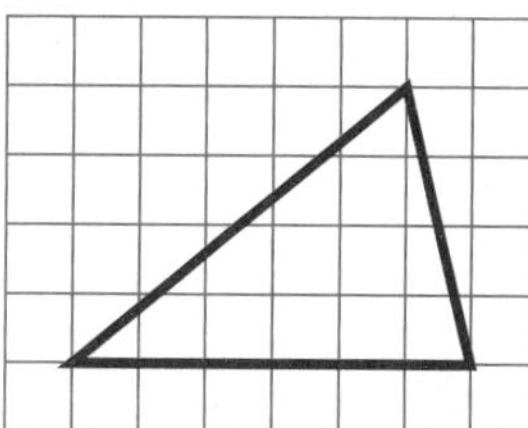

93.

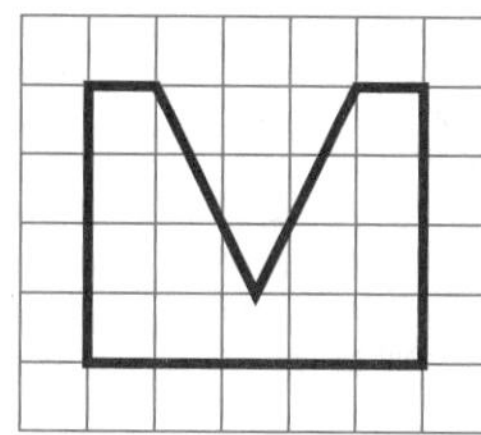

94.

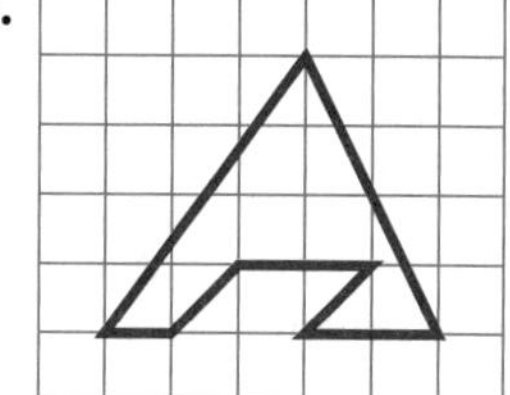

Building Number Skills

Choose the larger number in each pair.

95. 3.012 or 3.11

96. 0.0200 or 0.20

97. $\frac{5}{6}$ or 0.833

98. $\frac{5}{12}$ or 0.417

99. Is 4.789 closer to 4.78 or 4.79?

100. Is 0.0015 closer to 0.00 or 0.01?

Convert to a decimal.

101. $8\frac{1}{8}\%$

102. $16\frac{2}{3}\%$

Round each decimal to hundredths.

103. 2.0852

104. 3.998

105. 1.0047

106. 12.1463

Use your calculator to convert each number to decimal form. Round your answers to thousandths.

107. $\frac{11}{13}$

108. $\frac{7}{3}$

109. $2\frac{1}{6}$

110. $4\frac{3}{7}$

Use your calculator to compute. Write your answers with repeater bar notation.

111. $2\frac{3}{4} \times 3\frac{1}{3}$

112. $6\frac{1}{2} - 5\frac{4}{9}$

113. $20 \div 2\frac{2}{5}$

114. $1\frac{3}{8} + 4\frac{2}{3}$

CHAPTER

3

Signed Numbers

3.1 Negative Numbers

A. What Are Negative Numbers?

Fill in the following list with opposites. The first two are done for you.

Activity

1. 5 feet above ground	**5 feet below ground**
2. **Earn $20**	Spend $20
3. Gain 15 yards	____________
4. ____________	12° below 0
5. 6 pounds heavier	____________
6. ____________	3 weeks ago
7. Win $5	____________
8. ____________	4 miles behind
9. Climb 400 feet	____________
10. ____________	2 years younger

We can use a thermometer to visualize opposites. On the thermometer shown in Figure 3.1, the opposite of 5° *above* 0 is 5° *below* 0. We also call 5° below 0 negative 5°, which is written as $-5°$. The symbol $-$ means "negative" or "opposite of." Temperatures below 0° are represented by **negative numbers.** Temperatures above 0° are represented by **positive numbers.**

Figure 3.1

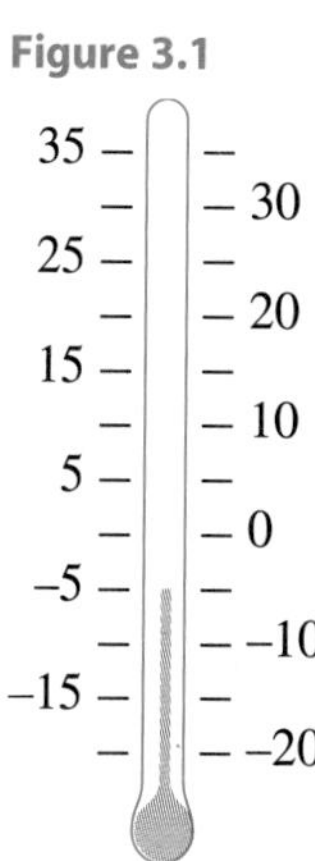

EXERCISE 1

Fill in the thermometer in Figure 3.3 to show $-4°$.

Figure 3.3

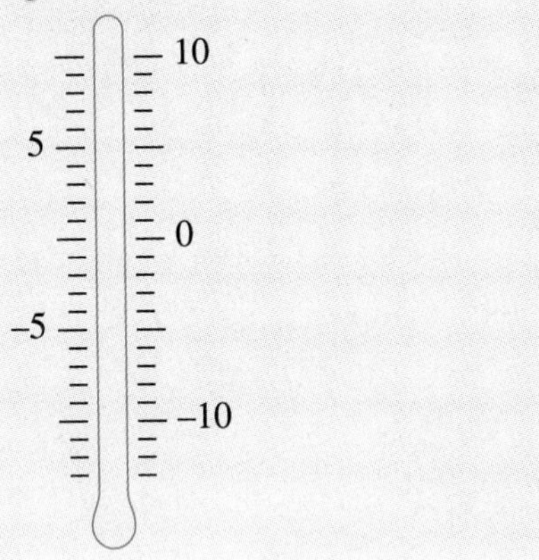

EXAMPLE 1

Sketch thermometers that show the indicated temperatures.

a. $14°$ **b.** $-8°$

Solution

Figure 3.2

a.

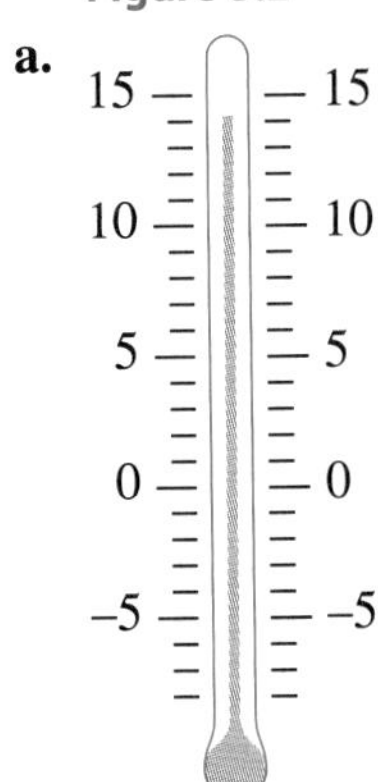

b.

10 — 10
5 — 5
0 — 0
-5 — -5
-10 — -10

EXERCISE 2

Saaid lost \$558 in a bad investment. Use a signed number to represent the change in Saaid's finances.

Elevations above sea level are given by positive numbers, and elevations below sea level are given by negative numbers. If your checking account is overdrawn, it has a negative balance. In general, positive numbers denote gains, increases, profits, and the like, and negative numbers denote losses or decreases.

The symbol + indicates a positive number. For example, $+3$ means "positive 3." However, we usually write positive 3 as just 3. When we deal with positive and negative numbers together, we call them **signed numbers.**

EXAMPLE 2

Use a signed number to describe each situation.

a. Death Valley is 273 feet below sea level.

Elevation of Death Valley: -273 feet

b. The state of Ohio ended 1992 with a deficit of \$8 million.

Net worth of Ohio: $-\$8$ million

EXAMPLE 3  

The temperature was $6°$ yesterday evening and fell $10°$ overnight. What was the temperature in the morning?

Solution

Illustrate $6°$ on the thermometer, as shown in Figure 3.4. Then count *down* $10°$ from there. This brings us to $-4°$. The temperature in the morning was $-4°$.

Figure 3.4

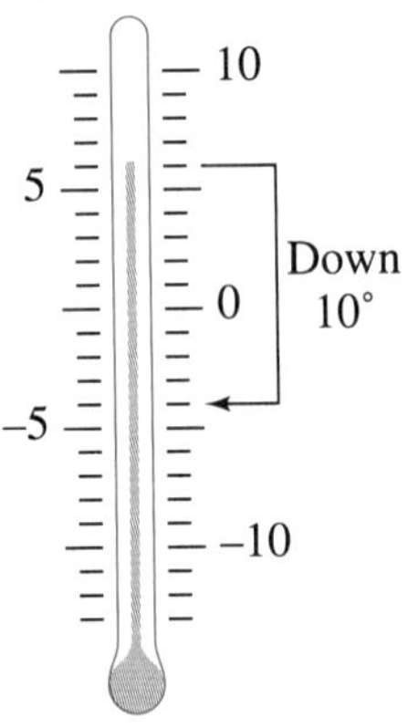

Bar Graphs

We can use a bar graph to display both positive and negative values. Bars that extend above the zero line indicate positive values, and bars that extend below the zero line indicate negative values.

EXAMPLE 4

The bar graph in Figure 3.6 shows normal low temperatures for each month in Bismarck, North Dakota.

Figure 3.6 Bismarck low temperatures

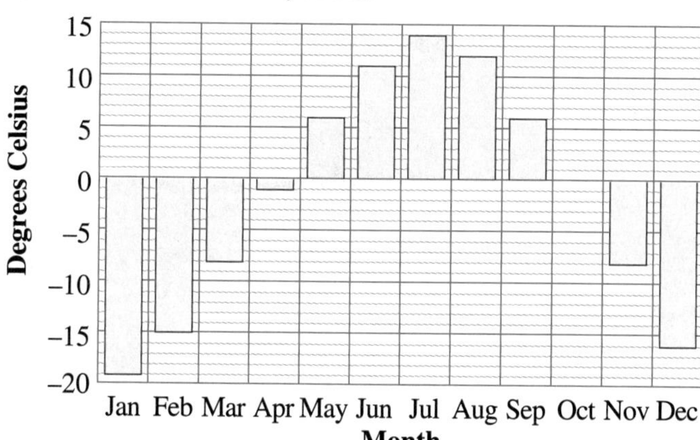

(Source: *American Almanac 1994–1995*)

a. What is normally the low temperature in December?

b. Which month normally has the coldest temperature? What is the coldest temperature?

Solution

a. The bar for December extends down to -16, so the normal low for December is $-16°$ Celsius (16°C).

b. January has the coldest temperature, -19°C.

EXERCISE 3

The temperature was 4° in the evening yesterday and dropped 11° overnight. What was the temperature this morning? Illustrate your calculation on the thermometer in Figure 3.5.

Figure 3.5

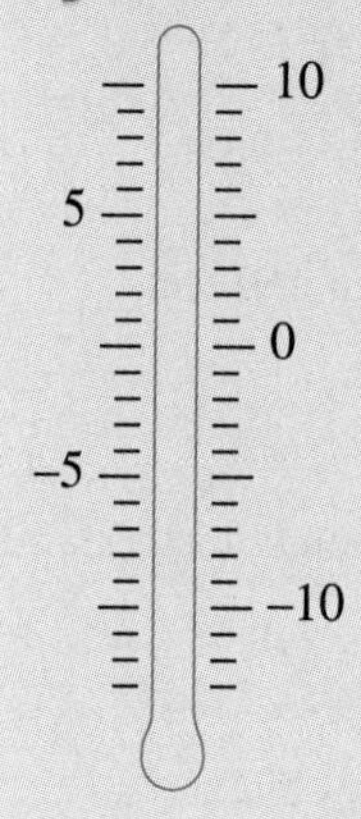

EXERCISE 4

a. Refer to the bar graph in Figure 3.6. What is the normal low temperature in Bismarck in February? Is it colder in Bismarck in February or in December?

b. Which two months have the same negative normal low temperature? What is that temperature?

ANSWERS TO 3.1A EXERCISES

1.

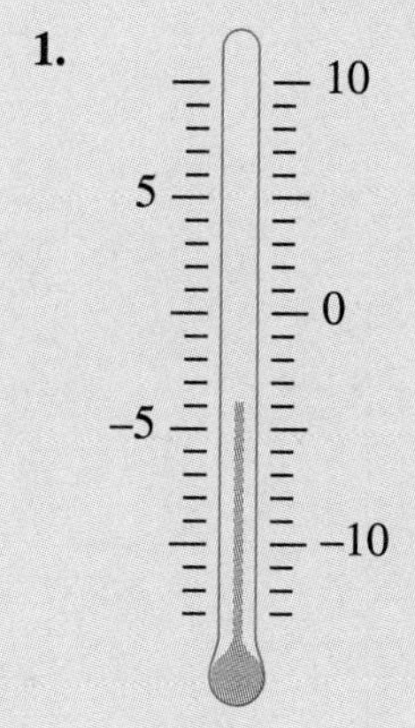

3. $-7°$

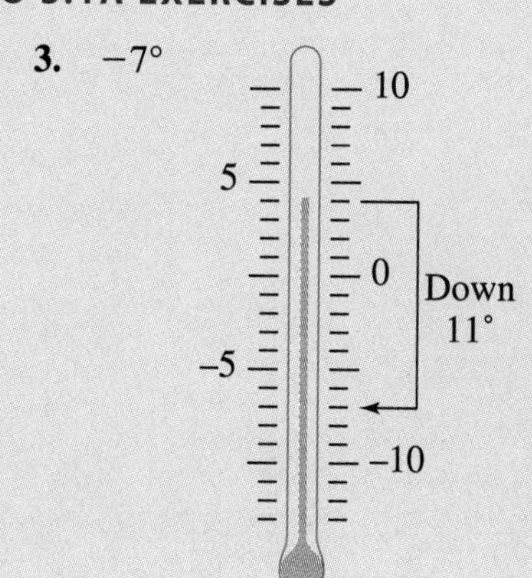

2. $-\$558$

4a. $-15°$; it is colder in December.

4b. March and November; -8°C

HOMEWORK 3.1A

Use a positive or negative number to describe each of the following situations.

1. Your checking account is overdrawn by $100.

2. The ski resort is 2000 feet above sea level.

3. Marty grew 2 inches this year.

4. The quarterback was sacked for a loss of 8 yards.

5. The plane descended 450 feet.

6. Hazel deposited $80 in her account.

7. Major Motors Company lost $3 million.

8. Test scores for eighth graders are up 23 points.

9. Aristotle was born in 384 B.C.

10. Barney lost 15 pounds.

Fill in the thermometers to show the indicated temperatures.

11. $-3°$

12. $-7°$

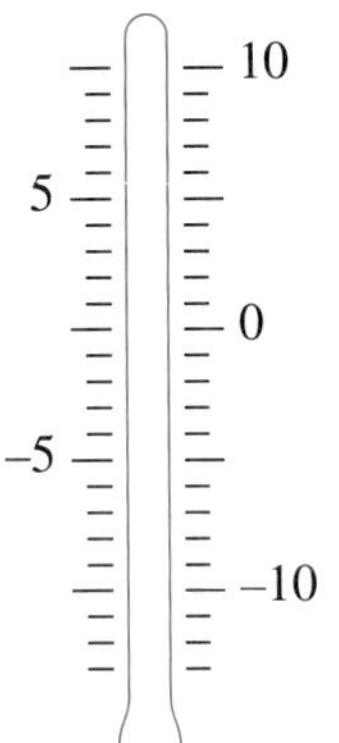

13. 0°

14. 6°

15. $-15°$

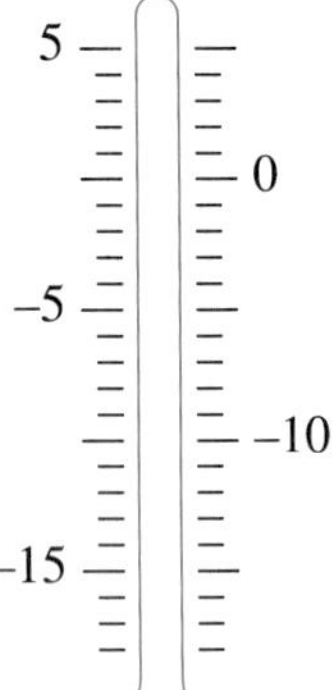

16. $-13°$

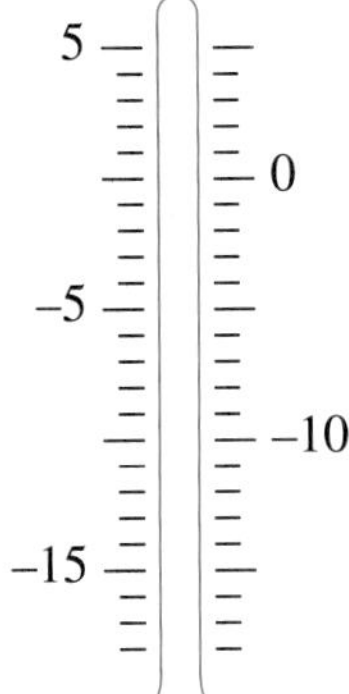

17. 2°

18. 10°

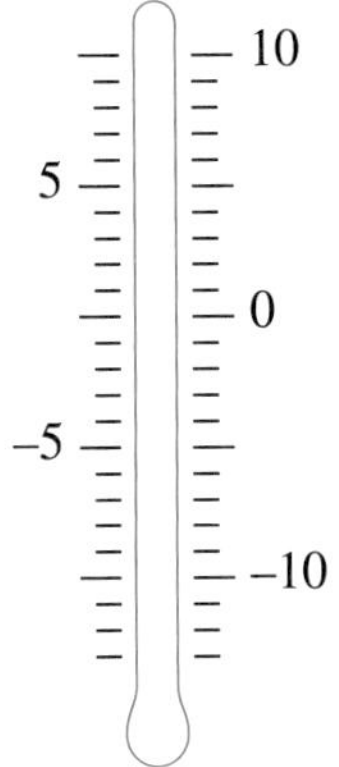

For Problems 19–24, use the thermometers to answer the questions.

19. The temperature was $-9°$ this morning and rose $12°$ by noon. What was the temperature at noon?

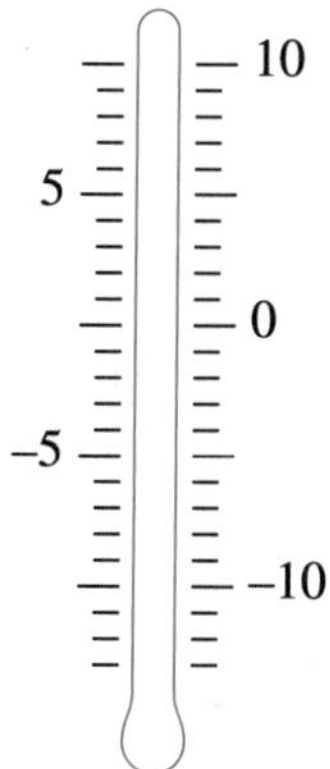

20. The temperature was $-2°$ yesterday afternoon and fell $6°$ by the evening. What was the temperature yesterday evening?

21. The temperature was $7°$ at noon and fell $15°$ in the next 4 hours. What was the temperature at 4 P.M.?

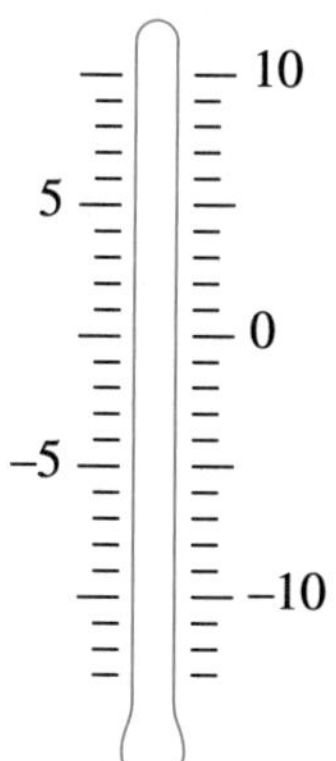

22. The temperature was $-12°$ last night and rose $8°$ overnight. What was the temperature this morning?

23. The temperature was $-3°$ at 6 A.M. and fell $10°$ over the next 6 hours. What was the temperature at noon?

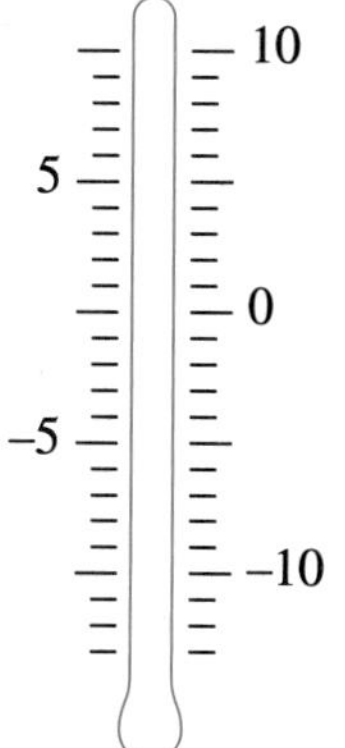

24. The temperature was $9°$ yesterday and fell $9°$ overnight. What was the temperature this morning?

25. The bar graph in Figure 3.7 shows normal low temperatures each month in Milwaukee, Wisconsin.

Figure 3.7 Milwaukee low temperatures

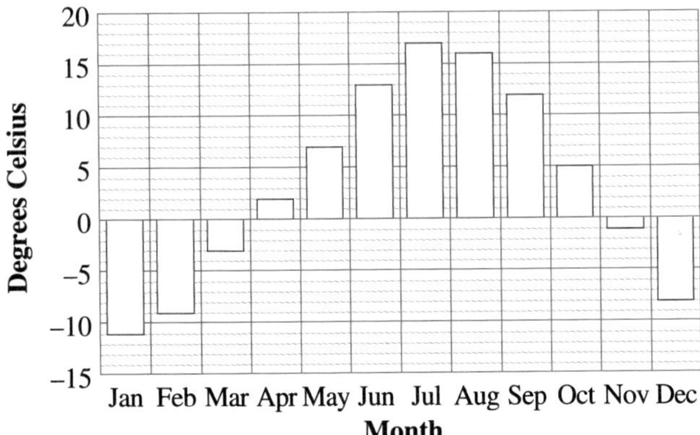

(Source: *American Almanac 1994–1995*)

a. What is normally the low temperature in March?

b. Which months have the coldest and warmest low temperatures, and what are those temperatures?

c. How much colder is the low temperature in February than it is in June?

26. The bar graph in Figure 3.8 shows the overnight low temperatures in four towns.

Figure 3.8 Low temperatures

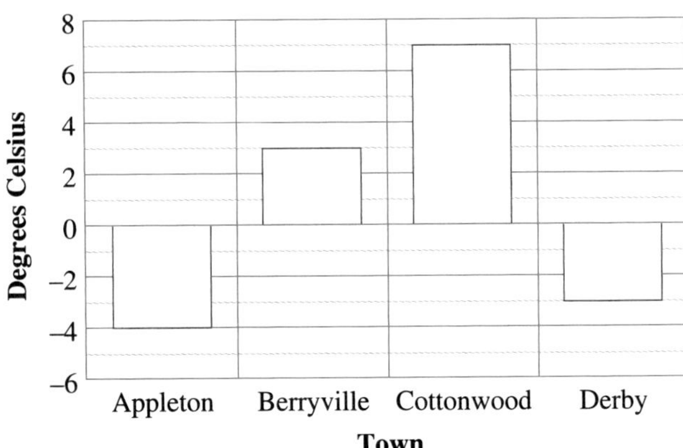

a. What was the overnight low in Derby?

b. Which town had the coldest temperature, and what was that temperature?

c. How much colder was it in Appleton than in Cottonwood?

27. The table gives the overnight low temperatures in four cities. Use the grid in Figure 3.9 to make a bar graph showing each town's overnight low temperature.

City	Mobile	Juneau	Peoria	Charlotte
Low (°C)	4	−7	−8	−1

Figure 3.9 Overnight low temperatures

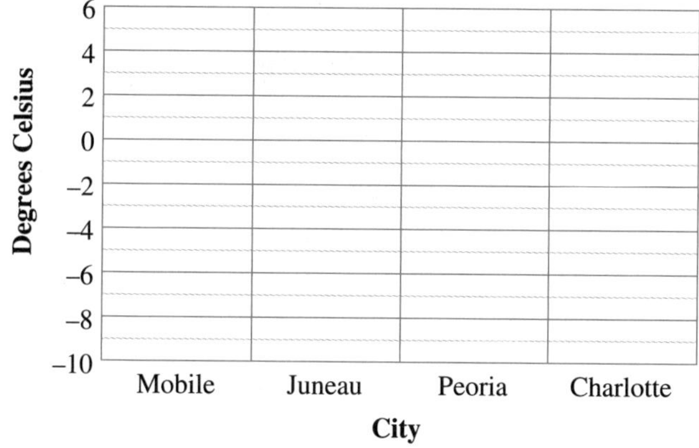

28. The table gives normal low temperatures for each month in Omaha, Nebraska. Use the grid in Figure 3.10 to make a bar graph showing each month's low temperature.

Month	Jan	Feb	Mar	Apr	May	Jun
Low (°C)	−12	−9	−2	4	11	16

Month	Jul	Aug	Sep	Oct	Nov	Dec
Low (°C)	19	17	12	5	−2	−9

(Source: *American Almanac 1994–1995*)

Figure 3.10 Omaha low temperatures

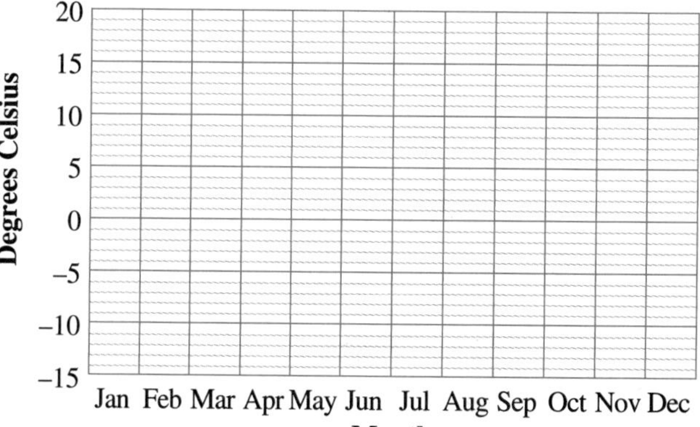

Decide whether you should calculate an area or a perimeter for each situation. (See Lesson 2.2 to review area and perimeter.)

29. You'd like to jog 3 miles by jogging laps around the hockey field.

30. You'd like to resurface your driveway.

31. You'd like to buy enough Christmas lights to border your roof.

32. You'd like to fence in your swimming pool area.

33. You'd like to plow enough land to plant corn for a family of five.

34. You'd like to apply water sealer to your redwood deck.

35. You'd like to know how many people can attend a lecture in the community hall.

36. You'd like to know how many books you can store if you build a shelf around the top of your dorm room.

37. You'd like to know how much shoreline is available for wild fowl around a lake.

38. You'd like to know how much water evaporates each day from the surface of the lake.

Building Number Skills

In this chapter we review some basic facts about fractions. The exercises will ask you to think about the *meaning* of fractions, rather than the rules for calculating with them.

The **denominator** of a fraction tells us how many equal pieces to divide the whole into. The **numerator** tells us how many pieces to take.

Sketch a diagram illustrating each fraction.

Example: $\frac{2}{3}$ Divide into three equal pieces. Shade two pieces.

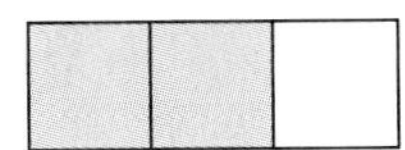

39. $\frac{3}{4}$

40. $\frac{2}{5}$

41. $\frac{3}{5}$

42. $\frac{5}{6}$

43. $\frac{5}{8}$

44. $\frac{3}{8}$

45. $\frac{1}{3}$

46. $\frac{1}{4}$

47. $\frac{8}{9}$

48. $\frac{4}{9}$

B. The Number Line; Inequality Symbols

The Number Line

We often use a scale like a thermometer to illustrate signed numbers. This scale, called a **number line,** is shown in Figure 3.11.

Figure 3.11

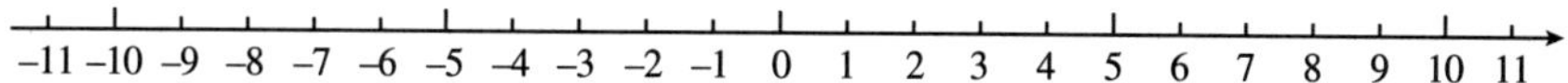

To draw a number line, begin by marking a position for zero in the middle of the line. Choose a length to represent 1 unit and mark off the counting numbers (1, 2, 3, etc.) to the *right* of zero. These marks represent positive numbers. Each positive number has its opposite or negative to the *left* of zero. For example, if we mark off 3 units to the left of zero, we come to the opposite of 3, or -3. The marks to the left of zero represent negative numbers.

The opposite of a negative number is a positive number. For example, the opposite of -6 is 6. We write the opposite of -6 as $-(-6)$, so

$$-(-6) = 6$$

The opposite of negative 6 is 6.

Together, the positive whole numbers, the negatives of the whole numbers, and zero make up a set of numbers called the **integers.**

If we want to indicate a particular number on the number line, we place a dot at its position. This is called the **graph** of the number.

EXAMPLE 5 

Graph the numbers -6, -2.5, 0, 4, and $5\frac{1}{4}$ on a number line.

Solution

We graph fractions by placing dots between the appropriate integers on the number line. For example, we graph $5\frac{1}{4}$ between 5 and 6. The completed graph is shown in Figure 3.12.

Figure 3.12

EXERCISE 5

Graph the numbers $-2\frac{3}{4}$, -0.2, and 1.6 on the number line in Figure 3.13.

Figure 3.13

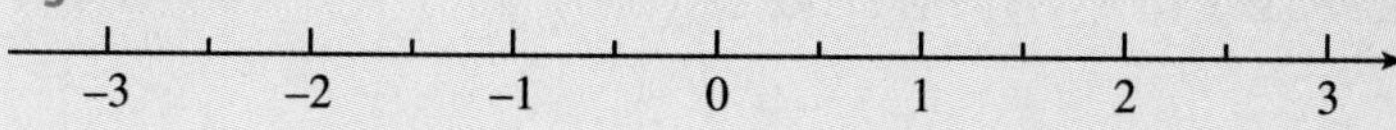

Inequality Symbols

Which is colder, a temperature of $-5°$ or a temperature of $-10°$? Since $-10°$ is colder, it makes sense to say that -10 is less than -5. On the number line, -10 lies to the left of -5, as shown in Figure 3.14.

Figure 3.14

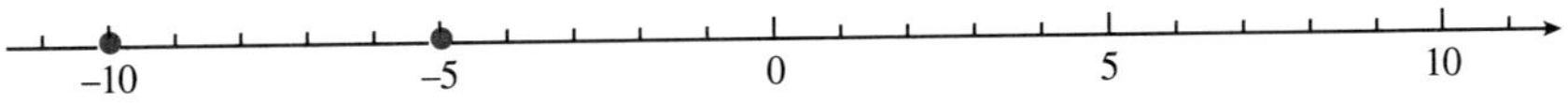

The numbers increase as we move from left to right on the number line. For example, -4 is less than -3, -2 is less than 0, and 3 is less than 6. In particular, every negative number is less than every positive number.

We use special symbols to denote *less than* and *greater than:*

> **$<$ means "is less than."**
> **$>$ means "is greater than."**

For example,

$-8 < -6$ means "-8 is less than -6."
$2 > -5$ means "2 is greater than -5."

These two symbols, $<$ and $>$, are called **inequality symbols,** to distinguish them from the equality symbol, $=$.

EXERCISE 6
Use an inequality symbol to write the phrase "-2 is greater than w."

EXAMPLE 6

Use inequality symbols to write the following phrases.

a. -9 is less than 4. **b.** -1 is greater than -12.
c. x is greater than -5. **d.** 6 is less than z.

Solution

a. $-9 < 4$ **b.** $-1 > -12$
c. $x > -5$ **d.** $6 < z$

EXERCISE 7
Choose a variable to represent the unknown quantity in the following situation and write an inequality:

The lottery jackpot is now more than $1,000,000.

EXAMPLE 7

Choose a variable to represent the unknown quantity in the following situation and write an inequality:

Aaron's brother can't vote yet; he is under 18 years old.

Solution
The unknown quantity is Aaron's brother's age:

Aaron's brother's age: *a*

Now use an inequality symbol to write that Aaron's brother's age is less than 18 years:

$$a < 18$$

ANSWERS TO 3.1B EXERCISES

5.

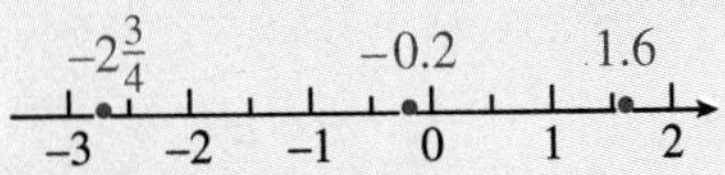

6. $-2 > w$

7. $J > 1{,}000{,}000$

HOMEWORK 3.1B

Graph each set of numbers on the number line provided.

1. $-3, 5, 1\frac{1}{2}, -2\frac{3}{4}, 3.5$

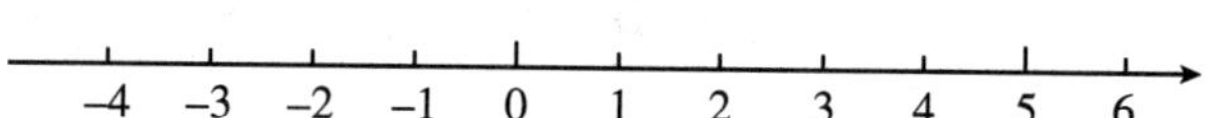

2. $2, -4.3, -6, 2\frac{1}{3}, -\frac{1}{2}$

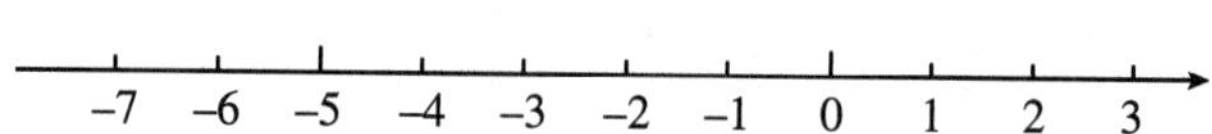

3. $3.5, -\frac{2}{3}, -2, 2, -3\frac{1}{2}$

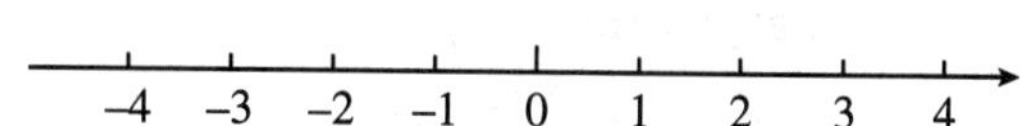

4. $-5, \frac{3}{4}, -1\frac{3}{4}, 2.25, -0.5$

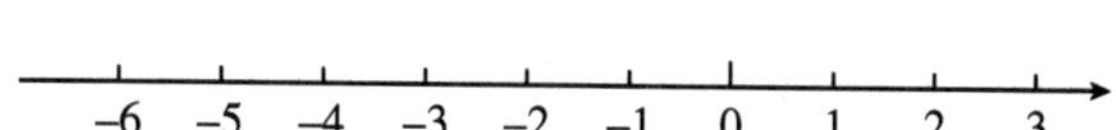

Use inequality symbols to write each phrase.

5. 6.5 is greater than 6.07.

6. 2.23 is less than 2.3.

7. -12 is less than -2.

8. -5 is greater than -9.

9. $-\frac{3}{4}$ is less than $-\frac{1}{2}$.

10. $-2\frac{1}{3}$ is greater than $-2\frac{2}{3}$.

Use a number line to help you decide whether each statement is true or false.

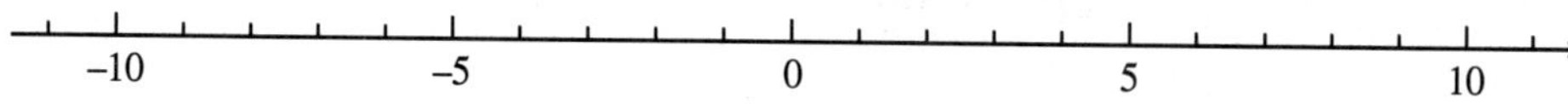

11. $-3 > -5$

12. $-8 < 2$

13. $-7 > 6$

14. $-9 < -10$

15. $-12 < -9$

16. $-1 > -8$

17. $-\frac{1}{2} < -1\frac{1}{2}$

18. $-\frac{3}{4} > \frac{1}{2}$

19. $0.1 < 0.01$

20. $0.099 > 0.9$

Give several values of the variable that make the inequality true.

21. $x > 2$

22. $z < 5$

23. $t < -3$

24. $v > -6$

25. $-4 > w$

26. $-1 < y$

Simplify each expression.

27. The opposite of 4

28. The opposite of $3\frac{1}{2}$

29. The opposite of -15

30. The opposite of -6.4

31. $-(-2)$

32. $-(-10)$

33. $-[-(-7)]$

34. $-[-(-25)]$

For Problems 35–42, choose a variable for the unknown quantity and write an inequality.

35. George's Aunt Ethel is a senior citizen; she's over 65 years old.

36. Jimmy gets into movies for half-price; he's under 12 years old.

37. The temperature of liquid copper exceeds its melting point of 1083°C.

38. The temperature of dry ice is below its evaporation point of −79°C.

39. My score on the physics test fell below the cutoff for an A, 85 points.

40. The rent on this apartment is more than my upper limit of $450.

41. The divers must stay above a depth of −50 feet.

42. Even if he answers the next question correctly, Marvin's score on *Jeopardy* will be worse than −400.

43. The bar graph in Figure 3.15 shows the number of miles of railroad in several countries in 1900 and in 2000. Complete the table and use signed numbers to show the change in miles of railroad for each country over the last century.

Figure 3.15 Number of miles of railroad, 1900 and 2000

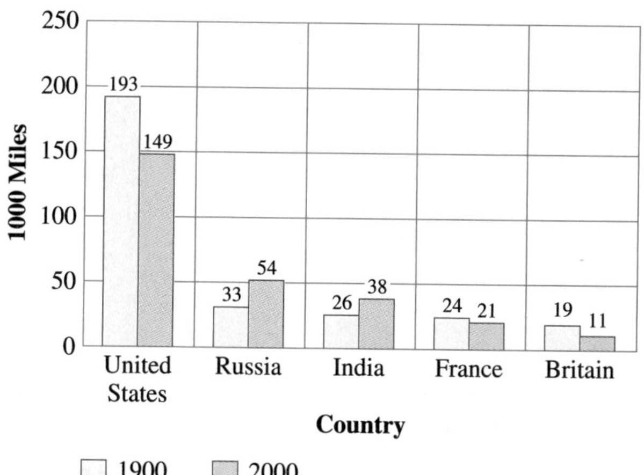

Country	Railroads in 1900 (thousands of miles)	Railroads in 2000 (thousands of miles)	Change in miles
U.S.			
Russia			
India			
France			
Britain			

44. The bar graph in Figure 3.16 shows the average airfare to various destinations from each city in 1998 and in 1999. Complete the table and use signed numbers to show the change in average fare for each city from 1998 to 1999.

(Source: *Consumer Reports* Travel Letter)

Figure 3.16 Cost of average airfare, 1998 and 1999

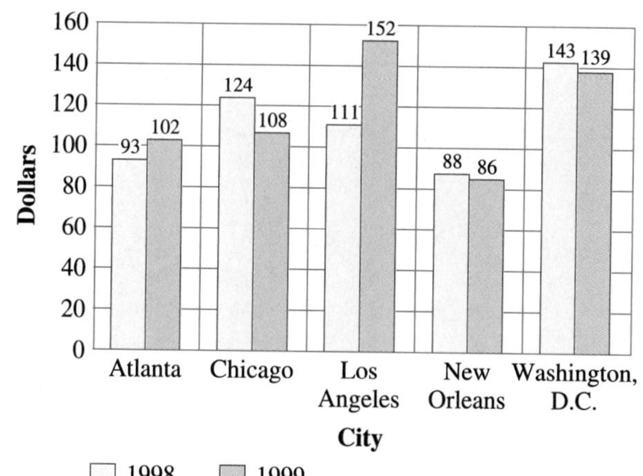

City	Average airfare in 1998	Average airfare in 1999	Change in airfare
Atlanta			
Chicago			
Los Angeles			
New Orleans			
Washington, D.C.			

Building Number Skills

An **improper fraction** is a fraction that is greater than or equal to 1. This means that its numerator is greater than or equal to its denominator.

Sketch a diagram illustrating each improper fraction.

Example: $\frac{6}{4}$

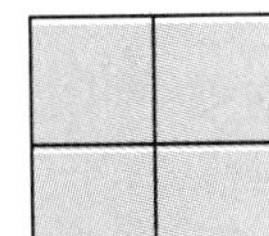

45. $\frac{4}{3}$

46. $\frac{3}{2}$

47. $\frac{7}{4}$

48. $\frac{5}{3}$

49. $\frac{5}{2}$

50. $\frac{8}{3}$

51. $\frac{8}{5}$

52. $\frac{9}{4}$

53. $\frac{9}{6}$

54. $\frac{10}{4}$

3.2 Adding Signed Numbers

A. How Do We Add Signed Numbers?

Adding Two Numbers with the Same Sign

We will use number lines to help us understand how to add signed numbers. First, let's see how to show the sum of two positive numbers on a number line.

EXAMPLE 1

Use a number line to illustrate the sum $(+4) + (+3)$

Solution

Plot the first number, $+4$, as shown on the number line in Figure 3.17. From that point, move 3 units in the positive direction, or to the right. This brings us to $+7$, so $(+4) + (+3) = +7$.

Figure 3.17

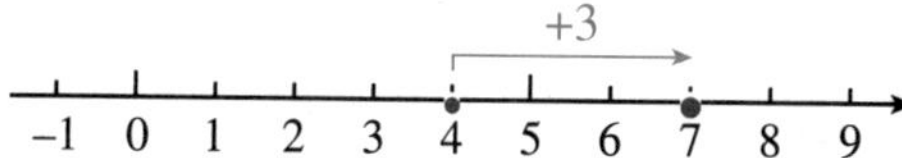

What about the sum of two negative numbers? For example, what is the sum of the following?

$$(-5) + (-3)$$

Think of negative numbers as representing debts. Then it makes sense that the sum of two negative numbers is another negative number: If you owe \$5 to one person and \$3 to another person, then you are \$8 in debt. We can illustrate this sum on a number line as shown in Example 2.

EXAMPLE 2 

Use a number line to illustrate the sum $(-5) + (-3)$

Solution

Plot the first number, -5, on the number line, as shown in Figure 3.18. From that point, move 3 units in the negative direction, or to the left. This brings us to -8, so $(-5) + (-3) = -8$.

Figure 3.18

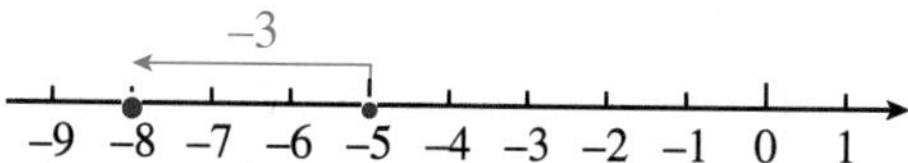

EXERCISES

Illustrate the following sums on number lines.

1. $(+2) + (+4)$

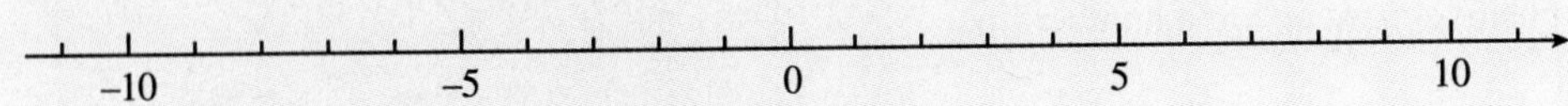

2. $(-1) + (-5)$

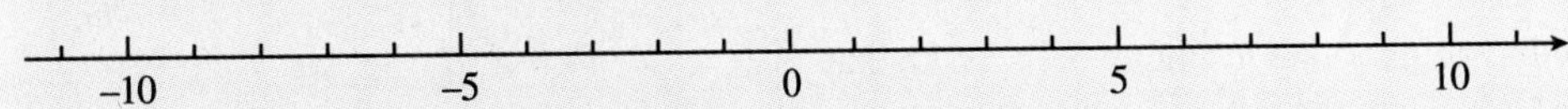

3. $(-4) + (-3)$

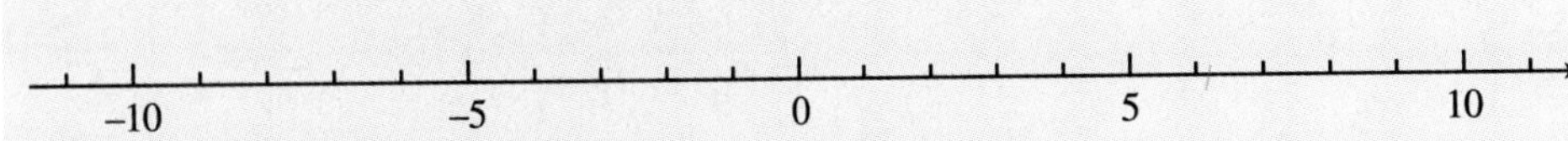

From Examples 1 and 2, we see that

1. *The sum of two positive numbers is positive.*
2. *The sum of two negative numbers is negative.*

Adding Two Numbers with Opposite Signs

Next we consider the sum of a positive number and a negative number—for example,

$$(+6) + (-8)$$

In terms of money, this could mean that you have \$6 but you owe someone \$8, so your net financial status is \$2 in debt. We illustrate this sum on a number line in Example 3.

EXAMPLE 3 

Use a number line to illustrate the sum $(+6) + (-8)$

Solution

Plot the first number, $+6$, as shown on the number line in Figure 3.19. From that point, move 8 units in the negative direction, or to the left. This brings us to -2, so $(+6) + (-8) = -2$.

Figure 3.19

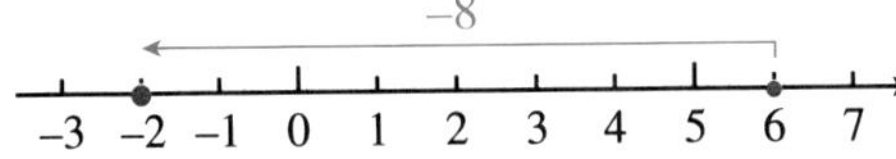

In Example 3 note that when your debts are greater than your assets, your net worth is negative. On the other hand, if your assets are greater than your debts, then your net worth is positive. Suppose you have \$10, but you owe someone \$6. We can represent this situation by the sum

$$(+10) + (-6)$$

Once you pay off your debt, you still have \$4 left. We illustrate this sum on a number line in Example 4.

EXAMPLE 4

Use a number line to illustrate the sum $(+10) + (-6)$

Solution

Plot the first number, $+10$, as shown on the number line in Figure 3.20. From that point, move 6 units in the negative direction, or to the left. This brings us to $+4$, so $(+10) + (-6) = +4$.

Figure 3.20

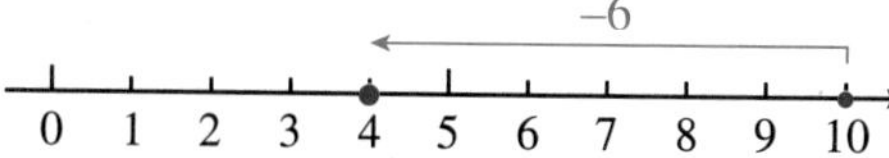

From Examples 3 and 4, we see that the sum of a positive number and a negative number can be either positive or negative. If the positive number is farther from zero than the negative number, the sum will be positive. If the negative number is farther from zero than the positive number, the sum will be negative.

EXERCISES

Illustrate the following sums on number lines.

4. $(-4) + (+8)$

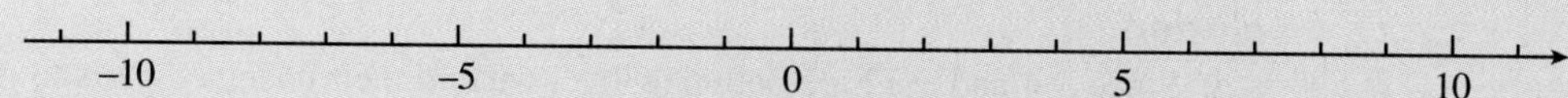

5. $(-9) + (+2)$

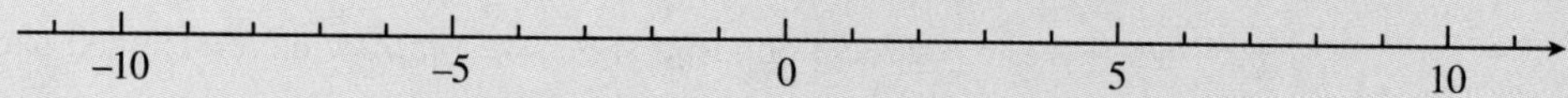

6. $(+7) + (-12)$

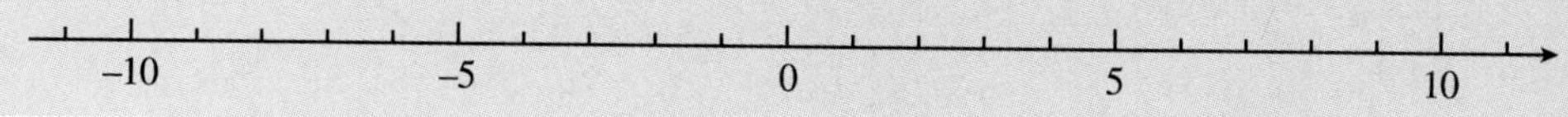

Rules for Addition

Once we know the sign of the answer, how do we find its actual value? Look again at the preceding examples. If we *subtract* the unsigned parts of the two numbers, we find the unsigned part of their sum. We can see that this is reasonable if we consider another example about money. Suppose you have a debt of \$15, but you earn \$20. You need \$15 of your earnings to pay off the debt, so we *subtract* \$15 from \$20. The remainder, \$5, is your net worth, as illustrated in Figure 3.21.

Figure 3.21

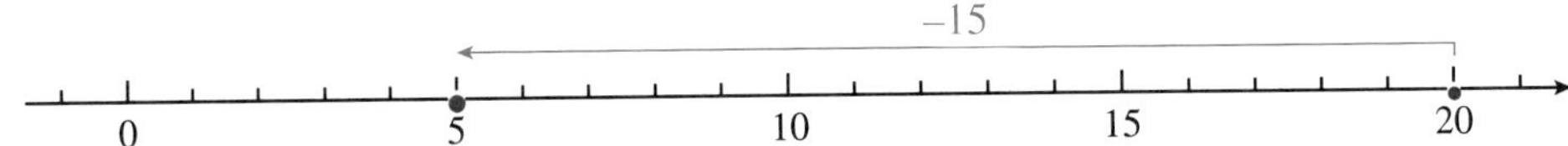

Let's summarize what we have learned about addition.

To Add Two Numbers with the Same Sign

1. Add the unsigned parts of the numbers.
2. Use the same sign for the sum.

To Add Two Numbers with Opposite Signs

1. Subtract the unsigned parts of the numbers.
2. If the positive number is farther from zero, the sum is positive. If the negative number is farther from zero, the sum is negative.

EXAMPLE 5

Use the rules for addition to find the following sums.

a. $(-4) + (-12)$ **b.** $(-8) + (+2)$

c. $(-7) + (+9)$

Solution

a. Since -4 and -12 are both negative, we *add* their unsigned parts to get

$$4 + 12 = 16$$

We use the same sign, $-$, for the answer. Thus,

$$(-4) + (-12) = -16$$

b. Since -8 and $+2$ have opposite signs, we *subtract* their unsigned parts to get

$$8 - 2 = 6$$

The negative number is farther from zero, so the answer is negative. Thus,

$$(-8) + (+2) = -6$$

c. Since -7 and $+9$ have opposite signs, we *subtract* their unsigned parts to get

$$9 - 7 = 2$$

The positive number is farther from zero, so the answer is positive. Thus,

$$(-7) + (+9) = +2$$

EXERCISE 7
Use the rules for addition to find the following sums:
a. $(-2) + (-2)$

b. $(-4) + (+3)$

c. $(+6) + (-8)$

Using a Calculator

For more complicated sums, you may want to use a calculator. To enter a negative number on a scientific calculator, press the [+/−] button *after* the number. For example, to enter -8, press

8 [+/−]

The [+/−] key is called the *change sign key.* Note that we do *not* use the [−] key to enter a negative number. The [−] key is used for subtraction.

EXAMPLE 6

Use a calculator to add $(-36.9) + (+53.2)$

Solution

Enter the sum as shown below:

36.9 [+/−] [+] 53.2 [=]

The calculator will return the answer, 16.3.

On some calculators, including most graphing calculators, we use the [(−)] key to enter a negative number. The sum in Example 6 would be entered as

[(−)] 36.9 [+] 53.2 [=]

EXERCISE 8
Use a calculator to add $(+41.8) + (-96.7)$

EXAMPLE 7

Seymour's stock gained 3 points on Tuesday but lost 7 points on Wednesday. What was the net change in the stock since Tuesday?

Solution

Use $+3$ to denote a 3-point gain and -7 to denote a 7-point loss. Then

$$\mathbf{(+3) + (-7) = -4}$$

The net change in Seymour's stock was a 4-point loss.

EXERCISE 9
When the ballots in a disputed district were recounted, Senator Dubya lost 58 votes, but he gained 42 votes when the absentee ballots were tallied. What was the net change in votes for Senator Dubya?

ANSWERS TO 3.2A EXERCISES

1.

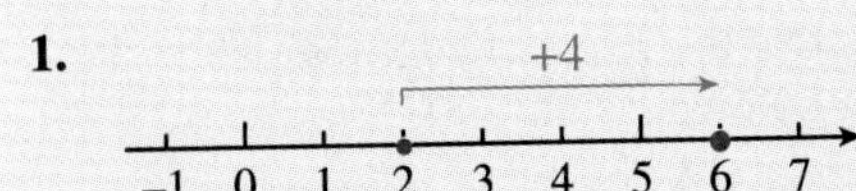

$(+2) + (+4) = 6$

2.

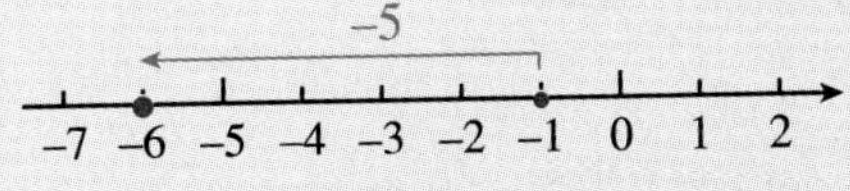

$(-1) + (-5) = -6$

3.

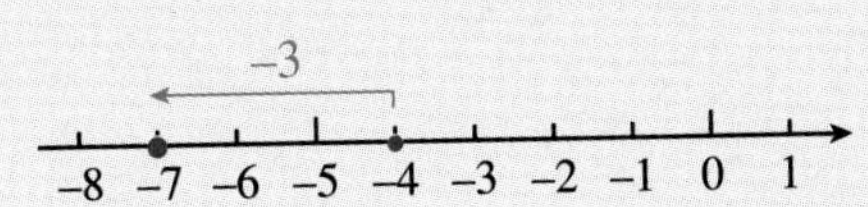

$(-4) + (-3) = -7$

4.

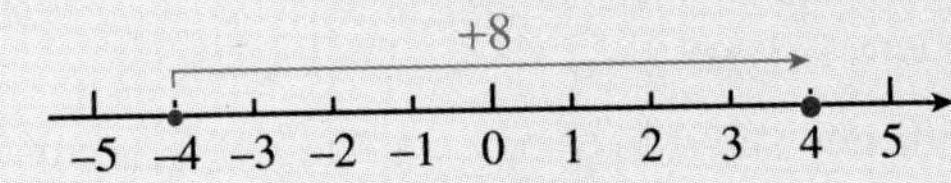

$(-4) + (+8) = 4$

5.

+2

−9 −8 −7 −6 −5 −4 −3 −2 −1 0 1

$(-9) + (+2) = -7$

6.

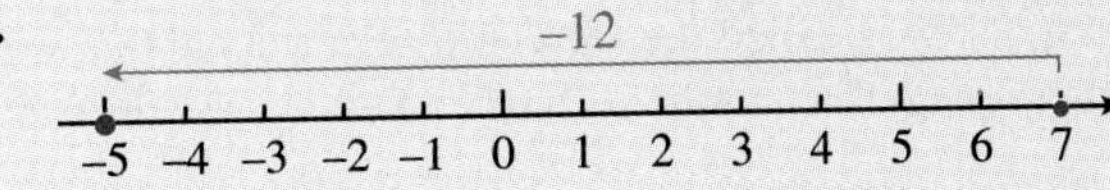

$(+7) + (-12) = -5$

7a. -4 **7b.** -1 **7c.** -2

8. -54.9

9. -16 votes

HOMEWORK 3.2A 

For Problems 1–10, use number lines to illustrate the following sums of signed numbers.

1. $(+4) + (+5)$

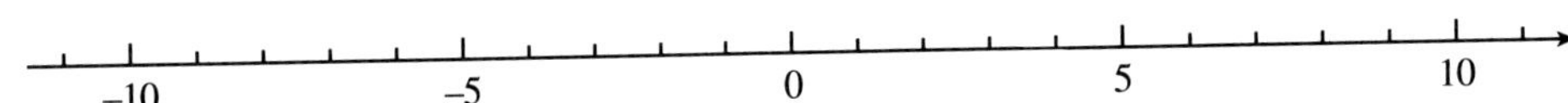

2. $(+3) + (+6)$

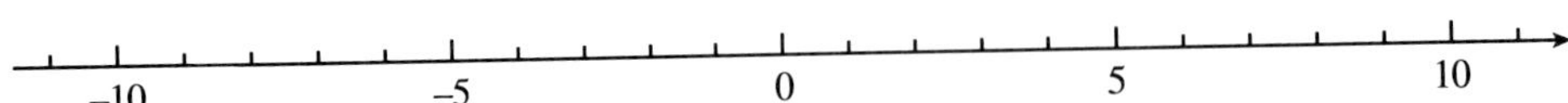

3. $(-2) + (-6)$

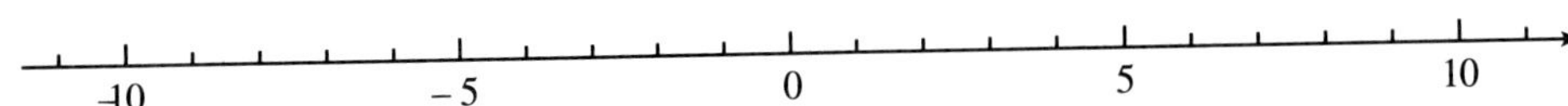

4. $(-7) + (-2)$

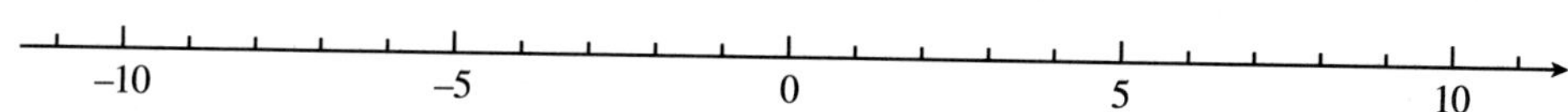

5. $(-3) + (+7)$

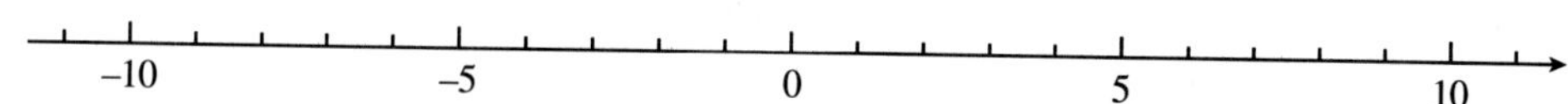

6. $(-9) + (+4)$

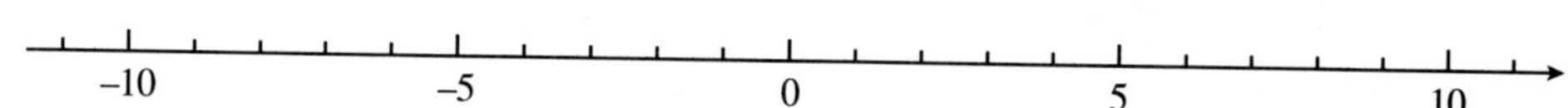

7. $(+10) + (-12)$

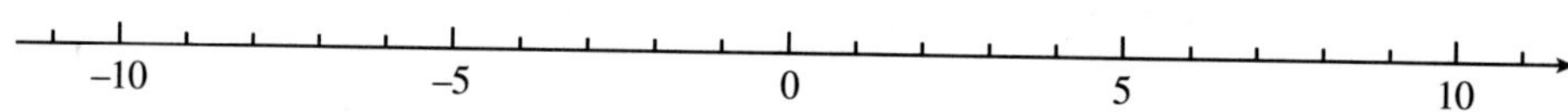

8. $(+8) + (-5)$

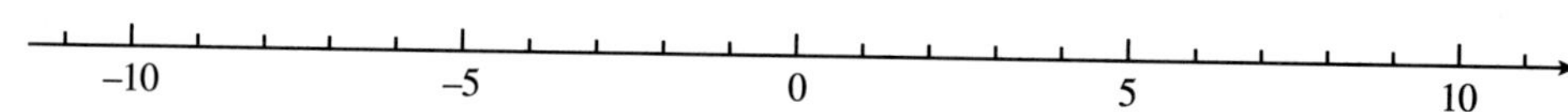

9. $(-8) + (+4)$

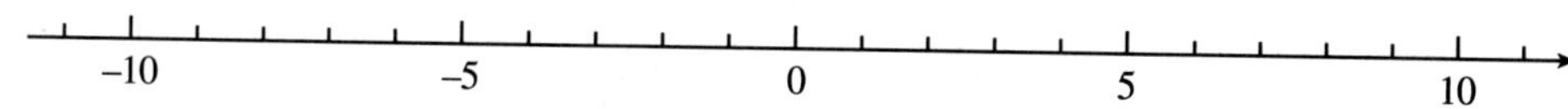

10. $(+1) + (-6)$

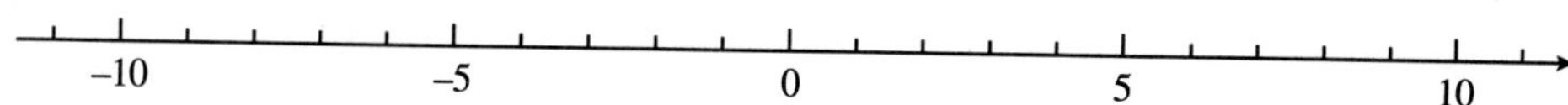

11. Explain in your own words why it makes sense that the sum of two negative numbers is negative.

12. Explain in your own words why it makes sense that the sum of two numbers with opposite signs has the sign of the number that is farther from zero.

13. Make up an example about money that shows how to add two numbers with opposite signs.

14. Make up an example about money that shows how to add two negative numbers.

Use the rules for addition to compute the following sums.

15. $(+7) + (+9)$

16. $(+4) + (+18)$

17. $(-5) + (-6)$

18. $(-3) + (-9)$

19. $(-12) + (-17)$

20. $(-15) + (-21)$

21. $(-8) + (-8)$

22. $(-15) + (-15)$

23. $-25 + (-50)$

24. $-40 + (-70)$

25. $(+12) + (-5)$

26. $(+14) + (-4)$

27. $20 + (-7)$

28. $15 + (-8)$

29. $(-16) + (+24)$

30. $(-20) + (+29)$

31. $(-6) + (+13)$

32. $(-7) + (+19)$

33. $(-5) + (+5)$

34. $(-2) + (+2)$

35. $-8 + (+2)$

36. $-10 + (-3)$

37. $-17 + (+4)$

38. $-30 + (+5)$

39. $(+9) + (-18)$

40. $(+4) + (-15)$

41. $21 + (-15)$

42. $17 + (-14)$

43. $-16 + (-18)$

44. $16 + (-18)$

45. $-29 + 7$

46. $-32 + (-4)$

Use a calculator to compute each sum.

47. $(-38) + (-76)$

48. $(-39) + (-84)$

49. $(-0.5) + (-0.3)$

50. $(-2.5) + (-3.5)$

51. $-136 + (-245)$

52. $318 + (-205)$

53. $13.8 + (-18.3)$

54. $-6.9 + 15.7$

55. $-2.05 + 6.54$

56. $-0.15 + (-1.23)$

Use a sum of signed numbers to answer each question.

57. Yariv lost 27 pounds on his diet but then gained back 6 pounds over the holidays. What is the net change in Yariv's weight?

58. The Spartans gained 16 yards on their first down but then lost 7 yards on their next play. What was their net change in position?

59. The Mountaineering Club hiked up 1200 feet to the top of a ridge and then descended into a valley 1400 feet below the summit of the ridge. What was their net change in elevation?

60. Bruno is trying to bake potatoes. He increased the oven's temperature by 125° when he grew impatient, but then decreased the temperature by 75° when the skins started to burn. What was the net change in the oven temperature?

61. The captain of Flight 386 descended 5000 feet to avoid turbulence in the upper atmosphere and then descended another 1800 feet on the approach to a small airport. What was the plane's net change in altitude?

62. The temperature is expected to drop 15° by this afternoon and then drop another 18° overnight. What will be the net temperature change by tomorrow morning?

Find the sum of each pair of opposites.

63. $(-9) + (+9)$

64. $(+4) + (-4)$

65. $(+63) + (-63)$

66. $(-81) + (+81)$

67. $(-0.2) + (+0.2)$

68. $(+1.3) + (-1.3)$

69. Use an example about money to explain why the sum of two opposites is always zero.

70. a. What number can you add to 18 to produce a sum of 0?

b. What number can you add to -12 to produce a sum of 0?

Building Number Skills

Any number divided by itself is 1; for example, $\frac{8}{8} = 1$. We can use this fact to subtract fractions from 1. What is left if we take $\frac{5}{8}$ from 1? Since 1 is the same as $\frac{8}{8}$, we have

$$1 - \frac{5}{8} = \frac{8}{8} - \frac{5}{8} = \frac{3}{8}$$

Rewrite 1 as a fraction and then do the subtraction.

71. $1 - \frac{1}{3}$

72. $1 - \frac{1}{4}$

73. $1 - \frac{2}{5}$

74. $1 - \frac{7}{8}$

75. $1 - \frac{5}{9}$

76. $1 - \frac{3}{10}$

77. $1 - \frac{7}{12}$

78. $1 - \frac{2}{7}$

79. $1 - \frac{3}{4}$

80. $1 - \frac{2}{3}$

B. Signed Numbers and Variables

Variables can have negative values, too. We can use a bar graph to display the values of a variable.

EXAMPLE 8

Delbert carried the football six times during his last game. The bar graph in Figure 3.22 shows the number of yards he gained on each rush. What was his rushing total for the game?

Figure 3.22 Yards gained per rush

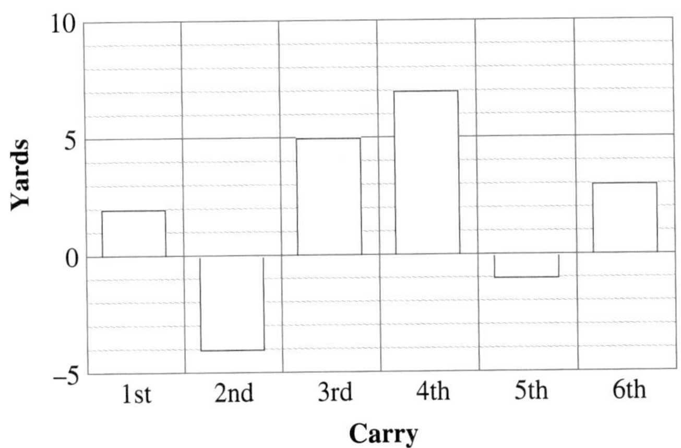

EXERCISE 10
Delbert carried the ball four times in the first half of the game. Refer to the bar graph in Figure 3.22 to find Delbert's rushing total for the first half.

Solution

We read the value of yards gained for each carry and add the numbers:

$$2 + (-4) + 5 + 7 + (-1) + 3 = 12$$

Delbert gained a total of 12 yards rushing.

Evaluation

When we substitute a negative number for a variable, it is a good idea to enclose the negative number in parentheses. The parentheses help us keep track of which signs indicate operations (adding or subtracting) and which signs indicate that a number is positive or negative.

EXAMPLE 9 

Evaluate the following sums for $a = -5.8$, $b = -3.2$, and $c = 6.1$.

a. $a + b$ **b.** $b + c + a$

Solution

a. Substitute -5.8 for a and -3.2 for b. Since both numbers are negative, we add their unsigned parts and make the answer negative:

$$a + b = (-5.8) + (-3.2) = -9$$

b. Substitute -3.2 for b, 6.1 for c, and -5.8 for a. Add the numbers in pairs from left to right:

$$\begin{aligned} b + c + a &= (-3.2) + 6.1 + (-5.8) \\ &= 2.9 + (-5.8) \\ &= -2.9 \end{aligned}$$

EXERCISE 11

Evaluate the following sums for $x = -5.1, \quad y = 2.3, \quad z = -1.7$

a. $x + y$

b. $x + y + z$

Solving Equations

Equations may involve negative numbers, and solutions to equations may be negative numbers.

EXAMPLE 10 

Solve the equations.

a. $x - 8 = -3$ **b.** $z - 12 = -18$

Solution

a. Because 8 is subtracted from the variable, add 8 to both sides of the equation:

$$\begin{aligned} x - 8 &= -3 \\ +8 \quad &\quad +8 \\ \hline x \quad &= \quad 5 \end{aligned}$$

The solution is 5.

b. Because 12 is subtracted from the variable, add 12 to both sides of the equation:

$$\begin{aligned} z - 12 &= -18 \\ +12 \quad &\quad +12 \\ \hline z \quad &= \quad -6 \end{aligned}$$

The solution is -6.

EXERCISE 12

Solve the equations.

a. $y - 4 = -9$

b. $w - 6 = -3$

Now consider the equation

$$-3 + x = 9$$

The left side of this equation says -3 added to x. Remember that our goal in solving equations is to isolate x on one side of the equation. Suppose we add $+3$ to both sides of the equation. Since

$$(-3) + (+3) = 0$$

we are left with $0 + x$, or just x, on the left side. On the right side we have

$$9 + 3, \quad \text{or} \quad 12$$

The solution looks like this:

$$\begin{aligned} -3 + x &= \quad 9 \\ +3 \qquad &\quad +3 \\ \hline x &= \quad 12 \end{aligned}$$

Thus, our knowledge of opposites gives us a new way to solve equations involving addition: We note what number is added to the variable and *add its opposite* to both sides of the equation.

EXERCISE 13

Solve the equations.

a. $-5 + b = -11$

b. $6 = -4 + w$

ANSWERS TO 3.2B EXERCISES

10. 10 yards

11a. -2.8

11b. -4.5

12a. -5

12b. 3

13a. -6

13b. 10

EXAMPLE 11

Solve the equations.

a. $-6 + a = 14$

b. $-8 = -1 + v$

Solution

a. Because -6 is added to the variable a, we add $+6$ to both sides of the equation:

$$\begin{array}{rcr} -6 + a &=& 14 \\ +6 \quad & & +6 \\ \hline a &=& 20 \end{array}$$

The solution is 20.

b. Because -1 is added to the variable v, we add $+1$ to both sides of the equation:

$$\begin{array}{rcl} -8 &=& -1 + v \\ +1 & & +1 \\ \hline -7 &=& \quad v \end{array}$$

The solution is -7.

HOMEWORK 3.2B 

Use a sum of signed numbers to answer each question.

1. The crew of a new submarine dove 700 feet to test their equipment and then rose 250 feet to their cruising depth. What was their net change in elevation?

2. Rocky lost \$450 betting on the horses this afternoon and then dropped \$245 at a poker game in the evening. What was the net change in his financial status for the day?

3. Morgan lost \$5600 on the stock market this morning and lost another \$3450 in the afternoon. What was the net change in the value of his portfolio?

4. Rachelle made \$568.25 daytrading on Tuesday but lost \$472.75 on Wednesday. What was the net change in her funds?

5. Wanda deposited her paycheck of \$1876.47 and then wrote checks totaling \$1595.62 to pay her bills. What was the net change in her bank account?

6. Carmella's Boutique made \$3019.28 on spring dresses this year but lost \$3752.91 on an unpopular line of sweater shorts. What was the boutique's net profit on the two items?

For Problems 7–12, use the rules for addition to compute the following sums.

7. $8 + (-6)$

8. $-8 + (-6)$

9. $-8 + 6$

10. $-20 + 30$

11. $-20 + (-30)$

12. $20 + (-30)$

13. Figure 3.23 shows the quarterly profits earned by Orinoco.com in 2001.

Figure 3.23 Quarterly profits

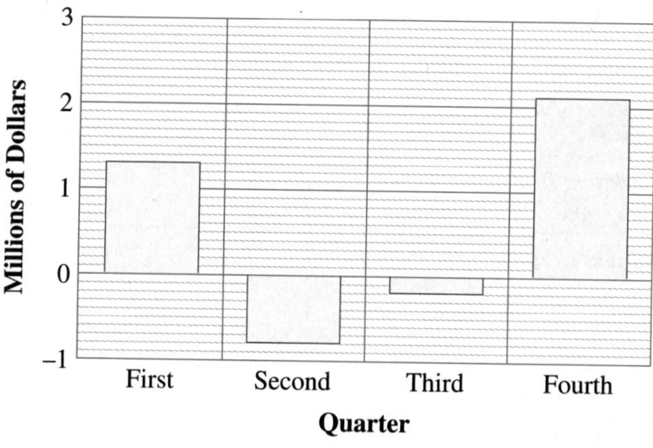

a. What are the profits for the four quarters shown?

b. What was Orinoco.com's total annual profit for 2001?

14. Figure 3.24 shows how much Zhila won in each of the six games of poker she played.

Figure 3.24 Winnings at poker

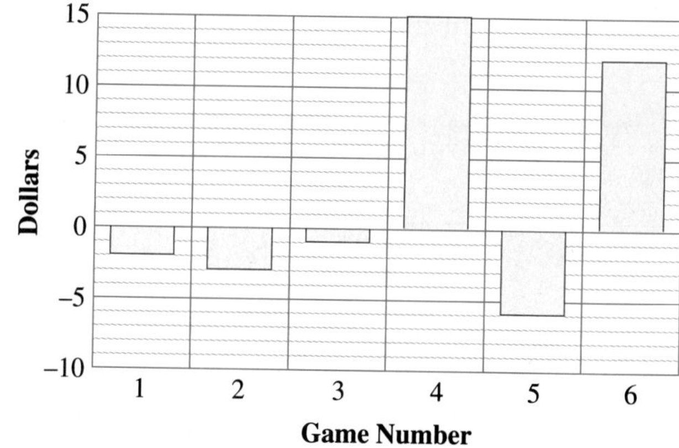

a. Write Zhila's poker record as a sum of signed numbers.

b. What was the financial outcome of Zhila's poker?

15. Figure 3.25 shows the water level in the reservoir, relative to its normal depth of 34 feet. For example, in January, when the level was 3 feet above normal, the water depth was $34 + 3 = 37$ feet.

Figure 3.25 Water level in reservoir

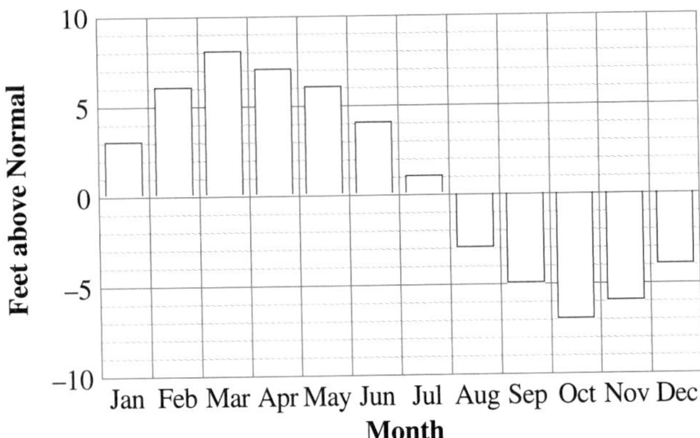

a. Using values from the bar graph, fill in the table:

Month	Jan	Feb	Mar	Apr	May	June
Relative water level (ft)	3					
Water depth (ft)	37					

Month	July	Aug	Sep	Oct	Nov	Dec
Relative water level (ft)						
Water depth (ft)						

b. Use Figure 3.26 to make a bar graph showing the actual depth of the water in the reservoir.

Figure 3.26 Actual water level in reservoir

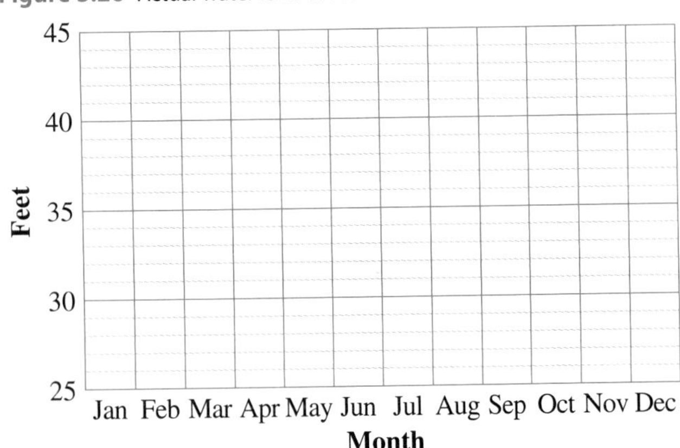

16. Figure 3.27 shows the annual rainfall in Los Angeles, relative to the average value of 15 inches over the previous century. For example, in 1987 rainfall was 6 inches below normal, or $15 - 6 = 9$ inches.

Figure 3.27 Los Angeles rainfall

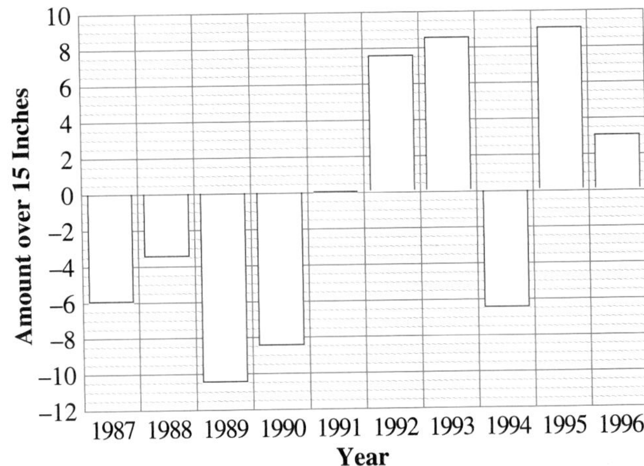

a. Using values from the bar graph, fill in the table:

Year	1987	1988	1989	1990	1991	1992
Deviation from average	−6					
Annual rainfall (in.)	9					

Year	1993	1994	1995	1996
Deviation from average				
Annual rainfall (in.)				

b. Use Figure 3.28 to make a bar graph showing the annual rainfall in Los Angeles over the decade from 1987 to 1996.

Figure 3.28 Los Angeles rainfall

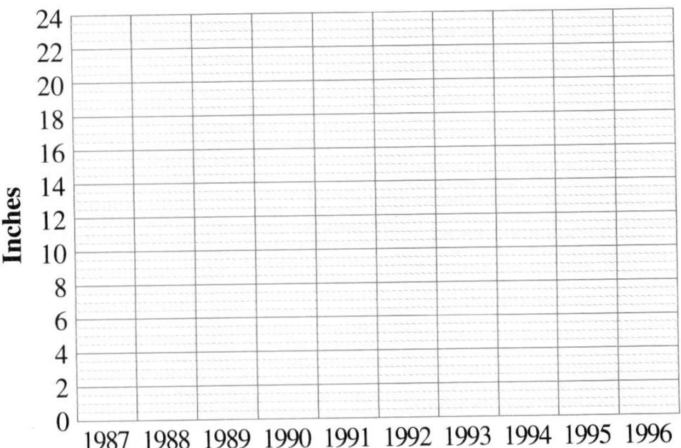

Evaluate each sum for $m = -8$, $p = -4$, and $t = 6$.

17. $m + p$

18. $m + m$

19. $t + m$

20. $p + t$

21. $m + p + t$

22. $p + p + p$

23. $t + p + t$

24. $m + p + p$

Solve.

25. $x - 9 = -2$

26. $w - 8 = -15$

27. $a - 4 = -18$

28. $b - 11 = -9$

29. $-5 + v = -14$

30. $-6 + y = -3$

31. $-12 + g = -12$

32. $-26 + h = -26$

33. $-5.7 + n = -8.4$

34. $-8.5 + k = -3.1$

35. $s - 83.2 = -12.6$

36. $d - 39.6 = -58.3$

37. $j - 0.04 = -0.6$

38. $c - 0.8 = -0.02$

Evaluate each sum for $x = 2.5, y = -7.2$, and $z = -4.8$.

39. $x + y$

40. $y + y$

41. $z + x$

42. $z + y$

43. $z + z + z$

44. $x + y + x$

45. $x + x + y$

46. $x + y + z$

Write an algebraic expression to answer each question. (See Lesson 1.4 to review algebraic expressions.)

47. Rachel bought six pairs of shorts at p dollars each. How much did she spend?

48. Daniel spent d dollars on a package of eight oil filters. How much did each oil filter cost?

49. Abby has \$1200 saved for vacation. If she is gone for v days, how much can she spend per day?

50. Brenda's department has an annual budget of \$5000 for supplies. If they spent b dollars in the first quarter, how much is left for the rest of the year?

51. Carol and Alex together spend \$1800 per year on car insurance. If Carol's insurance is c dollars, how much is Alex's insurance?

52. Bita needs t dollars for next year's tuition. She has saved \$80. How much more does she need?

53. Danit would like to read m pages of her economics text this weekend. She has read 28 pages already. How many more pages must she read?

54. Sunray Solar Energy Company made \$10,000 in dividends for its n stockholders. How much will each stockholder receive?

55. Edgar has to write a 5000-word essay. If his printer prints w words per page, how many pages will the essay cover?

56. The horses on the Pierce College farm eat q pounds of oats per month. If oats come in 50-pound sacks, how many sacks should the trainer order each month?

Building Number Skills

If the numerator of a fraction is bigger than the denominator, the fraction is bigger than 1, and it can be written as a mixed number. For example,

$$\frac{7}{4} = \frac{4+3}{4} = \frac{4}{4} + \frac{3}{4} = 1\frac{3}{4}$$

Write each improper fraction as a mixed number.

57. $\frac{9}{5}$

58. $\frac{13}{9}$

59. $\frac{11}{6}$

60. $\frac{11}{8}$

61. $\frac{10}{7}$

62. $\frac{5}{3}$

63. $\frac{5}{4}$

64. $\frac{8}{5}$

65. $\frac{13}{8}$

66. $\frac{12}{7}$

3.3 Subtracting Signed Numbers

A. How Do We Subtract Signed Numbers?

Subtracting a Positive Number

Compare the two number lines shown in Figure 3.29. The first number line illustrates the addition problem

$$(+8) + (-3)$$

The second number line illustrates the subtraction problem

$$(+8) - (+3)$$

Figure 3.29

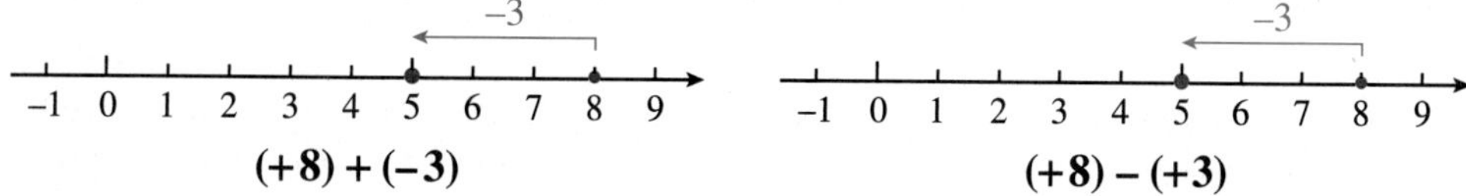

In each figure, we start at +8 on the number line and move 3 units to the *left*. Both problems have the same answer, +5; that is,

$$(+8) + (-3) = 5 \quad \text{and} \quad (+8) - (+3) = 5$$

This example illustrates an important fact:

Subtracting a positive number is the same as adding the corresponding negative number.

Of course, we don't need a number line to calculate simple subtractions like $8 - 3$. But number lines may help us with less familiar problems.

EXAMPLE 1  

Use number lines to illustrate the sum or difference.

a. $(-3) + (-5)$ **b.** $(-3) - (+5)$

Solution

a.

Figure 3.30a

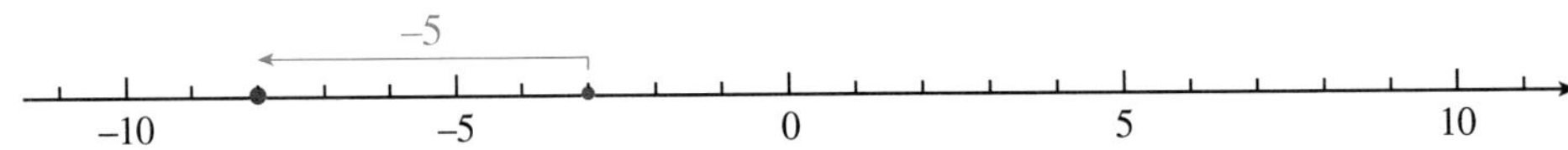

b.

Figure 3.30b

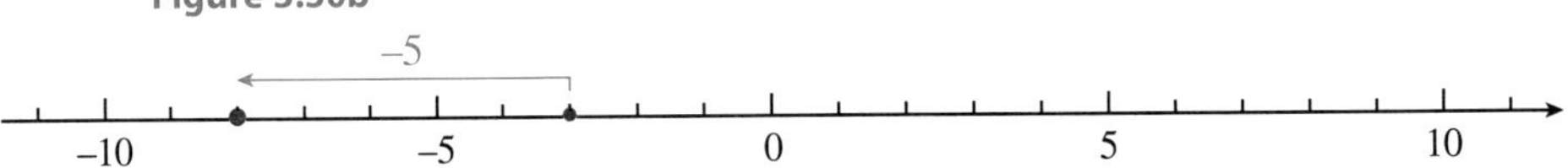

EXERCISES

Use number lines to illustrate each of the following problems. The answers for each pair should be the same! Keep in mind that when we subtract a positive number, we must move to the left on the number line.

1a. $(+2) + (-8)$ **b.** $(+2) - (+8)$

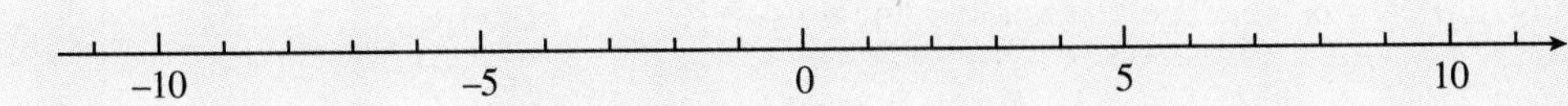

2a. $(+9) + (-4)$ **b.** $(+9) - (+4)$

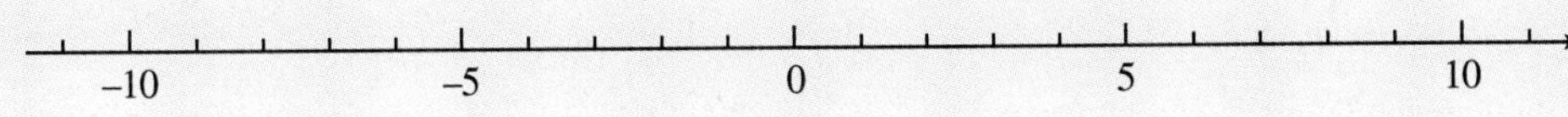

If you are convinced that subtracting a positive number is the same as adding a negative number, then you don't need to use a number line to compute differences. We can rewrite any subtraction problem as an equivalent addition problem and then use the rules for addition from Lesson 3.2.

EXAMPLE 2 

Rewrite each subtraction problem as an equivalent addition problem and then compute.

a. $6 - (+10)$ **b.** $-4 - (+3)$

Solution

a. Subtracting positive 10 is the same as adding negative 10, so

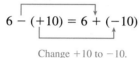

Change subtraction to addition.

$$6 - (+10) = 6 + (-10)$$

Change $+10$ to -10.

Now use the rules for addition: To add two numbers with *opposite* signs, we *subtract* their unsigned parts to get $10 - 6 = 4$. Since 10 is greater than 6, the answer is negative. Thus,

$$6 + (-10) = -4.$$

b. Subtracting positive 3 is the same as adding negative 3, so

Change subtraction to addition.

$$-4 - (+3) = -4 + (-3)$$

Change $+3$ to -3.

Now use the rules for addition. To add two numbers with the *same* sign, we *add* their unsigned parts to get $4 + 3 = 7$. Since both numbers are negative, the answer is negative. Thus,

$$-4 + (-3) = -7$$

EXERCISE 3
Rewrite each subtraction problem as an equivalent addition problem and then compute.
a. $2 - (+8)$

b. $-3 - (+5)$

Subtracting a Negative Number

What does it mean to subtract a negative number? Consider the problem

$$2 - (-5)$$

If we use a number line to help us, we begin by plotting the first number, 2. We should then move 5 units away, but in which direction? Recall that when we *add* -5, we move to the *left.* Since there are only two directions on the number line (right and left), when we *subtract* -5 we must move to the *right.* We finish the problem as shown in Figure 3.31, to obtain an answer of 7.

Figure 3.31

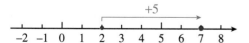

Subtracting -5 on the number line looks the same as adding $+5$. In both problems we move 5 units to the right, and in both problems the answer is 7; that is,

$$2 - (-5) = 7 \quad \text{and} \quad 2 + (+5) = 7$$

This makes sense if we think in terms of money, with negative numbers representing debts. Suppose you have a net worth of \$200, after writing a check for \$500 to repay a loan from your father. Then your father calls to say "As your graduation present, I've decided to cancel your \$500 debt, so keep the money." You now have \$700. Subtracting a debt of \$500 has the same effect as adding \$500 to your account!

The example above illustrates another important fact:

Subtracting a negative number is the same as adding the corresponding positive number.

EXAMPLE 3  

Use number lines to illustrate each sum or difference.

a. $6 - (-3)$ **b.** $6 + (+3)$

Solution

a.

Figure 3.32a

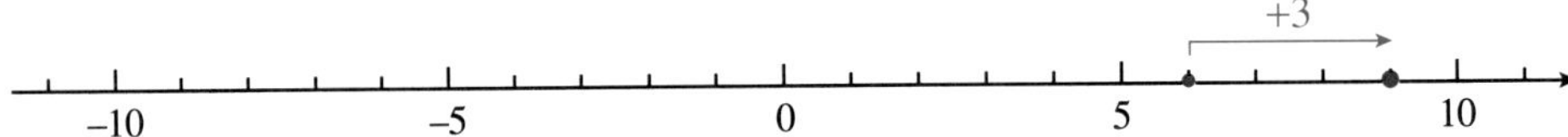

b.

Figure 3.32b

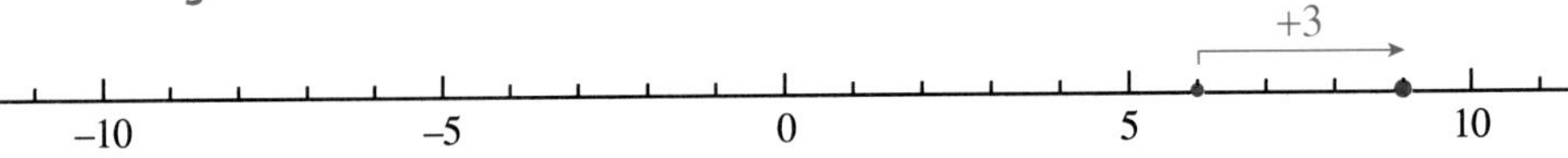

EXERCISES

Demonstrate the following subtraction problems on the number line and note that each has the same answer as the equivalent addition problem.

4a. $-8 - (-2)$ **b.** $-8 + (+2)$

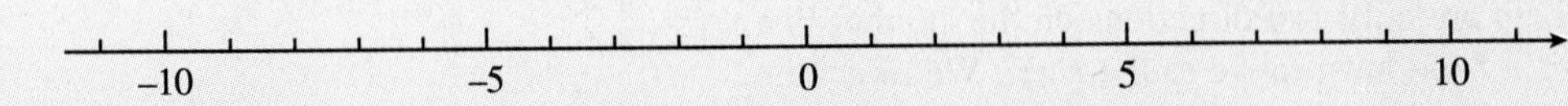

5a. $-4 - (-7)$ **b.** $-4 + (+7)$

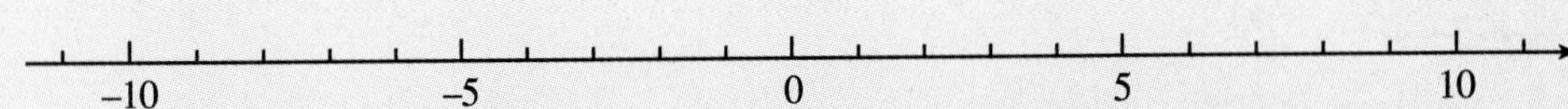

If you are convinced that every subtraction problem can be rewritten as an equivalent addition problem, you don't need to use number lines to compute differences.

EXAMPLE 4 

Rewrite each subtraction problem as an equivalent addition problem and then compute.

a. $-8 - (-12)$ **b.** $15 - (-20)$

Solution

a. Subtracting -12 is the same as adding $+12$, so

Change subtraction to addition.

$$-8 - (-12) = -8 + (+12)$$

Change -12 to $+12$.

Now use the rules for addition: Since -8 and $+12$ have opposite signs, we subtract their unsigned parts to get $12 - 8 = 4$. Since 12 is larger than 8, the answer is positive. Thus,

$$-8 + (+12) = 4$$

b. Subtracting -20 is the same as adding $+20$, so

Change subtraction to addition.

$$15 - (-20) = 15 + (+20)$$

Change -20 to $+20$.

We don't need the rules to add two positive numbers:

$$15 + 20 = 35$$

Rule for Subtraction

We don't really have any new rules to follow for subtraction. Instead, we change every subtraction problem to an equivalent addition problem and then follow the rules for addition.

Rules for Subtracting Signed Numbers

1. To subtract a positive number, add the corresponding negative number.
2. To subtract a negative number, add the corresponding positive number.

EXAMPLE 5 

Compute $-8 - 13$

Solution

We read this expression as negative 8 subtract 13. Subtracting a positive number is the same as adding a negative number, so we can rewrite the problem as

$$-8 + (-13)$$

Finally, to add two negative numbers, we add the numbers and make the answer negative:

$$-8 + (-13) = -21$$

EXERCISE 6
Rewrite each subtraction problem as an equivalent addition problem and then compute.

a. $3 - (-2)$

b. $-19 - (-16)$

Using a Calculator

We can use a calculator to perform subtraction problems involving negative numbers. Remember that the [−] key is used only for subtraction; the [+/−] key is used to enter a negative number.

EXERCISE 7
Use a calculator to compute
$5.7 - (-8.4)$

EXAMPLE 6

Use a calculator to compute $-4.8 - (-7.2)$

Solution

Enter the difference as shown below:

4.8 [+/−] [−] 7.2 [+/−] [=]

The calculator will return the answer 2.4.

Profit and Loss

In Lesson 2.3 we learned a formula for calculating profit:

$$\text{Profit} = \text{Revenue} - \text{Cost}$$

If a company's costs exceed its revenues, its profit will have a negative value, which means that the company will experience a loss.

EXAMPLE 7

Figure 3.33 shows a company's costs and revenues for the years 1997–2001.

Figure 3.33 Cost and revenue

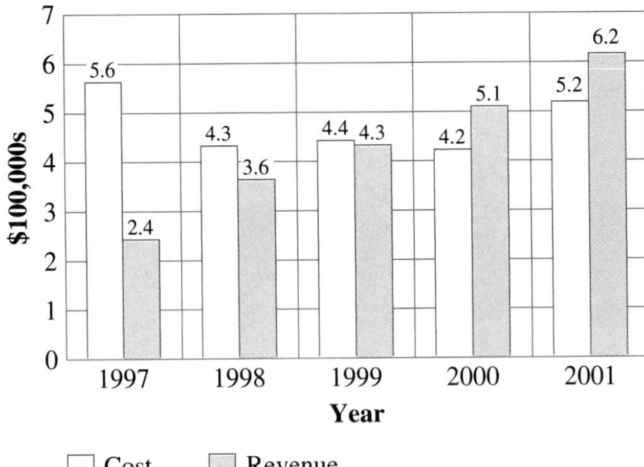

a. Fill in the table with the values of revenue and cost from the bar graph.

Year	1997	1998	1999	2000	2001
Revenue					
Cost					

b. Compute the company's profit for 1997.

Solution

a. Read the values of Cost and Revenue from the bar graph. Because the graph is scaled in units of \$100,000, we multiply each value from the graph by \$100,000. For example, the company's costs in 1997 were

$$5.6 \times \$100{,}000 = \$560{,}000$$

Year	1997	1998	1999	2000	2001
Revenue	240,000	360,000	430,000	510,000	620,000
Cost	560,000	430,000	440,000	420,000	520,000

b. To calculate the company's profit in 1997, use the formula

$$\begin{aligned} P &= R - C \\ &= 240{,}000 - 560{,}000 = -320{,}000 \end{aligned}$$

The company lost \$320,000 in 1997.

EXERCISE 8

a. Complete the table showing the company's profit for each of the 5 years.

Year	1997	1998	1999	2000	2001
Revenue	240,000	360,000	430,000	510,000	620,000
Cost	560,000	430,000	440,000	420,000	520,000
Profit					

b. In which of the 5 years did the company lose money?

ANSWERS TO 3.3A EXERCISES

1.

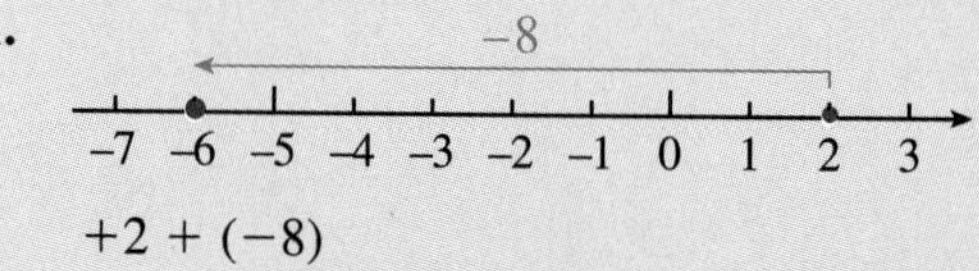

$+2 + (-8)$

2.

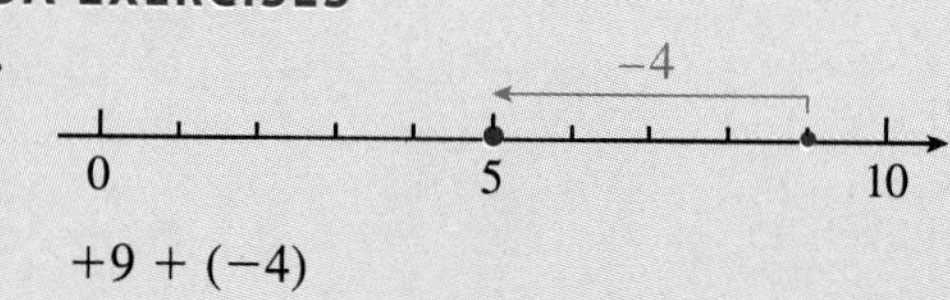

$+9 + (-4)$

3a. -6 **3b.** -8

4.

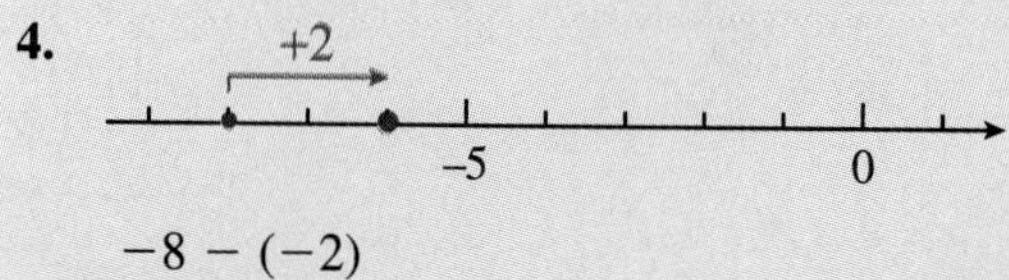

$-8 - (-2)$

5.

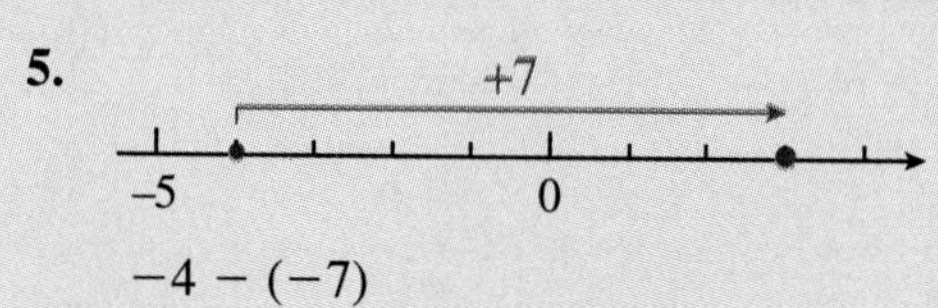

$-4 - (-7)$

6a. 5 **6b.** -3 **7.** 14.1

8a.

Year	1997	1998	1999	2000	2001
Revenue	240,000	360,000	430,000	510,000	620,000
Cost	560,000	430,000	440,000	420,000	520,000
Profit	−320,000	−70,000	−10,000	90,000	100,000

8b. 1997, 1998, and 1999

HOMEWORK 3.3A 

(a) Rewrite each subtraction problem as an equivalent addition problem.
(b) Illustrate each problem on a number line.

1. $8 - (-4)$

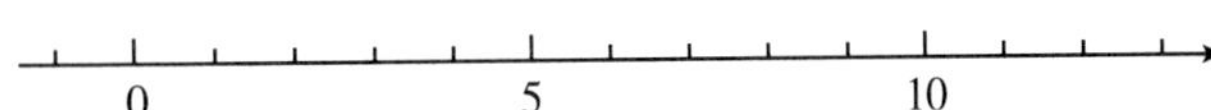

2. $9 - (+6)$

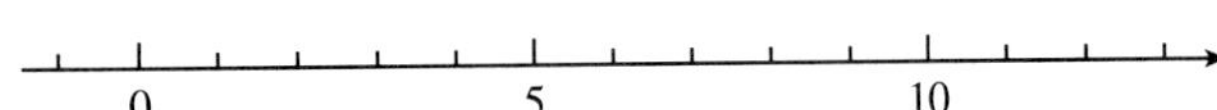

3. $-3 - (+8)$

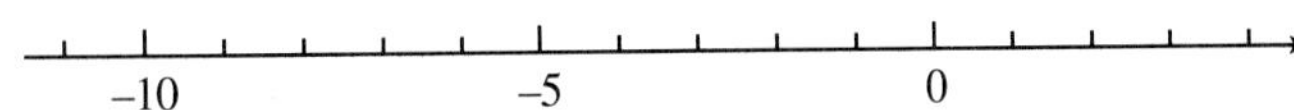

4. $-5 - (-7)$

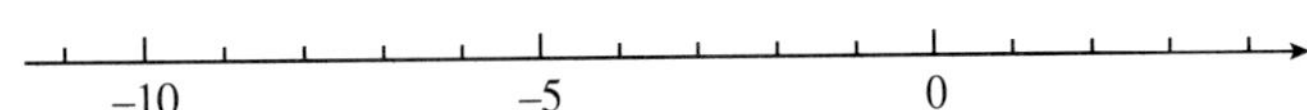

5. $11 - (+5)$

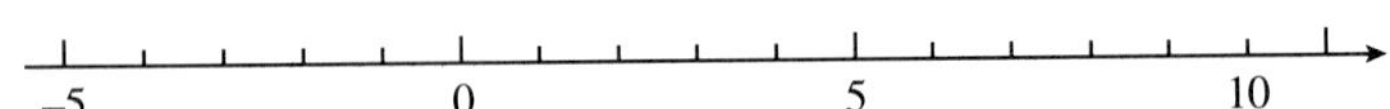

6. $8 - (-1)$

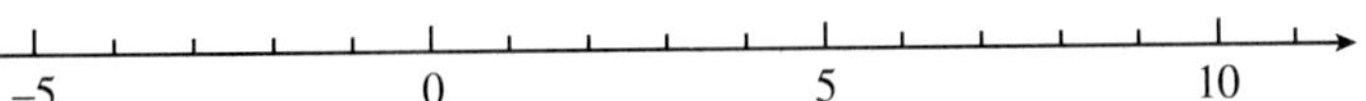

7. $-9 - (-2)$

8. $-4 - (-5)$

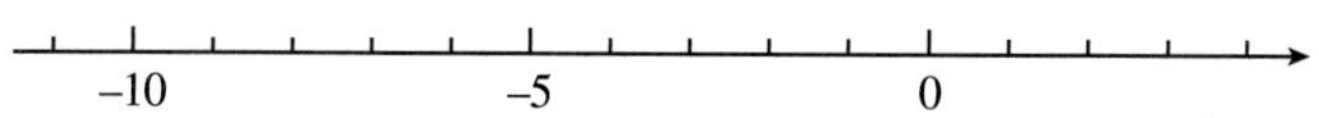

9. Use ideas about money to explain why subtracting a negative number is the same as adding a positive number.

10. Explain why we don't really need any new rules for subtraction of signed numbers.

Subtract. If necessary, use the rule for subtraction.

11. $12 - (+3)$

12. $8 - (+6)$

13. $5 - (+11)$

14. $7 - (+10)$

15. $-6 - (+9)$

16. $-4 - (+3)$

17. $-8 - (+4)$

18. $-1 - (+7)$

19. $-5 - (+5)$

20. $-9 - (+9)$

21. $7 - (+7)$

22. $2 - (+2)$

23. $9 - (-8)$

24. $7 - (-5)$

25. $17 - (-20)$

26. $11 - (-16)$

27. $-13 - (+9)$

28. $-10 - (+4)$

29. $-12 - (+19)$

30. $-18 - (+21)$

31. $-17 - (+17)$

32. $-32 - (+32)$

33. $-16 - (-5)$

34. $-9 - (-20)$

35. $-22 - (-28)$

36. $-25 - (-18)$

37. $-30 - (-30)$

38. $-14 - (-14)$

39. Delbert participated in a student math competition at his school. The competition consisted of three rounds of problem solving. Delbert earns points for correct answers, but points are subtracted from his score for each incorrect answer. Figure 3.34 shows Delbert's results after three rounds.

Figure 3.34 Contest points

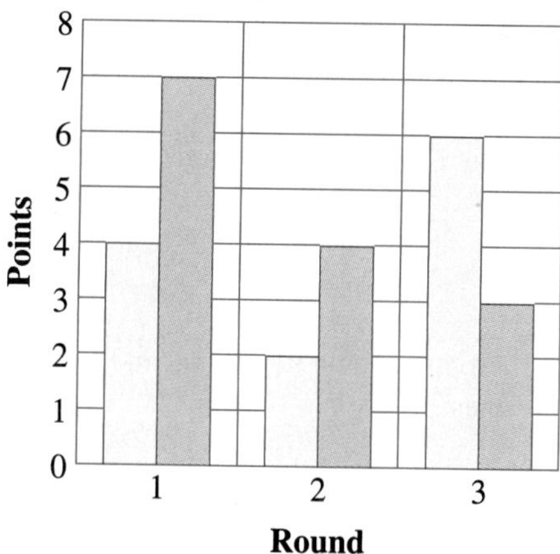

a. Find Delbert's score on each round.

b. The final score for each contestant is the sum of the scores on each round. Find Delbert's final score.

40. Figure 3.35 shows the quarterly revenue and cost figures for Infocom, a new internet service provider.

Figure 3.35 Cost and revenue by quarter

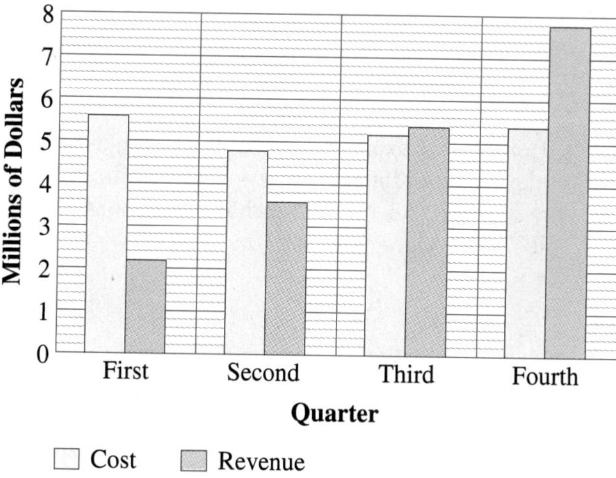

a. Compute Infocom's profit for each quarter of the year.

b. Compute Infocom's profit for the entire year (the sum of its profits for each quarter).

For Problems 41–48, use a calculator to find the following differences.

41. $23.8 - (-82.9)$

42. $-18.3 - (+74.1)$

43. $-249 - (-152)$

44. $-453 - (-826)$

45. $0.012 - (-0.39)$

46. $0.39 - (-0.012)$

47. $-1.05 - (-0.15)$

48. $-2.02 - (-0.2)$

49. a. Read each problem aloud in English and then compute the values:

$16 - (+4)$ $\quad 16 + (-4)$

b. Consider the expression $16 - 4$. Should this expression be interpreted as 4 subtracted from 16 or as -4 added to 16? Will both interpretations give the same value? Which do you prefer?

50. a. Read each problem aloud in English and then compute the values:

$8 - (+15)$ $\quad 8 + (-15)$

b. Consider the expression $8 - 15$. Should this expression be interpreted as 15 subtracted from 8 or as -15 added to 8? Will both interpretations give the same value? Which do you prefer?

51. a. Read each problem aloud in English and then compute the values:

$-3 - (+9)$ $\quad -3 + (-9)$

b. Consider the expression $-3 - 9$. Should this expression be interpreted as 9 subtracted from -3 or as -9 added to -3? Will both interpretations give the same value? Which do you prefer?

52. a. Read each problem aloud in English and then compute the values:

$-18 - (+6)$ $\quad -18 + (-6)$

b. Consider the expression $-18 - 6$. Should this expression be interpreted as 6 subtracted from -18 or -6 added to -18? Will both interpretations give the same value? Which do you prefer?

53. a. Write the expression $8 - 15$ as an equivalent addition problem and then as an equivalent subtraction problem.

b. In your opinion, is it easier to compute $8 - 15$ as a subtraction problem or as an addition problem?

54. a. Write the expression $-4 - 5$ as an equivalent addition problem and then as an equivalent subtraction problem.

b. In your opinion, is it easier to compute $-4 - 5$ as a subtraction problem or as an addition problem?

Simplify each expression by treating it as an addition.

55. a. $-5 - 7$

b. $2 - 9$

56. a. $8 - 11$

b. $-4 - 9$

Add or subtract as indicated.

57. $13 + (-5)$

58. $-16 + (-7)$

59. $-27 + (-13)$

60. $28 + (-15)$

61. $35 - 11$

62. $-15 - 20$

63. $-40 - 30$

64. $44 - 22$

65. $-18 + 53$

66. $-18 - 53$

67. $-3 - (-5)$

68. $-3 - 5$

69. $3 - 5$

70. $3 - (-5)$

Building Number Skills

You know that there are four-fourths in one whole. How many fourths are there in two wholes or in three wholes? In the following exercises, illustrate your answers with a sketch.

Example: How many thirds are in two wholes? **Six**

$2 = \frac{?}{3}$  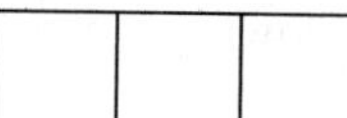 $\mathbf{2 = \frac{6}{3}}$

71. How many fourths are in two wholes?

72. How many halves are in three wholes?

73. How many thirds are in three wholes?

74. How many fifths are in two wholes?

75. How many sixths are in three wholes?

76. How many thirds are in four wholes?

77. How many halves are in five wholes?

78. How many sixths are in two wholes?

79. How many eighths are in three wholes?

80. How many ninths are in two wholes?

B. Signed Numbers and Variables

Evaluation

Keeping track of negative signs can be tricky when subtraction is involved. It is especially important to enclose negative numbers in parentheses when we are evaluating algebraic expressions.

EXAMPLE 8  

Evaluate the following expressions for $w = -6$, $x = -15$, and $z = 8$.

a. $w - x$ **b.** $x - z$

Solution

a. Substitute -6 for w and -15 for x. Enclose each value within parentheses:

$$w - x = (-6) - (-15) \quad \text{Rewrite as an addition.}$$
$$= -6 + 15 \quad \text{Follow the rules for addition.}$$
$$= 9$$

b. Substitute -15 for x and 8 for z. Enclose each value within parentheses:

$$\begin{aligned} x - z &= (-15) - (8) && \text{Rewrite as an addition.} \\ &= -15 + (-8) && \text{Follow the rules for addition.} \\ &= -23 \end{aligned}$$

EXERCISE 9

Evaluate the following expressions for $a = -7$, $b = 3$, and $c = -8$.

a. $a - b$

b. $c - a$

CAUTION!

Addition and subtraction operations should be performed *in order from left to right.* In the example below, notice that we get the wrong answer if we don't perform the calculations in order.

Correct Method	*Incorrect Method*
$20 - 12 - 5$	$20 - 12 - 5$
$= 8 - 5$	$= 20 - 7$
$= 3$ ← **Right answer**	$= 13$ ← **Wrong answer**

EXAMPLE 9

Simplify each expression.

a. $-15 - 6 + 8$ **b.** $8 - 6 - (-10)$

Solution

Perform the calculations in order from left to right.

a. $$\begin{aligned} -15 - 6 + 8 &= \mathbf{-15 - 6} + 8 && \text{Add } -6 \text{ to } -15. \\ &= -21 + 8 && \text{Add } -21 \text{ and } 8. \\ &= -13 \end{aligned}$$

b. $$\begin{aligned} 8 - 6 - (-10) &= \mathbf{8 - 6} - (-10) && \text{Subtract 6 from 8.} \\ &= 2 - (-10) && \text{Rewrite as an addition.} \\ &= 2 + 10 = 12 \end{aligned}$$

EXERCISE 10

Simplify each expression.

a. $10 - 16 + 3$

b. $-7 + 5 - (-3)$

Solving Equations

Sometimes we can simplify one or both sides of an equation before trying to solve.

EXAMPLE 10 

Solve.

a. $x - (-12) = 34$ **b.** $-5 + t = 7 - 15$

Solution

a. Simplify the left side of the equation and then solve.

$$\begin{aligned} x - (-12) &= 34 && \text{Rewrite the subtraction as addition.} \\ x + 12 &= 34 && \text{Subtract 12 from both sides.} \\ \mathbf{-12} &\quad \mathbf{-12} \\ \hline x &= 22 \end{aligned}$$

EXERCISE 11
Solve.

a. $w - (-12) = -15$

b. $-6 + p = -3 + (-5)$

EXERCISE 12
One year ago the water level in the Sweetwater Reservoir was 26.5 feet. Due to high usage and drought conditions, the water level is now 19.8 feet. What was the net change in the water level?

b. Simplify the right side of the equation and then solve.

$$\begin{aligned} -5 + t &= 7 - 15 && \text{Add 7 and } -15. \\ -5 + t &= -8 && \text{Add 5 to both sides.} \\ \underline{+5} \quad & \quad \underline{+5} \\ t &= -3 \end{aligned}$$

Net Change

We can compute the **net change** in a variable by subtracting its initial value from its final value:

$$\textit{net change} = \textit{final value} - \textit{initial value}$$

If the final value is larger than the initial value, the net change is positive. If the final value is smaller than the initial value, the net change is negative.

EXAMPLE 11

Figure 3.36 shows the prime-time Nielsen ratings for the major networks for the week ending 3 February 2001, compared to their previous totals.

Figure 3.36 TV ratings

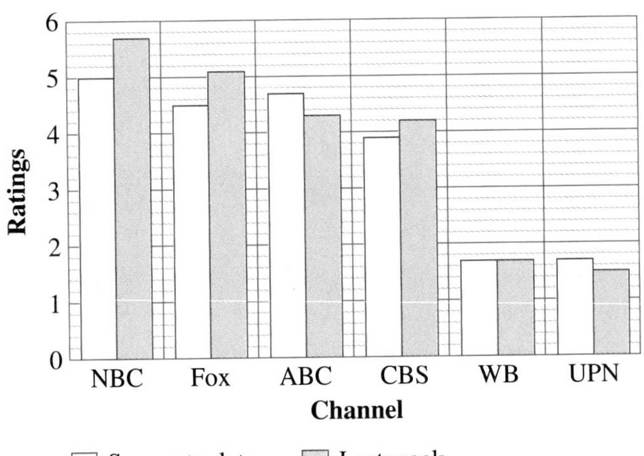

(Source: *Los Angeles Times*)

a. What was the net change in ABC's rating?

b. Make a table showing the net change in ratings for each of the networks.

Solution

a. The net change in ABC's rating is given by

$$\begin{aligned} \text{net change} &= \text{final value} - \text{initial value} \\ &= 4.3 - 4.7 = -0.4 \end{aligned}$$

ABC's rating dropped by 0.4 point.

b. Use the formula to compute the net change for each network.

Network	NBC	Fox	ABC	CBS	WB	UPN
Change in rating	0.7	0.6	−0.4	0.3	0	−0.2

If we know the starting value and the net change in a variable, we can calculate its final value.

EXAMPLE 12

Figure 3.37 shows the change in annual sales for Solaria.com over 6 years.

Figure 3.37 Change in annual sales

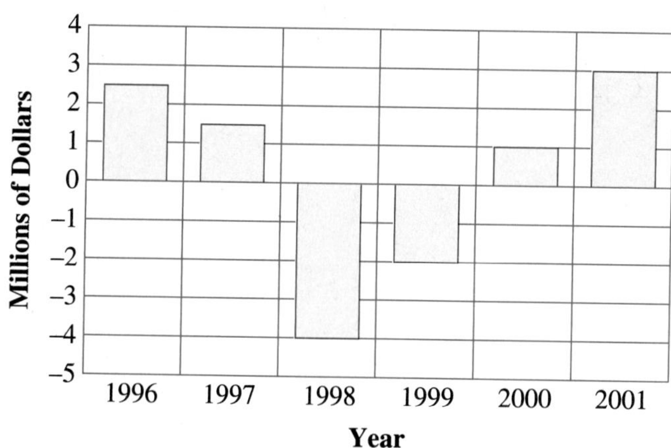

a. In 1995 Solaria posted \$4 million in sales. What was its sales total in 1996?

b. Make a table showing Solaria's sales totals for the next 5 years.

Solution

a. Solaria's sales were \$4 million in 1995 and sales increased by \$2.5 million in 1996, so their sales total in 1996 was

$$4 + 2.5 = 6.5 \text{ million dollars}$$

or \$6,500,000.

b. To compute the sales figure for each year, add the change in sales to the figure for the previous year:

$$\begin{aligned} &1997\text{:} && 6.5 + 1.5 = 8 \text{ million dollars} \\ &1998\text{:} && 8 - 4 = 4 \text{ million dollars} \\ &1999\text{:} && 4 - 2 = 2 \text{ million dollars} \\ &2000\text{:} && 2 + 1 = 3 \text{ million dollars} \\ &2001\text{:} && 3 + 3 = 6 \text{ million dollars} \end{aligned}$$

Thus,

Year	1997	1998	1999	2000	2001
Sales (million \$)	8	4	2	3	6

EXERCISE 13

If Solaria.com projects a net change of −\$2.5 million in sales for 2002 compared with 2001, what will be the total sales for 2002?

ANSWERS TO 3.3B EXERCISES

9a. −10

9b. −1

10a. −3

10b. 1

11a. $w = -27$

11b. $p = -2$

12. −6.7 feet

13. \$3.5 million

HOMEWORK 3.3B 

Simplify each expression.

1. $8 - 10 - 6$
2. $-16 + 7 - 12$
3. $-3 - (-18) - 9$
4. $14 - 20 + 3$
5. $6 - (-6) - 6$
6. $-5 - (-5) - 5$
7. $-2 + 3 - (-4) - 5$
8. $9 - 8 + 7 - (-6)$
9. $-12 - 12 - 12$
10. $15 - (-15) - (-15)$

For Problems 11–20, evaluate each expression for $a = -3$, $b = 8$, and $c = -7$.

11. $-a$
12. $-(-b)$
13. $-(-c)$
14. $a - c$
15. $-b - a$
16. $a + b - c$
17. $b - c - a$
18. $c - b - a$
19. $a - b + c$
20. $b - a + c$

21. Figure 3.38 shows the value of U.S. exports to Venezuela for the years 1994–1999.

Figure 3.38 U.S. exports to Venezuela

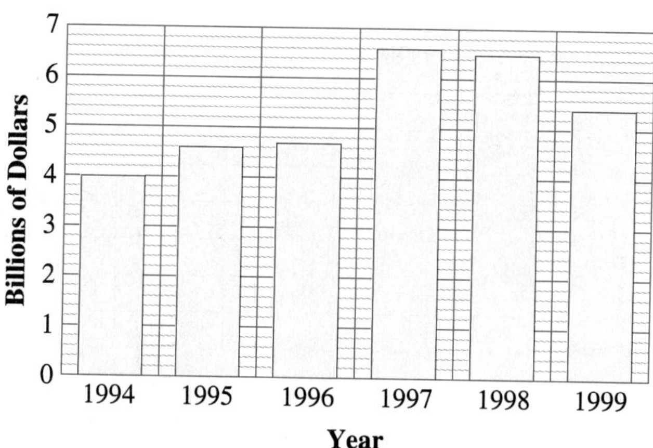

(Source: U.S. Department of Commerce)

a. By how much did the exports change in 1995 compared with 1994? In 1996 compared with 1995? Fill in the table:

Year	1995	1996	1997	1998	1999
Change in exports (billion $)					

b. Make a bar graph showing changes in exports on the grid in Figure 3.39. Use the table from part (a).

Figure 3.39 Changes in exports to Venezuela

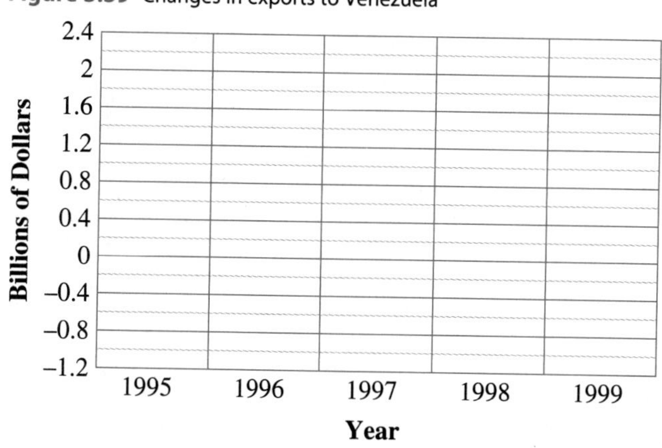

22. Figure 3.40 shows the value of U.S. exports to South Africa for the years 1994–1999.

Figure 3.40 U.S. exports to South Africa

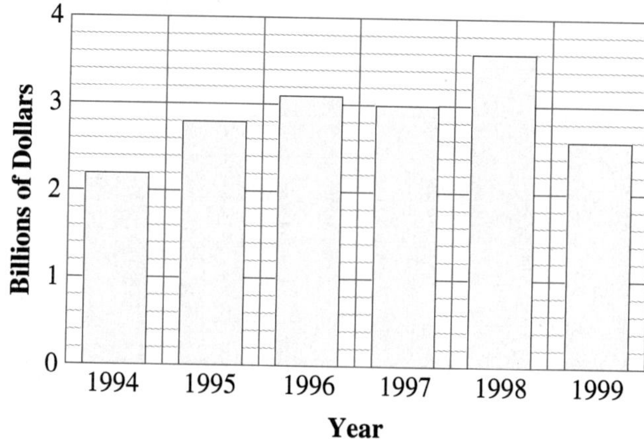

(Source: U.S. Department of Commerce)

a. By how much did the exports change in 1995 compared with 1994? In 1996 compared with 1995? Fill in the table:

Year	1995	1996	1997	1998	1999
Change in exports (billion $)					

b. Make a bar graph showing changes in exports on the grid in Figure 3.41. Use the table from part (a).

Figure 3.41 Changes in exports to South Africa

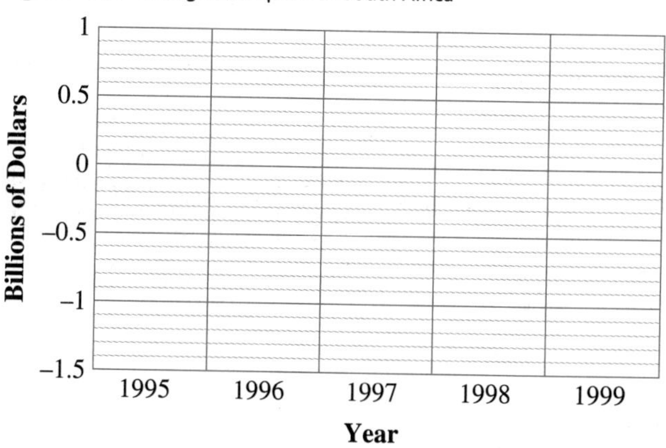

23. In 1995 Boeing Corporation filled 468 orders for planes from commercial airlines. Figure 3.42 shows the change in the number of orders for aircraft over the next few years.

Figure 3.42 Change in Boeing orders

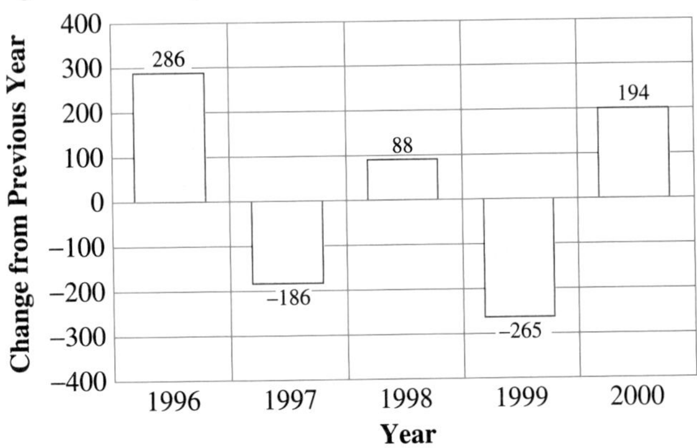

(Source: *Los Angeles Times*)

a. Complete the table showing the number of commercial aircraft built by Boeing in each year:

Year	1996	1997	1998	1999	2000
Aircraft					

b. In which year did Boeing build the most airplanes?

24. In 1995 Airbus Industrie filled 106 orders for planes from commercial airlines. Figure 3.43 shows the change in the number of orders for aircraft over the next few years.

Figure 3.43 Change in Airbus orders

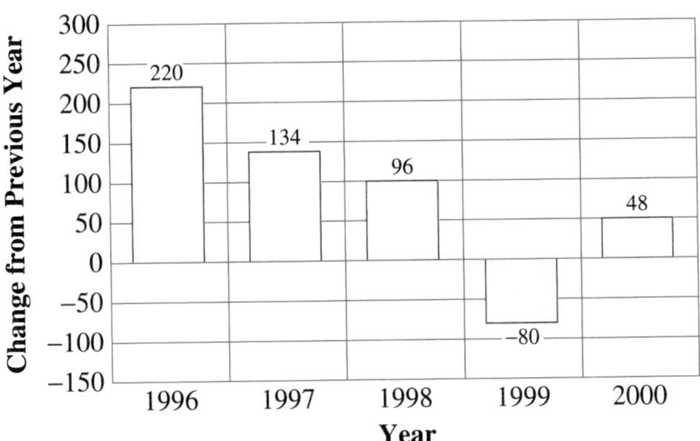

(Source: *Los Angeles Times*)

a. Complete the table showing the number of commercial aircraft built by Airbus in each year:

Year	1996	1997	1998	1999	2000
Aircraft					

b. In which year did Airbus first build more commercial airplanes than Boeing? (See Problem 23.)

Solve each equation. Check your solutions by substituting into the original equation to see if a true statement results.

25. $13 + a = 9$

26. $7 + b = -10$

27. $p + 12 = -6 - 4$

28. $q + 15 = -7 - 2$

29. $h + 16 = 12 - (-7)$

30. $k + 8 = 5 - (-11)$

31. $m - 6 = -11$

32. $n - 14 = -9$

33. $-36 + t = -22$

34. $-17 + s = -11$

35. $41 + u = -33 + 14$

36. $56 + v = 35 - 48$

37. $c - 6.8 = -3.9$

38. $d - 5.1 = -7.8$

39. $f - (-4.8) = -2.3$

40. $g - (-8.5) = 2.4$

Evaluate each expression for $w = -0.2$, $v = -1.4$, and $t = 2.1$.

41. $v - t$

42. $-v - w$

43. $v - t + w$

44. $-t - t - t$

45. $-w + t + v$

46. $t - v - w$

Evaluate the algebraic expression to complete the table.

47.

x	$x + 4$
−7	
−5	
−3	
−1	

48.

w	$9 + w$
−20	
−16	
−12	
−8	

49.

h	$h - 5$
−8	
−5	
−2	
2	

50.

m	$8 - m$
−10	
−5	
5	
10	

Write an algebraic expression for each quantity. (See Lesson 1.4 to review algebraic expressions.)

Example: How many inches are there in f feet?

There are 12 inches in 1 foot, so in f feet there are $12f$ inches.

51. How many feet are there in k yards?

52. How many ounces are there in p pounds?

53. How many pounds are there in t tons?

54. How many feet are there in m miles?

55. How many quarts are there in g gallons?

56. How many gallons are there in q quarts? (*Be careful:* Should the number of gallons be larger or smaller than the number of quarts?)

57. How many pounds are there in z ounces?

58. How many yards are there in h feet?

59. How many feet are there in n inches?

60. How many miles are there in y yards?

Building Number Skills

Write each improper fraction as a whole number. Illustrate with a sketch.

Example $\frac{6}{3}$: 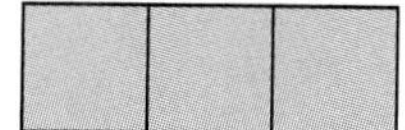 $\frac{6}{3} = 2$

How many groups of three thirds?

61. $\frac{8}{4}$

62. $\frac{6}{2}$

63. $\frac{9}{3}$

64. $\frac{16}{4}$

65. $\frac{8}{2}$

66. $\frac{12}{4}$

67. $\frac{5}{5}$

68. $\frac{6}{6}$

69. $\frac{12}{3}$

70. $\frac{12}{6}$

Do you see a way to work these exercises without using a diagram? Describe your method and give an example.

3.4 Multiplying and Dividing Signed Numbers

A. How Do We Multiply and Divide Signed Numbers?

Multiplication

Multiplication is really just a shortcut for repeated addition. For example,

$$5 \cdot 2 \quad \text{means} \quad \underbrace{2 + 2 + 2 + 2 + 2}_{\text{Five 2s added together}}$$

Similarly,

$$5 \cdot (-2) \quad \text{means} \quad \underbrace{(-2) + (-2) + (-2) + (-2) + (-2)}_{\text{Five } (-2)\text{s added together}}$$

Since the sum of five -2s added together is -10, the product $5(-2)$ is equal to -10 also. Think of this problem in terms of money: If you owe five different people \$2 each, then you are actually in debt by \$10. From these examples, the following rule seems reasonable.

The product of two numbers with opposite signs is negative.

It is also true that $(-2)(5) = -10$, because we can multiply two numbers in either order and get the same answer. Zero times any number, positive or negative, is zero.

EXERCISE 1
Multiply.
a. $5(-4)$

b. $(-7)(2)$

EXAMPLE 1 

a. $(-3)(6) = -18$ **b.** $8(-4) = -32$
c. $9(-0.3) = -2.7$ **d.** $(-13)(0) = 0$

Earlier we argued that erasing a debt of \$500 had the same effect as adding \$500 to your net worth. This example helped us understand that subtracting a negative number is the same as adding a positive number. Now imagine that *three* debts of \$500 are erased from your balance sheet. This is equivalent to *increasing* your net worth by \$1500. We express this mathematically as

$$-3(-500) = 1500$$

In general, we have the following rule.

The product of two negative numbers is positive.

EXERCISE 2
Multiply.
a. $(-9)(-7)$

b. $(-6)(-0.4)$

EXAMPLE 2  

a. $(-4)(-5) = 20$ **b.** $(-1.5)(-8) = 12$

We know that the product of two positive numbers is positive. We can summarize our findings as follows.

Rules for Multiplying Signed Numbers

1. The product of two numbers with the same sign is positive.
2. The product of two numbers with opposite signs is negative.
3. The product of any number and zero is zero.

We can also find products of signed numbers using a calculator.

EXAMPLE 3

Use a calculator to find the product $6.2(-3.4)$

Solution

Enter the product as shown:

6.2 [×] 3.4 [+/−] [=]

The calculator will return the answer -21.08.

EXERCISE 3

Use a calculator to find the product $(-8.25)(6.4)$

Division

Division is the opposite operation for multiplication. This means that every division fact can be rewritten as an equivalent multiplication fact. For example,

$$\frac{12}{4} = \mathbf{3} \qquad \text{because} \qquad 4 \cdot \mathbf{3} = 12$$

We can use this idea to investigate division of signed numbers. Each division problem below is rewritten as a multiplication problem:

$$\frac{-20}{5} = \boxed{?} \qquad \text{because} \qquad 5 \cdot \boxed{?} = -20$$

$$\frac{-14}{-2} = \boxed{?} \qquad \text{because} \qquad -2 \cdot \boxed{?} = -14$$

Consider the first multiplication problem, $5 \cdot \boxed{?}\ -20$. From what we have learned about multiplication of signed numbers, we know that the question mark must be replaced by a negative number. (Why?) In fact,

$$5 \cdot \boxed{\mathbf{-4}} = -20, \qquad \text{so} \qquad \frac{-20}{5} = \boxed{\mathbf{-4}}$$

From this problem, we see that

The quotient of a negative number and a positive number is negative.

In the second multiplication problem, $-2 \cdot \boxed{?}\ -14$, the question mark must be replaced by a positive number. (Why?) In fact,

$$-2 \cdot \boxed{7} = -14, \qquad \text{so} \qquad \frac{-14}{-2} = \boxed{7}$$

From this problem, we see that

The quotient of two negative numbers is positive.

EXERCISE 4

Divide.

a. $\dfrac{-56}{-7}$

b. $\dfrac{-12}{2}$

EXAMPLE 4  

a. $\dfrac{27}{-3} = -9$ b. $\dfrac{-28}{-7} = 4$

We can also use a calculator to compute quotients of signed numbers.

EXERCISE 5

Use a calculator to find the quotient $\dfrac{-25.2}{1.4}$

EXAMPLE 5

Use a calculator to find the quotient $\dfrac{-2.5}{-0.02}$

Solution

Enter the division problem as shown:

2.5 [+/−] [÷] 0.02 [+/−] [=]

The calculator will return the answer 125.

Quotients Involving Zero

What about quotients involving zero? First consider two examples where zero is divided by a number. For each example, we rewrite the quotient as a product:

$$\frac{0}{8} = \boxed{?} \quad \text{because} \quad 8 \cdot \boxed{?} = 0$$

$$\frac{0}{-5} = \boxed{?} \quad \text{because} \quad -5 \cdot \boxed{?} = 0$$

What number can we use to replace the question marks and make true statements? For both multiplication problems, the only number we can use is zero. Thus,

$$\frac{0}{8} = \boxed{0} \quad \text{because} \quad 8 \cdot \boxed{0} = 0$$

and

$$\frac{0}{-5} = \boxed{0} \quad \text{because} \quad -5 \cdot \boxed{0} = 0$$

In general, we can conclude that

The quotient of zero divided by any number (except zero) is zero.

Now consider two examples in which a number is divided by zero. Again, we rewrite the quotients as products:

$$\frac{15}{0} = \boxed{?} \quad \text{because} \quad 0 \cdot \boxed{?} = 15$$

$$\frac{-4}{0} = \boxed{?} \quad \text{because} \quad 0 \cdot \boxed{?} = -4$$

What number can we use to replace the question marks and make true statements? Since zero times any number is zero, we cannot find any number that will result in a product of 15 or a product of −4. There is no solution to either multiplication problem, and consequently there are no solutions to the division problems either. We say that division by zero is *undefined.*

The quotient of any number divided by zero is undefined.

EXAMPLE 6 

a. $\frac{0}{-5} = 0$ b. $\frac{-18}{0}$ is undefined. c. $\frac{0}{0}$ is undefined.

EXERCISE 6

Divide if possible.

a. $\frac{-83}{0}$

b. $\frac{0}{-83}$

We can summarize the rules for division as follows.

Rules for Dividing Signed Numbers

1. The quotient of two numbers with the same sign is positive.
2. The quotient of two numbers with opposite signs is negative.
3. Zero divided by any number (except zero) is zero.
4. The quotient of any number divided by zero is undefined.

ANSWERS TO 3.4A EXERCISES

1a. −20 **1b.** −14 **2a.** 63 **2b.** 2.4 **3.** −52.8 **4a.** 8 **4b.** −6 **5.** −18 **6a.** Undefined **6b.** 0

HOMEWORK 3.4A 

For Problems 1–24, multiply or divide, if possible.

1. 3(−6) **2.** −4(−8) **3.** −9(−2)

4. −6 · 8 **5.** −7 · 4 **6.** 5(−9)

7. −36 ÷ 4 **8.** −42 ÷ (−6) **9.** −56 ÷ (−8)

10. 70 ÷ (−10) **11.** 24 ÷ (−8) **12.** −63 ÷ 7

13. $0 \div (-12)$

14. $0(-12)$

15. $-12 \div 0$

16. $-12 \cdot 0$

17. $12 \div (-12)$

18. $12(-12)$

19. $\frac{-84}{-6}$

20. $\frac{-110}{5}$

21. $\frac{96}{-6}$

22. $\frac{-95}{-5}$

23. $\frac{0}{-7}$

24. $\frac{-7}{0}$

25. Use the idea of repeated addition to explain why the product of a positive number and a negative number is negative.

26. True or false:

a. The sum of two negative numbers is positive.

b. The difference of two negative numbers is positive.

c. The product of two negative numbers is positive.

d. The quotient of two negative numbers is positive.

Compare the expressions in (a) and (b) and simplify each.

27. **a.** $-6(-8)$ **b.** $-6 - 8$

28. **a.** $5 - (-9)$ **b.** $-5(-9)$

29. **a.** $12 - (-4)$ **b.** $12 \div (-4)$

30. **a.** $-36 \div 9$ **b.** $-36 - 9$

31. **a.** $-40 - (-8)$ **b.** $-40 \div (-8)$

32. **a.** $24 + (-6)$ **b.** $24 \div (-6)$

33. **a.** $-3 + (-8)$ **b.** $-3(-8)$

34. **a.** $-2(-13)$ **b.** $-2 - (-13)$

35. **a.** $-9(7)$ **b.** $-9 + 7$

36. **a.** $8 - 12$ **b.** $8(-12)$

Write each division problem as an equivalent multiplication problem.

37. $-40 \div 8 = -5$

38. $-80 \div (-16) = 5$

39. $\frac{56}{-4} = -14$

40. $\frac{-72}{-3} = 24$

41. $\frac{a}{-7} = 28$

42. $\frac{b}{12} = -6$

Use your calculator to multiply or divide. Round your answers to two decimal places if necessary.

43. $-18(-23)$

44. $374 \div (-22)$

45. $-434 \div (-14)$

46. $25(-19)$

47. $\dfrac{288}{-16}$

48. $\dfrac{-312}{-26}$

49. $(-0.3)(26.1)$

50. $(2.8)(-5.4)$

Choose the correct equation for each problem. (See Lesson 1.5 to review equations.)

51. If Ivory's car gets 26 miles to the gallon, how many gallons will she need in order to drive 380 miles?
a. $26g = 380$
b. $380 - 26 = g$
c. $\dfrac{g}{26} = 380$

52. Stanley made $36,000 in commissions, giving him an average earnings of $1125 per week. How many weeks did Stanley work?
a. $\dfrac{w}{36{,}000} = 1125$
b. $36{,}000w = 1125$
c. $1125w = 36{,}000$

53. The Lake Forest Little League will use 180 baseballs this season. How many packages of 1 dozen should the league organizer order?
a. $180p = 12$
b. $12p = 180$
c. $\dfrac{p}{12} = 180$

54. There are 1760 yards in 1 mile. How many miles are in 7920 yards?
a. $7920m = 1760$
b. $1760m = 7920$
c. $7920 - m = 1760$

55. How many pounds is 132 ounces equal to?
a. $16p = 132$
b. $132p = 16$
c. $\dfrac{p}{16} = 132$

56. Elsie and Jen share a paper route with 86 houses. If Elsie delivers papers to 47 houses, how many is Jen responsible for?
a. $47 - h = 86$
b. $47 + h = 86$
c. $h - 86 = 47$

Building Number Skills

Use a diagram to express each mixed number as an improper fraction.

2 wholes + $\frac{1}{3}$

Example: $2\frac{1}{3} =$ [diagram: three bars each divided into thirds; first two fully shaded, third with one third shaded] $= \frac{7}{3}$

Divide each whole into thirds.
Count the thirds.

57. $1\frac{3}{4} =$

58. $1\frac{2}{3} =$

59. $3\frac{1}{3} =$

60. $3\frac{1}{4} =$

61. $2\frac{3}{8} =$

62. $2\frac{5}{6} =$

63. $4\frac{1}{2} =$

64. $3\frac{1}{2} =$

65. $1\frac{5}{9} =$

66. $2\frac{2}{9} =$

Do you see a way to work these exercises without using a diagram? Describe your method and give an example.

B. Signed Numbers and Variables

Evaluation

Suppose we want to evaluate the expression $12b$ for $b = -8$. (Recall that $12b$ means 12 *times* b.) It is very important to enclose -8 in parentheses when we substitute for b, as follows:

$$12b = 12(-8) = -96$$

The parentheses tell us to multiply 12 times -8. If we forget the parentheses, the expression will look like $12 - 8$, a subtraction instead of a multiplication.

EXERCISE 7

Evaluate each expression for $a = -2$ and $b = -18$

a. $\frac{b}{a}$

b. $2ab$

EXAMPLE 7  

Evaluate each expression for $n = -3$ and $p = -6$.

a. $5np$

b. $\frac{n}{p}$

Solution

Substitute -3 for n and -6 for p.

a. $5np = 5(-3)(-6)$
$= -15(-6)$
$= 90$

b. $\frac{n}{p} = \frac{-3}{-6}$
$= 0.5$

EXERCISE 8

a. Evaluate $100p$ for Home Depot's stock.

b. What does this mean for Home Depot stock?

EXAMPLE 8

Figure 3.44 shows the change in price per share of several stocks after 1 day. If the price per share changes by p dollars, then 100 shares of the stock gain or lose $100p$ dollars in value.

Figure 3.44 Price change

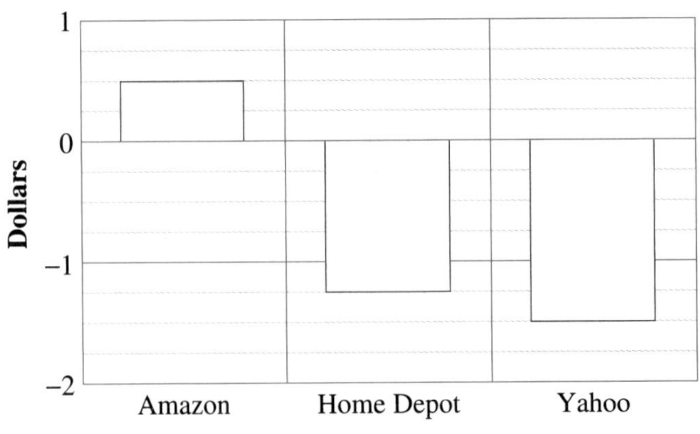

(Source: *Los Angeles Times*, January 27, 2001)

a. How much did 100 shares of Amazon stock change in value?

b. Evaluate $100p$ for the change in price of Yahoo. What does that value signify?

Solution

a. The price of Amazon stock gained half a dollar, or \$0.50 per share, so 100 shares gained $100(0.50) = \$50$ in value.

b. Yahoo's stock lost \$1.50 per share, so $p = -1.50$ and

$$100p = 100(-1.50) = -150$$

This means that 100 shares *lost* \$150 in value.

CAUTION!

Multiplications and divisions should be performed *in order from left to right.* In the example below, we get the wrong answer if we don't perform the calculations in order:

Correct Method	*Incorrect Method*
$120 \div (-6) \div (-2)$	$120 \div (-6) \div (-2)$
$= -20 \div (-2)$	$= 120 \div (3)$
$= 10$ ← **Right answer**	$= 40$ ← **Wrong answer**

EXERCISE 9

Simplify $-64 \div 8 \div (-2)$

EXAMPLE 9 

Simplify $-36 \div (-2)(-6) \div 3$

Solution

Perform the operations in order from left to right:

$$\mathbf{-36 \div (-2)}(-6) \div 3 \quad \text{The quotient of } -36 \text{ and } -2 \text{ is } 18.$$
$$= \mathbf{18(-6)} \div 3 \quad \text{The product of 18 and } -6 \text{ is } -108.$$
$$= \mathbf{-108 \div 3} = -36 \quad \text{The quotient of } -108 \text{ and 3 is } -36.$$

EXERCISE 10

Solve each equation.

a. $\dfrac{y}{-5} = 15$ **b.** $-8r = 24$

Solving Equations

The strategy we learned earlier for solving equations still applies: If the variable is multiplied by a number, we should divide both sides of the equation by that number. If the variable is divided by a number, we should multiply both sides of the equation by that number.

EXAMPLE 10 

Solve each equation.

a. $-6x = -102$ **b.** $\dfrac{w}{-7} = 21$

Solution

a. $\dfrac{-6x}{\mathbf{-6}} = \dfrac{-102}{\mathbf{-6}}$

$x = 17$

b. $\mathbf{-7}\left(\dfrac{w}{-7}\right) = 21(\mathbf{-7})$

$w = -147$

ANSWERS TO 3.4B EXERCISES

7a. 9
7b. 72
8a. -125
8b. They lost \$125 in value.
9. 4
10a. $y = -75$
10b. $r = -3$

HOMEWORK 3.4B 

Simplify each expression.

1. $(-3)(-4)(-6)$

2. $8(-2)(-5)$

3. $(-5)(-5)(-5)(-5)$

4. $(-2)(-2)(-2)(-2)(-2)$

5. $-80 \div 8 \div (-2)$

6. $-128 \div (-16) \div (-4)$

7. $12(-20) \div (-5)(-2)$

8. $-25(-30) \div 15(-5)$

9. $-10 \div 10(-10) \div (-10)$

10. $-8 \div 8(-8) \div 8$

Evaluate each expression for $h = -2$, $v = -12$, and $m = 8$.

11. $-3m$

12. $\dfrac{v}{-3}$

13. $\dfrac{26}{h}$

14. $-6hv$

15. $\dfrac{10h}{m}$

16. hmv

For Problems 17–28, evaluate each expression for $d = -4$, $q = -0.8$, and $k = -2.4$.

17. $d \cdot d \cdot d$

18. $q \cdot q$

19. $d + d + d$

20. $q + q$

21. $\frac{q}{8}$

22. $\frac{8}{q}$

23. $k \div q \div d$

24. $k \div d \div q$

25. $2qk$

26. $-9dq$

27. $\frac{0}{q}$

28. $\frac{k}{0}$

29. Figure 3.45 shows the change in price per share for five Dow-Jones stocks on 1 day of trading. If the price increase is p dollars, then 50 shares of the stock increase in value by $50p$ dollars.

Figure 3.45 Change in stock prices

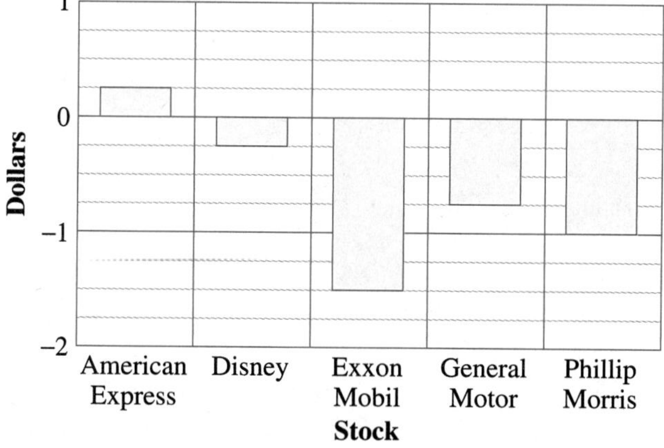

(Source: *Los Angeles Times,* 27 January 2001)

a. What was the change in the price of 1 share of Phillip Morris stock?

b. How did the value change for 50 shares of Phillip Morris?

c. Find the change in value of 50 shares of each of the other four stocks.

d. If you own 50 shares of each stock shown in the graph, what was the total change in the value of your portfolio?

30. Figure 3.46 shows changes in the price of Sony stock for 1 week. When Sony's price rises by p dollars, then 100 shares of the stock gain $100p$ dollars in value.

Figure 3.46 Change in Sony stock

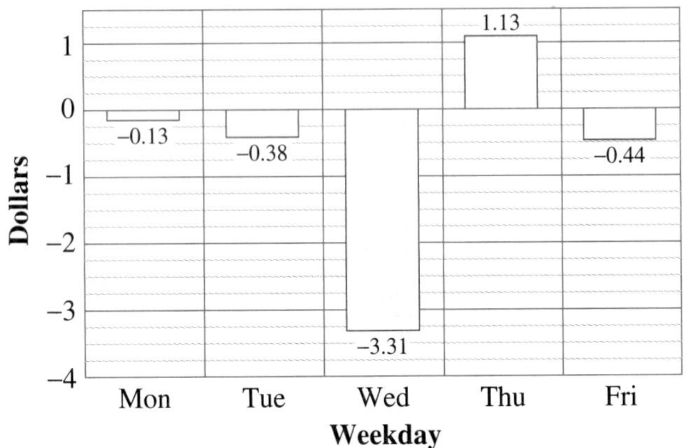

(Source: *Los Angeles Times*, 23–27 January 2001)

a. What was the change in price of the stock on Monday? What does this mean for Sony stockholders?

b. Compute the change in the value of 100 shares of Sony stock on Monday.

c. Compute the daily change in the value of 100 shares of Sony for the other 4 days.

d. What was the net change in the value of 100 shares of Sony for the week?

Solve each equation.

31. $8q = -24$

32. $7w = -49$

33. $-2r = -18$

34. $-5t = -60$

35. $-6y = 42$

36. $-3u = 45$

37. $\frac{p}{9} = -9$

38. $\frac{a}{4} = -12$

39. $\frac{s}{-5} = -15$

40. $\frac{d}{-11} = -3$

41. $\frac{f}{-4} = 18$

42. $\frac{g}{-20} = 5$

43. $\frac{h}{10} = -2.8$

44. $\frac{k}{-100} = -3.7$

45. $\frac{z}{-0.001} = 59$

46. $\frac{x}{0.1} = -82$

47. $1.5c = -90$

48. $-6.4v = 3.2$

49. $-0.3b = -0.12$

50. $-0.8n = -80$

Solve each equation. You may need to use any of the four operations to solve these equations.

51. $x + 8 = -3$

52. $-5 + x = 7$

53. $-3x = -24$

54. $7x = -42$

55. $x - 12 = -2$

56. $15 + x = 9$

57. $\frac{x}{-2} = -8$

58. $\frac{x}{15} = -3$

Solve each problem by writing and solving an equation. Follow the three steps on page 114.

59. The ratio of men to women majoring in chemistry at Baldwin State is 2.5, or $\frac{5}{2}$. If 18 women are chemistry majors, how many men are chemistry majors?

60. Staci divided a pad of drawing paper among the 48 children at day camp. If each child got three sheets of paper, how many sheets were on the pad?

61. The trail to the swimming hole is 1.4 miles longer than the trail to the lodge. If it is 4.1 miles to the swimming hole, how far is it to the lodge?

62. Wendell pays 12% of his salary in state taxes. If he pays $312 per month in state taxes, what is his salary?

63. Of the voters surveyed, 38% favor term limits. If 304 voters favored term limits, how many voters were surveyed?

64. A space traveler on Jupiter weighs 2.64 times what she would on Earth. If Astra weighs 330 pounds on Jupiter, what does she weigh on Earth?

Evaluate the algebraic expression to complete the table.

65.

b	$-4b$
-6	
-4	
4	
6	

66.

a	$\dfrac{a}{-4}$
-6	
-4	
4	
6	

67.

n	$\dfrac{-12}{n}$
-18	
-8	
8	
18	

68.

q	$12 - q$
-18	
-8	
8	
18	

Building Number Skills

Use a diagram to express each improper fraction as a mixed number.

3 wholes + $\frac{1}{2}$

Example: $\dfrac{7}{2} =$ 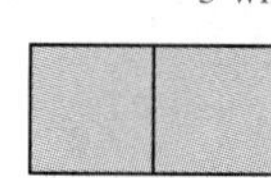 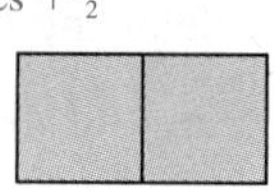 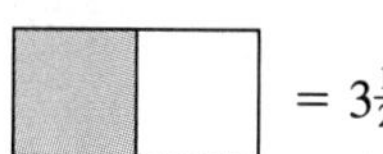 $= 3\frac{1}{2}$

Group the halves into wholes. How many halves left over?

69. $\dfrac{5}{3} =$

70. $\dfrac{7}{4} =$

71. $\dfrac{9}{4} =$

72. $\dfrac{8}{3} =$

73. $\dfrac{19}{8} =$

74. $\dfrac{17}{6} =$

75. $\dfrac{16}{3} =$

76. $\dfrac{14}{3} =$

77. $\dfrac{10}{4} =$

78. $\dfrac{14}{4} =$

Do you see a way to do these exercises without using a diagram? Describe your method and give an example.

3.5 Problem Solving

We can use equations to solve problems involving negative numbers.

EXAMPLE 1

Choose the correct equation to model each problem. Then solve the equation and answer the question in the problem.

$$-18 + n = 63 \qquad -63 + n = -18$$
$$-63 + n = 18 \qquad -18 + n = -63$$

a. After making a payment toward her phone bill, Lorrie's balance changed from −\$63 to −\$18. How much was her payment?

Step 1 The unknown quantity is Lorrie's payment:

Amount of Lorrie's payment: *n*

Step 2 Start with her original balance and add her payment to get her new balance:

$$\mathbf{-63 + n = -18}$$

Step 3 Now solve the equation:

$$\begin{array}{rcr} -63 + n &=& -18 \\ \mathbf{+63} & & \mathbf{+63} \\ \hline n &=& \mathbf{45} \end{array}$$

Lorrie's payment was \$45.

b. The average daytime temperature in Duluth changes from −18° in January to 63° in June. How much does the average temperature change in that time?

Step 1 The unknown quantity is the change in average temperature:

Change in average temperature: *n*

Step 2 Start with the average temperature in January and add the change in temperature to get the average temperature in June:

$$\mathbf{-18 + n = 63}$$

Step 3 Now solve the equation:

$$\begin{array}{rcr} -18 + n &=& 63 \\ \mathbf{+18} & & \mathbf{+18} \\ \hline n &=& \mathbf{81} \end{array}$$

The average temperature increased by 81°.

In Chapter 1 we solved applied problems by writing an equation about the problem. We used the three steps reviewed below.

EXERCISE 1

Choose the correct equation from the list in Example 1 to model the following problem. Then solve the equation and answer the question in the problem:

Öniz was scuba diving 63 feet below the water's surface when she began following a fish. She stopped when she was still 18 feet below the surface. How many feet did she rise?

Steps for Solving Applied Problems

Step 1 Identify the quantity you are asked to find, and choose a variable to represent it.

Step 2 Find some quantity that can be expressed in two different ways, and write an equation.

Step 3 Solve the equation, and answer the question in the problem.

EXAMPLE 2 

Jeremy's new restaurant ended the first quarter with a loss of $10,270, but he made a profit in the second quarter and ended his first 6 months only $3800 in debt. How much was his second-quarter profit?

Solution

Step 1 What are we asked to find in the problem?

The second-quarter profit

Choose a variable to represent the unknown quantity:

Second-quarter profit: *p*

Step 2 Find something in the problem that we can express in two *different* ways:

Start with the restaurant's financial status after the first quarter and add its earnings in the second quarter to get the total after 6 months:

$$-10{,}270 + p = -3800$$

Step 3 Solve the equation:

$$\begin{aligned} -10{,}270 + p &= -3{,}800 \\ +10{,}270 \quad &\quad +10{,}270 \qquad \text{Add 10,270 to both sides.} \\ p &= 6{,}470 \end{aligned}$$

Answer the question in the problem:

Jeremy made a profit of $6470 in the second quarter.

EXERCISE 2

Svetlana parachuted from a plane at an elevation of 2534 feet to the floor of Death Valley, at an elevation of −273 feet. What was her change in elevation?

EXERCISE 3

Refer to Figure 3.47. The ending prices for ExxonMobil and Wal-Mart stocks were \$79.44 and \$50.81 respectively. What were their starting prices?

EXAMPLE 3

Figure 3.47 shows the change in stock prices for three Dow-Jones companies over the week of January 15–19, 2001.

Figure 3.47 Change in stock prices

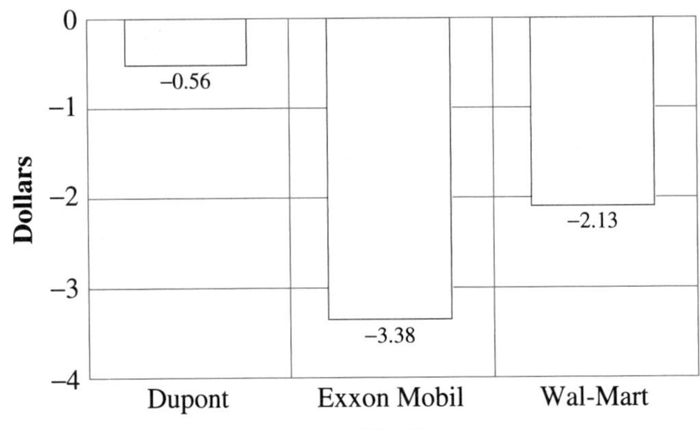

(Source: *Los Angeles Times,* 21 January 2001)

a. What was the change in the price of Dupont stock?

b. The ending price of Dupont was \$42.63 per share. What was the starting price?

Solution

a. The bar for Dupont stock is labeled with -0.56, so Dupont stock lost \$0.56 per share.

b. Step 1 Let s represent the starting price in dollars.

Step 2 We can represent the ending price of \$42.63 in two different ways.

The ending price is the result of adding the change to the starting price. Thus,

$$\underset{\text{Ending price}}{42.63} = \underset{\text{Starting price}}{s} + \underset{\text{Change in price}}{(-0.56)}$$

Step 3 Solve the equation. Add the opposite of -0.56 to both sides:

$$\begin{aligned} 42.63 &= s + (-0.56) \quad \text{Add 0.56 to both sides.} \\ +\mathbf{0.56} & \quad\;\; + \quad \mathbf{0.56} \\ \hline 43.19 &= s \end{aligned}$$

The starting price was \$43.19.

ANSWERS TO 3.5 EXERCISES

1. $-63 + n = -18$. She rose 45 feet.

2. Her change in elevation was -2807 feet; that is, her elevation decreased by 2807 feet.

3. ExxonMobil, \$82.82; Wal-Mart, \$52.94

HOMEWORK 3.5

Choose the correct equation to model the problems.

$-12 + n = -5$	$-12 + n = 5$
$n + 12 = -5$	$n + 12 = 5$

1. Last night the temperature rose 12°, and this morning the temperature is 5°. What was the temperature yesterday?

2. Marta's score was −12, but after the last hand her score rose to −5. How many points did she make on the last hand?

3. Bradley rode the elevator down 12 floors and emerged 5 floors below ground. What floor did he start on?

4. Orrin was \$12 in debt, but he did some yard work this weekend; after paying his debt, he has \$5. How much did he make?

Choose the correct equation to model the problems.

$3n = -30$	$\frac{n}{3} = -30$
$n + 3 = -30$	$n - 3 = -30$

5. Delbert lost 3 points on this round of Squibble, making his cumulative score −30. What was his score after the previous round?

6. The temperature has been dropping for 3 days, for a total change of −30°. What was the average temperature change per day?

7. A cyclist is biking down a long slow incline, and for three minutes her elevation changed by −30 yards per minute. What was the total change in her elevation?

8. Francine paid her \$3 dues to the philatelists' club, and now their account has a balance of −\$30. What was the balance before Francine paid her dues?

For each problem, (a) identify the unknown quantity and choose a variable to represent it, (b) find something that can be expressed in two different ways and write an equation that models the problem, and (c) solve the equation and answer the question in the problem.

9. From last night to this afternoon, the temperature in Iron Mountain had risen 17°, resulting in a temperature of −6°. What was the temperature last night?

10. The temperature in Cumberland dropped 26° from Wednesday to Thursday. If the temperature on Thursday was −11°, what was the temperature on Wednesday?

11. Lisa deposited $132 in her bank account today, giving her a new balance of $74. What was her balance yesterday?

12. Clyde wrote a check for $63 worth of art supplies, and his account is now overdrawn by $81. What was his account balance before purchasing art supplies?

13. Seung went hiking in Death Valley, and after lunch he climbed 248 feet. This brought him to an elevation of 109 feet. What was his elevation before lunch?

14. Gita went spelunking in Emerald Cave. She descended 47 feet to a popular cavern whose elevation is −35 feet. What was her starting elevation?

15. The Greek philosopher Socrates died in 399 B.C. at an age of 71. In what year was Socrates born?

16. Augustus Caesar died in A.D. 14, 58 years after the death of Julius Caesar. In what year did Julius Caesar die?

17. Thad's net yardage gain was -18 yards before Saturday's football game, and after the game his net gain was -26 yards. How many yards did Thad gain during the game?

18. Rowan's score was -75 at intermission on the College Bowl, but he finished the game with a score of 200. What was the change in Rowan's score after intermission?

19. The Yellow Submarine rose from -286 feet to -159 feet. What was its change in elevation?

20. Starting at the lowest level of the basement parking garage and riding the elevator to her office on the ninth floor, Briar's elevation changes from -45 feet to 108 feet. What is her change in elevation?

For Problems 21–30, simplify.

21. $-18 - 12$

22. $7(-15)$

23. $\frac{-72}{6}$

24. $-8 + 14$

25. $-21 - (-17)$

26. $-48 \div (-3)$

27. $-16(-4)$

28. $\frac{0}{-8}$

29. $-15 \div 0$

30. $13 - 19$

Figure 3.48 Price change in stock prices

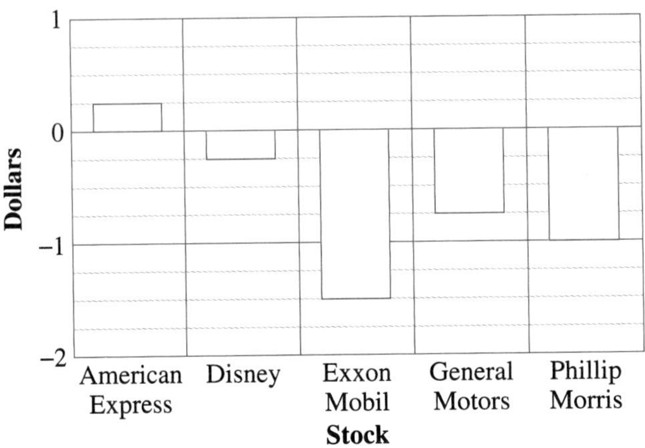

(Source: *Los Angeles Times*, 27 January 2001)

31. Figure 3.48 shows the change in price per share for five Dow-Jones stocks.

a. What was the change in the price of a share of ExxonMobil stock? What does the number mean?

b. The final price of ExxonMobil was \$81 per share. What was the starting price?

32. Use Figure 3.48 to answer the following questions:

a. What was the change in price of a share of General Motors stock? What does the number mean?

b. The final price in General Motors was \$53.75 per share. What was the starting price?

Building Number Skills

You know that there are 24 hours in a day. Use the diagram of 24 dots to help you compute the number of hours in each fraction of a day.

33. $\frac{2}{3}$ **34.** $\frac{3}{4}$ **35.** $\frac{5}{6}$ **36.** $\frac{1}{6}$ **37.** $\frac{5}{8}$

38. $\frac{7}{8}$ **39.** $\frac{1}{4}$ **40.** $\frac{1}{3}$ **41.** $\frac{1}{8}$ **42.** $\frac{3}{8}$

3 Summary

Lesson 3.1

Every **positive number** has an opposite called a **negative number.**

The positive and negative numbers together are called the **signed numbers.**

We use a **number line** to illustrate the signed numbers.

Together, the positive whole numbers, the negatives of the whole numbers, and zero make up a set of numbers called the **integers.**

To indicate a particular number on the number line, we place a dot at its position. This dot is called the **graph** of the number.

The two symbols, $<$ and $>$, are called **inequality symbols:**

$<$ means is "less than."

$>$ means is "greater than."

Lesson 3.2

Rules for Adding Signed Numbers

To Add Two Numbers with the Same Sign

1. Add the unsigned parts of the numbers.
2. Use the same sign for the sum.

To Add Two Numbers with Opposite Signs

1. Subtract the unsigned parts of the numbers.
2. If the positive number is farther from zero, the sum is positive. If the negative number is farther from zero, the sum is negative.

When we substitute a negative number for a variable in an algebraic expression, we enclose the negative number in parentheses.

Equations may involve negative numbers, and solutions to equations may be negative numbers.

Lesson 3.3

Rules for Subtracting Signed Numbers

1. To subtract a positive number, add the corresponding negative number.
2. To subtract a negative number, add the corresponding positive number.

Additions and subtractions should be performed in order from left to right.

Sometimes we can simplify one or both sides of an equation before trying to solve.

The **net change** in a variable is given by

$$\textit{net change} = \textit{final value} - \textit{initial value}$$

Lesson 3.4

Rules for Multiplying Signed Numbers

1. The product of two numbers with the same sign is positive.
2. The product of two numbers with opposite signs is negative.
3. The product of any number and zero is zero.

Rules for Dividing Signed Numbers

1. The quotient of two numbers with the same sign is positive.
2. The quotient of two numbers with opposite signs is negative.
3. Zero divided by any number (except zero) is zero.
4. The quotient of any number divided by zero is undefined.

Multiplications and divisions should be performed in order from left to right.

Lesson 3.5

Review the strategy for solving equations.

1. Ask yourself: Which operation has been performed on the variable?
2. Perform the opposite operation on both sides of the equation.
3. Check your solution.

Review the steps for solving applied problems.

1. Identify the quantity you are asked to find and choose a variable to represent it.
2. Find some quantity that can be expressed in two different ways and write an equation.
3. Solve the equation and answer the question in the problem.

CHAPTER 3 REVIEW 

Add the following words to your glossary, if you have not already done so.

negative number	signed number	number line	numerator
positive number	integer	graph	denominator
inequality symbol	improper fraction		

1. Name three different quantities you might measure using negative numbers.

2. Draw a number line and graph $-4, -2\frac{3}{4}, -2\frac{1}{4}, 0, \frac{1}{2}$, and 3.75

3. What is the difference between an *equation* and an *inequality*?

4. Give examples of how inequality symbols are used.

5. Explain in your own words how to add two signed numbers.

6. Explain how to subtract two signed numbers.

7. Explain how to multiply or divide two signed numbers.

8. The product of zero and any number is ________.

9. The quotient of any number divided by zero is ________.

10. The quotient of zero divided by any number is ________.

Replace the blank in each problem with the correct symbol: <, >, or =.

11. -3 ____ -5

12. -8 ____ -4

13. -2 ____ 2

14. 7 ____ -7

15. $-\frac{3}{2}$ ____ -1.5

16. -0.75 ____ $-\frac{3}{4}$

(a) Choose a variable for the unknown quantity and (b) write an inequality to describe the situation.

17. Zoe can only afford rent that is less than $650 per month.

18. Carl is over 6 feet tall.

19. Last night the temperature fell below $-15°$.

20. My score at Phizzbin never fell as low as -150 points.

Use a number line to illustrate the following sums.

21. $-2 + (-4)$

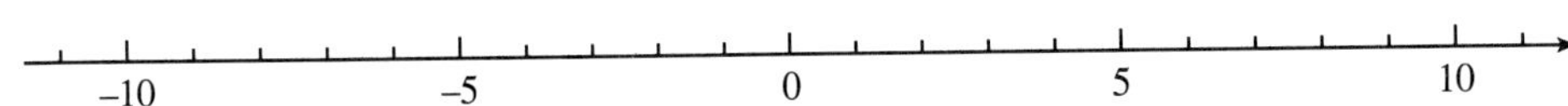

22. $-8 + 6$

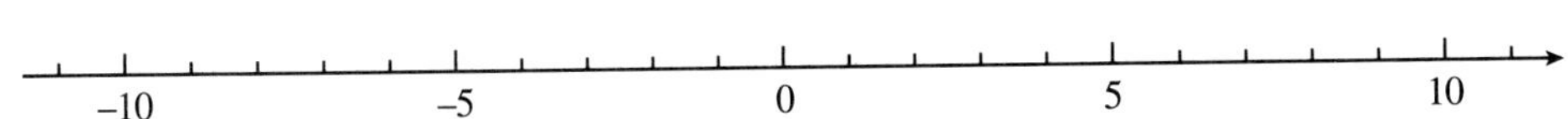

Write each subtraction problem as an equivalent addition problem.

23. $-5 - (-7)$

24. $-2 - 9$

Fill in the blank.

25. The sum of two negative numbers is a ________________ number.

26. The product of two negative numbers is a ________________ number.

27. The sum of a number and its opposite is ______________.

28. The quotient of a number and its opposite is ______________.

Compute.

29. $-8(-4)$

30. $10 - (-6)$

31. $\frac{-36}{4}$

32. $-19 + 12$

33. $-11 - 23$

34. $-48 \div (-3)$

35. $-15 - (-8)$

36. $5(-6)$

37. $18 - 20$

38. $\frac{0}{-9}$

39. $-12 + 12$

40. $-25 - 25$

Compute.

41. $-3 + (-8) - 14$

42. $6 - 10 - (-4)$

43. $-20 - 5 + 7$

44. $-1 - 1 + 2$

45. $-3(-9)(-2)$

46. $-42 \div 2(-3)$

Evaluate for $a = -4$, $b = 6$, and $c = -8$.

47. $a - c$

48. bc

49. $\frac{b}{a}$

50. $b + a - c$

51. $5ac$

52. $-a \div c$

53. Figure 3.49 shows the U.S. balance of trade for each year in the 1990s. The balance of trade is the difference between the amount of U.S. exports and imports for that year.

a. What does it mean when the balance of trade is negative?

b. What was the total U.S. trade balance for the decade of the 1990s?

Figure 3.49 U.S. international trade balance

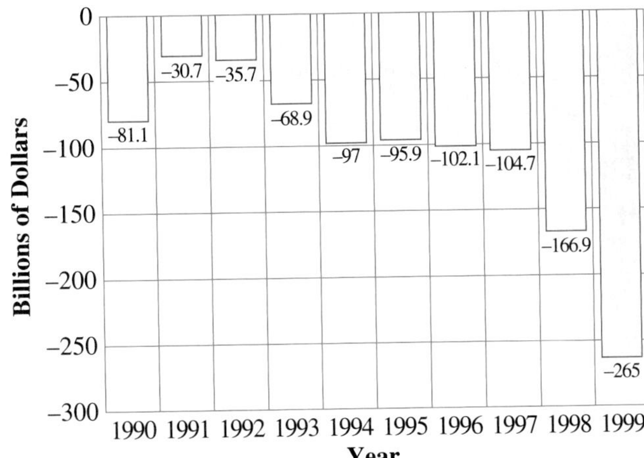

(Source: U.S. Department of Commerce)

c. Complete the table showing the net increase in the balance of trade for each year over the previous year.

Year	1991	1992	1993	1994	1995	1996	1997	1998	1999
Increase in trade balance (billion $)									

54. Figure 3.50 shows by how much the price of gasoline varied relative to the state average at six gas stations across town. The state average was $1.793 per gallon.

Figure 3.50 Price of gasoline (above the average price)

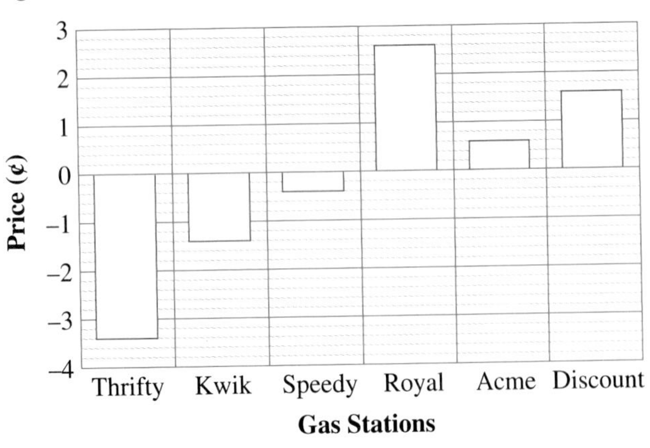

a. Using values from the bar graph, fill in the table:

	Thrifty	*Kwik*	*Speedy*	*Royal*	*Acme*	*Discount*
Deviation from average ($)						
Actual price per gallon ($)						

b. Use Figure 3.51 to make a bar graph of the prices of gasoline at the different stations.

Figure 3.51 Gasoline prices

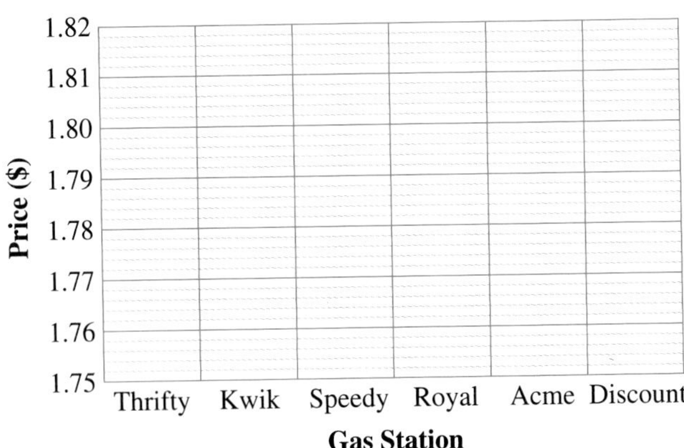

Solve each equation.

55. $m + 6 = -8$

56. $p - 12 = -15$

57. $-72 = -8n$

58. $-7q = 28$

59. $-16 + b = 20$

60. $-11 = -4 + a$

61. $\frac{h}{6} = -12$

62. $18 = \frac{k}{-4}$

63. $-2.9 = 0.4d$

64. $-1.6g = -15$

65. $6.7 = -4.8 + y$

66. $z - 0.04 = 0.04$

For Problems 67–70, (a) write an equation that models the problem and (b) solve the equation. Write your answer in a sentence.

67. Last year Imogene was deeply in debt, but she paid back \$2500 and now her net worth is −\$4800. What was her net worth last year?

68. Keesha just deposited her tax refund of \$820, and now her checking account balance is \$460. What was her balance before the refund?

69. During a special promotion, the Kwik Shop lost \$0.45 on each Super Soda they sold. At the end of the promotion, their profit on Super Sodas was −\$542.70. How many Super Sodas did they sell?

70. The temperature in Frozen Falls, South Dakota, dropped steadily for 15 hours, for a total net change of −39°. What was the temperature change per hour?

Figure 3.52 Weekly change in stock price

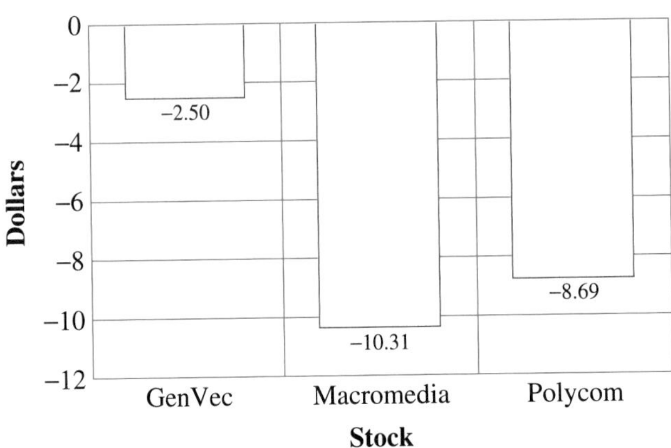

(Source: *Los Angeles Times,* 21 January 2001)

71. Figure 3.52 shows the change in stock prices of three companies for the week that ended on Friday, 19 January 2001.

a. The ending price of GenVec was \$7.50 per share. What was the starting price?

b. If you owned 100 shares of GenVec stock, how much was your investment worth at week's end?

72. Use Figure 3.52 to answer the following questions:

a. The ending price of Macromedia was \$31.63 per share. What was the starting price?

b. If you owned 100 shares of Macromedia stock, how much was your investment worth at week's end?

Building Number Skills

For Problems 73 and 74, sketch a diagram illustrating each fraction.

73. $\frac{5}{6}$

74. $\frac{8}{5}$

75. How many fourths are there in five wholes?

76. How many eighths are there in three wholes?

(a) Sketch a diagram illustrating each mixed number and (b) write each mixed number as an improper fraction.

77. $2\frac{1}{4}$

78. $1\frac{2}{3}$

(a) Sketch a diagram illustrating each improper fraction and (b) write each improper fraction as a mixed number.

79. $\frac{9}{2}$

80. $\frac{23}{6}$

Rewrite 1 as a fraction and subtract.

81. $1 - \frac{2}{9}$

82. $1 - \frac{1}{5}$

Make a diagram of 18 dots and use it to illustrate the following products.

83. $\frac{2}{3}(18)$

84. $\frac{5}{6}(18)$

CHAPTER

4

Order of Operations

4.1 Order of Operations with Positive Numbers

A. Which Operations Come First?

How do we simplify the following expression?

$$5 + 2 \cdot 4$$

Should we add $5 + 2$ first and then multiply by 4, or should we multiply $2 \cdot 4$ first and then add 5?

Option 1		*Option 2*	
$\mathbf{5 + 2} \cdot 4$	Add first.	$5 + \mathbf{2 \cdot 4}$	Multiply first.
$= 7 \cdot 4$	Then multiply.	$= 5 + 8$	Then add.
$= 28$		$= 13$	

The two options give us different answers, so they cannot both be right. You might decide to perform the operations from left to right in the order they appear, but this causes problems. (We would like $5 + 2 \cdot 4$ and $2 \cdot 4 + 5$ to have the same answer.) Instead, people who use mathematics have agreed upon the following rule:

Always perform multiplications and divisions before additions and subtractions.

By following this rule, everyone means the same thing when they write an algebraic expression, and everyone gets the same answers. So option 2 is correct:

$$
\begin{aligned}
&5 + 2 \cdot 4 && \text{Perform the multiplication first.}\\
&= 5 + 8 && \text{Now add.}\\
&= 13
\end{aligned}
$$

EXERCISE 1
Simplify each expression.

a. $22 - 2 \cdot 7$

b. $8 + 6 \div 2$

EXAMPLE 1

Use the preceding rule to simplify each expression.

a. $25 - 6 \cdot 3$ Multiply first.
$= 25 - 18$ Subtract.
$= 7$

b. $48 \div 10 - 2$ Divide first.
$= 4.8 - 2$ Subtract.
$= 2.8$

Longer Expressions

What about longer expressions? For example, how should we simplify the following expression?

$$10 + 3 \cdot 8 \div 2 - 4 \cdot 2$$

We should start with multiplications and divisions, working from left to right. It is helpful to underline all the multiplication and division operations before beginning, as shown below:

$10 + \underline{\mathbf{3 \cdot 8} \div 2} - \underline{4 \cdot 2}$ Multiply 3 times 8.
$= 10 + \underline{\mathbf{24 \div 2}} - \underline{4 \cdot 2}$ Divide 24 by 2.
$= 10 + 12 - \underline{\mathbf{4 \cdot 2}}$ Multiply 4 times 2.
$= 10 + 12 - 8$

After we have finished all the multiplications and divisions, we perform additions and subtractions in order from left to right:

$\mathbf{10 + 12} - 8$ Add 10 plus 12.
$= \mathbf{22} - 8$ Subtract 8 from 22.
$= 14$

We can combine these rules into a set of guidelines:

1. *First, perform all multiplications and divisions in order from left to right.*
2. *Then, perform all additions and subtractions in order from left to right.*

EXAMPLE 2

Use the preceding guidelines to simplify each expression.

a. $18 + 3(12) - 24 \div 8$ **b.** $5 + 5 \cdot 5 - 5 \div 5 - 5$

Solution

a. Underline all the multiplication and division operations:

$18 + \underline{3(12)} - \underline{24 \div 8}$ Multiply 3 times 12; then divide 24 by 8.
$= \mathbf{18 + 36} - 3$ Add 18 plus 36; then subtract 3.
$= 54 - 3 = 51$

b. Underline all the multiplication and division operations:

$$5 + \underline{5 \cdot 5} - \underline{5 \div 5} - 5 \quad \text{Perform the multiplication and the division.}$$
$$= \mathbf{5 + 25} - 1 - 5 \quad \text{Add and subtract from left to right.}$$
$$= \mathbf{30 - 1} - 5$$
$$= 29 - 5 = 24$$

EXERCISE 2

Simplify each expression.

a. $10 + 10 \div 2 \cdot 8$

b. $26 - 6 \div 3 + 4 \cdot 7$

Using a Calculator

A scientific calculator is designed to follow the order of operations. It performs multiplications and divisions before additions and subtractions, no matter which operations are entered first. To test whether you have a scientific calculator, evaluate the expression

$$3 + 5(2)$$

by entering

3 [+] 5 [×] 2 [=]

If your calculator returns the (correct) answer 13, you have a scientific calculator. If your calculator gives the answer 16, it did not follow the order of operations. In that case, you must be careful to enter calculations in the same order they should be performed. For example, you should enter the expression $3 + 5(2)$ as

5 [×] 2 [+] 3 [=]

ANSWERS TO 4.1A EXERCISES

1a. 8 **1b.** 11

2a. 50 **2b.** 52

HOMEWORK 4.1A

Simplify each expression. Use the guidelines given in the lesson.

1. $2(4) - 3$

2. $4(3) - 5$

3. $2 + 4 \cdot 3$

4. $9 - 3 \cdot 2$

5. $25 - 3(6.4)$

6. $7 + 2(18.9)$

7. $24 \div 6 + 2$

8. $40 \div 10 - 2$

9. $80 - 56 \div 8$

10. $70 - 35 \div 7$

11. $18 + \dfrac{18}{3}$

12. $12 - \dfrac{36}{4}$

Solve. (See Lessons 3.2B and 3.3B to review solving equations.)

13. $-4 + z = 5$

14. $-9 + w = -1$

15. $x - (-8) = 3$

16. $y - (-2) = -6$

17. $-12 = -5 + a$

18. $-4 = -10 + b$

Write and solve an equation to answer each question. (See Lesson 3.5 to review problem solving.)

19. Grace won \$15 in her first game of poker, but after the second game she was in debt by \$23. What was the outcome of her second game?

20. The high temperature today was 6°F, an increase of 11°F above last night's low temperature. What was the overnight low?

21. On Monday, stocks for Upstart.com sold for \$2.38 per share. On Friday, the stock was worth \$1.84 per share. What was the net change in the price per share?

22. On a hike in Death Valley, Rocio climbed 120 feet in altitude and is now 35 feet below sea level. What was her initial elevation?

Simplify each expression. Use the guidelines given in the lesson.

23. $45 - 24 \div 4\,(3)$

24. $100 - 75 \div 25(3)$

25. $84 - 2(5)(6)$

26. $5 + 3(4)(10)$

27. $18 \cdot 5 - 3 \cdot 12$

28. $24 \cdot 5 + 5 \cdot 13$

29. $3(8.2) - 6(2.1)$

30. $5(1.3) + 4(2.2)$

31. $2 + 3 \cdot 8 - 6 + 3$

32. $4 \cdot 9 - 3 \cdot 2 + 4$

33. $24 \div 6 + 2 \cdot 8 \div 4$

34. $48 \div 3 \cdot 4 - 20 - 8$

Compute each sum or difference without using a calculator. (See Lessons 3.2 and 3.3 to review sums and differences of signed numbers.)

35. $-7 - 9$

36. $-4 - (-12)$

37. $8 - (-5)$

38. $-12 + 16$

39. $-6 + 15$

40. $18 - 29$

41. $-15 - (-15)$

42. $13 - (-13)$

43. $-4 - 4$

44. $-9 + 9$

Calculate the areas of the pieces in each quilt block.

45. The pine tree is assembled on a 6×6 centimeter grid (Figure 4.1).

a. Find the area of the blue triangle. Find the area of the blue trapezoid. What is the total area of blue fabric?

b. Find the area of the red tree trunk.

c. Find the area of the yellow background fabric.

Figure 4.1

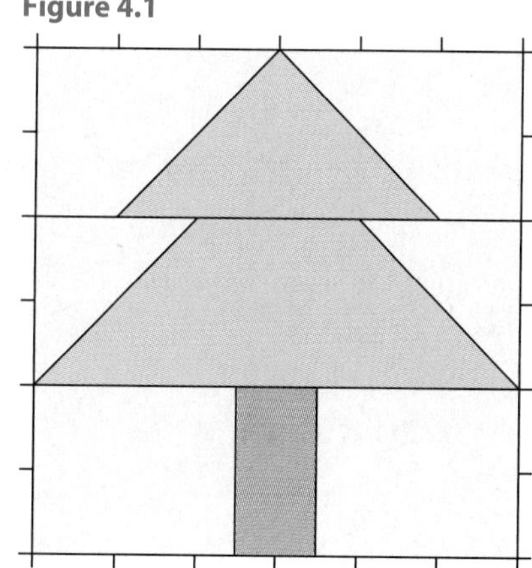

46. The pine tree is assembled on a 7×7 centimeter grid (Figure 4.2).

a. The tree includes eight blue trapezoids. Find the area of each. What is the total area of the blue fabric?

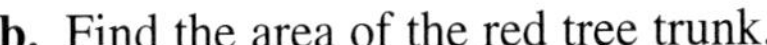

b. Find the area of the red tree trunk.

c. Find the area of the yellow background fabric.

Figure 4.2

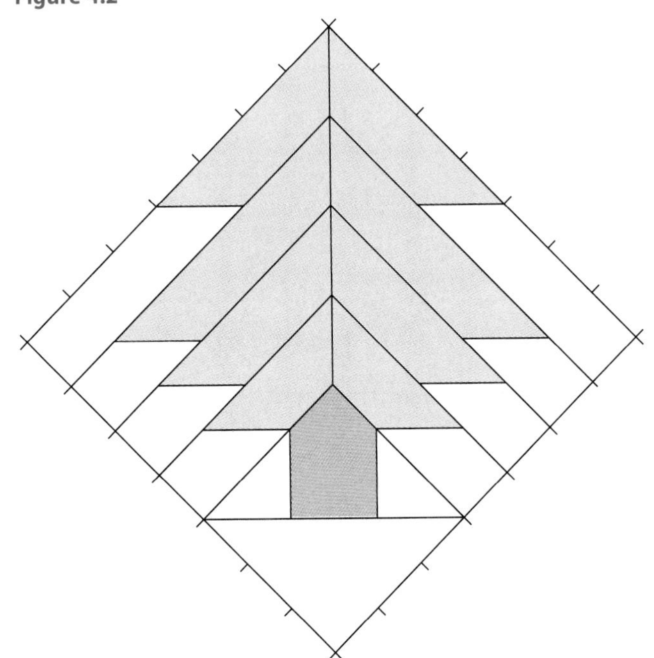

Building Number Skills

In this chapter we review place value and rounding decimal numbers and then use these skills in estimation and approximation. Try to do these exercises mentally—without a calculator or pencil and paper.

Mentally multiply or divide by a power of 10.

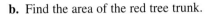

Example: $12.5 \times 1000 = 12500$ Move the decimal three places to the right.

47. 37×100

48. $286 \times 10{,}000$

49. 7.43×1000

50. 62.1×100

51. $8400 \div 10{,}000$

52. $52{,}300 \div 1000$

53. $36 \div 100{,}000$

54. $4956 \div 100$

55. $28.6 \div 100$

56. $8.03 \div 10$

B. Parentheses and Fraction Bars

Parentheses and Brackets

How can we write the following phrase in mathematical symbols?

Add 2 and 5 and then multiply the sum by 3.

If you do the computations as stated, you'll see that the result is 21. But neither of the following expressions gives the right answer:

Option 1	*Option 2*
$2 + 5 \cdot 3$	$3 \cdot 2 + 5$
$= 2 + 15 = 17$	$= 6 + 5 = 11$

According to our guidelines, multiplication comes before addition. We need an override mechanism to use when we really want addition to come first. For this purpose, we use parentheses:

$$\mathbf{3(2 + 5)}$$

The parentheses tell us to perform any operations inside first. Thus,

$$3(2 + 5) = 3(7) = 21$$

This rule is another guideline:

Perform any operations inside the parentheses first.

EXAMPLE 3

Simplify each expression.

a. $18 - 6(9 - 7)$ **b.** $24 \div (6 - 3) \cdot 4$

Solution

a. $18 - 6(\mathbf{9 - 7})$ Perform the subtraction inside the parentheses first.
$= 18 - \mathbf{6(2)}$ Next, perform the multiplication.
$= 18 - 12 = 6$

b. $24 \div (\mathbf{6 - 3}) \cdot 4$ Perform the subtraction inside the parentheses first.
$= \mathbf{24 \div 3} \cdot 4$ Perform multiplications and divisions in order.
$= 8 \cdot 4 = 32$

EXERCISE 3

Simplify each expression.

a. $6 \div (12 - 10) \cdot 3$

b. $7 + 3(5 + 2)$

Nested Parentheses

Parentheses are sometimes called *grouping devices* because they group together parts of an expression. Square brackets, [ ], and curly braces, { }, are also used as grouping devices.

EXERCISE 4
Simplify $14 - 4[5 - 3(1 + 2)]$

EXAMPLE 4 

Simplify according to the order of operations:

$$6 - 3[8 - 2(10 - 7)]$$

Solution

We begin with the innermost set of parentheses:

$6 - 3[8 - 2(\mathbf{10 - 7})]$	Subtract inside the parentheses.
$= 6 - 3[8 - \mathbf{2(3)}]$	Inside the square brackets, multiply first.
$= 6 - 3[\mathbf{8 - 6}]$	Subtract inside the square brackets.
$= 6 - \mathbf{3[2]}$	Multiply.
$= 6 - 6 = 0$	

Fraction Bars

A fraction bar is another kind of grouping symbol. Any operations that appear above or below a fraction bar should be completed first. For example, in the expression

$$\frac{16 + 14}{8 - 2}$$

we compute the sum $16 + 14$ and the difference $8 - 2$ *before* dividing:

$$\frac{16 + 14}{8 - 2} = \frac{30}{6} = 5$$

EXERCISE 5
Simplify $5 + \frac{7 + 9}{2 \cdot 4}$

EXAMPLE 5  

Simplify according to the order of operations: $16 - \frac{24}{2(4)}$

Solution

$16 - \frac{24}{2(4)}$	Multiply below the fraction bar.
$= 16 - \frac{\mathbf{24}}{\mathbf{8}}$	Divide.
$= 16 - 3 = 13$	

We combine all the rules in this section into a set of guidelines for simplifying algebraic expressions. These guidelines are called the **order of operations.**

Order of Operations

1. First, perform all operations inside the parentheses or above or below a fraction bar.
2. Next, perform all multiplications and divisions in order from left to right.
3. Finally, perform all additions and subtractions in order from left to right.

Using a Calculator

When we write division problems by hand, we can use either a fraction bar or a division symbol, ÷, to indicate a quotient. For example, 15 divided by 3 can be written either

$$\frac{15}{3} \quad \text{or} \quad 15 \div 3$$

The calculator does not have a fraction bar, so we must use the [÷] key to perform divisions. If there are operations to be performed above or below the fraction bar, *we must enclose those operations in parentheses.* For example, to simplify the expression

$$\frac{120}{4(5)}$$

we must enter

120 [÷] [(] 4 [×] 5 [)] [=]

[If your calculator does not have parentheses, you must perform the calculation in steps: First, compute 4(5) to get 20 and then divide 120 by 20 to get 6.] Can you explain why the keying sequence

120 [÷] 4 [×] 5

is incorrect?

EXAMPLE 6

Use a calculator to compute $\frac{4 + 12}{80}$

Solution

Because the calculator does not have a fraction bar, we must enclose $4 + 12$ in parentheses. Enter the expression as

[(] 4 [+] 12 [)] [÷] 80 [=]

The calculator returns the answer 0.2.

EXERCISE 6

Use a calculator to compute $\frac{24}{3 \cdot 4}$

CAUTION!

In Example 6, the keying sequence

4 [+] 12 [÷] 80

gives an incorrect answer. Can you explain why?

ANSWERS TO 4.1B EXERCISES

3a. 9

3b. 28

4. 30

5. 7

6. 2

HOMEWORK 4.1B 

Simplify. Note the difference between (a) and (b) in each pair of expressions.

1. a. $8 + 2 \cdot 5$

b. $(8 + 2)5$

2. a. $10 - 6 \div 2$

b. $(10 - 6) \div 2$

3. a. $\dfrac{24}{2 + 6}$

b. $\dfrac{24}{2} + 6$

4. a. $\dfrac{8 + 4}{2}$

b. $8 + \dfrac{4}{2}$

5. a. $(9 - 4) - 3$

b. $9 - (4 - 3)$

6. a. $27 \div (9 \div 3)$

b. $(27 \div 9) \div 3$

7. a. $6 \cdot 8 - 6$

b. $6(8 - 6)$

8. **a.** $24 \div 3 \cdot 2$

b. $24 \div (3 \cdot 2)$

9. **a.** $\dfrac{36}{6(3)}$

b. $\dfrac{36}{6}(3)$

10. **a.** $20 - 8 + 2$

b. $20 - (8 + 2)$

Write each phrase as a mathematical expression.

11. **a.** 3 times the sum of 5 and 8

b. The sum of 3 times 5 and 8

12. **a.** Find the sum of 18 and 30 and then divide by 6.

b. Add to 18 the quotient of 30 divided by 6.

13. **a.** Subtract the difference of 12 and 8 from 25.

b. Subtract 8 from the difference of 25 and 12.

14. **a.** Find the difference of 34 and 7 and then multiply by 3.

b. Multiply 7 times 3 and then subtract the product from 34.

Simplify each expression according to the order of operations.

15. $4(3 + 5)$

16. $3(7 - 4)$

17. $12 - (6 - 4)$

18. $32 \div (8 \div 4)$

19. $\dfrac{48}{4(6)}$

20. $\dfrac{90}{3(15)}$

21. $400 \div (100 \div 4)$

22. $20 - (20 - 20)$

23. $3 + 2(6 - 1)$

24. $5 + 3(2 + 3)$

25. $12 - 2(1 + 3)$

26. $20 - 3(6 - 4)$

27. $6 \cdot 10 - (2 + 7) \cdot 4$

28. $8 \cdot 9 + (7 - 5) \cdot 6$

29. $3(9 - 5) + 5(12 - 4)$

30. $7(5 + 6) - 4(9 - 5)$

31. $\frac{5 + 19}{4}$

32. $\frac{28 - 13}{5}$

33. $\frac{16}{20 - 12}$

34. $\frac{25}{15 - 10}$

35. $\frac{30 - 9}{12 - 9}$

36. $\frac{6 + 18}{6 + 6}$

Use a calculator to simplify each expression. Round your answers to two decimal places.

37. $\frac{12.8 + 24.6}{3.5}$

38. $\frac{45.2 - 16.3}{2.4}$

39. $\frac{38}{79 - 24}$

40. $\frac{56}{21 + 51}$

41. $\frac{156 - 36.7}{2.8(7.4)}$

42. $\frac{64.5 + 59.1}{0.2(43)}$

Solve. (See Lesson 3.4B to review solving equations.)

43. $-4 + x = -12$

44. $-4x = -12$

45. $\frac{x}{-4} = -12$

46. $x - (-4) = -12$

47. $18 = -9 + w$

48. $18 = \frac{w}{-9}$

49. $18 = -9w$

50. $18 = w - (-9)$

Building Number Skills

Mentally multiply by a decimal fraction.

Example: $386.2 \times 0.01 = 3.862$ Move the decimal two places to the left.

51. 59.6×0.1

52. 23.48×0.01

53. 17.65×0.001

54. 41.9×0.001

55. 563.7×0.001

56. 86.9×0.01

57. 1248×0.01

58. 3728×0.001

59. 8×0.0001

60. 4×0.1

4.2 Algebraic Expressions with Two Operations

A. Writing Algebraic Expressions

Many algebraic expressions involve two or more operations.

EXAMPLE 1

Emma can order tapes from a mail-order catalog for \$5 each. She pays a \$4 handling charge on her order.

a. Fill in the table below:

Number of tapes	2	3	5	8	10	12
Calculation	5(**2**) + 4	5(**3**) + 4	5(**5**) + 4	5(**8**) + 4	5(**10**) + 4	5(**12**) + 4
Total cost	**14**	**19**	**29**	**44**	**54**	**64**

b. Describe in words how to find the total cost of Emma's order:

Multiply the number of tapes by 5 and then add 4.

c. If t represents the number of tapes Emma orders, write an algebraic expression for the total cost of her order:

$$5t + 4$$

EXERCISE 1

Francisco is sharing a taxicab with three friends from the airport to the hotel. The four friends will share the fare equally, but Francisco will also tip the driver \$5 for handling his luggage.

a. Fill in the table:

Fare (\$)	10	16	20	30
Calculation				
Francisco's cost (\$)				

b. Describe in words how to compute Francisco's cost based on the fare.

c. If F represents the amount of the taxicab fare, write an algebraic expression for Francisco's share.

Some algebraic expressions involve more than one variable.

EXAMPLE 2 

Choose variables for each unknown quantity and write the following phrase as an algebraic expression:

The ratio of the interest earned to the amount invested.

Solution

The interest earned and the amount invested are unknown quantities.

Interest earned: I

Amount invested: P

The algebraic expression is a ratio: $\frac{I}{P}$

EXERCISE 2

Choose variables for each unknown quantity and write an algebraic expression for the following phrase:

The total of Ben's purchases and Emily's purchases.

EXAMPLE 3

Choose variables for each unknown quantity and write the following phrase as an algebraic expression:

12 times the difference of your monthly salary and your house note

Solution

Your monthly salary and your house note are unknown quantities.

Monthly salary: s

House note: n

The algebraic expression is $12(s - n)$.

EXERCISE 3

Choose variables for each unknown quantity and write an algebraic expression for the following phrase.

2 tablespoons times the number of cups, plus 1 extra tablespoon

ANSWERS TO 4.2A EXERCISES

1a.

Fare ($)	10	16	20	30
(Calculation)	$\frac{10}{4} + 5$	$\frac{16}{4} + 5$	$\frac{20}{4} + 5$	$\frac{30}{4} + 5$
Francisco's cost ($)	7.50	9	10	12.50

1b: Divide the fare by 4 and add 5 to that quotient.

1c. $\frac{F}{4} + 5$

2. Ben's purchases: B; Emily's purchases: E; $B + E$

3. Number of cups: C; $2C + 1$

HOMEWORK 4.2A 

1. The oven temperature started at 75° and is rising at 30° per minute. Fill in the following table:

Minutes elapsed	1	3	5	6	8	10
(Calculation)						
Oven temperature						

a. Describe in words how to find the oven temperature at any time.

b. If m represents the number of minutes elapsed since the oven was turned on, write an algebraic expression for the oven temperature.

2. Herman weighed 215 pounds when he went on a diet. He has been losing 3 pounds a week. Fill in the following table:

Weeks passed	2	4	6	9	12	15	18
(Calculation)							
Herman's weight							

a. Describe in words how to find Herman's weight at the end of any week.

b. If w represents the number of weeks that Herman has been dieting, write an algebraic expression for his weight.

3. Luisa's parents have agreed to pay her tuition ($800 per year) plus half her annual living expenses while she is in school. Fill in the following table:

Living expenses	2400	3000	3600	4000	4500	5000
(Calculation)						
Parents will pay						

a. Describe in words how to find the amount Luisa's parents will contribute to Luisa's support.

b. If a represents Luisa's annual living expenses, write an algebraic expression for the amount her parents will contribute to her support.

4. Aunt Charlotte is leaving $1000 to her cat, and the rest of her estate will be divided equally among her three nephews. Fill in the following tablet:

Aunt Charlotte's estate	10,000	16,000	25,000	40,000	100,000
(Calculation)					
Each nephew's share					

a. Describe in words how to find each nephew's share of the estate.

b. If e stands for the amount of Aunt Charlotte's estate, write an algebraic expression for the amount each nephew will inherit.

5. Mildred canned 80 pints of tomatoes. She kept some for herself and divided the rest equally among her four daughters. Fill in the following table:

Pints kept	4	8	12	20	28
(Calculation)					
Pints for each daughter					

a. Describe in words how to find each daughter's share of the canned tomatoes.

b. If M stands for the number of pints Mildred kept, write an algebraic expression for the number of pints she gave each daughter.

6. Station KPUB plans to add 8% of the revenue from this spring's pledge drive to its new-recordings budget. They already have $800 set aside for new recordings. Fill in the following table:

Pledge drive revenue	5000	10,000	12,000	16,000	25,000
(Calculation)					
New recordings budget					

a. Describe in words how to find the budget for new recordings.

b. If P stands for the pledge drive revenue, write an algebraic expression for the new recordings budget.

Choose the correct algebraic expression for each situation.

$2t + 12$ $12 - 2t$

$2(t + 12)$ $2t - 12$

7. Janine's history book has 12 chapters. If she studies 2 chapters a week, how many chapters will she have left after t weeks?

8. Arturo is 12 years older than twice the age of his nephew. If Arturo's nephew is t years old, how old is Arturo?

9. Rick made 12 fewer than twice as many phone calls as his roommate made this month. If Rick's roommate made t phone calls, how many calls did Rick make?

10. Every winter, the Civic Society knits mittens for the children of the county orphanage. This year there are 12 more children than last year. If there were t children last year, how many mittens will they need this year?

Choose the correct algebraic expression for each English phrase.

$\frac{m}{12} - 3$ $\frac{m - 3}{12}$

$\frac{12}{m - 3}$ $3 - \frac{12}{m}$

11. 12 divided by 3 less than m

12. 3 less than the quotient of m divided by 12

13. The quotient of 3 less than m divided by 12

14. Subtract from 3 the quotient of 12 and m

Write an algebraic expression for each situation.

15. Tuition at Woodrow University is $400 plus $30 per unit. How much tuition will you pay if you enroll in u units?

16. Moira's income is $50 more than one-third of her mother's income. If her mother's income is I, how much is Moira's income?

17. After paying their monthly expenses of $560, Hank and his three partners split the rest of their revenue. If this month's revenue is R, how much is Hank's share?

18. Otis buys 200 pounds of dog food at a time and uses 15 pounds a week for his dog Ralph. How much dog food does Otis have left after w weeks?

19. Getaway Tours offers a Caribbean cruise for $2000 per person if 12 people sign up. For each additional person who signs up, the price per person is reduced by $60. How much will you pay for the cruise if p additional people sign up?

20. Renee receives $600 for appearing in a corn chip commercial, plus a residual of $80 each time the commercial is aired. If the commercial plays t times, how much will Renee make?

Choose variables for each unknown quantity and write algebraic expressions for the following phrases.

21. The sum of the length and the width

22. The principal times the interest rate

23. The product of the base and the height

24. The sum of the principal and the interest

25. The ratio of the number of gallons of alcohol to the total volume

26. The weight of the copper divided by the total weight

Use two operations to write algebraic expressions for the following phrases.

27. 5 hours more than the product of 7 and Arnold's daily workout time

28. The ratio of school lunches needed to 12 cafeteria workers, increased by 8

29. The sum of rent and utilities, decreased by $60

30. 10 less than the difference of your budget and your expenses

31. One-half the sum of the radius and 3

32. 2 times the sum of 5 and the length

Simplify. Follow the order of operations.

33. $4000 - 400 - 40 - 4(4)$

34. $2 + 20 + 200 + 2000(2)$

35. $6 + 4 \cdot 10 + 5 \cdot 100 + 3 \cdot 1000$

36. $10{,}000 \cdot 8 + 100 \cdot 7 + 10 \cdot 6 + 5$

37. $13.1(0.8) - 12.4 \div 6.2$

38. $4.6 \div 0.4 + 0.6(9.3)$

39. $240 \div (6 + 3 \cdot 14)$

40. $120 \div (60 - 8 \cdot 7)$

41. $16 - 12 \div 4 + 3(20 - 2 \cdot 7)$

42. $28 + 6 \div 2 - 2(5 + 3 \cdot 2)$

Fill in the blank with the correct symbol: >, <, or =. (See Lesson 3.1B to review inequality symbols.)

43. $-5 - 3$ ____ -5

44. $-2 - 7$ ____ -7

45. $-4(-6)$ ____ 10

46. $-3(-5)$ ____ 15

47. $-8 - (-2)$ ____ -8

48. $-3 - (-9)$ ____ -6

49. $-7 - 7$ ____ 0

50. $4 - (-4)$ ____ 0

Building Number Skills

Mentally find each sum or difference.

Example: 800 + 700 = 1500 Add 7 + 8, followed by two zeros

51. 400 + 800

52. 1200 + 500

53. 1800 − 700

54. 2300 − 1200

55. 60 + 90

56. 30 + 180

57. 250 − 120

58. 300 − 160

59. 1500 + 1700

60. 2400 + 1300

B. Evaluating Algebraic Expressions

When we evaluate an algebraic expression, we follow the order of operations explained in Lesson 4.1A.

EXAMPLE 4

Evaluate $\frac{12}{x-2}$ for each of the values in the table:

x	6	10	20
$\frac{12}{x-2}$			

Solution

Substitute the given value for the variable. Then simplify: Begin by performing the subtraction below the fraction bar:

$$\text{For } x = 6: \frac{12}{x-2} = \frac{12}{\mathbf{6}-2} = \frac{12}{4} = 3$$

$$\text{For } x = 10: \frac{12}{x-2} = \frac{12}{\mathbf{10}-2} = \frac{12}{8} = \frac{3}{2}$$

$$\text{For } x = 20: \frac{12}{x-2} = \frac{12}{\mathbf{20}-2} = \frac{12}{18} = \frac{2}{3}$$

Fill in the table with the values of $\frac{12}{x-2}$:

x	6	10	20
$\frac{12}{x-2}$	3	$\frac{3}{2}$	$\frac{2}{3}$

EXAMPLE 5  

Evaluate $\frac{h-k}{2b}$ for $h = 20$, $k = 8$, and $b = 6$.

Solution

Substitute the given values for the variables and follow the order of operations.

$$\frac{h-k}{2b} = \frac{20-8}{2(6)} \quad \text{Perform operations above and below fraction bar.}$$

$$= \frac{12}{12} = 1$$

EXERCISE 4

Evaluate $3 + 2h$ for the values in the table:

h	1	4	10
$3 + 2h$			

EXERCISE 5

Evaluate $\frac{h(B+b)}{2}$ for $h = 5$, $B = 7$, and $b = 3$.

ANSWERS TO 4.2B EXERCISES

4a. 5

4b. 11

4c. 23

5. 25

HOMEWORK 4.2B 

Evaluate each expression to complete the table.

1.

z	0	3	10
$5z + 4$			

2.

y	2	5	7
$8y - 6$			

3.

b	4	6	9
$26 - 2b$			

4.

a	0	1	8
$12 + 3a$			

5.

h	13	7	24
$\frac{h - 5}{4}$			

6.

v	8	3	9
$\frac{v + 12}{10}$			

7.

d	3	6.5	0.2
$7(d + 1)$			

8.

g	6	4.5	5.6
$3(g - 4)$			

9.

m	3	12	0.5
$\frac{m}{3 + m}$			

10.

s	0	2	6.5
$\frac{s}{8 - s}$			

Evaluate each expression for the given values of the variables.

11. $4y - x$, for $x = 7$ and $y = 3$

12. $2y + x$, for $x = 8$ and $y = 9$

13. $7(s + t)$, for $s = 12$ and $t = 28$

14. $8(s - t)$, for $s = 27$ and $t = 15$

15. $5a - 6b$, for $a = 10$ and $b = 4$

16. $4a + 3b$, for $a = 8$ and $b = 7$

17. $\frac{3w + z}{z}$, for $w = 8$ and $z = 6$

18. $\frac{w + 2z}{z}$, for $w = 9$ and $z = 3$

19. $\frac{h}{g} - k$, for $h = 1.2$, $g = 0.6$, and $k = 0.8$

20. $v + \frac{u}{t}$, for $v = 0.3$, $u = 0.8$, and $t = 0.5$

For Problems 21–28, evaluate each formula for the given values of the variables.

The perimeter of a rectangle is given by the formula

$$P = 2l + 2w$$

where l is the length of the rectangle and w is its width.

21. The length of a rectangular rug is 7.4 feet, and its width is 4.8 feet. How long a fringe is needed to border the entire rug?

22. Mario's vegetable garden is 25 meters long and 15 meters wide. How much chicken wire does he need to enclose the garden?

The area of a trapezoid is given by the formula

$$A = \frac{h}{2}(b + c)$$

where h is the height of the trapezoid and b and c are the upper and lower bases (Figure 4.3).

Figure 4.3

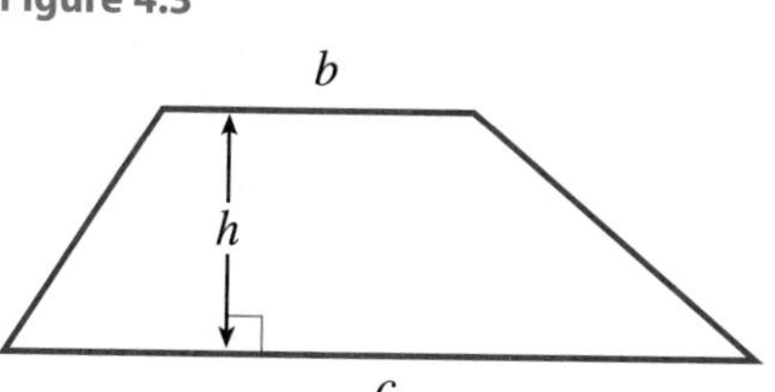

23. One section of the roof on Emery's garage has the shape of a trapezoid (Figure 4.4). The top of the roof is 30 feet long, and the bottom of the roof is 40 feet long. The distance between the top and bottom is 15 feet. What is the area of this section of the roof?

Figure 4.4

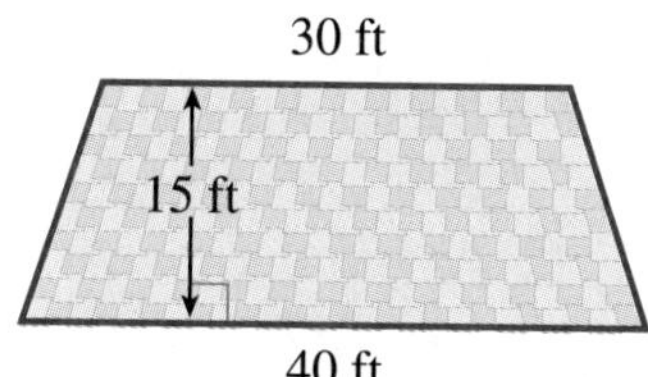

24. The glass cooktop on Joy's modern high-tech stove has the shape of a trapezoid (Figure 4.5). It is 36 inches wide at the back and 28 inches wide in front. The distance from front to back is 20 inches. What is the area of the cooktop?

Figure 4.5

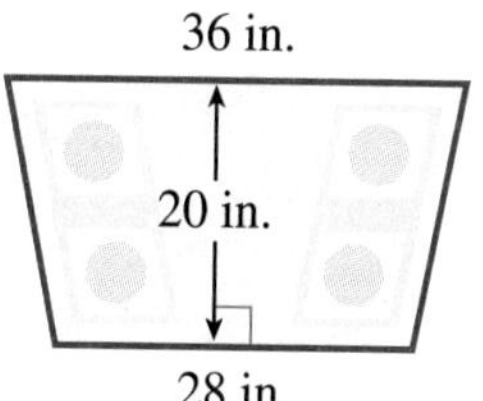

The temperature in degrees Celsius is given by

$$C = \frac{5F - 160}{9}$$

where F is the temperature in degrees Fahrenheit.

25. Normal body temperature is 98.6°F. What is normal body temperature in degrees Celsius?

26. Water freezes at 32°F. What is the freezing point of water in degrees Celsius?

If you deposit P dollars in an account earning simple interest rate r, then after t years the amount of money in the account is given by

$$A = P + Prt$$

27. Marla deposits \$50,000 at an interest rate of 6% for 5 years. How much is in her account at the end of that time?

28. Clyde pulls off a bank heist and gets away with \$800,000. Before he is apprehended, he manages to deposit the money in a Swiss bank account that pays 5.5% simple annual interest. When he gets out of jail 7 years later, Clyde closes his account and relocates to Rio de Janeiro. How much money does Clyde withdraw from the bank?

Compute without using a calculator. (See Lessons 3.2 through 3.4 to review operations on signed numbers.)

29. $3(-8)$

30. $-6(-5)$

31. $3 - 8$

32. $-6 - 5$

33. $20 \div (-5)$

34. $-20 - (-5)$

35. $-20(-5)$

36. $5 - 20$

37. $5 - (-20)$

38. $5(-20)$

39. Figure 4.6 shows normal low temperatures each month in Milwaukee, Wisconsin.

Figure 4.6 Milwaukee low temperatures

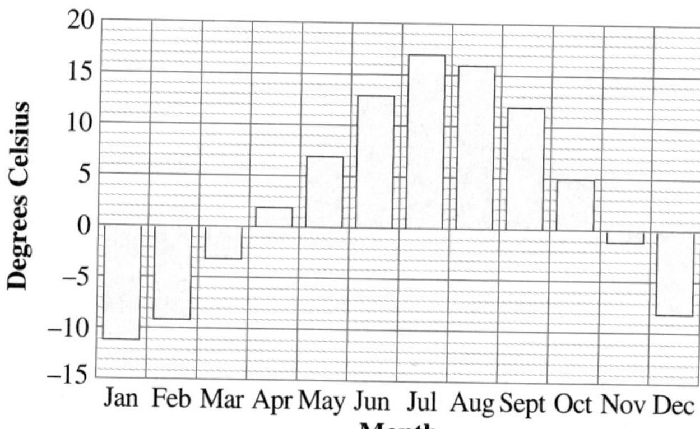

(Source: *American Almanac 1994–1995*)

a. Compute the 20 − T, the difference between a comfortable room temperature and Milwaukee's normal low temperature for each month. Fill in the table:

Month	Jan	Feb	Mar	Apr	May	June
20 − T(°C)						

Month	July	Aug	Sep	Oct	Nov	Dec
20 − T(°C)						

b. Use the grid in Figure 4.7 to make a bar chart showing the values in the table.

Figure 4.7 Change needed for comfortable temperature

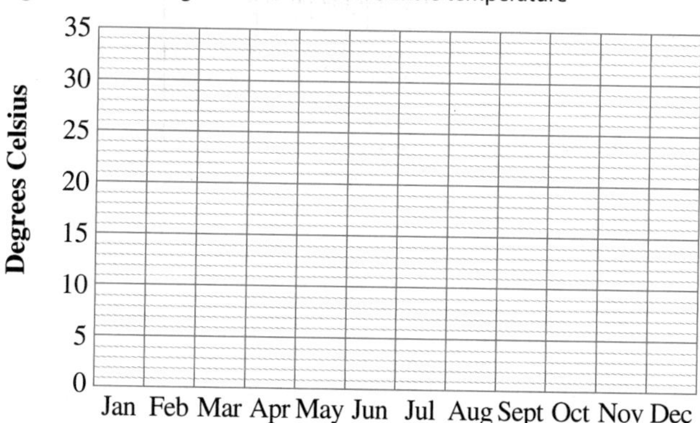

40. Online Grocer is a new grocery-delivery service. Figure 4.8 shows their costs and revenues for 1999–2001, in millions of dollars.

Figure 4.8 Online Grocer's costs and revenue

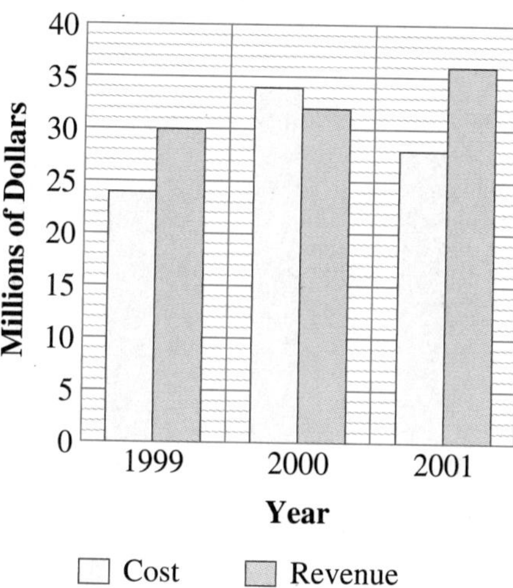

a. Compute the company's profit, $P = R - C$, for each of the 3 years. Fill in the table:

Year	1999	2000	2001
Profit (million $)			

b. Use the grid in Figure 4.9 to make a bar chart showing the company's profits in each year.

Figure 4.9 Online Grocer's profit

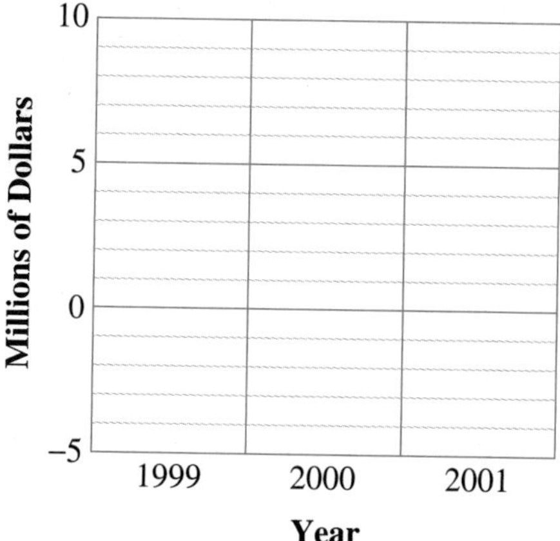

Building Number Skills

Mentally find each product or quotient.

Example: $20 \times 300 = (2 \times 3)$ followed by three zeros $= 6000$

41. 40×80

42. 120×30

43. 500×7000

44. 900×300

45. $\dfrac{2400}{40}$

46. $\dfrac{7200}{800}$

47. $35{,}000 \div 700$

48. $680{,}000 \div 200$

49. 160×4000

50. 1500×500

4.3 Equations with Two Operations

In Lesson 2.1 we solved equations by undoing the operation performed on the variable. For example, in the equation

$$4x = 28$$

the variable x has been *multiplied* by 4 to get 28, so we *divide* both sides of the equation by 4:

$$\frac{4x}{\mathbf{4}} = \frac{28}{\mathbf{4}} \quad \text{Divide both sides by 4.}$$

$$x = 7$$

The solution of the equation is 7. We can check this by substituting **7** for x in the equation:

Check: $4(\mathbf{7}) = 28$ True

Since this statement is true, 7 is indeed the solution.

In this lesson we learn how to solve equations with two or more operations.

EXAMPLE 1

Emma can order tapes from a mail-order catalog for \$5 each. She pays a \$4 handling charge on her order.

a. Write an algebraic expression for the cost of ordering t tapes.

b. Write an equation for the number of tapes Emma can order with \$34.

c. Describe in words how to find the number of tapes Emma can order with \$34.

Solution

a. Each tape costs \$5, so we *multiply* t by 5 and then *add* \$4 for the handling charge:

$$\text{Cost of ordering } t \text{ tapes: } \mathbf{5t + 4}$$

b. We set the cost of ordering t tapes equal to \$34:

$$5t + 4 = 34$$

c. Emma must pay the \$4 handling charge on any order, so if we first *subtract* \$4 from her \$34, we see that she has \$30 left to spend on tapes. Next, since each tape costs \$5, we *divide* \$30 by \$5 to see how many tapes she can buy: $30 \div 5 = 6$, so Emma can buy six tapes.

In Example 1c we have actually described a method for solving the equation $5t + 4 = 34$:

$$\begin{aligned} 5t + 4 &= 34 && \text{First, subtract 4 from both sides.} \\ \mathbf{-4} \quad &\quad \mathbf{-4} && \\ 5t &= 30 && \\ \frac{5t}{\mathbf{5}} &= \frac{30}{\mathbf{5}} && \text{Next, divide both sides by 5.} \\ t &= 6 && \end{aligned}$$

To check our solution, we substitute $t = \mathbf{6}$ into the original equation:

$$\begin{aligned} \textbf{Check: } 5(\mathbf{6}) + 4 &= 34 \\ 30 + 4 &= 34 \quad \text{True} \end{aligned}$$

The solution checks. Emma can order six tapes with \$34.

EXERCISE 1

Phoebe is stacking books on a 30-centimeter footstool. Each book is 3 centimeters thick.

a. Write an algebraic expression for the height of the top of the stack of n books.

b. Write an equation for the number of books Phoebe needs for the stack to be 54 centimeters high.

c. Describe how to find the number of books needed to make the stack 54 centimeters high.

A Strategy for Solving Equations

Suppose you want to give a bracelet as a surprise gift to a friend for her birthday. First, you wrap the bracelet in a small box and then you wrap the small box in a large box (Figure 4.10). To open the gift, your friend first unwraps the large box and then unwraps the small box. In other words, she *undoes the wrappings in reverse order.* We use this same idea to solve equations.

Figure 4.10

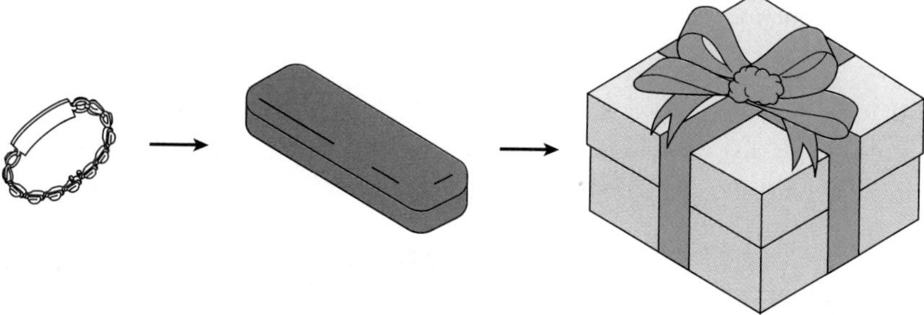

Consider the equation

$$2x - 5 = 13$$

What operations have been performed on the variable x? To answer this question, imagine that you have a value for x and you want to evaluate the expression $2x - 5$. Following the order of operations, you would first *multiply* the value of x by 2 and then you would *subtract* 5 from the result.

To solve the equation, we must undo these operations *in the reverse order.* Whatever operation was performed last, we must undo first. Thus, we should first *add* 5 to both sides of the equation and then *divide* both sides by 2.

Operations Performed on x	*Steps for Solution*
1. Multiplied by 2	**1.** Add 5
2. Subtracted 5	**2.** Divide by 2

Now let's use our plan to solve the equation:

$$\begin{aligned} 2x - 5 &= 13 \\ +5 \quad & +5 \qquad \text{Add 5 to both sides of the equation.} \\ 2x &= 18 \\ \frac{2x}{\mathbf{2}} &= \frac{18}{\mathbf{2}} \qquad \text{Divide both sides of the equation by 2.} \\ x &= 9 \end{aligned}$$

The solution of the equation is 9. To check the solution, we substitute **9** for x in the original equation:

Check: $2(\mathbf{9}) - 5 = 13$

$18 - 5 = 13$ True

The solution checks.

Here is a summary of our strategy for solving equations:

To Solve an Equation:

1. List the operations performed on the variable *in order.*
2. Undo those operations *in reverse order.*

EXERCISE 2
Solve the equation $3b - 4 = 14$

EXAMPLE 2 

Solve the equation $4 + \frac{a}{3} = 9$

Solution

Ask yourself what operations were performed on the variable a. Then make a plan for solving the equation by undoing those operations in reverse order.

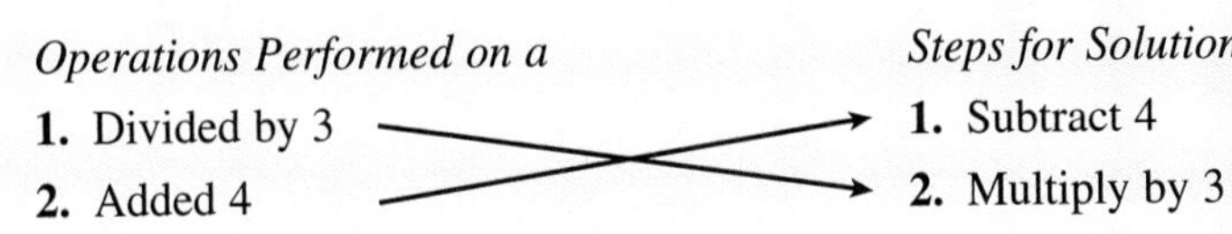

(continued)

Now follow the steps in the plan:

$$\begin{aligned} 4 + \frac{a}{3} &= 9 \\ \mathbf{-4} \quad &\quad \mathbf{-4} \end{aligned}$$ Subtract 4 from both sides of the equation.

$$\frac{a}{3} = 5$$

$$\mathbf{3}\left(\frac{a}{3}\right) = \mathbf{3}(5)$$ Multiply both sides by 3.

$$a = 15$$

The solution is 15.

$$\textbf{Check:} \quad 4 + \frac{\mathbf{15}}{3} = 9$$

$$4 + 5 = 9$$ True

Solving Equations

As usual, it is a good idea to simplify both sides of the equation before beginning to solve.

EXAMPLE 3

Solve the equation $300(0.06)t = 250 = 340$

Solution

First, simplify each side of the equation. We can multiply 300(0.06) to get 18, which gives us

$$18t + 250 = 340$$

Next, analyze the operations performed on the variable t. Solve the equation by undoing those operations in reverse order:

Operations Performed on t	*Steps for Solution*
1. Multiplied by 18	**1.** Subtract 250
2. Added 250	**2.** Divide by 18

Now follow the steps in the plan:

$$\begin{aligned} 18t + 250 &= 340 \\ \mathbf{-250} \quad &\quad \mathbf{-250} \end{aligned}$$ Subtract 250 from both sides.

$$18t = 90$$

$$\frac{18t}{\mathbf{18}} = \frac{90}{\mathbf{18}}$$ Divide both sides by 18.

$$t = 5$$

The solution is 5.

$$\textbf{Check:} \quad 300(0.06)(\mathbf{5}) + 250 = 340$$

$$90 + 250 = 340$$ True

EXERCISE 3

Solve the equation $\frac{r}{5} + 2 = 7 - 3$

EXERCISE 4

Solve the equation $\frac{w+3}{2} - 5 = 12$

ANSWERS TO 4.3 EXERCISES

1a. $3n + 30$

1b. $3n + 30 = 54$

1c. Subtract 30 from 54 to get 24, the height of the stack without the footstool, and divide the result by 3 to get 8, the number of books in the stack.

2. $b = 6$

3. $r = 10$

4. $w = 31$

In the next example we use three steps to solve the equation.

EXAMPLE 4

Solve the equation $\frac{5t - 3}{6} = 17$

Solution

Analyze the order of operations performed on the variable t. Then make a plan for solving the equation by undoing those operations in the reverse order:

Operations Performed on t	*Steps for Solution*
1. Multiplied by 5	**1.** Multiply by 6
2. Subtracted 3	**2.** Add 3
3. Divided by 6	**3.** Divide by 5

Now follow the steps in the plan:

$$\mathbf{6}\left(\frac{5t-3}{6}\right) = \mathbf{6}(17) \quad \text{Multiply both sides by 6.}$$

$$5t - 3 = 102$$

$$\underline{\mathbf{+3}} \quad \underline{\mathbf{+3}} \quad \text{Add 3 to both sides.}$$

$$5t = 105$$

$$\frac{5t}{\mathbf{5}} = \frac{105}{\mathbf{5}} \quad \text{Divide both sides by 5.}$$

$$t = 21$$

The solution is 21.

Check: $\frac{5(\mathbf{21}) - 3}{6} = 17$

$$\frac{105 - 3}{6} = 17$$

$$\frac{102}{6} = 17 \quad \text{True}$$

HOMEWORK 4.3

Solve each equation. Show the steps in your solution.

1. $2x + 5 = 27$

2. $3y + 4 = 25$

3. $4a - 6 = 14$

4. $5b - 1 = 19$

5. $17 = 7t - 4$

6. $29 = 8w - 3$

7. $21 = 6h + 9$

8. $30 = 9k + 3$

9. $\frac{m}{4} - 5 = 2$

10. $\frac{n}{7} - 4 = 5$

11. $8 + \frac{v}{3} = 10$

12. $2 + \frac{u}{6} = 8$

13. $\frac{4p}{5} = 8$

14. $\frac{7q}{2} = 14$

15. $5(z + 4) = 35$

16. $4(s - 6) = 16$

17. $36 = 9(f - 7)$

18. $28 = 7(3 + g)$

Use a calculator as needed to help you solve the equations.

19. $5m - 0.35 = 2.4$

20. $6r + 2.8 = 7.6$

21. $0.08d + 55.5 = 103.5$

22. $0.12v - 0.36 = 0.72$

23. $10.2 = \frac{w}{1.8} + 2.64$

24. $41.3 = \frac{c}{3.5} - 18.7$

Choose variables for each unknown quantity and write algebraic expressions.

25. The rebate deducted from the sale price

26. The distance traveled divided by the time elapsed

27. The ratio of field goals to attempts

28. The difference between the height of the roof and the height of the tree

29. The sum of the escrow fees and 1.5% of the selling price

30. The product of the number of students and $3 more than last year's entrance fee

31. 62% of the sum of your test average and your homework average

32. The sum of your business expenses and 29% of your medical expenses

(a) Complete the table of values and (b) use the table to solve the equation.

33. a.

x	2	4	6	8	10	12
$3x - 6$						

b. $3x - 6 = 12$

34. a.

x	2	4	6	8	10	12
$3(x - 6)$						

b. $3(x - 6) = 12$

35. a.

x	4	6	8	10	12	14	16
$\frac{4 + 2x}{3}$							

b. $\frac{4 + 2x}{3} = 12$

36. a.

x	4	6	8	10	12	14	16
$4 + \frac{2x}{3}$							

b. $4 + \frac{2x}{3} = 12$

Simplify each side of the equation and then solve.

37. $298 = \frac{1}{2}(18)h + 46$

38. $316 = \frac{1}{2}(16)g + 4(15)$

39. $1080 = 600 + 600(0.04)t$

40. $1187.50 = 950 + 950(0.05)t$

41. $5w - 3(6.4) = 120 + 0.2(38)$

42. $12 + 50r = 2(173) + 4(85)$

Solve each equation by using three steps.

43. $\frac{2g}{7} + 12 = 16$

44. $\frac{5d}{3} - 5 = 10$

45. $11 = \frac{9z}{4} - 7$

46. $21 = \frac{3c}{8} + 15$

47. $\frac{3y - 4}{5} = 4$

48. $\frac{2x + 5}{7} = 3$

49. $2(5b + 1) = 32$

50. $3(4r - 9) = 45$

Building Number Skills

Convert units without using a calculator. (See the inside front cover for metric conversions.)

51. 57 meters = _______ centimeters

52. 15 grams = _______ centigrams

53. 4.25 kilograms = _______ grams

54. 6.2 kilometers = _______ meters

55. 8.2 liters = _______ milliliters

56. 1.38 meters = _______ millimeters

57. 78 centigrams = _______ gram

58. 347.2 centimeters = _______ meters

59. 5456.2 meters = _______ kilometers

60. 96.5 grams = _______ kilogram

4.4 Problem Solving

A. Problem Solving with Formulas

Many problems can be solved by using an appropriate formula. You may want to review the formulas we studied in Lessons 2.3 and 2.4.

EXAMPLE 1

Garden snails travel at about 2.6 feet per minute. At that rate, how long will it take a snail to cross a 10-foot patio?

Solution

We use the formula $d = rt$, with $d = 10$ feet and $r = 2.6$ feet per minute:

$$10 = 2.6t$$

Solve the equation for t:

$$\frac{10}{\mathbf{2.6}} = \frac{2.6t}{\mathbf{2.6}} \quad \text{Divide both sides by 2.6.}$$

$$3.85 \approx t$$

It will take the snail approximately 3.85 minutes to cross the patio.

In Example 1 the symbol ≈ means "is approximately equal to." We use ≈ instead of = when we round an answer.

EXERCISE 1

How long will it take the snail to cross a 13-foot patio?

Formulas with Two or More Operations

The formula

$$F = \frac{9}{5}C + 32$$

is used to convert temperatures in degrees Celsius to degrees Fahrenheit. It requires two operations: First, we multiply C by $\frac{9}{5}$ and then we add 32. We can also use the formula to convert from degrees Fahrenheit to degrees Celsius.

EXAMPLE 2  

Normal body temperature for people is 98.6°F. Use the formula to convert normal temperature to degrees Celsius.

Solution

Substitute 98.6 for F in the formula:

$$98.6 = \frac{9}{5}C + 32$$

Solve the equation for C. We will undo the operations in three steps:

$$\begin{aligned} 98.6 &= \frac{9}{5}C + 32 \\ \mathbf{-32} & \quad\quad\quad \mathbf{-32} \quad \text{Subtract 32 from both sides.} \\ 66.6 &= \frac{9}{5}C \\ \mathbf{5}(66.6) &= \mathbf{5}\left(\frac{9C}{5}\right) \quad \text{Multiply both sides by 5.} \\ 333 &= 9C \\ \frac{333}{\mathbf{9}} &= \frac{9C}{\mathbf{9}} \quad \text{Divide both sides by 9.} \\ 37 &= C \end{aligned}$$

Normal body temperature is 37°C.

EXERCISE 2

The boiling temperature of water is 212°F. Use the formula in Example 2 to find the boiling temperature of water in degrees Celsius.

ANSWERS TO 4.4A EXERCISES

1. 5 minutes

2. 100°C

HOMEWORK 4.4A

Choose the correct formula and solve the problem. When necessary, round your answers to two decimal places.

$d = rt \qquad I = Prt \qquad A = lw \qquad P = rW \qquad A = \frac{S}{n}$

1. If a rectangular garden plot is 12 feet wide, how long must it be in order to provide 180 square feet of gardening space?

2. The official record for swimming the English Channel, a distance of 18 miles, is held by Penny Dean, who completed her swim in 7 hours 40 minutes. What was her average speed, in miles per hour? (*Hint:* Write 7 hours and 40 minutes in hours. What fraction of an hour is 40 minutes?)

3. The average weight of the 18 students in Jeanine's health class is 146 pounds. What is the sum of their weights?

4. Professor Kwon claims that 60% of his physics students pass his classes. If 75 students passed the professor's class this year, how many students did he start with?

5. Katrin invested $2000 in a friend's business venture, and after 1 year she received a dividend of $170. What percent rate of return did her investment earn?

6. Over the past 15-week semester, Mercedes spent $840 on food. How much did she spend each week, on the average?

Use the formula given in the problem to answer the question.

7. The perimeter of a rectangle is given by $P = 2l + 2w$. Find the length of a rectangle whose perimeter is 86 meters and whose width is 18 meters.

8. The area of a triangle is given by $A = \frac{1}{2}bh$. Find the height of a triangle whose area is 35 square feet and whose base is 14 feet.

9. The cost of renting a car for 1 day is given by $C = d + pm$, where d is the daily fee, p is the price per mile, and m is the number of miles driven. Suppose the daily fee is \$15 and the price per mile is \$0.20. If your rental bill was \$75 for 1 day, how far did you drive?

10. The cost of belonging to a health club is given by $C = I + pm$, where I is the initiation fee, p is the monthly fee, and m is the number of months. Amber paid \$425 for a 1-year membership at Sports Palace. If the initiation fee is \$125, how much is the monthly fee?

11. If you deposit P dollars in an account earning simple interest rate r, then after t years the amount of money in the account is given by $A = P + Prt$. Three years ago, Floyd deposited \$600 in his account, and he now has \$708. What is the interest rate on his account?

12. If you deposit \$5000 in an account that pays 7.5% interest, how long will it be before you have \$9500? (Use the formula in Problem 11.)

13. The area of a trapezoid is given by $A = \frac{h}{2}(b + c)$, where h is the height of the trapezoid and its bases are b and c. The area of a trapezoid is 54 square inches, and its height is 6 inches. If one of the bases is 8 inches long, how long is the other base?

14. The acceleration of a car is given by $a = \frac{f - s}{t}$, where f is the final speed, s is the starting speed, and t is the time to reach the final speed. Find f if $s = 10$, $t = 5$, and $a = 7$.

(a) Write an algebraic expression and fill in the table.
(b) Write an equation and use the table from part (a) to solve it.

15. a. 9 more than the product of 12 and the average monthly rainfall m:

m	0.5	0.75	1	1.25	2

b. The total rainfall this year was 24 inches, 9 inches more than last year. What was the average monthly rainfall last year?

16. a. The product of 5 and the dog-walking fee d reduced by 50:

d	28	30	31	35	37

b. Micah spent \$50 advertising his new dog-walking service. He charges \$5 to walk your dog, and so far his profit is \$135. How many dogs has he walked?

17. a. The sum of the score s and 156, divided by 3:

s	82	84	86	88	90

b. Corinna earned 156 points on her last two tests. What score does she need on the third test to have an average of 80?

18. a. The product of 5 and 0.25 less than the regular price p:

p	0.95	1.10	1.20	1.25	1.35

b. Latrisha had five coupons for \$0.25 off the price of a large can of dog food. She used the coupons on five cans of dog food and paid \$4.75. What is the regular price of a can of dog food?

The quilt designs in Problems 19 and 20 are not built on grids. Use the dimensions given to find the areas.

19. The quilt block in Figure 4.11 depicts a lighthouse. The block is 9 inches wide and 16 inches tall. The door of the lighthouse is $1\frac{1}{2}$ inches wide and $2\frac{1}{2}$ inches tall. The housing for the light is $4\frac{1}{2}$ inches tall and 6 inches wide at its widest point. The light itself measures 3 inches by $1\frac{1}{2}$ inches. Find the area of the red pieces that make up the main part of the building.

Figure 4.11

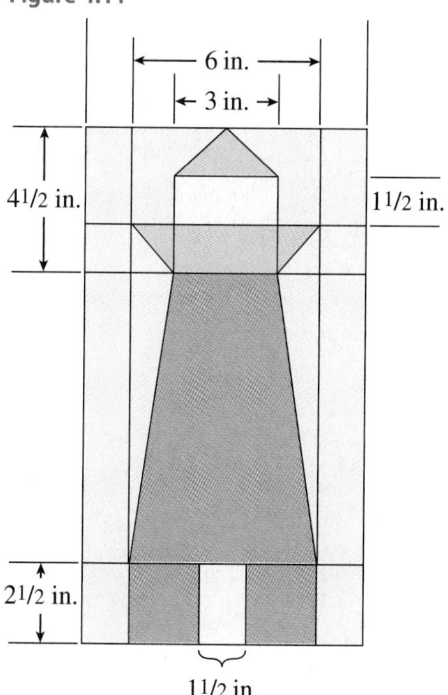

20. The quilt block in Figure 4.12 depicts a church. The block is 8 inches wide and 12 inches tall. The walls of the church and the triangular part of the steeple are both $3\frac{1}{2}$ inches tall. The horizontal strip, which includes the base of the steeple, is pieced together from four squares of fabric. Find the area of the dark blue pieces that make up the church roof and steeple.

Figure 4.12

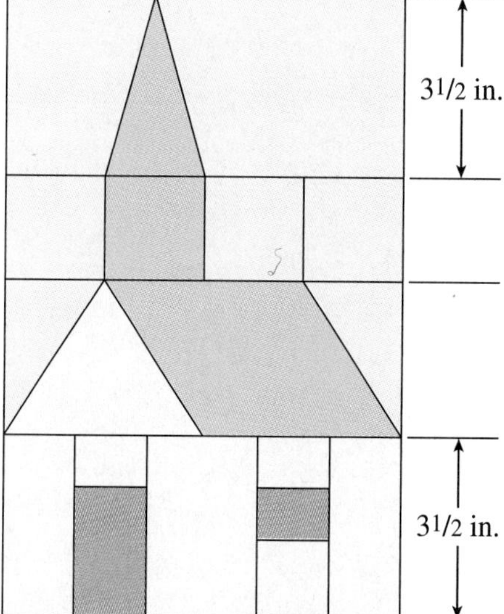

Solve and show the steps in your solution.

21. $6(x - 4) = 114$

22. $8 + \frac{x}{6} = 20$

23. $\frac{9x + 2}{5} = 13$

24. $\frac{4x}{7} - 9 = 7$

25. $\frac{2}{3}(x + 8) = 12$

26. $\frac{3}{5}x - 6 = 15$

Simplify by following the order of operations. (See Lesson 4.1 to review the order of operations.)

27. $18 - 4(3)$

28. $(18 - 4) \cdot 3$

29. $5 \cdot 6 - 2$

30. $5(6 - 2)$

31. $3 \cdot 12 - 2 \cdot 4$

32. $3(12 - 2) \cdot 4$

33. $3(12 - 2 \cdot 4)$

34. $(3 \cdot 12 - 2) \cdot 4$

35. $35 - (3 \cdot 15 - 2 \cdot 6)$

36. $(35 - 3)(15 - 2 \cdot 6)$

Compute without using a calculator.

37. $-17 - (-12)$

38. $-7(-8)$

39. $-15 - 6$

40. $-14 - (-13)$

41. $9(-11)$

42. $7 - 18$

43. $8 - (-9)$

44. $-6 - (-6)$

45. $-5(-5)$

46. $-2 - 2$

Building Number Skills

Mentally estimate each product or quotient by rounding one of the numbers.

Example: $6 \times 383 \approx 6 \times 400 = 2400$ Round 383 to hundreds.

47. 6×53 (Round to tens.)

48. 8×59 (Round to tens.)

49. 4×865 (Round to hundreds.)

50. 9×217 (Round to hundreds.)

51. $1493 \div 5$ (Round to hundreds.)

52. $3575 \div 6$ (Round to hundreds.)

53. $\frac{21{,}440}{7}$ (Round to thousands.)

54. $\frac{32{,}058}{4}$ (Round to thousands.)

55. $8 \times 29{,}874$ (Round to thousands.)

56. $5 \times 79{,}899$ (Round to thousands.)

B. Writing Equations

Consider the problem in Example 3.

EXAMPLE 3

Myron is buying a new computer system. He made a down payment of \$500 and agreed to make 18 equal monthly payments. If the total price of the system is \$6350, how much will each payment be?

There are many ways to solve this problem. Here are three methods that do not involve algebra.

Method 1 The guess-and-check method. Guess a value for the amount of each payment and check to see whether that value will yield the correct total price for the computer system. For example, guess that the payment is \$200. Myron would then pay

$$\underbrace{500}_{\text{Down payment}} + \underbrace{18(200)}_{\text{Monthly payments}} = 4100$$

Since \$4100 is too low, we should try a larger value for our next guess at the monthly payment—maybe \$250. We would continue in this way, adjusting our guesses, until we closed in on the correct amount.

Method 2 Make a table. This is a more organized version of the guess-and-check method. We made tables of values most recently in Lesson 4.2. A table of values for this problem might look like this:

Monthly payment	100	150	200	250	300	350
Total price	2300	3200	4100	5000	5900	6800

Since the total price of the computer system is between \$5900 and \$6800, we can narrow our search to monthly payments between \$300 and \$350. We could make another table with smaller increments between possible monthly payments. However, this is still a tedious method for solving the problem.

Method 3 Use arithmetic skills. Since Myron paid \$500 down on the system, we can **subtract 500** from the total price to find out how much he has left to pay in monthly installments:

$$6350 - 500 = 5850$$

Since the balance will be paid in 18 equal installments, we can **divide 5850 by 18** to find the amount of each installment:

$$\frac{5850}{18} = 325$$

Each monthly payment will be \$325.

Method 3 solves the problem efficiently. We can solve this problem using arithmetic because we understand exactly how the payments should work. However, some problems are too difficult for us to see exactly what to do. For more difficult problems, algebraic equations are often a useful tool. If we can describe the problem by an equation, we can work backward to the solution by solving the equation.

Using Algebra to Solve Problems

In Lesson 2.5 we solved problems using algebraic equations by following a three-step strategy.

Steps for Solving Problems

Step 1 Identify the unknown quantity and choose a variable to represent it.

Step 2 Find some quantity that can be expressed in two different ways and write an equation that models the problem.

Step 3 Solve the equation and answer the question in the problem.

Here is a solution to Example 3 that uses an algebraic equation.

Solution

Step 1a What are we asked to find in the problem?

The amount of each monthly payment

Step 1b Choose a variable to represent the unknown quantity.

Amount of monthly payment: *p*

Step 2a Find something in the problem that we can express in two *different* ways:

The total price of the system

First way: We know that the total price of the system is **$6350.**

Second way: The total price is the sum of the down payment and all the monthly payments.

Total price of system: $\underbrace{\mathbf{500}}_{\text{Down payment}} + \underbrace{\mathbf{18}p}_{\text{Monthly payments}}$

Step 2b Write an equation using the expressions from Step 2a.

$$\mathbf{500 + 18}p\ \mathbf{= 6350}$$

Step 3a Solve the equation:

$$\begin{aligned} 500 + 18p &= 6350 \\ \mathbf{-500} \quad &\quad \mathbf{-500} \qquad \text{Subtract 500 from both sides.} \\ 18p &= 5850 \\ \frac{18p}{\mathbf{18}} &= \frac{5850}{\mathbf{18}} \qquad \text{Divide both sides by 18.} \\ p &= 325 \end{aligned}$$

Step 3b Answer the question in the problem:

The monthly payments will be $325.

EXERCISE 3

Suppose that Myron's computer system goes on sale for $5540 just before he buys it. If he still makes 18 monthly payments after paying $500 down, how much will each of the monthly payments be?

The two steps used to solve the equation (subtract 500 and divide by 18) are the same steps we used on page 262 in Method 3, the arithmetic solution.

EXERCISE 4
On the same quiz Lucy scored 18 out of 20, which was 2 points higher than her average over the previous six quizzes. How many points had Lucy accumulated over the first six quizzes?

EXAMPLE 4  

Lloyd scored only 13 out of 20 on his last quiz. That was 5 points lower than his average score over the previous six quizzes. How many quiz points did Lloyd accumulate over six quizzes?

Solution

Step 1a What are we asked to find in the problem?

Lloyd's total quiz points over six quizzes

Step 1b Choose a variable to represent the unknown quantity:

Total quiz points: q

Step 2a Find something in the problem that we can express in two *different* ways:

Lloyd's score on the last quiz

First way: We know Lloyd's last score was **13.**

Second way: We can subtract **5** points from his average score.

Last quiz score: $\underbrace{\frac{q}{6}}_{\text{Average score}} - \underbrace{5}_{\text{5 points lower}}$

Step 2b Write an equation using the expressions from Step 2a:

$$\frac{q}{6} - 5 = 13$$

Step 3a Solve the equation:

$$\frac{q}{6} - 5 = 13$$
$$\underline{+5 \quad +5} \quad \text{Add 5 to both sides.}$$
$$\frac{q}{6} = 18$$
$$6\left(\frac{q}{6}\right) = 6(18) \quad \text{Multiply both sides by 6.}$$
$$q = 108$$

Step 3b Answer the question in the problem:

Over six quizzes, Lloyd accumulated 108 points.

ANSWERS TO 4.4B EXERCISES
3. \$280
4. 96 points

HOMEWORK 4.4B 

Choose the correct equation for each problem and solve.

$$3y + 12 = 30 \qquad 3y - 12 = 30$$

$$\frac{y}{3} + 12 = 30 \qquad \frac{y}{3} - 12 = 30$$

1. Zora bought three boxes of Christmas cards and used 12 cards. She has 30 cards left. How many cards were in each box?

2. Denton read one-third of his history assignment plus 12 pages of economics, for a total of 30 pages. How long is the history assignment?

3. Cindy planned to bicycle to Sagebrush State Park in three equal daily segments, but a thunder shower came up and she rode only 30 miles the first day, 12 miles shorter than she planned. How far is it to the park?

4. Margaret makes \$3 an hour baby-sitting. If she already has \$12, how many hours must she work in order to buy a \$30 gift for her mother?

Choose the correct equation for each problem and solve.

$$\frac{x + 20}{4} = 80 \qquad \frac{x}{4} + 20 = 80$$

$$4x + 20 = 80 \qquad 4(x - 20) = 80$$

5. Delbert bought four compact discs and a \$20 book for a total of \$80. How much did each compact disc cost?

6. If Barbara makes just 20 points on the next round of Castles and Kingdoms, her average score over four rounds will be 80 points per round. How many points does she have now?

7. Francine and her three roommates shared a winning lottery ticket. Adding her share to her $20 savings gives Francine $80. How much did the lottery ticket win?

8. Four students bought concert tickets with their $20 student-discount coupons. They paid a total of $80. How much do the tickets sell for regularly?

For each problem,
(a) identify the unknown quantity and choose a variable to represent it,
(b) use the hint to help you write an equation, and
(c) solve the equation. Write your answer in a sentence.

9. Irwin bought some tapes through the mail at a cost of $7 each. He also paid $2 for shipping and handling. If the total cost of Irwin's order was $30, how many tapes did he buy? (*Hint:* Write an expression in terms of your variable for the total cost of Irwin's order.)

10. Cassandra ordered some photo albums as gifts. Each album cost $12, and the total shipping cost was $4. If Cassandra's order totaled $76, how many albums did she order? (*Hint:* Write an expression in terms of your variable for the total cost of Cassandra's order.)

11. Fran wants to buy a $58 radio. She already has $30. If she earns $4 per hour as a cashier, how many hours must she work to earn the money she needs? (*Hint:* Write an expression in terms of your variable for the amount of money Fran has saved.)

12. David is saving to buy a pair of roller blades for $111. He already has $27. If he saves $12 per week, how long will it take to save the money he needs? (*Hint:* Write an expression in terms of your variable for the amount of money David has saved.)

13. The cost of a stereo is \$269. That is \$9 more than 4 times the cost of a camera. How much does the camera cost? (*Hint:* Write an expression in terms of your variable for the cost of the stereo.)

14. If you divide the cost of a TV set by 5 and subtract \$13, you get the cost of a tape recorder. The tape recorder costs \$39. How much does the TV cost? (*Hint:* Write an expression in terms of your variable for the cost of the tape recorder.)

15. This year Georgia paid \$2840 in state taxes. That was \$600 plus 7% of her adjusted income. What was Georgia's adjusted income this year? (*Hint:* Write an expression in terms of your variable for Georgia's taxes.)

16. Gilbert's speech instructor will add 5% of his score in the State Debating Contest to his homework total. If Gilbert would like to raise his total from 652 points to 675 points, what must he score in the debate? (*Hint:* Write an expression in terms of your variable for Gilbert's homework score.)

For Problems 17–26, compute without using a calculator. (See Lessons 3.2 through 3.4 to review operations on signed numbers.)

17. $-12 - 8 + 6$

18. $20 - 10 - 5$

19. $36 - (-20) - 16$

20. $-28 - 12 - (-10)$

21. $48 \div (-4)(-2)$

22. $-45 \div 9(-5)$

23. $-100 \div 20 \div (-5)$

24. $80 \div (-8) \div (-2)$

25. $-3(-3)(-3)(-3)$

26. $-3 - 3 - 3 - 3$

27. The quilt block shown in Figure 4.13 is called Grandmother's Flower Garden. The center of the flower and each petal are hexagons. To find their area, we can divide each hexagon into two equal trapezoids, with the longer base twice the length of the shorter base. The block is 25 centimeters wide and 26 centimeters tall.
    a. Find the dimensions (short base, long base, and height) of one trapezoid.

    b. Find the total area of the blue quilt pieces.

Figure 4.13

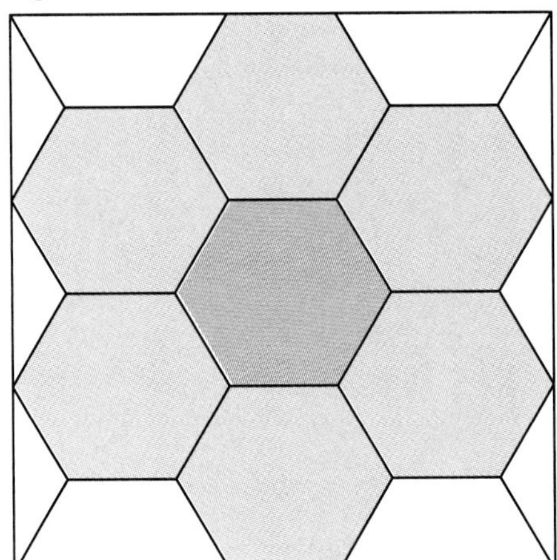

28. The quilt block shown in Figure 4.14 is an eight-pointed star. The star is made up of eight parallelograms. The block is a 24-centimeter square, and the inner square, shown as a dotted line, is 10 centimeters by 10 centimeters.
    a. Find the dimensions (base and height) of one parallelogram.

    b. Calculate the area of the blue quilt pieces.

Figure 4.14

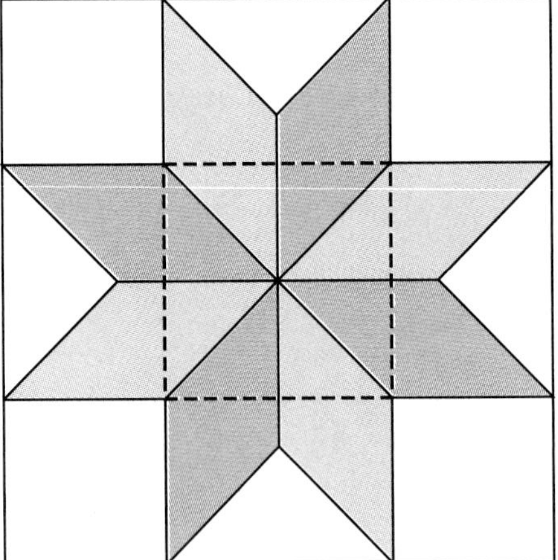

Building Number Skills

Mentally estimate each sum or difference by rounding the given numbers.

Example: $1208 + 413 \approx 1200 + 400 = 1600$ Round each number to hundreds.

Round to hundreds.

29. $758 + 635$

30. $941 + 617$

31. $2378 + 1327$

32. $2219 + 5292$

33. $883 - 504$

34. $789 - 288$

Round to tens.

35. $43 + 28$

36. $66 + 92$

37. $256 - 138$

38. $862 - 139$

4.5 Order of Operations with Signed Numbers

A. Simplifying Expressions

In Lesson 4.1 we studied the order of operations. In this lesson we will simplify expressions that involve negative numbers.

Sums and Differences

An expression like $-6 - 8$ can be interpreted as either an addition, $-6 + (-8)$, or as a subtraction, $-6 - (+8)$. Both give the same answer, but the addition $-6 + (-8)$ is easier to compute. (Can you explain why?)

EXAMPLE 1

Simplify each expression.

a. $-6 - 8$

b. $5 - 12$

Solution

a. Think of $-6 - 8$ as *adding* two numbers: -6 and -8. The sum of -6 and -8 is -14, so $-6 - 8 = -14$.

b. Think of $5 - 12$ as *adding* 5 and -12. Since 5 and -12 have opposite signs, we subtract 5 from 12 and make the answer negative. Thus, $5 - 12 = -7$.

EXERCISE 1

Simplify each expression.

a. $8 - 17$

b. $-4 - 5$

In an addition problem, the + and − symbols between numbers always indicate the sign of the next number. A subtraction such as $3 - (-9)$ can be converted into an addition, $3 + 9$. Remember to perform additions and subtractions in order from left to right.

EXERCISE 2
Simplify $-2 + 6 - 5 - (-7)$

EXAMPLE 2

Simplify $2 - 7 + 3 - (-9)$

Solution
Interpret operations as addition wherever possible.

$$\begin{aligned} &\mathbf{2 - 7} + 3 - (-9) && \text{Add 2 and } -7 \text{ to get } -5. \\ &= \mathbf{-5 + 3} - (-9) && \text{Add } -5 \text{ and 3 to get } -2. \\ &= -2 - (-9) && \text{Rewrite as an addition.} \\ &= -2 + 9 && \text{Add } -2 \text{ and 9.} \\ &= 7 \end{aligned}$$

It may seem strange at first to read $2 - 7$ as 2 added to −7, but this method will be very helpful when we simplify more complicated expressions. However, you will get the same answer if you think of $2 - 7$ as 2 subtract 7, so you can use whichever approach makes most sense to you.

Products and Quotients

Because we use minus signs to indicate negative numbers as well as subtraction, we must be careful to distinguish between these two uses.

EXERCISE 3
Compare the two expressions and simplify each.

a. $-2(-7)$

b. $-2 - (-7)$

EXAMPLE 3  

Compare the two expressions and simplify each.

a. $-8 - (-5)$ **b.** $-8(-5)$

Solution
Note the placement of parentheses in each expression. The first expression says −8 subtract −5, and the second expression says −8 times −5.

a. In the first expression, the parentheses merely enclose −5, and the highlighted minus sign indicates that the operation is subtraction:

$$-8 - (-5)$$

This minus sign tells us to subtract.

Thus, we simplify the first expression as follows:

$$\begin{aligned} -8 - (-5) &= -8 + 5 && \text{Rewrite as an addition.} \\ &= -3 && \text{Add } -8 \text{ and } +5. \end{aligned}$$

b. In the second expression, the parentheses between the two numbers tell us to multiply:

$$-8(-5)$$

These parentheses tell us to multiply.

We simplify the second expression as follows:

$-8(-5) = 40$ The product of two negative numbers is positive.

Combined Operations

Of course, we will also encounter expressions that involve both multiplication and addition or subtraction. The order of operations tells us to perform multiplications (and divisions) before additions and subtractions, so it is a good idea to begin by breaking up the expression into its terms. (See Lesson 1.2 to review the distinction between factors and terms.)

EXAMPLE 4

Simplify $-6 - 3(-5)$

Solution

First, study the expression and analyze the operations involved:

$$-6 - 3(-5)$$

These parentheses indicate multiplication.

This sign indicates subtraction.

The expression is composed of two terms: $\underline{-6} - \underline{3(-5)}$. We must simplify each term before attempting to combine them. We begin with the multiplication $3(-5)$:

$$\begin{aligned} -6 - 3(-5) &= -6 - \mathbf{3(-5)} && \text{The product of 3 and } -5 \text{ is } -15. \\ &= -6 - (-15) && \text{Rewrite as an addition.} \\ &= -6 + 15 && \text{Add } -6 \text{ and } 15. \\ &= 9 \end{aligned}$$

EXERCISE 4
Simplify $-5 - 2(-8)$

Remember that multiplications and divisions should be performed in order from left to right.

EXAMPLE 5

Simplify $5 - 2(-8) + 6 \div 3(-12)$

Solution

First, underline the terms of the expression. Each term is separated from the next one by an addition or subtraction symbol:

$$\underline{5}\ \boxed{-}\ \underline{2(-8)}\ \boxed{+}\ \underline{6 \div 3(-12)}$$

(continued)

EXERCISE 5
Simplify $8 - 6 \div (-3) + (-4)(7)$

ANSWERS TO 4.5A EXERCISES

1a. −9

1b. −9

2. 6

3a. 14

3b. 5

4. 11

5. −18

This expression has three terms. Simplify each term by performing multiplications and divisions *in order from left to right.* And—this is important—don't lose the + or − symbols separating the terms!

$$\underline{5} \boxed{-} \underline{2(-8)} \boxed{+} \underline{6 \div 3(-12)}$$ To simplify the last term: $6 \div 3(-12) = 2(-12) = -24$

$$= 5 \boxed{-} (-16) \boxed{+} (-24)$$

Notice that we inserted parentheses around −16 and around −24. This helps identify the minus symbols in front of 16 and 24 as negative symbols, not subtraction symbols. Finally, rewrite the subtraction as an addition and add in order from left to right:

$$5 - (-16) + (-24)$$ Change subtraction to addition.

$$= 5 + 16 + (-24)$$ $5 + 16 = 21$

$$= 21 + (-24) = -3$$

HOMEWORK 4.5A 

Rewrite each expression as an addition of signed numbers and then compute the sum.

1. $6 + 2 - 12 + 3$

2. $-9 + 4 - 3 + 2$

3. $-3 - 4 - (-2)$

4. $7 - 10 - (-5)$

5. $6 - 8 - (-9) - 1$

6. $-7 - 2 - 3 - (-8)$

7. $-28 - (-35) - 63$

8. $57 - 82 - 75$

9. $-12.8 + 21.7 - 19.2 + 17.5$

10. $-24.1 - 18.9 + 13.7 - (-20.6)$

Simplify.

11. **a.** $12 - 20$ **b.** $12(-20)$ **c.** $12 - (-20)$

12. **a.** $-12 - 20$ **b.** $-12(-20)$ **c.** $-12 - (-20)$

13. **a.** $4 - 6 - 8$ **b.** $4(-6)(-8)$

14. **a.** $4 - 6(-8)$ **b.** $4(-6) - 8$

15. **a.** $-5 + 9(-16)$ **b.** $-5(9) - 16$

16. **a.** $-5(9)(-16)$ **b.** $-5 + 9 - 16$

17. **a.** $3 - 3 - 3 + 3$ **b.** $3 - 3(-3)(3)$

18. **a.** $3(-3) + 3(-3)$ **b.** $3(-3) - 3(-3)$

For each problem,
(a) identify the unknown quantity and choose a variable to represent it,
(b) use the hint to help you write an equation, and
(c) solve the equation. Write your answer in a sentence.

19. Alida takes her chocolate pudding out of the freezer, but must let it come to room temperature before eating it. The pudding comes out of the freezer at 15°, and room temperature is 75°. If the pudding's temperature rises on average 6° per minute, how long must Alida wait? (*Hint:* Write an expression in terms of your variable for the temperature of the pudding.)

20. Hilda's new puppy weighed 4.4 pounds when she brought it home, and it should gain weight at approximately 0.8 pound per week. How many weeks will it take for the puppy to reach its adult weight of 38 pounds? (*Hint:* Write an expression in terms of your variable for the puppy's weight.)

21. When Pak enrolled at Carver College, he gave up his full-time job and now makes $500 more than one-third of his previous annual salary. If he makes $8500 a year now, what was his old salary? (*Hint:* Write an expression in terms of your variable for Pak's current salary.)

22. After paying the inheritance tax of $5000, JP Tycoon's six children split his legacy equally. If each offspring received $428,000, what was JP's total bequest? (*Hint:* Write an expression in terms of your variable for each child's share.)

(a) How many terms are there in each expression? Underline the terms.
(b) Simplify each expression. Compute multiplications before additions or subtractions.

23. $3 - 9(-2)$

24. $-5 + 2(-7)$

25. $-4(-2)(-3)$

26. $6(-5) \cdot 8$

27. $-2(-4) + 3(-6)$

28. $6(-7) + (-8)(-4)$

29. $12 - 8(-3) + 5 - 8$

30. $-3 + 9(-3) - 5 - 7$

31. $-240 + 8 \cdot 20 - 4 \cdot 10 \cdot 5$

32. $30 \cdot 5 - 3 \cdot 50 \cdot 2 - 0 \cdot 7$

33. $-5 - 2(-4)(9) + 7 - 2$

34. $20 - 4 - 6(-3)(-2)$

Simplify. Perform multiplications and divisions in order from left to right.

35. $-120 \div 6(-2)$

36. $80 \div (-4)(-5)$

37. $90 \div (-6) \div (-3) \cdot 2$

38. $-144 \div (-12) \div 12(-3)$

39. $-10(-18) \div 6(-3)$

40. $16(-20) \div (-4)(-5)$

Building Number Skills

Mentally estimate each sum or difference by rounding the given numbers.

Example: $-38 - 53 \approx -40 - 50 = -90$ Round each number to tens.

Round to tens.

41. $-72 - 51$

42. $-96 - 78$

43. $-48 + 67$

44. $-81 + 22$

45. $-77 + 43$

46. $-29 + 58$

Round to hundreds.

47. $-567 - 583$

48. $-815 - 638$

49. $-739 + 273$

50. $-482 + 871$

EXERCISE 6

Compare the expressions and simplify each.

a. $-2(-3) - 5$

b. $-2(-3)(-5)$

c. $(-2 - 3)(-5)$

B. Parentheses and Fraction Bars

Parentheses are often used to denote multiplication, but they can also be used to enclose part of an expression. If there are operations *inside* (between) the parentheses, we should perform those operations first.

EXAMPLE 6

Compare the three expressions and simplify each.

a. $-4(6 - 10)$ **b.** $-4(6)(-10)$ **c.** $-4(6) - 10$

Solution

a. The first expression has an operation inside the parentheses, $6 - 10$, which must be performed first:

$$\begin{aligned} &-4(6 - 10) && \text{Add 6 and } -10 \text{ inside the parentheses.} \\ &= -4(-4) && \text{Multiply.} \\ &= 16 \end{aligned}$$

b. The second expression is a product of three factors:

$$\begin{aligned} &-4(6)(-10) && \text{Multiply from left to right.} \\ &= -24(-10) \\ &= 240 \end{aligned}$$

c. The third expression consists of two terms, $-4(6)$ and 10:

$$\begin{aligned} &-4(6) - 10 && \text{Multiply } -4 \text{ times 6.} \\ &= -24 - 10 && \text{Add } -24 \text{ and } -10. \\ &= -34 \end{aligned}$$

Nested Parentheses

If an expression involves nested parentheses (one set of parentheses inside another), we start with the innermost set of parentheses and work out.

EXAMPLE 7 

Simplify $-3 - 2[10 + 2(5 - 8)]$

Solution

In this expression, the square brackets are the outside grouping symbols, and the curved parentheses are nested inside. Start with the operation inside the parentheses, $5 - 8$:

$-3 - 2[10 + 2(\mathbf{5 - 8})]$	Add 5 and -8.
$= -3 - 2[10 + \mathbf{2(-3)}]$	Multiply $2(-3)$ inside square brackets.
$= -3 - 2[\mathbf{10 - 6}]$	Add $10 - 6$ inside square brackets.
$= -3 - \mathbf{2[4]}$	Multiply $-2(4)$.
$= -3 - 8$	Add $-3 - 8$.
$= -11$	

EXERCISE 7

Simplify $-5 + 2[3 - 7(8 - 11)]$

Fraction Bars

A fraction bar is a division symbol, but it is also a grouping symbol like parentheses. You should simplify any expressions above or below a fraction bar before dividing the bottom into the top.

EXAMPLE 8 

Simplify $\dfrac{4 - 40}{4 + 8}$

Solution

Do *not* try to divide 4 by 4, or -40 by 8! We first simplify the expressions above and below the fraction bar and *then* divide:

$$\frac{4 - 40}{4 + 8} = \frac{-36}{12} = -3$$

EXERCISE 8

Simplify $\dfrac{6 - 14}{3 - 7}$

We repeat here for reference the rules for the order of operations.

Order of Operations

1. First, perform all operations inside the parentheses or above or below a fraction bar.
2. Next, perform all multiplications and divisions in order from left to right.
3. Finally, perform all additions and subtractions in order from left to right.

Evaluating Algebraic Expressions

When substituting values into an algebraic expression, remember to enclose any negative values in parentheses.

EXERCISE 9

Evaluate $\dfrac{V - v}{t}$ for $V = 4$, $v = -6$, and $t = 5$.

EXAMPLE 9 

Evaluate $6a - 3ab$ for $a = -2$ and $b = -5$.

Solution

Substitute the given values for the variables. Enclose each in parentheses. The original expression had two terms, so there are still two terms after substituting the numerical values:

$$6a - 3ab = \underline{6(-2)} - \underline{3(-2)(-5)}$$ Simplify each term by multiplying: $6(-2) = -12;\ 3(-2)(-5) = 30$

$$= -12 - 30$$ Add.

$$= -42$$

ANSWERS TO 4.5B EXERCISES

6a. 1 **6b.** −30 **6c.** 25

7. 43

8. 2

9. 2

HOMEWORK 4.5B 

Simplify. Pay close attention to the placement of parentheses.

1. a. $4 - 7$ **b.** $4(-7)$ **c.** $4 - (-7)$ **d.** $-(4 - 7)$

2. a. $-4 - 7$ **b.** $-4(-7)$ **c.** $-4 - (-7)$ **d.** $-(-4 - 7)$

3. $-8 - 3(-2)$ **4.** $-8(-3) - 2$ **5.** $(-8)(-3)(-2)$

6. $(-8 - 3)(-2)$ **7.** $-8(-3 - 2)$ **8.** $-8 - (3 - 2)$

9. $-8 - 3 - 2$

10. $(-8 - 3) - 2$

11. $-(8 - 3 - 2)$

12. $8 - (-3) - 2$

13. $8 - (-3 - 2)$

14. $8 - (-3)(-2)$

Simplify.

15. $10 - (20 - 30)$

16. $-18 - (12 - 8)$

17. $-120 \div (30 \div 6)$

18. $90 \div (-15 \div 3)$

19. $16[20 \div (-5)]$

20. $-64 \div [-2(-8)]$

21. $-9 + 12 \div (-3)(8 - 2)$

22. $40 \div (-4) + 12(-6 - 6)$

23. $-10 - (6 - 9) + 2(8 - 13)$

24. $-3(12 - 6) - (4 - 7) + 8$

Simplify. Perform operations above or below fraction bars first.

25. $\dfrac{36}{6 - 10}$

26. $\dfrac{-20 - 10}{6}$

27. $\dfrac{-30 + 15}{3 - 8}$

28. $\dfrac{9 - 15}{3 - 5}$

29. $\dfrac{8 - 2(-17)}{4 \cdot 6 - 3}$

30. $\dfrac{-27 + 6(3)}{-9 + 3(4)}$

31. $\dfrac{-4(15 - 33)}{6(5 - 11)}$

32. $\dfrac{(-36 + 6)(-3)}{(-15 + 12 \cdot 2)}$

Use the order of operations to evaluate each expression for the values in the table.

33.

z	-2	0	-5
$15 - 5z$			

34.

w	-9	0	-2
$-3w + 7$			

35.

a	-4	4	2
$-2(3a - 8)$			

36.

b	-3	2	10
$-4(10 - 7b)$			

37.

c	-7	-2	2
$(c + 3)(c - 4)$			

38.

d	-6	1	4
$(d - 2)(d - 6)$			

39.

u	-4	3	5
$\dfrac{3u}{u - 5}$			

40.

v	-8	0	4
$\dfrac{-8v}{8 - v}$			

41.

r	-6	0	2
$\frac{r-2}{r+2}$			

42.

s	-4	2	6
$\frac{3s-2}{s-6}$			

Evaluate each expression for the given values of the variables.

43. $-3 - 4pq$, for $p = -3$ and $q = -5$

44. $7 - 9wz$, for $w = -4$, and $z = -10$

45. $4 - b(2 + 3c)$, for $b = 6$ and $c = -8$

46. $(5 - w)(-2y) - 3$, for $w = -3$ and $y = -4$

Write and solve an equation for each problem. (See Lesson 2.4 to review geometric formulas.)

47. Denecia wants the sail on her model sailboat to have an area of 720 square centimeters. If she makes a triangular sail with a base of 40 centimeters, what is the sail's height?

48. A paint manufacturer claims that a gallon of its paint will cover an area of 250 square feet. If Assad is painting a wall that is 8 feet tall from floor to ceiling, how wide an area can he paint?

49. If C stands for the temperature in degrees Celsius, then the temperature in degrees Fahrenheit is given by $F = 1.8C + 32$. Red wine should be stored at 56°F. What is this temperature in degrees Celsius?

50. The temperature in degrees Celsius is given by $C = \frac{5F - 160}{9}$, where F is the temperature in degrees Fahrenheit. You are vacationing on the French Riviera and the weather report predicts a high of 33°C. What is this temperature in degrees Fahrenheit?

Building Number Skills

Estimate each product or quotient by rounding one of the numbers.

Example: $\frac{-2683}{3} \approx \frac{-2700}{3} = 900$ Round to hundreds.

51. $-4(678)$

52. $-8(214)$

53. $-5(-394)$

54. $-3(-918)$

55. $\frac{-2625}{2}$

56. $\frac{-3579}{9}$

57. $\frac{-4758}{-8}$

58. $\frac{-5535}{-11}$

59. $\frac{11{,}748}{-6}$

60. $\frac{28{,}319}{-4}$

4 Summary

Lesson 4.1

Order of Operations

1. First, perform all operations inside the parentheses or above or below a fraction bar.
2. Next, perform all multiplications and divisions in order from left to right.
3. Finally, perform all additions and subtractions in order from left to right.

Lesson 4.2

Algebraic expressions may involve two or more operations.

When we evaluate an algebraic expression, we follow the order of operations.

Lesson 4.3

To solve an equation that has more than one operation:

1. List the operations performed on the variable *in order.*
2. Undo those operations *in reverse order.*

Lesson 4.4

The symbol $\approx$ means "approximately equal to."

Steps for Solving Problems

Step 1 Identify the unknown quantity and choose a variable to represent it.

Step 2 Find some quantity that can be expressed in two different ways and write an equation that models the problem.

Step 3 Solve the equation and answer the question in the problem.

Lesson 4.5

Because we use minus signs to indicate negative numbers as well as subtraction, we must be careful to distinguish between these two uses.

It is a good idea to begin simplifying an expression by breaking it up into its terms.

If an expression involves nested parentheses (one set of parentheses inside another), we start with the innermost set of parentheses and work out.

When substituting values into an algebraic expression, remember to enclose any negative values in parentheses.

CHAPTER 4 REVIEW 

For Problems 1–6, answer true or false.

1. Operations should be performed in the order they are written.

2. Always perform multiplications before divisions.

3. Start by performing all operations inside parentheses.

4. Perform any operations below a fraction bar before doing the division.

5. When simplifying nested parentheses, always start with the square brackets.

6. When solving an equation, always subtract before dividing.

7. Where should you place parentheses in the expression
$$2 + 5 \cdot 12 \div 3$$
   a. If the division should be performed first?

   b. If the addition should be performed first?

8. Where should you place parentheses in the expression
$$6 + 16 \div 2 \cdot 4$$
   a. If the division should be performed first?

   b. If the multiplication should be performed first?

9. Explain what the parentheses mean in each expression and then simplify.
   a. $-8 - 3(-4)$

   b. $-8 - (3 - 4)$

10. Find the error in the following calculation and correct it: Evalute $2x - 8$, for $x = -3$.
$$2 - 3 - 8 = -1 - 8 = -9$$

Simplify each expression according to the order of operations.

11. $-6 + 4 - (-9) - 2$

12. $5 - 8 - 4 - (-7)$

13. $13 - 8(6 - 2)$

14. $13 - 8(6)(-2)$

15. $-3 + 8(-6 - 3)$

16. $10 - 4(8 - 12)$

17. $48 \div (-6) \div (-2)$

18. $48 \div (-6 - 2)$

19. $\dfrac{-8 + 12}{-4 + 2}$

20. $\dfrac{-18}{-18 + 6}$

Evaluate each expression for the given values of the variables.

21. $mx + b$, for $m = -3$, $x = -4$, and $b = -6$

22. $\dfrac{s}{t} - v$, for $s = -54$, $t = 18$, and $v = 12$

23. $-a(3b - c)$, for $a = 12$, $b = -9$, and $c = -4$

24. $(x - 2y)(3x + y)$, for $x = -6$ and $y = -2$

25. $\dfrac{2mp}{p - 2m}$, for $m = 3$ and $p = -18$

26. $\dfrac{h + d}{h - d}$, for $h = -4$ and $d = -12$

For Problems 27–30, use a calculator to simplify each expression.

27. $-0.5(-2.4) - 1.8(0.2)$

28. $3.6 - 4.2 - 2.4(-2.4)$

29. $\dfrac{4.7 - 4(3 - 7.7)}{-6.8 - 2.6}$

30. $\dfrac{(-4.2 - 2.4)(-3.2)}{1.6(-8)}$

31. Use the formula $A = P + Prt$ to find the account balance after 5 years if you invest \$2500 at 7.5% interest rate.

32. Use the formula $P = 2l + 2w$ to find the perimeter of a rug that measures $38\frac{1}{4}$ inches $\times$ $64\frac{3}{4}$ inches.

33. Use the formula $A = \dfrac{h}{2}(b + c)$ to find the area of a trapezoid that is 9 centimeters tall and whose upper and lower bases are 15 centimeters and 20 centimeters.

34. Use the formula $C = \dfrac{5F - 160}{9}$ to find the Celsius temperature when the Fahrenheit temperature is $-13°$.

(a) Write an algebraic expression in terms of the variable,
(b) evaluate your expression for the given value of the variable, and
(c) write an equation and solve it.

35. **a.** Claire's checking account had \$400 in January, and she deposits \$80 per month. If she doesn't write any checks, what will be the account balance after *m* months?

b. What will be Claire's balance after 9 months?

c. How many months will it take for Claire's balance to reach \$1440?

36. **a.** Delbert's telephone service charges a \$2 access fee for a long-distance call to London plus \$0.80 per minute. How much will it cost Delbert to talk to Francine in London for *m* minutes?

b. What is the cost of a 15-minute phone call to London?

c. How long can Delbert talk to Francine in London for \$26?

37. a. Every school day (5 days a week), each member of the cross-country team runs a certain distance d, depending on her ability level, plus 0.5 mile around the track. What distance does each team member run in 1 week?

b. Evelyn runs 6 miles before working out on the track. How far does she run in 1 week?

c. Greta runs 22.5 miles per week. What is her assigned distance d?

38. a. Corey is planning a 4-day rafting trip for a group of tourists. He must plan on 2 gallons of water for each tourist, as well as for each of the three guides. If there are t tourists in the group, how much water should Corey bring?

b. If there are 15 tourists, how much water will the expedition need?

c. If the supply raft can carry at most 24 gallons of water, how many tourists can go on the expedition?

39. a. Digby bought stereo equipment by making a \$50 down payment and arranging to pay the balance in ten equal installments. If the total cost of the equipment is C, how much is each installment payment?

b. If the total cost of Digby's stereo equipment was \$670, how much is each payment?

c. Darla bought stereo equipment under the same plan, and her payments are \$79 each. What was the total cost of Darla's stereo?

40. a. For English class, Earlene has to read a novel from an approved list plus a 30-page essay in 9 days. If the novel she chooses is p pages long and she reads an equal amount each day, how many pages should she read per day?

b. Earlene chooses to read *A Tale of Two Cities,* which is 357 pages long. How many pages should she read per day?

c. Shawn figures that he can finish 60 pages per day. How long a novel can he read?

41. Describe a general strategy for solving an equation that involves two or more operations.

42. Describe a three-step process for solving problems algebraically.

Solve.

43. $5x - 8 = 16 + 2$

44. $11 - 8 = 4y - 13$

45. $57 = 3z + 16 - 4$

46. $9 + 6w - 1 = 40$

47. $2(0.7)a - 4.3 = 8.3$

48. $6.5 + 0.8(3)b = 35.3$

49. $3 = \dfrac{4h - 17}{5}$

50. $\dfrac{9 + 5k}{8} = 8$

51. $21.6 + \dfrac{v}{2.5} = 28.8$

52. $\dfrac{t}{3.4} - 2.8 = 13.2$

Solve each problem by following the three steps on page 263.

53. The Tree People planted new trees in an area that was burned by brush fires. Of the seeds planted, 60% sprouted, but gophers ate 38 of the new sprouts. That left 112 new saplings. How many seeds were planted?

54. Of the freshman class at State College, 70% achieved sophomore standing. Eighteen sophomores transferred to another school, which left 458 from the original freshman class. How large was the freshman class?

55. During their annual sale, Sierra Sportsware sold 8 kayaks at $30 off their original price. The kayaks brought in $1072. What was the regular price of a kayak?

56. The treasurer for the Botanical Society would like to raise the annual dues by $15. That would give the Society $1760 in dues next year. If the Society has 32 members, what are the dues this year?

Building Number Skills

Mentally multiply or divide.

57. 3.57×1000

58. $3.57 \div 1000$

59. 5640×0.001

60. 1.89×0.01

Mentally compute.

61. $380 - 260$

62. $1200 + 1500$

63. 600×300

64. $\frac{3600}{40}$

Convert units.

65. 2350 milliliters = _______ liters

66. 4.7 kilometers = _______ meters

Use rounding to estimate the answer.

67. $228 + 593$

68. $9312 - 5817$

69. 4×3854

70. $\frac{1596}{4}$

71. $\frac{-2448}{6}$

72. $-8 \times (-290)$

73. $-718 - 798$

74. $384 - 879$

CHAPTER

5

Exponents and Roots

5.1 Exponents

A. What Is an Exponent?

Squares and Cubes

The area of a rectangle can be calculated using the formula $A = lw$, where l stands for the length of the rectangle and w for its width (see Figure 5.1). The area tells us how many square tiles, 1 unit on a side, will fit inside the rectangle.

Figure 5.1

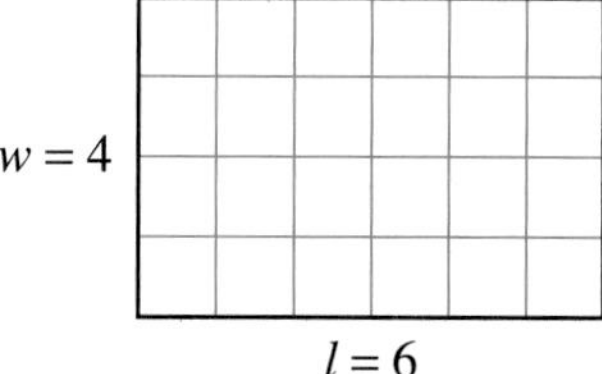

Figure 5.2

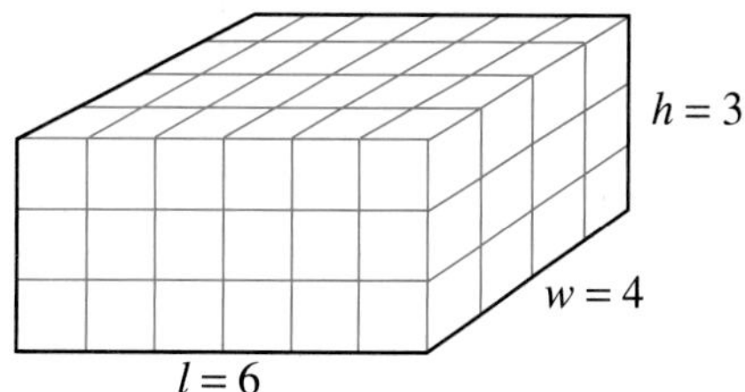

Similarly, the volume of a box (measured in cubic units) can be calculated using the formula $V = lwh$, where l, w, and h stand for the length, width, and height of the box. The volume tells us how many blocks, 1 unit on a side, will fit inside the box. Thus, the volume of the box in Figure 5.2 is

$$\begin{aligned} V &= lwh \\ &= (6)(4)(3) = 72 \text{ cubic inches} \end{aligned}$$

A square, shown in Figure 5.3, is a rectangle whose length and width are equal. We will use *s*, for side, to denote both the length and width of the square. The area of the square is then

$$A = s \cdot s$$

A cube, shown in Figure 5.4, is a box whose length, width, and height are all equal, so its volume is given by

$$V = s \cdot s \cdot s$$

Figure 5.3

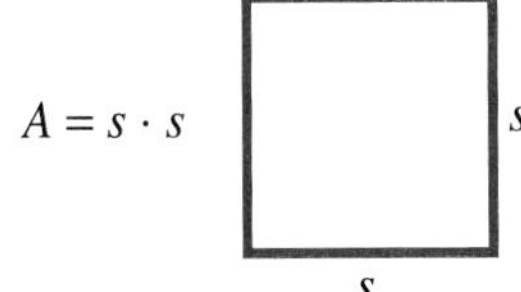

Figure 5.4

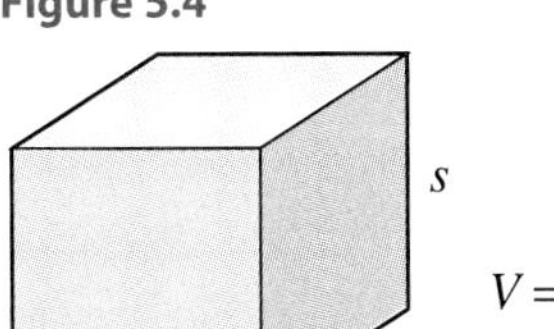

EXERCISE 1

a. Find the area of a square whose side is 3 centimeters long.

b. Find the volume of a cube whose side is 5 inches long.

EXAMPLE 1

a. Find the area of a square whose side is 6 feet long.

b. Find the volume of a cube whose side is 4 meters long.

Solution

a. $A = s \cdot s$
$= 6(6)$
$= 36$ square feet

b. $V = s \cdot s \cdot s$
$= 4(4)(4)$
$= 64$ cubic meters

Repeated Multiplication

Finding the area of a square or the volume of a cube involves *repeated multiplication* by the same number. We indicate repeated multiplication with a symbol called an exponent. The **exponent** tells us how many times to use the **base** as a factor. For example,

5^3 means $5 \cdot 5 \cdot 5$, or 125.

A number or variable with an exponent is called a **power** of its base. We read the symbol 5^3 as "five raised to the third power," or simply as "five to the third."

EXAMPLE 2  

a. $2^5 = \underbrace{2 \cdot 2 \cdot 2 \cdot 2 \cdot 2}_{\text{Five factors of 2}} = 32$ Read: "Two to the fifth power."

b. $8^2 = \underbrace{8 \cdot 8}_{\text{Two factors of 8}} = 64$ Read: "Eight to the second power."

c. $10^4 = \underbrace{10 \cdot 10 \cdot 10 \cdot 10}_{\text{Four factors of 10}} = 10{,}000$ Read: "Ten to the fourth power."

CAUTION!

Note that 5^3 does *not* mean 5 times 3 or 15; we already have a way to write 5(3). Remember that an exponent or power indicates *repeated* multiplication of the base.

EXERCISE 2
Compute the power: 12^3

The formulas for the area of a square and the volume of a cube can be written with exponents as

$$A = s^2 \qquad \text{and} \qquad V = s^3$$

The calculations in Example 1 then look like

$$A = s^2 \qquad\qquad V = s^3$$
$$= 6^2 = 36 \qquad\qquad = 4^3 = 64$$

Because the exponents 2 and 3 are used frequently, they are given special names:

6^2 is read "six squared."
4^3 is read "four cubed."

Powers with larger exponents do not have special names.

Powers of Negative Numbers

We must be careful when using exponents with negative numbers. Compare the following two expressions:

$$(-2)^4 \qquad \text{and} \qquad -2^4$$

The first expression is the fourth power of negative 2, which means

$$(-2)^4 = (-2)(-2)(-2)(-2) = 16$$

The second expression is the negative of the fourth power of 2, which means

$$-2^4 = -\ 2 \cdot 2 \cdot 2 \cdot 2 = -16$$

In -2^4 the exponent 4 applies *only to the base,* 2; the negative sign is applied *after* the power is computed. Consequently, we can state the following rule:

When we raise a negative number to a power, we must enclose the negative number in parentheses.

EXAMPLE 3 

a. $-9^2 = -9 \cdot 9 = -81$ Only 9 is squared.
b. $(-9)^2 = (-9)(-9) = 81$ -9 is squared.

EXERCISE 3
Compute.
a. $(-3)^4$

b. -3^4

Using a Calculator

You can always compute a power as a repeated multiplication. For example, to compute 7^4 you would enter

7 [×] 7 [×] 7 [×] 7 [=]

and the calculator displays the answer 2401.

Scientific calculators usually have a key labeled [x^y] or [y^x], called the power key, for computing powers. To compute 7^4 using the power key, we enter

7 [y^x] 4 [=]

Graphing calculators have a caret key, [^], for entering powers. On a graphing calculator we enter 7^4 as

7 [^] 4 [ENTER]

Also, many calculators have a key labeled [x^2] for computing squares of numbers.

EXAMPLE 4

Use a calculator to compute the powers.

a. $(1.2)^3$ **b.** $(-12)^4$

Solution

a. Enter the following operations:

1.2 [y^x] 3 [=] or 1.2 [^] 3 [ENTER]

to find that $(1.2)^3 = 1.728$.

b. If your calculator has a [+/−] key, enter the following operations:

12 [+/−] [y^x] 4 [=]

to find that $(-12)^4 = 20{,}736$. On a graphing calculator, enter

[( ] [(−)] 12 [ )] [^] 4

Some calculators will not raise a negative number to a power; they return an error message. In that case, you can raise the positive number to the desired power and then decide on the correct sign for the answer.

EXERCISE 4

Use a calculator to compute each power.

a. $(-1.1)^4$

b. $(-2.3)^3$

ANSWERS TO 5.1A EXERCISES

1a. 9 square centimeters

1b. 125 cubic inches

2. 1728

3a. 81

3b. −81

4a. 1.4641

4b. −12.167

HOMEWORK 5.1A 

1. a. Find the area of a square that measures 8 inches on a side.

b. Find the volume of a cube whose base is the square in part (a).

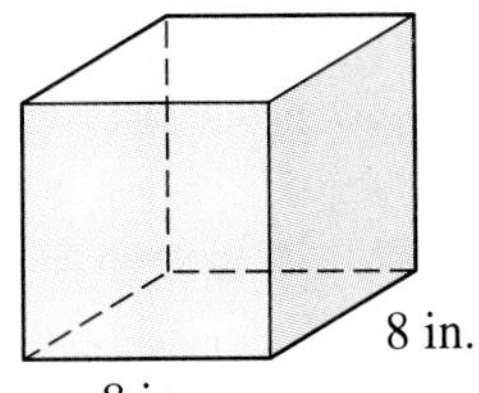

2. a. Find the area of a square that measures 0.1 meter on a side.

b. Find the volume of a cube whose base is the square in part (a).

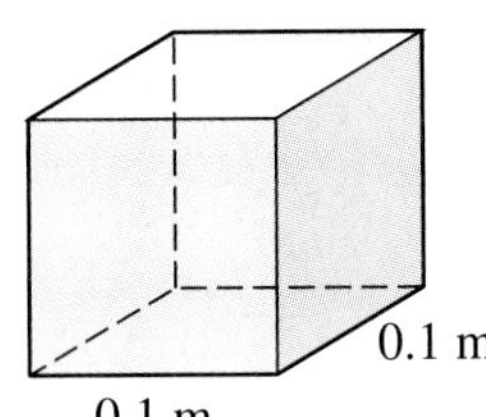

3. How many 1-centimeter cubes will fit inside a box that measures 16 centimeters by 20 centimeters by 8 centimeters?

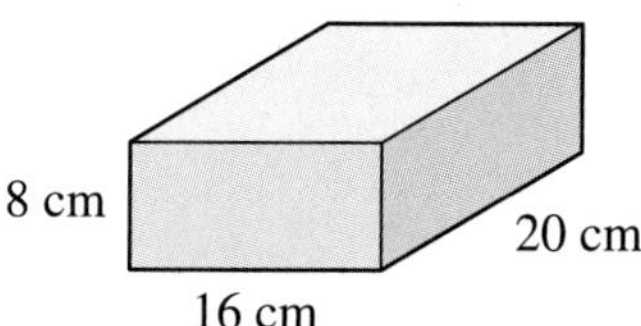

4. How many cubic feet of dirt came out of a hole that is 12.5 feet wide, 15 feet long, and 8.2 feet deep?

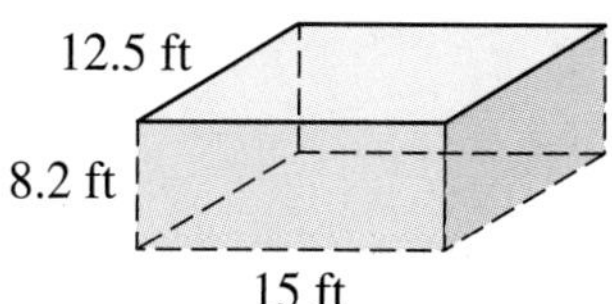

5. The side of a cube is 20 centimeters long.

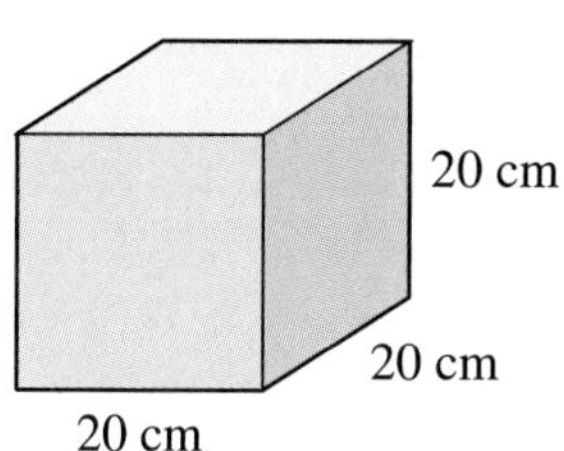

a. Find the area of one face of the cube.

b. How many faces does the cube have? Find the total surface area (the sum of the areas of all the faces) of the cube.

c. Find the volume of the cube.

6. a. How many square feet are in a square yard? (Draw a picture.)

b. How many square inches are in a square foot?

c. How many cubic feet are in a cubic yard?

d. How many cubic inches are in a cubic foot?

7. Add the terms *exponent* and *base* to your glossary.

8. What is a *power* of a number?

9. Explain the difference between $(-5)^2$ and -5^2.

10. What does the expression "five cubed" mean?

Write each product using exponents. (You do not have to compute the powers.)

11. $4 \cdot 4 \cdot 4 \cdot 4 \cdot 4 \cdot 4 \cdot 4 \cdot 4$

12. $9 \cdot 9 \cdot 9 \cdot 9 \cdot 9 \cdot 9$

13. $(0.5)(0.5)(0.5)$

14. $(1.7)(1.7)$

15. $(-1)(-1)$

16. $(-3)(-3)(-3)$

17. $a \cdot a \cdot a \cdot a \cdot a \cdot a$

18. $z \cdot z \cdot z \cdot z$

19. $2 \cdot 2 \cdot 2 \cdot 2 \cdot q \cdot q \cdot q$

20. $6 \cdot 6 \cdot 6 \cdot 6 \cdot 6 \cdot w \cdot w \cdot w \cdot w$

21. a. Compute the squares of all the integers from 1 to 10.

1^2	2^2	3^2	4^2	5^2	6^2	7^2	8^2	9^2	10^2

b. Compute the cubes of all the integers from 1 to 10.

1^3	2^3	3^3	4^3	5^3	6^3	7^3	8^3	9^3	10^3

22. a. Compute all the powers of 2 up to 2^{10}.

2^1	2^2	2^3	2^4	2^5	2^6	2^7	2^8	2^9	2^{10}

b. Compute all the powers of 3 up to 3^{10}.

3^1	3^2	3^3	3^4	3^5	3^6	3^7	3^8	3^9	3^{10}

23. a. Compute all the powers of 10 up to 10^5.

b. Describe how to write down the number 10^{24}.

24. a. Explain the difference between and 4(3) and 3^4.

b. Explain the difference between $4x$ and x^4.

Compute each power.

25. a. 2^3 **b.** 5^2 **c.** 7^2 **d.** 6^1

26. a. 3^4 **b.** 3^3 **c.** 2^5 **d.** 5^3

27. a. 2^4 **b.** 2^1 **c.** 4^3 **d.** 1^3

28. a. 1^4 **b.** 6^2 **c.** 8^2 **d.** 4^4

29. **a.** 9^2 **b.** 6^3 **c.** 10^1 **d.** 12^2

30. **a.** 10^2 **b.** 11^2 **c.** 13^1 **d.** 7^3

Compute each power.

31. **a.** $(-3)^2$ **b.** $(-2)^3$

32. **a.** $(-1)^3$ **b.** $(-1)^4$

33. **a.** $(-5)^3$ **b.** $(-9)^2$

34. **a.** $(-4)^2$ **b.** $(-4)^3$

35. **a.** -3^2 **b.** -2^3

36. **a.** -1^4 **b.** -1^5

37. **a.** -4^2 **b.** -4^3

38. **a.** -8^2 **b.** -3^3

Use your calculator to compute.

39. **a.** 2.6^2 **b.** 3.2^2 **c.** 4.9^3 **d.** 0.3^2

40. **a.** 8.5^4 **b.** 7.3^3 **c.** 2.1^4 **d.** 0.8^3

Solve. (See Lesson 4.3 to review solving equations.)

41. $\frac{x}{2} + 5 = 3 - 10$

42. $2x - 3 = 5 - 7$

43. $\frac{2x - 3}{-5} = -7$

44. $\frac{2x}{5} - 3 = -7$

45. $2(x - 3) - 5 = -7$

46. $-5(2x - 3) = -10$

47. **a.** Multiply 35 by 100.

b. Multiply 35 by 1000.

c. Multiply 35 by 10,000.

48. **a.** Multiply 0.263 by 100.

b. Multiply 0.263 by 1000.

c. Multiply 0.263 by 10,000.

49. Compute:
a. 0.074×10^2 **b.** 0.074×10^3 **c.** 0.074×10^4

50. Compute:
a. 2.65×10^2 **b.** 2.65×10^3 **c.** 2.65×10^4

Building Number Skills

We can reduce a fraction if we can arrange the shaded and unshaded pieces into larger groups of equal size. For example, to reduce $\frac{12}{18}$ we form groups of six pieces each:

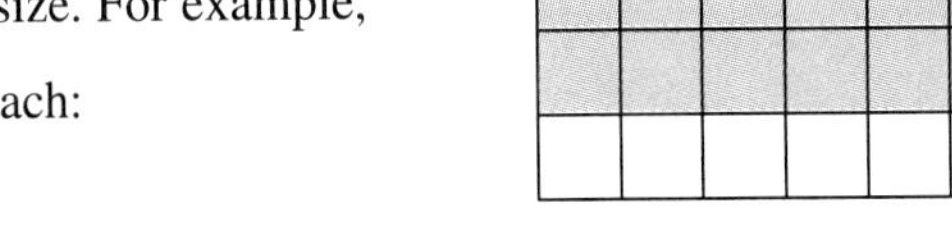

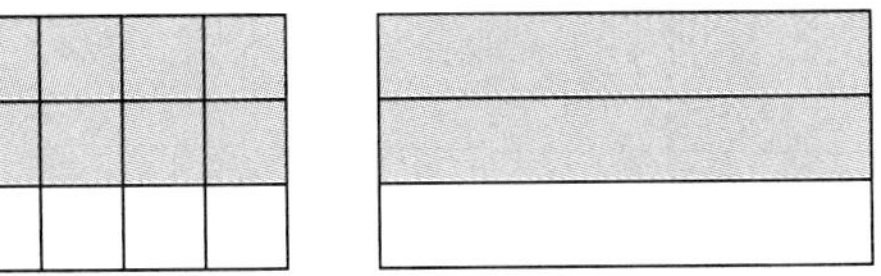

$$\frac{12}{18} = \frac{2}{3}$$

Use the diagrams to help you reduce each fraction.

51. $\frac{3}{6} =$

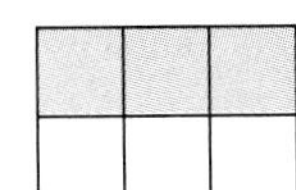

52. $\frac{6}{8} =$

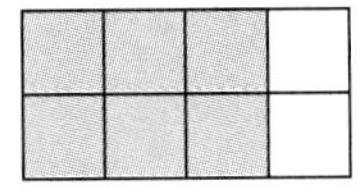

53. $\frac{6}{9} =$

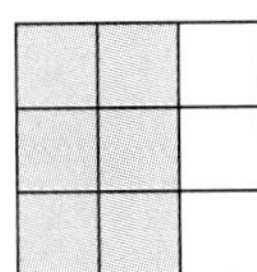

54. $\frac{4}{6} =$

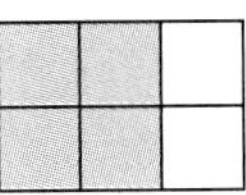

55. $\frac{2}{8} =$

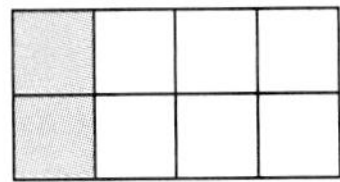

56. $\frac{3}{9} =$

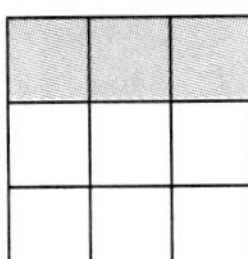

57. $\frac{15}{24} =$

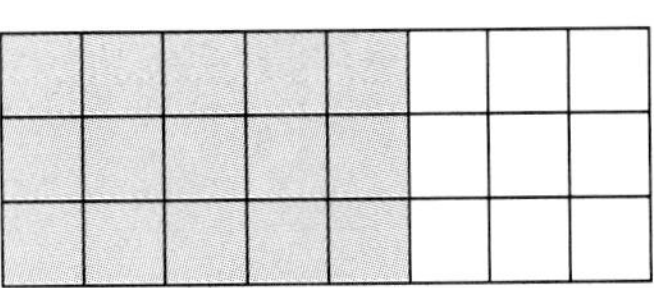

58. $\frac{8}{16} =$

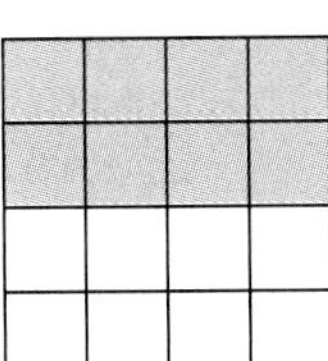

59. $\frac{10}{12} =$

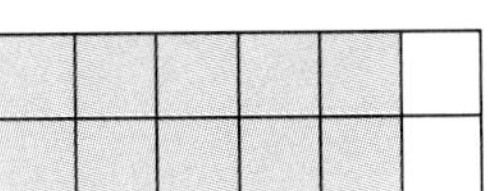

60. $\frac{12}{15} =$

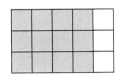

Do you see a way to do these exercises without using a diagram? Describe your method and give an example.

B. Scientific Notation

People who deal with information often encounter very large numbers in their work. For instance, the nearest star to our solar system is Proxima Centauri, a faint, reddish star at a distance of approximately 24,800,000,000,000 miles. Economists work with figures like the national debt, which has surpassed $5,656,000,000,000.

Calculations with such large numbers can be difficult. Most calculators have only an eight- or ten-digit display, so numbers in the trillions cannot even be entered. To get around this problem, we write numbers with a special notation that uses powers of 10.

Powers of 10

We can mentally multiply a number by a power of 10 by moving the decimal point to the right. For example, you can check that

$$3.62 \times 10^5 = 3\ 62\ 000. = 362{,}000$$

Move the decimal five places to the right.

We move the decimal point the same number of places as the exponent on 10.

EXAMPLE 5 

a. $28 \times 10^8 = 2\ 8\ 00\ 000\ 000.$ Move the decimal eight places to the right.

$= 2{,}800{,}000{,}000$

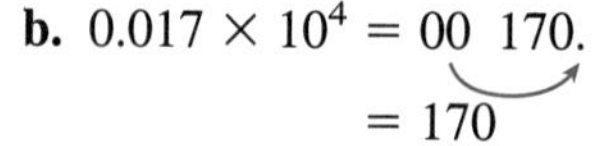

b. $0.017 \times 10^4 = 00\ 170.$ Move the decimal four places to the right.

$= 170$

EXERCISE 5

Simplify.

a. 1.23×10^5

b. 1.23×10^2

We can also reverse the process and write numbers with powers of 10. For example,

$$800 = 8 \times 100 = 8 \times 10^2 \qquad \text{and} \qquad 2000 = 2 \times 1000 = 2 \times 10^3$$

We figure out what power of 10 to use by counting backward to the new location of the decimal.

EXAMPLE 6 

Determine the correct power of 10.

a. $358{,}000 = 35.8 \times 10$___ **b.** $19{,}760{,}000 = 1.976 \times 10$___

Solution

a. Counting backward to the new location of the decimal, we see that we must move the decimal point four places:

$$35\ 8\ 000 \rightarrow 35.8$$

Four places

Therefore, we need an exponent of 4 for 10. Thus,

$$358{,}000 = 35.8 \times 10^4$$

b. Counting backward to the new location of the decimal, we see that we need an exponent of 7 for 10:

1 9 760 000 → 1.976

Seven places

Thus,

$$19{,}760{,}000 = 1.976 \times 10^7$$

Very Large Numbers

How would you compare several very large numbers? Let's consider three numbers:

50,000,000,000   5,000,000,000,000   500,000,000,000

Begin by counting the zeros in each number. We can write the numbers as

5×10^{10} 5×10^{12} 5×10^{11}

The number with the highest power of 10 is the largest.

Now compare the two numbers

37,000,000,000,000   and   2,174,925,800,000,000

Counting zeros is not helpful here. We need to compare the factors in front of the power of 10 as well. To do this, we *choose powers of* 10 *so that each number begins with a factor between* 1 *and* 10. Move the decimal places as shown:

3.7 000 000 000 000   and   2. 174 925 800 000 000

Counting backward to the decimal, we find the correct powers of 10:

3.7×10^{13} and 2.1749258×10^{15}

Now it is easy to see that the second number is larger.

This method for writing large numbers is called **scientific notation.** Regular numbers are said to be in *standard notation.*

To Write a Number in Scientific Notation

1. Move the decimal point so that there is only one nonzero digit to the left of the decimal.
2. Multiply by a power of 10, where the exponent on 10 is the number of places you moved the decimal.

EXAMPLE 7  

Write in scientific notation.

a. 245.3   **b.** 38,200,000,000

Solution

a. $245.3 = 2.453 \times 10^2$ Move the decimal two places.

b. $38{,}200{,}000{,}000 = 3.82 \times 10^{10}$ Move the decimal ten places.

EXERCISE 6

Determine the correct power of 10.

a. $1200 = 1.2 \times$ ______

b. $12{,}000{,}000 = 1.2 \times$ ______

EXERCISE 7

Write in scientific notation:

a. 98,700,000,000

b. 98.7

Using a Calculator

You can use scientific notation to enter very large numbers in your calculator. First, look for a key labeled EE or EXP. To enter the number 2.61×10^5, enter 2.61 as usual. Then press the EXP key. The calculator display should change to

2.61 00

Now enter *just the exponent* on the power of 10; for our example, we enter 5. The calculator should then display

2.61 05

This is the way most calculators display a number in scientific notation. You are supposed to know that 05 is the power of 10.

By entering the numbers in scientific notation, you can perform calculations with very large numbers.

EXAMPLE 8

Use your calculator to compute $(-32{,}700{,}000{,}000{,}000) \div (1{,}200{,}000{,}000)$.

Solution

First, convert each number to scientific notation:

$$-32{,}700{,}000{,}000{,}000 = -3.27 \times 10^{13} \quad \text{Move the decimal 13 places.}$$
$$1{,}200{,}000{,}000 = 1.2 \times 10^{9} \quad \text{Move the decimal 9 places.}$$

Enter the quotient with the following key strokes.

3.27 +/− EXP 13 ÷ 1.2 EXP 9 =

(We press the change sign key after 3.27 to make that number negative.) The calculator returns the quotient as

−2.725 04

which we interpret as -2.725×10^4, or $-27{,}250$.

EXERCISE 8

Use a calculator to compute

$$\frac{12{,}000{,}000{,}000{,}000}{30{,}000{,}000{,}000}$$

ANSWERS TO 5.1B EXERCISES

5a. 123,000

5b. 123

6a. 10^3

6b. 10^7

7a. 9.87×10^{10}

7b. 9.87×10^1

8. 400

HOMEWORK 5.1B

Write each number in standard notation.

1. 6×10^4

2. 9×10^2

3. 275×10^3

4. 93×10^5

5. 48.623×10^2

6. 17.837×10^3

7. 7.28×10^{12}

8. 6.259×10^{9}

9. 0.74×10^{8}

10. 0.35892×10^{4}

11. 0.0000006×10^{6}

12. 0.00094×10^{11}

Determine the correct power of 10.

13. 113,000 = 113 × ______

14. 2,720,000 = 272 × ______

15. 5,492,000 = 54.92 × ______

16. 60,940,000 = 60.94 × ______

17. 2,697.283 = 2.697283 × ______

18. 591.57 = 5.9157 × ______

19. 643,000,000,000,000,000 = 6.43 × ______

20. 1,386,700,000,000,000,000,000,000 = 1.3867 × ______

Write in scientific notation.

21. 3,500,000

22. 48,900

23. 27.6

24. 38.92

25. 67,520

26. 83,600,000

27. 7,920,000,000,000,000,000,000

28. 60,777,000,000,000,000

Use your calculator and scientific notation to perform the calculations.

29. (23,500,000,000)(187,000,000)

30. (9,478,000,000)(510,000,000,000)

31. (982,000,000,000,000) ÷ (4,000,000,000)

32. (260,000,000,000,000) ÷ (52,000,000,000,000)

33. $\frac{(36{,}000{,}000{,}000)(4{,}800{,}000)}{(14{,}400{,}000{,}000{,}000)}$

34. $\frac{(750{,}000{,}000{,}000)(160{,}000{,}000{,}000)}{(240{,}000{,}000{,}000{,}000{,}000)}$

35. $\frac{7.2 \times 10^{24}}{(1.8 \times 10^{15})(3.2 \times 10^{7})}$

36. $\frac{2.16 \times 10^{32}}{(6 \times 10^{14})(1.5 \times 10^{12})}$

37. $(8 \times 10^{12})^3$

38. $(5 \times 10^{9})^4$

Round your answers to two decimal places if necessary.

39. Light travels at a speed of 186,000 miles per second. How many seconds does it take light to reach Earth from Proxima Centauri? (See the lesson to find the distance to Proxima Centauri.) How many days is that?

40. The population of the United States is approximately 284,500,000. If we divide up the national debt equally among the population, how much is your share? (See the lesson to find the national debt.)

41. The mass of Earth is 13,200,000,000,000,000,000,000,000 pounds and its volume is 38,250,000,000,000,000,000,000 cubic feet. What is the density of Earth, in pounds per cubic foot? (Density is mass divided by volume.)

42. A 1-mile tall stack of bills contains about 19,008,000 bills. In November 1923 the circulation of Reichsbank marks in Germany was 400,338,326,350,700,000,000. How tall a stack of bills would that make?

43. a. Explain the difference between 3.2^4 and 3.2×10^4. Which number is expressed in scientific notation?

b. Simplify each of the numbers in part (a).

44. Are 6.5^2 and 6.5×10^2 the same number? Explain why or why not.

45. Choose the best number for each.

3.65×10^2	5.28×10^3	4.32×10^4
6.048×10^5	5.256×10^6	2.788×10^7

a. The number of feet in 1 mile.

b. The number of days in 1 year.

c. The number of seconds in 1 week.

d. The number of square feet in 1 square mile.

46. Choose the best approximation for each.

5×10^2	5×10^3	5×10^4
5×10^5	5×10^6	5×10^7

a. The population of the United States

b. The height of the tallest mountains, in feet

c. The distance from New York to Los Angeles, in miles

d. The number of hours in a century

Simplify. (See Lesson 4.5 to review order of operations on signed numbers.)

47. $-4(-6) - 3(-4)$

48. $-4(-6 - 3) - 4$

49. $15 - 3(8) - 11(-2)$

50. $(15 - 3)(8 - 11) - 2$

51. $6 - 3[-2(-5) - 3]$

52. $6 - 3[-2 - 5(-3)]$

Evaluate for $a = -4$ and $b = -3$.

53. $(a - b)(a + b)$

54. $a - ab + b$

55. $a(ab - b)$

56. $a - b[a - b(a - b)]$

Building Number Skills

Convert each percent to a fraction and reduce.

Example: $35\% = \frac{35}{100} = \frac{7}{20}$

57. 80%

58. 60%

59. 45%

60. 85%

61. 24%

62. 72%

63. 38%

64. 74%

65. 96%

66. 8%

5.2 Order of Operations

A. Simplifying Expressions

How should we simplify the following expression?

$$3 \cdot 5^2$$

Should we multiply 3 times 5 and then square the result, or should we square 5 first and then multiply by 3? In other words, which option is correct?

Option 1		*Option 2*	
$3 \cdot 5^2$	Multiply first.	$3 \cdot 5^2$	Square first.
$\rightarrow 15^2$	Then square.	$\rightarrow 3 \cdot 25$	Then multiply.
$= 225$		$= 75$	

Because an exponent applies only to its base, we choose option 2.

Powers should be computed before multiplications (or divisions.)

If we really want to multiply before computing the power, we can use parentheses around the product. Compare the two calculations in Example 1.

EXAMPLE 1

Simplify.

a. $3 \cdot 5^2 = 3 \cdot 25$ Compute the power first.
$= 75$ Multiply.

b. $(3 \cdot 5)^2 = 15^2$ Multiply inside the parentheses.
$= 225$ Compute the power.

EXERCISE 1
Simplify.
a. $7 \cdot 2^3$

b. $(7 \cdot 2)^3$

From Example 1, we see that operations inside parentheses come before raising to powers. With these new guidelines, our revised order of operations looks like this:

Order of Operations

1. First, perform all operations inside parentheses or above or below a fraction bar.
2. Next, compute all powers.
3. Next, perform all multiplications and divisions in order from left to right.
4. Finally, perform all additions and subtractions in order from left to right.

EXAMPLE 2  

Simplify.

a. $10 - 2 \cdot 3^2$ **b.** $(8 - 5^2)(8 - 5)^2$

Solution

a. $10 - 2 \cdot \mathbf{3^2}$ Compute the power first.
$= 10 - 2 \cdot 9$ Multiply.
$= 10 - 18$ Subtract.
$= -8$

b. $(8 - 5^2)(\mathbf{8 - 5})^2$ Subtract 8 − 5 inside parentheses.
$= (8 - 5^2)(3)^2$ Compute powers.
$= (8 - 25)(9)$ Subtract 8 − 25 inside parentheses.
$= (-17)(9)$ Multiply.
$= -153$

EXERCISE 2
Simplify $5 + 4 \cdot 3^2$

EXERCISE 3

Use a calculator to simplify

$$4(3.14)(1.5)^2$$

EXAMPLE 3

Use a calculator to simplify the expression $\frac{4}{3}(3.14)(4.5)^3$.

Solution

A scientific calculator is programmed to compute any powers before it computes products and quotients. Thus, if you have a scientific calculator, you can enter the expression just as it is written,

4 [÷] 3 [×] 3.14 [×] 4.5 [y^x] 3 [=]

and the calculator returns the answer 381.51. (On a graphing calculator, enter [^] instead of [y^x].)

If you do not have a scientific calculator, you must be careful to enter the operations in the correct order. First, compute the power $(4.5)^3$ and then multiply the result by $\frac{4}{3}(3.14)$.

Evaluating Expressions

Compare the two expressions

$$2x^3 \qquad \text{and} \qquad (2x)^3$$

In the first expression, the exponent 3 applies *only* to x. To evaluate the expression, we first raise x to the third power and then multiply by 2. On the other hand, the parentheses in the second expression tell us that the exponent 3 applies to $(2x)$, so we first compute $2x$, then cube the result.

EXERCISE 4

Evaluate each expression for $v = 6$.

a. $\frac{1}{2}v^2$

b. $\left(\frac{1}{2}v\right)^2$

EXAMPLE 4

Evaluate each expression for $x = 5$.

a. $2x^3$ **b.** $(2x)^3$

Solution

Substitute 5 for x in each expression.

a. $2x^3 = 2(5)^3$ First, compute the power.

$= 2(125)$ Multiply.

$= 250$

b. $(2x)^3 = (2 \cdot 5)^3$ Multiply inside the parentheses.

$= (10)^3$ Compute the power.

$= 1000$

CAUTION!

Compare the expressions

$$-z^4 \qquad \text{and} \qquad (-z)^4$$

In the first expression, the exponent 4 applies *only* to z, not to the minus sign! To evaluate $-z^4$ we first compute the power z^4, then multiply the result by -1. In other words,

$$-z^4 = -1 \cdot z^4$$

EXAMPLE 5

Evaluate each expression for $z = 3$.

a. $-z^4$ **b.** $(-z)^4$

Solution

Substitute 3 for z in each expression.

a. $-z^4 = -1 \cdot (3)^4$ First, compute the power.

$= -1 \cdot 81$ Multiply by -1.

$= -81$

b. $(-z)^4 = (-3)^4$ Raise -3 to the fourth power.

$= (-3)(-3)(-3)(-3)$

$= 81$

EXERCISE 5

Evaluate each expression for $k = 4$.

a. $-k^3$

b. $(-k)^3$

Remember that when we substitute a negative value for a variable, we should enclose the value in parentheses.

EXAMPLE 6  

Evaluate $5n^3 - n^2$, for $n = -3$.

Solution

Substitute -3 for n. Enclose -3 in parentheses:

$5n^3 - n^2 = 5(-3)^3 - (-3)^2$ Compute powers.

$= 5(-27) - 9$ Multiply.

$= -135 - 9$ Add.

$= -144$

EXERCISE 6

Evaluate $-16t^2 + 56t$, for $t = -2$.

ANSWERS TO 5.2A EXERCISES

1a. 56 **1b.** 2744 **2.** 41 **3.** 28.26

4a. 18 **4b.** 9 **5a.** -64 **5b.** -64 **6.** -176

HOMEWORK 5.2A

Evaluate. Compare the order of operations in each pair of expressions.

1. a. -4^2 **b.** $(-4)^2$

2. a. $5 \cdot 2^3$ **b.** $(5 \cdot 2)^3$

3. a. $6(-4)^2$ **b.** $(6 - 4)^2$

4. a. $3^2 + 4^2$ **b.** $(3 + 4)^2$

5. a. $12 - 4^2$ **b.** $12(-4)^2$

6. a. $2(-5^2)$ **b.** $2(-5)^2$

Use the order of operations to simplify.

7. $4 \cdot 2^3$ **8.** $5(-3)^2$ **9.** $10(-2)^4$ **10.** $6 \cdot 5^2$

11. $(3 \cdot 4)^2$ **12.** $(-2 \cdot 5)^3$ **13.** $(2 + 8)^2$ **14.** $2^2 + 3^2$

15. $3^2 - 6^2$ **16.** $(7 - 3)^2$ **17.** $3 + 5^2$ **18.** $12 - 2^3$

Choose the correct meaning for each expression.

19. a. $3x$ means $x + x + x$ or $x \cdot x \cdot x$.

b. x^3 means $x + x + x$ or $x \cdot x \cdot x$.

20. a. $(-x)^4$ means $-x - x - x - x$ or $(-x)(-x)(-x)(-x)$.

b. $4(-x)$ means $-x - x - x - x$ or $(-x)(-x)(-x)(-x)$.

Use the order of operations to simplify.

21. $4(2 - 3^2)$

22. $6(1 - 3)^2$

23. $4^2 + 8^2 \div 2^3$

24. $6^2 - 3^4 \div 3^3$

25. $(-2)(3) - 2^2$

26. $3(-5) - 2^3$

27. $14 - (-3)^2$

28. $18 - (-4)^2$

29. $5 - 4 \cdot 3^2$

30. $-6 + 2 \cdot 4^2$

31. $4 - (2 \cdot 3)^2$

32. $-(3 \cdot 2)^2 - 5$

For Problems 33–42, evaluate for $x = -2$ and $y = -3$.

33. x^4

34. y^3

35. $2y^4$

36. $5x^3$

37. $(-x)^2$

38. $-y^2$

39. $6 - 3x^2$

40. $1 + 3x^3$

41. $-2x^3 - x$

42. $-2y^2 + y$

43. Explain the difference between $2x$ and x^2. Illustrate with a numerical example.

44. Explain the difference between $3x$ and x^3. Illustrate with a numerical example.

Write an equation and solve it to answer the question. (See Lesson 4.4A to review problem solving with formulas.)

45. A roll of carpet contains 400 square feet of carpet. If the roll is 16 feet wide, how long is the piece?

46. The peregrine falcon is the fastest flying creature. At a 45° angle of descent, it can cover 312 yards in 3 seconds. What is the falcon's speed in yards per second? In miles per hour?

47. Beth's monthly income is $1800, and her rent is $576 per month. What percent of her monthly income does Beth spend on rent?

48. Andrea made a down payment of $1776 on her new car. If this was 12% of the price of the car, how much did the car cost?

49. In the 50 states, the average area devoted to state parks is 222,960 acres of land. How many acres of state park are there in the whole United States?

50. Latrisha invested $6500 in a certificate of deposit for 1 year and earned $308.75 interest. What rate of interest did the certificate earn?

Building Number Skills

Add the fractions and write your answer as a mixed number. (To review adding fractions, see Appendix A.3.)

Example: $\frac{5}{8} + \frac{7}{8} = \frac{12}{8}$ Add the numerators; keep the same denominator.

$= \frac{3}{2} = 1\frac{1}{2}$ Reduce the answer.

51. $\frac{2}{5} + \frac{4}{5}$

52. $\frac{5}{7} + \frac{4}{7}$

53. $\frac{7}{12} + \frac{11}{12}$

54. $\frac{9}{10} + \frac{7}{10}$

55. $\frac{7}{9} + \frac{8}{9}$

56. $\frac{8}{15} + \frac{11}{15}$

57. $\frac{5}{6} + \frac{5}{6}$

58. $\frac{5}{8} + \frac{5}{8}$

59. $\frac{9}{16} + \frac{11}{16}$

60. $\frac{19}{20} + \frac{17}{20}$

B. Circles and Spheres

Circumference of a Circle

The distance from the center of a circle to any point on the circle itself is called the **radius** of the circle. The **diameter** of a circle is the length of a line segment joining two points on the circle and passing through the center. The diameter of a circle is thus twice the radius (Figure 5.5).

Figure 5.5

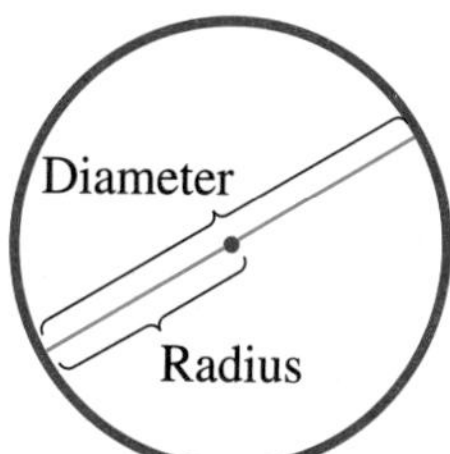

The perimeter of a circle is called its **circumference.** The longer the diameter of a circle, the bigger the circle will be and the larger its circumference. But is there a formula for finding the circumference of a circle in terms of its diameter?

If you measure the diameters and circumferences of various circles, you will find that the circumference is always a little more than 3 times the diameter. In algebraic language we say that

$$C = \pi d$$

where C stands for the circumference, d stands for the diameter, and the Greek letter π (pi, read "pie") stands for a number a little more than 3. Several ancient cultures, including the Egyptians and Babylonians, had reasonable approximations for the number π. However, its exact value cannot be written down as a fraction, and its decimal representation never ends. The first few digits of the number π are shown below:

$$\pi \approx 3.141592654...$$

Most scientific calculators have a key for π that you can use in calculations. If your calculator does not have a key for π, you can use the approximation $\pi \approx 3.14$.

The **circumference** C of a circle is given by

$$C = \pi d$$

where d is the **diameter** of the circle.

EXERCISE 7
Find the distance around the edge of a 10 peso coin with diameter 28 millimeters.

EXAMPLE 7  

Find the distance around the edge of a circular skating rink whose diameter is 50 feet.

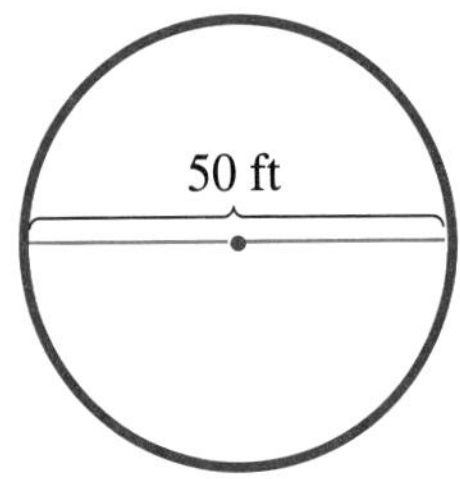

Solution

The distance around the edge is the circumference of the circle. Thus, we are looking for the circumference of a circle whose diameter is 50 feet:

$$C = \pi d = \pi(50)$$

Using a calculator to evaluate π yields $C = 157.07963$ feet, or approximately 157.08 feet. If we use the approximation 3.14 for π, we find

$$C = (3.14)(50) = 157 \text{ feet}$$

Since the diameter of a circle is twice the radius, or $d = 2r$, we can also express the circumference in terms of the radius:

The **circumference** of a circle is given by

$$C = 2\pi r$$

where r is the **radius** of the circle.

EXERCISE 8
A tree trunk has a circumference of 2 meters. What is the radius of the trunk?

EXAMPLE 8

The distance around a circular pond at the arboretum is approximately 500 yards. How far is it from the edge of the pond to the gazebo at its center?

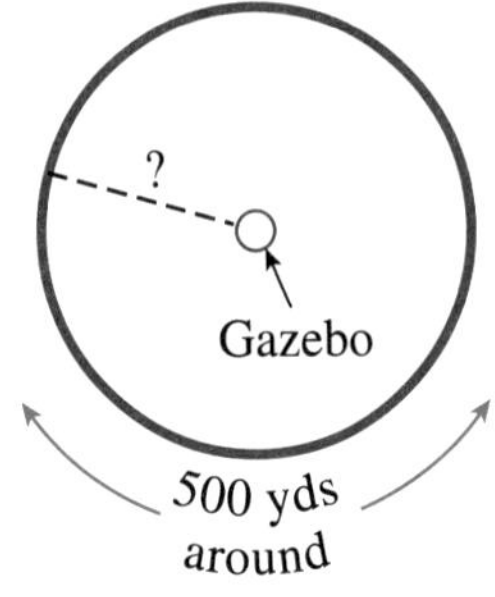

Solution

We are given the circumference of a circle and asked to find its radius. We use the formula $C = 2\pi r$ and substitute the value of C, 500 yards. We then solve the formula for r.

$$\begin{aligned} C &= 2\pi r && \text{Substitute 500 for } C. \\ 500 &= 2\pi r && \text{Divide both sides by 2.} \\ 250 &= \pi r && \text{Divide both sides by } \pi. \\ \frac{250}{\pi} &= r \end{aligned}$$

If you use your calculator's value for π, you will find $r \approx 79.58$ yards. If you use the approximation 3.14 for π, you will find $r \approx 79.62$ yards.

CAUTION!

In Example 8 you can solve the equation in one step by dividing both sides by 2π. Be careful, though; the following calculator keying sequence is *incorrect*. (Do you see why?)

500 [÷] 2 [×] [π] [=]

Area of a Circle

There is also a formula for finding the area of a circle if you know its radius.

> The **area** A of a circle is given by
>
> $$A = \pi r^2$$
>
> where r is the **radius** of the circle.

EXAMPLE 9 

Find the area of the skating rink in Example 7.

Solution

The diameter of the skating rink was 50 feet, so its radius is half that, or 25 feet. We substitute 25 for r in the formula for area:

$$\begin{aligned} A &= \pi r^2 \\ &= \pi(25)^2 && \text{Compute the power first.} \\ &= \pi(625) && \text{Multiply.} \\ &= 1963.50 \end{aligned}$$

The area of the skating rink is approximately 1963.50 square feet.

EXERCISE 9

If a slab from the tree trunk in Exercise 8 is used as a table top, what is the area of the table top?

Figure 5.6

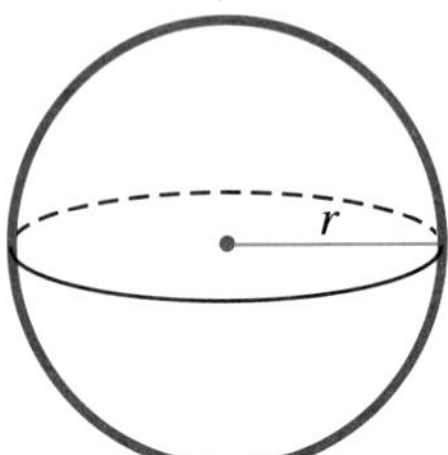

Volume of a Sphere

We can find the volume of a sphere if we know its radius.

The **volume** of a sphere is given by

$$V = \frac{4}{3}\pi r^3$$

where r is the **radius** of the sphere (Figure 5.6).

The volume of a figure measures how much space is inside the figure and is given in cubic units. It may seem difficult to measure the inside of a round object like a sphere in cubic units, but you can imagine filling the sphere with liquid and then pouring the liquid into a box to measure its volume.

EXERCISE 10

Find the volume of a basketball with a radius of 4.75 inches.

EXAMPLE 10  

Find the volume of a spherical water tank whose radius is 20 yards.

Solution

Substitute 20 for r in the formula for the volume of a sphere:

$$V = \frac{4}{3}\pi r^3 = \frac{4}{3}\pi(20)^3 \quad \text{Compute the power first.}$$

$$= \frac{4}{3}\pi(8000) \quad \text{Multiply 8000 by } \pi.$$

$$\approx \frac{4}{3}(25132.741) \quad \text{Multiply by 4 and divide by 3.}$$

$$\approx 33510.322$$

The volume is approximately 33,510.32 cubic yards.

ANSWERS TO 5.2B EXERCISES

7. $28\pi \approx 88$ millimeters

8. Approximately 0.32 meter

9. Approximately 0.32 square meter

10. Approximately 448.9 cubic inches

HOMEWORK 5.2B 

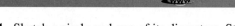

1. Sketch a circle and one of its diameters. State a formula for the circumference of a circle in terms of its diameter.

2. Sketch a circle and one radius. State a formula for the circumference of a circle in terms of its radius.

3. Sketch a circle and one radius. State a formula for the area of a circle in terms of its radius.

4. Sketch a sphere and one radius. State a formula for the volume of a sphere in terms of its radius.

Find the circumference of each circle. Round your answers to two decimal places.

5. $r = 8$ inches

6. $r = 3$ feet

7. $r = 6$ feet

8. $r = 5$ inches

9. $d = 1.2$ meters

10. $d = 10$ centimeters

11. $d = 25$ centimeters

12. $d = 3.8$ meters

Find the radius of each circle whose circumference is given. Round your answers to two decimal places.

13. $C = 60$ yards

14. $C = 8.7$ feet

15. $C = 42.5$ inches

16. $C = 100$ meters

Find the area of each circle. Round your answers to two decimal places.

17. $r = 4$ feet

18. $r = 7$ inches

19. $r = 0.2$ centimeter

20. $r = 0.5$ meter

21. $d = 25$ yards

22. $d = 13$ yards

Find the volume of each sphere. Write your answers in scientific notation.

23. $r = 2$ meters

24. $r = 6$ centimeters

25. $r = 0.01$ centimeter

26. $r = 0.03$ inch

27. $r = 520$ feet

28. $r = 250$ meters

Each problem involves an area, a circumference, or a volume. Decide which measure is appropriate and then solve the problem.

29. A circular pizza pan has a diameter of 14 inches. How much pizza dough is needed to cover it?

30. Francine wants to install a circular window in the stairwell in her house. The window has a diameter of 30 inches. How much weather stripping will Francine need to seal the window?

31. To find the diameter of a large tree in his yard, Delbert wraps a string around the trunk and then measures its length. If the string is 72.25 inches long, what is the diameter of the tree?

32. A rotating lawn sprinkler waters a circle whose edge is 8 feet from the sprinkler. How much lawn does the sprinkler water?

33. The first solo transatlantic balloon crossing was completed in 1984 in a helium-filled balloon called *Rosie O'Grady*. The diameter of the balloon was 58.7 feet. Assuming that the balloon was approximately spherical, calculate its volume.

34. The circumference of a spherical candle is 15.7 inches. What is the volume of wax in the candle?

35. Evaporation from a body of water depends upon the exposed area of the surface. To calculate the area of a large circular fish pond, Graham measures the distance around the edge of the pond as 75.4 feet. Calculate the area of the pond.

36. Francine has 200 yards of fencing material. How much space can she enclose for her horse if she makes a circular pen?

37. The diameter of the Sun is approximately 1,391,400 kilometers.

a. Write the diameter of the Sun in scientific notation. What is the radius of the Sun?

b. What is the volume of the Sun? Write your answer in scientific notation and in standard notation.

38. The distance from Earth to the Sun is approximately 93,000,000 miles.

a. Write the distance from Earth to the Sun in scientific notation.

b. Assuming that Earth revolves around the Sun in a circular orbit, calculate the distance that Earth travels around the Sun in 1 year. Write your answer in scientific notation and in standard notation.

Find the area and perimeter of each figure. Round your answers to two decimal places.

39. 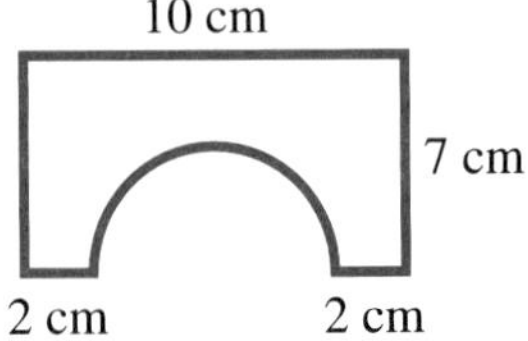

40. 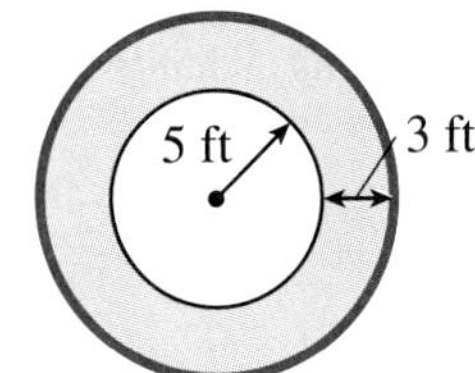

41. 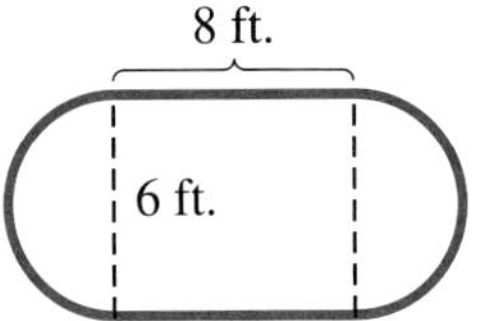

42. 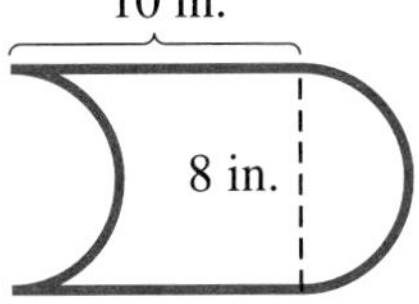

43. 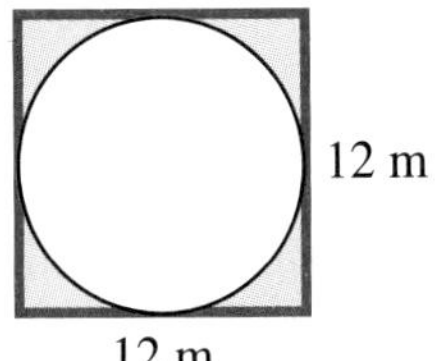

44. 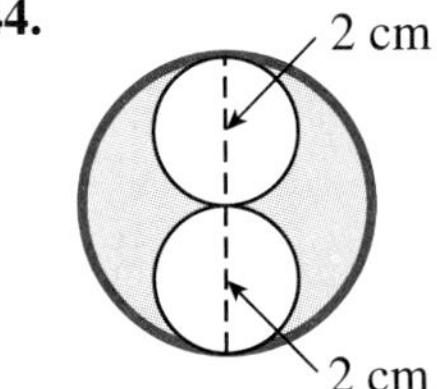

45. The quilt pattern in Figure 5.7 is called Drunkard's Path. The design is drawn on a four-patch block that is 30-centimeters square. The radius of each quarter-circle is four-fifths of the side of the patch it sits in. Find the total area of the blue pieces.

Figure 5.7

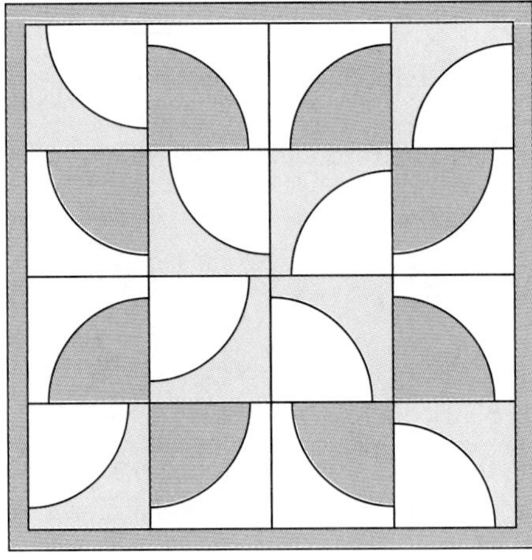

46. The miniature quilt in Figure 5.8 is called Birdhouse Row. Each birdhouse is a square $6\frac{1}{2}$ inches on a side. The blue roof is $1\frac{1}{2}$ inches wide, and the small patch of yellow at the bottom of the birdhouse is an isosceles triangle with two sides of length $1\frac{1}{2}$ inches. The door is a circle of diameter $1\frac{1}{2}$ inches. Find the total area of the red pieces in the quilt.

Figure 5.8

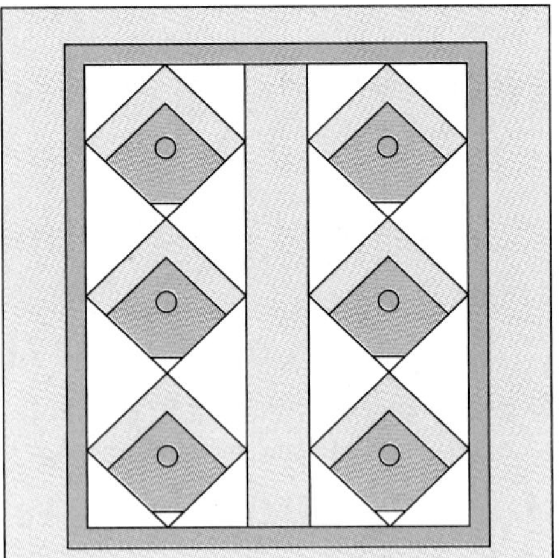

47. Angela has fenced off a 15-foot-square vegetable garden in the corner of her backyard. Her pet goat, Capri, is tethered by a 15-foot rope to the corner of the garden fence, as shown in Figure 5.9. What is the area of grass that Capri can reach?

Figure 5.9

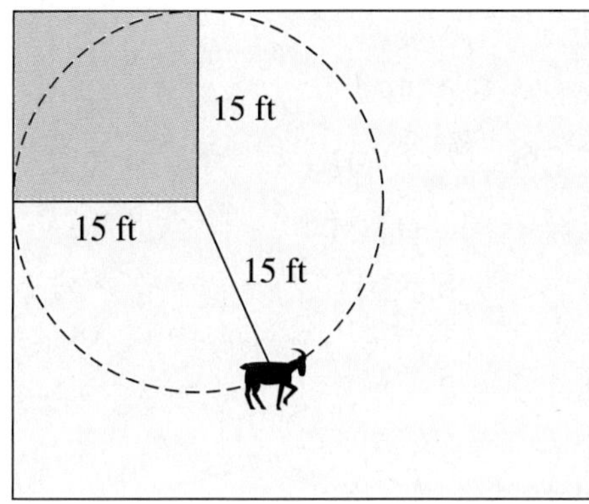

48. Delbert's front lawn is 40 feet long and 20 feet wide. He installs sprinklers around the perimeter, as shown in Figure 5.10. If the sprinklers are adjusted so that there is no overlap in their range, what is the area of the lawn that will be missed?

Figure 5.10

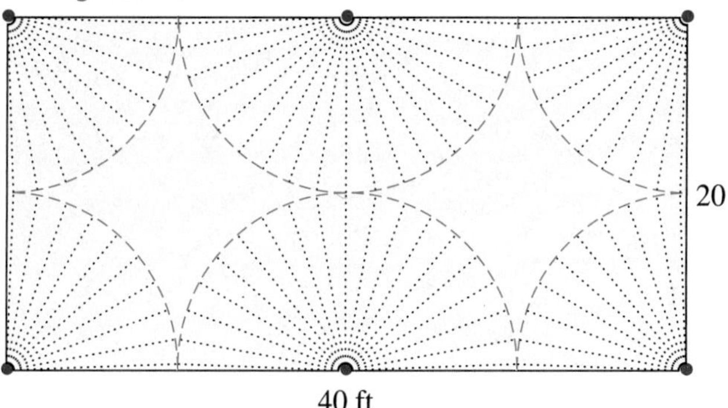

Simplify.

49. $3^2 - (2 - 4)^2$

50. $-4^2(-3 - 2)^2$

51. $-2 - 3(-3)^3 - 2$

52. $3 - 4(-2)^3(-3)$

53. $4^2 - (-4^2 - 4)^2$

54. $2^3 - (2 - 2^3)^3$

Evaluate for $x = -3$ and $y = -5$.

55. $x^2(x - 2y)$

56. $y(x^2 + y)$

57. $-x^2 - xy^2$

58. $-xy(x - y)^2$

59. $x - y(x^2 + y)^2$

60. $xy^3 + (xy)^3$

Building Number Skills

Add or subtract the mixed numbers by combining the whole numbers and then combining the fractions.

Example: $6\frac{3}{5} - 4\frac{2}{5} = (6 - 4) + \left(\frac{3}{5} - \frac{2}{5}\right)$ Combine the whole numbers and then combine the fractions.

$= 2 + \frac{1}{5} = 2\frac{1}{5}$

61. $2\frac{1}{9} + 5\frac{4}{9}$

62. $3\frac{2}{7} + 1\frac{3}{7}$

63. $8\frac{2}{3} - 5\frac{1}{3}$

64. $7\frac{8}{11} - 6\frac{3}{11}$

65. $1\frac{3}{8} + 6\frac{3}{8}$

66. $5\frac{2}{9} + 2\frac{4}{9}$

67. $9\frac{5}{6} - 4\frac{1}{6}$

68. $4\frac{9}{10} - 1\frac{7}{10}$

69. $3\frac{2}{5} + 6\frac{3}{5}$

70. $2\frac{5}{8} + 2\frac{3}{8}$

5.3 Like Terms

A. Simplifying Expressions

Like Terms

You know from arithmetic that you can combine quantities of the same type. For example, if you have three cats and your roommate has two cats, then together you have (3 + 2) cats, or 5 cats. Or, suppose tuition at your college is $20 per unit. If you are enrolled in eight units this semester and plan to enroll in ten units next semester, then you can figure your tuition for the year by adding the fees for each semester:

8($20) + **10**($20), or **18**($20)

Either calculation gives your total tuition for the year, $360.

Now suppose that next year the tuition at your college is going up to some unknown figure, x dollars per unit. If you take the same number of units next year, your tuition will be

$$8x + 10x, \quad \text{or} \quad 18x \text{ dollars}$$

We can add up multiples of x dollars exactly the same way we added up multiples of $20. In this expression, $8x$ and $10x$ are called **like terms** because the variable part of each term, x, is the same. The numbers multiplied by the variable, 8 and 10 in this example, are called the **coefficients** of the variable.

To Add or Subtract Like Terms

1. Add or subtract the coefficients.
2. Do not change the variable part of the terms.

EXAMPLE 1

Add like terms.

a. $9m - 4m$ **b.** $-6st - 8st$

Solution

In each case, add the coefficients of the terms. The variable part of the terms remains the same.

a. $9m - 4m = 5m$ because $9 - 4 = 5$.

b. $-6st - 8st = -14st$ because $-6 - 8 = -14$.

EXERCISE 1

Add like terms: $-4ab + 11ab$

Note that since

$$x = 1 \cdot x \quad \text{and} \quad -x = -1 \cdot x,$$

the coefficient of x is 1 and the coefficient of $-x$ is -1. Also note that since zero times any number is zero, we have

$$0 \cdot x = 0$$

EXERCISE 2
Combine like terms: $pq - 5pq$

EXAMPLE 2

Combine like terms.

a. $3x + x = 4x$ because $3 + 1 = 4$.
b. $-w + 3w = 2w$ because $-1 + 3 = 2$.

Equivalent Expressions

Combining like terms gives us a simpler expression that is *equivalent* to the original. **Equivalent expressions** have the same value when we substitute a number for the variable. It is always a good idea to replace complicated expressions with simpler ones if we can.

We *cannot* combine *unlike* terms. (If you have three cats and your roommate has two canaries and you try to combine them, you will still have three cats and two canaries, if you're lucky.) If the variable parts of the terms are not identical, then they are not like terms, and they cannot be combined. Here are some examples of *unlike* terms:

$3x$ and $5y$ are *not* like terms.
$6a$ and $-2ab$ are *not* like terms.
$-4z$ and -8 are *not* like terms.

So, expressions such as

$$3x + 5y \quad \text{or} \quad 6a - 2ab \quad \text{or} \quad -4z - 8$$

cannot be rewritten in any simpler way; they just stay the way they are.

If an expression has some like terms and some unlike terms, we can combine the like terms to simplify the expression.

EXERCISE 3
Simplify by combining like terms:
$5rs - 2r + 3 - 2rs + r - 1$

EXAMPLE 3  

Simplify by combining like terms: $2w - 7 + 6wz - 8w - 5z + wz - 2$

Solution

Rearrange the terms of the expression so that like terms are together. Remember that a + or − sign applies to the term that comes *after* it.

$$2w - 8w - 7 - 2 + 6wz + wz - 5z$$

Now combine like terms by adding their coefficients:

$$2w - 8w = -6w$$
$$-7 - 2 = -9$$
$$6wz + wz = 7wz$$

Because $-5z$ is the only z term, we cannot combine it with any other term. The simplified expression is

$$-6w - 9 + 7wz - 5z$$

None of these terms are like terms, so the expression cannot be simplified any further. The order of the terms in the answer does not matter.

Constant Multiples of Terms

Recall that $3x$ means 3 times x, or

$$x + x + x$$

because multiplication is really just repeated addition. Similarly, $3(2x)$ means 3 times $2x$, or

$$2x + 2x + 2x$$

By combining like terms, we see that $2x + 2x + 2x = 6x$. Thus,

$$3(2x) = 6x$$

Thus, to multiply a term by a constant, we multiply the coefficient of the term by the constant.

EXAMPLE 4

The members of the Concert Choir agree that each member of the choir should sell five tickets to the spring concert.

a. If each ticket sells for x dollars, how much will each member make from ticket sales?

Each member will earn $5x$ dollars.

b. If the choir has 20 members, how much will they earn from ticket sales?

The choir will earn $20(5x)$, or $100x$ dollars.

The same procedure applies if the constant or the coefficient is a negative number.

EXAMPLE 5

a. $-6(7a) = -42a$ **b.** $-2.5(-6.4q) = 16q$

Order of Operations

When we simplify algebraic expressions involving variables, we follow the order of operations.

EXAMPLE 6 

Simplify $8b + 6 - 4(5b) + 9(11b - 7b)$

Solution

Follow the order of operations:

$$8b + 6 - 4(5b) + 9(\mathbf{11b - 7b}) \quad \text{Subtract inside parentheses.}$$
$$= 8b + 6 - \mathbf{4(5b)} + \mathbf{9(4b)} \quad \text{Multiply.}$$
$$= 8b + 6 - 20b + 36b \quad \text{Add: Combine like terms.}$$
$$= 24b + 6$$

EXERCISE 4

Every staff member will carry home two sets of new regulations from the meeting. Each set of regulations weighs r ounces.

a. How many ounces will each staff member carry home?

b. If there are nine staff members, what is the total weight of regulations they carry home?

EXERCISE 5

Simplify $3(-14w)$

EXERCISE 6

Simplify $1 + 5(m - 4m) - 6(-2m)$

ANSWERS TO 5.3A EXERCISES

1. $7ab$

2. $-4pq$

3. $3rs - r + 2$

4a. $2r$ ounces

4b. $18r$ ounces

5. $-42w$

6. $1 - 3m$

HOMEWORK 5.3A 

1. Explain what *like terms* are. Give an example of like terms and an example of unlike terms.

2. What is a *coefficient*? Give an example.

True or false.

3. Like terms must have the same coefficient.

4. A plus (+) or minus (−) sign applies to the term that follows it.

5. The coefficient of the term ab is 1.

6. When we combine like terms, the variable parts of the terms are altered as well as the coefficients.

Choose a value for each variable and show that the two expressions are the same for that value. (Do *not* choose 0 or 1.)

7. $5y + 3y$; $8y$

8. $12c - 10c$; $2c$

9. $-9g + 2g$; $-7g$

10. $-4v - 6v$; $-10v$

Add or subtract like terms.

11. $4y + 2y$

12. $3y + 4y$

13. $-6x + 2x$

14. $7x - 9x$

15. $-8b + 8b$

16. $4b - 4b$

17. $-3pq + 12pq$

18. $-8st + 11st$

19. $-32W - 47W$

20. $-29K - 28K$

21. $-7.6a - 5.2a$

22. $-6.1a - 4.3a$

23. $-12.7x - (-3.3x)$

24. $4.2x - (-5.5x)$

25. $3x - 4x + 2x$

26. $-2x + 6x - x$

27. $-ab + 5ab - (-3ab)$

28. $3bc - (-4bc) - 8bc$

Combine like terms.

29. $6t + 3 - 4t$

30. $2 + 5t - 8t$

31. $3 + 4y - (-8y) - 7$

32. $7 - 4y + 6y - (-8)$

33. $-2st + 5s - 6st - (-4s)$

34. $-5u - 6uv + 8uv + 9u$

Simplify.

35. $6(5d)$

36. $10(7v)$

37. $-8(-3h)$

38. $-9(3r)$

39. $0.25(16a)$

40. $0.30(800x)$

41. $20(0.50m)$

42. $100(0.75n)$

43. **a.** Explain the difference between $4x + 7x$ and $4(7x)$.

b. Evaluate both of the expressions in part (a) for $x = -3$.

44. **a.** Explain the difference between $-3a - 2a$ and $-3(-2a)$.

b. Evaluate both of the expressions in part (a) for $a = 4$.

Simplify. Follow the order of operations.

45. $25s - 3(5s)$

46. $-18u + 3(9u)$

47. $16 - 6(4y)$

48. $32 - 8(9c)$

49. $-7(2h) - (6h - 9h)$

50. $4(-3g) - (8g + 5g)$

51. $-4 - 4(4t - 4t) - 4t$

52. $6 - 6(-6z) + 6z - 6$

53. $(5 - 9)a - 5 - 9a - 5(-9a)$

54. $2w - (-2w - 2w) - 2(w - 2w)$

Simplify. (See Lesson 5.2A to review the order of operations with exponents.)

55. $8(-6)^2$

56. $8(-6^2)$

57. $12 - 8(-6)^2$

58. $12 - (8 - 6)^2$

59. $(12 - 8) - 6^2$

60. $12(-8)(-6^2)$

61. $-2^3(-2)^2 - 2^4$

62. $(-2^3 - 2)^2 - (-2)^4$

Building Number Skills

You know that there are 60 minutes in 1 hour. Use the diagram to help you compute the number of minutes in each fraction of an hour.

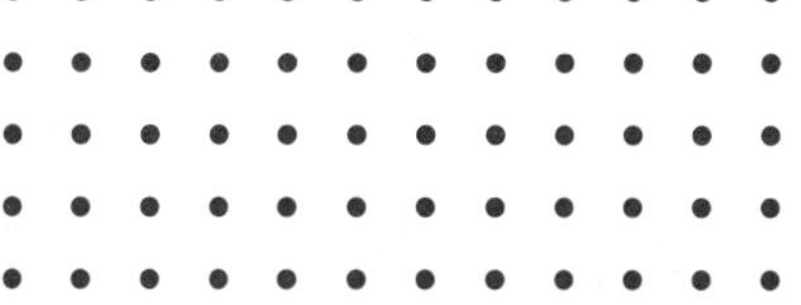

63. $\frac{1}{3}$

64. $\frac{1}{4}$

65. $\frac{1}{5}$

66. $\frac{1}{12}$

67. $\frac{5}{6}$

68. $\frac{7}{10}$

69. $\frac{5}{12}$

70. $\frac{3}{5}$

71. $\frac{8}{15}$

72. $\frac{11}{15}$

B. Terms with Exponents

Combining Like Terms

In *like terms* the variable parts of the terms must be exactly the same. This applies to any exponents involved as well. For example,

$$5x^2 \quad \text{and} \quad -2x^2 \quad \text{are like terms.}$$
$$xy^3 \quad \text{and} \quad 4xy^3 \quad \text{are like terms.}$$

However,

$$6x \quad \text{and} \quad 3x^2 \quad \text{are } not \text{ like terms.}$$

To add or subtract like terms, we add or subtract their coefficients and leave the variable part of the terms unchanged.

EXERCISE 7
Combine like terms: $-7r^3 + 8r^3$

EXAMPLE 7  

Combine like terms.

a. $4x^2 + 3x^2$ **b.** $-7a^2b + 5a^2b$

Solution

a. Add the coefficients, $4 + 3$, and leave the variable part, x^2, unchanged:

$$4x^2 + 3x^2 = (4 + 3)x^2 = 7x^2$$

b. Add the coefficients, $-7 + 5$, and leave the variable part, a^2b, unchanged:

$$-7a^2b + 5a^2b = (-7 + 5)a^2b = -2a^2b$$

CAUTION!

When adding like terms, we do not add the exponents. For example,

$$4x^2 + 3x^2 = 7x^4 \text{ is } incorrect.$$

We cannot add or subtract *unlike* terms.

EXERCISE 8
Simplify if possible: $3m^2n + 5mn^2$

EXAMPLE 8  

a. $3t + 5t^3$ cannot be simplified.
b. $-2wz^2 - 2w^2z$ cannot be simplified.

Multiplying Variable Expressions

What about multiplying variable expressions? We will need one fact from arithmetic: If we want to multiply several factors, we can rearrange them in any order and get the same product. For example,

$$(2)(3)(4)(5) = (6)(4)(5)$$
$$= (24)(5) = 120$$

Now let's rearrange the factors and multiply again:

$$(4)(3)(5)(2) = (12)(5)(2)$$
$$= (60)(2) = 120$$

The product is the same.

This fact is called the **commutative law,** and it works with variables, too. (You can review the laws of arithmetic in Appendix A.10.)

EXAMPLE 9 

Multiply $(2a)(3a)$

Solution

Rearrange the factors with the coefficients together and the variable factors together:

$$(2a)(3a) = 2 \cdot 3 \cdot a \cdot a$$

Now we can multiply $2 \cdot 3$ to get 6 and $a \cdot a$ to get a^2. Thus,

$$(2a)(3a) = 2 \cdot 3 \cdot a \cdot a = 6a^2$$

EXERCISE 9

Multiply $(7q)(-2q)$

To multiply two powers with the same base, we need only recall what exponents mean.

EXAMPLE 10  

Multiply $b^3(b^4)$

Solution

Recall that b^3 means $b \cdot b \cdot b$ and b^4 means $b \cdot b \cdot b \cdot b$. Consequently, their product is

$$b^3 \cdot b^4 = b \cdot b \cdot b \cdot b \cdot b \cdot b \cdot b, \quad \text{or} \quad b^7$$

Thus, $b^3 \cdot b^4 = b^7$.

EXERCISE 10

Multiply $a^2 \cdot a^5$

On the other hand, we *cannot* simplify a product of two powers with *different bases.* For instance, the product $a^3 \cdot b^4$ cannot be simplified; we simply write a^3b^4.

ANSWERS TO 5.3B EXERCISES

7. r^3

8. Cannot be simplified.

9. $-14q^2$

10. a^7

CAUTION!

It is important to distinguish between multiplying and adding when using algebraic expressions. Remember that

$$2x \quad \text{means} \quad x + x \quad \text{Addition}$$

whereas

$$x^2 \quad \text{means} \quad x \cdot x \quad \text{Multiplication}$$

For example, if $x = 5$,

$$2x = 5 + 5 = 10 \quad \text{and} \quad x^2 = 5 \cdot 5 = 25$$

HOMEWORK 5.3B 

Simplify by combining like terms.

1. $6w^3 - 9w^3$
2. $-4s^2 + 7s^2$
3. $-8bc + 5bc$
4. $uv - 2uv$
5. $-pq^2 - pq^2$
6. $-3u^3v - 3u^3v$

Simplify by combining like terms. If there are no like terms, write "cbs" ("cannot be simplified").

7. $2a + 6a^2 - 3a - 5a^2$
8. $-d^3 - d^2 + 2d^3 - d^2$
9. $-4yz + 3yz^2 + 4yz + 2y^2z$
10. $6pq + 2pq^2 - 2p^2q - 6pq$
11. $5x^2 - 6x + 2$
12. $2y^2 + 4y - 1$
13. $3m + m^3 + 3m^3 + 3$
14. $2z^4 - 4z^2 - 4z^4 - 2z^2$
15. $b + b^2 - ab - ab^2$
16. $2hk - 4h + 4k - 2hk$

Find the product.

17. $m \cdot m^2$

18. $n^2 \cdot n^2$

19. $k^3 \cdot k^3$

20. $h^4 \cdot h$

21. $(6x^2)(-5x^3)$

22. $(-3y^2)(-4y^4)$

23. $-4p(3q^2)$

24. $2g^2(-3h^2)$

25. $s^3t^2(st^2)$

26. $ac^3(a^3c)$

27. $-6x^3z^2(-3zx^2)$

28. $-8w^4p(3p^2w^2)$

Simplify each expression if possible. If the expression cannot be simplified, write "cbs" ("cannot be simplified").

29. a. $x + x$ **b.** $x \cdot x$

30. a. $x^2 + x^2$ **b.** $x^2(x^2)$

31. a. $x + x^2$ **b.** $x(x^2)$

32. a. $2x + 3x$ **b.** $2x(3x)$

33. a. $x^2 + x^3$ **b.** $x^2(x^3)$

34. a. $x + y$ **b.** $x \cdot y$

35. **a.** $x + xy$ **b.** $x(xy)$

36. **a.** $xy + xy$ **b.** $xy(xy)$

37. **a.** $2x + 3y$ **b.** $2x(3y)$

38. **a.** $x^2y + xy^2$ **b.** $x^2y(xy^2)$

Evaluate each group of expressions for the given value. Which pair of expressions in each group is equivalent?

39. For $a = 3$: **a.** $a^2 + a^2$ **b.** $2a^2$ **c.** a^4

40. For $b = 2$: **a.** $b^2(b^3)$ **b.** b^5 **c.** b^6

41. For $v = 5$: **a.** $2v(3v)$ **b.** $6v$ **c.** $6v^2$

42. For $t = 4$: **a.** $2t + 3t$ **b.** $5t^2$ **c.** $5t$

43. For $w = 2$: **a.** $5w^2(-2w^2)$ **b.** $3w^2$ **c.** $-10w^4$

44. For $q = 4$: **a.** $-6q^2 + 3q^2$ **b.** $-3q^2$ **c.** -3

Find the perimeter of each figure.

45.

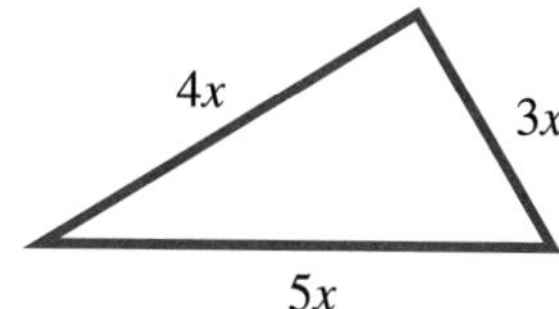

46.

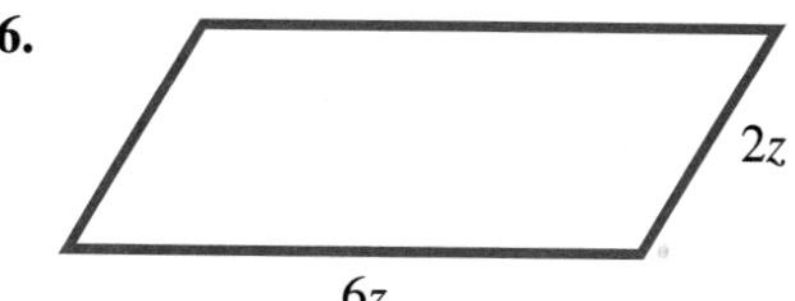

Find the area of each figure.

47.

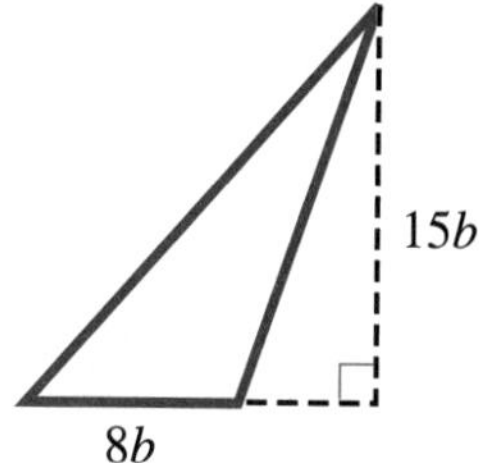

48.

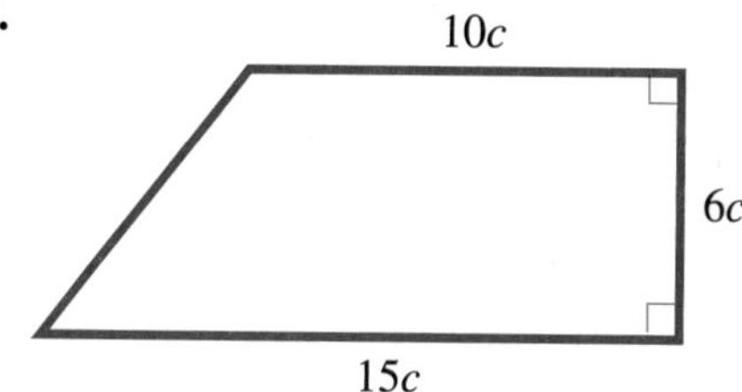

Find the area and perimeter of each rectangle.

49.

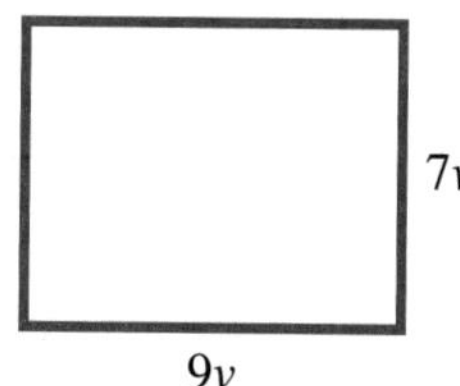

50.

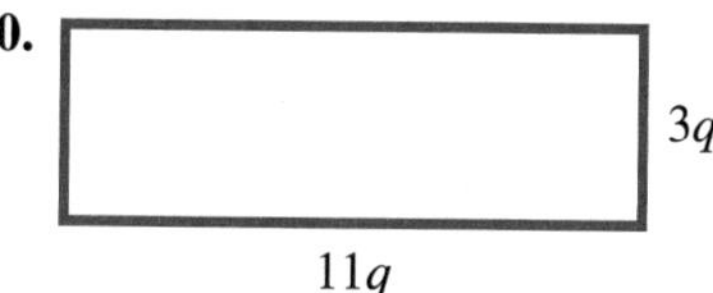

Building Number Skills

Use a diagram to find each product.

Example: $\frac{2}{3}\left(\frac{5}{8}\right) = \mathbf{\frac{10}{24}}$ (Can this product be reduced?)

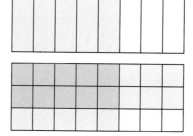

1. First draw $\frac{5}{8}$ vertically.
2. Divide into three parts horizontally and shade two of them.
3. What fraction is shaded both ways? Ten shaded out of 24 total pieces.

51. $\frac{1}{3}\left(\frac{4}{5}\right)$

52. $\frac{1}{2}\left(\frac{3}{5}\right)$

53. $\frac{3}{4}\left(\frac{2}{3}\right)$

54. $\frac{3}{4}\left(\frac{5}{6}\right)$

55. $\frac{2}{3}\left(\frac{2}{5}\right)$

56. $\frac{2}{3}\left(\frac{3}{8}\right)$

57. $\frac{3}{8}\left(\frac{1}{3}\right)$

58. $\frac{5}{8}\left(\frac{4}{5}\right)$

59. $\frac{2}{5}\left(\frac{5}{8}\right)$

60. $\frac{3}{5}\left(\frac{5}{6}\right)$

Do you see a way to do these exercises without using a diagram? Describe your method and give an example.

5.4 Square Roots

A. What Is a Square Root?

The area of the square in Figure 5.11 is 25 square inches. What is the length s of each side of the square? Since the area of a square is given by the formula $A = s^2$, we are looking for a number s whose square is 25. Or, as an equation,

$$s^2 = 25$$

You can probably see that the number we need is 5, or $s = 5$, since $5^2 = 25$. The number 5 is called the square root of 25.

Figure 5.11

$A = 25$ $s = ?$

s is called a **square root** of b if $s^2 = b$.

EXAMPLE 1

a. 3 is a square root of 9 because $3^2 = 9$.

b. 8 is a square root of 64 because $8^2 = 64$.

c. $\frac{2}{3}$ is a square root of $\frac{4}{9}$ because $\left(\frac{2}{3}\right)^2 = \frac{4}{9}$.

EXERCISE 1
What positive number is a square root of 16?

You may have noticed that 5 is not the only square root of 25. In fact, -5 is another square root of 25 because

$$(-5)^2 = (-5)(-5) = 25$$

Every positive number has two square roots, one positive and one negative.

For most applications, only the positive square root makes sense. (For instance, we cannot have a square whose length is -5 inches.) To distinguish between the two square roots of a number, we have a special symbol called a **radical sign** to denote the *positive* square root of a number. Thus,

$\sqrt{16}$ means the *positive* square root of 16

so

$$\sqrt{16} = 4$$

If we want the *negative* square root of a number, we write $-\sqrt{16} = -4$.

EXAMPLE 2 

a. $\sqrt{144} = 12$

b. $-\sqrt{36} = -6$

c. $-\sqrt{1} = -1$

d. $\sqrt{17^2} = 17$

EXERCISE 2
Evaluate $\sqrt{4}$

Square Roots of Negative Numbers

What about the square root of a negative number? What is the square root of -9? Let's check the only reasonable possibilities, 3 and -3:

$$3^2 = 9 \quad \text{and} \quad (-3)^2 = 9$$

Because the square of *any* number is positive (or zero), we cannot find a number whose square is -9. This means that -9 has no square root, and neither does any other negative number.

The square root of a negative number is undefined.

CAUTION!

Be careful when using radical notation:

$-\sqrt{9}$ means the opposite of the square root of 9,

but

$\sqrt{-9}$ means the square root of negative 9.

Thus,

$$-\sqrt{9} = -3 \quad \text{but} \quad \sqrt{-9} \quad \text{is undefined.}$$

Order of Operations

Square roots occupy the same position in the order of operations as exponents do: after parentheses but before multiplications and divisions.

Order of Operations

1. First, perform all operations inside parentheses, above or below a fraction bar, or inside a radical.
2. Next, compute all powers and roots.
3. Next, perform all multiplications and divisions in order from left to right.
4. Finally, perform all additions and subtractions in order from left to right.

EXAMPLE 3

Simplify.

a. $4 + 3\sqrt{49}$

b. $\dfrac{6 + \sqrt{64}}{9 - \sqrt{4}}$

Solution

a. $4 + 3\sqrt{49}$ Compute the square root.

$= 4 + 3(7)$ Multiply.

$= 4 + 21$ Add.

$= 25$

b. $\dfrac{6 + \sqrt{64}}{9 - \sqrt{4}}$ Compute the square roots.

$= \dfrac{6 + 8}{9 - 2}$ Add and subtract above and below the fraction bar.

$= \dfrac{14}{7} = 2$

EXERCISE 3

Simplify $\sqrt{9 + 16}$

ANSWERS TO 5.4A EXERCISES

1. 4
2. 2
3. 5

HOMEWORK 5.4A 

For Problems 1–4, find two square roots for each number.

1. a. 36 **b.** 49 **c.** 1

2. a. 4 **b.** 25 **c.** 100

3. a. 64 **b.** 81 **c.** 144

4. a. 121 **b.** 169 **c.** 225

5. What is the difference between the square of a number and the square root of a number? Give examples.

6. If $a = 16$ and $b = 4$, which of the following is true?
a. $a^2 = b$ **b.** $b^2 = a$
c. $a = \sqrt{b}$ **d.** $b = \sqrt{a}$

7. What is a radical sign, and what does it mean?

8. Explain the difference between $-\sqrt{16}$ and $\sqrt{-16}$.

9. Since $(-5)^2 = 25$, does $\sqrt{25} = -5$? Explain why or why not.

10. If $a^2 = b^2$, is it necessarily true that $a = b$?

Evaluate each radical if possible. If the radical is undefined, say so.

11. a. $\sqrt{49}$ **b.** $\sqrt{169}$

12. a. $\sqrt{0}$ **b.** $\sqrt{9}$

13. a. $-\sqrt{121}$ **b.** $\sqrt{-25}$

14. a. $-\sqrt{144}$ **b.** $\sqrt{-64}$

15. a. $-\sqrt{1}$ **b.** $\sqrt{-1}$

16. a. $\sqrt{400}$ **b.** $\sqrt{900}$

17. a. $\sqrt{\frac{1}{16}}$ **b.** $\sqrt{\frac{1}{25}}$

18. a. $-\sqrt{\frac{9}{49}}$ **b.** $-\sqrt{\frac{16}{81}}$

19. a. $-\sqrt{\frac{196}{25}}$ **b.** $\sqrt{\frac{4}{289}}$

20. a. $\sqrt{\frac{256}{121}}$ **b.** $-\sqrt{\frac{169}{324}}$

Simplify. Follow the order of operations.

21. $7\sqrt{16}$

22. $8\sqrt{25}$

23. $3 + 5\sqrt{81}$

24. $4 + 6\sqrt{64}$

25. $12 - \sqrt{225}$

26. $6 - \sqrt{100}$

27. $-1 - 2\sqrt{9}$

28. $2 - 5\sqrt{36}$

29. $\dfrac{6 + \sqrt{144}}{3}$

30. $\dfrac{20 + \sqrt{225}}{5}$

(a) Write an algebraic expression and (b) write and solve an equation. (See Lessons 4.2 and 4.4 to review these topics.)

31. **a.** At 8 A.M. the temperature was 72°, and it has been rising by 6° every hour. At this rate, what will be the temperature after h hours?

b. When was the temperature 96°?

32. **a.** Avram has typed 480 words of his term paper and is still typing at a rate of 30 words per minute. How many words will Avram have typed after m minutes?

b. When will Avram have typed 1230 words?

33. **a.** For her mother's retirement party, Daniella spent \$50 on a gift plus her share of the cost of the party. If the party costs P dollars and 12 people are contributing (including Daniella), how much did Daniella spend?

b. If Daniella spent \$68, how much did the party cost?

34. **a.** Delbert shares a house with four roommates. He pays \$200 rent per month, plus an equal share of the utilities. If the utilities cost U dollars this month, how much does Delbert owe?

b. Delbert paid \$263 last month. What was the utility bill for the house?

35. a. Brian reserved \$60 from the office Sunshine Fund for a party and used the rest to buy new desk sets for the 15 people in the office. If the sunshine fund originally had S dollars, how much can Brian spend on each desk set?

b. If each desk set cost \$23, how much was originally in the sunshine fund?

36. a. Vincent's school provided him with C colored pencils to distribute to the students in his class, and he bought 8 more pencils so that each student would get the same number. If there are 23 students in Vincent's class, how many pencils does each get?

b. If each student got four colored pencils, how many pencils did the school provide?

37. a. Raylyn's auto registration fee is \$20 plus 2% of the value of her car. If Raylyn's car is worth B dollars, how much will her registration fee be?

b. If Raylyn paid \$340 for registration, how much is her car worth?

38. a. Maryam has to carry a dummy weighing 50 pounds plus 60% of her own weight for 20 yards as part of the test to become a lifeguard. If Maryam weighs W pounds, how much must she carry?

b. If Maryam must carry 134 pounds, how much does she weigh?

Simplify each pair of expressions and compare the results. If the expression cannot be simplified, write "cbs" ("cannot be simplified").

39. a. $-\sqrt{225}$

b. $\sqrt{-225}$

40. a. $\sqrt{9+16}$

b. $\sqrt{9}+\sqrt{16}$

41. a. $4\sqrt{25}$

b. $\sqrt{4 \cdot 25}$

42. a. $\sqrt{9}\sqrt{36}$

b. $\sqrt{9 \cdot 36}$

43. a. $5 + 2\sqrt{64}$

b. $(5 + 2)\sqrt{64}$

44. a. $\sqrt{\dfrac{16}{49}}$

b. $\dfrac{\sqrt{16}}{\sqrt{49}}$

45. a. $\sqrt{100} - \sqrt{64}$

b. $\sqrt{100 - 64}$

46. a. $\dfrac{\sqrt{81}}{4}$

b. $\sqrt{\dfrac{81}{4}}$

47. a. $10 - 3\sqrt{144}$

b. $(10 - 3)\sqrt{144}$

48. a. $\sqrt{4\sqrt{81}}$

b. $\sqrt{4}\sqrt{81}$

Building Number Skills

The quotient $4 \div \frac{1}{3}$ means "How many thirds are there in 4?" We can use a diagram to illustrate the division.

Use a diagram to find each quotient.

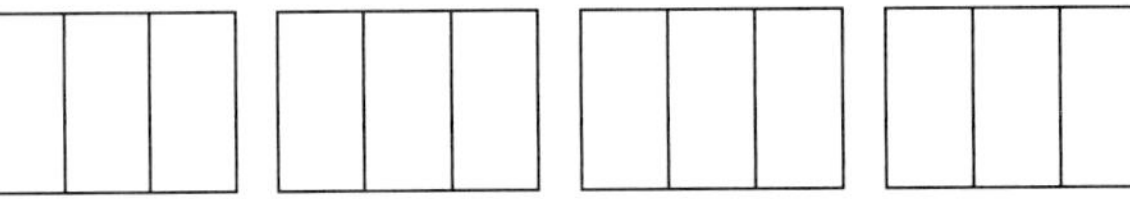

Example: $4 \div \frac{1}{3} = \mathbf{12}$

Draw four wholes and divide each into thirds. How many thirds?

49. $3 \div \frac{1}{2}$

50. $2 \div \frac{1}{3}$

51. $3 \div \frac{1}{3}$

52. $2 \div \frac{1}{2}$

53. $2 \div \frac{1}{4}$

54. $3 \div \frac{1}{4}$

55. $1 \div \frac{1}{8}$

56. $1 \div \frac{1}{6}$

57. $4 \div \frac{1}{6}$

58. $5 \div \frac{1}{8}$

Do you see a way to do these exercises without using a diagram? Describe your method and give an example.

B. Expressions and Equations

By following the order of operations, we can evaluate expressions that involve radicals.

EXAMPLE 4 

Evaluate $5 - 2\sqrt{v+3}$ for $v = 13$.

Solution

Substitute 13 for v in the expression:

$$
\begin{aligned}
5 - 2\sqrt{v+3} &= 5 - 2\sqrt{\mathbf{13} + 3} && \text{Add 13 + 3 inside the radical.}\\
&= 5 - 2\sqrt{16} && \text{Compute the radical}\\
&= 5 - 2(4) && \text{Multiply.}\\
&= 5 - 8 && \text{Subtract.}\\
&= -3
\end{aligned}
$$

EXERCISE 4

Evaluate $\sqrt{169 - r^2}$ for $r = 5$.

Solving Equations with x^2

Taking a square root is the opposite of squaring a number. For example, if we start with any number, say,

$$n = 3$$

and then square it, we get

$$n^2 = 9$$

If we then take the square root of the result, we find

$$\sqrt{n^2} = \sqrt{9}$$

or, after simplifying,

$$n = 3$$

Performing the two operations of squaring and taking square roots brings us back to our original number, 3. (Actually, we have to be a little bit careful, because every positive number has two square roots, but we'll worry about that in a minute.)

Thus, we can undo the squaring operation by performing its opposite operation, taking square roots. This will enable us to solve equations like

$$x^2 = 64$$

where the variable is squared. Saying that x^2 equals 64 is the same as saying that x is a square root of 64. You can probably guess that 8 is a solution to the equation, but can you see that -8 is also a solution? Both 8 and -8 are square roots of 64. However, the symbol $\sqrt{64}$ denotes only the *positive* root. If we want to indicate both square roots of 64, we use the notation

$$\pm\sqrt{64}$$

where the symbol $\pm$ is read "plus or minus." Thus, $\pm\sqrt{64}$ denotes both solutions, 8 and -8.

In general, an equation involving the square of the variable has two solutions.

EXERCISE 5

Solve $n^2 = \frac{4}{9}$

EXAMPLE 5

Solve the equation $x^2 = 225$

Solution

Since the variable x is squared, we take the square root of both sides:

$$\begin{aligned} x^2 &= 225 \\ x &= \pm\sqrt{225} && \text{Take the square root of both sides.} \\ &= \pm 15 && \text{Simplify the square root.} \end{aligned}$$

The equation has two solutions, 15 and -15.

Equations with More Than One Operation

Recall that if more than one operation has been performed on the variable, we undo the operations in reverse order.

EXERCISE 6

Solve $12 + \frac{1}{2}v^2 = 30$

EXAMPLE 6

Solve the equation $3x^2 + 16 = 64$

Solution

First, analyze the equation and plan the solution:

Operations Performed on x	*Steps for Solution*
1. Squared	**1.** Subtract 16
2. Multiplied by 3	**2.** Divide by 3
3. Added 16	**3.** Take square roots

Carry out the plan:

$$\begin{aligned} 3x^2 + 16 &= 64 \\ -16 \quad & \quad -16 && \text{Subtract 16 from both sides.} \\ 3x^2 &= 48 \\ \frac{3x^2}{3} &= \frac{48}{3} && \text{Divide both sides by 3.} \\ x^2 &= 16 \\ x &= \pm\sqrt{16} && \text{Take square roots of both sides.} \\ &= \pm 4 && \text{Simplify the square root.} \end{aligned}$$

The solutions are 4 and -4.

ANSWERS TO 5.4B EXERCISES

4. 12

5. $\pm\frac{2}{3}$

6. $v = \pm 6$

HOMEWORK 5.4B 

Evaluate each expression to complete the table.

1.

k	0	1	4
$8\sqrt{k} + 12$			

2.

j	0	1	4
$2\sqrt{j} - 10$			

3.

h	7	10	42
$3\sqrt{h - 6}$			

4.

g	5	21	60
$5\sqrt{g + 4}$			

5.

b	0	9	400
$2b - 10\sqrt{b}$			

6.

d	1	16	100
$-4d + 4\sqrt{d}$			

7.

z	121	144	256
$(12 - \sqrt{z})^2$			

8.

y	169	225	324
$(\sqrt{y} - 15)^2$			

9.

x	36	81	225
$-3x + 2x\sqrt{x}$			

10.

w	64	144	289
$8w - 4w\sqrt{w}$			

Solve each equation.

11. $a^2 = 25$

12. $b^2 = 49$

13. $y^2 = 196$

14. $z^2 = 225$

15. $3s^2 = 48$

16. $5t^2 = 405$

17. $-8p^2 = -72$

18. $-4q^2 = -144$

19. $9m^2 = 9$

20. $6n^2 = 0$

21. $7v^2 = 1183$

22. $4u^2 = 1296$

23. $5x^2 - 7 = 73$

24. $4y^2 + 6 = 150$

25. $105 - 2w^2 = 7$

26. $32 - 3z^2 = 5$

27. $9 + \frac{Z^2}{4} = 25$

28. $144 + \frac{W^2}{9} = 169$

Follow the order of operations to simplify.

29. $\sqrt{16 + 4(12)}$

30. $\sqrt{81 - 2(16)}$

31. $(2 + \sqrt{25})(5 - \sqrt{64})$

32. $(6 + \sqrt{9})(3 - \sqrt{49})$

33. $\dfrac{\sqrt{36} - 10}{\sqrt{36} - 2}$

34. $\dfrac{\sqrt{16} - 8}{4 - \sqrt{64}}$

35. $\sqrt{4 + \sqrt{25}}$

36. $\sqrt{16 + \sqrt{81}}$

For Problems 37–44, simplify each expression without using pencil, paper, or calculator.

37. $(\sqrt{144})^2$

38. $(\sqrt{121})^2$

39. $\sqrt{9^2}$

40. $\sqrt{8^2}$

41. $\sqrt{81^2}$

42. $\sqrt{64^2}$

43. $(\sqrt{225})(\sqrt{225})$

44. $(\sqrt{169})(\sqrt{169})$

45. a. Complete the table for the given values of a and b:

a	b	$a+b$	a^2	b^2	a^2+b^2	$\sqrt{a^2+b^2}$
3	4					
5	12					
2	6					

b. Is it true that $\sqrt{a^2+b^2} = a + b$?

46. a. Complete the table for the given values of a and b:

a	b	$a+b$	$\sqrt{a+b}$	$\sqrt{a}$	$\sqrt{b}$	$\sqrt{a}+\sqrt{b}$
9	16					
36	64					
4	25					

b. Is it true that $\sqrt{a+b} = \sqrt{a} + \sqrt{b}$?

47. a. Complete the table for the given values of a and b:

a	b	$a+b$	$(a+b)^2$	a^2	b^2	a^2+b^2
2	3					
3	5					
4	7					

b. Is it true that $(a+b)^2 = a^2 + b^2$?

48. a. Complete the table for the given values of a and b:

a	b	$a + b$	$\sqrt{a}$	$\sqrt{b}$	$\sqrt{a} + \sqrt{b}$	$(\sqrt{a} + \sqrt{b})^2$
4	9					
16	25					
1	4					

b. Is it true that $(\sqrt{a} + \sqrt{b})^2 = a + b$?

Building Number Skills

Use a diagram to find each quotient.

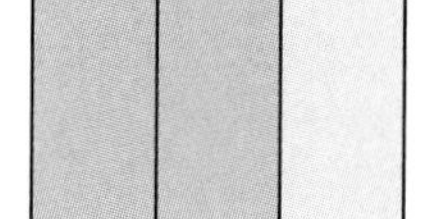
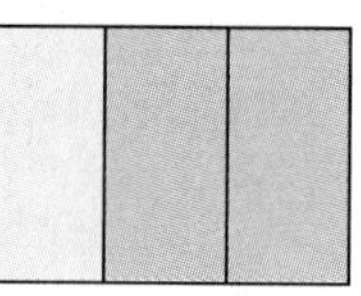
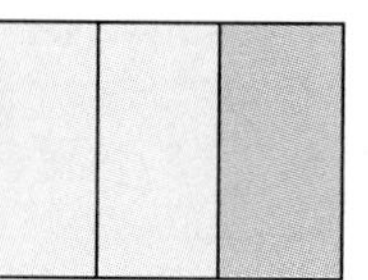
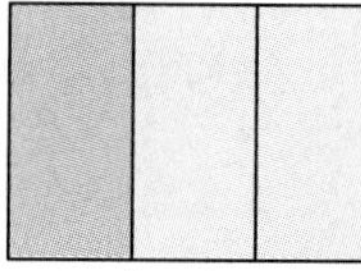

Example: $4 \div \frac{2}{3} = \mathbf{6}$

Draw four wholes and divide each into thirds.
How many groups of two-thirds?

49. $3 \div \frac{3}{4}$

50. $2 \div \frac{2}{3}$

51. $4 \div \frac{4}{3}$

52. $3 \div \frac{3}{2}$

53. $6 \div \frac{3}{8}$

54. $6 \div \frac{3}{4}$

55. $6 \div \frac{2}{3}$

56. $9 \div \frac{3}{8}$

57. $9 \div \frac{3}{2}$

58. $10 \div \frac{5}{8}$

5.5 Applications of Square Roots

A. Approximating Square Roots

Numbers such as 16 and 25 are called *perfect squares* because they are the squares of whole numbers:

$$16 = 4^2 \qquad \text{and} \qquad 25 = 5^2$$

Their square roots are easy to find. In fact, you have probably memorized the square roots of several perfect squares by now. But how can we find the square roots of numbers like 3 or 5, which are not perfect squares?

Activity Let's try to find the square root of 5 by guessing and correcting:

Because $2^2 = 4$, $\sqrt{5}$ must be bigger than 2.
And because $3^2 = 9$, $\sqrt{5}$ must be smaller than 3.
So $\sqrt{5}$ is a number between 2 and 3.

Suppose we guess 2.5 for $\sqrt{5}$. To check our guess, we can square 2.5 and see if we get 5:

$$(2.5)^2 = 6.25$$

so 2.5 is too big for the square root of 5.

Check the squares of several other numbers between 2 and 2.5:

$$(2.1)^2 = 4.41$$
$$(2.2)^2 = 4.84$$
$$(2.3)^2 = 5.29$$

From these values, we see that $\sqrt{5}$ must be between 2.2 and 2.3. Use your calculator to fill in the following list:

$(2.21)^2 =$ ______________ $(2.26)^2 =$ ______________

$(2.22)^2 =$ ______________ $(2.27)^2 =$ ______________

$(2.23)^2 =$ ______________ $(2.28)^2 =$ ______________

$(2.24)^2 =$ ______________ $(2.29)^2 =$ ______________

$(2.25)^2 =$ ______________

You should have discovered that $(2.23)^2$ is too small and $(2.24)^2$ is too big. Therefore, $\sqrt{5}$ is a number between 2.23 and 2.24. Use your calculator to fill in the following list:

$(2.231)^2 =$ ______________ $(2.236)^2 =$ ______________

$(2.232)^2 =$ ______________ $(2.237)^2 =$ ______________

$(2.233)^2 =$ ______________ $(2.238)^2 =$ ______________

$(2.234)^2 =$ ______________ $(2.239)^2 =$ ______________

$(2.235)^2 =$ ______________

You should have discovered that $(2.236)^2$ is too small and that $(2.237)^2$ is too big. Therefore, $\sqrt{5}$ is a number between 2.236 and 2.237.

Maybe you think that 2.236 is close enough, and you don't want to fill in any more lists. But maybe you are curious to know how far we have to go to find the exact decimal value of $\sqrt{5}$. It turns out that no matter how many decimal places we find, *we will never obtain an exact decimal form for* $\sqrt{5}$. However, the accuracy improves with each decimal place, so we can find a decimal *approximation* to $\sqrt{5}$ as accurate as we like.

Using a Calculator for Square Roots

Because square roots like $\sqrt{5}$ are time-consuming to approximate, most scientific calculators give values for square roots to as many digits as their displays allow. Look for a key labeled $\boxed{\sqrt{x}}$, and enter

5 

(On a graphing calculator, enter the radical symbol first: $\boxed{\sqrt{\ }}$ 5.) The calculator will return an approximation for $\sqrt{5}$—for example, 2.236068. You can then round off the number to as many decimal places as you need.

EXAMPLE 1

Use a calculator to find approximate values for each square root. Round your answers to three decimal places.

a. $\sqrt{3}$ **b.** $\sqrt{243}$ **c.** $\sqrt{0.02}$

Solution

a. Enter 3 $\boxed{\sqrt{x}}$ and the calculator displays 1.7320508. Rounding to three decimal places gives 1.732.

b. Enter 243 $\boxed{\sqrt{x}}$ and the calculator displays 15.588457. Rounding to three decimal places gives 15.588.

c. Enter 0.02 $\boxed{\sqrt{x}}$ and the calculator displays 0.1414214. Rounding to three decimal places gives 0.141.

EXERCISE 1

Use a calculator to approximate $\sqrt{226}$ to three decimal places.

Expressions and Equations

You can also use your calculator to find approximations for expressions that involve radicals or for solutions of equations that involve the square of the variable. Let's agree to round all approximate values to three decimal places, unless otherwise specified.

EXAMPLE 2

Simplify $5 - 2\sqrt{7}$

Solution

The order of operations tells us how to perform the calculations:

1. Compute $\sqrt{7}$.
2. Multiply $\sqrt{7}$ times -2.
3. Add 5 to the result.

(continued)

EXERCISE 2
Approximate $1 - \sqrt{5}$ to three decimal places.

If we enter the calculations in exactly this order, we will be sure to get the correct result. Enter:

7 [$\sqrt{x}$] [×] 2 [+/−] [+] 5 [=]

The calculator displays the result −0.2915026. Rounding to three places gives −0.292. However, most scientific calculators know the order of operations and will automatically perform the operations in the correct order. Test your calculator by entering

5 [−] 2 [×] 7 [$\sqrt{x}$] [=]

If your calculator knows the order of operations, you should get the same result as before.

To solve an equation involving x^2, we first isolate x^2 and then take square roots.

EXERCISE 3
a. Solve $3y^2 + 2 = 23$

b. Find decimal approximations for the solutions, rounded to thousandths.

EXAMPLE 3  

a. Solve $4x^2 - 19 = 33$.
b. Find decimal approximations for the solutions.

Solution

a. We undo the operations in reverse order:

Operations Performed on x	*Steps for Solution*
1. Squared	**1.** Add 19
2. Multiplied by 4	**2.** Divide by 4
3. Subtracted 19	**3.** Take square roots

Carry out the plan:

$$4x^2 - 19 = 33$$
$$\underline{+19 \quad +19} \qquad \text{Add 19 to both sides.}$$
$$4x^2 = 52$$
$$\frac{4x^2}{4} = \frac{52}{4} \qquad \text{Divide both sides by 4.}$$
$$x^2 = 13$$
$$x = \pm\sqrt{13} \qquad \text{Take the square root of both sides.}$$

To check that $\sqrt{13}$ and $-\sqrt{13}$ are indeed the solutions, substitute either one back into the original equation.

Check: $4(\sqrt{13})^2 - 19 = 33$
$4(13) - 19 = 33$
$52 - 19 = 33$ True; the solutions check.

b. Finally, use your calculator to obtain an approximate value for $\sqrt{13}$. Enter 13 [$\sqrt{x}$], and the calculator displays 3.6055513. Rounding to three decimal places gives us two approximate solutions, 3.606 and −3.606.

CAUTION!

In Example 3, the *exact* solutions of the equation are $\sqrt{13}$ and $-\sqrt{13}$. When we use a calculator to evaluate $\sqrt{13}$, we are finding *approximations* for the solutions.

ANSWERS TO 5.5A EXERCISES
1. 15.033
2. −1.236
3a. $y = \pm 7$
3b. ±2.646

HOMEWORK 5.5A 

1. Use your calculator to make a table showing the whole numbers between 1 and 10 and their square roots. Round your answers to three decimal places.

1	2	3	4	5	6	7	8	9	10

2. Use your calculator to make a table showing the whole numbers between 11 and 20 and their square roots. Round your answers to three decimal places.

11	12	13	14	15	16	17	18	19	20

Use your calculator to approximate the square roots. Round your answers to three decimal places.

3. a. $\sqrt{48.6}$ **b.** $\sqrt{29.3}$ **4. a.** $\sqrt{56.8}$ **b.** $\sqrt{73.9}$

5. a. $\sqrt{1.4}$ **b.** $\sqrt{1.8}$ **6. a.** $\sqrt{0.6}$ **b.** $\sqrt{0.3}$

7. a. $\sqrt{419}$ **b.** $\sqrt{836}$ **8. a.** $\sqrt{2498}$ **b.** $\sqrt{5612}$

Each number below is approximately the square root of a whole number. Find the whole number.

9. 9.220 **10.** 6.557 **11.** 12.961

12. 23.937 **13.** 63.891 **14.** 39.281

Find an approximation for each expression. Round your answers to thousandths.

15. $-2 + 6\sqrt{8}$

16. $-5 - 3\sqrt{6}$

17. $\dfrac{8 - 2\sqrt{12}}{4}$

18. $\dfrac{-9 + 3\sqrt{18}}{6}$

19. $\dfrac{5 - \sqrt{5}}{-3 + \sqrt{13}}$

20. $\dfrac{2 + \sqrt{32}}{8 - \sqrt{52}}$

Solve each equation and check your solutions. Then round your answers to three decimal places.

21. $2z^2 = 10$

22. $5y^2 = 15$

23. $3h^2 - 8 = 13$

24. $4k^2 + 7 = 47$

25. $8p^2 + 6 = 18$

26. $6q^2 - 9 = 1$

27. $23.6 - 5t^2 = 4.4$

28. $17.9 - 2s^2 = 8.1$

For Problems 29–34, evaluate each expression to complete the table. Round your answer to three decimal places.

29.

m	-3	2	9
$9\sqrt{m+5}$			

30.

p	8	10	16
$-4\sqrt{p-6}$			

31.

q	3	4	-6
$\sqrt{q^2 - 4}$			

32.

n	2	5	-4
$\sqrt{n^2 + 2}$			

33.

a	2	5	12
$\sqrt{a} - \sqrt{a + 3}$			

34.

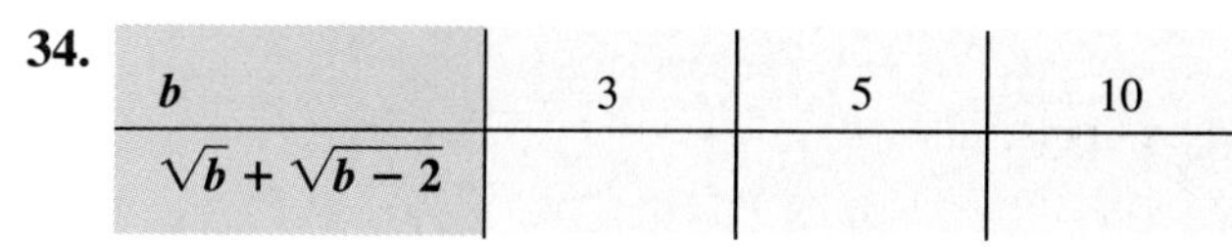

b	3	5	10
$\sqrt{b} + \sqrt{b - 2}$			

35. The distance m in miles that you can see on a clear day from a height of h miles is given by the formula $m = 89.4\sqrt{h}$. How far can you see from an airplane flying at an altitude of 4.7 miles?

36. Insurance investigators use the length in feet, d, of a car's skid marks to estimate its speed in miles per hour just before braking. The speed is given by the formula $v = \sqrt{24d}$. How fast was a car traveling if it left skid marks 200 feet long?

37. The period of a pendulum (the time it takes to complete one full swing) is given in seconds by the formula $T = 6.28\sqrt{\dfrac{L}{32}}$, where L is the length of the pendulum in feet. What is the period of the Foucault pendulum in the United Nations headquarters in New York, whose length is 75 feet?

38. If an object falls from a height h in meters, the time it will take to reach the ground is given in seconds by the formula $t = \sqrt{\dfrac{h}{4.9}}$. How long would it take for a pebble to fall from the Sears Tower in Chicago, which is 443.2 meters tall?

39. Find the diameter of a circular window if its area should be approximately 16 square feet.

40. A chocolate cake recipe fits into a rectangular pan measuring 8 by 10 inches. If you want to bake the same cake in a circular pan of the same size, what should be its radius?

Simplify each expression without using pencil, paper, or calculator.

41. $(\sqrt{17})^2$

42. $(\sqrt{23})^2$

43. $\sqrt{453^2}$

44. $\sqrt{872^2}$

45. $\sqrt{39}\sqrt{39}$

46. $\sqrt{2}\sqrt{2}$

For Problems 47–52, evaluate each expression for the given value.

47. $6x^2 + 2$, for $x = \sqrt{3}$

48. $4a^2 - 5$, for $a = \sqrt{5}$

49. $-9 - 2b^2$, for $b = \sqrt{11}$

50. $-7 + 3w^2$, for $w = \sqrt{14}$

51. $12 - m^2$, for $m = \sqrt{12}$

52. $-n^2 + 10$, for $n = \sqrt{11}$

53. If you start with any positive number and square it and then take the square root of the result, what do you end up with?

54. If you start with any positive number and take its square root and then square the result, what do you end up with?

Building Number Skills

Add or subtract, and reduce your answer if possible.

55. $\frac{5}{8} + \frac{1}{8}$

56. $\frac{7}{8} - \frac{5}{8}$

57. $7\frac{7}{9} - 2\frac{4}{9}$

58. $4\frac{3}{10} + 1\frac{1}{10}$

59. $\frac{8}{12} + \frac{6}{12}$

60. $\frac{10}{12} - \frac{8}{12}$

61. $5\frac{12}{16} - 3\frac{6}{16}$

62. $1\frac{10}{16} + \frac{14}{16}$

63. $\frac{13}{8} - \frac{3}{8}$

64. $\frac{5}{8} + \frac{7}{8}$

B. Pythagorean Theorem

A triangle in which one of the angles is a right angle, or 90°, is called a **right triangle.** The side opposite the right angle is the longest side of the triangle and is called the **hypotenuse.** The other two sides of the triangle are called the **legs** (Figure 5.12).

Figure 5.12

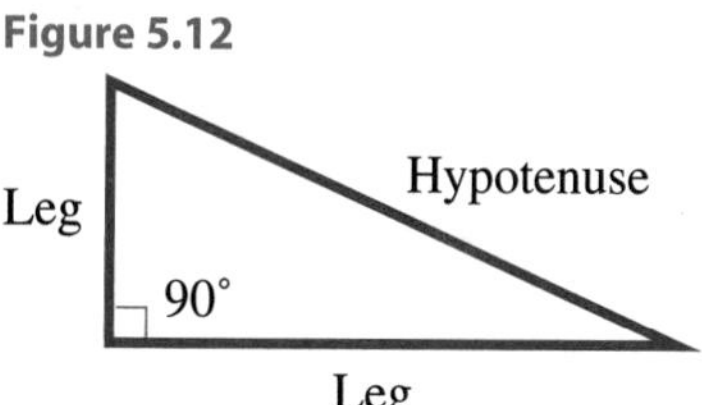

If we know the lengths of any two sides of a right triangle, we can find the third side by using a formula called the Pythagorean theorem. Several ancient civilizations—including the Egyptians, Babylonians, and Chinese—discovered this formula, but it is named in honor of the Greek mathematician Pythagoras, who gave a proof of the formula. We will use the variable c for the length of the hypotenuse, and the lengths of the legs will be denoted by a and b (it doesn't matter which is which).

Pythagorean Theorem

$$a^2 + b^2 = c^2$$

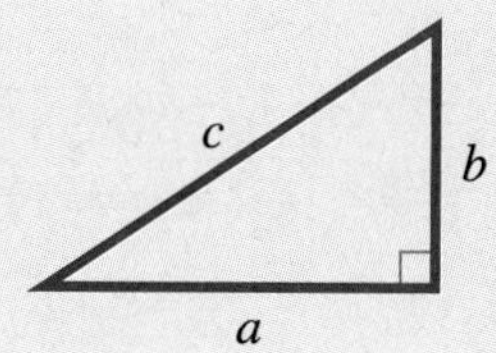

In words, the Pythagorean theorem says that the square of the hypotenuse of a right triangle is equal to the sum of the squares of the two legs.

EXAMPLE 4

The two legs of a right triangle are 6 inches and 8 inches long. How long is the hypotenuse?

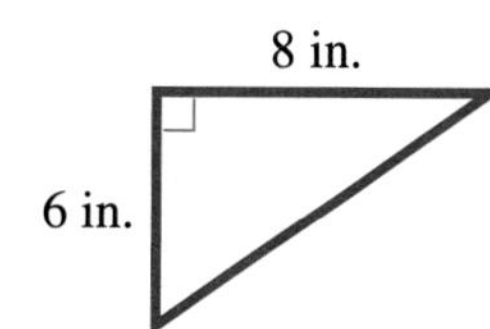

EXERCISE 4

The two short sides of a right triangle are 9 meters and 40 meters. What is the hypotenuse?

Solution

Substitute the lengths of the two legs into the Pythagorean theorem for a and b. We are looking for the length of the hypotenuse, c, so it remains a variable:

$$a^2 + b^2 = c^2 \quad \text{Substitute 6 for } a \text{ and 8 for } b.$$
$$6^2 + 8^2 = c^2 \quad \text{Compute the powers.}$$
$$36 + 64 = c^2 \quad \text{Simplify the left side.}$$
$$100 = c^2 \quad \text{Take square roots.}$$
$$\pm\sqrt{100} = c \quad \text{Simplify the radical.}$$
$$\pm 10 = c$$

The length of the hypotenuse cannot be -10 inches, so we discard that solution. The hypotenuse is 10 inches long.

EXAMPLE 5  

The length of a rectangle is 17 centimeters, and its diagonal is 20 centimeters long. What is the width of the rectangle?

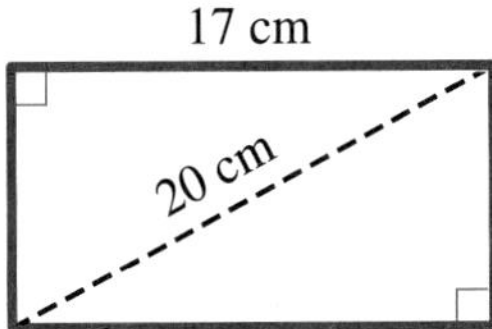

EXERCISE 5

A baseball diamond is a square whose sides are 90 feet. How far is it from first base to third base (which are on opposite corners of the square)?

Solution

The diagonal of the rectangle is the hypotenuse of a right triangle, as shown in the figure. We are looking for the width of the rectangle, which forms one of the legs of the right triangle. We use the Pythagorean theorem, substituting 20 for c and 17 for b:

$$a^2 + b^2 = c^2$$
$$a^2 + (17)^2 = (20)^2 \quad \text{Substitute 20 for } c \text{ and 17 for } b.$$

Now solve the equation for a. Begin by computing the powers:

$$a^2 + 289 = 400 \quad \text{Subtract 289 from both sides.}$$
$$\underline{\ \mathbf{-289} \quad \mathbf{-289}}$$
$$a^2 = 111 \quad \text{Take square roots of both sides.}$$
$$a = \pm\sqrt{111}$$

Use a calculator to evaluate the square root. The width of the rectangle is approximately 10.536 centimeters. (Because a represents the width of a rectangle, we discard the negative solution.)

ANSWERS TO 5.5B EXERCISES

4. 41 meters

5. $\sqrt{16{,}200} \approx 127.28$ feet

HOMEWORK 5.5B 

For Problems 1–12, find the length of the missing side in each right triangle.

1. 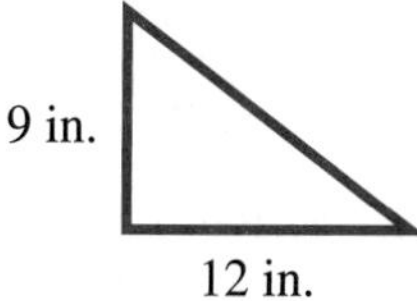

2. 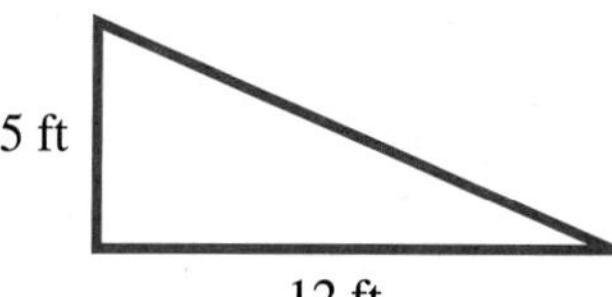

3. 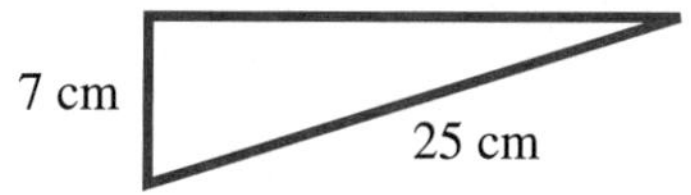

4. 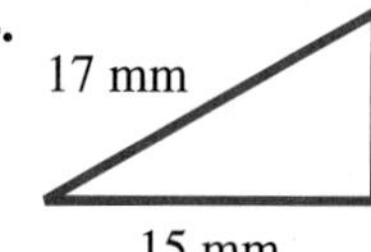

5. 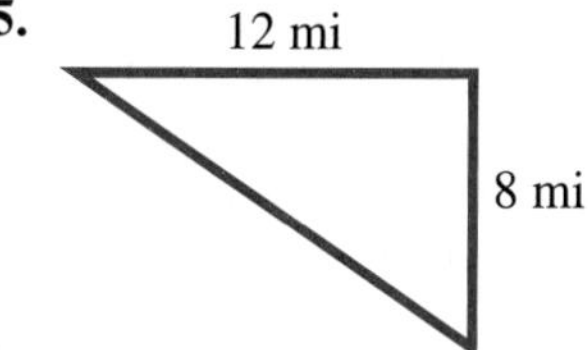

6. 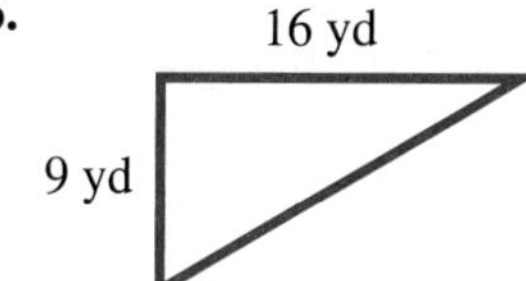

7. 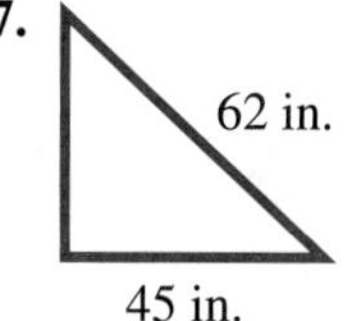

8. 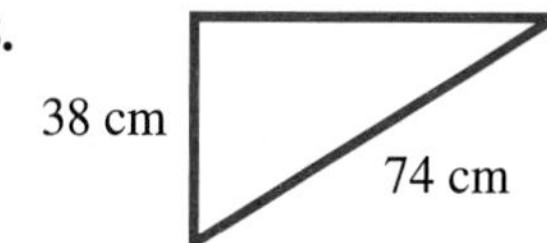

9. 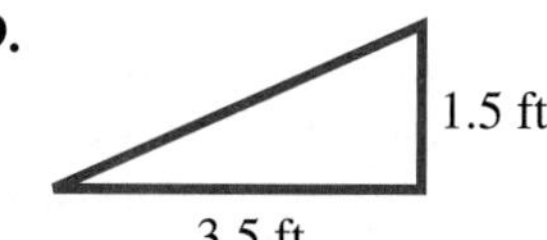

10. 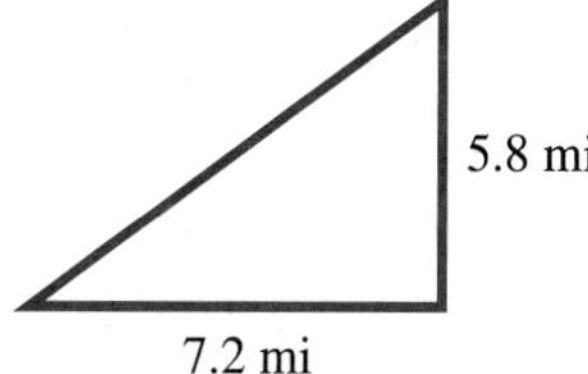

11. 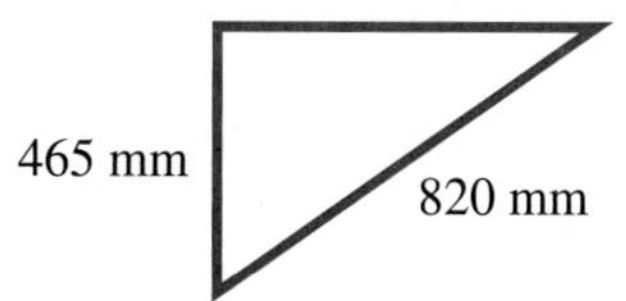

12. 

13. Juliet's window is 24 feet above the ground, and there is a 10-foot moat at the base of the wall. How long a ladder will Romeo need to reach the window?

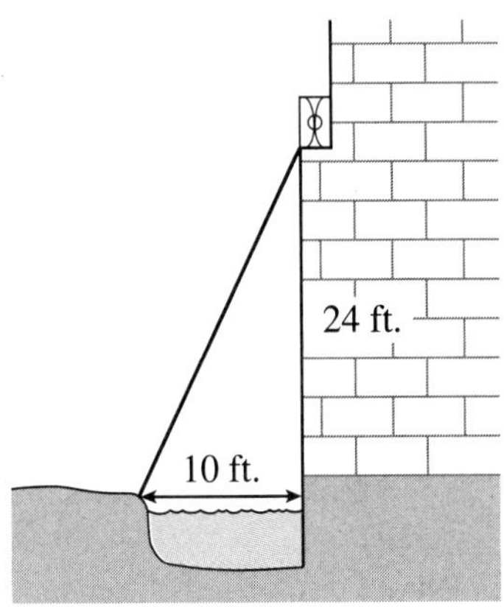

14. It is 16 miles from the highway to Sunrise and 30 miles from the junction to Conway. How far is it along the back road from Sunrise to Conway?

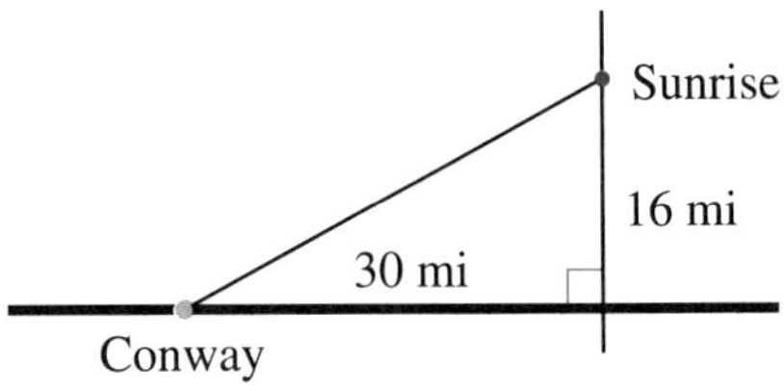

15. Clark is assembling a swing set at the neighborhood park. The slide is 16 feet long, and the platform at the top is 8 feet high. How far should the sand pit at the bottom of the slide be positioned from the base of the ladder?

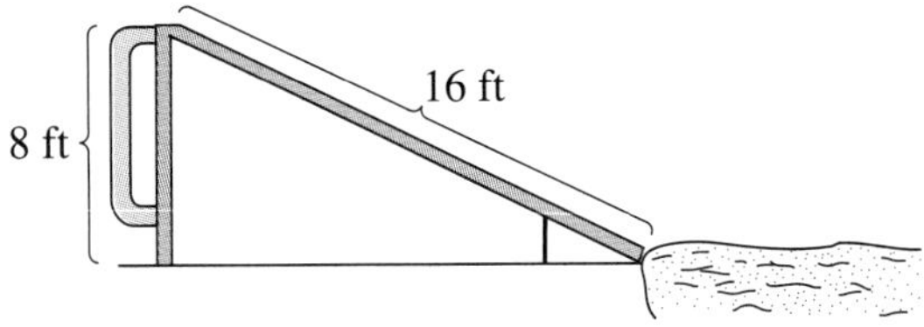

16. Marlene visited the Quetzalcoatl pyramid near Mexico City last summer. She measured the base and found it is about 1400 feet on each side. She unrolled a ball of string as she climbed the face of the pyramid, and it was about 722 feet to the top. How tall is the pyramid?

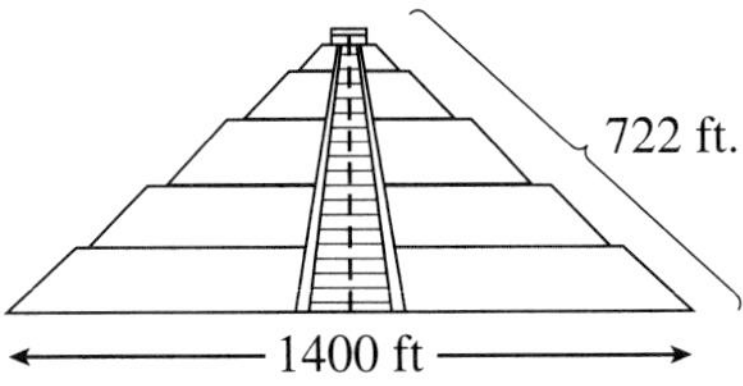

17. A pup tent is made out of two pieces of oil cloth sewn together along the sides. The floor piece is 15 feet wide, and the piece for the top is 17 feet wide. Poles are erected along the center line to hold up the roof. How long should the poles be?

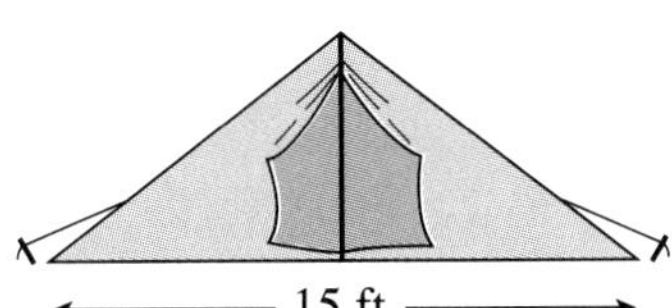

18. A surveyor would like to know the distance across the lake shown. She picks a spot *P* on a line perpendicular to the width of the lake, and measures the two distances shown. How wide is the lake?

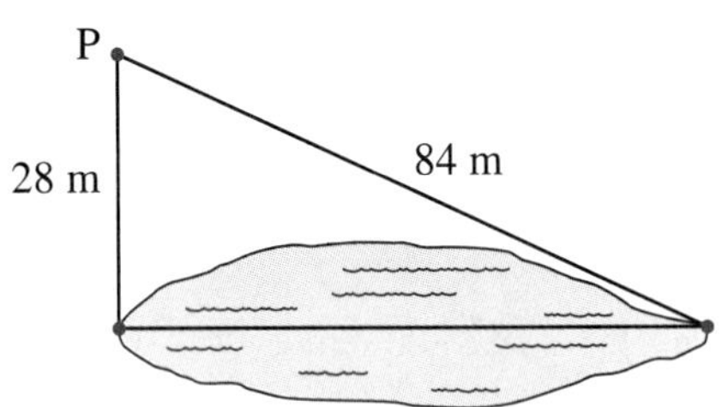

19. Find the length d of the diagonal of the box shown. (*Hint:* Find x first.)

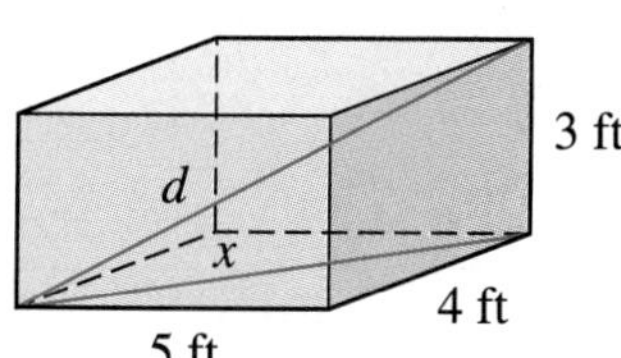

20. Find the length of each hypotenuse in the figure.

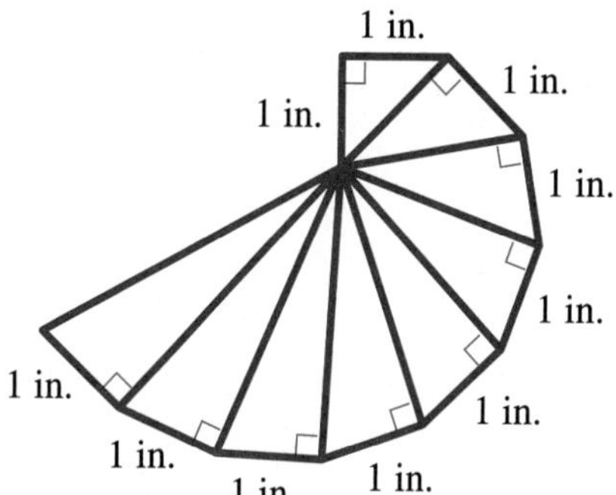

Explain why each of these is an incorrect application of the Pythagorean theorem.

21.

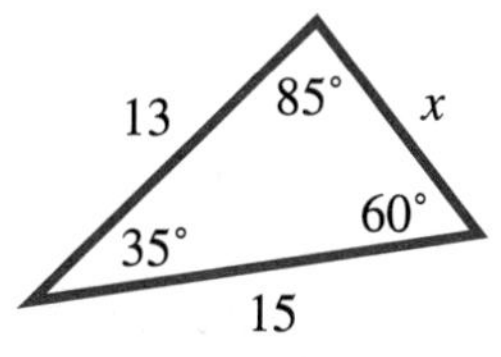

$x^2 + 13^2 = 15^2$

22.

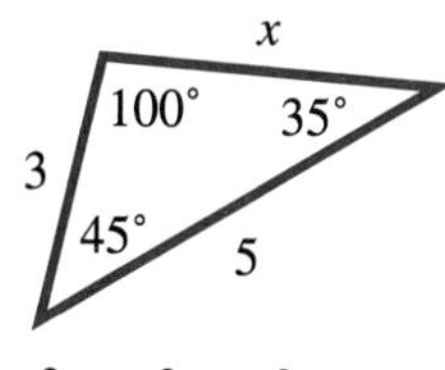

$3^2 + x^2 = 5^2$

23.

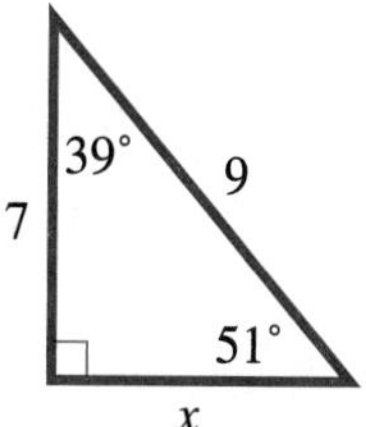

$7 + x = 9$

24.

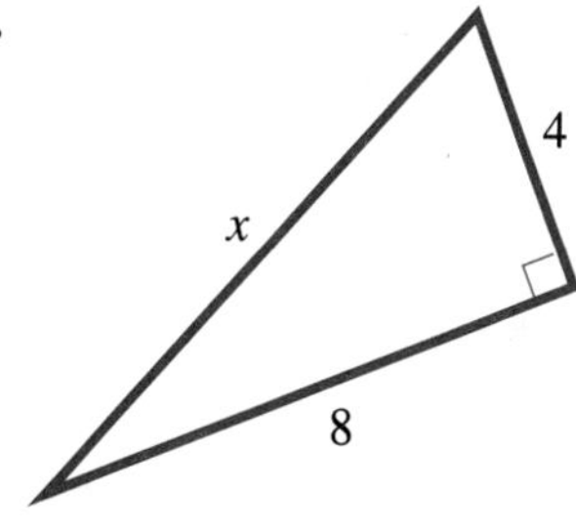

$8 + 4 = x$

25.

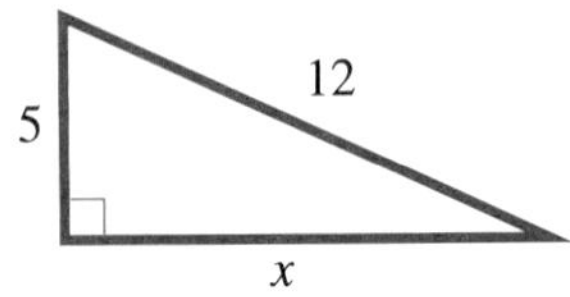

$5^2 + 12^2 = x^2$

26.

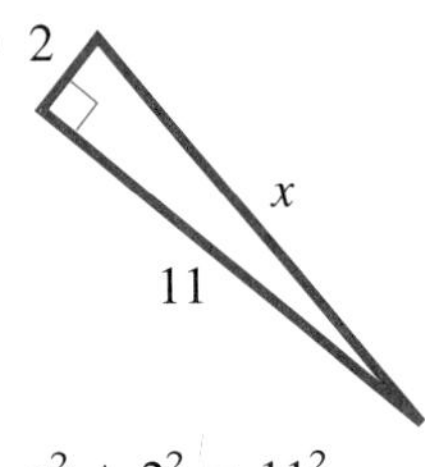

$x^2 + 2^2 = 11^2$

27.

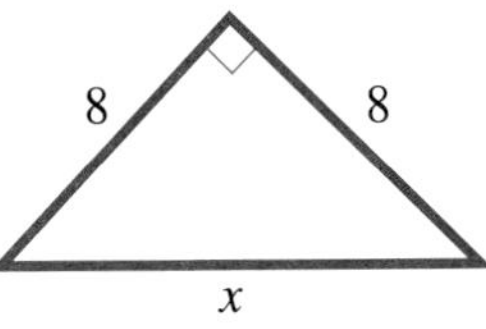

$x = 8^2 + 8^2$

28.

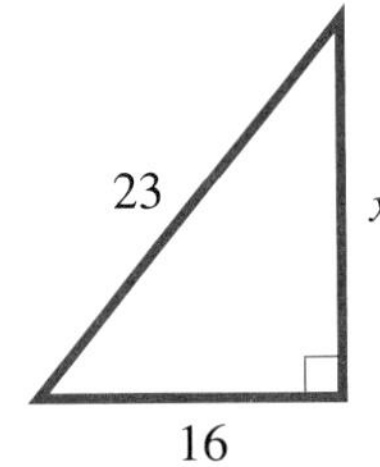

$x^2 + 16 = 25$

Simplify each expression if possible. If the expression cannot be simplified, write "cbs" ("cannot be simplified").

29. **a.** $4(5x)$ **b.** $4x(5x)$ **c.** $4x + 5x$ **d.** $4 + 5x$

30. **a.** $-3x(-2x)$ **b.** $-3 - 2x$ **c.** $-3x - 2x$ **d.** $-3(-2x)$

31. **a.** $x^2(x^3)$ **b.** $x^2 + x^3$ **c.** $x^3 + x^3$ **d.** $x^3(x^3)$

32. **a.** $4x^2 - 6x^2$ **b.** $6x^3 - 4x^2$ **c.** $6x^2(-4x^2)$ **d.** $4x^3(-6x^2)$

Explain the difference between each pair.

33. $2x^3$ and $(2x)^3$

34. $-\sqrt{a}$ and $\sqrt{-a}$

35. Square of x and square root of x

36. Square feet and cubic feet

37. -1^2 and $(-1)^2$

38. Area and volume

39. $\sqrt{9+16}$ and $\sqrt{9}+\sqrt{16}$

40. $(\sqrt{4}+\sqrt{9})^2$ and $(\sqrt{4})^2+(\sqrt{9})^2$

Building Number Skills

Multiply or divide. Reduce your answer if possible.

41. $\frac{3}{5}(30)$

42. $\frac{5}{8}(32)$

43. $\frac{3}{4}\left(\frac{8}{9}\right)$

44. $\frac{2}{3}\left(\frac{9}{16}\right)$

45. $3 \div \frac{1}{3}$

46. $2 \div \frac{1}{2}$

47. $8 \div \frac{4}{5}$

48. $10 \div \frac{5}{2}$

49. $2 \div \frac{4}{3}$

50. $2 \div \frac{4}{5}$

Challenge Problems

Here are two challenging problems that use what you have learned about scientific notation, circles and spheres, and the Pythagorean theorem.

1. When the railroad lays a new track, they always leave a small amount of space between the ends of adjacent rails. This is because the rails expand on hot days, and without some room between rails the tracks would buckle. Suppose a 15-mile section of rail expands 1 inch, causing the track to buckle as shown in the figure. (Not to scale!) How high will the rail rise in the middle? First, choose your best guess:
   a. 1 centimeter **b.** 2 inches
   c. 3 feet **d.** 15 yards

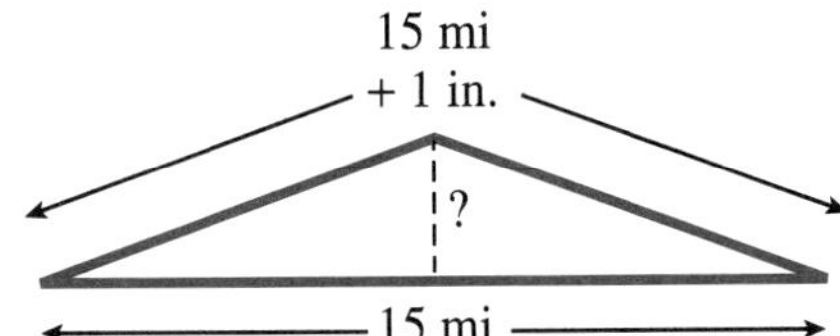

Now work out the answer. Be careful with units!

2. The radius of Earth is approximately 3963 miles. Suppose you have a string stretched tight around Earth's equator.
   a. How long is the string?
   b. Now suppose you add 10 feet to the length of the string. How long is it now (in feet)?
   c. Stretch out your new string so that it forms a circle a little bit bigger than the equator. What is the radius of the new circle?
   d. How high above the surface of Earth does the new circle of string sit?

5 Summary

Lesson 5.1

An **exponent** tells us how many times to use the **base** as a factor in the multiplication. A number or variable with an exponent is called a **power** of its base.

The area of a square of side s is given by $A = s^2$.

The volume of a cube of side s is given by $V = s^3$.

An exponent applies only to its base. When we raise a negative number to a power, we must enclose the negative number in parentheses.

To multiply a number by a power of 10 we move the decimal point to the right the same number of places as the exponent on 10.

To write a number in **scientific notation:**

1. Move the decimal point so that there is only one nonzero digit to the left of the decimal.
2. Multiply by a power of 10, where the exponent on 10 is the number of places you moved the decimal.

Lesson 5.2

Order of Operations

1. First, perform all operations inside parentheses or above or below a fraction bar.
2. Next, compute all powers.
3. Next, perform all multiplications and divisions in order from left to right.
4. Finally, perform all additions and subtractions in order from left to right.

The distance from the center of a circle to any point on the circle itself is called the **radius** of the circle. The **diameter** of a circle is the length of a line segment joining two points on the circle and passing through the center. The diameter of a circle is thus twice the radius.

The perimeter of a circle is called its **circumference.**

The circumference C of a circle is given by $C = \pi d$, where d is the diameter of the circle and $\pi \approx 3.14159$.

The circumference can also be expressed in terms of the radius: $C = 2\pi r$.

The area of a circle is given by $A = \pi r^2$, where r is the radius of the circle.

The volume of a sphere is given by $V = \frac{4}{3}\pi r^3$, where r is the radius of the sphere.

Lesson 5.3

Like terms are terms in which the variable parts are identical. The constant factor in each term is called the **coefficient.**

To add or subtract like terms:

1. Add or subtract the coefficients.
2. Do not change the variable part of the terms.

We cannot combine unlike terms.

Equivalent expressions have the same value when evaluated at the same number.

To multiply a term by a constant, we multiply the coefficient of the term by the constant.

When adding like terms, we do not add the exponents.

If we want to multiply together several factors, we can rearrange the factors in any order and get the same product. This is called the **commutative law** for multiplication.

We can multiply together two powers with the same base. However, we *cannot* simplify a product of two powers with *different bases.*

It is important to distinguish between multiplying and adding when using algebraic expressions.

Lesson 5.4

A number s is called a **square root** of b if $s^2 = b$.

Exery positive number has two square roots, one positive and one negative.

We use a special symbol called a **radical sign** to denote the *positive* square root of a number.

The square root of a negative number is undefined.

Order of Operations

1. First, perform all operations inside parentheses, above or below a fraction bar, or inside a radical.
2. Next, compute all powers and roots.
3. Next, perform all multiplications and divisions in order from left to right.
4. Finally, perform all additions and subtractions in order from left to right.

By following the order of operations, we can evaluate expressions that involve radicals.

Taking a square root is the opposite of squaring a number.

To solve an equation involving x^2, take the square root of both sides of the equation.

An equation involving the square of the variable has two solutions.

If more than one operation has been performed on the variable, we undo the operations in reverse order.

Lesson 5.5

A number that is the square of a whole number is called a perfect square.

We cannot find an exact decimal value for the square root of a number that is not a perfect square. However, we can use a calculator to find an approximation for the square root.

A **right triangle** is a triangle in which one of the angles is a right angle, or 90°. The side opposite the right angle is the longest side of the triangle, and is called the **hypotenuse.** The other two sides of the triangle are called the **legs.**

If we know the lengths of any two sides of a right triangle, we can find the third side by using the Pythagorean theorem: $a^2 + b^2 = c^2$, where c is the hypotenuse.

CHAPTER 5 REVIEW 

Add the following words to your glossary, if you have not done so already.

exponent	base	squared	cubed
power	coefficient	like terms	scientific notation
radius	diameter	circumference	pi
square root	radical sign	perfect square	Pythagorean theorem
right triangle	hypotenuse	legs	

1. What is a power? Give an example.

2. In the order of operations, when should powers be computed?

3. Draw a circle and label a diameter and a radius.

4. Draw a sphere and label its radius.

5. If the dimensions of a rectangle are given in inches, then its area will have units of ______________.

6. If the dimensions of a box are given in meters, then its volume will have units of ______________.

7. In the expression 5^3, 5 is called the ______________, and 3 is called the ______________.

8. Explain the difference between the expressions $3x$ and x^3.

9. Explain the difference between -4^2 and $(-4)^2$.

10. If the y^x key is not working on your calculator, how can you compute $(2.6)^4$?

True or false.

11. Like terms must have the same variable parts.

12. Two like terms are added or subtracted by combining their coefficients.

13. Powers should be computed right after multiplications.

14. The square root of a negative number is negative.

15. The volume of a sphere is measured in cubic units.

16. The circumference of a circle is measured in square units.

17. The Pythagorean theorem holds for all triangles.

18. The equation $x^2 = 7$ has no solution.

Write each expression as a product without using exponents.

19. 7^3a^5

20. $(-2)^3z^3$

21. Explain an easy way to multiply 38.762 times 10,000,000.

22. When you write a number in scientific notation, where should you locate the decimal point? How do you know what power of 10 you need?

23. Your calculator display shows as an answer [8.6 12]. What does this mean?

24. Explain how you would compute $\frac{32.5}{(2.8)(3.6)}$ using your calculator.

25. Explain the difference between 5.6^3 and 5.6×10^3.

26. Explain how to complete each of the two problems and then find the answers.

a. Evaluate x^2 for $x = 18$.

b. Solve $x^2 = 18$.

Simplify each pair of expressions according to the order of operations.

27. a. $3(-5)^2$ **b.** $3 - 5^2$

28. a. $\dfrac{12 - 2^3}{8 - 4^2}$ **b.** $\dfrac{(12 - 2)^3}{(8 - 4)^2}$

29. a. $(-4)(-3)^2$ **b.** $(-4 - 3)^2$

30. a. $16 - 8(-2)^2$ **b.** $16 - (8 - 2^2)$

Evaluate each expression for $a = -4$ and $b = -5$.

31. a. $3a^2$ **b.** $(3a)^2$

32. a. $-a^2 - a - 4$ **b.** $-ab^2(a - ab)$

Combine like terms.

33. $-7.4t - 5.1 + 7.4t$

34. $-3 + 13w - 7w$

35. $4x^2 - 3x - 2x^2$

36. $2z - 2z^2 - 2z^3 - 2z^2$

For Problems 37–44, the statement is *incorrect*. Find the error and correct the statement. If the statement cannot be simplified, write "cbs" ("cannot be simplified").

37. $5x^3 + 2x^3 = 7x^6$

38. $9v^4 - 3v^4 = 6$

39. $2c^5 - 8c^2 = -6c^3$

40. $3z^2 + 6z^5 = 9z^7$

41. $p^3(p^5) = p^{15}$

42. $h^4(h^4) = 2h^4$

43. $a^2b^3 = (ab)^5$

44. $y^8(-2y^3) = -2y^5$

45. The diameter of Jupiter is about 85,000 miles.

a. What is the distance around the equator of Jupiter, to the nearest mile?

b. What is the volume of Jupiter?

46. The Barringer meteor crater in Arizona is circular in shape and 4150 feet in diameter. Find its area.

Write your answers in scientific notation.

47. The distance from Earth to the Sun is approximately 92,900,000 miles. Assume that Earth's orbit is a circle.

a. Approximately what is the area (in square miles) of the region enclosed by Earth's orbit?

b. Approximately what is the distance that Earth travels in each trip around the Sun?

c. Approximately how long would it take a spaceship to follow the path of Earth's orbit if the ship traveled at 1000 miles per hour?

d. If the Sun were to expand until its outer edge reached Earth's orbit, approximately what would be the volume of the Sun?

48. Earth is roughly a sphere with radius approximately equal to 3960 miles.

a. Approximately what is the volume (in cubic miles) of Earth?

b. Approximately what is the distance in *feet* around Earth at the equator? (*Note:* 1 mile = 5280 feet.)

c. If a train could travel around the world along the equator, approximately how long would it take to circle Earth if it traveled at 100 feet per second?

d. Approximately what is the area (in square miles) of a circle with the same radius as Earth's?

For Problems 49–52, find the area and the perimeter of each figure. Round your answers to hundredths.

49.

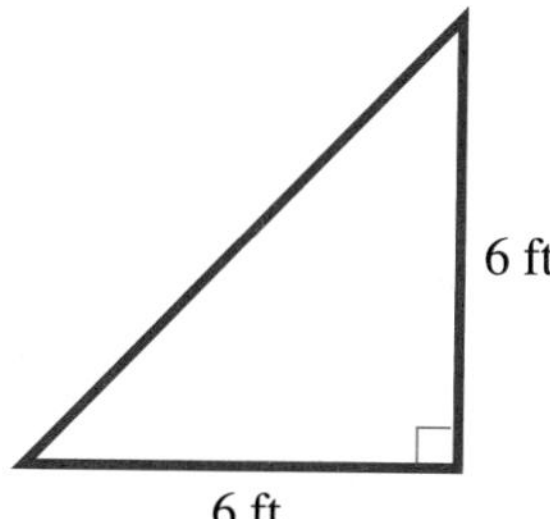

50.

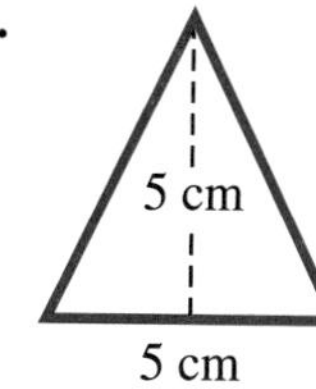

51.

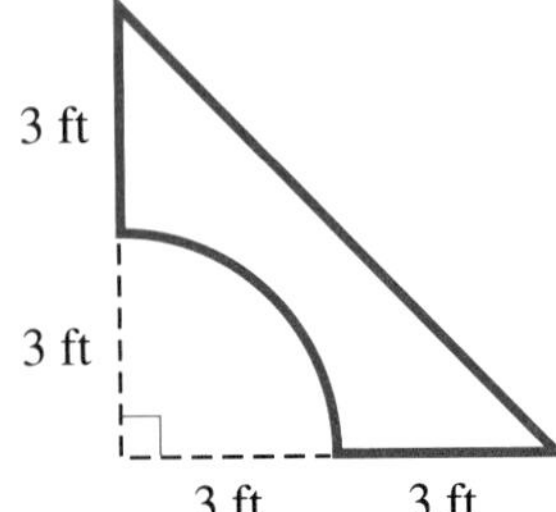

52.

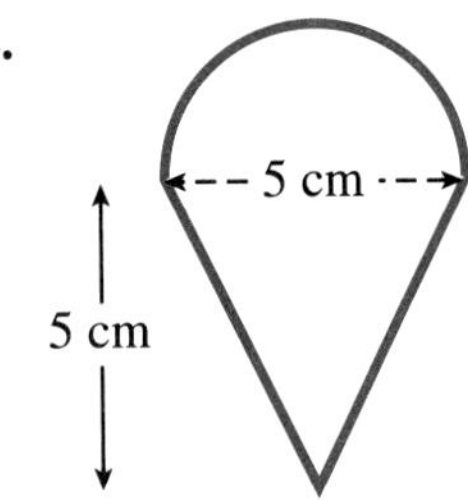

53. If s is the square root of b, then $s^2 =$ ________.

54. a. How many square roots does a positive number have?

b. How many square roots does a negative number have?

55. Explain the difference between $\sqrt{-16}$ and $-\sqrt{16}$.

56. Explain how to simplify $(\sqrt{5})^2$ without a calculator.

57. a. 2.25 is the square of what number?

b. 2.25 is the square root of what number?

58. Is there any number that is equal to its own square?

Simplify each expression according to the order of operations. Round your answers to two decimal places if necessary.

59. $8 - 2\sqrt{25}$

60. $\sqrt{8^2 - 5^2}$

61. $\dfrac{-6 + \sqrt{36 - 12}}{6}$

62. $16\sqrt{16\sqrt{16}}$

For Problems 63–66, solve each equation.

63. $z^2 = 81$

64. $2m^2 = 450$

65. $4w^2 - 8 = 2$

66. $\dfrac{k^2}{2.7} = 8.1$

67. If an object falls from a height of h centimeters, then the number of seconds t that the object takes before striking the ground is given by the formula $t = \dfrac{1}{7}\sqrt{\dfrac{h}{10}}$. How long would it take a marble to fall from a height of 1000 centimeters? Give an exact answer and an approximation rounded to three decimal places.

68. Pizza Hatch makes a 10-inch (diameter) pizza. The manager at Little Guy's Pizzeria wants to make a square pizza with the same area. How long must she make each edge of the square pizza? Give an exact answer and an approximation rounded to three decimal places.

69. Which of the following are the angles of a right triangle?

a. 60°, 60° 60°

b. 100°, 40°, 40°

c. 90°, 30°, 60°

d. 80°, 60°, 40°

70. How do you know which side of a right triangle is the hypotenuse?

71. Explain the Pythagorean theorem and illustrate your explanation with a sketch.

72. Is it possible that the sides of a right triangle could be 10 centimeters, 15 centimeters, and 20 centimeters in length? Why or why not?

73. Two of the sides of a right triangle are each 6 inches long. How long is the third side?

74. A path that runs along the diagonal of a rectangular park is about 1830 yards long. The width of the park is 500 yards. How long is the park? Round your answer to the nearest whole number.

Building Number Skills

For Problems 75 and 76, use a diagram to illustrate each function.

75. Reduce $\frac{8}{12}$

76. Reduce $\frac{12}{16}$

Write as a fraction in reduced form.

77. 28%

78. 45%

Use a diagram to find each product or quotient.

79. $\frac{1}{3}(21)$

80. $\frac{1}{6}(24)$

81. $\frac{3}{4}$ of $\frac{16}{9}$

82. $\frac{2}{3}$ of $\frac{3}{8}$

83. $6 \div \frac{1}{6}$

84. $2 \div \frac{1}{3}$

85. $10 \div \frac{5}{4}$

86. $5 \div \frac{5}{6}$

CHAPTER

6

Graphs

6.1 Line Graphs

A good way to organize and display information is to use a graph. Graphs are especially useful for illustrating the relationship between two variables. In Lesson 1.1 we considered bar graphs. The height of the bar illustrated the value of the variable. However, we don't really need the whole bar to convey that information; we could instead place a dot at the top of each bar. If we connect the dots with line segments, we have created a **line graph.**

EXAMPLE 1 

The bar graph in Figure 6.1a shows the percent of all flights that experienced late departures from U.S. airports.

a. Make a line graph that gives the same information.

b. Describe the overall trend in late departures over the time period 1995–2000.

Figure 6.1 Late departures

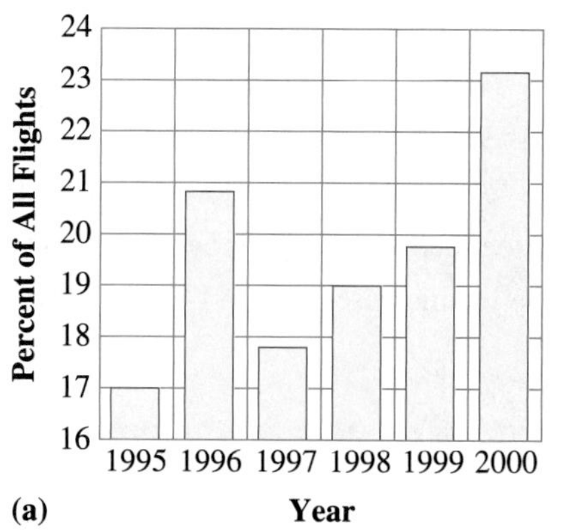

(a)

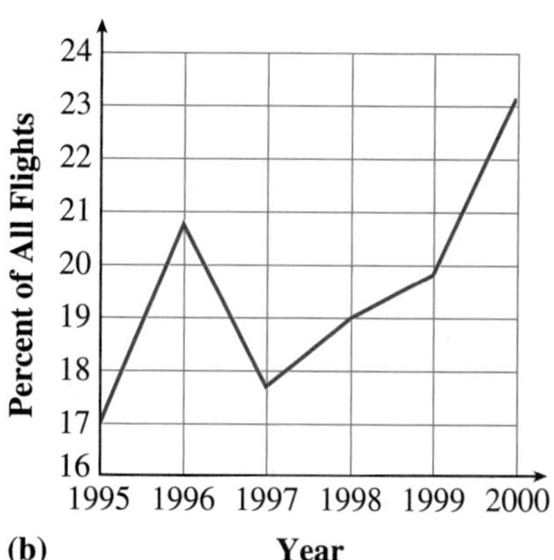

(b)

(Source: Department of Transportation)

EXERCISE 1

The bar graph in Figure 6.2a shows the percent of canceled flights at U.S. airports.

Figure 6.2 Canceled flights

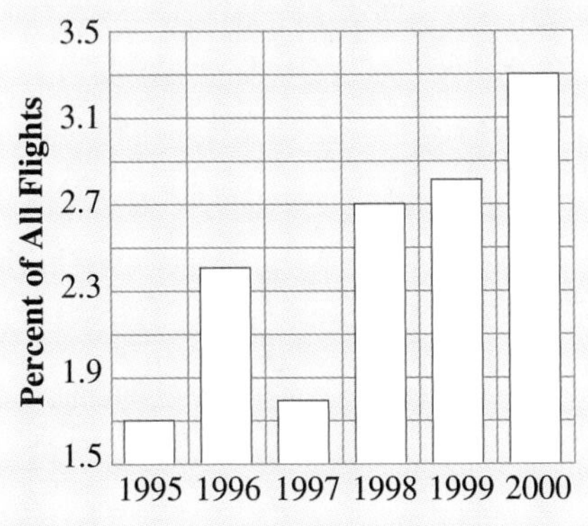

(a)

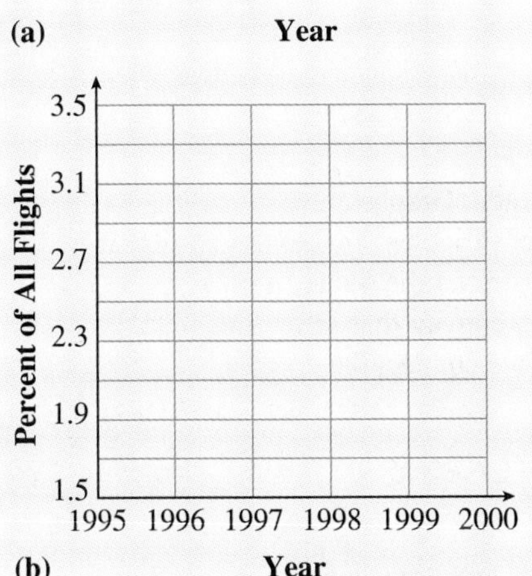

(b)

a. In Figure 6.2b, make a line graph that gives the same information.

b. In which year was there a decline in canceled flights?

Solution

a. Place a dot at the top of each bar and connect the dots with line segments, as shown in Figure 6.1b.

b. The percent of late departures increased over the 5-year period. There was a spike in 1996, returning to previous levels in the following year, and a steady increase from 1997 to 2000.

Try making your own line graph in Exercise 1.

We often use line graphs to illustrate trends in data over time.

EXAMPLE 2  

Figure 6.3 shows the annual fees for attending the University of California (UC) and the California State University (CSU) from 1984 to 1994.

Figure 6.3 University student annual fees

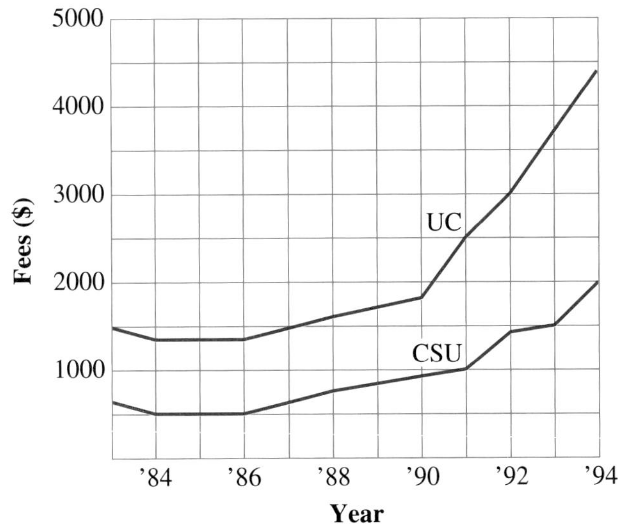

a. What was the annual fee at UC in 1985?

b. When did the annual fee at CSU first reach $1000?

c. How much did the annual fee at UC increase from 1992 to 1993?

d. How much did the annual fee at CSU increase over the same time period?

Solution

a. The annual fee at UC was approximately $1300.

b. The annual fee at CSU first reached $1000 in 1991.

c. At UC the fee increased from about $3000 to $3700, or by about $700.

d. At CSU the fee increased from $1400 to $1500, or about $100.

EXERCISE 2

The bar graph in Figure 6.4a shows the total annual U.S. sales of motorboats and sailboats over the decade from 1991 to 2000.

Figure 6.4 Boat sales

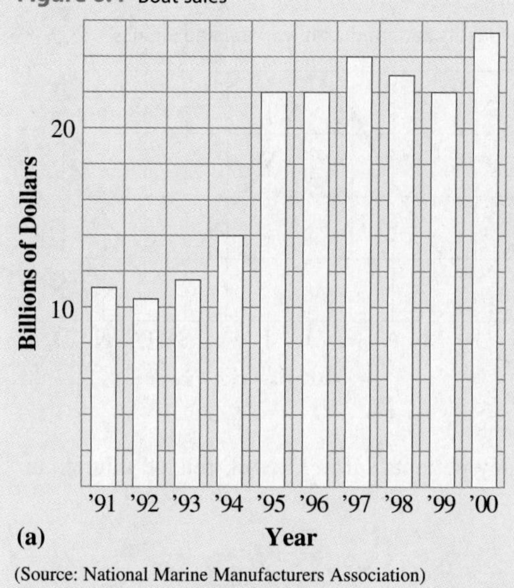

(a)

(Source: National Marine Manufacturers Association)

Billions of Dollars
20
10
'91 '92 '93 '94 '95 '96 '97 '98 '99 '00
Year

(b)

a. Use the grid in Figure 6.4b to convert the bar graph into a line graph.

b. How much did annual sales increase over the decade?

c. In which years did annual sales decrease over the previous year?

d. In which year did the greatest increase in sales occur?

ANSWERS TO 6.1 EXERCISES

1a.

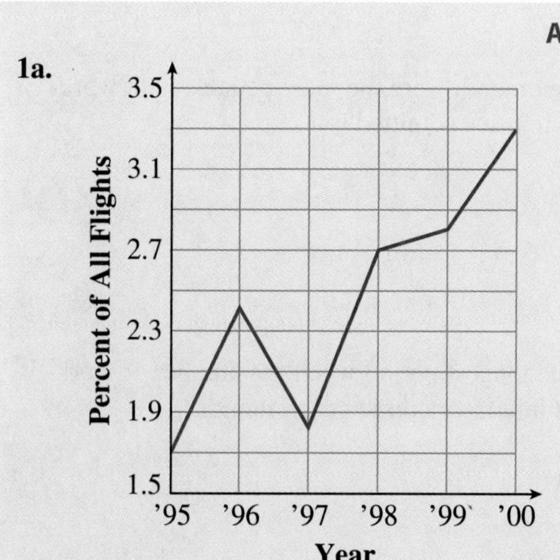

1b. 1997

2a.

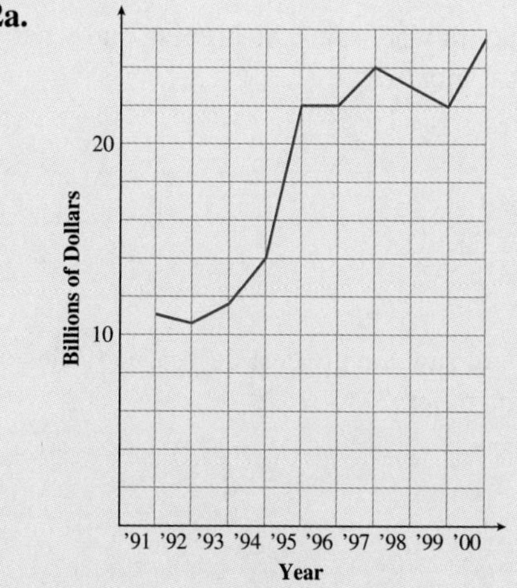

2b. 14 billion dollars

2c. 1992, 1998, and 1999

2d. 1995

HOMEWORK 6.1 

1. Figure 6.5 shows monthly sales of roller-blading equipment at the Sports Exchange last year.

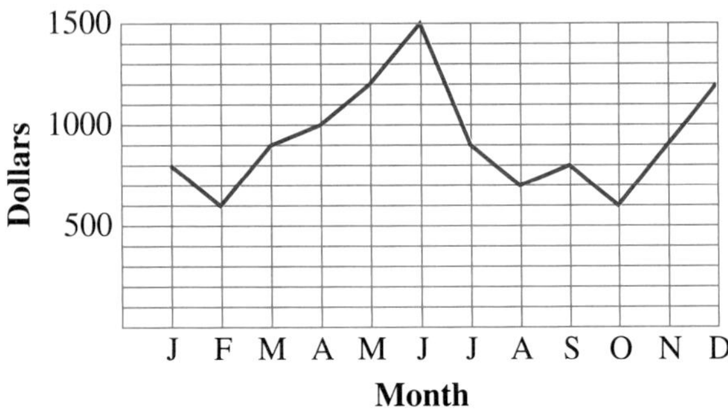

Figure 6.5 Sales of roller-blading equipment

a. How much roller-blading equipment was sold in April?

b. In which month(s) did the Sports Exchange sell $900 worth of equipment?

c. In which month were sales the highest, and what were the sales that month?

d. Which month saw the greatest decrease in sales over the previous month?

2. Figure 6.6 shows the number of patients admitted to Mercy Hospital with work-related injuries last year.

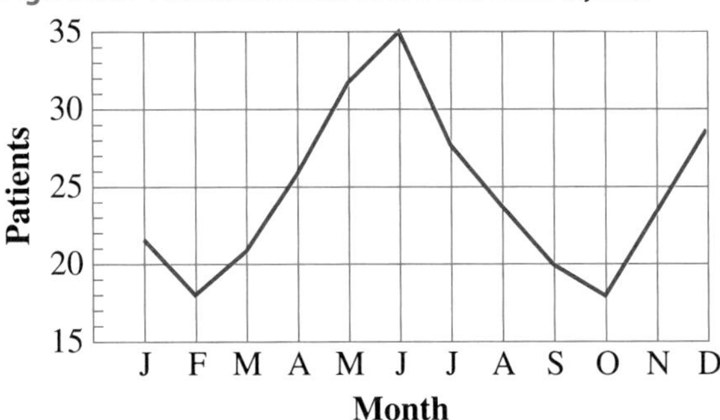

Figure 6.6 Patients admitted with work-related injuries

a. How many patients suffered work-related injuries in March?

b. In which month(s) were the fewest patients with work-related injuries admitted to the hospital?

c. In which month were the most patients with work-related injuries admitted?

d. Find the longest period when the number of work-related injuries increased each month.

3. Figure 6.7 shows the number of students enrolled part-time or full-time at 2-year colleges over a 25-year period.

Figure 6.7 Total enrollment in 2-year colleges

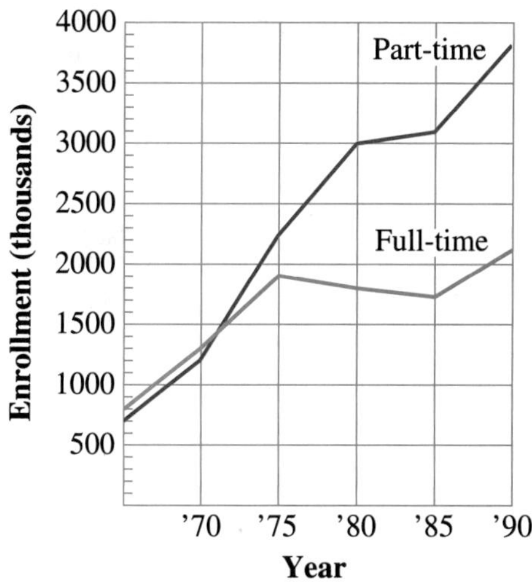

a. How many students were enrolled part-time at 2-year colleges in 1970?

b. In what year were approximately 1,700,000 students enrolled full-time?

c. What was the total enrollment (part-time and full-time) in 1990?

d. By how much did part-time enrollment increase from 1975 to 1980?

4. Figure 6.8 shows the number of cases of AIDS diagnosed in women in Los Angeles County over several years.

Figure 6.8 Incidence of AIDS among women

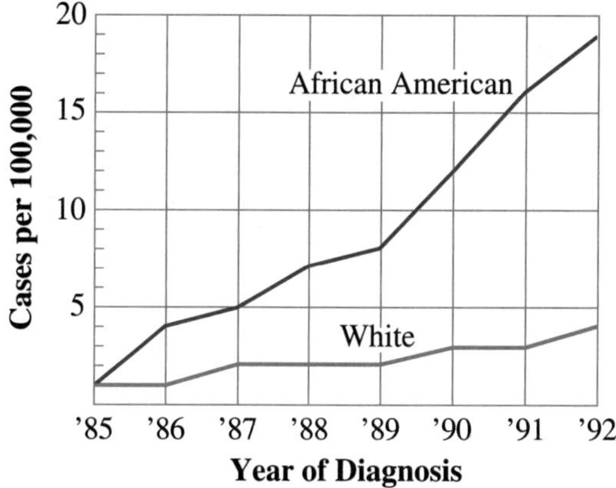

a. How many cases of AIDS were diagnosed among African-American women in 1986?

b. How many cases were diagnosed among white women in 1991?

c. What was the total number of cases per 100,000 diagnosed in 1989?

d. What was the increase in the number of cases among African-American women from 1989 to 1990?

5. a. The table shows women's annual earnings as a percent of men's from 1980 to 1989. Use the data to create a line graph on the grid provided.

Year	1980	1981	1982	1983	1984
Women's earnings (% of men's)	60	59	61.5	63.5	63.5

Year	1985	1986	1987	1988	1989
Women's earnings (% of men's)	64.5	64	65	66	68.5

(Source: Bureau of Labor Statistics)

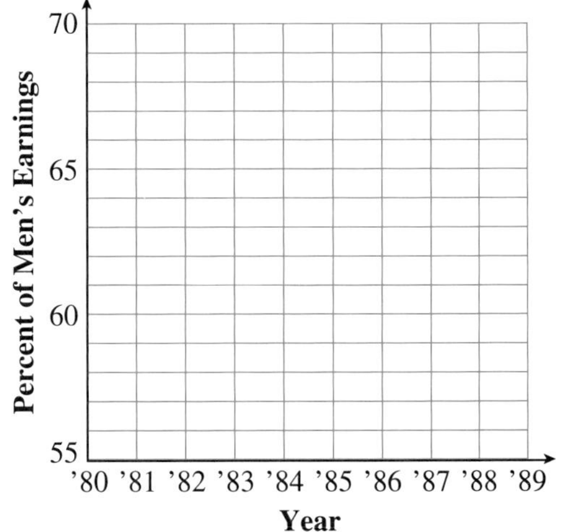

b. In which year(s) in the 1980s did women's earnings make the greatest gain, relative to men's?

c. In which year(s) did women's earnings decline, relative to men's?

d. What was the net gain in women's earnings (as a percent of men's) over the decade of the 1980s?

6. a. The table shows women's annual earnings as a percent of men's from 1990 to 1999. Use the data to create a line graph on the grid provided.

Year	1990	1991	1992	1993	1994
Women's earnings (% of men's)	71.5	69.5	70.5	71.5	72

Year	1995	1996	1997	1998	1999
Women's earnings (% of men's)	71.5	74	74.5	73	73

(Source: Bureau of Labor Statistics)

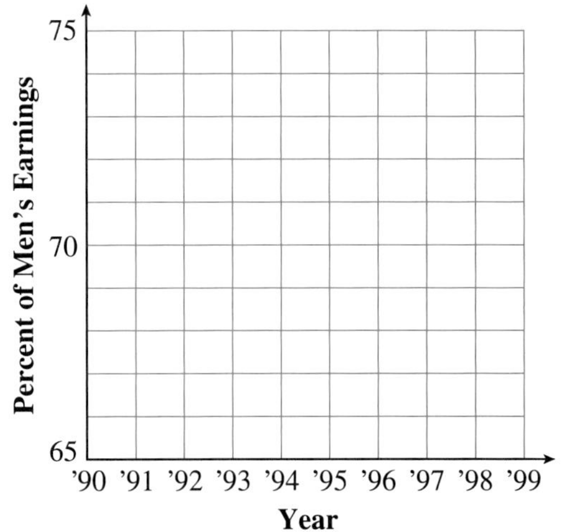

b. In which year in the 1990s did women's earnings make the greatest gain, relative to men's?

c. In which year(s) did women's earnings decline, relative to men's?

d. What was the net gain in women's earnings (as a percent of men's) over the decade of the 1990s? How does this compare to progress in the 1980s?

7. The bar graph in Figure 6.9 shows the projected increase in jobs in southern California through 2025.

Figure 6.9 Jobs

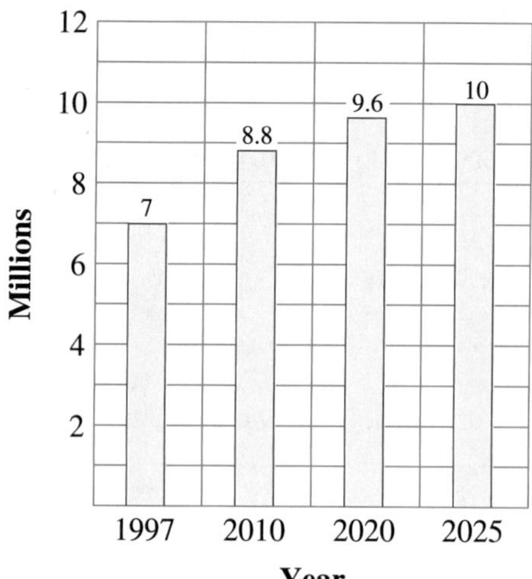

(Source: *Los Angeles Times,* 15 December 2000)

a. Use the grid provided to create a line graph for the same information.

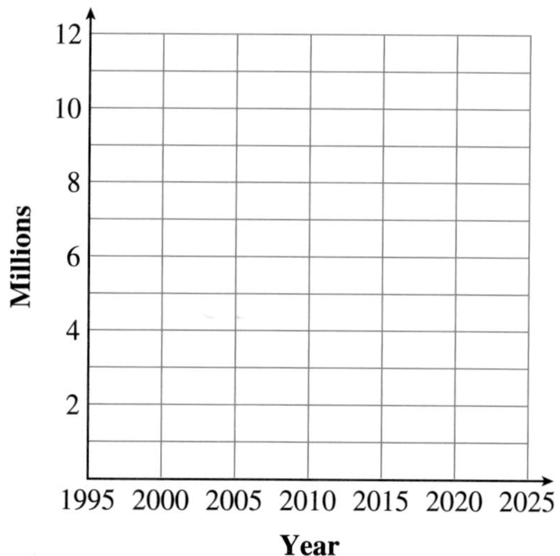

b. What feature of the original bar graph could be considered misleading?

8. The bar graph in Figure 6.10 shows the projected increase in the number of households in southern California through 2025.

Figure 6.10 Households

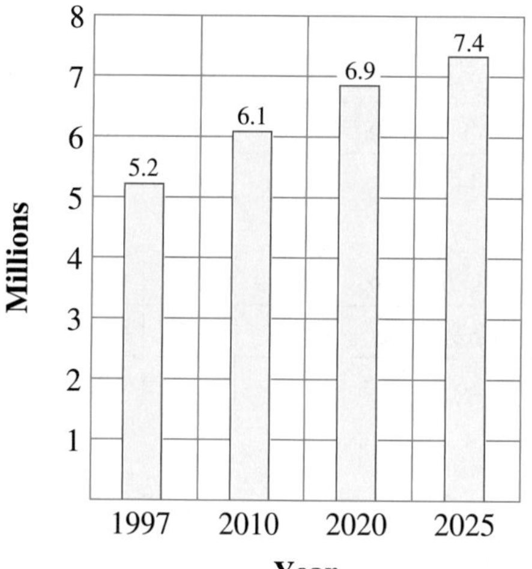

(Source: *Los Angeles Times* 15 December 2000)

a. Use the grid provided to create a line graph for the same information.

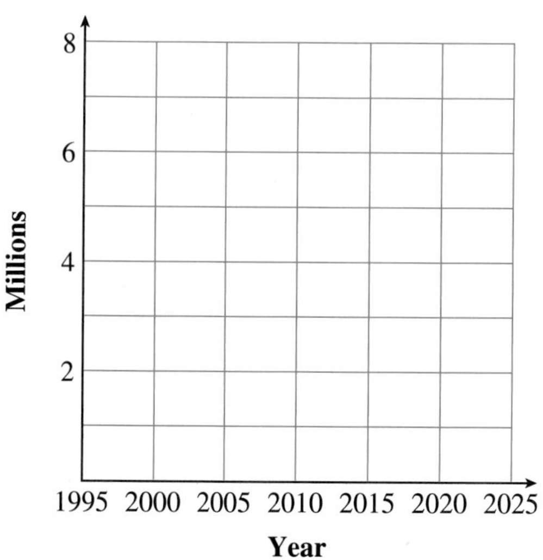

b. What feature of the original bar graph could be considered misleading?

9. A national opinion poll asked voters the question "Do you agree or disagree that most men are better suited emotionally for politics than women?" The results are shown in the table. (Note that the question was not asked every year.)

Year	1974	1975	1977	1978	1983	1985
Disagree (%)	49	48	47	55	63	61
Agree (%)	44	48	46	43	35	38

Year	1986	1988	1989	1990	1991	1993
Disagree (%)	62	66	67	69	70	75
Agree (%)	36	33	28	26	25	20

(Source: National Opinion Research Center)

a. On the grid provided, make two line graphs illustrating the data.

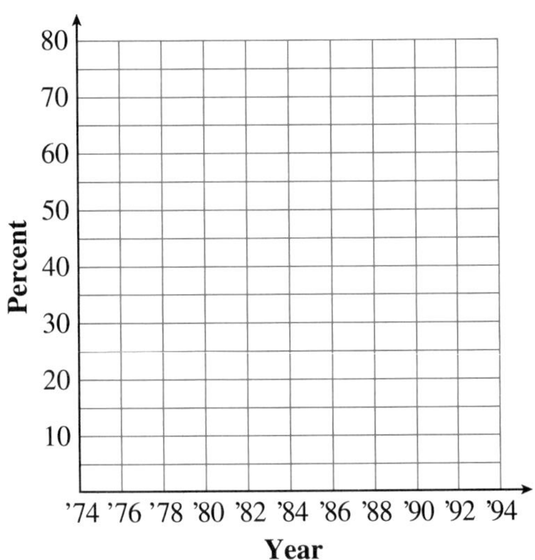

b. What trends in public opinion do you see in the graphs?

c. Why don't the percent of people who agree and the percent who disagree in each year add up to 100%?

10. As the "baby boomers" age, more people in their 40s and 50s are riding motorcycles. The table shows the number of motorcycle fatalities during the 1990s.

Year	1994	1995	1996	1997	1998	1999
Fatalities (under age 35)	1500	1440	1280	1150	1240	1170
Fatalities (35 and over)	820	800	900	950	1050	1300

(Source: National Highway Traffic Safety Administration)

a. On the grid provided, make two line graphs illustrating the data.

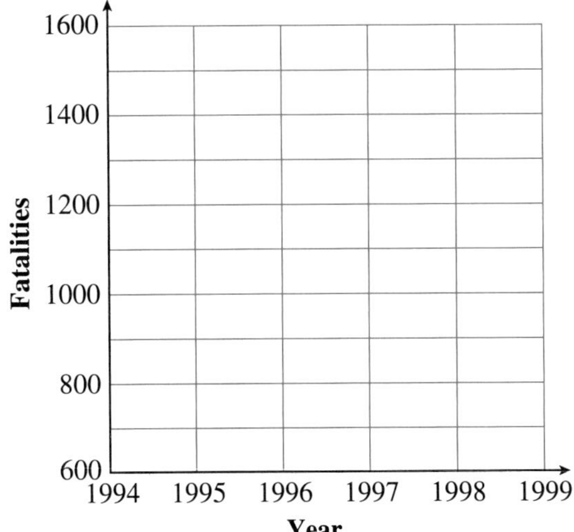

b. What trends do you see in fatalities among motorcycle drivers?

c. During which year did the number of fatalities among older drivers first exceed the number of fatalities among younger drivers?

11. The UCLA Bruins played the Ball State Cardinals during the NCAA basketball tournament in 2000. The table shows the Bruins' lead by the minute in the first half of the game:

Minute	0	1	2	3	4	5	6	7	8	9	10
Bruins' lead	0	0	3	3	−1	1	5	5	3	6	5

Minute	11	12	13	14	15	16	17	18	19	20
Bruins' lead	7	9	4	4	2	4	2	2	−1	−4

a. Make a line graph on the grid provided.

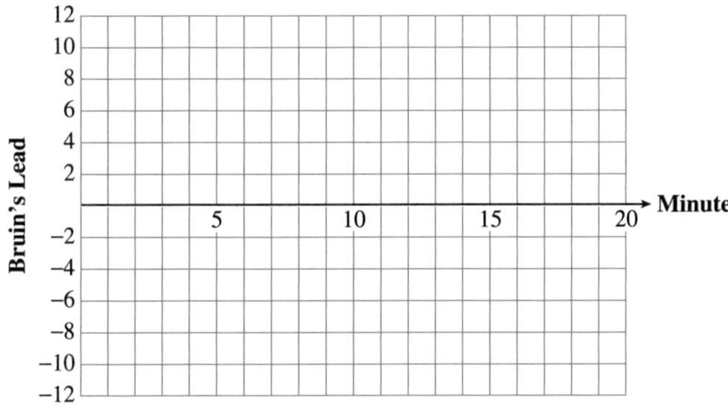

b. When did the Bruins have their biggest lead over the Cardinals?

c. What was the Cardinals' net gain over the Bruins during the last 8 minutes of the first half?

d. Who was winning at the end of the first half? By how many points?

12. The UCLA Bruins played the Ball State Cardinals during the NCAA basketball tournament in 2000. The table shows the Bruins' lead by the minute in the second half of the game:

Minute	0	1	2	3	4	5	6	7	8	9	10
Bruins' lead	−4	−6	−4	−4	2	1	3	6	9	12	11

Minute	11	12	13	14	15	16	17	18	19	20
Bruins' lead	7	10	10	12	10	9	4	5	5	8

a. Make a line graph on the grid provided.

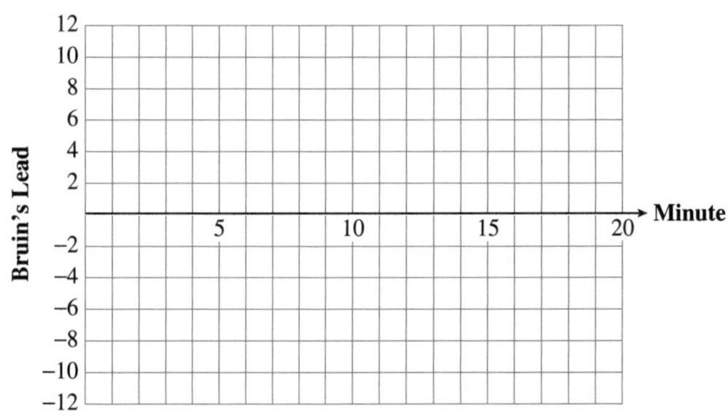

b. When did the Cardinals have their biggest lead over the Bruins?

c. When did the Cardinals' biggest net gain over the Bruins occur, and how many points did they gain?

d. Who won the game? By how many points?

Simplify each pair of expressions. (See Lesson 5.2 to review the order of operations.)

13. a. $\dfrac{5 - 2^3}{5 - 2}$ **b.** $\dfrac{5(-2)^3}{5(-2)}$

14. a. $\sqrt{5^2 + 12^2}$ **b.** $\sqrt{5^2} + \sqrt{12^2}$

15. a. $\sqrt{(4)(9)}$ **b.** $(\sqrt{4})(\sqrt{9})$

16. a. $(\sqrt{5 \cdot 7})^2$ **b.** $(\sqrt{5})^2 \cdot (\sqrt{7})^2$

17. a. $\sqrt{15^2 - 12^2}$ **b.** $\sqrt{15^2} - \sqrt{12^2}$

18. a. $(\sqrt{1} + \sqrt{4})^2$ **b.** $(\sqrt{1})^2 + (\sqrt{4})^2$

19. a. $\dfrac{18 + \sqrt{81}}{3}$ **b.** $\dfrac{18\sqrt{81}}{3}$

20. a. $\sqrt{\dfrac{100}{25}}$ **b.** $\dfrac{\sqrt{100}}{\sqrt{25}}$

Building Number Skills

Every fraction can be expressed in many different but equivalent ways. For example, $\frac{5}{10}$ is the same portion of one whole as $\frac{1}{2}$, so the two fractions are equal.

Use a diagram to build each fraction. (See Appendix A.1 to review building fractions.)

Example: $\frac{2}{3} = \frac{?}{6}$

Illustrate two-thirds.
Divide the whole into six equal pieces.
How many are shaded?

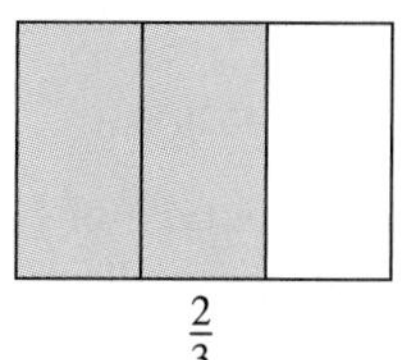

21. $\frac{3}{4} = \frac{?}{8}$

22. $\frac{1}{2} = \frac{?}{6}$

23. $\frac{2}{3} = \frac{?}{9}$

24. $\frac{2}{3} = \frac{?}{12}$

25. $\frac{3}{4} = \frac{?}{12}$

26. $\frac{1}{2} = \frac{?}{8}$

27. $\frac{3}{8} = \frac{?}{16}$

28. $\frac{5}{6} = \frac{?}{12}$

29. $\frac{5}{8} = \frac{?}{24}$

30. $\frac{3}{8} = \frac{?}{32}$

Do you see a way to do these exercises without using a diagram? Describe your method and give an example.

6.2 Display of Data

Researchers in many fields collect large quantities of information. They hope to learn more about their subjects by analyzing trends in the data or finding connections between variables. Their first task is to organize the data in useful ways. We consider two types of graphs used by statisticians, histograms and boxplots.

A. Histograms

A **histogram** is a type of bar graph that shows how frequently each value of a variable occurs in a collection of data.

EXAMPLE 1

In her sociology class, Alida was asked to investigate how often American families change their place of residence. She asked 15 classmates how many different homes they lived in before coming to college. Here are the results of her survey:

Aaron	2	Hillary	1	Steffi	6
Barbara	2	Juana	6	Tung	3
David	6	Mariel	3	Valerie	4
Elisa	3	Paula	3	Will	3
Geraldo	3	Sean	12	Xavier	4

a. To analyze the data, we begin by making a **line plot.** Use a ruler to mark off the values from 1 to 12 on a number line. Then place an X over the appropriate number for each data value. For example, Aaron lived in two different places, so we place an X above 2 on our number line. Continue until all the data values are displayed, being careful to make each X the same height. The completed line plot is shown in Figure 6.11.

Figure 6.11 Number of homes before starting college

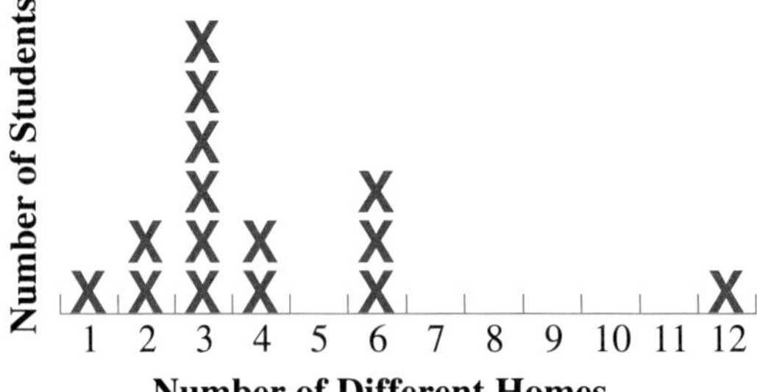

b. The largest and smallest values on the line plot are called the **extremes** of the data. For Alida's survey, the extremes are 1 and 12. The difference between the largest and smallest data values is called the **range** of the data. The range of this collection of data is

$$\begin{aligned}\textbf{range} &= \textbf{high value} - \textbf{low value}\\ &= 12 - 1 = 11\end{aligned}$$

c. Next we'll use our line plot to construct a histogram. Over each value on the number line, we construct a bar with the same height as the column of Xs. The histogram is shown in Figure 6.12.

Figure 6.12 Number of homes before starting college

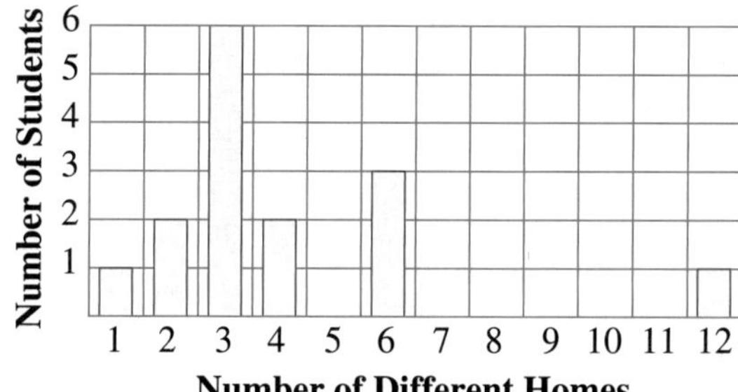

EXERCISE 1

Etna conducted a poll of 27 households asking how many pets they kept. Figure 6.13a shows a line plot for the results of her poll. Make a histogram for the data on the grid in Figure 6.13b.

Figure 6.13 Pets in household

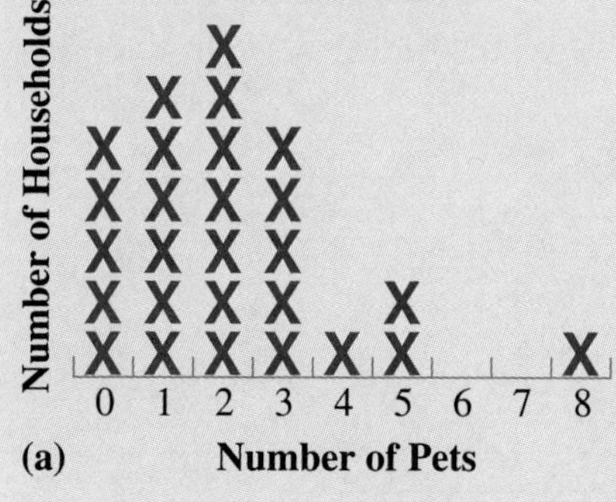

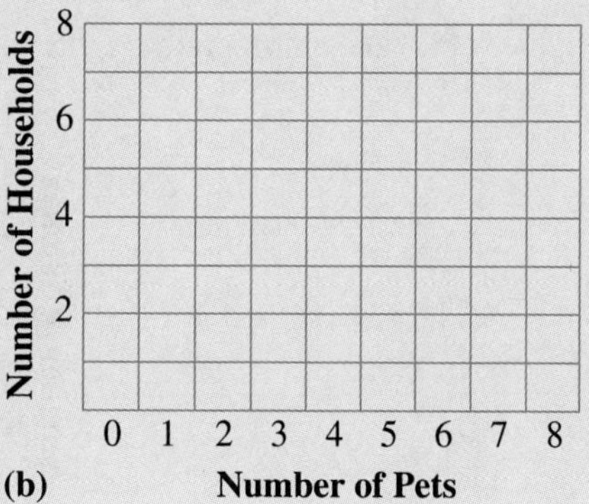

Using Intervals

If the data extend over a wide range, we can divide the line plot into intervals.

EXAMPLE 2

Each student in a health class counted his or her pulse for 1 minute. The results are recorded below:

49	54	63	63	66	67	68	69	69
71	72	73	74	76	77	78	81	94

Make a histogram for the data.

Solution

We first make a line plot, using intervals of length 5, as shown in Figure 6.14a. The first interval includes all values from 45 to 49, the next will include all values from 50 to 54, and so on. The corresponding histogram is shown in Figure 6.14b.

Figure 6.14

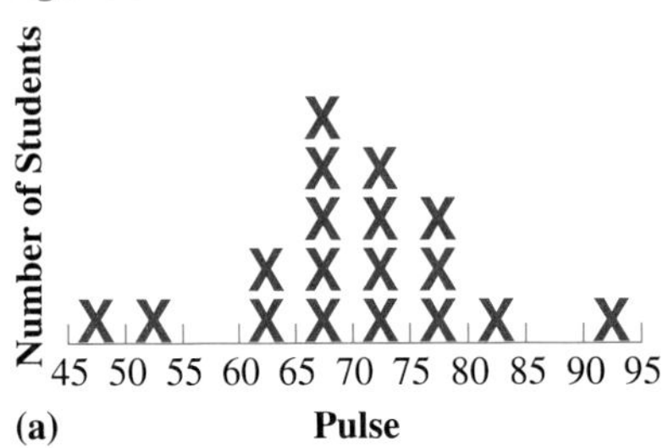

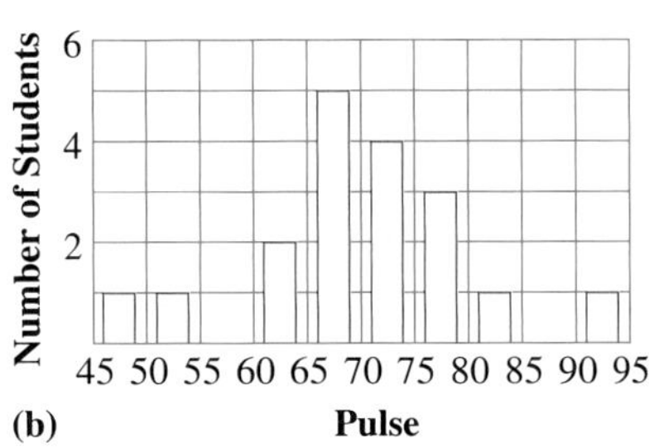

EXERCISE 2

The members of a women's soccer team were weighed as part of their annual physical exams. The weights, in pounds, are as follows:

112	115	119	122	124	126
131	132	134	136	138	139
139	140	144	146	152	164

On the grids, make a line plot (Figure 6.15a) and a histogram (Figure 6.15b) for the data, using intervals of length 10.

Figure 6.15

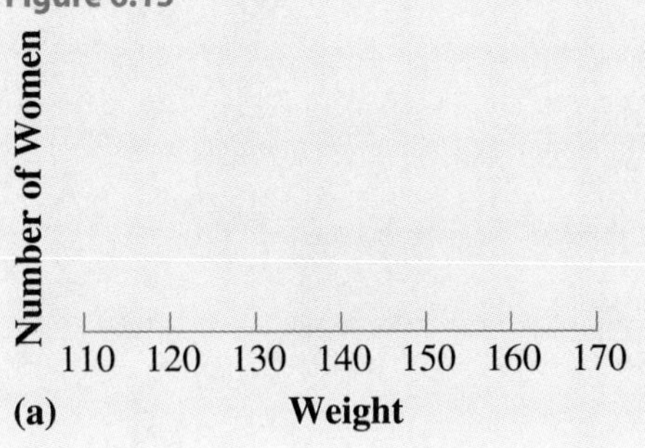

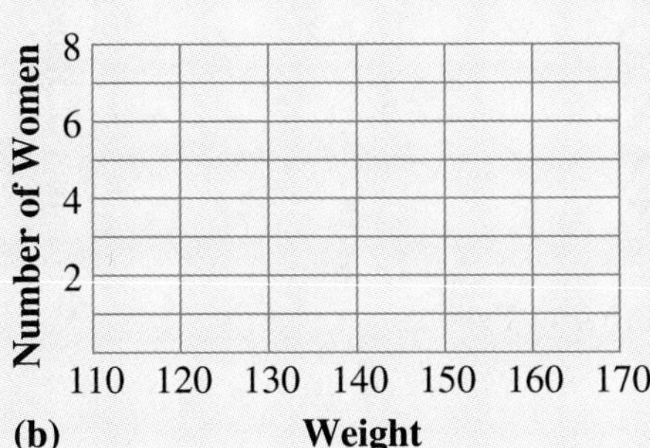

The Mean and the Mode

The value that occurs most frequently is called the **mode** of a data set. For the survey in Example 1, the mode is 3 because more students lived in three different places. The mode is one of three different average values considered by statisticians.

A second average value is called the **mean.** The mean is the average value familiar to most people. We have computed this average in previous lessons using the formula

$$\textbf{mean} = \frac{S}{n}$$

where S is the sum of all the data values and n is the number of data values. You can use your calculator to check that the mean of the survey data in Example 1 is 4.

Which average value is better, the mean or the mode? Both are useful in different situations. The mode is useful if one score occurs more frequently than the others. In some sense it is the most popular data value. The mean can be strongly affected by the presence of a few "outlying" data values. In Example 1, most of the data values are 6 or below, with one much larger value, 12. This value affected the calculation of the mean, causing it to be slightly higher than the typical response from the people surveyed.

The Median

A third type of average is called the **median.** The median is the middle score in a collection of data.

EXAMPLE 3

Compute the mean, the median, and the mode for the following ages of students in a history seminar:

17 18 19 19 19 20 21 22 22 46

Solution

The mean is

$$\frac{S}{n} = \frac{17 + 18 + 19 + 19 + 19 + 20 + 21 + 22 + 22 + 46}{10}$$

$$= \frac{233}{10} = 22.3$$

The median is the middle value when the ages are arranged in order. In this example, there is an even number of students in the seminar, so we take the age halfway between the middle two ages:

$$\text{median} = \frac{19 + 20}{2} = 19.5$$

The median reflects more accurately the typical age of the students than the mean does. This is because the one atypical age, 46, affects the calculation of the mean more than it affects the median.

The mode is 19, because there are more students (three of them) of age 19 than of other ages.

EXERCISE 3

The students in a media arts class were asked how many VCRs they owned. Compute the mean, the median, and the mode for the data:

0 1 1 1 1 2 2 2 3 5

The mean, the median, and the mode are called **measures of central tendency** because they usually fall somewhere in the middle of the range of data. Each of these average values summarizes the data by giving the value of a typical data point.

ANSWERS TO 6.2A EXERCISES

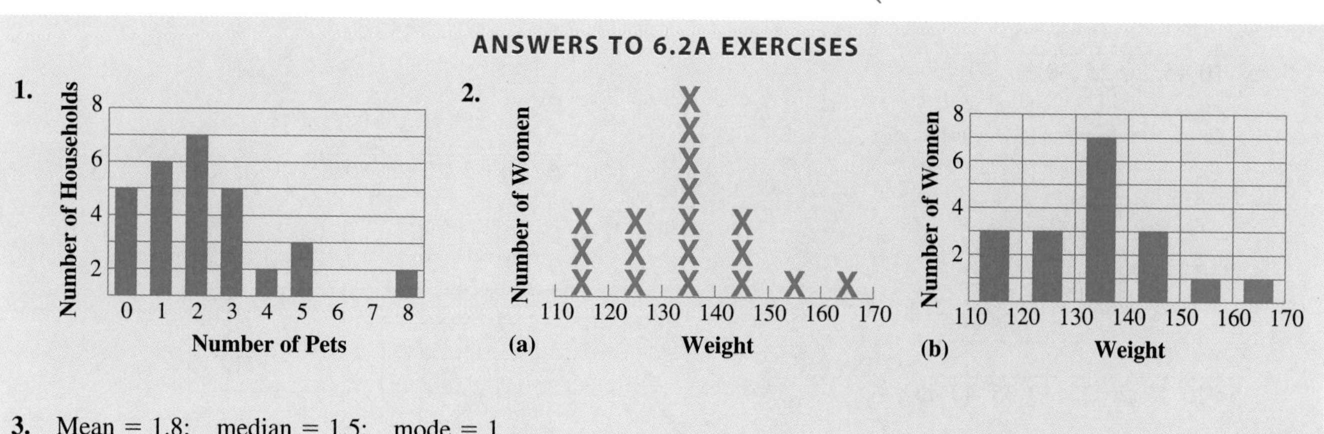

3. Mean = 1.8; median = 1.5; mode = 1

HOMEWORK 6.2A 

For the data sets in Problems 1–6: (a) Find the extremes and compute the range; (b) make a line plot of the data and then construct a histogram; and (c) compute the mean, median, and mode of the data.

1. Wendy surveyed the students in her geology class to see how many credit hours each is enrolled in this semester. She recorded the following responses:

5	6	12	18	12	11	8	15	12
16	9	12	15	14	9	17	13	

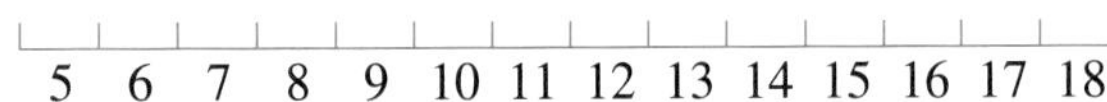

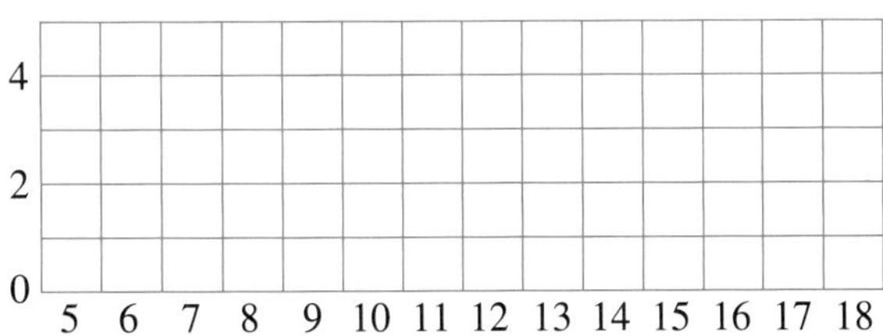

2. Dawn asked each of the workers in her office to keep track of how many books they read last year. She recorded the following results:

10	3	7	4	6	9	2	4	8
12	6	7	2	6	3	5	6	

2 3 4 5 6 7 8 9 10 11 12

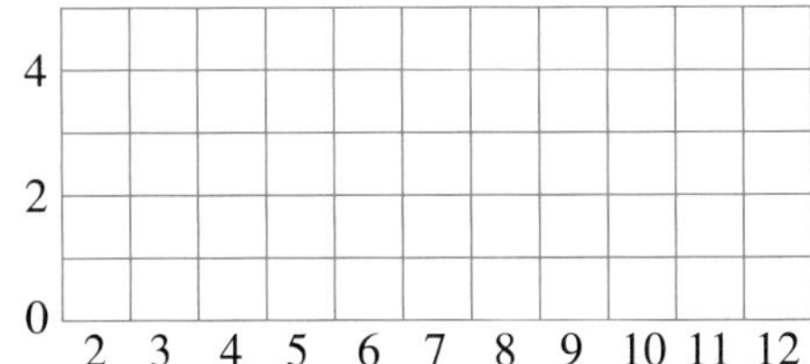

3. Professor Jennings asked the students in her philosophy class to fill out a survey about their career goals. She recorded the following responses to the question "How many hours do you work per week?" Use intervals of 5 hours to make your line plot.

25	20	18	32	6	10	40
12	24	20	35	16	12	15
30	22	12	15	12	26	18

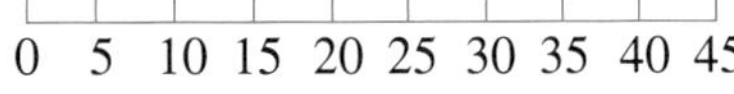

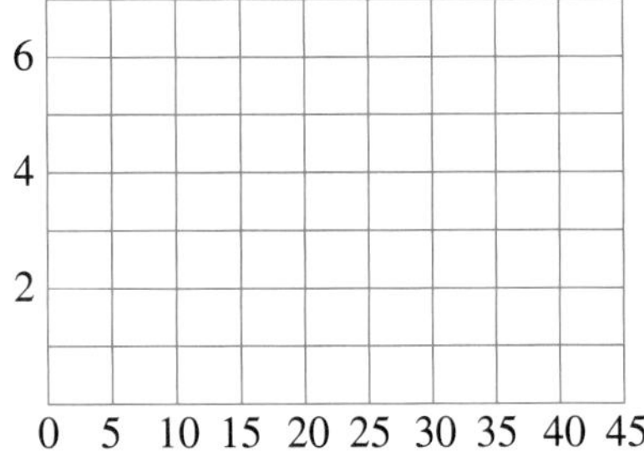

4. A representative of the Transit District surveyed students in the cafeteria to determine how many minutes they spend commuting one way to school. Each student agreed to record his or her commuting time each day for a week and report the average time. The representative collected the following results, in minutes. Use intervals of 5 minutes to make your line plot.

10	18	25	8	25	15	45
42	6	22	18	22	12	16
28	24	33	37	18	16	32

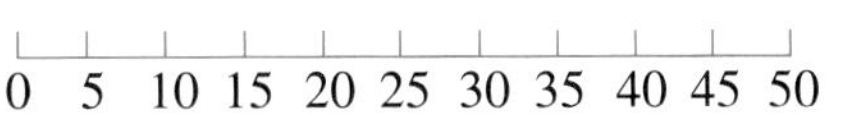

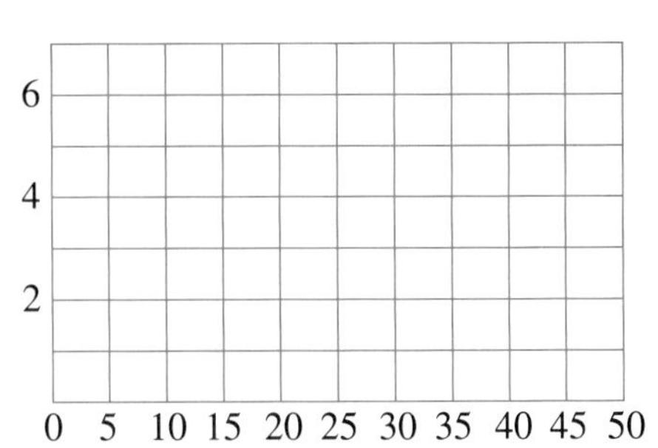

5. The following figures give the average beginning salaries offered to new graduates from various majors. Use intervals of $1000 to make your line plot. (*Source:* 1994–1995 *American Almanac*)

Accounting	$27,493
Business	$24,555
Marketing	$24,361
Civil engineering	$29,211
Chemical engineering	$39,482
Computer engineering	$33,963
Electrical engineering	$34,313
Mechanical engineering	$34,460
Nuclear engineering	$34,755
Petroleum engineering	$38,387
Engineering technology	$29,236
Chemistry	$28,002
Mathematics	$26,524
Physics	$26,835
Humanities	$24,373
Social sciences	$22,684
Computer science	$31,329

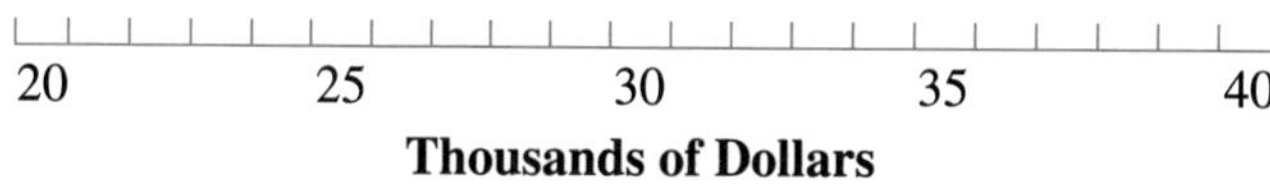

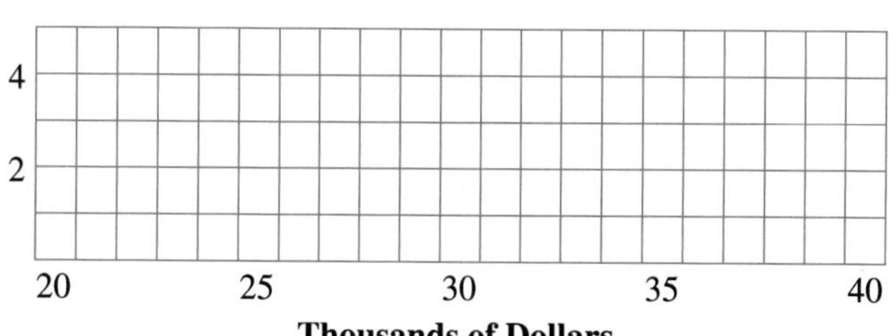

6. The following figures show the number of Americans, in millions, who participate in selected sports. Use intervals of 0.5 million to make your line plot. (*Source:* 1994–1995 *American Almanac*)

Aerobic exercising	5.8	Golf	3.0
Backpacking	1.5	Hiking	2.6
Baseball	1.9	Hunting	2.6
Basketball	4.9	Racquetball	2.0
Bicycle riding	5.5	Running or jogging	4.5
Bowling	7.0	Skiing	2.9
Camping	5.3	Soccer	1.3
Exercise walking	6.3	Softball	2.9
Exercising with equipment	6.9	Swimming	7.7
Fishing	6.0	Tennis	3.9
Football	2.8	Volleyball	5.1

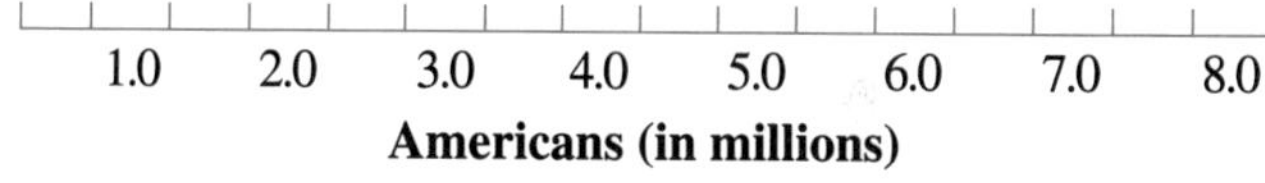

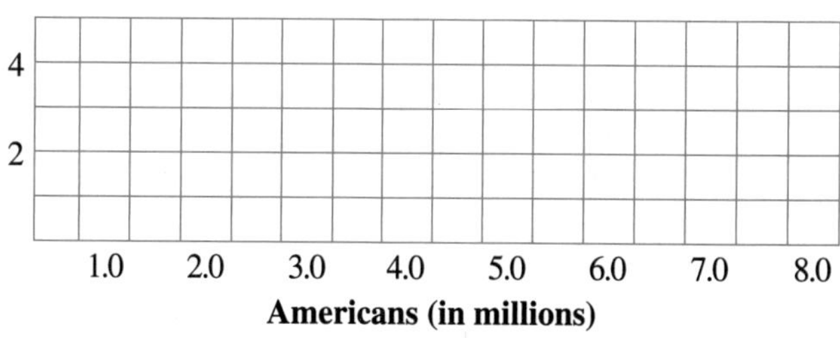

7. Seven houses are for sale in a particular neighborhood. Their listed prices are shown in the table.

House	*Price ($)*
A	190,000
B	100,000
C	110,000
D	2,500,000
E	100,000
F	170,000
G	120,000

a. Find the mean, median, and mode of the house prices.

b. Explain the advantages and disadvantages of each as an average value.

c. If you are a contractor planning to build a dozen similar houses in this neighborhood, which average would be most useful to you in planning what kind of house to build? Why?

8. There are only seven acting roles in a new movie. The salaries of the seven actors are shown in the table.

Actor	*Salary ($)*
A	1,200,000
B	100,000
C	20,000
D	10,000
E	7,000
F	7,000
G	7,000

a. Find the mean, median, and mode of the salaries.

b. You are summarizing the costs of the movie to potential producers. Which average salary do you report?

c. You are writing a brochure to recruit new actors to the studio. Which average salary do you report?

9. A researcher polled 100 families with children in her city in order to estimate the average number of children per family. She obtained the information shown in the following **frequency table**:

Number of children	1	2	3	4	5	6
Number of families	28	32	20	12	6	2

a. Make a histogram for the data.

Number of Families: 10, 20, 30

1 2 3 4 5 6

Number of Children

b. Calculate the mean of the data. [*Hint:* Use the formula mean $= \frac{S}{n}$. To find S, multiply each data value (number of children) times the frequency of its occurrence (number of families), and add the results.]

10. Grant asked 80 members of his health club how many hours per week they exercised. He recorded the information in the following frequency table:

Number of hours	2	3	4	5	6	8	10
Number of people	10	18	21	12	8	8	3

a. Make a histogram for the data.

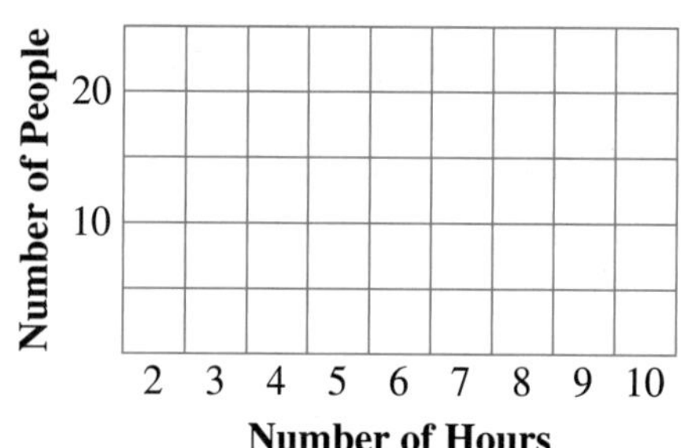

b. Calculate the mean of the data. [*Hint:* Use the formula mean $= \frac{S}{n}$. To find S, multiply each data value (number of hours) times the frequency of its occurrence (number of people), and add the results.]

11. Figure 6.16 shows the results of a survey asking people how many servings of vegetables they eat per week.

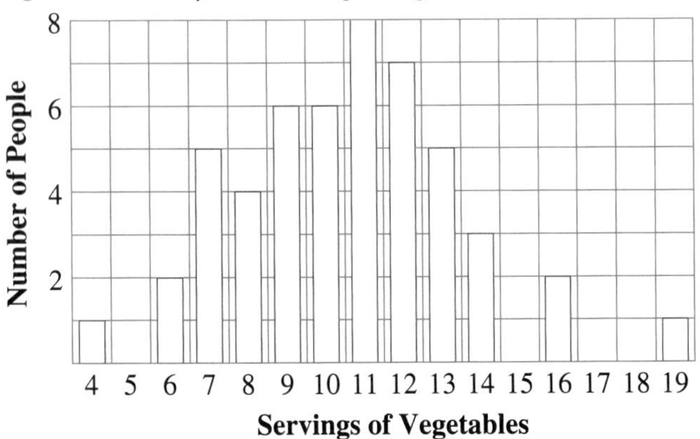

Figure 6.16 Survey results: servings of vegetables

a. Find the extremes of the data and calculate the range.

b. What is the mode of the data?

c. How many people eat 12 or more servings of vegetables per week?

d. How many people were included in the survey?

e. Find the median of the data.

f. What percent of the people surveyed eat 7 servings of vegetables per week?

g. Calculate the mean of the data. (*Hint:* Use the information in the histogram to make a frequency table like the one in Problem 9.)

12. Figure 6.17 shows the results of a survey asking people how many hours of television they watch per week.

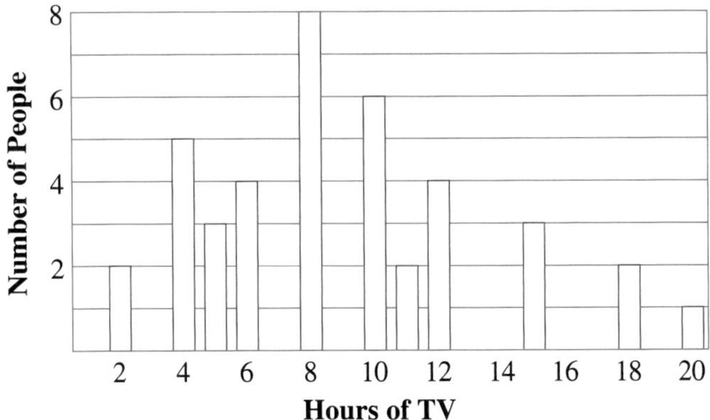

Figure 6.17 Survey results: hours of television

a. Find the extremes of the data and calculate the range.

b. What is the mode of the data?

c. How many people watch 10 or more hours of television per week?

d. How many people were included in the survey?

e. Find the median of the data.

f. What percent of the people surveyed watch 8 hours of television per week?

g. Calculate the mean of the data. (*Hint:* Use the information in the histogram to make a frequency table like the one in Problem 10.)

Simplify each expression if possible. If the expression is not defined or cannot be simplified, explain why not.

13. $\dfrac{2x}{0}$

14. $\dfrac{-0}{5}$

15. -6^3

16. -3^6

17. $a^2b + 2a^2b$

18. $a^2b(2a^2b)$

19. $2mn(3m^2n)$

20. $2m + 3n$

21. $\sqrt{121}$

22. $\sqrt{\dfrac{49}{81}}$

Building Number Skills

Build each pair of fractions to the same denominator.

Example: $\dfrac{1}{2} = \dfrac{\mathbf{3}}{6}$; $\dfrac{2}{3} = \dfrac{\mathbf{4}}{6}$

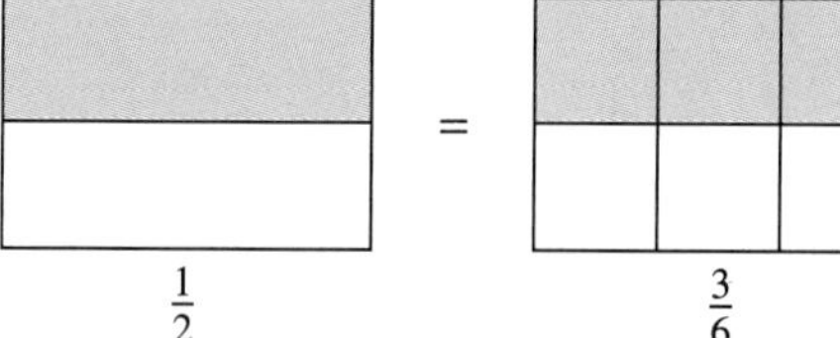

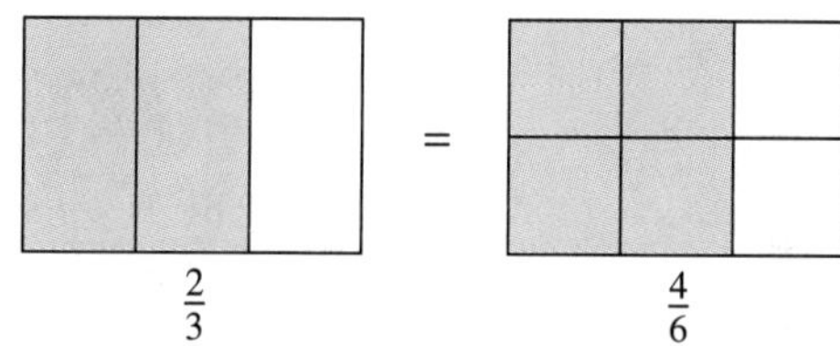

23. $\dfrac{1}{3} = \dfrac{?}{12}$; $\dfrac{3}{4} = \dfrac{?}{12}$

24. $\dfrac{1}{4} = \dfrac{?}{12}$; $\dfrac{2}{3} = \dfrac{?}{12}$

25. $\dfrac{3}{2} = \dfrac{?}{10}$; $\dfrac{4}{5} = \dfrac{?}{10}$

26. $\dfrac{3}{5} = \dfrac{?}{10}$; $\dfrac{5}{2} = \dfrac{?}{10}$

27. $\dfrac{3}{4} = \dfrac{?}{12}$; $\dfrac{5}{6} = \dfrac{?}{12}$

28. $\dfrac{4}{9} = \dfrac{?}{18}$; $\dfrac{5}{6} = \dfrac{?}{18}$

29. $\dfrac{1}{6} = \dfrac{?}{24}$; $\dfrac{3}{8} = \dfrac{?}{24}$

30. $\dfrac{5}{8} = \dfrac{?}{24}$; $\dfrac{5}{6} = \dfrac{?}{24}$

31. $\dfrac{5}{2} = \dfrac{?}{6}$; $\dfrac{5}{3} = \dfrac{?}{6}$

32. $\dfrac{2}{3} = \dfrac{?}{15}$; $\dfrac{2}{5} = \dfrac{?}{15}$

B. Quartiles

Statisticians are also interested in whether the data are all clustered around the average value or if they are spread out over a wide range of values. The three histograms in Figure 6.18 show data sets that have the same mean. In histogram (a), the data values are clustered around the mean; in histogram (b), the data are widely spread; and in histogram (c), the data are evenly distributed over their range.

Figure 6.18

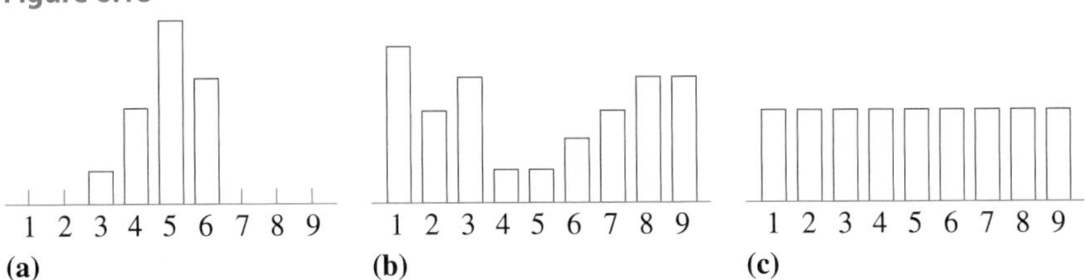

A simple way to measure the spread in a collection of data is to compute the upper and lower **quartiles.** The median divides the data into an upper half and a lower half. The median of the upper half is called the **upper quartile,** and the median of the lower half is called the **lower quartile.** (We use the term *quartile* because these three points divide the data roughly into quarters.)

EXAMPLE 4

The 21 students in Professor Ikkanda's biology class earned the following scores on the midterm exam. Compute median and the upper and lower quartiles.

Alia	53	Florence	72	Rosaura	86
Arman	96	Golnoosh	91	Sang Mee	66
Chris	64	Lloyd	24	Shervin	77
David	73	Margaret	76	Viken	79
Douglas	71	Olumide	88	Wissam	55
Eduardo	82	Paul	85	Yeuk-Cheung	46
Farida	77	Rita	12	Yusuf	60

Solution

First arrange the scores in order. The median score is 73 because ten scores are higher than 73 and ten scores are lower than 73.

median
↓
12 24 46 53 55 60 64 66 71 72 |73| 76 77 77 79 82 85 86 88 91 96

To compute the lower quartile we consider the ten scores below the median. The median of these scores is halfway between the fifth and sixth scores:

12 24 46 53 55 ↑ 60 64 66 71 72

Thus, the lower quartile is

$$LQ = \frac{55 + 60}{2} = 57.5$$

The upper quartile is the median of the scores above 73, or halfway between 82 and 85:

76 77 77 79 82 85 86 88 91 96

The upper quartile is

$$UQ = \frac{82 + 85}{2} = 83.5$$

EXERCISE 4

Compute the median and the upper and lower quartiles of the following bowling scores:

81	89	98	105	112
122	135	140	156	164
169	172	175	177	179
188	195	196	210	

Try Exercise 4 now.

The difference between the upper and lower quartiles is called the **interquartile range,** or **IQR.** Approximately half the data values lie within the interquartile range. For the biology scores, the interquartile range is

$$\begin{aligned} \text{IQR} &= \text{upper quartile} - \text{lower quartile} \\ &= 83.5 - 57.5 = 26 \end{aligned}$$

In other words, about half of the test scores lie within a 26-point spread around the median value of 73. The interquartile range gives us an idea of how spread out the data values are.

Boxplots

We use a diagram called a **boxplot** to display the median and the quartiles of a collection of data. We start by drawing a number line that includes the extremes of the data set. For the biology scores of Example 4 we can use a number line from 0 to 100.

Next, we draw a rectangle or box from the lower to the upper quartile. We indicate the median by drawing a vertical line through the box at that value. Finally, we draw two "whiskers" from either side of the box out to the extremes of the data. The boxplot for the biology test scores is shown in Figure 6.19:

Figure 6.19

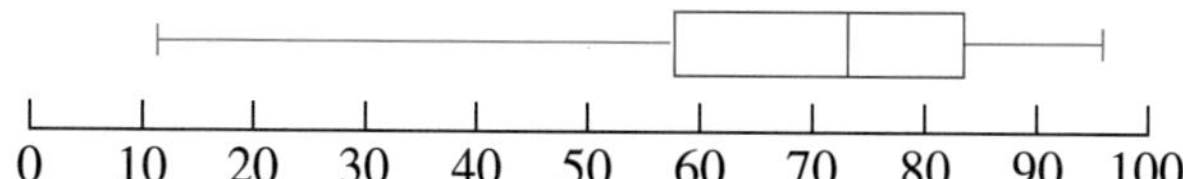

EXAMPLE 5

The 11 members of the West Hills Cycling Club have the following weights:

107 114 128 136 145 162 165 168 176 179 183

a. Make a histogram for the data.

b. Compute the mean, median, and mode of the data.

c. Compute the upper and lower quartiles and the interquartile range.

d. Construct a boxplot for the data.

Figure 6.20

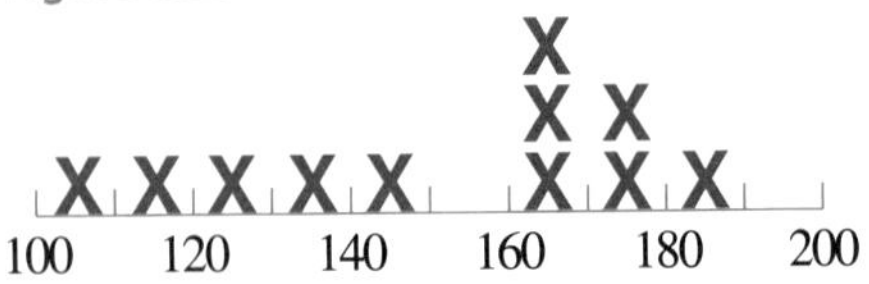

Figure 6.21

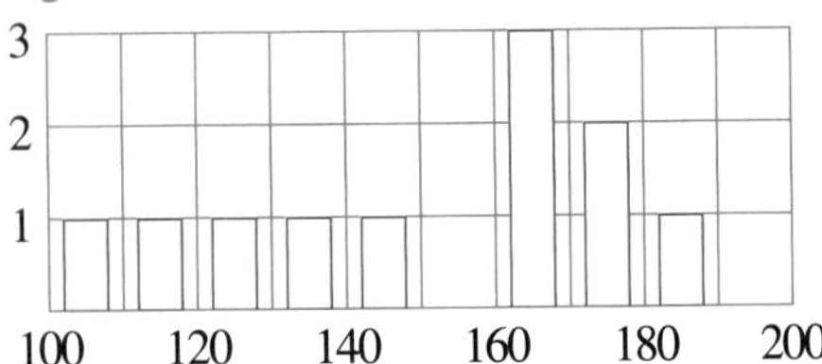

Solution

a. Begin by noting the extremes of the data, 107 and 183. Construct a number line that includes the extremes; a scale from 100 to 200 with intervals of length 10 is a good choice. Next, indicate each data point by placing an X above the appropriate interval, as shown in Figure 6.20.

Finally, draw in a vertical bar over each interval to show the height of each column of Xs. The completed histogram is shown in Figure 6.21.

b. The mean of the data is

$$\text{mean} = \frac{S}{n}$$

$$= \frac{107 + 114 + 128 + 136 + 145 + 162 + 165 + 168 + 176 + 179 + 183}{11}$$

$$= \frac{1663}{11} \approx 151.18$$

The median is the middle value, 162. There is no mode for this data set, because no data value occurs more than once.

c. The upper quartile is the median of the weights from 165 to 183. The middle value of these data is 176. The lower quartile is the middle value of the weights from 107 to 145, or 128. The interquartile range is thus

$$\text{IQR} = 176 - 128 = 48$$

d. To construct the boxplot, draw a rectangle from the lower quartile, 128, to the upper quartile, 176. Draw in a vertical line at the median, 162. Then draw in the whiskers from the edges of the box to the extremes, 107 and 183. Figure 6.22 shows the completed boxplot.

Figure 6.22

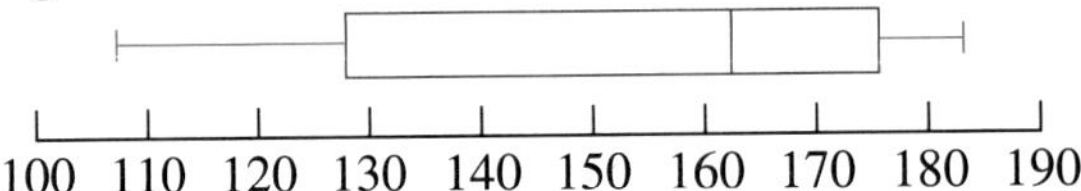

Now try Exercise 5.

EXERCISE 5

Construct a boxplot for the bowling scores from Exercise 4.

ANSWERS TO 6.2B EXERCISES

4. Median: 164; lower quartile: 112; upper quartile: 179

5.

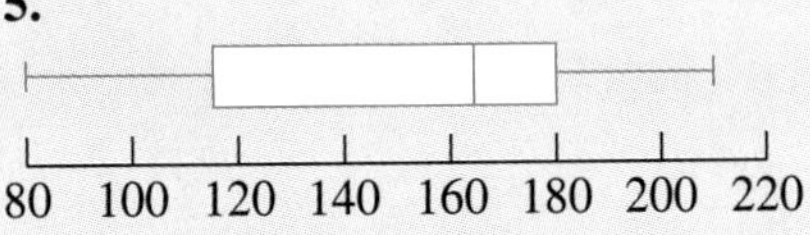

HOMEWORK 6.2B 

For Problems 1–6, (a) find the median and the upper and lower quartiles, (b) compute the interquartile range, and (c) construct a boxplot for the data.

1. In an aerobics class, the participants were asked to find their heart rates immediately after exercising. They recorded the following results:

136	142	156	148
165	152	168	174
127	148	144	153
135	145	166	151

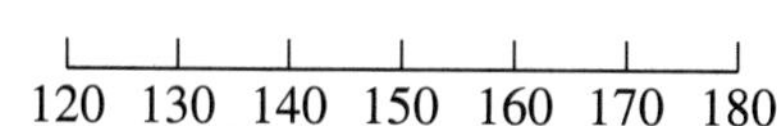

2. A group of runners were asked how many miles they ran, on average, each week. They gave the following responses:

25	60	24	42
48	30	36	56
35	21	44	32
52	48	45	35

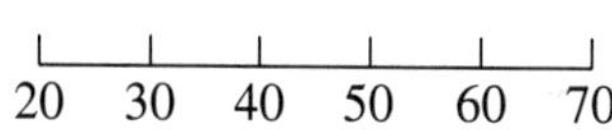

3. The state police set up a radar speed trap on the highway outside of town and clocked cars traveling at the following speeds, in miles per hour:

68	56	65	72
60	82	67	58
64	76	68	74
80	78	64	66

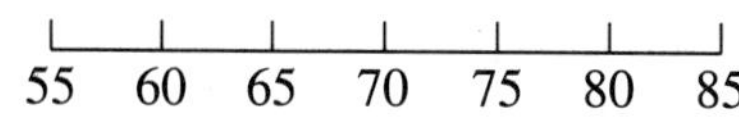

4. The gas company conducted a survey to see what temperature people maintain in their homes. They recorded the following thermostat readings during a telephone survey:

72	70	66	70
68	65	71	74
67	68	75	62
68	61	69	73

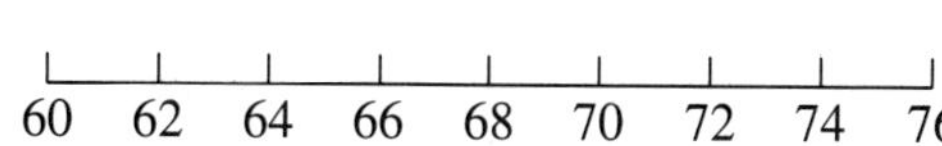

5. The following figures show the birthrates in 1991 for each of the 50 states and the District of Columbia. The rate is given as the number of births for every 1000 people. (*Source:* 1994–1995 *American Almanac*)

Maine	13.6	North Carolina	15.2
New Hampshire	14.8	South Carolina	16.2
Vermont	14.0	Georgia	16.7
Massachusetts	14.7	Florida	14.6
Rhode Island	14.7	Kentucky	14.6
Connecticut	14.8	Tennessee	15.0
New York	16.2	Alabama	15.4
New Jersey	15.6	Mississippi	16.7
Pennsylvania	14.1	Arkansas	15.0
Ohio	15.2	Louisiana	17.0
Indiana	15.3	Oklahoma	15.1
Illinois	16.8	Texas	18.3
Michigan	16.0	Montana	14.2
Wisconsin	14.5	Idaho	16.2
Minnesota	15.1	Wyoming	14.6
Iowa	13.9	Colorado	15.9
Missouri	15.3	New Mexico	18.0
North Dakota	14.0	Arizona	18.2
South Dakota	15.6	Utah	20.4
Nebraska	15.1	Nevada	17.2
Kansas	15.2	Washington	15.9
Delaware	16.5	Oregon	14.5
Maryland	16.3	California	20.1
Dist. of Columbia	19.7	Alaska	20.5
Virginia	15.5	Hawaii	17.6
West Virginia	12.5		

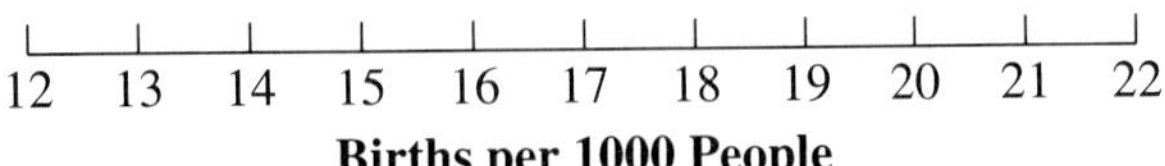

6. The following figures show the percent of the population in 1991 who had attained a bachelor's degree or higher for each of the 50 states and the District of Columbia. (*Source:* 1994–1995 *American Almanac*)

Maine	25.6	North Carolina	17.4
New Hampshire	18.8	South Carolina	16.6
Vermont	24.4	Georgia	19.3
Massachusetts	24.3	Florida	18.3
Rhode Island	21.3	Kentucky	13.6
Connecticut	27.2	Tennessee	16.0
New York	23.1	Alabama	15.7
New Jersey	24.9	Mississippi	14.7
Pennsylvania	17.9	Arkansas	13.3
Ohio	17.0	Louisiana	16.1
Indiana	15.6	Oklahoma	17.8
Illinois	21.0	Texas	20.3
Michigan	17.4	Montana	19.8
Wisconsin	17.7	Idaho	17.7
Minnesota	21.8	Wyoming	18.8
Iowa	16.9	Colorado	27.0
Missouri	17.8	New Mexico	20.4
North Dakota	18.1	Arizona	20.3
South Dakota	17.2	Utah	22.3
Nebraska	18.9	Nevada	15.3
Kansas	21.1	Washington	22.9
Delaware	21.4	Oregon	20.6
Maryland	26.5	California	23.4
Dist. of Columbia	33.3	Alaska	23.0
Virginia	24.5	Hawaii	22.9
West Virginia	12.3		

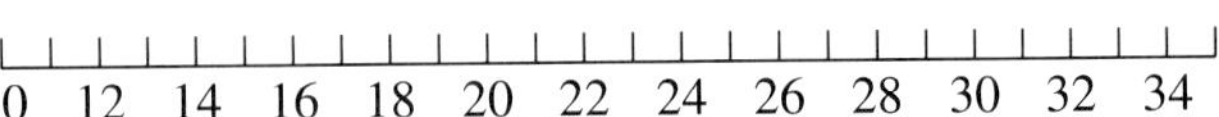

7. Figure 6.23 shows the spread of ages among children in two different neighborhoods, A and B.

Figure 6.23

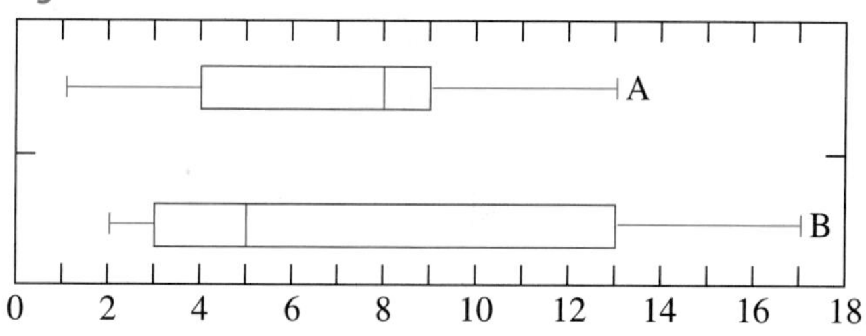

a. Which neighborhood has the youngest child?

b. Which neighborhood has the oldest child?

c. Which neighborhood has the wider range of children's ages?

d. Which neighborhood has the larger median age?

e. Suppose both neighborhoods have the same total number of children. Stefanie wants to have lots of play friends within a year of her age. Stefanie is 4 years old. Which neighborhood seems to be a better choice? Explain.

8. Figure 6.24 shows the price range of bicycles at three different bicycle shops, X, Y, and Z.

Figure 6.24

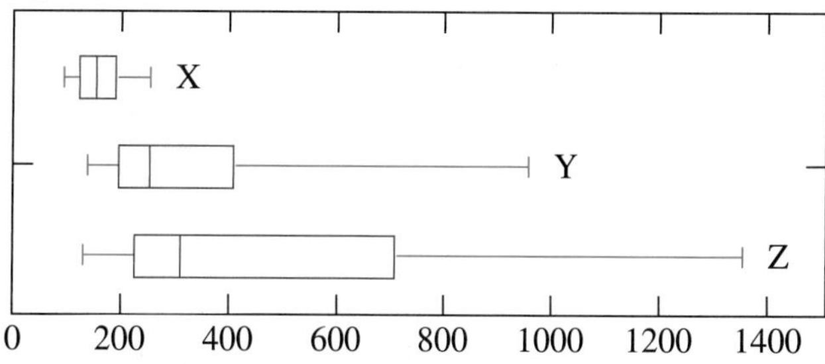

a. Which shop has the least expensive bicycle?

b. Which shop has the widest range of prices?

c. Which shop has the greatest fraction of its bicycles at prices below $200?

d. Which shop has the greatest fraction of it prices above $700?

e. Suppose the three stores have the same total number of bicycles. Which seems to have the most bicycles available in the range from $200 to $400? Explain.

Write an equation that expresses the second variable in terms of the first variable.

9.

x	y
2	−8
4	−16
5	−20
8	−32

10.

x	z
−2	6
−1	3
2	−6
4	−12

11.

p	q
−5	−2
−2	1
−1	2
2	5

12.

s	t
−10	−6
−7	−3
−5	−1
−4	0

13.

m	r
−6	−8
−3	−5
−2	−4
1	−1

14.

h	k
1	−5
3	−3
8	2
9	3

15.

a	b
−8	4
−3	1.5
4	−2
6	−3

16.

v	w
−8	−2
−4	−1
0	0
2	0.5

17.

n	p
1	2
2	5
3	8
4	11

18.

d	g
1	−3
2	−5
3	−7
4	−9

Simplify each expression if possible. If the expression is not defined or cannot be simplified, explain why not. (See Lesson 5.2 to review order of operations with radicals.)

19. $\frac{6 + 2^3}{7}$

20. $\frac{(5 + 3)^2}{4}$

21. $-\sqrt{-0}$

22. $-\sqrt{-9}$

23. $\sqrt{16 - 9}$

24. $\sqrt{(-3)^2}$

25. $\dfrac{\sqrt{4}-2}{3}$

26. $\dfrac{5k}{1-\sqrt{1}}$

27. $\dfrac{-w}{6-2\sqrt{9}}$

28. $\dfrac{19q}{3-3\sqrt{4}}$

Building Number Skills

Build both fractions to the same denominator and then add or subtract. (See Appendix A.3 to review adding fractions.)

Example: $\dfrac{1}{2}+\dfrac{2}{3}=\dfrac{\mathbf{3}}{\mathbf{6}}+\dfrac{\mathbf{4}}{\mathbf{6}}=\dfrac{7}{6}$

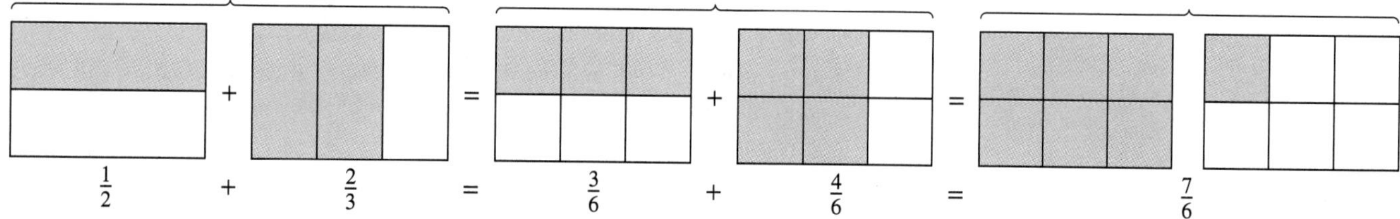

$\frac{1}{2} + \frac{2}{3} = \frac{3}{6} + \frac{4}{6} = \frac{7}{6}$

29. $\dfrac{1}{4}+\dfrac{2}{3}=\dfrac{?}{12}+\dfrac{?}{12}$

30. $\dfrac{1}{3}+\dfrac{3}{4}=\dfrac{?}{12}+\dfrac{?}{12}$

31. $\dfrac{3}{5}+\dfrac{5}{2}=\dfrac{?}{10}+\dfrac{?}{10}$

32. $\dfrac{3}{2}+\dfrac{4}{5}=\dfrac{?}{10}+\dfrac{?}{10}$

33. $\dfrac{4}{3}+\dfrac{3}{2}=\dfrac{?}{6}+\dfrac{?}{6}$

34. $\dfrac{5}{2}+\dfrac{5}{3}=\dfrac{?}{6}+\dfrac{?}{6}$

35. $\dfrac{2}{5}+\dfrac{3}{4}=\dfrac{?}{20}+\dfrac{?}{20}$

36. $\dfrac{4}{5}+\dfrac{1}{4}=\dfrac{?}{20}+\dfrac{?}{20}$

37. $\dfrac{2}{3}+\dfrac{2}{5}=\dfrac{?}{15}+\dfrac{?}{15}$

38. $\dfrac{1}{3}+\dfrac{3}{5}=\dfrac{?}{15}+\dfrac{?}{15}$

6.3 Graphs of Equations

A. Equations Relating Variables

By now you undoubtedly realize that algebra is the study of variables. We use algebraic expressions to describe and understand relationships between two (or more) variable quantities. Consider the following examples:

- A plumber charges \$30 for a house call, plus \$20 an hour. The cost C of the plumber's visit is

$$C = 30 + 20h$$

  where h is the number of hours she works.

- The area A of a square is given by

$$A = s^2$$

  where s is the length of one side of the square.

- If a car brakes suddenly while traveling at a high speed, its tires will leave skid marks on the road. The velocity v of the car just before braking is given in miles per hour by

$$v = 4.9\sqrt{d}$$

  where d is the length of the skid marks, in feet.

Making a Table of Values

In each of the preceding examples, an algebraic equation relates the two variables. To understand the relationship, we begin by making a table of values for the equation.

EXAMPLE 1 

Make a table of values for the cost of the plumber's visit,

$$C = 30 + 20h$$

What conclusions can you draw from the table?

Solution

Notice the form of the equation:

$$C = \text{(an expression involving } h\text{)}$$

The equation is a formula for finding the value of C, if we know the value of h. We say that C is expressed *in terms of h,* or that *C depends upon h.* We can choose values for h and use the equation $C = 30 + 20h$ to calculate the corresponding values of C. We make a table listing the values of *h first,* in the left column of the table. We have chosen the values 1 through 6 for the variable h. You can check the calculations to get the values of C in Table 1.

Table 1

h	C	$C = 30 + 20h$
1	50	$C = 30 + 20(\mathbf{1}) = 50$
2	70	$C = 30 + 20(\mathbf{2}) = 70$
3	90	$C = 30 + 20(\mathbf{3}) = 90$
4	110	$C = 30 + 20(\mathbf{4}) = 110$
5	130	$C = 30 + 20(\mathbf{5}) = 130$
6	150	$C = 30 + 20(\mathbf{6}) = 150$

From the table, we see that as the values of h increase, so do the values of C. The more hours the plumber works, the higher is her fee.

EXAMPLE 2

Make a table of values for the area of a square,

$$A = s^2$$

Solution

The area A of a square depends upon s, the length of the side. So we choose various values for s and use the equation $A = s^2$ to calculate the corresponding values of A.

Table 2

s	A	$A = s^2$
1	1	$A = (\mathbf{1})^2 = 1$
2	4	$A = (\mathbf{2})^2 = 4$
3	9	$A = (\mathbf{3})^2 = 9$
4	16	$A = (\mathbf{4})^2 = 16$
5	25	$A = (\mathbf{5})^2 = 25$
6	36	$A = (\mathbf{6})^2 = 36$

As we might expect, Table 2 shows that as the sides of a square increase in length, its area increases also.

Now try Exercise 1.

In all three of these examples, when we increase the values of the first variable, the second variable increases also. But there is more that we can learn from the algebraic expressions relating the variables. If a table of values helps us visualize the behavior of the variables, then a graph is even better.

Creating a Graph

In Lesson 6.1 we learned how to read values and interpret information from graphs. In this lesson we construct our own graphs from equations. We'll begin with Example 1,

$$C = 30 + 20h$$

EXERCISE 1

Make a table of values for the velocity of a car that left skid marks of length d, where

$$v = 4.9\sqrt{d}$$

Table 3

d	v	$v = 4.9\sqrt{d}$
100		$v = 4.9\sqrt{100}$
200		$v = 4.9\sqrt{200}$
300		$v = 4.9\sqrt{300}$
400		$v = 4.9\sqrt{400}$
500		$v = 4.9\sqrt{500}$
600		$v = 4.9\sqrt{600}$

Figure 6.25

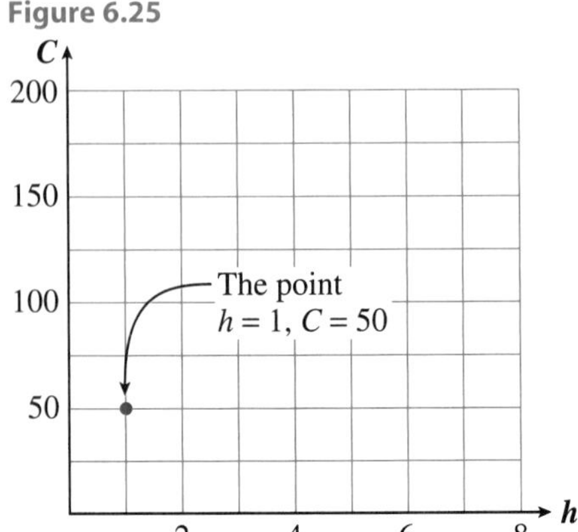

Because the values of C depend on the values of h, we say that C is the **dependent variable** and h is the **independent variable.** We use the horizontal axis to display the values of the independent variable and the vertical axis for the dependent variable, as shown in Figure 6.25.

Next, we mark off appropriate scales on the two axes. By consulting the table of values for Example 1, we see that the values of h run from 1 to 6, so we mark off the horizontal axis in whole number units. The values of C run from 50 to 150, so we use units of 10 on the vertical axis.

Now we are ready to plot the values in the table. Each pair of values will be recorded as a single point on the graph. The first pair is $h = 1$, $C = 50$. We locate $h = 1$ on the horizontal axis and then move *up* from that location until we are at the same height as $C = 50$ on the vertical axis. (A piece of graph paper helps keep things lined up.) Put a dot at this location, as shown in Figure 6.25.

EXAMPLE 3 

a. Complete the graph of the equation $C = 30 + 20h$.

b. How much will the plumber charge for $3\frac{1}{2}$ hours of work?

Solution

Table 1

h	C
1	50
2	70
3	90
4	110
5	130
6	150

a. Plot one point for each pair of values for h and C in Table 1. You should find that the points lie on a straight line. (Note that the point $h = 0$, $C = 30$ also satisfies the equation.) Draw a line through the points to complete the graph, as shown in Figure 6.26.

b. To find out how much the plumber charges for $3\frac{1}{2}$ hours work, we can use the equation, or we can use the graph. To use the equation, substitute $3\frac{1}{2}$ or **3.5** for h and solve for C:

$$C = 30 + 20(\mathbf{3.5}) = 100$$

To use the graph, first locate 3.5 on the horizontal or h axis. Then move straight up to the point P shown on the graph. This point represents the pair of values with $h = 3.5$. Finally, move horizontally back to the C axis to find the C value for point P. You should find $C = 100$, the same value the equation gave. Thus, the plumber charges \$100 for 3.5 hours of work.

Figure 6.26

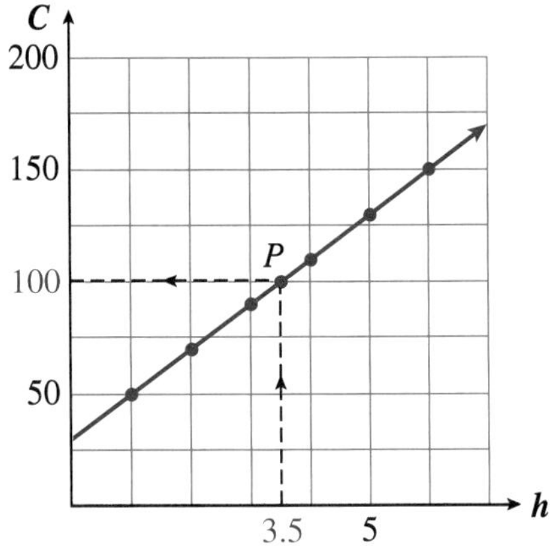

Here are the steps we used to create the graph.

Steps for Graphing an Equation

1. Make a table of values. Choose values for the independent variable and use the equation to find the values of the dependent variable.
2. Choose appropriate scales and label the axes.
3. Plot the points from the table and connect them with a smooth curve.

EXAMPLE 4

a. Graph the equation $A = s^2$.

b. If the area of a square is 20 square inches, what is the length of its side?

Solution

a. The length of one side of the square, s, is the independent variable, and A, the area of the square, is the dependent variable. The horizontal axis will represent s and the vertical axis will represent A. Because of the range of values for A displayed in the table, we scale the vertical axis in units of 5. Plot one point for each pair of values in the table and then connect them with a smooth curve, as shown in Figure 6.27.

s	A
1	1
2	4
3	9
4	16
5	25
6	36

b. To find the side of a square whose area is 20 square inches, we can use the equation or we can use the graph. To use the equation, substitute **20** for A and solve for s:

$$\mathbf{20} = s^2$$

$$\sqrt{20} = s \quad \text{Take square roots of both sides.}$$

$$4.47 \approx s \quad \text{Approximate the square root.}$$

To use the graph, first locate $A = 20$ on the vertical axis. Move horizontally to the point Q shown on the graph. This is the point that corresponds to an A value of 20. To find the s value that goes with $A = 20$, drop down directly from Q to the s axis. Although we cannot read the value of s exactly from our graph, we can see that it is approximately 4.5. Thus, a square whose side is approximately 4.5 inches long has an area of 20 square inches.

Figure 6.27

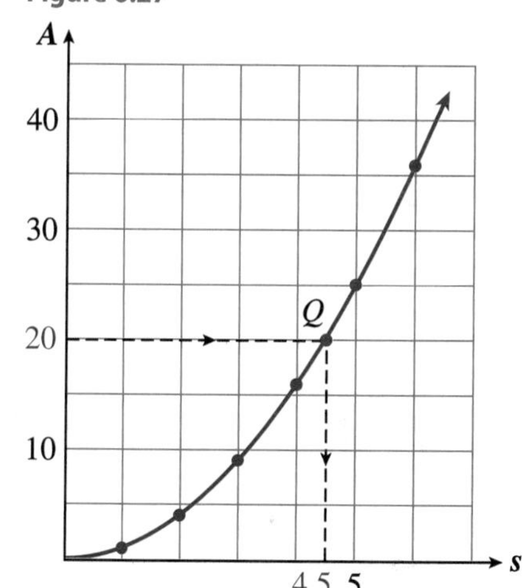

Now try Exercise 2.

All three of the graphs in the examples above are increasing, but in different ways. The graph in Example 3 increases at a constant rate; for each increase of 1 hour, the plumber's fee increases by $20. The graph in Example 4 bends upward, whereas the graph in Exercise 2 bends downward.

EXERCISE 2

a. Graph the equation $v = 4.9\sqrt{d}$ on the grid in Figure 6.28.

Plot the points from the table you made in Exercise 1.

d	v
100	49
200	69
300	85
400	98
500	110
600	120

Figure 6.28

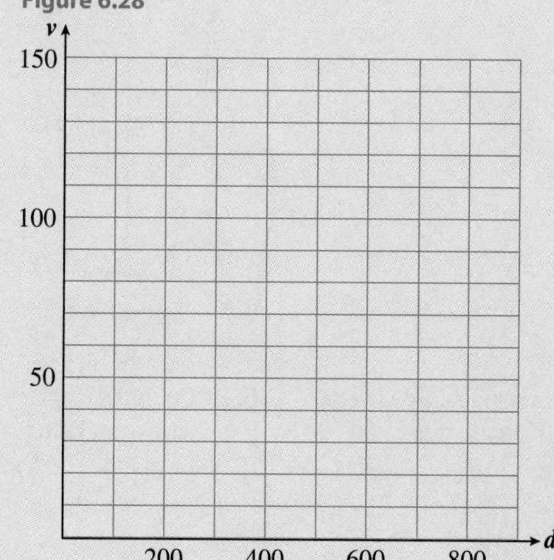

b. If a car leaves skid marks 800 feet long, how fast was it going when it braked? Verify your answer on the graph.

Evaluate v for $d = 800$:

$v =$

Locate the point on the graph with $d = 800$.

ANSWERS TO 6.3A EXERCISES

1.

d	v
100	49
200	69
300	85
400	98
500	110
600	120

2a.

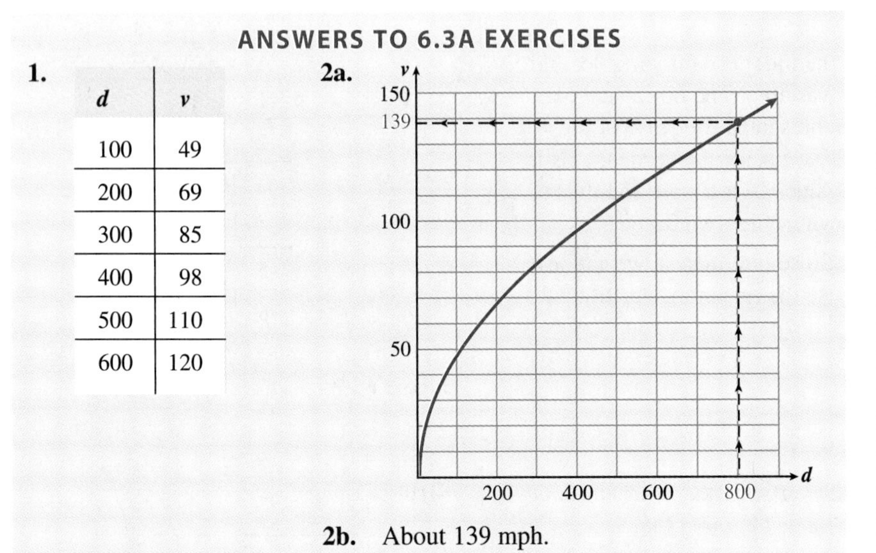

2b. About 139 mph.

HOMEWORK 6.3A 

For Problems 1–8, complete the table of values and plot the graph. Then use the graph to answer the questions.

1. Uncle Herb's diet allows him to eat a total of 1200 calories for lunch and dinner. If his lunch contains c calories, then the number of calories he can have for dinner is given by

$$d = 1200 - c$$

c	d
200	
400	
600	
800	
1000	
1200	

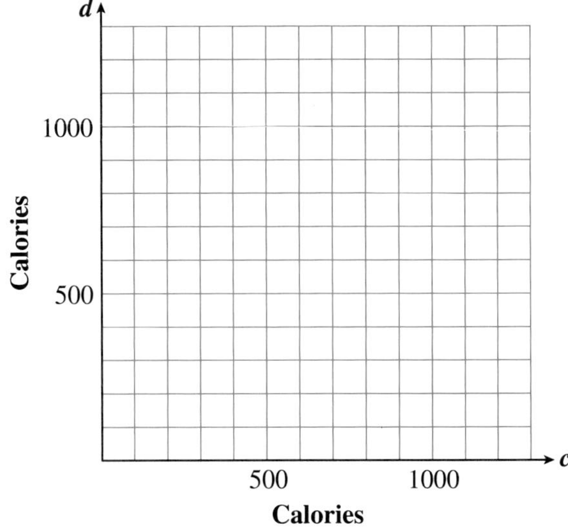

a. If Uncle Herb consumes 550 calories for lunch, how many calories can he have for dinner? Locate this point on your graph.

b. If Uncle Herb plans a 750-calorie dinner, how many calories can he have for lunch? Locate this point on your graph.

2. Softek bought a new photocopying machine for 20 thousand dollars ($20,000). Every year, the copier depreciates in value by 2 thousand dollars ($2000). Its value V after t years is given by

$$V = 20 - 2t$$

where V is measured in thousands of dollars.

t	V
0	
2	
4	
6	
8	
10	

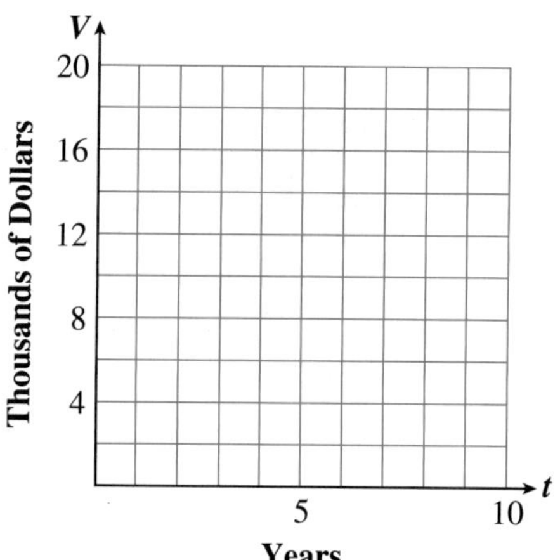

a. What is the value of the copier after 3 years? Locate this point on your graph.

b. When will the copier's value be $8000? Locate this point on your graph.

3. At the Custom Pizza shop, you can buy their special smoked-chicken pizza in any size you like. The cost C of the pizza in dollars is given by

$$C = \frac{1}{2}r^2$$

where r is the radius of the pizza.

r	C
2	
3	
4	
5	
6	
7	

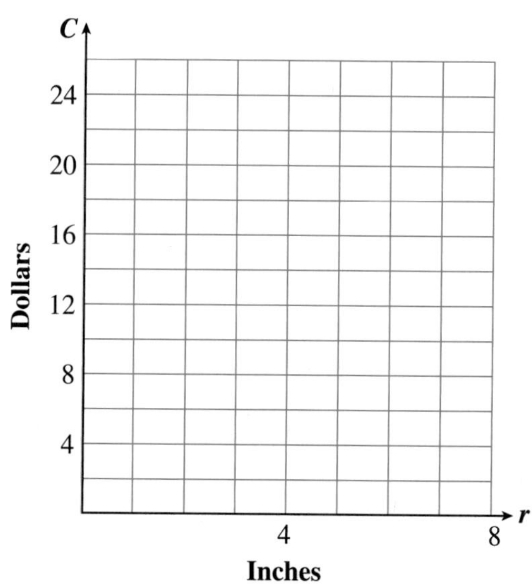

a. How much does a pizza of radius 3 inches cost? Locate this point on your graph.

b. How big a pizza can you buy for $18? Locate this point on your graph.

4. While hiking, Ryan drops a stone off the edge of a 400-foot cliff. After falling for t seconds, the height h of the stone above the base of the cliff is given in feet by

$$h = 400 - 16t^2$$

t	h
0	
1	
2	
3	
4	
5	

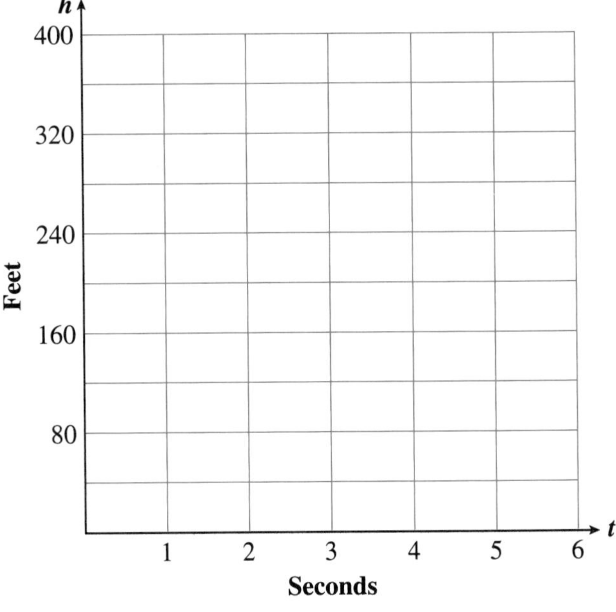

a. What is the height of the stone 3 seconds after being dropped? Locate this point on your graph.

b. How long will it take the stone to hit the ground? Locate this point on your graph.

5. Meryl wants to travel 600 miles to a mountain resort on her vacation. If she travels at an average speed of v miles per hour, then it will take her t hours to reach the resort, where t is given by

$$t = \frac{600}{v}$$

v	t
30	
50	
60	
100	
120	
150	
200	
300	

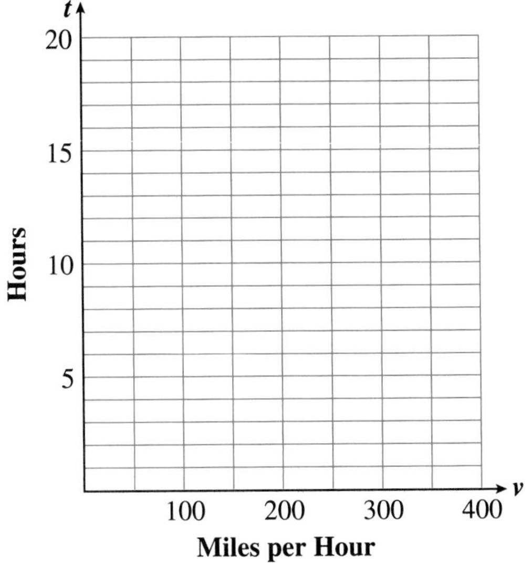

a. How long will it take Meryl to reach the resort if she averages 50 miles per hour? Locate this point on your graph.

b. How fast will Meryl have to travel if she wants to reach the resort in 3 hours? Locate this point on your graph.

6. Aisha plans to crochet a scarf that is 60 inches long. If she crochets n inches per day, it will take her d days to finish the scarf, where d is given by

$$d = \frac{60}{n}$$

n	d
2	
5	
10	
12	
15	
20	
30	

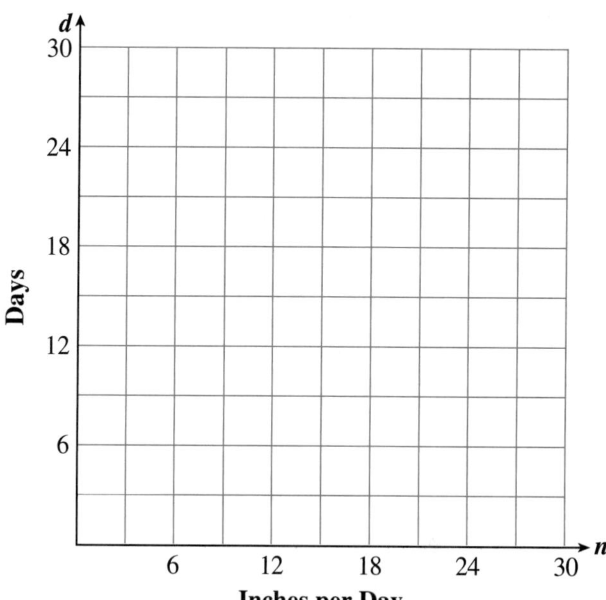

a. If Aisha crochets 2 inches per day, how long will it take to finish the scarf? Locate this point on your graph.

b. If Aisha wants to finish the scarf in 5 days, how much must she crochet each day? Locate this point on your graph.

7. If you are flying in an airplane at an altitude of h feet, on a clear day you can see for a distance of d miles, where d is given by

$$d = 1.22\sqrt{h}$$

h	d
5,000	
10,000	
15,000	
20,000	
25,000	
30,000	

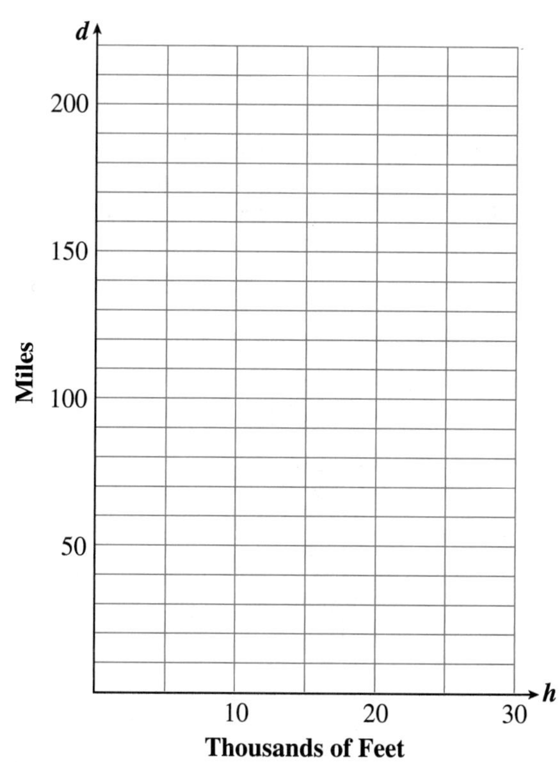

a. How far can you see from an altitude of 10,000 feet? Locate this point on your graph.

b. How high do you have to be to see 100 miles? Locate this point on your graph.

8. If an object falls from a height of h meters, its velocity v when it strikes the ground is given by

$$v = 4.4\sqrt{h}$$

in meters per second.

h	v
50	
100	
150	
200	
250	
300	
350	

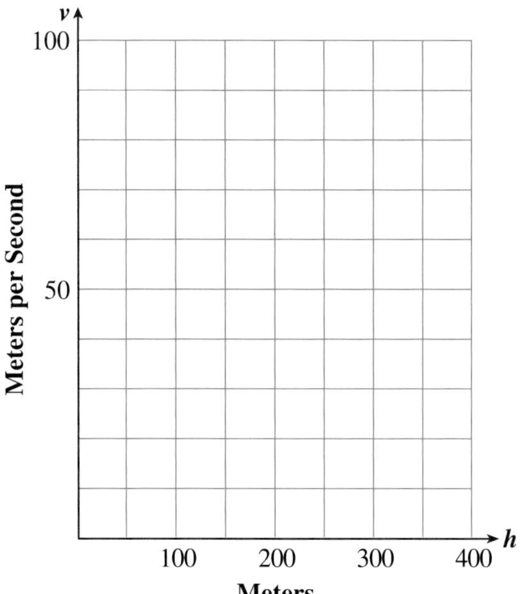

a. If a penny falls off the Washington Monument, 170 meters high, what will be its velocity when it hits the ground? Locate this point on your graph.

b. A rock dropped from the Royal Gorge bridge in Colorado will hit the water below at a velocity of 80 meters per second. How high is the bridge? Locate this point on your graph.

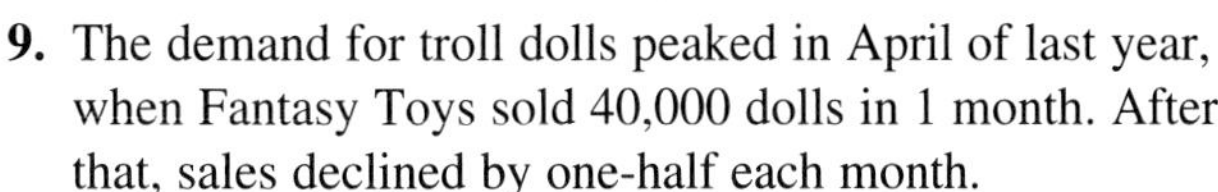

9. The demand for troll dolls peaked in April of last year, when Fantasy Toys sold 40,000 dolls in 1 month. After that, sales declined by one-half each month.

a. Make a table showing the number of troll dolls sold each month for the next 6 months.

Month	*Sales (thousands)*
April ($m = 0$)	
May ($m = 1$)	
June ($m = 2$)	
July ($m = 3$)	
August ($m = 4$)	
September ($m = 5$)	
October ($m = 6$)	

b. Graph the data in your table.

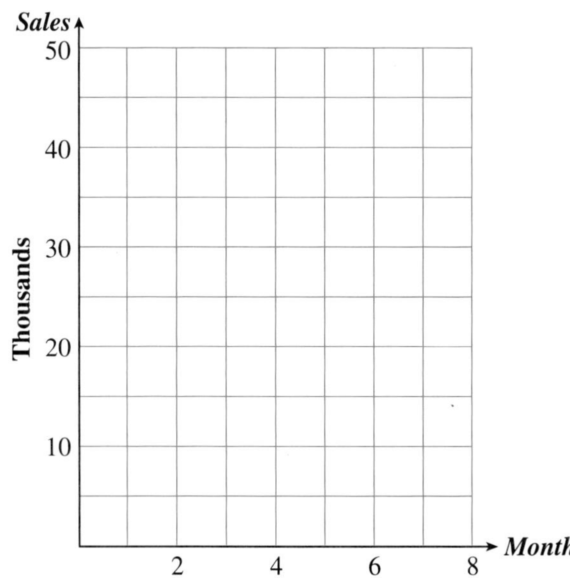

10. A biology student is culturing a certain bacteria. She starts with 100 bacteria, and under optimum conditions the number of bacteria will double every day.

a. Make a table showing the number of bacteria in the culture every day for one week.

Day	*Bacteria (hundreds)*
0	
1	
2	
3	
4	
5	
6	
7	

b. Graph the data in your table.

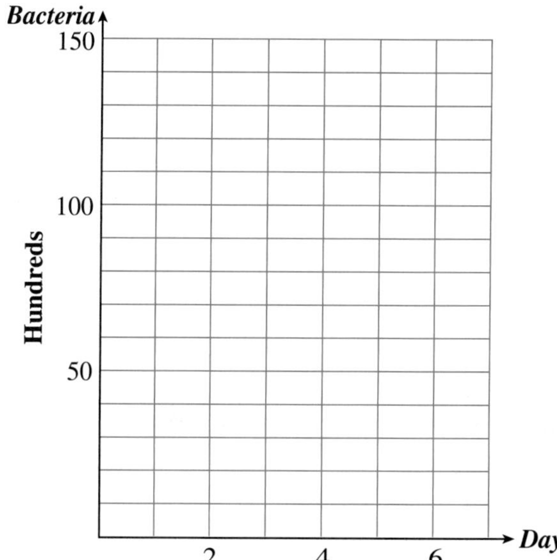

Simplify. (See Lesson 4.5 to review the order of operations.)

11. $2 - 3(4)$

12. $3 - 4(5)$

13. $\frac{114 - 48}{-6}$

14. $\frac{72 - 117}{9}$

15. $\frac{12 - 4}{6 - 8}$

16. $\frac{64 - 96}{8 - 4}$

17. $8 - 3(2 - 5)(3 - 8)$

18. $6 - 2(8 - 4)(4 - 8)$

19. $49 - (18 - 3 \cdot 4)(1 + 2 \cdot 3)$

20. $(34 - 7 \cdot 4)(3 + 6 \cdot 5) \div 2$

21. $\dfrac{6 - 3(5) + 9}{6 - 6(-3)}$

22. $\dfrac{4 - (-2)(-2)}{8 - (2 - 2)}$

Write an equation that expresses the second variable in terms of the first variable.

23.

x	y
2	−8
4	−16
5	−20
8	−32

24.

x	z
−2	6
−1	3
2	−6
4	−12

25.

p	q
−5	−2
−2	1
−1	2
2	5

26.

s	t
−10	−6
−7	−3
−5	−1
−4	0

27.

m	r
−6	−8
−3	−5
−2	−4
1	−1

28.

h	k
1	−5
3	−3
8	2
9	3

29.

a	b
−8	4
−3	1.5
4	−2
6	−3

30.

v	w
−8	−2
−4	−1
0	0
2	0.5

Building Number Skills

Rewrite each whole number as a fraction and then add or subtract. (See Appendix A.4 to review mixed numbers and improper fractions.)

Example: $2 + \frac{3}{5} = \frac{\mathbf{10}}{\mathbf{5}} + \frac{3}{5} = \frac{13}{5}$ How many fifths in 2 wholes? $2 = \frac{10}{5}$

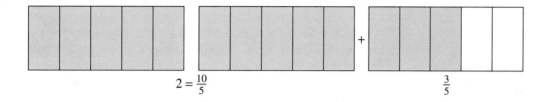

31. $3 + \frac{3}{4}$

32. $4 + \frac{2}{3}$

33. $2 + \frac{3}{8}$

34. $2 + \frac{1}{6}$

35. $4 - \frac{5}{6}$

36. $5 - \frac{7}{8}$

37. $2 - \frac{3}{5}$

38. $1 - \frac{2}{5}$

39. $1 + \frac{1}{3}$

40. $4 + \frac{1}{4}$

B. The Cartesian Coordinate System

We will often need graphs that display negative values as well as positive values. To simplify our work in this lesson, we will use x for the independent variable and y for the dependent variable; later we can use any names for the variables.

In Chapter 3 we used a number line to keep track of operations on signed numbers. To display the values of *two* related variables, we need *two* number lines. We construct a second number line perpendicular to the first one, as shown in Figure 6.29.

Figure 6.29

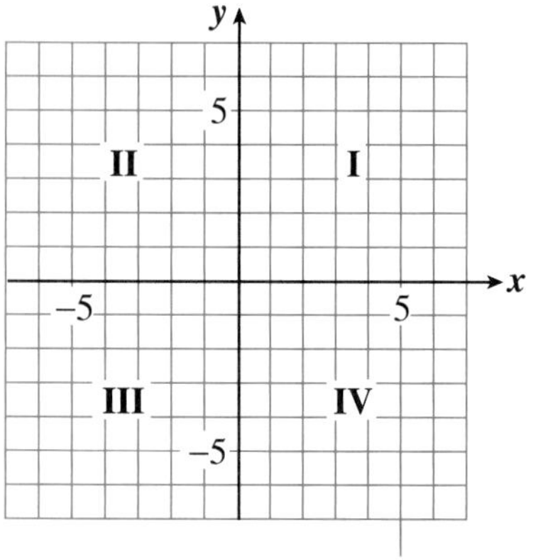

The values on this second number line increase from bottom to top, and the two lines intersect at the zero points of each. The horizontal number line is called the ***x* axis,** and the vertical number line is the ***y* axis.** The two axes divide the plane into four regions called **quadrants,** which are numbered as shown. The point where the two axes intersect is called the **origin.** This system for locating points in the plane was inspired by the French philosopher and mathematician René Descartes, who lived in the seventeenth century. Figure 6.29 is called a **Cartesian coordinate system** in his honor.

Plotting Points

Now when we graph an equation, we can use both positive and negative values for the variables. First let's see how to plot individual points. Each pair of values in the following table gives us a point on the graph:

x	y	
2	6	(2, 6)
−3	4	(−3, 4)
−5	−1	(−5, −1)
4	−3	(4, −3)
0	−2	(0, −2)
−3	0	(−3, 0)

Figure 6.30

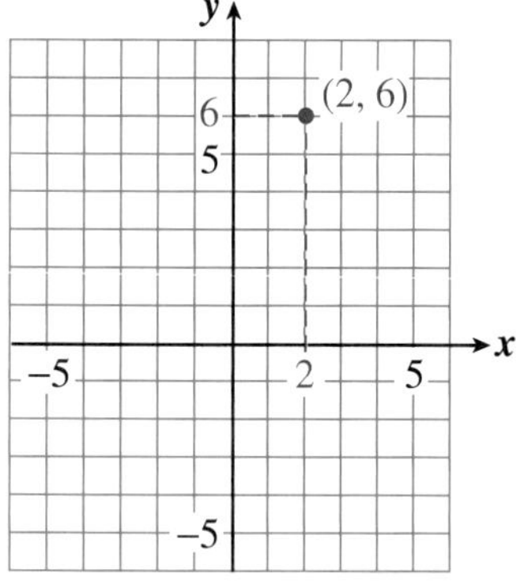

The first pair of values is $x = 2$ and $y = 6$, which we write as (2, 6). We call the numbers 2 and 6 the **coordinates** of the point; 2 is the ***x* coordinate,** and 6 is the ***y* coordinate.** The coordinates of a point describe its location in the plane, in much the same way that your address gives the location of your house. To plot this point, we first find 2 on the x axis (to the *right* of the origin) and from there move 6 units directly *up,* as shown in Figure 6.30.

The expression (2, 6) is called an **ordered pair** because the order of the two numbers is important. The first number in the pair is always the value of x, and the second number is the value of y.

The location of a point in the plane is given by the **ordered pair *(a, b)*,** where a is the ***x* coordinate** of the point and b is the ***y* coordinate.**

Now let's see how to plot the other points in the table. The signs of the coordinates tell us which direction to move when plotting the point.

EXAMPLE 5

Plot the points $(-3, 4)$, $(-5, -1)$, and $(4, -3)$.

Solution

From the origin we move along the x axis to the *right* if the x coordinate is positive and to the *left* if the x coordinate is negative. From this location on the x axis, we move *up* if the y coordinate is positive and *down* if the y coordinate is negative. For example, to plot $(-3, 4)$, we move 3 units to the left and then 4 units up. Figure 6.31 shows how to plot all three points.

Figure 6.31

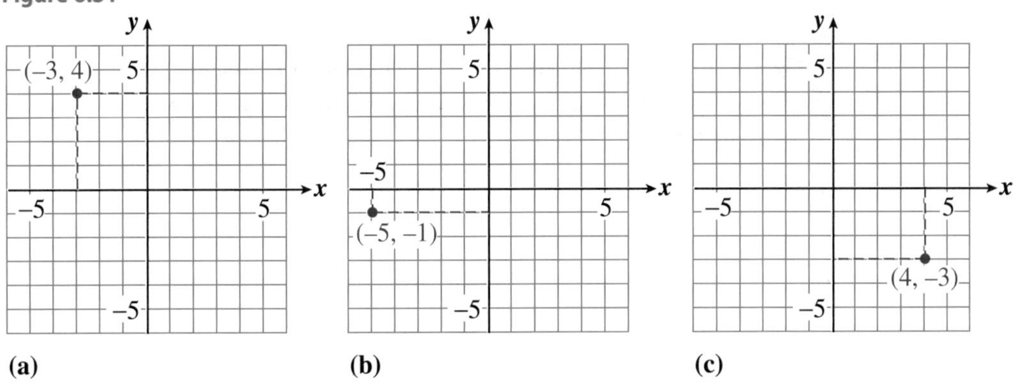

The signs of the coordinates determine in which of the four quadrants the point will appear. A point lies *on* one of the two axes only if one (or both) of its coordinates is zero.

EXAMPLE 6

Plot the points $(0, -2)$ and $(-3, 0)$.

Solution

If the x coordinate is 0, the point lies on the x axis, like the point $(0, -2)$ in Figure 6.32. If the y coordinate is 0, the point lies on the x axis, like the point $(-3, 0)$.

Figure 6.32

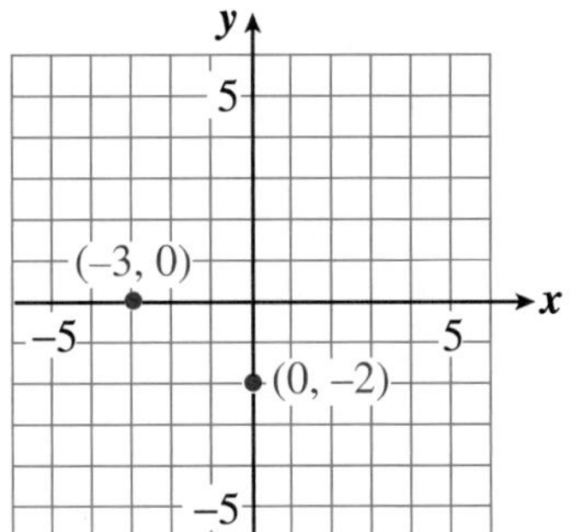

Now try Exercise 3.

EXERCISE 3

Plot the following points on the grid in Figure 6.33.

Figure 6.33

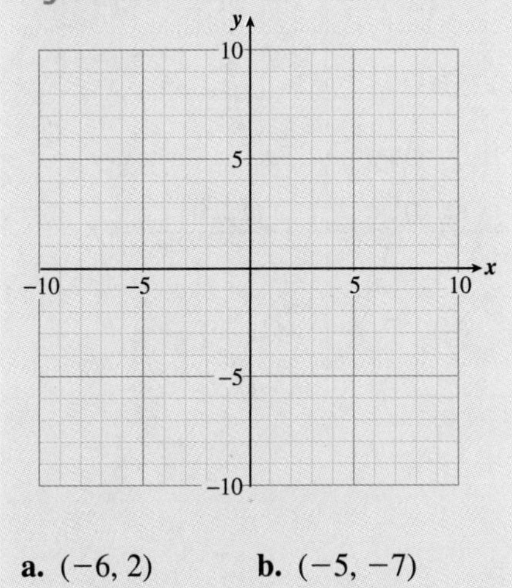

a. $(-6, 2)$ **b.** $(-5, -7)$

c. $(8, -4)$ **d.** $(1, 7)$

e. $(0, -9)$ **f.** $(-2, 0)$

Graphing an Equation

We can use a Cartesian coordinate system to graph an equation in two variables. Each ordered pair in the table of values is called a **solution** of the equation. The **graph** of the equation is just a picture of all the solutions to the equation.

EXAMPLE 7  

Graph the equation $y = x^2 - 4$.

Solution

Follow the three steps on page 412.

Step 1 Begin by choosing several values for the independent variable x. Choose both positive and negative values. Then use the equation to compute the y value for each x value.

x	y		
−4	12	$y = (\mathbf{-4})^2 - 4 = 12$	$(-4, 12)$
−3	5	$y = (\mathbf{-3})^2 - 4 = 5$	$(-3, 5)$
−2	0	$y = (\mathbf{-2})^2 - 4 = 0$	$(-2, 0)$
−1	−3	$y = (\mathbf{-1})^2 - 4 = -3$	$(-1, -3)$
0	−4	$y = (\mathbf{0})^2 - 4 = -4$	$(0, -4)$
1	−3	$y = (\mathbf{1})^2 - 4 = -3$	$(1, -3)$
2	0	$y = (\mathbf{2})^2 - 4 = 0$	$(2, 0)$
3	5	$y = (\mathbf{3})^2 - 4 = 5$	$(3, 5)$
4	12	$y = (\mathbf{4})^2 - 4 = 12$	$(4, 12)$

Step 2 Next, draw the x and y axes and mark off appropriate scales on each. Looking at the x values in our table, we see that the scale on the x axis should run from −5 to 5. On the y axis, we should label a scale from, say, −5 to 15.

Step 3 Finally, plot each of the points in the table. You can see that the points do *not* lie on a straight line. Draw a smooth curve through the points *in order,* from the smallest x coordinates to the largest. The completed graph is shown in Figure 6.34.

Figure 6.34

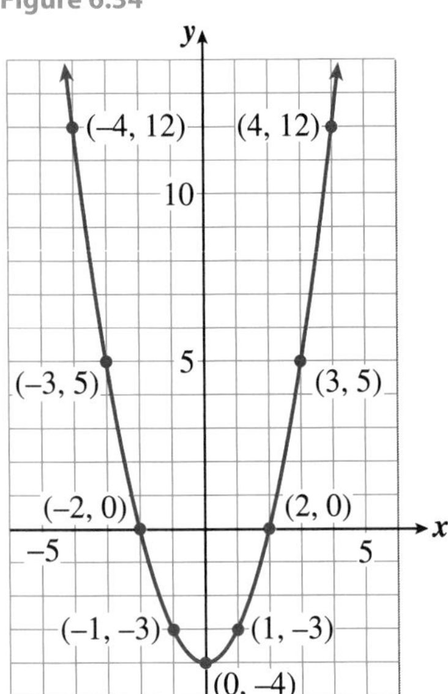

Now try Exercise 4.

EXERCISE 4

a. Complete the table for the equation $y = x^2 - 6$.

b. Graph the equation on the grid in Figure 6.35.

x	y
−4	
−3	
−2	
−1	
0	
1	
2	
3	
4	

Figure 6.35

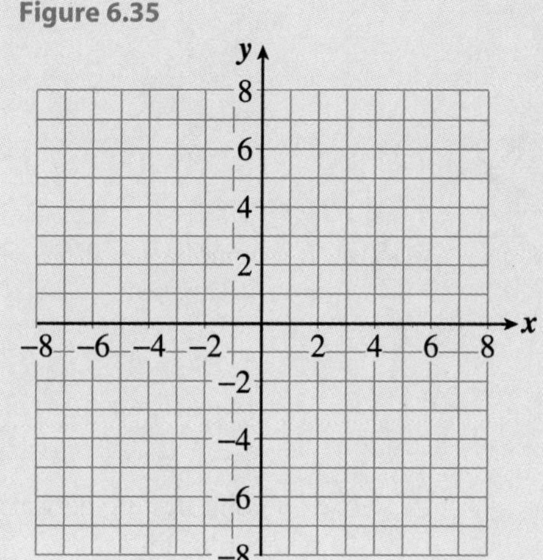

ANSWERS TO 6.3B EXERCISES

3.

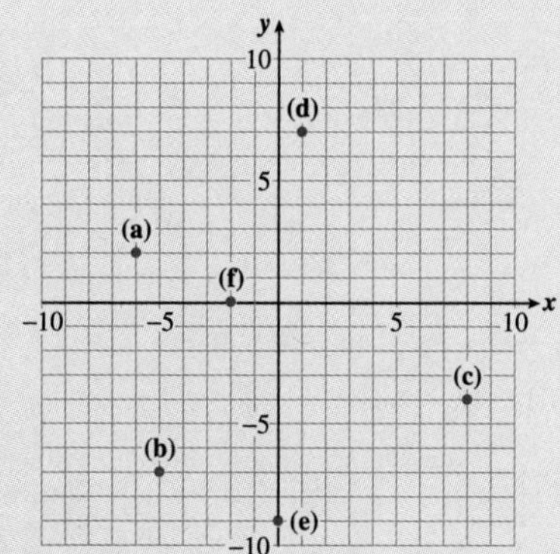

4a.

x	y
−4	10
−3	3
−2	−2
−1	−5
0	−6
1	−5
2	−2
3	3
4	10

4b.

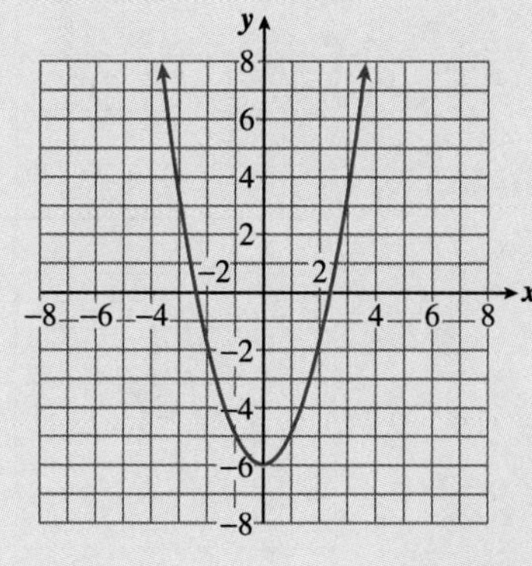

HOMEWORK 6.3B 

1. Write a paragraph describing the Cartesian coordinate system. Include in your description the x and y axes, the origin, and the quadrants.

2. Explain how to plot a point when you know its coordinates.

3. **a.** If a point lies in the third quadrant, what do you know about its coordinates?

   b. If a point lies in the second quadrant, what do you know about its coordinates?

4. **a.** If a point lies on the x axis, what do you know about its coordinates?

   b. If a point lies on the y axis, what do you know about its coordinates?

5. Plot all the points whose y coordinate is 3. What do you get?

6. Plot all the points whose x coordinate is -3. What do you get?

Plot each set of points on a Cartesian coordinate system.

7. (2, 3), (2, -3), (-2, 3), (-2, -3)

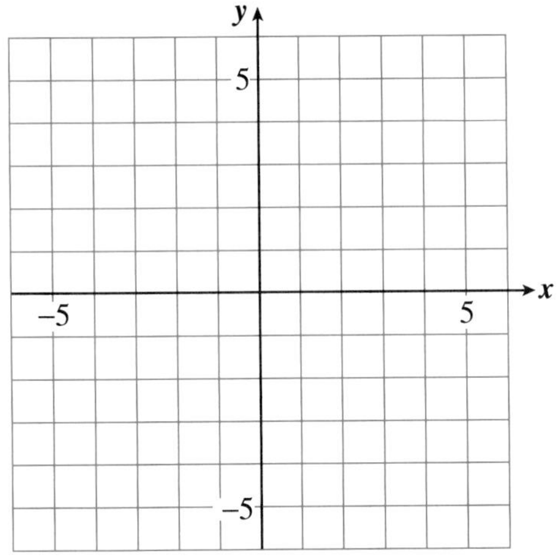

8. (0, 4), (4, 0), (-4, 0), (0, -4)

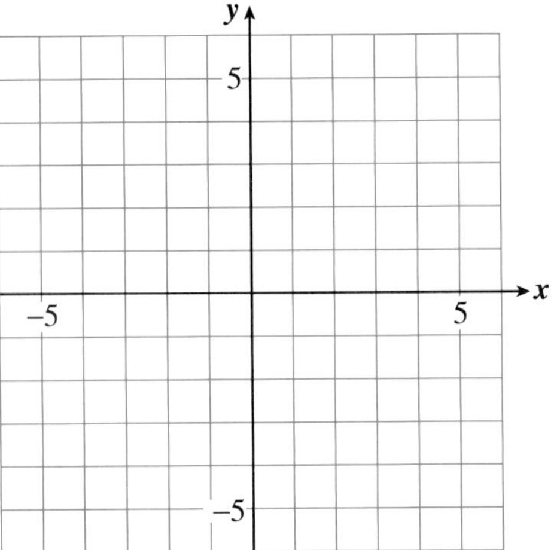

9. (1, 5), (5, 1), (−5, 1), (−1, 5)

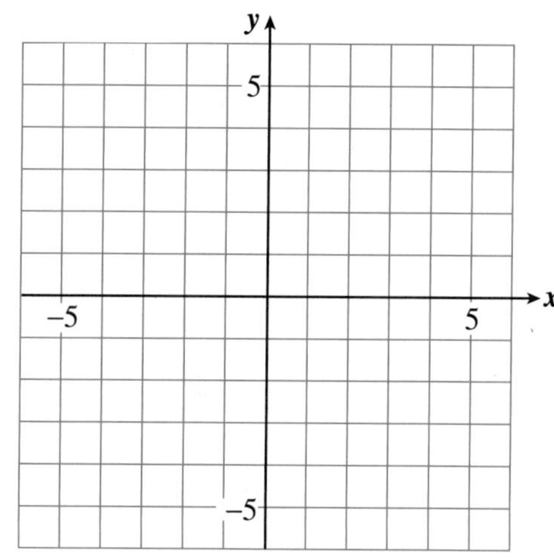

10. (2, 2), (6, 6), (−2, −2), (−6, −6)

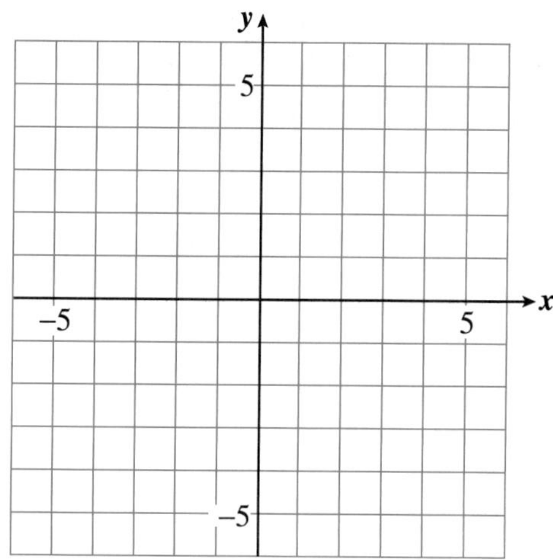

Complete the table of values and graph the equation on a Cartesian coordinate system.

11. $y = x^3 + 1$

x	-2	-1	$\frac{-1}{2}$	0	$\frac{1}{2}$	1	2
y							

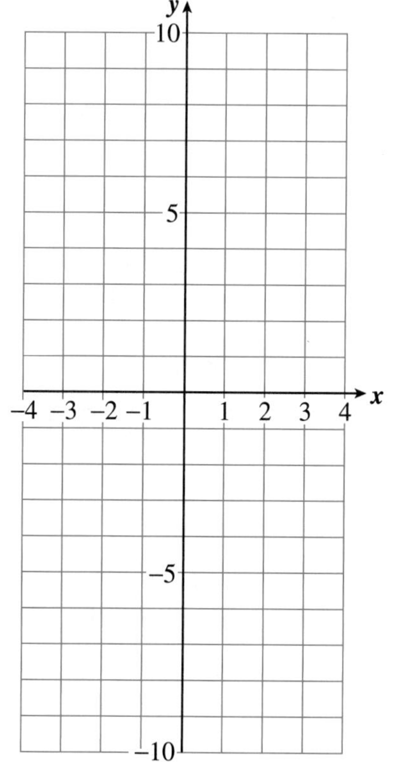

12. $y = x^3 - 1$

x	-2	-1	$\frac{-1}{2}$	0	$\frac{1}{2}$	1	2
y							

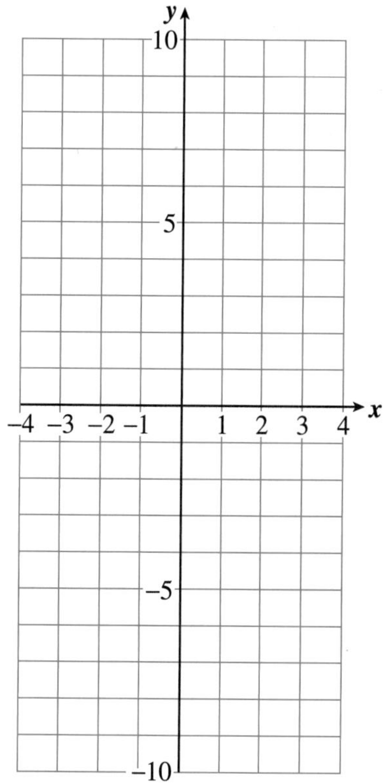

13. $y = \frac{1}{x}$

x	-4	-3	-2	-1	$\frac{-1}{2}$	$\frac{-1}{4}$
y						

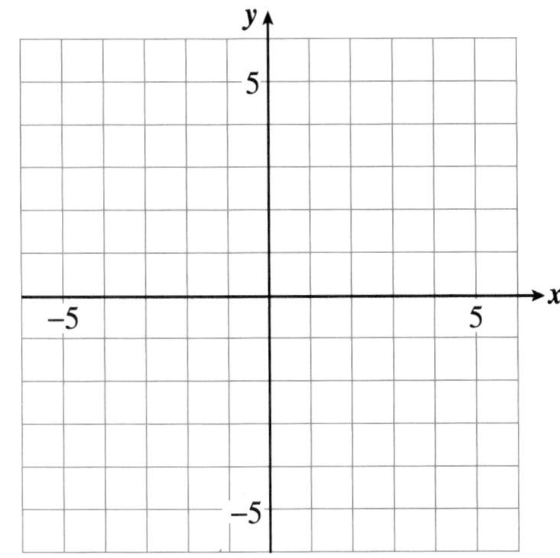

14. $y = \frac{1}{x}$

x	$\frac{1}{4}$	$\frac{1}{2}$	1	2	3	4
y						

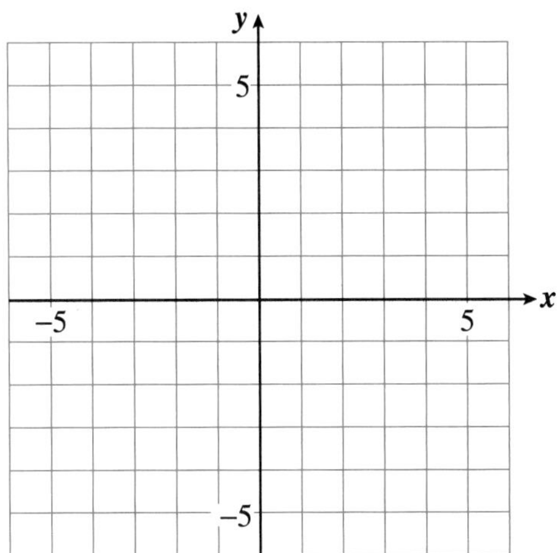

15. $y = \sqrt{x - 4}$

x	4	5	6	7	8	13	20
y							

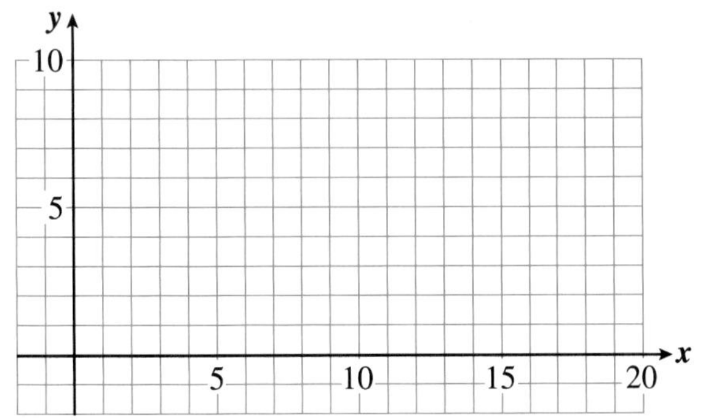

16. $y = \sqrt{x + 4}$

x	-4	-3	-2	-1	0	5	12
y							

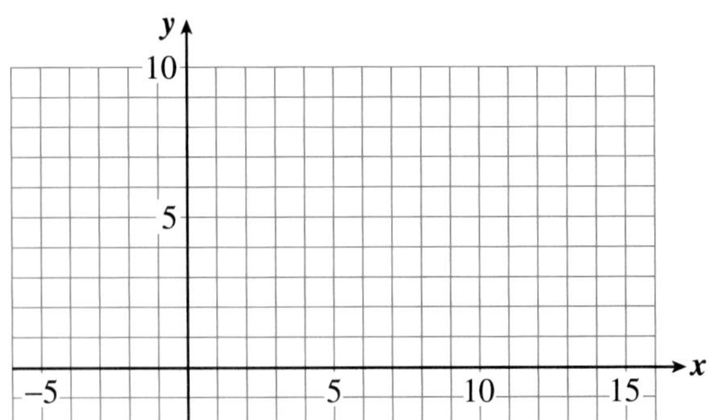

17. $y = 9 - x^2$

x	−4	−3	−2	−1	0	1	2	3	4
y									

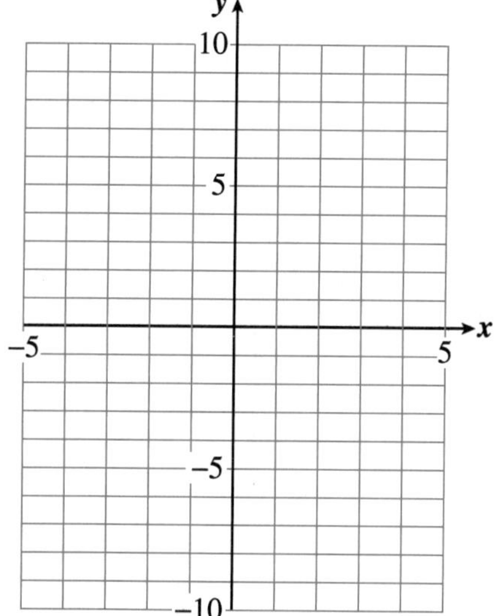

18. $y = x^2 + 2$

x	−4	−3	−2	−1	0	1	2	3	4
y									

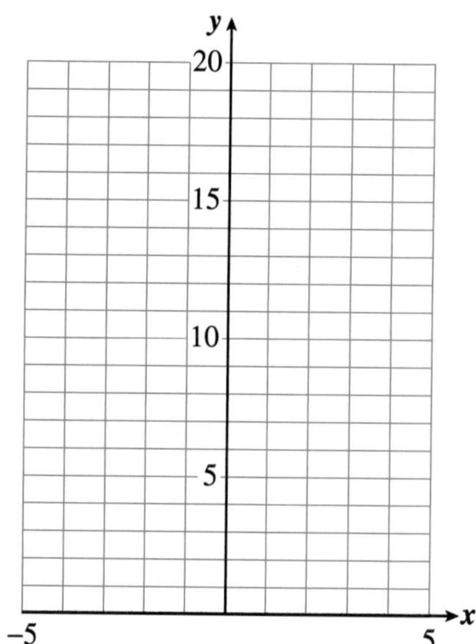

19. $y = x^2 + 2x$

x	−4	−3	−2	−1	0	1	2
y							

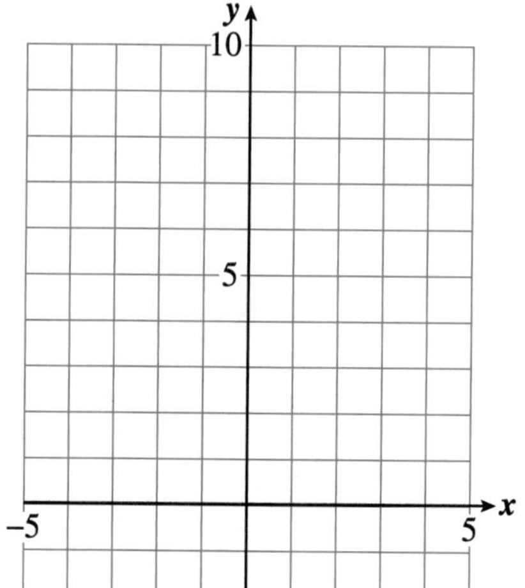

20. $y = x^2 - 2x$

x	−2	−1	0	1	2	3	4
y							

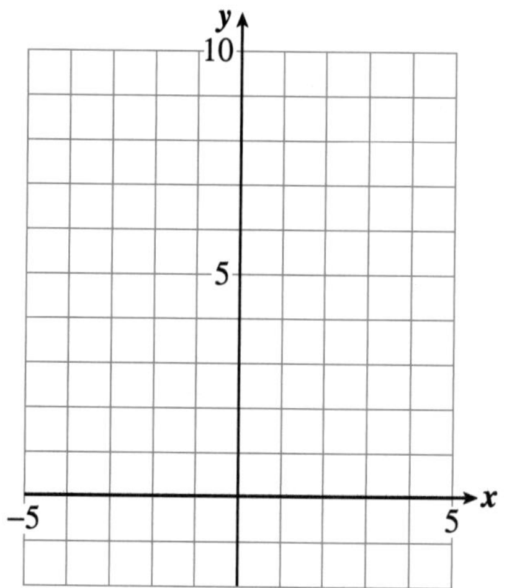

Fill in the table to go with each graph. Then, from the list of equations given, choose the correct equation for the graph.

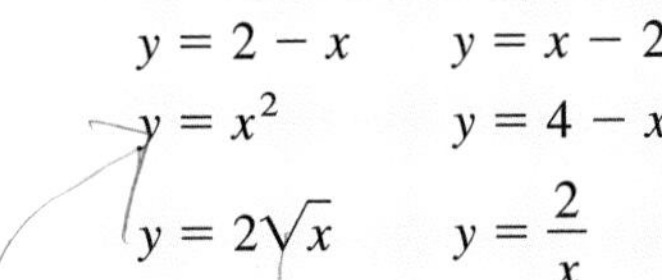

$y = 2 - x$ $\quad y = x - 2$

$y = x^2$ $\quad y = 4 - x^2$

$y = 2\sqrt{x}$ $\quad y = \dfrac{2}{x}$

21.

x	y
-2	
-1	
$\frac{1}{2}$	
1	

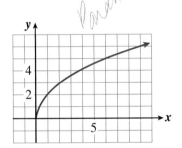

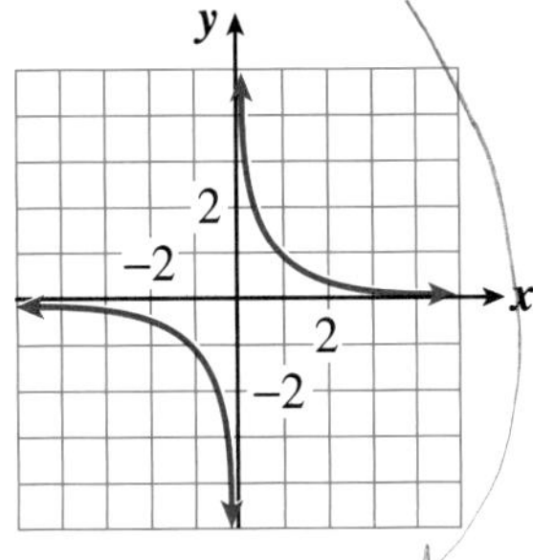

22.

x	y
-3	
0	
1	
3	

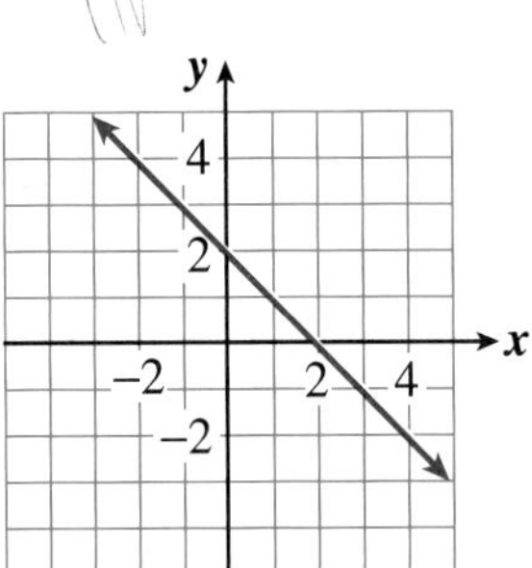

23.

x	y
0	
1	
4	
9	

24.

x	y
-2	
0	
1	
2	

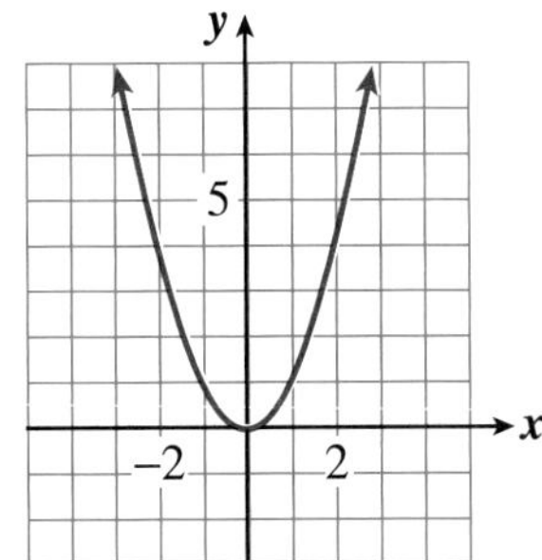

25.

x	y
5	
3	
0	
-1	

26.

x	y
-2	
0	
1	
3	

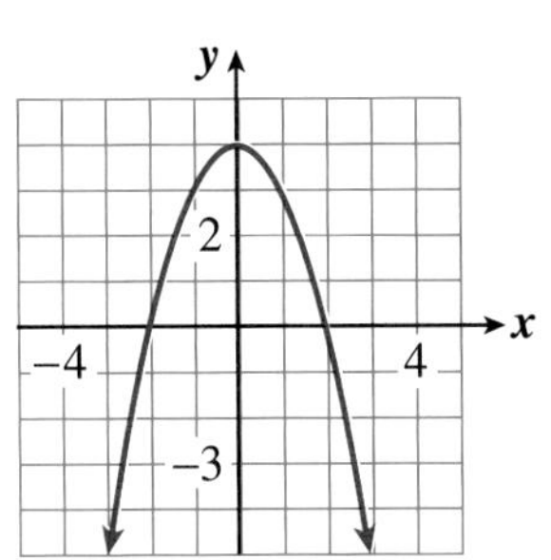

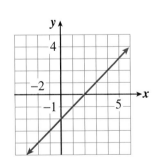

Choose the correct equation to model each problem and solve. (See Lesson 4.4 to review problem solving.)

$5(x + 25) = 75$	$\dfrac{x + 25}{5} = 75$
$\dfrac{x}{5} + 25 = 75$	$5x + 25 = 75$

27. Valerie needs to score only 25 points on her fifth exam for her average on the five tests to be 75. What is the sum of her first four exam scores?

28. Toan bought five tickets to an art show and paid an additional \$25 to buy a related book. If the total cost was \$75, how much was each ticket?

29. Joowon and his four friends split evenly the cost of the food bill at a nice restaurant, but Joowon paid the full \$25 for the bar tab. Joowon's total came to \$75. How much was the food bill for the meal?

30. Going into the bonus round of a game show, Kira's 2-day total consisted of yesterday's score plus the 25 points she earned so far today. If she answers the next question correctly, her score will be multiplied by 5, which would give her a new score of 75. What was Kira's score yesterday?

Building Number Skills

Build both fractions to the same denominator and then add or subtract. (See Appendix A.3 to review adding fractions.)

Example: $\dfrac{1}{2} + \dfrac{3}{8} = \dfrac{4}{8} + \dfrac{3}{8} = \dfrac{7}{8}$

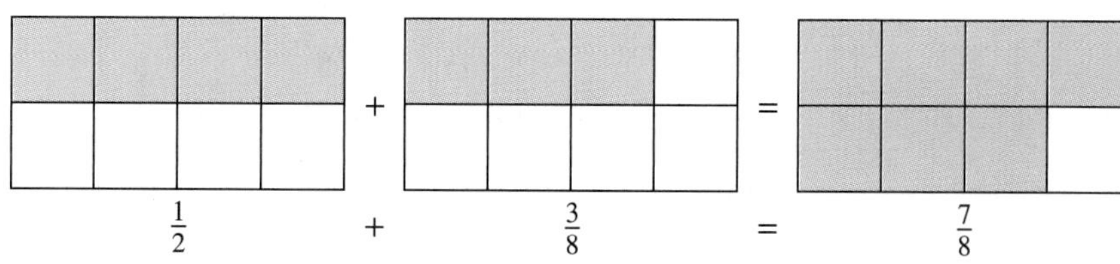

$\dfrac{1}{2} + \dfrac{3}{8} = \dfrac{7}{8}$

31. $\dfrac{1}{3} + \dfrac{5}{6} = \dfrac{?}{6} + \dfrac{?}{6}$

32. $\dfrac{2}{3} + \dfrac{4}{9} = \dfrac{?}{9} + \dfrac{?}{9}$

33. $\dfrac{3}{4} + \dfrac{5}{16} = \dfrac{?}{16} + \dfrac{?}{16}$

34. $\frac{1}{4} + \frac{7}{8} = \frac{?}{8} + \frac{?}{8}$

35. $\frac{7}{12} - \frac{1}{3} = \frac{?}{12} - \frac{?}{12}$

36. $\frac{1}{2} - \frac{1}{6} = \frac{?}{6} - \frac{?}{6}$

37. $\frac{2}{3} - \frac{4}{15} = \frac{?}{15} - \frac{?}{15}$

38. $\frac{7}{10} - \frac{2}{5} = \frac{?}{10} - \frac{?}{10}$

39. $\frac{5}{18} + \frac{1}{6} = \frac{?}{18} + \frac{?}{18}$

40. $\frac{3}{5} + \frac{6}{25} = \frac{?}{25} + \frac{?}{25}$

6.4 Lines

The simplest and most useful type of equation relating two variables has a very simple and useful graph. These equations are called **linear equations** because their graphs are straight lines. Any equation that looks like

$$y = ax + b$$

where a and b are constants, is a linear equation.

EXAMPLE 1

Graph the equation $y = -2x + 6$

Solution

Begin by making a table of values. Choose both positive and negative values for the independent variable x. Calculate the y value for each x value by substituting the x value into the equation.

x	y		
−3	12	$y = -2(\mathbf{-3}) + 6 = 12$	(−3, 12)
−2	10	$y = -2(\mathbf{-2}) + 6 = 10$	(−2, 10)
−1	8	$y = -2(\mathbf{-1}) + 6 = 8$	(−1, 8)
0	6	$y = -2(\mathbf{0}) + 6 = 6$	(0, 6)
1	4	$y = -2(\mathbf{1}) + 6 = 4$	(1, 4)
2	2	$y = -2(\mathbf{2}) + 6 = 2$	(2, 2)
3	0	$y = -2(\mathbf{3}) + 6 = 0$	(3, 0)
4	−2	$y = -2(\mathbf{4}) + 6 = -2$	(4, −2)

Figure 6.36

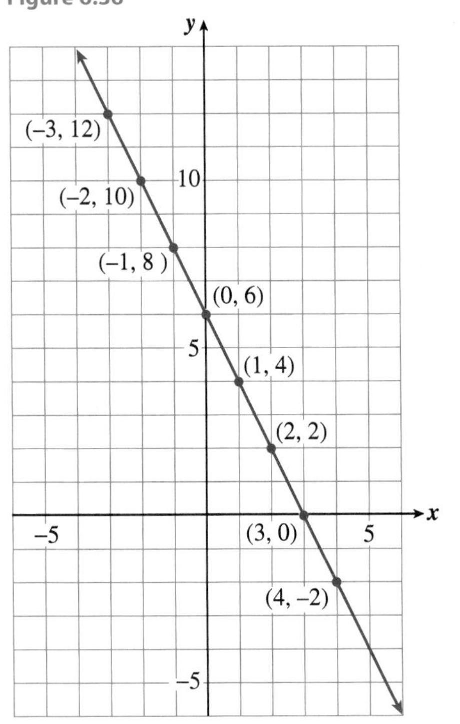

Next, sketch a Cartesian coordinate system with appropriate scales on the x and y axes. For this graph, we choose scales from −5 to 5 on the x axis and from −5 to 15 on the y axis. Plot each of the points in the table of values. You should find that all points lie on a straight line. The completed graph is shown in Figure 6.36.

In Example 1, if we choose more x values and plot more points, we will find that the line extends forever in either direction. To indicate this on the graph, we draw an arrowhead at each end of the line, as shown in the figure.

Now try Exercise 1.

EXERCISE 1

Graph the equation $y = 3x - 9$

Step 1 Complete the table.

x	y
−1	
0	
1	
2	
3	
4	

Step 2 Label the scales on the axes.

Step 3 On the grid in Figure 6.37, plot the points from the table and connect them with a straight line.

Figure 6.37

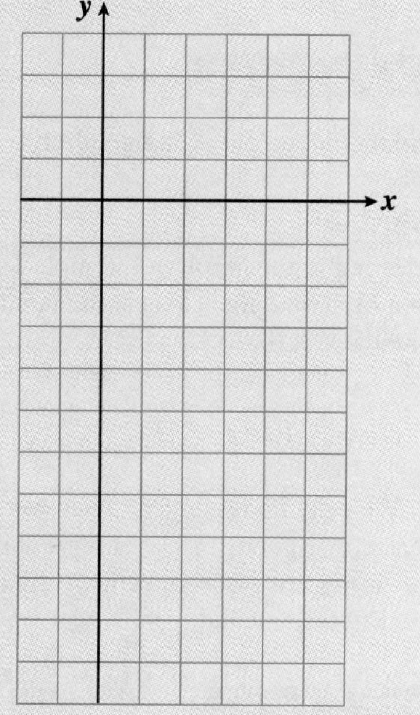

How many points do we need to plot, to draw an accurate graph? Perhaps you can see that to graph a line we really only need two points (and a ruler or straightedge). It is a good idea to plot three points, though, as a check.

EXAMPLE 2  

Graph the equation $y = x - 4$

Solution

Choose three different values for x and find the corresponding y values. Record your values in a table.

x	y		
−1	−5	$y = \mathbf{-1} - 4 = -5$	$(-1, -5)$
0	−4	$y = \mathbf{0} - 4 = -4$	$(0, -4)$
2	−2	$y = \mathbf{2} - 4 = -2$	$(2, -2)$

Plot the points and draw a line through them. (If all three points do not lie on the same line, there must be an error in your calculations!) The completed graph is shown in Figure 6.38.

Figure 6.38

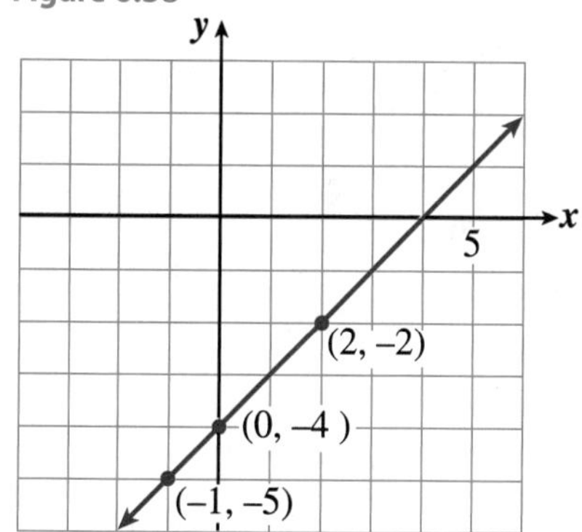

Now try Exercise 2.

EXERCISE 2

Graph the equation $y = 2x + 4$

Step 1 Complete the table:

x	y
−3	
0	
1	

Step 2 Label the scales on the axes.

Step 3 On the grid in Figure 6.39, plot the points from the table and connect them with a straight line.

Figure 6.39

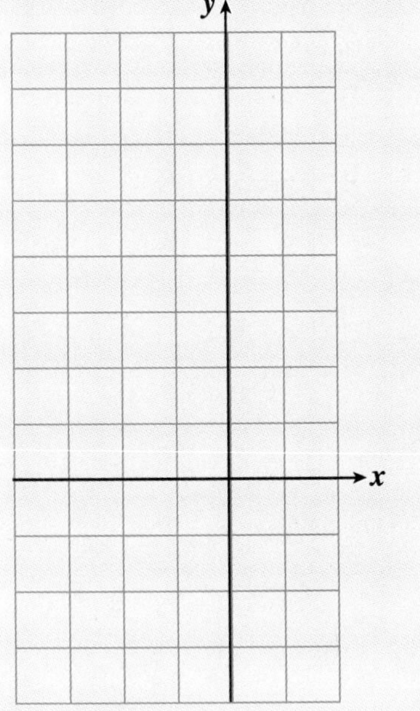

EXERCISE 3

Find the intercepts of the graph in Exercise 1, $y = 3x - 9$

Intercepts

The graph of a linear equation does not end at the last point we plot; it continues forever in both directions. Every line (unless it is horizontal or vertical) will cross both the x axis and the y axis eventually. The two intersection points are called the ***x* intercept** and the ***y* intercept,** respectively. In Example 2 the x intercept is the point (4, 0), and the y intercept is the point (0, −4). The intercepts are important both for applications and for understanding the graph itself. When graphing a line, we should show enough of the line to include both intercepts.

EXAMPLE 3

Find the intercepts of the graph in Example 1, $y = -2x + 6$.

Solution

Referring to the graph in Example 1, we see that the line crosses the x axis at the point (3, 0) and the y axis at the point (0, 6). Thus, the x intercept is (3, 0) and the y intercept is (0, 6).

Now try Exercise 3.

We can choose any x values we like when we calculate points for a graph. It is usually a good idea to choose one negative x value and one positive x value, and zero is always a convenient choice. If the coefficient of x is a fraction, we can simplify the calculations by choosing x values with the fraction in mind.

EXAMPLE 4 

a. Graph the equation $y = \frac{1}{3}x + 3$

b. Find the x and y intercepts of the line.

Solution

a. Because the coefficient of x is $\frac{1}{3}$, we can avoid working with fractions by choosing x values that are divisible by 3. For this example we choose −6, 0, and 3.

x	y
−6	1
0	3
3	4

$y = \frac{1}{3}(\mathbf{-6}) + 3 = -2 + 3 = 1$ (−6, 1)

$y = \frac{1}{3}(\mathbf{0}) + 3 = 0 + 3 = 3$ (0, 3)

$y = \frac{1}{3}(\mathbf{3}) + 3 = 1 + 3 = 4$ (3, 4)

Plot the points and draw the line through them. Extend the line far enough to cross both axes. The completed graph is shown in Figure 6.40.

b. The graph crosses the x axis at the point $(-9, 0)$ and the y axis at the point $(0, 3)$. Thus, the x intercept is $(-9, 0)$, and the y intercept is $(0, 3)$.

Figure 6.40

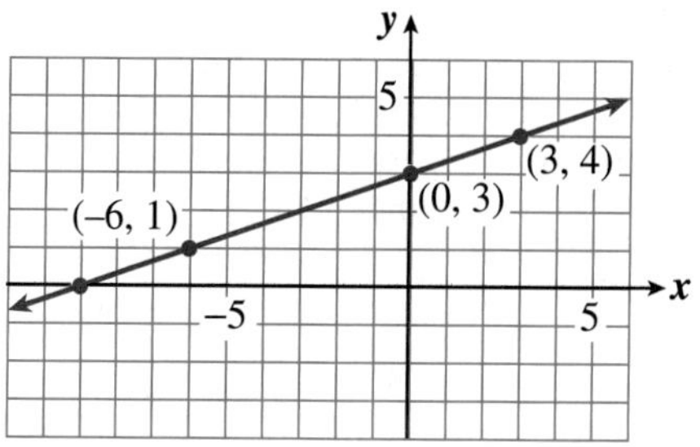

Now try Exercise 4.

EXERCISE 4

Graph the equation $y = \frac{3}{4}x - 2$.

Step 1 Complete the table:

x	y
-4	
0	
4	

Step 2 Label the scales on the axes.

Step 3 On the grid in Figure 6.41, plot the points from the table and connect them with a straight line.

Figure 6.41

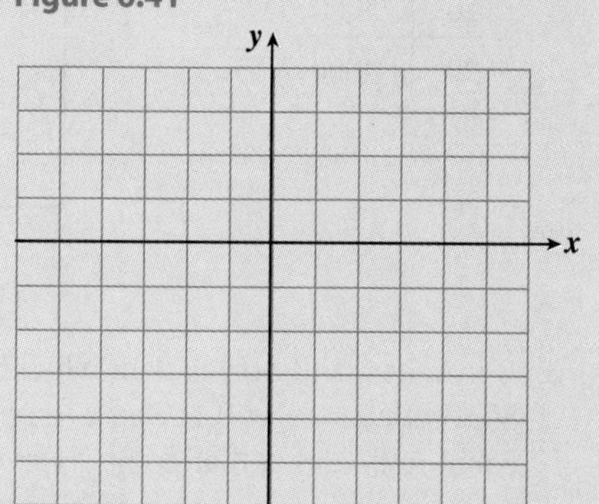

Applications of Linear Equations

There are many examples of variable relationships that lead to linear equations.

EXAMPLE 5

Byron borrowed \$6000 from his uncle to help pay for his college education. Now that he has graduated and has a job, he is paying back the loan at \$100 per month. (His uncle is not charging Byron any interest on the loan.)

a. Write an equation showing the amount of money, y, that Byron still owes his uncle after x months.

b. Graph your equation.

c. Use your graph to find out how long will it take Byron to pay off his debt.

EXERCISE 5

Chloe's All-Night Diner buys coffee in 50-pound containers. They use 2 pounds of coffee per day.

a. Write an equation for the number of pounds of coffee, C, left d days after they open a new container.

b. Graph your equation on the grid in Figure 6.43.

Figure 6.43

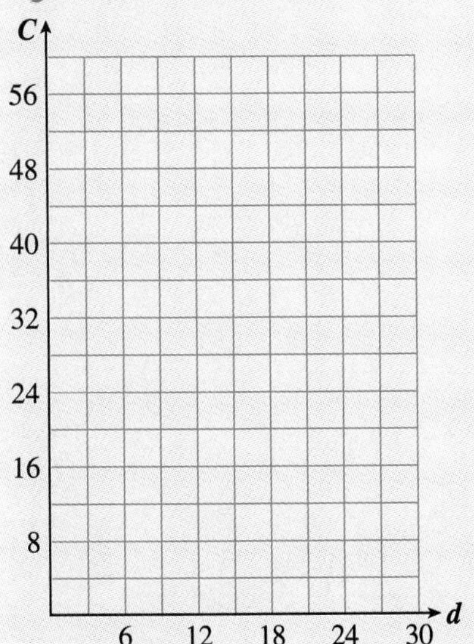

c. Chloe's orders more coffee when they have 8 pounds left in the container. Use your graph to find out how long that will take.

Solution

a. Byron subtracts \$100 from his debt each month, so

$$y = 6000 - 100x$$

b. Choose values for x and make a table showing several points on the graph. Then plot the points and connect them with a straight line, as shown in Figure 6.42.

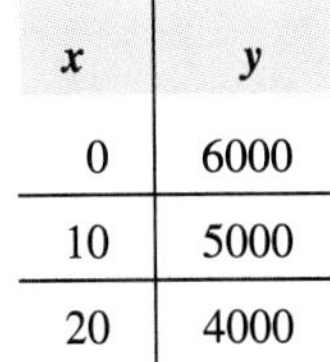

x	y
0	6000
10	5000
20	4000

Figure 6.42

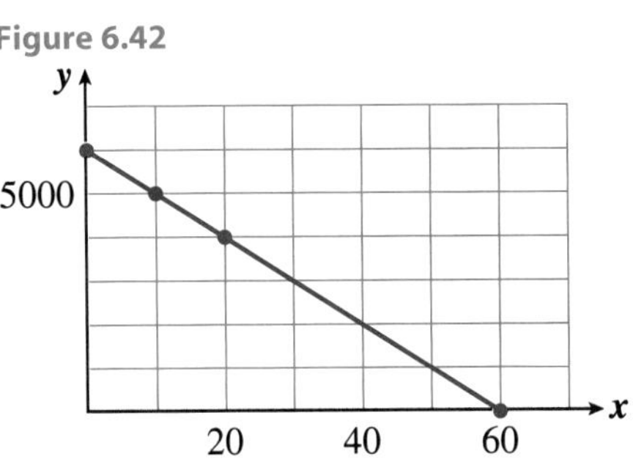

c. Byron has paid off his debt when the amount he still owes his uncle is \$0, or when $y = 0$. The x intercept of the graph has y coordinate equal to 0, and at that point $x = 60$. Thus, it will take Byron 60 months, or 5 years, to repay his uncle.

Now try Exercise 5.

ANSWERS TO 6.4 EXERCISES

1.

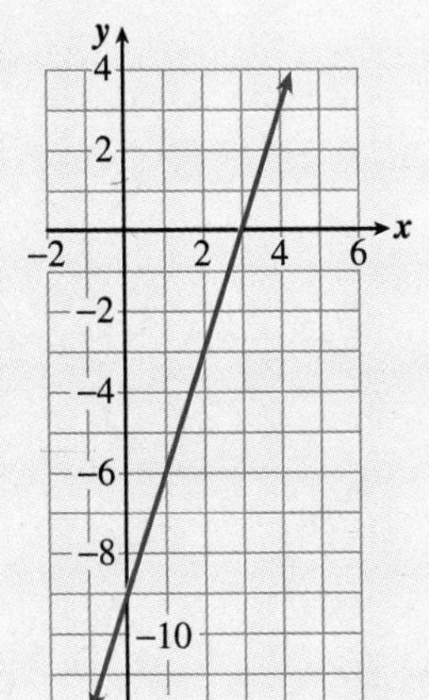

2.

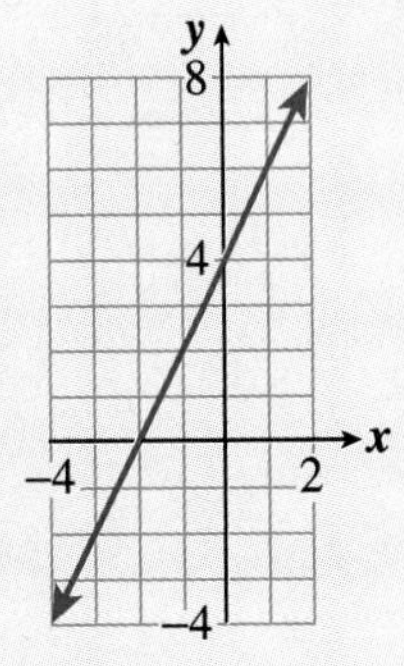

3. $(0, -9)$ and $(3, 0)$

4.

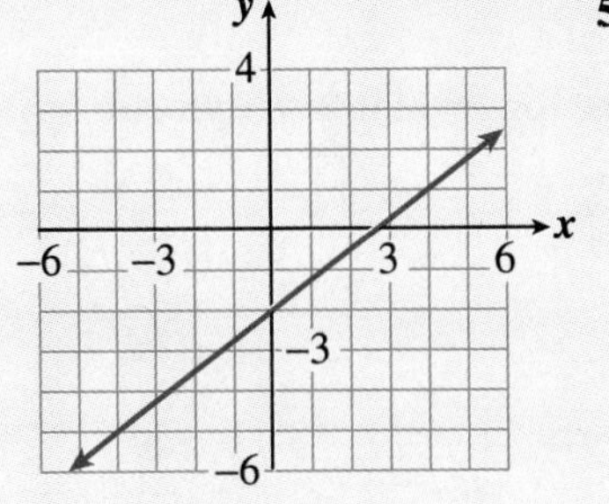

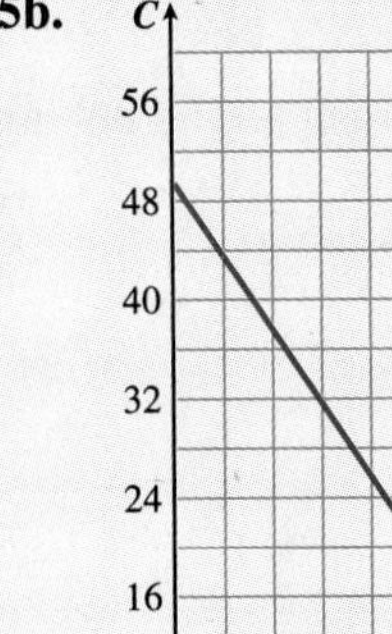

5a. $C = 50 - 2d$ **5c.** 21 days

5b.

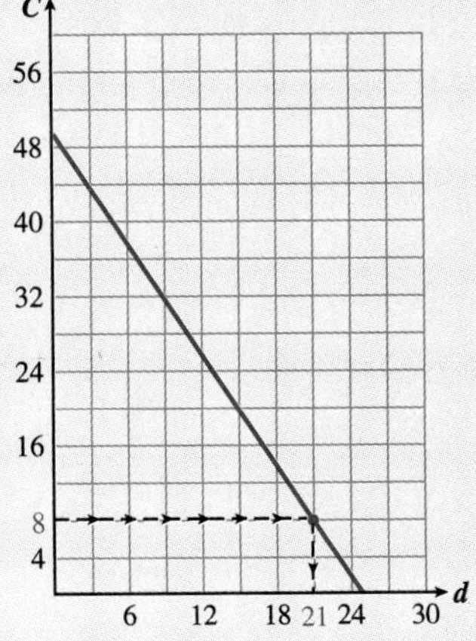

HOMEWORK 6.4 

1. a. Give an example of an equation whose graph is a straight line.

   b. Give an example of an equation whose graph is *not* a straight line.

2. What is a linear equation in two variables? Give an example.

3. What is the x intercept of a graph? What is the y intercept?

4. a. What can you say about the coordinates of the x intercept?

   b. What can you say about the coordinates of the y intercept?

For Problems 5–16, (a) make a table showing three solutions to the equation, (b) graph the equation, and (c) find the x and y intercepts of the graph.

5. $y = x$

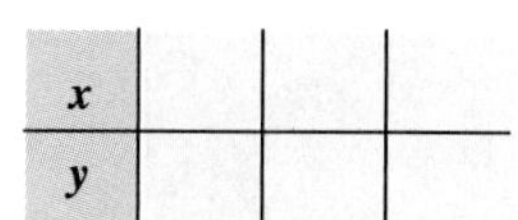

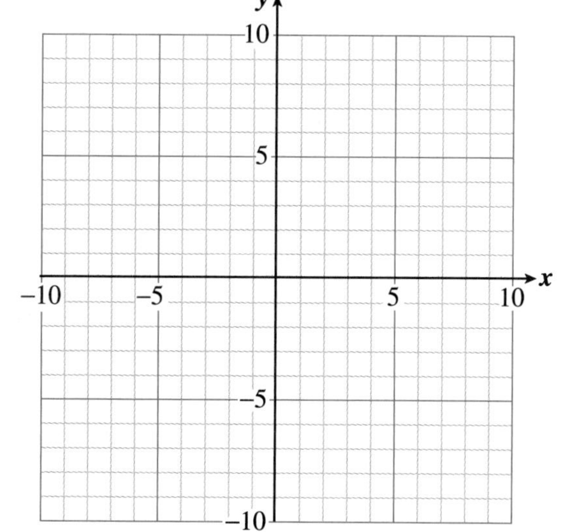

6. $y = 2x$

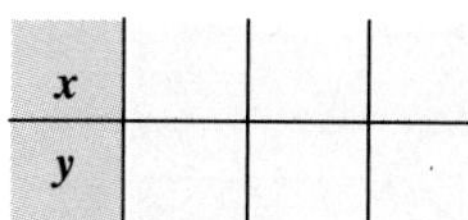

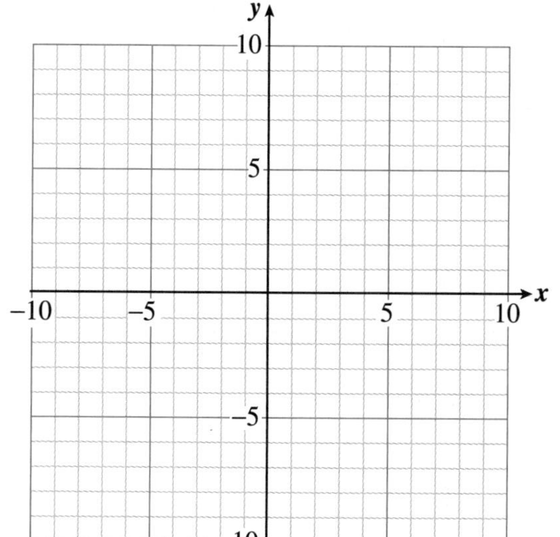

7. $y = x - 1$

x		
y		

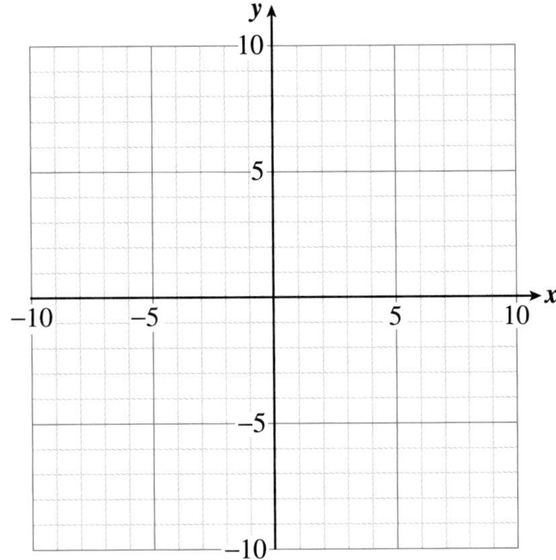

8. $y = x + 2$

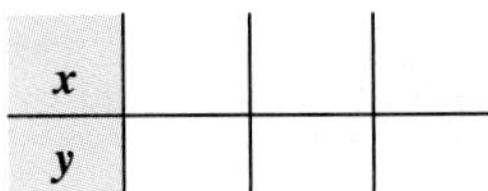

x		
y		

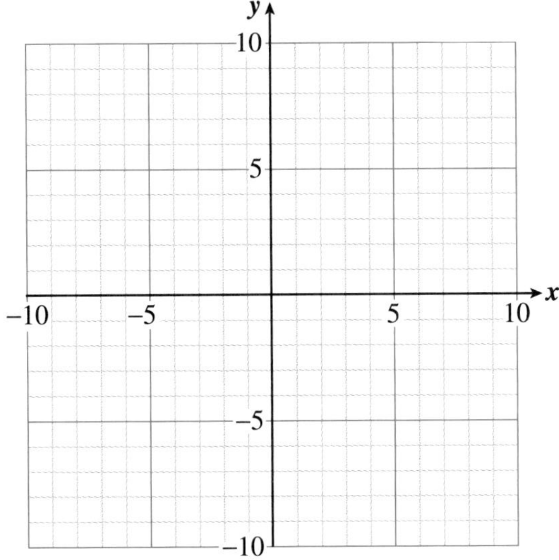

9. $y = -2x$

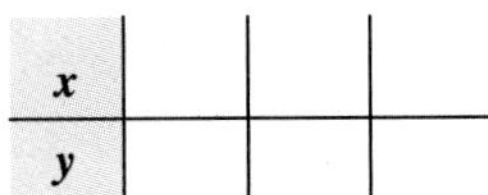

x		
y		

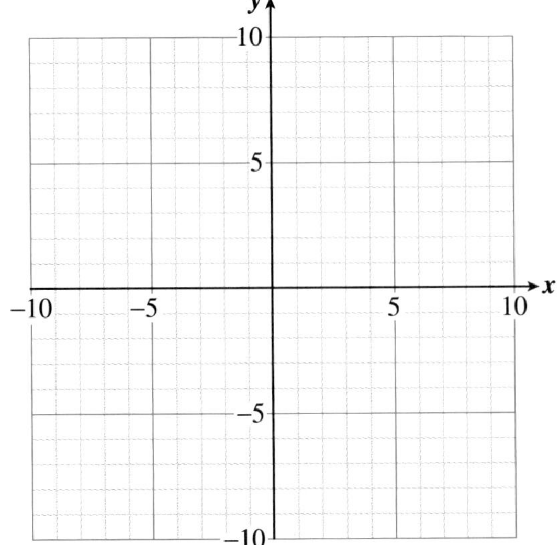

10. $y = -3x$

x		
y		

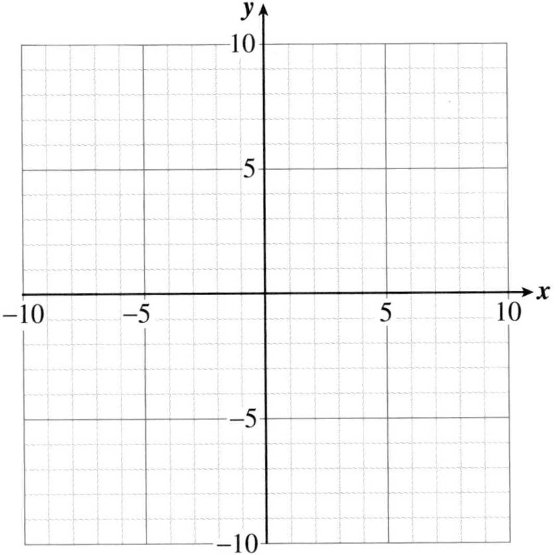

11. $y = -x - 5$

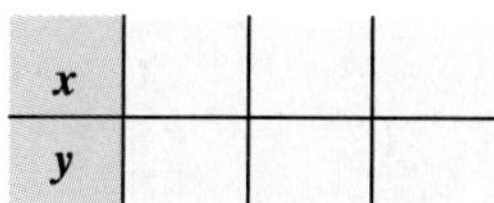

x			
y			

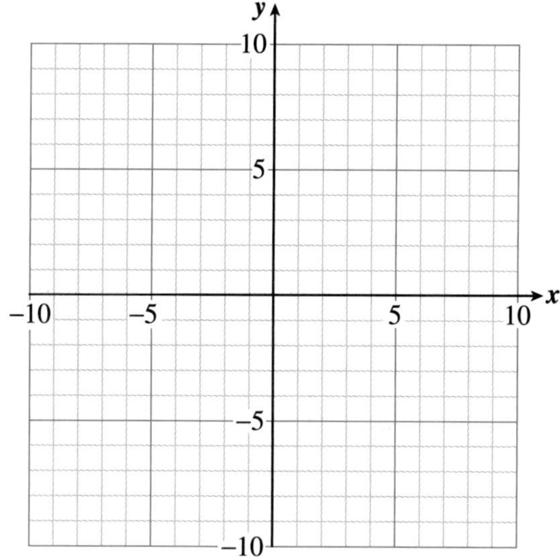

12. $y = -x + 4$

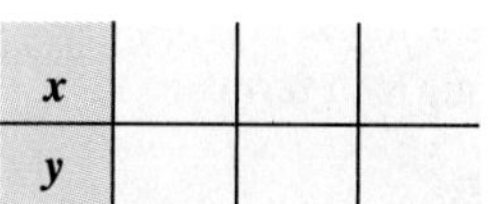

x			
y			

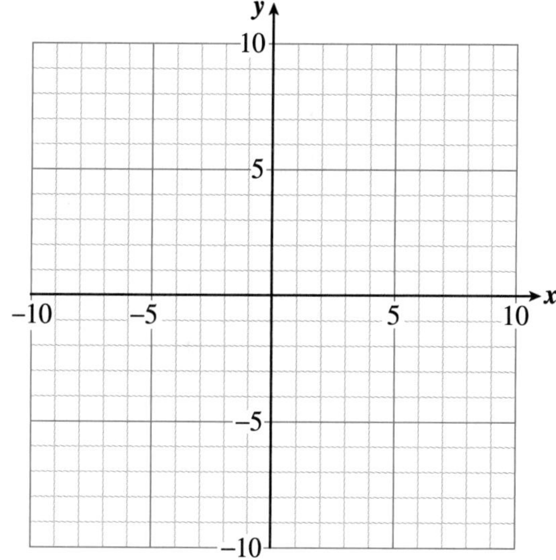

13. $y = 2x + 3$

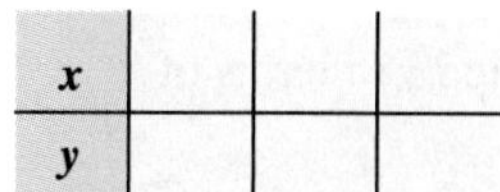

x			
y			

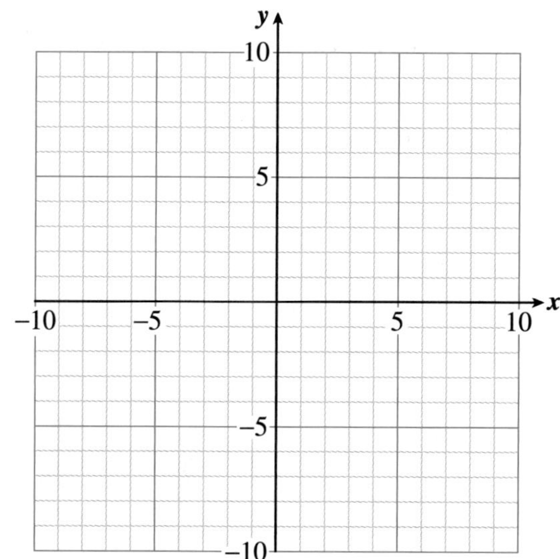

14. $y = -3x - 1$

x			
y			

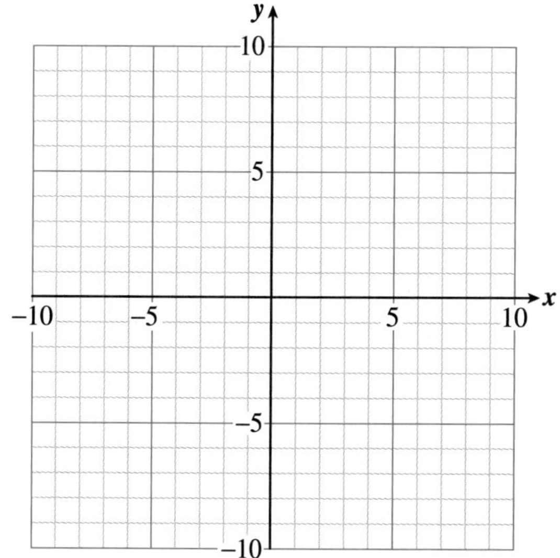

15. $y = \frac{1}{2}x - 4$

x		
y		

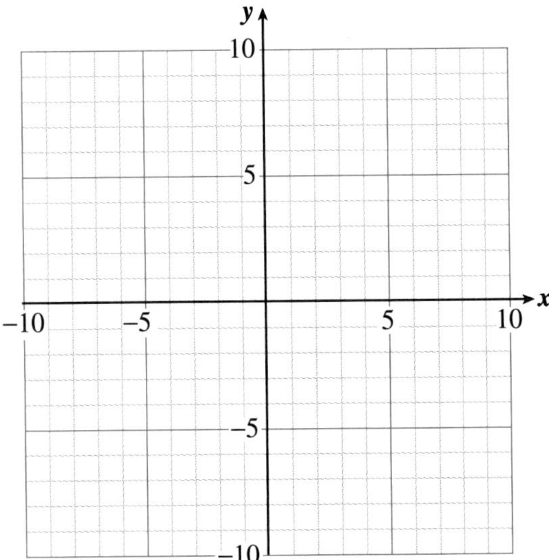

16. $y = \frac{2}{3}x + 2$

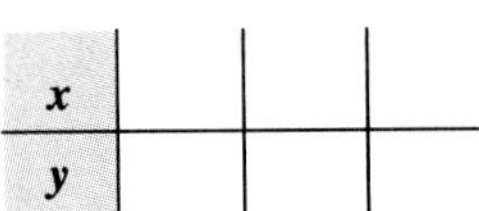

x		
y		

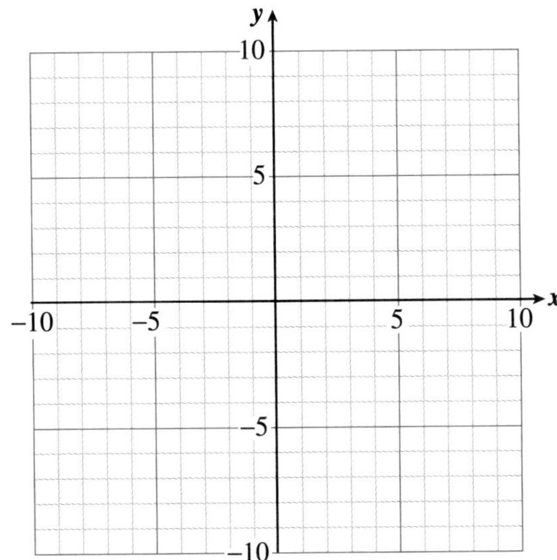

17. Eloise makes $20 an hour as a computer programmer. Let x represent the number of hours Eloise works in a week and let y represent her earnings.

a. Write an equation expressing y in terms of x.

b. Complete the table and graph your equation.

x	y
0	
10	
20	
40	

18. Anatole sells hand-tooled leather belts for $15 each. Let x represent the number of belts he sells in a week and let y represent his revenue.

a. Write an equation expressing y in terms of x.

b. Complete the table and graph your equation.

x	y
0	
5	
10	
20	

19. Stuart invested \$800 in a computer and word processor and now makes \$5 a page typing research papers. Let x represent the number of pages Stuart has typed and let y represent his profit from his business venture. (What is his "profit" when $x = 0$?)

a. Write an equation expressing y in terms of x.

b. Complete the table and graph your equation.

x	y
0	
50	
100	
200	

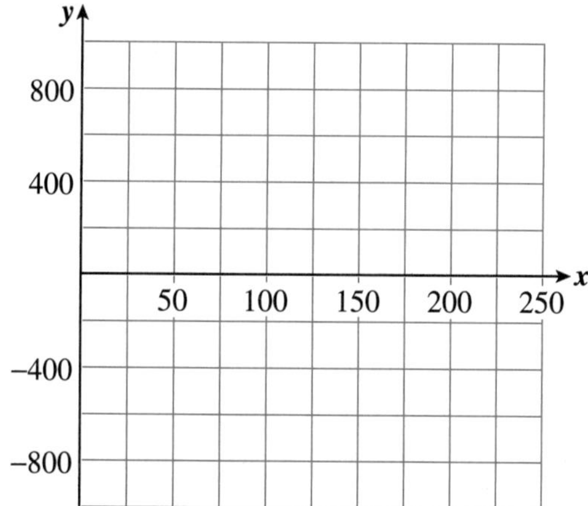

c. How many pages must Stuart type in order to break even? Write and solve an equation to answer this question and then verify the answer on your graph.

20. The Hammel Beach Light Opera ended last season \$1200 in debt, but they are selling tickets to their summer benefit performance at \$20 each. Let x represent the number of tickets they sell and let y represent their financial balance. (What is the balance when $x = 0$?)

a. Write an equation expressing y in terms of x.

b. Complete the table and graph your equation.

x	y
0	
20	
40	
100	

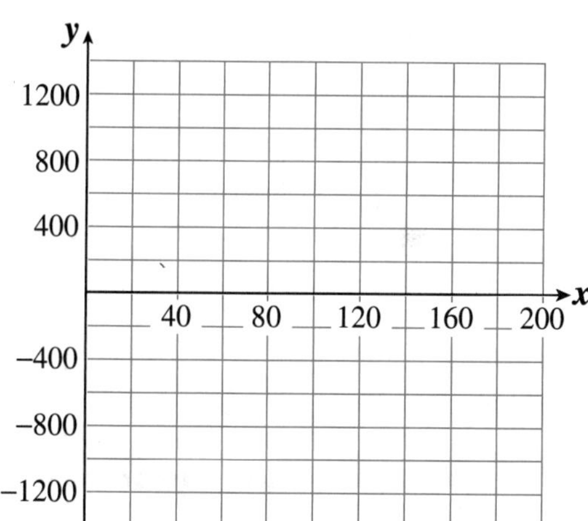

c. How many tickets must they sell in order to erase their debt? Write and solve an equation to answer this question and then verify the answer on your graph.

Solve each equation and check your solutions. (See Lessons 4.3 and 5.4 to review solving equations.)

21. $2x - 6 = -10$

22. $3x + 13 = 4$

23. $3c^2 - 7 = 5$

24. $\frac{d^2}{3} - 2 = 10$

25. $3h^2 = 15$

26. $\frac{k^2}{3} = 2$

27. $\frac{x^2}{2} + 3 = 6$

28. $5y^2 + 1 = 16$

Building Number Skills

Build both fractions to the same denominator and then add or subtract. (See Appendix A.3 to review adding fractions.)

Example: $\frac{1}{6} + \frac{3}{10} = \frac{5}{30} + \frac{9}{30} = \frac{14}{30} = \frac{7}{15}$

29. $\frac{5}{6} - \frac{4}{9} = \frac{?}{18} - \frac{?}{18}$

30. $\frac{5}{6} - \frac{3}{4} = \frac{?}{12} - \frac{?}{12}$

31. $\frac{3}{4} - \frac{3}{10} = \frac{?}{20} - \frac{?}{20}$

32. $\frac{9}{10} - \frac{5}{6} = \frac{?}{30} - \frac{?}{30}$

33. $\frac{3}{10}+\frac{7}{15}=\frac{?}{30}+\frac{?}{30}$

34. $\frac{1}{6}+\frac{4}{15}=\frac{?}{30}+\frac{?}{30}$

35. $\frac{5}{6}-\frac{5}{8}=\frac{?}{24}-\frac{?}{24}$

36. $\frac{3}{8}-\frac{1}{6}=\frac{?}{24}-\frac{?}{24}$

37. $\frac{1}{8}+\frac{7}{10}=\frac{?}{40}+\frac{?}{40}$

38. $\frac{7}{10}+\frac{7}{8}=\frac{?}{40}+\frac{?}{40}$

6.5 Equations Again

In Chapter 3 we learned that an expression such as

$$10 - 6 + 3$$

can be interpreted as the *sum* of three terms: 10, -6, and $+3$. The + and − symbols tell us whether the number that follows is positive or negative. This viewpoint will be very helpful when we solve equations involving negative numbers.

Solving Equations

Recall our strategy for solving equations: We undo the operations performed on the variable, in their reverse order. In the equation

$$8 - 2x = -6$$

we think of $8 - 2x$ as the *sum* of two terms, 8 and $-2x$. Then the operations performed on the variable are

Operations Performed on x

1. Multiplied by -2

2. Added 8

To solve the equation, we undo these two operations in reverse order, as shown in Example 1.

EXAMPLE 1

Solve $8 - 2x = -6$

Solution

Our plan for solution looks like this:

Operations Performed on x	*Steps for Solution*
1. Multiplied by -2	**1.** Subtract 8
2. Added 8	**2.** Divide by -2

Now we carry out the plan:

$$\begin{aligned} 8 - 2x &= -6 \\ \mathbf{-8} \quad & \quad \mathbf{-8} \\ \hline -2x &= -14 \\ \frac{-2x}{\mathbf{-2}} &= \frac{-14}{\mathbf{-2}} \\ x &= 7 \end{aligned}$$

Subtract 8 from both sides of the equation.

Divide both sides by -2.

The solution is 7.

As always, it is a good idea to check the solution in the original equation. Remember to enclose the solution in parentheses when you substitute into the equation.

CAUTION!

In Example 1 be sure to divide both sides by -2, not by 2. To reverse the operation "multiply by -2" we divide by the *same* number, -2.

Now try Exercise 1.

EXERCISE 1
Solve $10 - 4x = 2$

EXAMPLE 2

Solve $-4.4 = -6.8 - 0.2m$

Solution

First, we analyze the equation and make a plan for its solution:

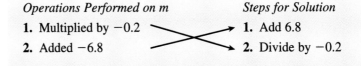

Operations Performed on m	*Steps for Solution*
1. Multiplied by -0.2	**1.** Add 6.8
2. Added -6.8	**2.** Divide by -0.2

Now we carry out the plan:

$$-4.4 = -6.8 - 0.2m$$

$$\underline{\mathbf{+6.8}} \quad \underline{\mathbf{+6.8}} \qquad \text{Add 6.8 to both sides.}$$

$$2.4 = \quad -0.2m$$

$$\frac{2.4}{\mathbf{-0.2}} = \frac{-0.2m}{\mathbf{-0.2}} \qquad \text{Divide both sides by } -0.2.$$

$$-12 = m$$

The solution is -12. (You should check the solution.)

CAUTION!

In Example 2 we reversed the operation "added -6.8" by adding the opposite, 6.8. (See Lesson 3.2B to review solving equations by adding the opposite of a number.)

Now try Exercise 2.

In Example 3 we use three steps to solve the equation.

EXAMPLE 3 

Solve $15 = \frac{-3x}{4} - 9$

Solution

Analyze the equation and formulate a plan for its solution:

Operations Performed on x	*Steps for Solution*
1. Multiplied by -3	**1.** Add 9
2. Divided by 4	**2.** Multiply by 4
3. Subtracted 9	**3.** Divide by -3

Now carry out the plan:

$$15 = \frac{-3x}{4} - 9$$

$$\underline{\mathbf{+9}} \quad \underline{\qquad \mathbf{+9}} \qquad \text{Add 9 to both sides.}$$

$$24 = \frac{-3x}{4}$$

$$\mathbf{4}(24) = \left(\frac{-3x}{4}\right)\mathbf{4} \qquad \text{Multiply both sides by 4.}$$

$$96 = -3x$$

$$\frac{96}{\mathbf{-3}} = \frac{-3x}{\mathbf{-3}} \qquad \text{Divide both sides by 3.}$$

$$-32 = x$$

The solution is -32. (You should check the solution.)

Try Exercise 3.

EXERCISE 2

Solve $-3.6 - 2.8t = -45.6$

EXERCISE 3

Solve $\frac{5 - 2x}{6} = -3$

ANSWERS TO 6.5A EXERCISES

1. 2

2. 15

3. 11.5

HOMEWORK 6.5 

1. **a.** What operations have been performed on x in the expression $2 - 5x$?

   b. How would you undo the operations in part (a)?

2. **a.** What operations have been performed on y in the expression $-1 - 6y$?

   b. How would you undo the operations in part (a)?

3. **a.** What operations have been performed on z in the expression $(z/3) + 4$?

   b. How would you undo the operations in part (a)?

4. **a.** What operations have been performed on w in the expression $(w - 7)/2$?

   b. How would you undo the operations in part (a)?

Solve each equation. Show your work.

5. $3 - 9x = -15$

6. $-6x - 5 = 19$

7. $-3x + 7 = -26$

8. $-2 - 8x = 38$

9. $\dfrac{-7x}{2} + 14 = 18$

10. $\dfrac{5x}{3} - 9 = -11$

11. $7 + \dfrac{x}{-2} = -2$

12. $8 + \dfrac{x}{-9} = -1$

13. $-17 = \frac{-6x}{5} - 5$

14. $15 = -3 + \frac{2x}{-3}$

15. $\frac{x + 8}{-4} = -3$

16. $\frac{x - 5}{8} = -7$

17. $\frac{6 - 5x}{8} = -3$

18. $\frac{-3 - 4x}{2} = -1$

For Problems 19–24, (a) identify the unknown quantity and choose a variable to represent it, (b) find something in the problem that can be described in two ways and write an equation (see Lesson 4.4 to review writing equations), and (c) solve the equation and answer the question in the problem.

19. Starting at a depth of −45 feet, a diver begins descending at a rate of −15 feet per minute. How long will it take him to reach a depth of −150 feet?

20. At noon the temperature was 16°, and it has been growing colder at a rate of 3° per hour. How long will it be before the temperature reaches −26°?

21. Alida has to let her chocolate pie cool to room temperature before cutting it. The pie comes out of the oven at 350°, and room temperature is 75°. If the pie cools on average 11° per minute, how long must Alida wait?

22. Dean is cycling down Mt. Whitney and descends at a rate of 80 feet per minute. If he started at an elevation of 9600 feet, how long will it be before he reaches an elevation of 1200 feet?

23. Lois spent Saturday afternoon playing hand after hand of Fizbin with her sister. At 3 o'clock Lois had a score of 60 points, but by 5 o'clock her score was -96. If Lois lost 12 hands after 3 o'clock and didn't win any, how much is each hand worth?

24. Jean-Paul started a tab at The Common Grounds coffee house by paying the owner \$50. One month later, Jean-Paul owed the owner \$72.40. If a cup of coffee costs \$0.85, how many cups did Jean-Paul drink?

25. An archaeological expedition gathers at a remote village in the hills, at an altitude of 750 feet. They begin an arduous descent on foot into the valley below. They find that they can descend approximately 120 feet in elevation per hour. Let x represent the number of hours they have been hiking and let y represent their elevation.
   a. Write an equation expressing y in terms of x.

   b. Complete the table and graph your equation.

x	y
0	
2	
5	
7	

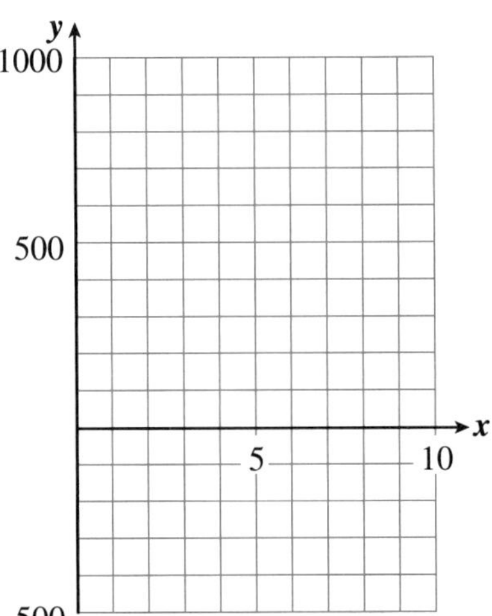

   c. How long will it take them to reach their site, which lies 210 feet below sea level on the valley floor? Write and solve an equation to answer this question and then verify the answer on your graph.

26. A group of adventurers would like to set a new elevation record for a hot-air balloon flight. The first layer of Earth's atmosphere is called the troposphere, and in this layer the atmospheric temperature decreases by about 8°C per kilometer above Earth's surface. The temperature on the ground is presently 24°C. Let x represent the balloon's altitude in kilometers and let y represent the temperature.
   a. Write an equation expressing y in terms of x.

   b. Complete the table and graph your equation.

x	y
0	
2	
5	
10	

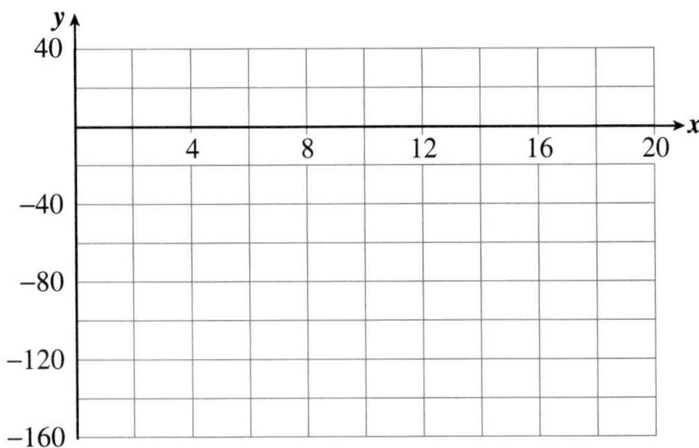

   c. What is the altitude at the top of the troposphere, where the temperature is -120°C? Write and solve an equation to answer this question, then verify the answer on your graph.

27. The equation that relates the temperature in degrees Fahrenheit to the temperature in degrees Celsius is

$$F = 1.8C + 32$$

Complete the table and graph this equation.

C	F
−20	
−10	
0	
10	
20	

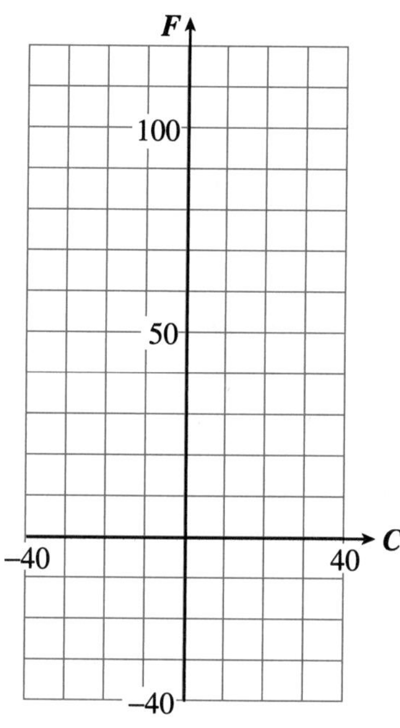

28. On the planet Kaldor in the Gamma Quadrant, there are no seasons. The temperature at noon on any day is given by the formula

$$T = 50 + 0.2h$$

where h is your elevation in Kaldorian feet and T is the temperature in Kaldorian degrees. Complete the table and graph this equation.

h	T
−1000	
−500	
0	
1000	
1500	

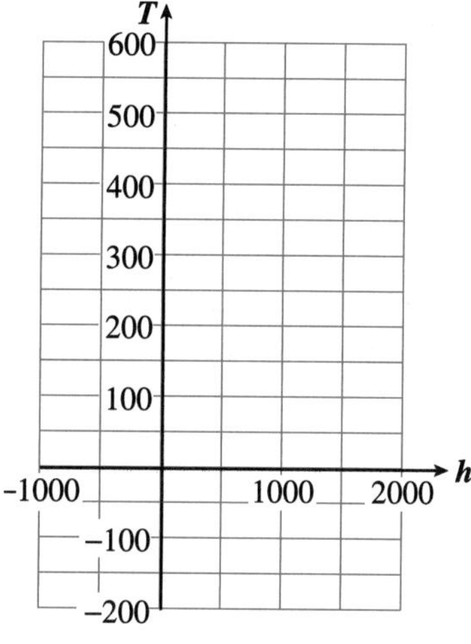

Evaluate each expression for the given value of the variables. (See Lesson 4.5 to review the order of operations.)

29. $-a$, for $a = -8.2$

30. $-(-b)$, for $b = -1.4$

31. $c - d$, for $c = -3$ and $d = -5$

32. $-k + l$, for $k = -7$ and $l = 4$

33. $-3m - 2n$, for $m = -5$ and $n = -2$

34. $\dfrac{24}{x} - 7y$, for $x = -3$ and $y = -2$

35. $\dfrac{3 - p}{-9 - q}$, for $p = -3$ and $q = -9$

36. $\dfrac{5 - r}{-15 - s}$, for $r = 5$ and $s = 15$

37. $-1 - 2(u + 7)$, for $u = -5$

38. $6 - 3(2 - v)$, for $v = -1$

Building Number Skills

To add unlike fractions, we build each fraction to the same denominator, called the LCD (lowest common denominator). The LCD is the smallest number that all the denominators divide into evenly. We can find the LCD by listing multiples of each denominator.

Find the LCD for each set of fractions.

Example: $\frac{1}{8}, \frac{1}{3}, \frac{1}{6}$

Multiples of 8: 8 16 [24] 32 40 48

Multiples of 3: 3 6 9 12 15 18 21 [24]

Multiples of 6: 6 12 18 [24] 30 36

The LCD is the first number that appears in all three lists. **LCD: 24**

39. $\frac{1}{9}, \frac{1}{12}$

40. $\frac{1}{8}, \frac{1}{10}$

41. $\frac{1}{10}, \frac{1}{15}$

42. $\frac{1}{15}, \frac{1}{20}$

43. $\frac{1}{12}, \frac{1}{15}$

44. $\frac{1}{12}, \frac{1}{16}$

45. $\frac{1}{4}, \frac{1}{6}, \frac{1}{10}$

46. $\frac{1}{6}, \frac{1}{9}, \frac{1}{12}$

47. $\frac{1}{4}, \frac{1}{8}, \frac{1}{16}$

48. $\frac{1}{3}, \frac{1}{9}, \frac{1}{81}$

6 Summary

Lesson 6.1

Graphs are useful for illustrating the relationship between two variables.

We can convert a bar graph into a **line graph** by replacing each bar with a point at the top of the bar and then connecting the points.

Line graphs can be used to show trends over time.

Lesson 6.2

A **histogram** is a bar graph that shows how frequently each value of a variable occurs.

The largest and smallest values on the histogram are called the **extremes** of the data.

The difference between the largest and smallest data values is called the **range** of the data.

There are three types of average value, or **measures of central tendency:**

- The value that occurs most frequently is called the **mode** of the data.
- The **mean** $= \dfrac{S}{n}$, where S is the sum of all the data values and n is the number of data values.
- The **median** is the middle score in a collection of data.

The upper and lower **quartiles** measure the spread in a collection of data. The median divides the data into an upper half and a lower half. The median of the upper half is called the **upper quartile,** and the median of the lower half is called the **lower quartile.**

The difference between the upper and lower quartiles is called the **interquartile range,** or **IQR.**

A **boxplot** displays the median and the quartiles of a collection of data.

Lesson 6.3

When making a graph of an equation, we use the horizontal axis to display the values of the **independent variable** and the vertical axis for the **dependent variable.**

Steps for Graphing an Equation

1. Make a table of values. Choose values for the independent variable and use the equation to find the values of the dependent variable.
2. Choose scales for the axes.
3. Plot the points from the table and connect them with a smooth curve.

A **Cartesian coordinate system** consists of a horizontal number line, called the ***x* axis,** and a vertical number line called the ***y* axis.**

The two axes divide the plane into four regions called **quadrants,** and the point where the two axes intersect is called the **origin.**

The location of a point in the plane is given by the **ordered pair** (a, b), where a is the ***x* coordinate** of the point and b is the ***y* coordinate.**

A **solution** of an equation in two variables is an ordered pair that makes the equation true.

The **graph** of the equation is a picture of all the solutions to the equation.

Lesson 6.4

A **linear equation** looks like

$$y = ax + b$$

where a and b are constants.

The graph of a linear equation is a straight line.

The ***x* intercept** of a line is the point where the line crosses the x axis. The ***y* intercept** is the point where the line crosses the y axis.

Lesson 6.5

To solve an equation, undo the operations performed on the variable, in reverse order.

CHAPTER 6 REVIEW

Add the following words to your glossary, if you have not already done so.

line graph	histogram	extremes	range
mean	median	mode	quartiles
interquartile range	boxplot	independent variable	dependent variable
x axis	*y* axis	quadrants	origin
ordered pair	coordinates	solution	graph of an equation
linear equation	*x* intercept	*y* intercept	

1. Explain how to convert a bar graph into a line graph.

2. What does the height of each bar on a histogram tell you?

3. Name three measures of central tendency.

4. Explain how to compute the upper and lower quartiles.

5. What can you compute to get an idea of how spread out the data are?

6. Do values of the dependent variable go in the first or second column of a table of values?

7. Which variable is plotted along the horizontal axis?

8. What are the coordinates of the origin?

9. What does the solution of an equation in two variables look like?

10. What is the graph of an equation in two variables?

11. Figure 6.44 shows the number of mountain lions shot in the suburbs of Los Angeles from 1971 to 1988.

Figure 6.44 Number of lions shot

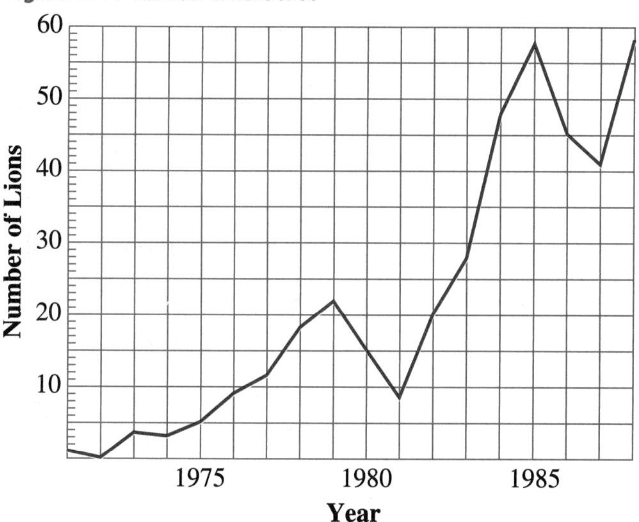

a. How many lions were shot in 1979?

b. In which year were 11 lions shot?

c. What was the increase in the number of lions shot between 1981 and 1985?

d. In which years was there a decline in the number of lions shot?

12. Figure 6.45 shows the percent of high school graduates between the ages of 16 and 24 who enroll in college each year.

Figure 6.45 College enrollment rates

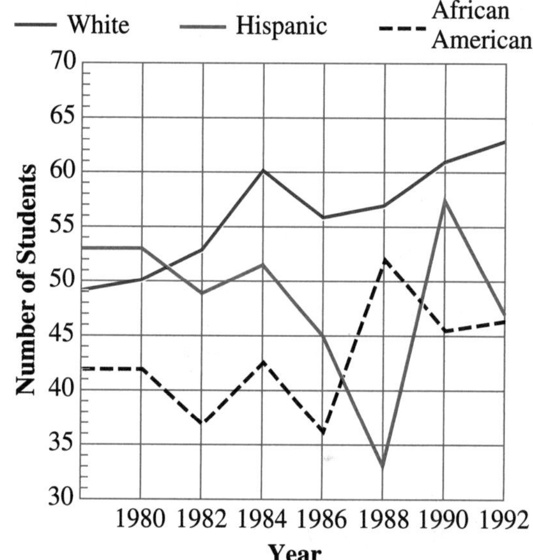

a. What percent of African-American high school graduates enrolled in college in 1982?

b. In what year did 53% of white high school graduates enroll in college?

c. In what year did the smallest percent of Hispanic high school graduates enroll in college?

d. How much did the percent of African-American high school graduates enrolling in college increase from 1986 to 1987? How did the enrollment of white and Hispanic high school graduates change during the same time period?

For the data sets (a) make a histogram for the data; (b) calculate the mean, median, and mode of the data; (c) find the upper and lower quartiles and the interquartile range; and (d) make a boxplot for the data.

13. The data show the number of women holding office in the state legislatures of each of the 50 states. Use intervals of ten to make your histogram. (Source: 1994–1995 *American Almanac*)

State	Women	State	Women
Alabama	8	Montana	30
Alaska	13	Nebraska	10
Arizona	30	Nevada	17
Arkansas	14	New Hampshire	142
California	27	New Jersey	15
Colorado	35	New Mexico	22
Connecticut	47	New York	35
Delaware	9	North Carolina	31
Florida	28	North Dakota	24
Georgia	41	Ohio	28
Hawaii	18	Oklahoma	13
Idaho	32	Oregon	25
Illinois	41	Pennsylvania	25
Indiana	29	Rhode Island	37
Iowa	22	South Carolina	22
Kansas	48	South Dakota	21
Kentucky	7	Tennessee	16
Louisiana	11	Texas	29
Maine	59	Utah	14
Maryland	46	Vermont	61
Massachusetts	46	Virginia	17
Michigan	30	Washington	58
Minnesota	55	West Virginia	22
Mississippi	19	Wisconsin	36
Missouri	37	Wyoming	22

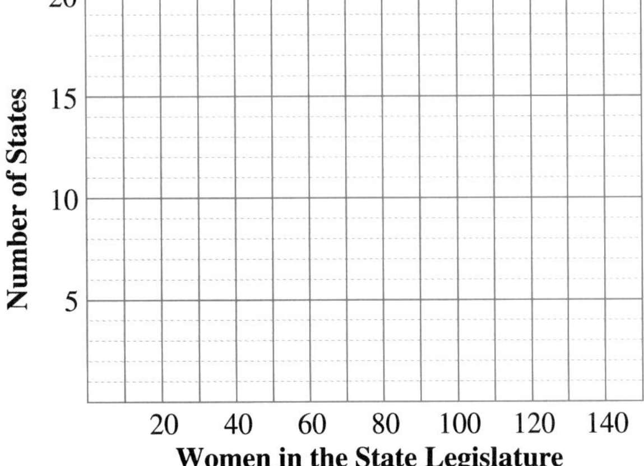

0 20 40 60 80 100 120 140

Women in the State Legislature

14. The data show 25 of the occupations with the longest occupational tenure, in years. Use intervals of whole years when making your histogram. (Source: 1994–1995 *American Almanac*)

Occupation	Years	Occupation	Years
Airplane pilots	12.2	Musicians	15.2
Automobile repairers	12.1	Pharmacists	12.7
Barbers	27.2	Plumbers	13.1
Brickmasons	12.6	Public transportation attendants	12.5
Chemical engineers	12.6	Supervisors, firefighters	15.0
Civil engineers	13.2	Supervisors, police	12.9
Dental technicians	12.9	Teachers, elementary	12.0
Dentists	15.1	Teachers, secondary	14.1
Electricians	12.3	Telephone repairers	12.9
Farmers	21.8	Timber cutting	11.5
Geologists	12.3	Tool and die makers	15.1
Locomotive operators	19.8	Veterinarians	14.0
Mechanics	13.7		

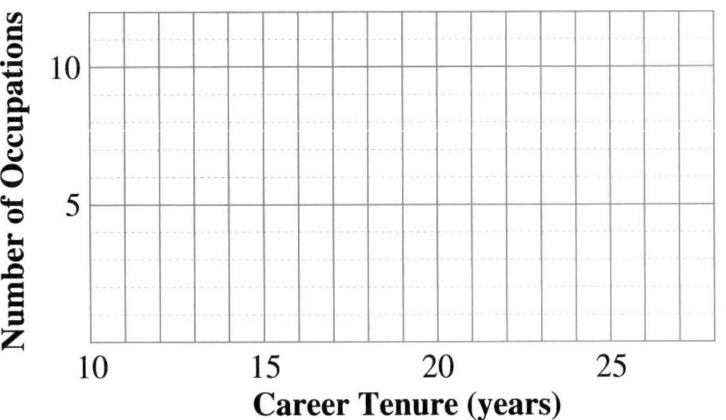

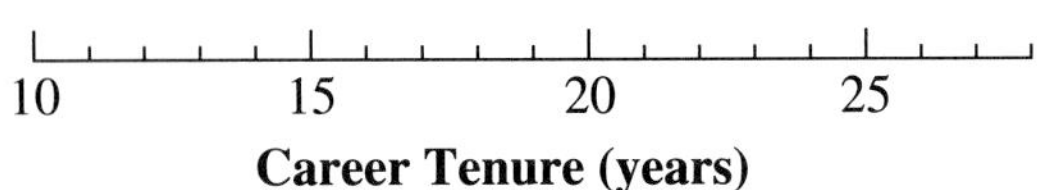

Career Tenure (years)

Complete the table of values and plot the graph. Then answer the questions.

15. The distance d in feet that an object falls in t seconds is given by

$$d = 16t^2$$

t	*(Computation)*	d
0	$16(0)^2$	0
$\frac{1}{4}$		
$\frac{1}{2}$		
$\frac{3}{4}$		
1		
$\frac{3}{2}$		

a. How far will the object fall in $\frac{5}{4}$ seconds? Locate the corresponding point on your graph.

b. How long will it take the object to fall 64 feet? Locate the corresponding point on your graph.

16. If an object is dropped from a height of h feet, the time in seconds it takes to hit the ground is given by

$$t = \frac{\sqrt{h}}{4}$$

h	*(Computation)*	t
1	$\frac{\sqrt{1}}{4}$	$\frac{1}{4}$
4		
9		
16		
25		

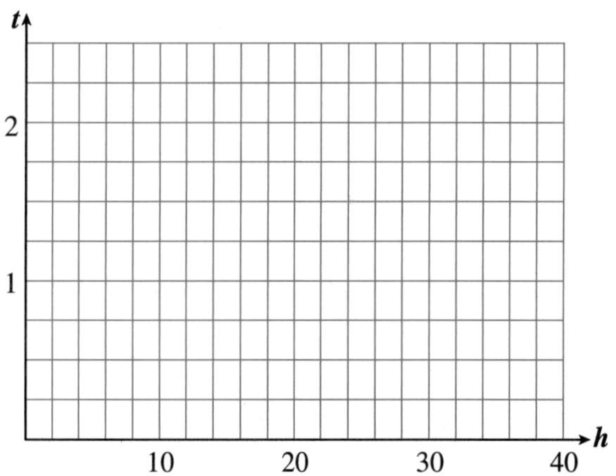

a. How long will it take the object to fall 20 feet? Locate the corresponding point on your graph.

b. If the object hits the ground in 1.5 seconds, how far did it fall? Locate the corresponding point on your graph.

Complete the table and graph the equation.

17. $y = 5 - x^2$

x	−4	−3	−2	−1	0	1	2	3	4
y									

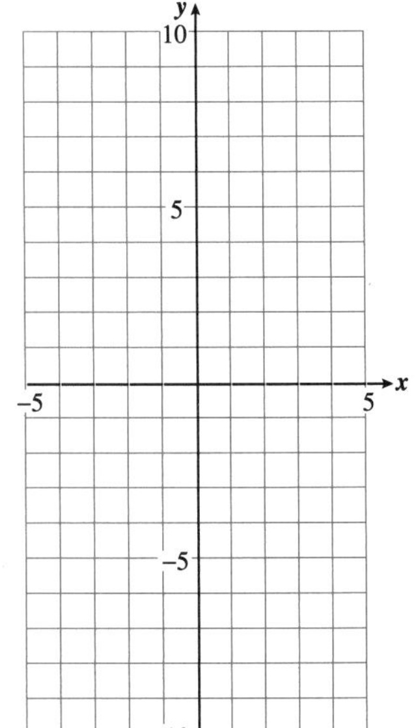

18. $y = \frac{6}{x}$, positive x values only

x	$\frac{1}{2}$	1	2	3	4	6	9	12
y								

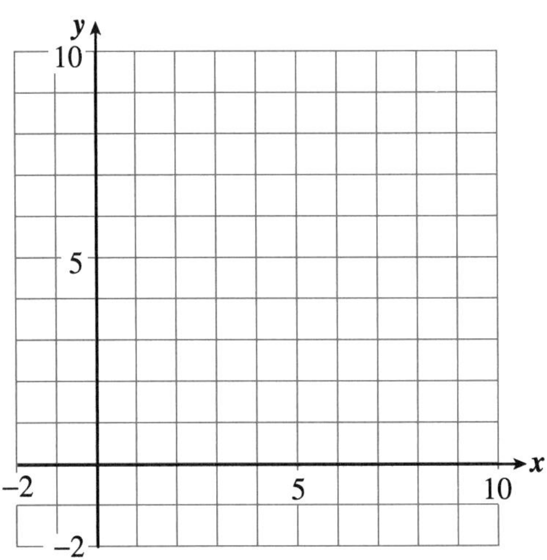

Fill in the table for each graph. Then, from the list of equations given, choose the correct equation for the graph.

$y = 3x + 1$ $\qquad y = \frac{4}{x}$ $\qquad y = 1 - x$

$y = \frac{1}{2}x^2 - 1$ $\qquad y = \sqrt{x + 8}$ $\qquad y = 8 - x^2$

19.

x	y
−8	
−7	
−4	
1	
8	

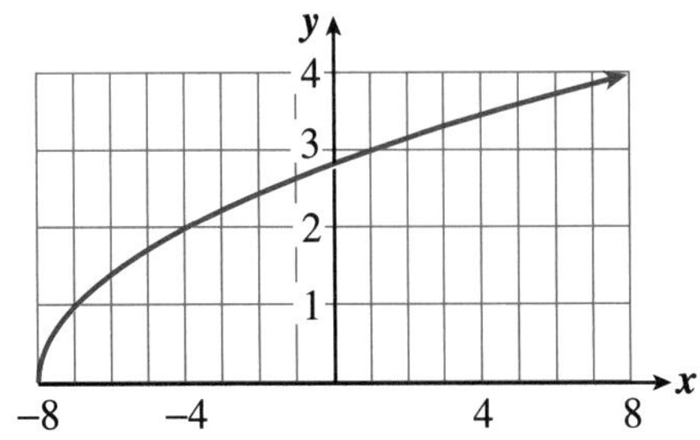

20.

x	y
−4	
−2	
−1	
1	
2	
4	

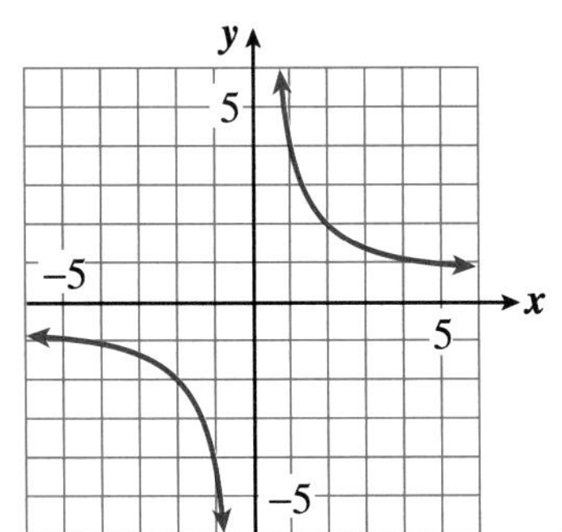

(a) Make a table showing three solutions to the equation, (b) graph the equation, and (c) find the *x* and *y* intercepts of the graph.

21. $y = 3x - 6$

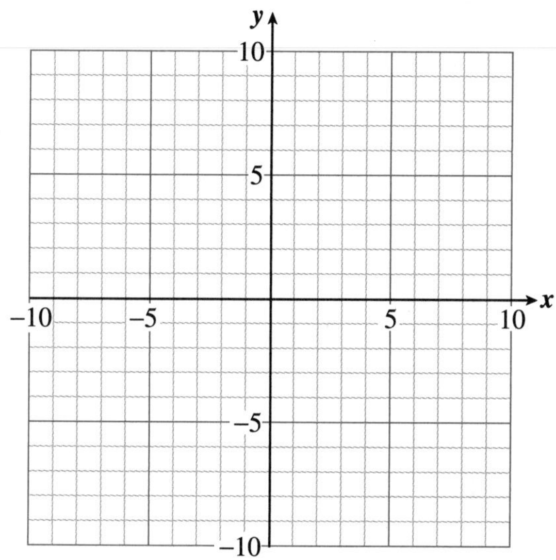

22. $y = -2x + 4$

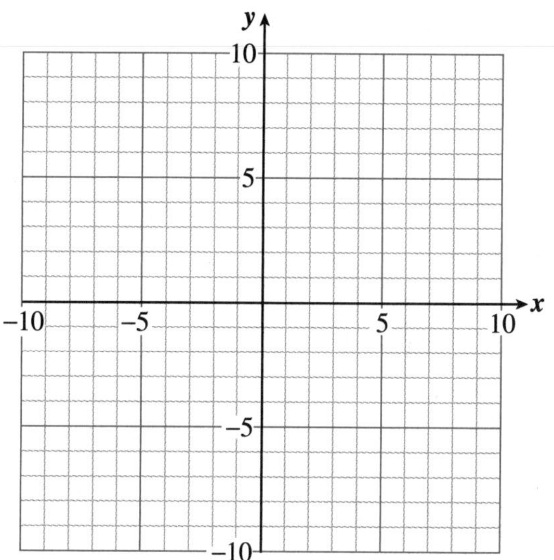

23. $y = \frac{2}{3}x - 2$

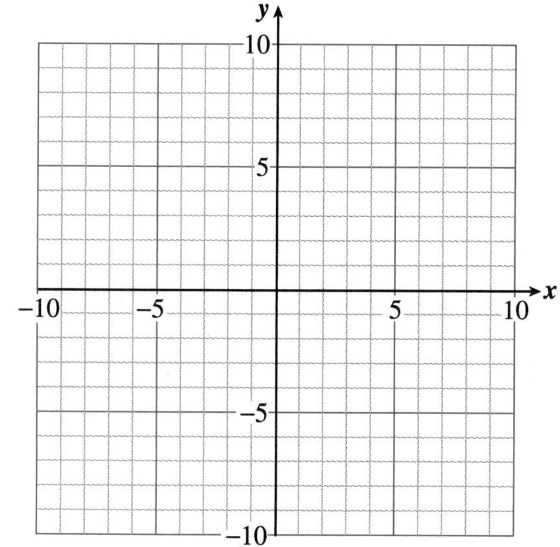

24. $y = \frac{-1}{2}x + 3$

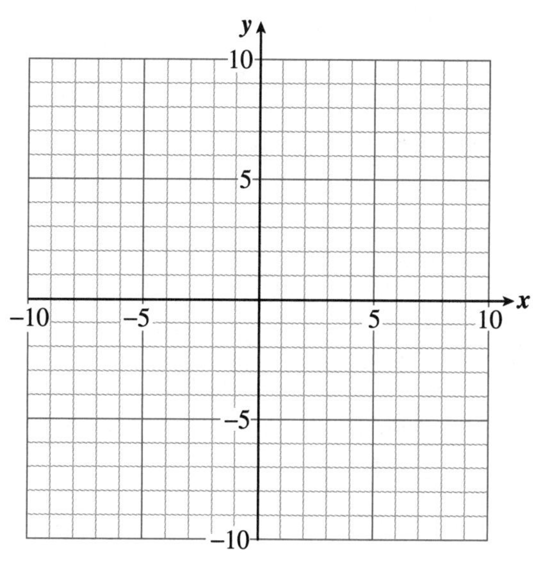

Solve.

25. $5 - 3s = 8$

26. $9 - \frac{b}{4} = 17$

27. $6 - \frac{2a}{3} = -18$

28. $-8 - \frac{t}{-5} = -3$

29. $\frac{2 - 7w}{4} = -3$

30. $\frac{-1 - 3z}{8} = -2$

(a) For part (a), write an algebraic expression; (b) for part (b), write an equation and solve.

31. Gil's math notebook has 300 pages, and he uses on average 6 pages per day for his notes and homework.

a. How many pages will Gil have left after *d* days?

b. When will Gil have 126 pages left?

32. Francine saved $5000 to go to school full-time. She spends $150 per week on food and rent.

a. How much of her savings will be left after *w* weeks?

b. When will Francine have $2450 left?

Complete the table and graph the equation. Then answer the questions.

33. Gina is 4 years older than Lisa.

Lisa's age	*Gina's age*
1	5
2	
5	
6	
x	

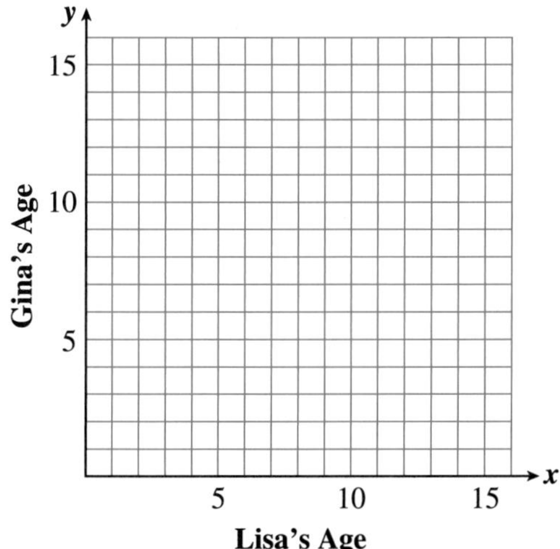

a. How old is Gina if Lisa is x years old?

b. What are the coordinates of the y intercept?

c. What does the y intercept say about the ages of Lisa and Gina?

34. Helen and Roberta are going to split $20 that they found.

Roberta's share	*Helen's share*
2	18
4	
10	
16	
x	

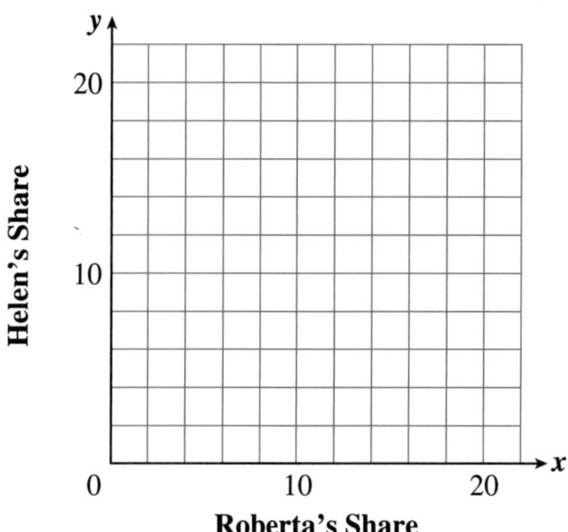

a. What is Helen's share if Roberta's share is x dollars?

b. What does the x intercept say about Roberta's and Helen's shares?

c. What does the y intercept say about Roberta and Helen's shares?

35. Ludmilla earns a commission of 5% of her real estate sales. Let x represent her sales, in thousands of dollars, and let y represent the commission she earns from her sales, in thousands of dollars.

a. Write an equation for y in terms of x.

b. Complete the table and graph the equation.

x	y
250	
600	
800	

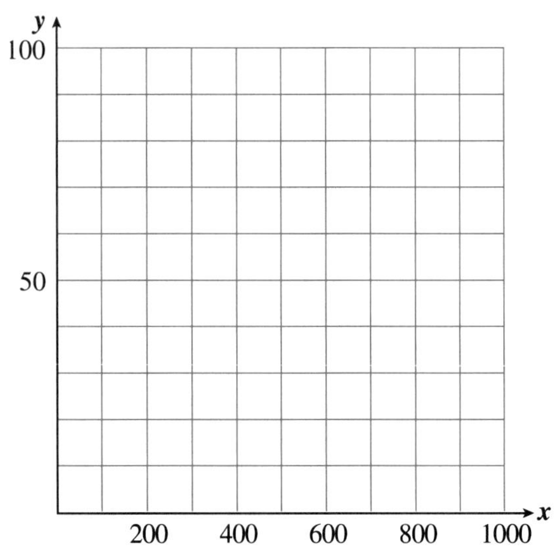

c. If Ludmilla earned $34,000 last month, how much were her sales?

36. Mojdeh rents a car from a company that charges $15.00 per day plus an additional $0.25 for each mile driven. She will be using the car for 1 day. Let x represent the number of miles she drives and let y represent the fee that the company charges her.

a. Write an equation for y in terms of x.

b. Complete the table and graph the equation.

x	y
50	
100	
200	

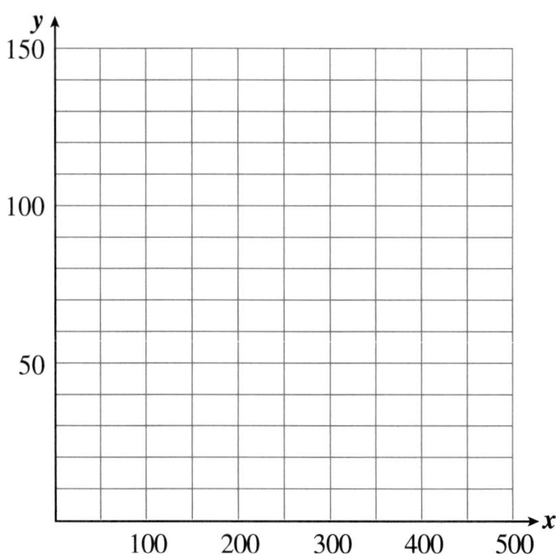

c. If the fee comes to a total of $75.00, how many miles did Mojdeh drive?

Building Number Skills

Use a diagram to build each fraction.

37. $\frac{3}{8} = \frac{?}{16}$

38. $\frac{2}{3} = \frac{?}{9}$

39. $\frac{2}{3} = \frac{?}{12}$

40. $\frac{5}{6} = \frac{?}{18}$

Rewrite each whole number as a fraction and then add or subtract.

41. $4 + \frac{2}{5}$

42. $2 + \frac{1}{3}$

43. $3 - \frac{3}{4}$

44. $4 - \frac{5}{6}$

Build both fractions to the same denominator and then add or subtract.

45. $\frac{1}{2} + \frac{1}{3}$

46. $\frac{1}{2} - \frac{1}{3}$

47. $\frac{2}{3} - \frac{3}{5}$

48. $\frac{3}{4} + \frac{3}{5}$

49. $\frac{2}{3} + \frac{2}{9}$

50. $\frac{3}{4} - \frac{5}{8}$

51. $\frac{8}{9} - \frac{5}{6}$

52. $\frac{3}{4} + \frac{1}{6}$

Find the LCD for each set of fractions.

53. $\frac{3}{4}, \frac{1}{6}, \frac{2}{9}$

54. $\frac{5}{12}, \frac{5}{18}, \frac{5}{8}$

CHAPTER

7

Fractions

7.1 Algebraic Fractions

A. What Is an Algebraic Fraction?

Review of Arithmetic Fractions

Suppose you have two ice cream sandwiches and you want to divide them equally among your three nephews, who have dropped by unexpectedly. How would you do it? Maybe you would divide each sandwich into three equal pieces, as shown in Figure 7.1. Then, because there are two sandwiches, each nephew gets two pieces. Thus, each nephew gets two-thirds of an ice cream sandwich.

Figure 7.1

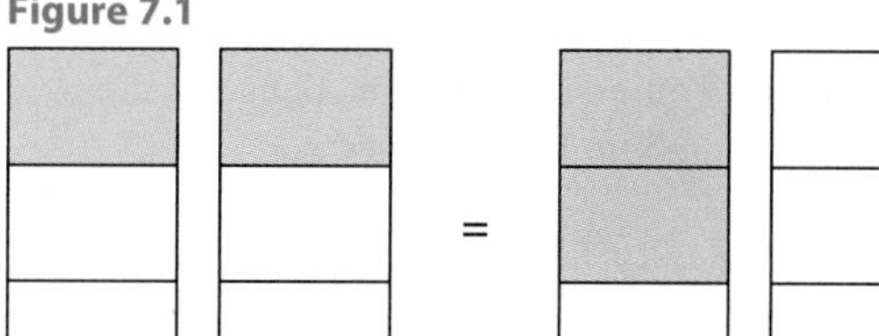

The fraction $\frac{2}{3}$ describes a portion of a whole amount (in this case, an ice cream sandwich). The **denominator,** 3, tells us how many *equal* pieces to divide the whole into, and the **numerator,** 2, tells us how many of the pieces each nephew gets. The fraction bar is a division symbol: 2 ice cream sandwiches divided between 3 nephews. In other words,

Numerator → 2

$$\frac{2}{3} \quad \text{means} \quad 2 \div 3$$

Denominator → 3

If the numerator is smaller than the denominator, the fraction is called a **proper fraction.** It represents a number smaller than 1. Examples of proper fractions are

$$\frac{3}{4}, \quad \frac{7}{10}, \quad \text{and} \quad \frac{5}{16}$$

If the numerator is bigger than or equal to the denominator, the fraction is called an **improper fraction.** It represents a number bigger than or equal to 1. Examples of improper fractions are

$$\frac{7}{4}, \quad \frac{25}{20}, \quad \text{and} \quad \frac{12}{12}$$

(If you have forgotten how to convert an improper fraction to a mixed number, you should review Section A.4 in the appendix.)

EXERCISE 1
Ray answered 5 questions correctly and the other 7 questions incorrectly on his quiz. What fraction of the questions did he answer correctly?

EXAMPLE 1

Write fractions to represent each situation.

a. There are 8 women and 5 men in Anna's yoga class. What fraction of the class is women?

b. At the company picnic, blueberry pie was served for dessert. Each pie was divided into 5 slices, and 32 people had a slice of pie. How many pies were eaten?

Solution

a. There are 13 people enrolled in the yoga class, and 8 of them are women. The class is $\frac{8}{13}$ women.

b. Each person ate $\frac{1}{5}$ of a pie, so $\frac{32}{5}$ pies were eaten—that is, $6\frac{2}{5}$ pies.

Now try Exercise 1.

Negative Fractions

We can also have a fraction of a negative number. Suppose the temperature drops 4° over 5 hours. To find the average change in temperature per hour, we divide the total change by the number of hours:

$$\frac{-4 \text{ degrees}}{5 \text{ hours}} \quad \text{or} \quad \frac{-4}{5} \text{ degrees per hour}$$

There are three ways to write a negative fraction. We can write the negative sign either in front of the fraction, in the numerator, or in the denominator. For example,

$$-\frac{4}{5} = \frac{-4}{5} = \frac{4}{-5}$$

(Use your calculator to verify that all three forms of the fraction are equal to -0.8.) However, if *both* the numerator and the denominator are negative, the fraction is positive. This is because the quotient of two negative numbers is positive. Thus,

$$\frac{-4}{-5} = \frac{4}{5}$$

EXAMPLE 2

Plot the following fractions on a number line:

$$\frac{5}{4}, \quad \frac{-3}{4}, \quad \frac{4}{-4}, \quad \frac{-1}{-4}, \quad \frac{0}{4}$$

Solution

Divide each unit on the number line into fourths. Notice which fractions are positive and which are negative and then plot, as shown in Figure 7.2.

Figure 7.2

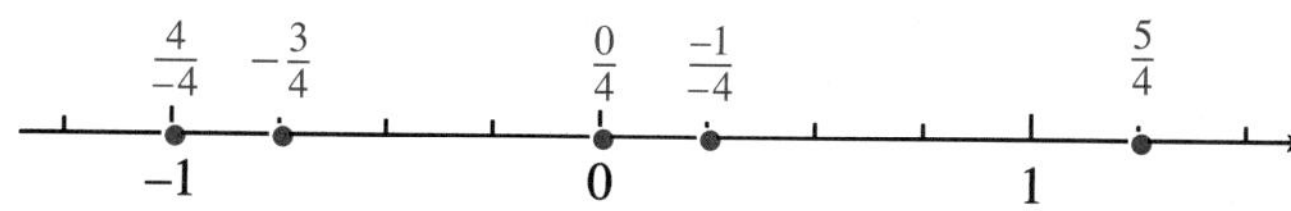

Now try Exercise 2.

EXERCISE 2

Plot $\frac{-4}{3}$, $\frac{1}{-3}$, $\frac{2}{3}$, and $\frac{-4}{-3}$ on the number line.

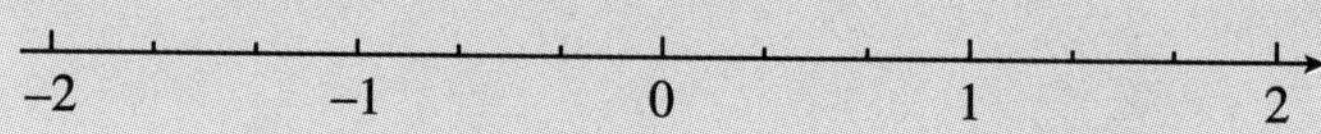

The positive and negative fractions and zero are called the **rational numbers.** Positive and negative whole numbers are included in the rational numbers, because every whole number can be written as a fraction with a denominator of 1. Examples of rational numbers are

$$-6, \quad \frac{-13}{5}, \quad 0, \quad \frac{2}{3}, \quad 11$$

Some square roots, like $\sqrt{4}$ and $\sqrt{121}$; are really rational numbers because they are equal to whole numbers. Other square roots, like $\sqrt{3}$ and $\sqrt{15}$, are not equal to any whole number or fraction, so they are not rational numbers.

Variables in Fractions

Variables may occur in the numerator or the denominator of a fraction, or both. Fractions that contain variables are called **algebraic fractions.**

EXAMPLE 3  

Choose variables and write an algebraic fraction for each situation.

a. The 5 people in Tom's study group order Chinese food and agree to split the bill equally. How much will Tom's share be?

The amount of the bill is unknown.

Total bill: b

Tom's share: $\frac{b}{5}$

b. Nurit has $800 to carpet a square bedroom in her house. What price per square foot can she afford?

The dimensions of the room are unknown.

Length of bedroom: s

Price per square foot: $\frac{800}{s^2}$

c. How long does it take an ocean liner to make a transatlantic voyage?

The distance and the speed of the liner are unknown.

Distance: D **Speed:** R

Time for voyage: $\dfrac{D}{R}$

Now try Exercise 3.

EXERCISE 3

The chess club will have 4 men and an unknown number w of women. What fraction of the club will be men?

ANSWERS TO 7.1A EXERCISES

1. $\dfrac{5}{12}$

2. 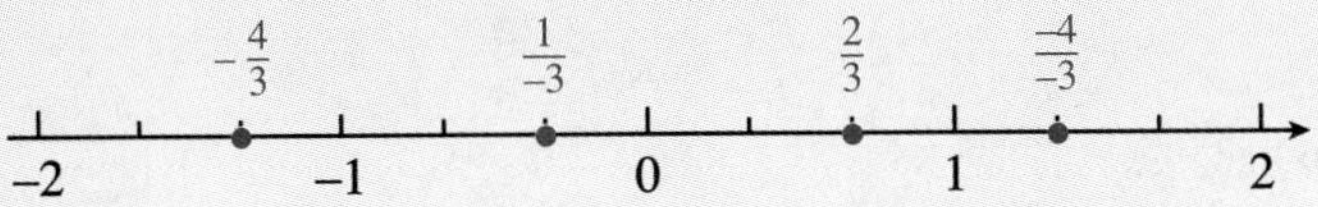

3. $\dfrac{4}{4+w}$

HOMEWORK 7.1A

1. What does the denominator of a fraction tell us? What does the numerator tell us?

2. What is a proper fraction? What is an improper fraction?

3. What is a rational number? Give some examples.

4. What is an algebraic fraction?

True or false

5. If two fractions have the same denominator, then the one with the larger numerator is larger. Give an example.

6. If two fractions have the same numerator, then the one with the larger denominator is larger. Give an example.

7. Every integer is also a rational number.

8. A square root cannot be a rational number.

9. The fraction $\dfrac{-3}{-7}$ is a negative number.

10. The fractions $\dfrac{-4}{9}$, $\dfrac{4}{-9}$, and $-\dfrac{4}{9}$ are all equal.

What fraction of each figure is shaded?

11.

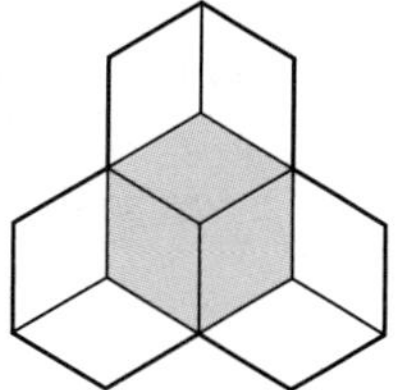

12.

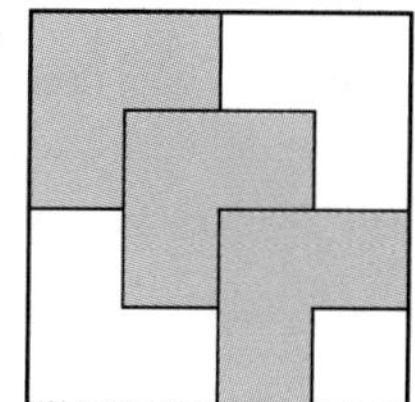

13.

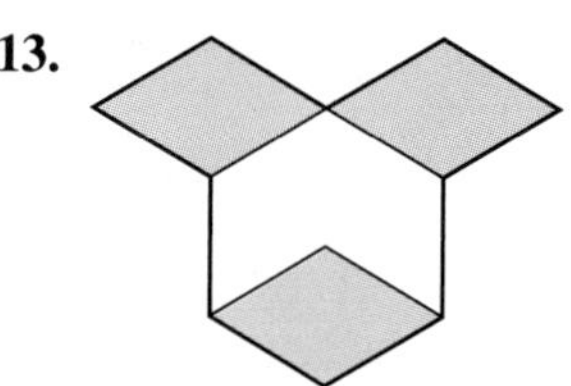

14.

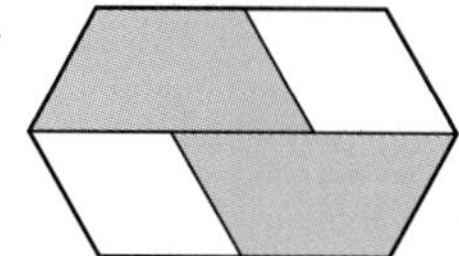

15.

16.

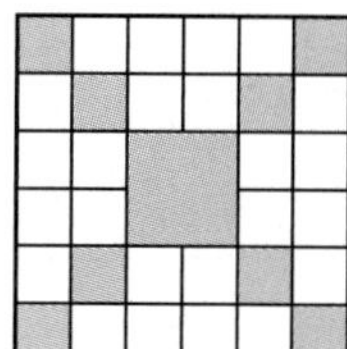

17.

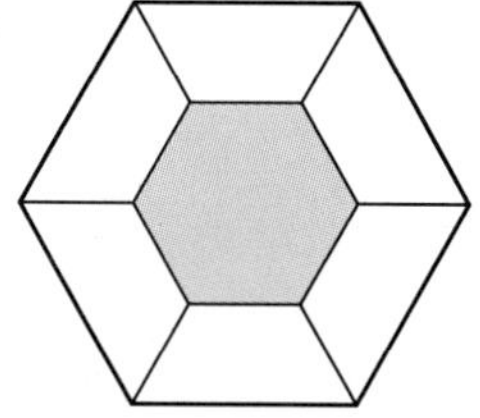

18.

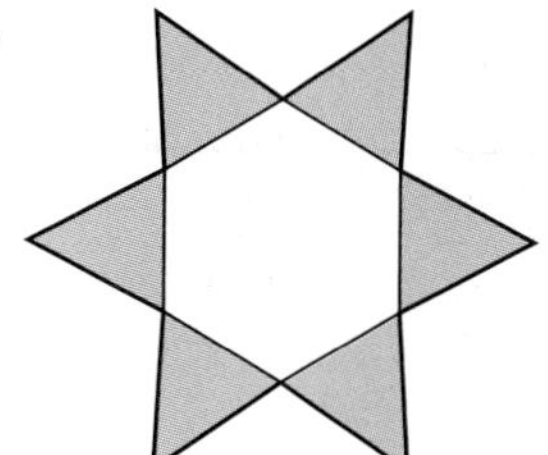

Write fractions to represent each situation.

19. Manuel's farm has 18 plum trees, and 7 of them show signs of wood borers. What fraction of the trees are infected with borers?

20. The City Bus Line has 58 bus stops throughout the city. This winter they will equip 13 of the bus stops with heaters. What fraction of the stops will have heaters?

21. In December, Marlene's kindergarten class was composed of 18 five-year-olds, 9 six-year-olds, and 2 seven-year-olds. What fraction of the class was five-year-olds?

22. Giuliano's serves three types of pizza. Last night they sold 15 pepperoni pizzas, 18 vegetarian pizzas, and 12 deluxe pizzas. What fraction of the pizzas sold were vegetarian?

23. Violet makes $1200 per month. She spends $300 a month on food and $750 a month on rent. What fraction of her salary does she spend on rent?

24. Longview College has 1600 students. There are 350 journalism majors and 650 majors in graphics arts. What fraction of the students major in graphics arts?

25. Sinbad the Wonder Cat eats $\frac{1}{3}$ of a large economy size can of cat food every day. How many cans of cat food does Sinbad eat in 2 weeks?

26. A dessert recipe calls for $\frac{1}{4}$ cup of raspberries for each serving. How many cups of raspberries are needed for nine servings?

Plot each set of fractions on a number line.

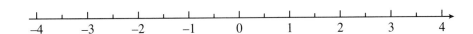

27. $\frac{-1}{3}, \frac{2}{3}, \frac{5}{3}, \frac{-6}{3}, \frac{0}{3}$

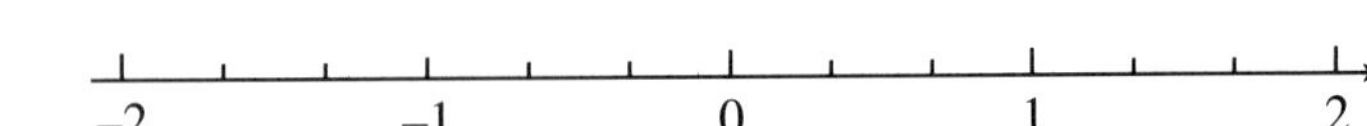

28. $\frac{3}{2}, \frac{-7}{2}, \frac{0}{2}, \frac{-2}{2}, \frac{8}{2}$

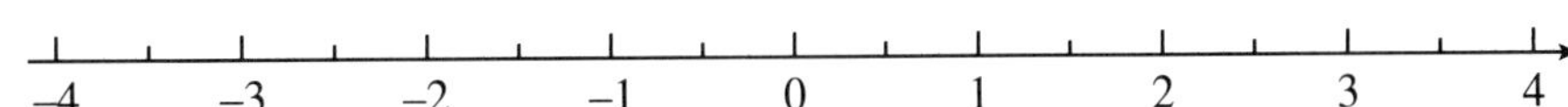

29. $\frac{3}{8}, \frac{9}{8}, \frac{-4}{8}, \frac{-7}{8}, \frac{15}{8}$

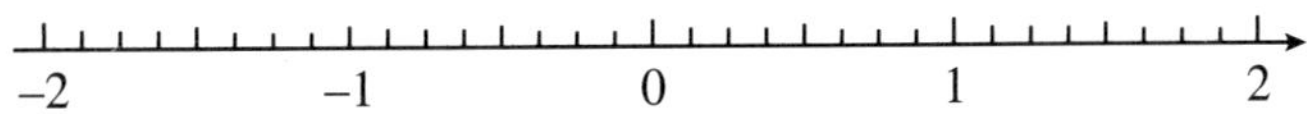

30. $\frac{-5}{6}, \frac{1}{6}, \frac{4}{6}, \frac{-9}{6}, \frac{6}{6}$

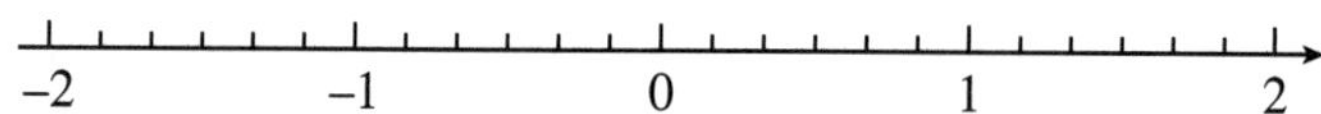

31. $\frac{3}{12}, \frac{6}{12}, \frac{9}{12}, \frac{1}{2}, \frac{1}{4}, \frac{3}{4}$

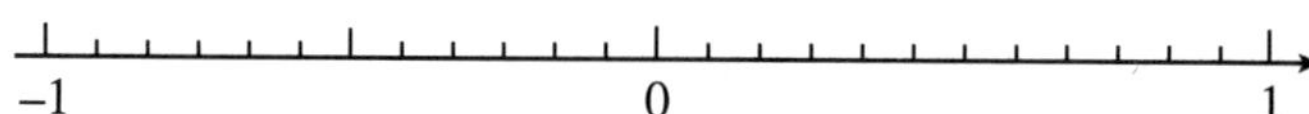

32. $\frac{2}{12}, \frac{4}{12}, \frac{8}{12}, \frac{1}{6}, \frac{1}{3}, \frac{2}{3}$

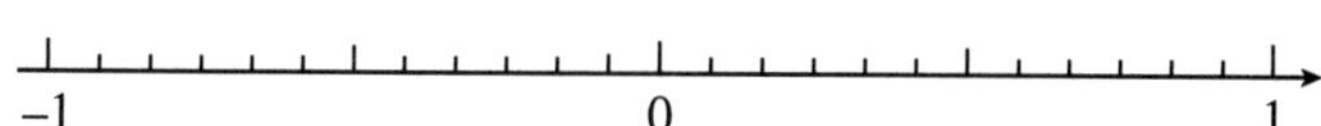

Choose variables to represent each unknown quantity and then write algebraic fractions.

33. There are 84 first-graders at Bonnair Elementary School. What fraction of the first-graders have been immunized against German measles?

34. There are 34 Sunfish sailboats registered at the South Shore Yacht Club. What fraction of the Sunfish are signed up for the Memorial Day Regatta?

35. Of the employees at Digitronics, 138 enrolled in a bonus incentive plan. What fraction of the employees enrolled?

36. Two hundred twelve 1998 model Satellite sports coupes were found to have defective brakes. What fraction of the sports coupes had defective brakes?

37. Gladys filled her car with gas three times while driving to Yellowstone National Park, a distance of 480 miles. What was her car's fuel efficiency during the trip? (See Lesson 2.3 for an appropriate formula.)

38. Julio bought a dozen oil filters on sale at a discount auto-parts store. What was the unit price of the oil filters? (See Lesson 2.3 for an appropriate formula.)

39. Write a fraction that gives the length of a rectangle in terms of its area and its width.

40. Write a fraction that gives your score on a test if you know the total number of points and the number of points you earned.

Solve. (See Lesson 6.5 to review solving equations.)

41. $-3x - 8 = 7$

42. $5y + 6 = -29$

43. $-15 = 12 - 9w$

44. $25 = -7 - 4t$

45. $\frac{c}{8} + 6 = -2$

46. $18 - \frac{b}{5} = -2$

47. $14 - \frac{2v}{5} = -4$

48. $\frac{3u}{8} + 1 = -5$

49. $-2 + \frac{8K}{3} = -12$

50. $9 - \frac{4H}{3} = 2$

Building Number Skills

Use rounding to choose the best answer. (See Appendix A.6 to review rounding.) Do *not* use a calculator.

Example: $489 - 196 \approx 500 - 200 = 300$

51. $38.2 + 21.7$ is about
a. 600 **b.** 60 **c.** 6

52. $297.5 + 208.4$ is about
a. 500 **b.** 5000 **c.** 50

53. $8.6 - 2.9$ is about
a. 0.6 **b.** 6 **c.** 60

54. $88 - 27.2$ is about
a. 600 **b.** 60 **c.** 0.6

55. $201 + 412.9$ is about
a. 60,000 **b.** 6000 **c.** 600

56. $302 + 528.47$ is about
a. 800 **b.** 8000 **c.** 80,000

57. $497 - 26.7$ is about
a. 500 **b.** 50 **c.** 5

58. $8091 - 126.5$ is about
a. 800 **b.** 8000 **c.** 80,000

59. $209 + 142 + 48.64$ is about
a. 400 **b.** 4000 **c.** 40,000

60. $48.5 + 19.7 + 3.27$ is about
a. 7 **b.** 70 **c.** 700

B. Equivalent Fractions

The Fundamental Principle of Fractions

A fraction can be written in many equivalent forms. For example, the fractions $\frac{2}{3}$ and $\frac{8}{12}$ represent the same amount, as you can see in Figure 7.3.

In Figure 7.3a, two out of three equal pieces are shaded. If we cut each of the pieces evenly into four smaller pieces, we get the picture in Figure 7.3b. There are now 3×4, or 12, equal pieces, of which 2×4, or 8, pieces are shaded.

If we multiply the numerator and denominator of $\frac{2}{3}$ by 4, we get the equivalent fraction $\frac{8}{12}$:

$$\frac{2 \cdot \mathbf{4}}{3 \cdot \mathbf{4}} = \frac{8}{12}$$

Or, we can start with $\frac{8}{12}$ and *divide* top and bottom by 4 to get $\frac{2}{3}$:

$$\frac{8 \div \mathbf{4}}{12 \div \mathbf{4}} = \frac{2}{3}$$

This example illustrates the fundamental principle of fractions.

Figure 7.3

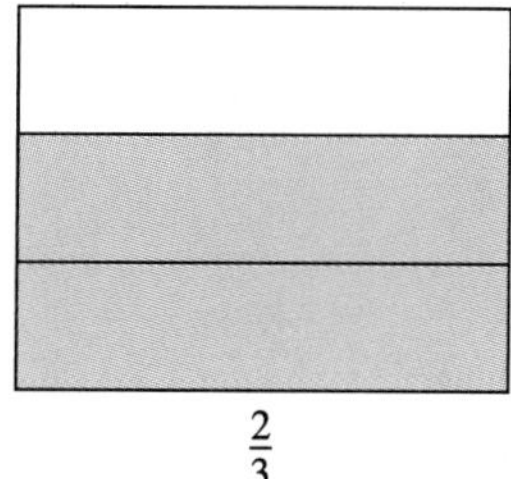

(a)

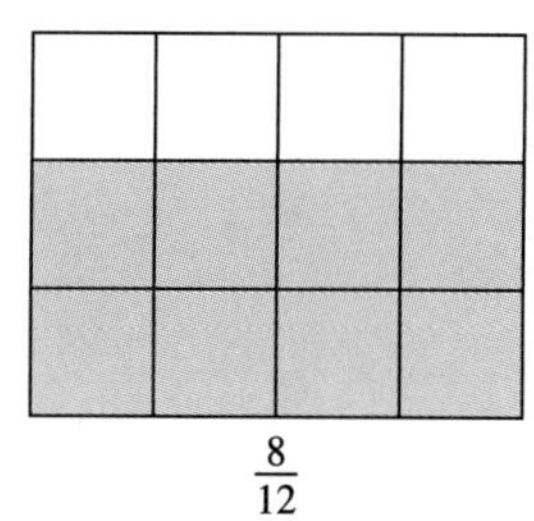

(b)

Fundamental Principle of Fractions

1. We can multiply the numerator and denominator of a fraction by the same (nonzero) number and the result will be an equivalent fraction.
2. We can divide the numerator and denominator of a fraction by the same (nonzero) number and the result will be an equivalent fraction.

We use the first part of the fundamental principle to *build* fractions and the second part to *reduce* fractions. (You may want to review Appendix A.1 on reducing and building arithmetic fractions.)

Reducing Fractions

When we **reduce a** fraction, we replace it by an equivalent fraction whose numerator and denominator are smaller numbers or simpler expressions. We will use a method that may look slightly different from the way you learned to reduce fractions. Our method is no harder than the old method, and it generalizes better to algebraic fractions. First, let's consider a fraction without variables, $\frac{18}{24}$.

We look for a number that divides evenly into both the numerator and the denominator of the fraction. It is most efficient if we find the largest such number. For the fraction $\frac{18}{24}$, the largest number is 6. Now write the numerator and denominator as products in which 6 is one of the factors, as shown below:

$$\frac{18}{24} = \frac{6 \cdot 3}{6 \cdot 4}$$

This step is called **factoring** the numerator and denominator, and 6 is called the **common factor.**

Finally, divide the numerator and denominator by their common factor 6. The fundamental principle of fractions tells us that the new fraction is equivalent to the old one. We show the division by drawing a slash through the common factors:

$$\frac{18}{24} = \frac{\not{6} \cdot 3}{\not{6} \cdot 4} = \frac{3}{4}$$

The reduced fraction is $\dfrac{3}{4}$.

EXAMPLE 4 

Reduce $\frac{16}{20}$

Solution

The largest common factor for numerator and denominator of $\frac{16}{20}$ is 4. Factor the numerator and denominator and divide each by the common factor.

$$\frac{16}{20} = \frac{\not{4} \cdot 4}{\not{4} \cdot 5} = \frac{4}{5}$$

The reduced fraction is $\frac{4}{5}$.

Now try Exercise 4.

EXERCISE 4

Reduce $\frac{56}{48}$

To Reduce a Fraction

1. Look for common factors in the numerator and denominator. If necessary, factor the numerator and denominator completely.
2. Write the numerator and denominator in factored form.
3. Divide the numerator and denominator by the common factors.

Reducing Algebraic Fractions

Now let's consider an algebraic fraction.

EXAMPLE 5 

Reduce the fraction $\frac{42ab^2}{35ab^3}$

Solution

First, consider the numerical part of the fraction: Look for the largest common factor of 42 and 35. This factor is 7, so we write 42 and 35 in factored form:

$$\frac{42ab^2}{35ab^3} = \frac{7 \cdot 6\,ab^2}{7 \cdot 5\,ab^3}$$

Next, write the variable parts of the numerator and denominator in factored form. Remember that b^2 means $b \cdot b$ and b^3 means $b \cdot b \cdot b$:

$$\frac{42ab^2}{35ab^3} = \frac{7 \cdot 6\,a\,b\,b}{7 \cdot 5\,a\,b\,b\,b}$$

Finally, divide any common factors from the numerator and denominator:

$$\frac{42ab^2}{35ab^3} = \frac{\not{7} \cdot 6\,\not{a}\,\not{b}\,\not{b}}{\not{7} \cdot 5\,\not{a}\,\not{b}\,\not{b}\,b} = \frac{6}{5\,b}$$

The reduced fraction is $\frac{6}{5b}$.

Now try Exercise 5.

EXERCISE 5

Reduce $\frac{20s^3t^2}{12st^3}$

CAUTION!

When reducing algebraic fractions, remember that we can only *divide* top and bottom by the same number or variable; we cannot subtract. Consider the following examples:

Does $\frac{6}{15} = \frac{3 \cdot 2}{3 \cdot 5} = \frac{2}{5}$? Yes.

To undo the multiplications we **divide** top and bottom by 3. This results in an equivalent fraction.

Does $\frac{5}{8} = \frac{3+2}{3+5} = \frac{2}{5}$? No!

We cannot **subtract** 3 from top and bottom. This does *not* result in an equivalent fraction.

The same rules apply to algebraic fractions: We can divide out common *factors,* but we cannot divide out common *terms.*

Does $\frac{2a}{5a} = \frac{2}{5}$? Yes.

a is a common **factor;** it can be divided out.

Does $\frac{2+a}{5+a} = \frac{2}{5}$? No!

a is a common **term;** it cannot be divided out.

Building Fractions

Fractions are easier to use when they are reduced to simplest form. However, we will need to **build** unlike fractions to higher terms before we can add or subtract them.

It is not hard to build a fraction; we multiply numerator and denominator by the same factor. For example,

$$\frac{3}{8} = \frac{3 \cdot 9}{8 \cdot 9} = \frac{27}{72}$$

According to the fundamental principle of fractions, $\frac{27}{72}$ is equivalent to the original fraction $\frac{3}{8}$. The number 9 in the example is called the **building factor.**

Usually we want to build a fraction so that it will have a specific denominator. For example, find the correct numerator so that

$$\frac{3}{4} = \frac{?}{20}$$

First ask yourself what building factor gives the new denominator 20. For this problem it is easy to see that the building factor is 5. Multiply the numerator by the building factor to obtain the answer:

$$\frac{3}{4} = \frac{3 \cdot 5}{4 \cdot 5} = \frac{15}{20}$$

Building Algebraic Fractions

If we cannot immediately see the building factor, we can factor the old and new denominators. Example 6 shows how to build an algebraic fraction.

EXAMPLE 6

Build $\frac{5}{4a} = \frac{?}{12a^2b}$

Solution

We first look for the building factor. Factor the new denominator:

$$12a^2b = 4 \cdot 3\,a\,a\,b = 4a \cdot 3ab$$

By comparing the new denominator with the old one, we see that we need a building factor of $3ab$. Multiply the numerator and denominator of the old fraction by the building factor:

$$\frac{5}{4a} = \frac{5 \cdot 3ab}{4a \cdot 3ab} = \frac{15ab}{12a^2b}$$

To Build a Fraction

1. Factor the old denominator and the new denominator, if necessary.
2. Find the building factor by comparing the old denominator with the new denominator. The extra factors in the new denominator are the building factor.
3. Multiply the numerator and denominator of the old fraction by the building factor.

EXERCISE 6

Build $\frac{w}{3} = \frac{?}{18w^2}$

CAUTION!

For a building problem, you should not reduce your answer! Because building is the opposite of reducing, you would get back to the original problem by reducing your answer. However, you can always *check* a building problem by reducing the new fraction to see if you get the original one back.

Now try Exercise 6.

Here is another example of building a fraction.

EXAMPLE 7 

Find the building factor and build $\frac{1}{15v} = \frac{?}{30v^3t}$

Solution

Finding a building factor is like filling in the blank in a multiplication problem:

$$15v = 15v \cdot \underline{\ ?\ } = 30v^3t$$

What numbers and variables should we multiply times $15v$ to get $30v^3t$? To find out, we factor both denominators and look for the missing factors:

$$\text{Old denominator:} \quad 15v = 3 \cdot 5v$$
$$\text{New denominator:} \quad 30v^3t = 2 \cdot 3 \cdot 5vvvt$$

The new denominator has all the same factors as the old denominator, but some extra ones as well: $\mathbf{2vvt}$, or $\mathbf{2v^2t}$. This is the building factor. If we multiply the old denominator by this building factor, we should get the new denominator:

$$15v \cdot 2v^2t = 30v^3t$$

Our building factor checks, so we can finish building. Multiply numerator and denominator of the old fraction by the building factor:

$$\frac{1}{15v} = \frac{1 \cdot 2v^2t}{15v \cdot 2v^2t} = \frac{2v^2t}{30v^3t}$$

Now try Exercise 7.

EXERCISE 7
Find the building factor and build:
$$\frac{7}{4r} = \frac{?}{24r^3s}$$

Building When the Denominator Is 1

We can build an expression that is not a fraction by rewriting it with a denominator of 1.

EXERCISE 8
Find the building factor and build:
$$5y = \frac{?}{8y}$$

EXAMPLE 8 

Find the building factor and build $3x = \dfrac{?}{2x}$

Solution

To make it look like a fraction, write $3x$ with a denominator of 1. The problem then becomes

$$\frac{3x}{1} = \frac{?}{2x}$$

Remember that *the numerators have nothing to do with finding the building factor.* As before, we compare the old denominator to the new denominator:

$$\text{Old denominator:} \quad 1$$
$$\text{New denominator:} \quad \mathbf{2x}$$

The extra factors in the new denominator are 2 and x, so the building factor is $\mathbf{2x}$. We multiply numerator and denominator of the old fraction by this building factor:

$$\frac{3x}{1} = \frac{3x \cdot 2x}{1 \cdot 2x} = \frac{6x^2}{2x}$$

Now try Exercise 8.

ANSWERS TO 7.1B EXERCISES

4. $\dfrac{7}{6}$
5. $\dfrac{5s^2}{3t}$
6. $\dfrac{6w^3}{18w^2}$
7. Building factor $6r^2s$; $\dfrac{42r^2s}{24r^3s}$
8. Building factor $8y$; $\dfrac{40y^2}{8y}$

HOMEWORK 7.1B 

1. What is the fundamental principle of fractions used for?

2. Which of the following fractions are equivalent to $\frac{6}{8}$?

a. $\frac{6+2}{8+2}$ **b.** $\frac{6-2}{8-2}$

c. $\frac{6 \cdot 2}{8 \cdot 2}$ **d.** $\frac{6 \div 2}{8 \div 2}$

3. When reducing a fraction, we can divide numerator and denominator by a common ______________, but not by a common ______________.

4. To build a fraction to an equivalent one with a given denominator, we first need to find the ______________.

Reduce each fraction.

5. $\frac{15}{12}$ **6.** $\frac{8}{6}$ **7.** $\frac{-56}{63}$ **8.** $\frac{40}{-45}$

9. $\frac{48}{-36}$ **10.** $\frac{-72}{40}$ **11.** $-\frac{120}{240}$ **12.** $-\frac{8800}{7700}$

Reduce each algebraic fraction.

13. $\frac{15}{3x}$ **14.** $\frac{8}{4a}$ **15.** $\frac{6}{2w}$ **16.** $\frac{18}{8r}$

17. $\frac{24b}{14}$ **18.** $\frac{32n}{12}$ **19.** $\frac{-5z}{6z}$ **20.** $\frac{-11s}{7s}$

21. $\frac{t}{17t}$ **22.** $\frac{y}{2y}$ **23.** $\frac{5u}{120uv}$ **24.** $\frac{7p}{21pq}$

25. $\dfrac{16ab}{-10ab}$

26. $\dfrac{28rt}{-21rt}$

27. $\dfrac{3a^2}{27a}$

28. $\dfrac{15x^2}{60x}$

29. $\dfrac{-9y^3z}{42yz}$

30. $\dfrac{-4mn^4}{14m^2n}$

31. $\dfrac{8u^3v^2}{12v^2w}$

32. $\dfrac{36bc^5d}{20bc^5}$

Reduce each fraction if possible. If the fraction cannot be reduced, state the reason.

33. a. $\dfrac{a + 4}{a + 5}$ b. $\dfrac{a \cdot 4}{a \cdot 5}$

34. a. $\dfrac{3 \cdot x}{7 \cdot x}$ b. $\dfrac{3 + x}{7 + x}$

35. a. $\dfrac{2 \cdot m}{4 \cdot n}$ b. $\dfrac{2 + m}{4 + n}$

36. a. $\dfrac{9 + p}{3 + q}$ b. $\dfrac{9 \cdot p}{3 \cdot q}$

37. a. $\dfrac{z - 3}{z + 9}$ b. $\dfrac{z(-3)}{z(+9)}$

38. a. $\dfrac{r(+8)}{r(-4)}$ b. $\dfrac{r + 8}{r - 4}$

39. a. $\dfrac{u(-v)}{u(v)}$ b. $\dfrac{u - v}{u + v}$

40. a. $\dfrac{r + s}{r - s}$ b. $\dfrac{r(s)}{r(-s)}$

41. a. $\dfrac{3 + x + y}{2 + x + y}$ b. $\dfrac{3xy}{2xy}$

42. a. $\dfrac{4pq}{6pq}$ b. $\dfrac{4 + p + q}{6 + p + q}$

(a) Find the building factor and (b) build the given fraction to an equivalent fraction with the new denominator.

43. $\frac{-5}{6} = \frac{?}{30}$

44. $\frac{-3}{8} = \frac{?}{24}$

45. $3 = \frac{?}{9}$

46. $12 = \frac{?}{2}$

47. $\frac{1}{2z} = \frac{?}{4z}$

48. $\frac{3}{5a} = \frac{?}{20a}$

49. $\frac{8}{m} = \frac{?}{11m}$

50. $\frac{2}{y} = \frac{?}{9y}$

51. $\frac{-4}{19} = \frac{?}{19b}$

52. $\frac{-23}{6} = \frac{?}{6n}$

53. $\frac{p}{7q} = \frac{?}{7pq}$

54. $\frac{v}{13u} = \frac{?}{13uv}$

55. $\frac{-9d}{w} = \frac{?}{2dw}$

56. $\frac{-14t}{s} = \frac{?}{5st}$

57. $\frac{x}{18} = \frac{?}{18x}$

58. $\frac{c}{5} = \frac{?}{5c^2}$

59. $\frac{-3g}{4} = \frac{?}{12g}$

60. $\frac{-13r}{3} = \frac{?}{30r}$

61. $\frac{-1}{ab} = \frac{?}{ab^2}$

62. $\frac{22}{y^2z} = \frac{?}{y^2z^2}$

63. $5n = \frac{?}{2n}$

64. $-4e = \frac{?}{3e}$

65. $\frac{-3w}{2r} = \frac{?}{24rw^3}$

66. $\frac{9k}{8l^4} = \frac{?}{32k^3l^4}$

For Problems 67 and 68, see Lesson 6.1 to review line graphs.

67. The *Centerville Gazette* puts out both a morning paper and an evening edition. Figure 7.4 shows the circulation of both editions from 1950 to 1990.

Figure 7.4 Circulation of morning and evening papers

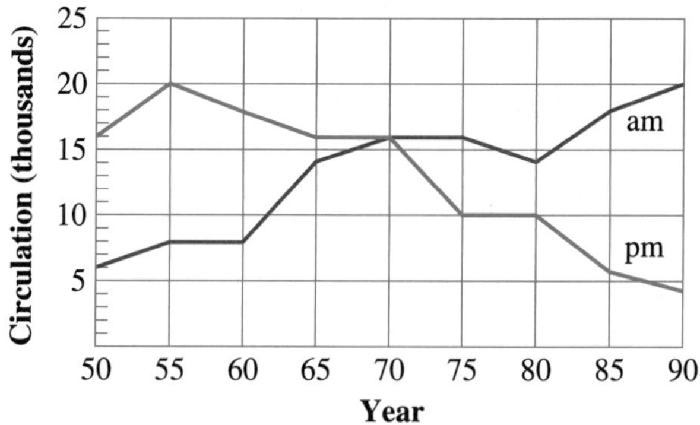

a. What was the circulation of the evening paper in 1980?

b. In what year did the circulation of the morning paper reach 14,000?

c. In what year did the circulation of the evening paper begin to decline?

d. In what year was the circulation of the two editions equal?

68. Murray's Meat Market has noticed an increase in the amount of chicken they sell each week and a decline in the amount of beef. Figure 7.5 shows their weekly sales of each type of meat over the past few years.

Figure 7.5 Average weekly sales of meat

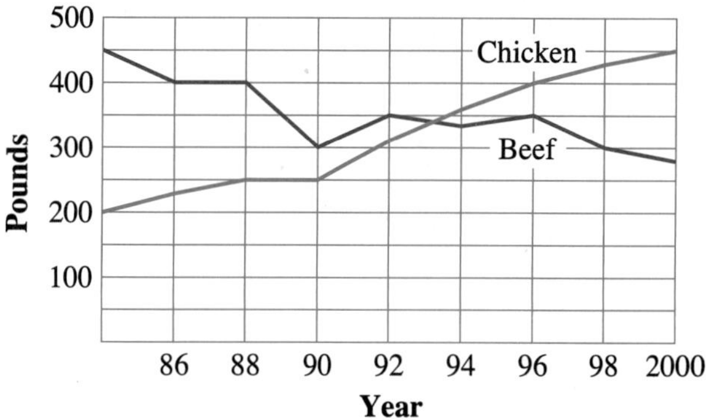

a. How many pounds of beef did Murray's sell each week in 1988?

b. In what year did Murray's sell 350 pounds of chicken each week?

c. In what year did Murray's first sell more chicken than beef?

d. What was the overall decrease in the amount of beef Murray's sells each week?

Complete the table and graph the equation and then answer the questions. (See Lesson 6.3 to review graphing equations.)

69. Ernestine wants to make a 12-mile trip on her bicycle. If her trip takes a total of t hours, then her average speed will be v miles per hour, where v is given by

$$v = \frac{12}{t}$$

t	v
0.75	
1	
1.2	
1.8	
2	
2.5	
3	
4	

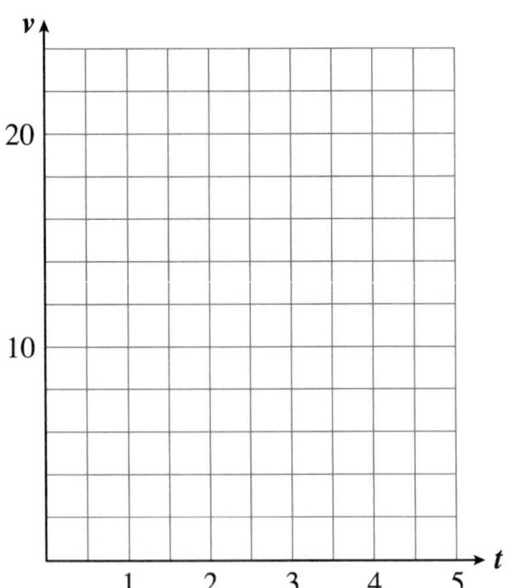

a. What will be Ernestine's average speed if the trip takes her 2 hours? Locate the corresponding point on your graph.

b. How long will Ernestine have to finish the trip if she wishes to maintain an average speed of 18 miles per hour? Locate the corresponding point on the graph.

70. The Administration Building at Rockville College is 64 feet high. If Delbert drops a water balloon from the top of the building, its height after t seconds is given by

$$h = 64 - 16t^2$$

t	h
0	
0.25	
0.5	
1	
1.5	
1.75	
2	

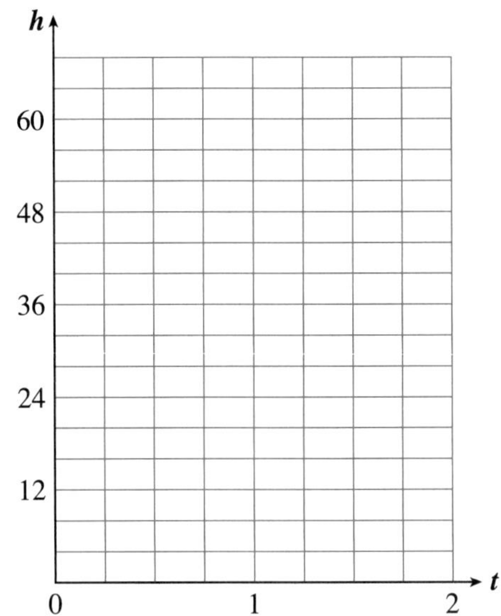

a. What is the height of the ballon after 1 second? Locate the corresponding point on your graph.

b. When will the balloon be 39 feet high? Locate the corresponding point on your graph.

Add or subtract like terms. (See Lesson 5.3 to review like terms.)

71. $-8z + 6z - (-5z) - 3z$

72. $5d - 11d - (-2d) + 3d$

73. $6y - 8y - (-2y) - 3y$

74. $-3y + 7y - 6y - 4y$

75. $6 + 2H - 9H - 6 - (-6H)$

76. $8 - 15R - 14 + 15R - (-5)$

77. $4 - 3a + 6ab - (-8a) - 10 - 9ab + 2a$

78. $-9 + 2c - 8cd - (-12) - (-5cd) + 3c - 9c$

Building Number Skills

Use rounding to choose the best answer. (See Appendix A.6 to review rounding.) Do *not* use a calculator.

Example: $29.2 + 51.4 \approx 30 + 50 = 80$

79. 48.1 + 31.6 is about
a. 70 **b.** 80 **c.** 90

80. 196.4 + 311.2 is about
a. 500 **b.** 600 **c.** 700

81. 9.4 − 2.3 is about
a. 6 **b.** 7 **c.** 8

82. 78 − 19.1 is about
a. 50 **b.** 60 **c.** 70

83. 311 + 292.4 is about
a. 500 **b.** 600 **c.** 700

84. 818 + 194.7 is about
a. 800 **b.** 900 **c.** 1000

85. 6265 − 236.6 is about
a. 4000 **b.** 5000 **c.** 6000

86. 723.9 − 18.76 is about
a. 500 **b.** 600 **c.** 700

87. 28.1 + 31.4 + 1.93 is about
a. 60 **b.** 70 **c.** 80

88. 206 + 391 + 26.9 is about
a. 500 **b.** 600 **c.** 700

7.2 Multiplying and Dividing Fractions

Figure 7.6

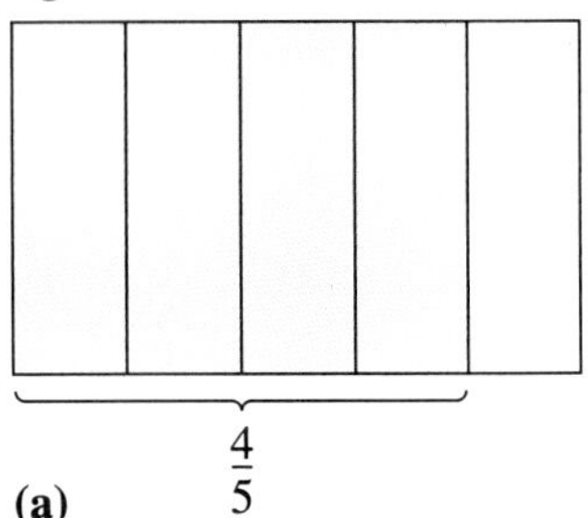

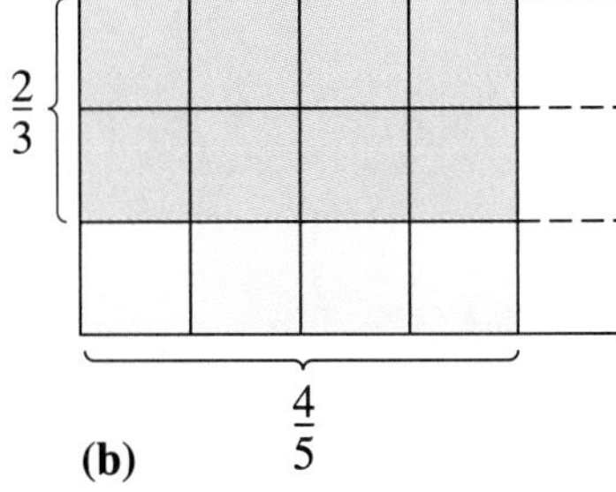

A. Multiplying Fractions

In arithmetic you learned the following rule for multiplying fractions.

To Multiply Two Fractions

1. Multiply the numerators together.
2. Multiply the denominators together.
3. Reduce the result, if necessary.

(You may want to review Appendix A.2 on multiplying and dividing arithmetic fractions.) To see why this rule works, consider an example. Suppose you make a chocolate cake for the Math Club bake sale, but at the end of the day $\frac{4}{5}$ of the cake is left in the pan. The remaining cake is shown in Figure 7.6a.

You decide to take $\frac{2}{3}$ of the remaining cake home and give the rest to your math teacher. You cut the remaining cake in thirds, as shown in Figure 7.6b. How much of the original cake are you taking home? Your share is

$$\frac{2}{3} \text{ of } \frac{4}{5} \quad \text{or} \quad \frac{2}{3} \times \frac{4}{5}$$

If you look at Figure 7.6b, you can see that you are taking home 8 pieces of cake and that the original cake would have had 15 pieces of the same size. This means that your share is $\frac{8}{15}$ of the original cake.

We get the same answer if we use the rule above:

$$\frac{2}{3} \times \frac{4}{5} = \frac{2 \times 4}{3 \times 5} = \frac{8}{15} \quad \begin{array}{l}\text{Multiply numerators together.}\\ \text{Multiply denominators together.}\end{array}$$

There is a shortcut for multiplying fractions that allows us to reduce the answer before we multiply. We do this by dividing out any common factors that appear in a numerator and a denominator, as shown in Example 1.

EXAMPLE 1  

Multiply $\frac{12}{35} \cdot \frac{28}{9}$

Solution

Notice that 7 divides evenly into one of the denominators (35) and one of the numerators (28). Write the fractions with the 7s factored out:

$$\frac{12}{\mathbf{7} \cdot 5} \cdot \frac{\mathbf{7} \cdot 4}{9}$$

Also, 12 and 9 are both divisible by 3, so we write them in factored form as well:

$$\frac{\mathbf{3} \cdot 4}{\mathbf{7} \cdot 5} \cdot \frac{\mathbf{7} \cdot 4}{\mathbf{3} \cdot 3}$$

Now divide the common factors from the numerators and denominators:

$$\frac{\not{\mathbf{3}} \cdot 4}{\not{\mathbf{7}} \cdot 5} \cdot \frac{\not{\mathbf{7}} \cdot 4}{\not{\mathbf{3}} \cdot 3}$$

(This step is often called *canceling*. Just remember that in this context canceling means dividing.) Finally, multiply together the remaining factors in the numerator and multiply the remaining factors in the denominator:

$$\frac{4}{5} \cdot \frac{4}{3} = \frac{16}{15}$$

You may prefer the following familiar shortcut for working out the product:

$$\frac{\overset{4}{\cancel{12}}}{\underset{5}{\cancel{35}}} \cdot \frac{\overset{4}{\cancel{28}}}{\underset{3}{\cancel{9}}} = \frac{4 \cdot 4}{5 \cdot 3} = \frac{16}{15}$$

Now try Exercise 1.

EXERCISE 1

Multiply $\frac{12}{7} \cdot \frac{14}{9}$

We can revise our rule for multiplying fractions to include canceling.

To Multiply Two or More Fractions Together

1. Divide out any common factors that appear in a numerator and a denominator.
2. Multiply together the remaining factors in the numerator.
   Multiply together the remaining factors in the denominator.
3. Reduce the product, if necessary.

If we divide out *all* the common factors in step 1, there will be no need to reduce the product. However, it is a good idea to check for common factors in the product in case we missed any.

Multiplying Algebraic Fractions

Here is an example of multiplying algebraic fractions.

EXAMPLE 2 

Multiply $\frac{ab}{6} \cdot \frac{3a}{2b}$

Solution

We can divide out common factors of b and 3:

$$\frac{a\cancel{b}}{2 \cdot \cancel{3}} \cdot \frac{\cancel{3}a}{2\cancel{b}}$$

Multiply together the remaining factors in the numerators and the remaining factors in the denominator:

$$\frac{a}{2} \cdot \frac{a}{2} = \frac{a^2}{4}$$

Because we divided out all the common factors in the first step, the product does not need to be reduced.

Now try Exercise 2.

EXERCISE 2

Multiply $\frac{2m}{5n} \cdot \frac{7mn^2}{6}$

EXERCISE 3

Simplify $\left(\frac{-t}{2}\right)^3$

ANSWERS TO 7.2A EXERCISES

1. $\frac{8}{3}$
2. $\frac{7m^2n}{15}$
3. $\frac{-t^3}{8}$

Power of a Fraction

An exponent tells us how many factors of the base to multiply together. (See Lesson 5.1 to review exponents.) The base of a power may be a fraction. For example,

$$\left(\frac{2a}{5}\right)^3 = \frac{2a}{5} \cdot \frac{2a}{5} \cdot \frac{2a}{5} = \frac{8a^3}{125}$$

EXAMPLE 3  

Simplify $\left(\frac{-4z}{3x}\right)^2$

Solution

Multiply the fraction times one copy of itself:

$$\left(\frac{-4z}{3x}\right)^2 = \frac{-4z}{3x} \cdot \frac{-4z}{3x} = \frac{16z^2}{9x^2}$$

Now try Exercise 3.

HOMEWORK 7.2A 

1. When we are reducing fractions, *cancel* is another name for which operation: add, subtract, multiply, or divide?

2. Explain in words how to multiply two fractions together.

3. The city park in Haverford covers $3\frac{3}{4}$ acres. The city council wants to maintain $\frac{3}{5}$ of the park as open space. How many acres of open space is that?

4. Of the employees at Softek, $\frac{5}{8}$ have college degrees, and $\frac{2}{3}$ of those with college degrees have advanced degrees. What fraction of all the employees have advanced degrees?

Multiply. Reduce your answers to lowest terms.

5. $\frac{1}{3} \cdot \frac{2}{9}$
6. $\frac{4}{5} \cdot \frac{3}{7}$
7. $\frac{8}{3} \cdot \frac{1}{8}$
8. $\frac{1}{4} \cdot \frac{4}{7}$
9. $\frac{-2}{d} \cdot \frac{c}{2}$
10. $\frac{-7}{a} \cdot \frac{b}{7}$
11. $\frac{8}{v} \cdot \frac{-u}{6}$
12. $\frac{-w}{12} \cdot \frac{15}{x}$

13. $9 \cdot \frac{x}{6}$

14. $\frac{y}{8} \cdot 4$

15. $\frac{w}{8} \cdot \frac{7}{z}$

16. $\frac{-p}{5} \cdot \frac{2}{s}$

17. $\frac{-r}{3} \cdot \frac{17}{r}$

18. $\frac{-9}{q} \cdot \frac{2q}{5}$

19. $\frac{-2}{a} \cdot \frac{3d}{4a}$

20. $\frac{8b}{3} \cdot \frac{9b}{7a}$

21. $-5c \cdot \frac{3}{20}$

22. $\frac{-8}{21} \cdot 7y$

23. $\frac{5}{6m^2} \cdot 2m$

24. $7n \cdot \frac{8}{63n^3}$

25. $\frac{12s}{5r} \cdot \frac{2r}{3s}$

26. $\frac{5b}{8d} \cdot \frac{3d}{10b}$

27. $\frac{-k^2}{14j} \cdot \frac{7j}{2k}$

28. $\frac{-5f}{18g} \cdot \frac{9g^2}{20f}$

29. $\frac{24u^3}{7v} \cdot \frac{21v}{8u}$

30. $\frac{14x}{5y} \cdot \frac{15y^4}{2x}$

31. $\frac{21r^2}{4rs} \cdot \frac{16s}{5r}$

32. $\frac{-18h}{6k} \cdot \frac{-k^2}{21h}$

Raise to the indicated power.

33. $\left(\frac{2}{3z}\right)^2$

34. $\left(\frac{5a}{4}\right)^2$

35. $\left(\frac{-n}{7}\right)^2$

36. $\left(\frac{-1}{9y}\right)^2$

37. $\left(\frac{5c}{2d}\right)^2$

38. $\left(\frac{7p}{3q}\right)^2$

39. $\left(\frac{-h}{3k}\right)^3$

40. $\left(\frac{-2x}{y}\right)^3$

For Problems 41–44, see Lesson 6.4 to review graphing lines. (a) make a table showing three solutions to the equation, (b) graph the equation, and (c) find the *x* and *y* intercepts of the graph.

41. $y = 4 - x$

x		
y		

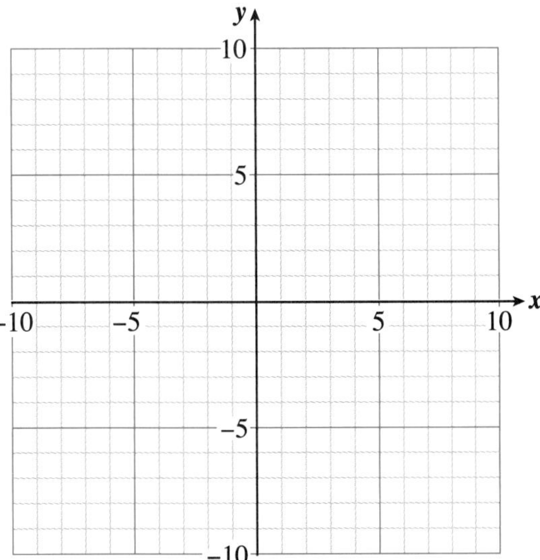

42. $y = 2x - 4$

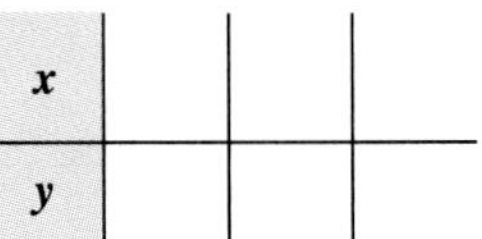

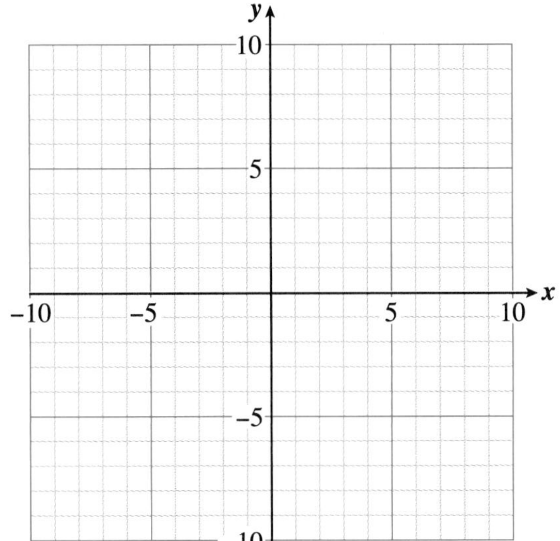

43. $y = \frac{1}{2}x - 1$

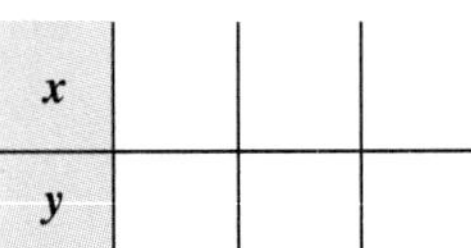

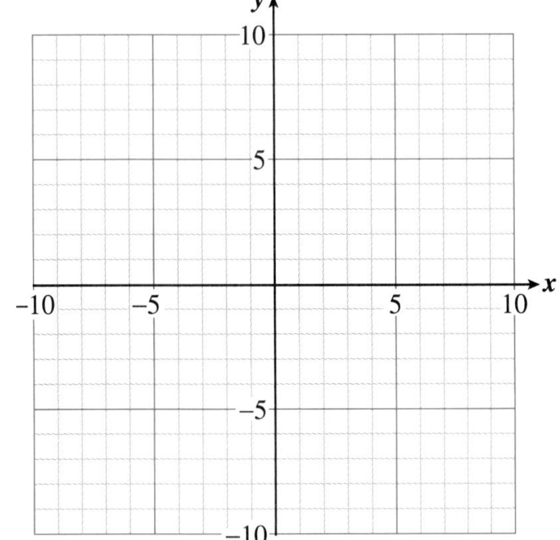

44. $y = \frac{-1}{3}x + 2$

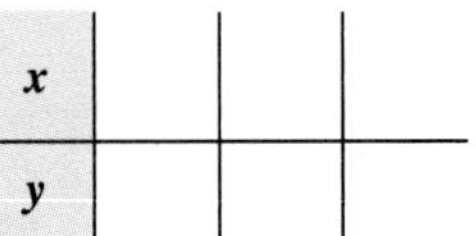

Building Number Skills

Use rounding to choose the best answer. (See Appendix A.6 to review rounding.) Do *not* use a calculator.

Example: $2.8(58) \approx 3(60) = 180$

45. 2.1(9.7) is about
a. 2 **b.** 20 **c.** 200

46. 3.9(3.2) is about
a. 1.2 **b.** 12 **c.** 120

47. 31(19) is about
a. 600 **b.** 6000 **c.** 60,000

48. 49(41) is about
a. 200 **b.** 2000 **c.** 20,000

49. 8.2(29) is about
a. 240 **b.** 2400 **c.** 24,000

50. 2.1(492) is about
a. 100 **b.** 1000 **c.** 10,000

51. 0.8(618) is about
a. 48 **b.** 480 **c.** 4800

52. 0.3(6890) is about
a. 210 **b.** 2100 **c.** 21,000

53. 0.2(0.82) is about
a. 0.16 **b.** 1.6 **c.** 16

54. 0.9(0.78) is about
a. 0.072 **b.** 0.72 **c.** 7.2

B. Dividing Fractions

You may recall a rule for dividing fractions: Invert and multiply. This is a short way of saying take the reciprocal of the second fraction and then multiply this reciprocal by the first fraction. (See Appendix A.2 to review dividing arithmetic fractions.)

The **reciprocal** of a fraction is obtained by interchanging its numerator and denominator. Some examples of reciprocals are shown in the margin.

Fraction	*Reciprocal*
$\frac{4}{7}$	$\frac{7}{4}$
$\frac{9}{2}$	$\frac{2}{9}$
$\frac{1}{3}$	$\frac{3}{1}$, or 3
5, or $\frac{5}{1}$	$\frac{1}{5}$

We will use the word *reciprocal* instead of *inverse* when discussing fractions because inverse has several other meanings in mathematics. Thus, if we want to divide $\frac{3}{4}$ by $\frac{2}{3}$, we write

Change to multiplication

$$\frac{3}{4} \div \frac{2}{3} = \frac{3}{4} \cdot \frac{3}{2} = \frac{9}{8}$$

Reciprocal

Take the reciprocal of the second fraction. Multiply by the second fraction.

Why does this rule for division work? Consider a simple example, $6 \div \frac{1}{3}$. This means how many one-thirds are there in 6? Draw six whole units as shown in Figure 7.7 and split each into thirds.

Figure 7.7

Because there are three thirds in each whole unit, we would expect 6×3, or 18 thirds in six units. The answer is 18. But 3 is the reciprocal of $\frac{1}{3}$, so using the reciprocal rule, we also find

$$6 \div \frac{1}{3} = 6 \times 3 = 18$$

For a more complicated example, you might try using Figure 7.7 to show that

$$6 \div \frac{2}{3} = \frac{6}{1} \times \frac{3}{2} = 9$$

(*Hint:* How many groups of two-thirds can you make from six whole units?)

We can state our rule for division as follows:

To Divide Two Fractions

1. Replace the second fraction by its reciprocal.
2. Multiply this reciprocal by the first fraction.
3. Reduce your answer if necessary.

Dividing Algebraic Fractions

Here are some examples of dividing algebraic fractions. We will show a separate step each time we divide out a common factor, but you can write these as a single step.

EXAMPLE 4 

Divide $\frac{15u}{4v} \div \frac{10u}{8}$

Solution

Replace the second fraction by its reciprocal and multiply:

$$\frac{15u}{4v} \div \frac{10u}{8} = \frac{15u}{4v} \cdot \frac{8}{10u}$$

$$= \frac{15u}{\cancel{4}v} \cdot \frac{\overset{2}{\cancel{8}}}{10u} \quad \text{Divide numerator and denominator by 4.}$$

$$= \frac{\overset{3}{\cancel{15u}}}{v} \cdot \frac{2}{\underset{2}{\cancel{10u}}} \quad \text{Divide numerator and denominator by } 5u.$$

$$= \frac{3}{v} \cdot \frac{2}{2} \quad \text{Divide numerator and denominator by 2. Multiply numerators; multiply denominators.}$$

$$= \frac{3}{v}$$

Now try Exercise 4.

EXERCISE 4

Divide $\frac{7p}{4} \div \frac{2}{p}$

EXAMPLE 5

Divide $\frac{-12b}{5} \div 6b$

Solution

Write $6b$ with a denominator of 1 to make it look like a fraction:

$$\frac{-12b}{5} \div \frac{6b}{1} = \frac{-12b}{5} \cdot \frac{1}{6b}$$ Replace second fraction by its reciprocal. Change division to multiplication.

$$= \frac{-\overset{2}{\cancel{12b}}}{5} \cdot \frac{1}{\underset{1}{\cancel{6b}}}$$ Divide numerator and denominator by $6b$. Multiply.

$$= \frac{-2}{5}$$

Now try Exercise 5.

EXERCISE 5

Divide $\frac{15r}{2} \div 6r^2$

ANSWERS TO 7.2B EXERCISES

4. $\frac{7p^2}{8}$

5. $\frac{5}{4r}$

HOMEWORK 7.2B 

1. How do we find the reciprocal of a fraction or whole number?

2. Explain in words how to divide one fraction by another fraction.

3. A real estate developer buys $12\frac{1}{2}$ acres of land and plans to divide it into lots for houses. Each lot will be $\frac{5}{6}$ of an acre. How many lots will there be?

4. The county road crew can paint the lines on a stretch of new road at a rate of $\frac{7}{10}$ miles per hour. How long will it take them to paint an $8\frac{3}{4}$ mile section of newly repaired county road?

Divide. Reduce your answers to lowest terms.

5. $\frac{-16}{9} \div \frac{4}{3}$

6. $\frac{5}{7} \div \frac{-20}{21}$

7. $\frac{-15}{4} \div \frac{-3}{8}$

8. $\frac{-28}{3} \div \frac{-7}{9}$

9. $\frac{14}{21} \div -4$

10. $\frac{-15}{10} \div 9$

11. $\frac{5}{3a} \div \frac{15}{a}$

12. $\frac{9m}{2} \div \frac{3m}{8}$

13. $\frac{24}{5h} \div \frac{-8}{5}$

14. $\frac{-18}{35} \div \frac{10}{7k}$

15. $\frac{-9}{2p} \div -36p$

16. $\frac{-6}{11v} \div -12v$

17. $\frac{-z^2}{2} \div \frac{z}{4}$

18. $\frac{4}{t} \div \frac{-16}{t^2}$

19. $\frac{24x^3}{7} \div \frac{8x^4}{21}$

20. $\frac{-1}{14b^2} \div \frac{-5}{7b^3}$

21. $\frac{-15}{c^5} \div \frac{20}{9c^3}$

22. $\frac{k^4}{12} \div \frac{-3k^2}{8}$

23. $\frac{a}{b} \div \frac{c}{d}$

24. $\frac{a}{b} \div \frac{b}{c}$

25. $\frac{a}{c} \div \frac{b}{c}$

26. $\frac{a}{b} \div \frac{a}{c}$

27. $2A^2 \div \frac{6A}{B^2}$

28. $\frac{M}{4P^2} \div 2M^2$

29. $\frac{3T^2}{4K^3} \div 9T^3$

30. $\frac{1}{5GH} \div \frac{G^2}{H^2}$

31. $\frac{4m}{3k} \div \frac{-8}{9mk}$

32. $\frac{-6v}{35t} \div \frac{27t}{21v}$

33. $\frac{-2p^2r^2}{p} \div \frac{-4r^3}{p^2r}$

34. $\frac{3qd^3}{qd^2} \div \frac{qd^2}{2q^2d}$

Write an equation that expresses the second variable in terms of the first variable.

35.

p	w
1	2
2	5
3	10
4	17

36.

n	b
1	0
2	3
3	8
4	15

37.

m	k
−9	6
−6	4
3	−2
6	−4

38.

t	H
−10	4
−5	2
10	−4
15	−6

Simplify. (See Lesson 5.3 to review adding and multiplying variable expressions.)

39. $-4a^3 + 6a^3$

40. $5z^2 - 8z^2$

41. $-8p - 9p^2 - 3p$

42. $10q^3 - 2q - 7q^3$

43. $3b(-4b^2)$

44. $-7y^2(-3y)$

45. $W^3 + W^3$

46. $T^2 - (-T^2)$

47. **a.** $2m + 2m + 2m + 2m$

b. $(2m)(2m)(2m)(2m)$

48. **a.** $-5n - 5n - 5n$

b. $(-5n)(-5n)(-5n)$

Building Number Skills

Use rounding to choose the best answer. (See Appendix A.6 to review rounding.) Do *not* use a calculator.

Example: $295 \div 5.9 \approx 300 \div 6 = 5$

49. $8.8 \div 2.9$ is about
a. 0.3 **b.** 3 **c.** 30

50. $5.9 \div 3.1$ is about
a. 0.2 **b.** 2 **c.** 20

51. $\dfrac{391}{21}$ is about
a. 2 **b.** 20 **c.** 200

52. $\dfrac{918}{32}$ is about
a. 3 **b.** 30 **c.** 300

53. $\dfrac{809}{3.9}$ is about
a. 2 **b.** 20 **c.** 200

54. $\dfrac{1206}{3.1}$ is about
a. 40 **b.** 400 **c.** 4000

55. $\dfrac{15{,}898}{19.5}$ is about
a. 8 **b.** 80 **c.** 800

56. $\dfrac{18{,}112}{21.3}$ is about
a. 90 **b.** 900 **c.** 9000

57. $120 \div 0.4$ is
a. 3 **b.** 30 **c.** 300

58. $3600 \div 0.9$ is
a. 40 **b.** 400 **c.** 4000

7.3 Adding and Subtracting Fractions

A. Like Fractions

We can easily add fractions with the same denominator. Remember that the denominator of a fraction tells us what kind of fraction it is: thirds, fourths, and so on. For example,

$$\frac{1}{5} + \frac{2}{5} = \frac{3}{5}$$

Figure 7.8

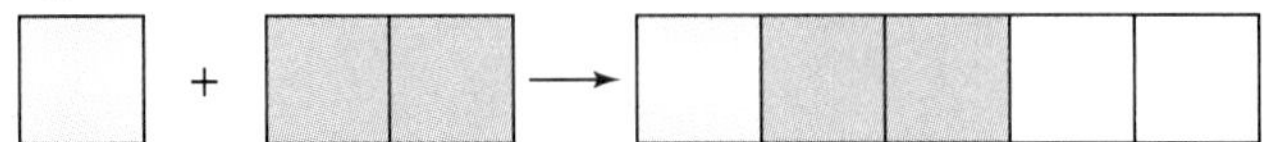

(see Figure 7.8). The *kind* of fraction doesn't change, so we don't change the denominator. The numerators tell us how many fifths we have, so we add the numerators. Subtraction works the same way. (See Appendix A.3 to review adding and subtracting arithmetic fractions.)

Fractions with the same denominator are called **like fractions;** fractions with different denominators are called **unlike fractions.** In this lesson we learn how to add and subtract like fractions. (We consider unlike fractions in Lesson 7.4.)

When working with algebraic fractions, keep in mind that the denominators of like fractions must be *exactly* the same. Thus,

$\frac{1}{2b}$ and $\frac{-5}{2b}$ are like fractions, but

$\frac{2}{3a}$ and $\frac{2}{3a^2}$ are *not* like fractions.

We can state our rule for adding or subtracting like fractions as follows:

To Add or Subtract Like Fractions

1. Combine the numerators over the same denominator.
2. Simplify the numerator if possible.

Adding and Subtracting Algebraic Fractions

Here are some examples of adding and subtracting like algebraic fractions.

EXAMPLE 1  

Subtract $\frac{3}{2x} - \frac{7}{2x}$

Solution

$$\frac{3}{2x} - \frac{7}{2x} = \frac{3-7}{2x} \quad \text{Combine the numerators.}$$

$$= \frac{-4}{2x} \quad \text{Simplify the numerator.}$$

$$= \frac{-2}{x} \quad \text{Reduce.}$$

Now try Exercise 1.

EXERCISE 1

Subtract $\frac{7}{10t} - \frac{3}{10t}$

The rules for adding and subtracting signed numbers still apply when we add or subtract fractions. You may want to review those rules from Lessons 3.2 and 3.3.

EXAMPLE 2  

Add $\left(\frac{-4c}{ab}\right) + \left(\frac{-2c}{ab}\right)$

Solution

$$\left(\frac{-4c}{ab}\right) + \left(\frac{-2c}{ab}\right) = \frac{-4c + (-2c)}{ab} \quad \text{Combine the numerators.}$$

$$= \frac{-6c}{ab} \quad \text{Simplify the numerator.}$$

Now try Exercise 2.

EXERCISE 2

Add $\frac{-3xy}{14} + \frac{5xy}{14}$

EXERCISE 3

Subtract $\frac{1}{a} - \frac{-2b}{a}$

EXAMPLE 3

Add $\frac{-3}{5} + \frac{2b}{5}$

Solution

$$\frac{-3}{5} + \frac{2b}{5} = \frac{-3 + 2b}{5} \quad \text{Combine the numerators.}$$

The numerator cannot be simplified, because -3 and $2b$ are not like terms. We cannot go any further with this problem, so the sum is $\frac{(-3 + 2b)}{5}$.

Now try Exercise 3.

More Adding Like Fractions

We can also add or subtract like fractions if there are two or more terms in the denominator. Remember that the denominator doesn't change when we add or subtract like fractions.

EXERCISE 4

Subtract $\frac{11n}{n - 5} - \frac{6n}{n - 5}$

EXAMPLE 4

Add $\frac{3w}{w + 2} + \frac{2w}{w + 2}$

Solution

$$\frac{3w}{w + 2} + \frac{2w}{w + 2} = \frac{3w + 2w}{w + 2} \quad \text{Combine the numerators.}$$

$$= \frac{5w}{w + 2} \quad \text{Simplify the numerator.}$$

Now try Exercise 4.

CAUTION!

In Example 4, the answer cannot be reduced. We cannot cancel the *w*s because *w* is not a factor of the denominator. (For the same reason, the original fractions,

$$\frac{3w}{w + 2} \quad \text{and} \quad \frac{2w}{w + 2}$$

cannot be reduced, either.)

If the *numerators* contain more than one term, we must be careful when combining them. Because a fraction bar is a grouping symbol, it acts like parentheses in the order of operations. Consequently, a subtraction sign applies to *all* the terms of the second numerator, not just to the first term.

EXAMPLE 5 

Subtract $\dfrac{2m-5}{m}-\dfrac{m-2}{m}$

Solution

Be careful when subtracting the numerators: We must subtract *both* terms of the second numerator:

$$\frac{2m-5}{m}-\frac{m-2}{m}=\frac{2m-5-(m-2)}{m}$$

Notice how we enclose the second numerator in parentheses to show that the subtraction sign applies to both terms. Now simplify the numerator by removing parentheses and combining like terms:

$$\frac{2m-5-(m-2)}{m}=\frac{2m-5-m+2}{m}$$ Change signs of *both* terms of second numerator.

$$=\frac{m-3}{m}$$ Combine like terms

The result cannot be reduced. We cannot cancel the *m*s because m is a term, not a factor, of the numerator.

Now try Exercise 5.

EXERCISE 5

Subtract $\dfrac{3w+3}{3w}-\dfrac{2w+1}{3w}$

ANSWERS TO 7.3A EXERCISES

1. $\dfrac{2}{5t}$
2. $\dfrac{xy}{7}$
3. $\dfrac{1+2b}{a}$
4. $\dfrac{5n}{n-5}$
5. $\dfrac{w+2}{3w}$

HOMEWORK 7.3A 

1. What are like fractions?

2. Is it reasonable that $\frac{1}{2}+\frac{1}{3}$ could equal $\frac{1}{5}$? Why or why not?

3. Explain why adding $\frac{2}{9}+\frac{5}{9}$ is a lot like adding $2x+5x$.

4. Explain why trying to add $\frac{2}{3}+\frac{3}{5}$ is a lot like trying to add $2x+3y$.

Add or subtract the like fractions. Give your answers in simplest form.

5. $\frac{2}{9} + \frac{5}{9}$

6. $\frac{3}{7} + \frac{1}{7}$

7. $\frac{11}{12} - \frac{5}{12}$

8. $\frac{9}{10} - \frac{3}{10}$

9. $\frac{3}{c} + \frac{4}{c}$

10. $\frac{2}{w} + \frac{6}{w}$

11. $\frac{5}{3q} - \frac{1}{3q}$

12. $\frac{11}{5b} - \frac{3}{5b}$

13. $\frac{-7a}{6} + \frac{11a}{6}$

14. $\frac{-3v}{8} - \frac{3v}{8}$

15. $\frac{-5s}{4k} - \frac{3s}{4k}$

16. $\frac{-9x}{2y} + \frac{3x}{2y}$

17. $\frac{-5r}{8} + \frac{5r}{8}$

18. $\frac{4f}{7} - \frac{4f}{7}$

19. $\frac{6p}{5} - \frac{2}{5}$

20. $\frac{-7n}{4} + \frac{3}{4}$

21. $\frac{10}{3v} - \frac{4h}{3v}$

22. $\frac{-3}{14g} - \frac{9m}{14g}$

23. $\frac{-4v^3}{9} - \frac{4v^3}{9}$

24. $\frac{4v^3}{9} - \frac{4v^2}{9}$

25. $\frac{8}{ab} + \left(\frac{-9}{ab}\right)$

26. $\frac{-12}{mk} - \left(\frac{-9}{mk}\right)$

27. $\frac{-3}{x^2} - \left(\frac{-1}{x^2}\right)$

28. $\frac{-9}{z^2} + \left(\frac{-2}{z^2}\right)$

Add or subtract the like fractions.

29. $\dfrac{5}{b+3} + \dfrac{2}{b+3}$

30. $\dfrac{8}{a-5} - \dfrac{4}{a-5}$

31. $\dfrac{3c}{n-2} - \dfrac{8c}{n-2}$

32. $\dfrac{7p}{g+1} - \dfrac{9p}{g+1}$

33. $\dfrac{-9m}{2m-3} + \dfrac{5m}{2m-3}$

34. $\dfrac{-2s}{3s+4} - \dfrac{-4s}{3s+4}$

35. $\dfrac{4w}{w+3} - \dfrac{3}{w+3}$

36. $\dfrac{3k}{k-2} + \dfrac{4}{k-2}$

37. $\dfrac{2x-1}{x} + \dfrac{2x+4}{x}$

38. $\dfrac{b-7}{2b} + \dfrac{3b-8}{2b}$

39. $\dfrac{u+5}{3u} - \dfrac{2u-1}{3u}$

40. $\dfrac{2j-9}{j} - \dfrac{4j-5}{j}$

For Problems 41–44, compute.

41. a. $\dfrac{1}{5} + \dfrac{3}{5}$ b. $\dfrac{1}{5} - \dfrac{3}{5}$ c. $\dfrac{1}{5} \cdot \dfrac{3}{5}$ d. $\dfrac{1}{5} \div \dfrac{3}{5}$

42. a. $\dfrac{2}{x} + \dfrac{5}{x}$ b. $\dfrac{2}{x} - \dfrac{5}{x}$ c. $\dfrac{2}{x} \cdot \dfrac{5}{x}$ d. $\dfrac{2}{x} \div \dfrac{5}{x}$

43. a. $\dfrac{a}{3} + \dfrac{b}{3}$ b. $\dfrac{a}{3} - \dfrac{b}{3}$ c. $\dfrac{a}{3} \cdot \dfrac{b}{3}$ d. $\dfrac{a}{3} \div \dfrac{b}{3}$

44. a. $\dfrac{p}{q} + \dfrac{r}{q}$ b. $\dfrac{p}{q} - \dfrac{r}{q}$ c. $\dfrac{p}{q} \cdot \dfrac{r}{q}$ d. $\dfrac{p}{q} \div \dfrac{r}{q}$

45. Write a few sentences describing the differences between the methods for multiplying fractions, dividing fractions, and adding (or subtracting) like fractions.

46. Correct each statement.

a. We can only multiply two fractions if they have the same denominator.

b. To divide fractions, take the reciprocal of the second fraction and multiply by the reciprocal of the first fraction.

c. To add two like fractions, add their denominators and keep the same numerator.

d. If a fraction is negative, both its numerator and its denominator must represent negative numbers

For Problems 47–50, see Lesson 6.4 to review graphing lines. (a) Make a table showing three solutions to the equation, (b) graph the equation, and (c) find the *x* and *y* intercepts of the graph.

47. $y = 5 - x$

x			
y			

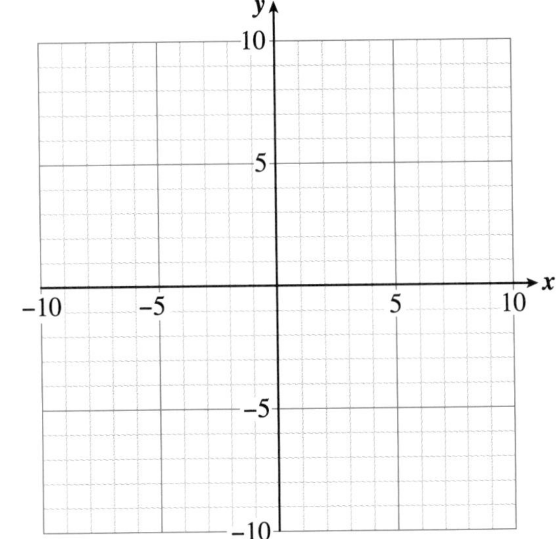

48. $y = x + 3$

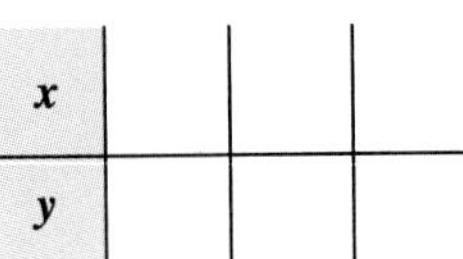

x			
y			

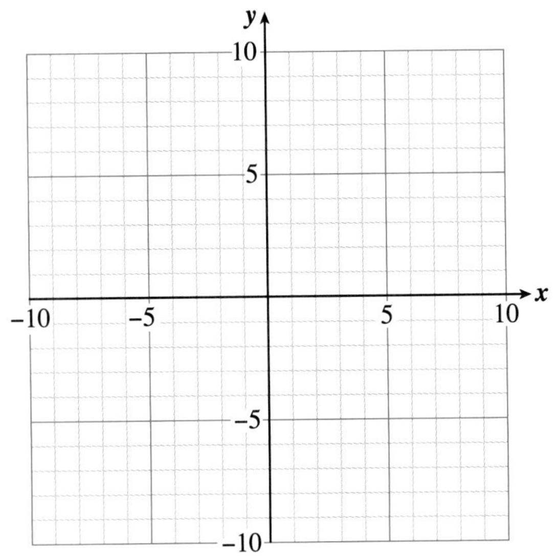

49. $y = \frac{1}{2}x + 1$

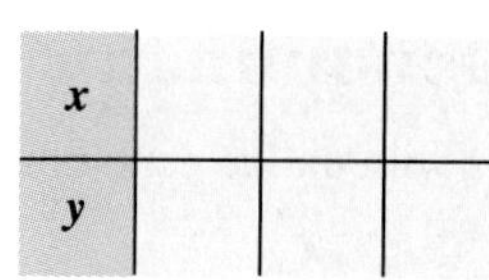

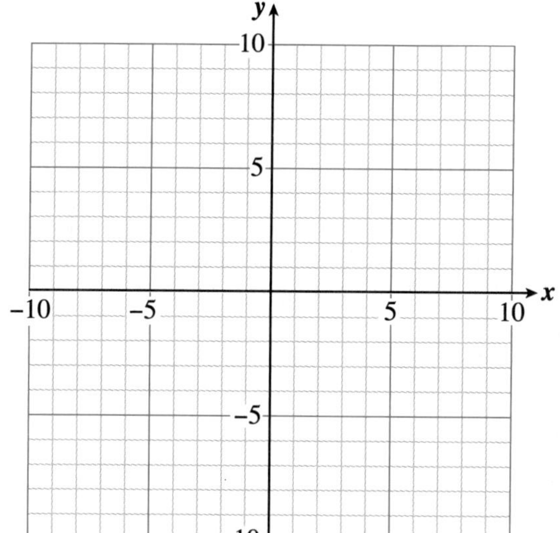

50. $y = 3 - \frac{1}{2}x$

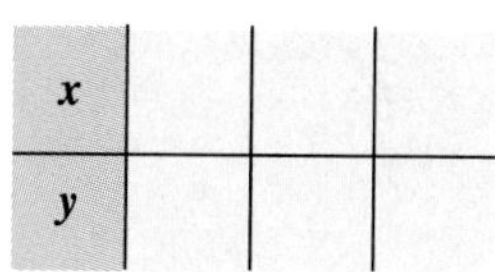

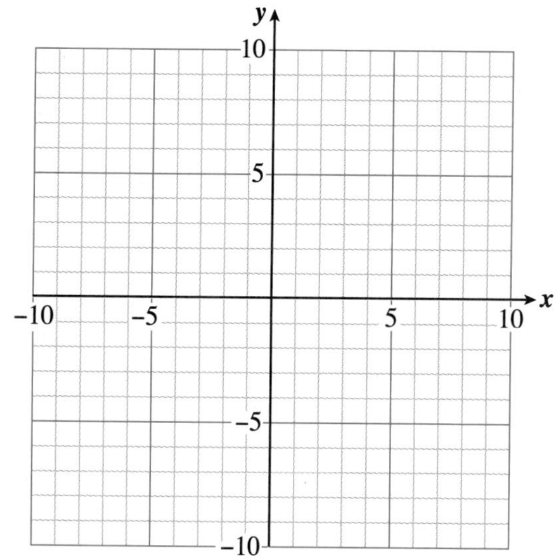

Building Number Skills

Place the decimal point so that each answer is correct. Do *not* use pencil and paper or a calculator.

Example: $27.2 \times 3.5 = 952$

Answer: 95.2

51. $55.6 + 28.73 = 8433$

52. $62.19 + 74.7 = 13689$

53. $946.2 - 627.53 = 31867$

54. $782.1 - 735.4 = 467$

55. $6.4 \times 37.6 = 24064$

56. $4.4 \times 21.39 = 94116$

57. $257.4 \div 46.8 = 55$

58. $756.8 \div 236.5 = 32$

59. $621.7 \times 0.8 = 49736$

60. $585.3 \times 0.4 = 23412$

B. Lowest Common Denominator

How can we add or subtract fractions with unlike denominators? As an example, consider

$$\frac{1}{3}+\frac{1}{4}$$

Figure 7.9

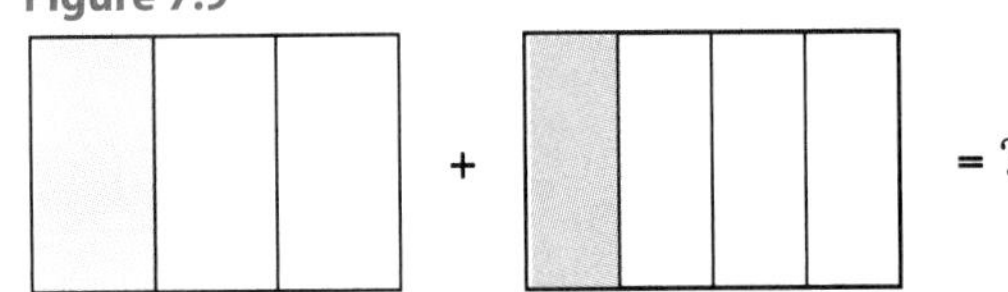

(see Figure 7.9). It is not clear what kind of fraction the sum will be. (Notice in particular that the sum cannot be $\frac{2}{7}$, because $\frac{2}{7}$ is smaller than $\frac{1}{3}$.)

Our strategy will be to build $\frac{1}{3}$ and $\frac{1}{4}$ into like fractions. Recall that to build a fraction we multiply its numerator and denominator by the same number. Let's look at some possibilities:

$$\frac{1}{3}=\frac{2}{6}=\frac{3}{9}=\mathbf{\frac{4}{12}}=\frac{5}{15}=\ldots$$

$$\frac{1}{4}=\frac{2}{8}=\mathbf{\frac{3}{12}}=\frac{4}{16}=\ldots$$

The smallest number that both fractions can have as denominator is 12. (These two fractions have other common denominators, such as 24 and 36, but the smallest denominator is the easiest to use.) We call 12 the **lowest common denominator** (or LCD) for $\frac{1}{3}$ and $\frac{1}{4}$. It is the smallest number that 3 and 4 both divide into evenly. (See Appendix A.3 to review finding the LCD for arithmetic fractions.)

Finding the LCD is the first step in adding or subtracting unlike fractions. In this lesson we concentrate on the first step.

Finding the LCD

How can we find the LCD in an efficient way? Sometimes it is easy to see the LCD—you can probably see right away that 12 is the LCD for 6 and 4. However, sometimes it is not so easy to find the LCD. Here is a method to use when you cannot see the LCD immediately.

Suppose we would like to find the LCD for $\frac{5}{18}$ and $\frac{13}{24}$. We want the smallest number that 18 and 24 both divide into evenly. We begin by factoring each denominator completely. If the numbers are large, we can keep track of our work with a *factor tree,* as shown in Figure 7.10.

Figure 7.10

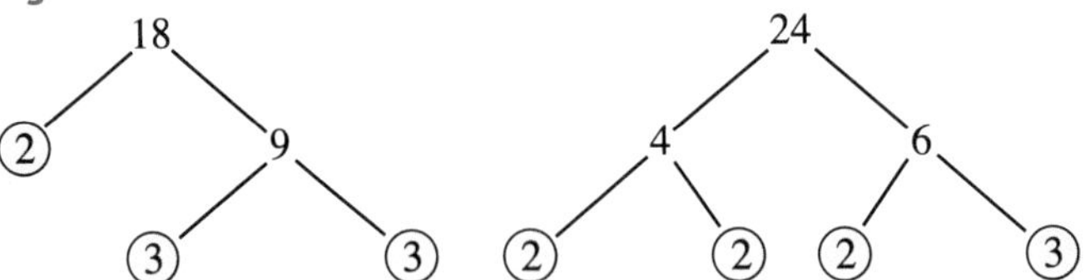

To make the factor tree, find any number that is a factor of the denominator and keep factoring until you can't go any further. The final factors are shown in color. It doesn't matter how you start; you will always come down to the same factored form. (Try starting the factorization of 24 as 3 times 8—you will end up with the same final factors.) Thus,

$$18 = 2 \cdot 3 \cdot 3 \qquad \text{and} \qquad 24 = 2 \cdot 2 \cdot 2 \cdot 3$$

Because each denominator must divide evenly into the LCD, the LCD must have enough factors to cover each denominator. Our LCD will be made up of 2s and 3s, but how many of each do we need? We need two 3s to cover 18, and we need three 2s to cover 24. Thus,

$$\text{LCD} = 2 \cdot 2 \cdot 2 \cdot 3 \cdot 3 = 72$$

You can check that 18 and 24 both divide evenly into 72, and no number smaller than 72 will work.

Here are the steps for finding the LCD.

To Find the LCD of Two or More Fractions

1. Factor each denominator completely and arrange the factors in order.
2. For each factor,
   a. Which denominator has the most copies of that factor? Circle them. (If there is a tie, either denominator will do.)
   b. Include all the circled factors in the LCD.
3. Multiply together the factors of the LCD.

EXAMPLE 6

Find the lowest common denominator of $\frac{3}{10}, \frac{5}{12},$ and $\frac{4}{15}.$

Solution

First, factor each denominator completely:

$$10 = 2 \cdot 5$$
$$12 = 2 \cdot 2 \cdot 3$$
$$15 = 3 \cdot 5$$

The factors involved are 2, 3, and 5. The factorization of 12 has the most 2s, so we circle those. At most one 3 appears in any factorization, so we can circle either one, and likewise we circle either appearance of the factor 5:

$$10 = 2 \cdot 5$$
$$12 = \boxed{2 \cdot 2} \cdot 3$$
$$15 = \boxed{3} \cdot \boxed{5}$$

Finally, we include all the circled factors in the LCD, and multiply:

$$\text{LCD} = 2 \cdot 2 \cdot 3 \cdot 5 = 60$$

The LCD is 60.

Now try Exercise 6.

EXERCISE 6

Find the lowest common denominator of

$$\frac{1}{4}, \frac{5}{6}, \text{ and } \frac{1}{9}$$

EXERCISE 7
Find the lowest common denominator of

$$\frac{m}{12}, \ \frac{5}{m^2n}, \ \text{and} \ \frac{3m}{8n}.$$

Algebraic Fractions

Our method for finding the LCD also works for algebraic fractions.

EXAMPLE 7  

Find the lowest common denominator for $\frac{1}{2x}$, $\frac{3}{4x^2y}$, and $\frac{1}{8xy^2}$.

Solution

Factor each denominator completely:

$$2x = 2 \cdot x$$

$$4x^2y = 2 \cdot 2 \cdot \boxed{xx}y \qquad 4x^2y \text{ has the most } xs.$$

$$8xy^2 = \boxed{2 \cdot 2 \cdot 2} \cdot x\boxed{yy} \qquad 8xy^2 \text{ has the most 2s, and the most } ys.$$

The factors involved are 2, x, and y. Find the denominator in which each factor occurs most often, and circle the factors as shown. Include all the circled factors in the LCD.

$$\text{LCD} = 2 \cdot 2 \cdot 2 \cdot xxyy = 8x^2y^2$$

The LCD is $8x^2y^2$.

Now try Exercise 7.

ANSWERS TO 7.3B EXERCISES

6. 36

7. $24m^2n$

HOMEWORK 7.3B 

1. The LCD of two fractions is the ______________ number that both denominators will ______________.

2. Answer each of the following questions. If your answer is no, explain why not. If your answer is yes, explain when it could happen.

a. Is it possible for the LCD to be smaller than one of the two denominators?

b. Is it possible for the LCD to be equal to one of the two denominators?

3. Find the first eight multiples of each denominator (the first two are done for you). What is the smallest number that is a multiple of *both* denominators?

$\frac{5}{8}$ Multiples of 8: 8, 16, . . .

$\frac{7}{10}$ Multiples of 10: 10, 20, . . .

4. Repeat problem 3 for the fractions $\frac{11}{18}$ and $\frac{4}{15}$.

Is the given LCD correct? Explain why or why not.

5. $\frac{3}{8}$ and $\frac{9}{16}$; LCD: 16

6. $\frac{7}{24}$ and $\frac{17}{36}$; LCD: 12

7. $\frac{1}{8xy^2}$ and $\frac{1}{12x^2y}$; LCD: $4xy$

8. $\frac{2}{9a^2}$ and $\frac{5}{27a^3}$; LCD: $27a^3$

Use a factor tree to factor each whole number completely.

9. 180

10. 336

11. 360

12. 294

Find the smallest number that each denominator will divide into evenly.

13. $\frac{2}{15}, \frac{7}{9}$

14. $\frac{3}{20}, \frac{13}{18}$

15. $\frac{5}{14}, \frac{8}{35}$

16. $\frac{23}{24}, \frac{11}{16}$

17. $\frac{3}{10}, \frac{5}{6}, \frac{5}{12}$

18. $\frac{7}{6}, \frac{7}{8}, \frac{7}{9}$

19. $\frac{13}{15}, \frac{13}{24}, \frac{13}{35}$

20. $\frac{7}{24}, \frac{19}{54}, \frac{11}{36}$

Find the lowest common denominator.

21. $\frac{1}{3x}, \frac{2}{x}$

22. $\frac{4}{b}, \frac{5}{2b}$

23. $\frac{7}{2g}, \frac{5}{6}$

24. $\frac{9}{10}, \frac{6}{5k}$

25. $\frac{25}{24t}, \frac{5}{18t}$

26. $\frac{11}{27v}, \frac{1}{12v}$

27. $\frac{3}{4n}, \frac{5}{2n^2}$

28. $\frac{2}{9m^2}, \frac{8}{3m}$

29. $\frac{b}{5a}, \frac{a}{3b}$

30. $\frac{x}{7y}, \frac{y}{2x}$

31. $\frac{6z}{w^3}, \frac{z}{w}$

32. $\frac{4p}{c^2}, \frac{4}{c^3}$

33. $\frac{2}{3xy}, \frac{9}{4y^2}$

34. $\frac{5}{6a^2b}, \frac{6}{5ab}$

35. $\frac{3}{8uv^2}, \frac{1}{12u^2v}$

36. $\frac{17}{36hk}, \frac{19}{30hk^2}$

37. $\frac{1}{5p}, \frac{1}{25p^2}, \frac{1}{125p^3}$

38. $\frac{2}{3q}, \frac{2}{9q^2}, \frac{2}{27q^3}$

39. $\frac{2}{qr}, \frac{3}{rs}, \frac{4}{sq}$

40. $\frac{b}{cd^2}, \frac{c}{bd^2}, \frac{d}{bc^2}$

Combine the fractions. (See Lessons 7.2 and 7.3A to review operations on fractions.)

41. a. $\frac{3}{5a} + \frac{b}{5a}$ **b.** $\frac{3}{5a} \cdot \frac{b}{5a}$

42. a. $\frac{1}{6z} - \frac{5w}{6z}$ **b.** $\frac{1}{6z} \div \frac{5w}{6z}$

43. a. $\frac{3r}{4p^2} \div \frac{9r}{2p}$ **b.** $\frac{3r}{4p^2} \cdot \frac{9r}{2p}$

44. a. $\frac{12m}{25hk} \cdot \frac{5h^2}{8mk}$ **b.** $\frac{12m}{25hk} \div \frac{5h^2}{8mk}$

Simplify. (See Lesson 5.4 to review square roots.)

45. $\sqrt{36 + 64}$

46. $\sqrt{100 - 36}$

47. $5\sqrt{81}$

48. $3\sqrt{121}$

49. $7 - 3\sqrt{169}$

50. $10 - 6\sqrt{49}$

51. $\sqrt{\frac{225}{16}}$

52. $\sqrt{\frac{400}{81}}$

53. $\frac{8 + \sqrt{16}}{4}$

54. $\frac{12 + \sqrt{9}}{3}$

Building Number Skills

Choose the true statement. Do not use a calculator.

55. a. $2 < \sqrt{6} < 3$ **b.** $3 < \sqrt{6} < 4$ **c.** $\sqrt{6} > 4$

56. a. $2 < \sqrt{10} < 3$ **b.** $3 < \sqrt{10} < 4$ **c.** $\sqrt{10} > 4$

57. a. $2 < \sqrt{20} < 3$ **b.** $3 < \sqrt{20} < 4$ **c.** $\sqrt{20} > 4$

58. a. $4 < \sqrt{43} < 5$ **b.** $5 < \sqrt{43} < 6$ **c.** $\sqrt{43} > 6$

59. a. $8 < \sqrt{73} < 9$ **b.** $9 < \sqrt{73} < 10$ **c.** $\sqrt{73} > 10$

60. a. $8 < \sqrt{82} < 9$ **b.** $9 < \sqrt{82} < 10$ **c.** $\sqrt{82} > 10$

61. a. $\sqrt{98} < 9$ **b.** $9 < \sqrt{98} < 10$ **c.** $\sqrt{98} > 10$

62. a. $\sqrt{108} < 9$ **b.** $9 < \sqrt{108} < 10$ **c.** $\sqrt{108} > 10$

63. a. $2 < \sqrt{17.8} < 3$ **b.** $3 < \sqrt{17.8} < 4$ **c.** $\sqrt{17.8} > 4$

64. a. $3 < \sqrt{26.2} < 4$ **b.** $4 < \sqrt{26.2} < 5$ **c.** $\sqrt{26.2} > 5$

7.4 Adding and Subtracting Unlike Fractions

Using the LCD

How does the lowest common denominator help us add or subtract unlike fractions? Recall our example from Lesson 7.3, $\frac{1}{3} + \frac{1}{4}$, shown in Figure 7.11.

Figure 7.11

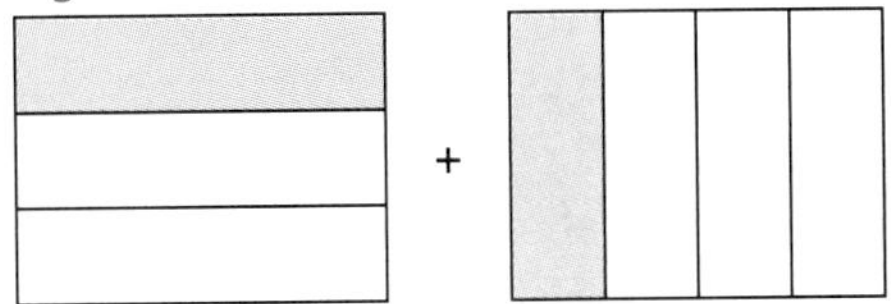

Thirds and fourths are not the same kind of fraction—they are not pieces of the same size. But if we can break up $\frac{1}{3}$ and $\frac{1}{4}$ into smaller pieces that *are* the same size, then we can combine all the pieces. This is what the LCD is for. We can break up both thirds and fourths into twelfths, as shown in Figure 7.12.

Figure 7.12

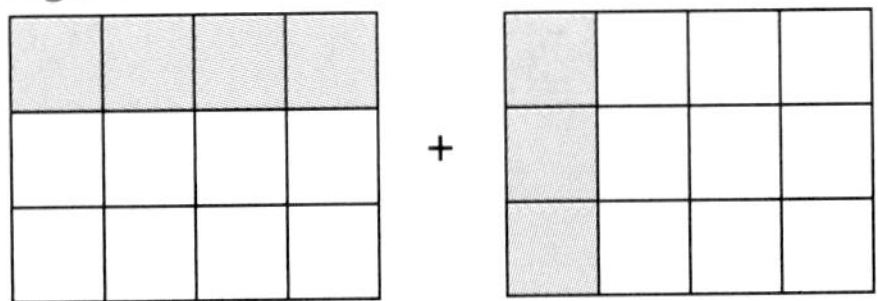

We can see from the figure that $\frac{1}{3} = \frac{4}{12}$ and $\frac{1}{4} = \frac{3}{12}$. (This is what building fractions is all about.) Once we have written the fractions with the same denominator, they are like fractions, and we can add them:

$$\frac{1}{3} + \frac{1}{4} = \frac{4}{12} + \frac{3}{12} = \frac{7}{12}$$

We can describe this method for adding and subtracting unlike fractions in three steps:

To Add or Subtract Unlike Fractions

1. Find the LCD of the fractions.
2. Build the fractions so that each has the LCD for its denominator.
3. Add (or subtract) the like fractions that result.

Here is an example using numbers only. (You may want to review adding and subtracting unlike arithmetic fractions in Appendix A.3.)

EXAMPLE 1

Add $\frac{3}{8} + \frac{1}{6}$

Solution

Step 1 Find the lowest common denominator. Factor each denominator completely:

$$8 = 2 \cdot 2 \cdot 2$$
$$6 = 2 \cdot 3$$

The LCD must include three 2s and one 3:

$$\text{LCD} = 2 \cdot 2 \cdot 2 \cdot 3 = 24$$

Step 2 Build each fraction to an equivalent one with the LCD as its denominator:

$$\frac{3}{8} = \frac{?}{24} \quad \text{and} \quad \frac{1}{6} = \frac{?}{24}$$

The building factor for the first fraction is 3, so

$$\frac{3}{8} = \frac{3 \cdot 3}{8 \cdot 3} = \frac{9}{24}$$

The building factor for the second fraction is 4, so

$$\frac{1}{6} = \frac{1 \cdot 4}{6 \cdot 4} = \frac{4}{24}$$

Step 3 Add the resulting like fractions by combining their numerators:

$$\frac{3}{8} + \frac{1}{6} = \frac{9}{24} + \frac{4}{24}$$
$$= \frac{9 + 4}{24} = \frac{13}{24}$$

The sum cannot be reduced.

Now try Exercise 1.

EXERCISE 1

Subtract $\frac{2}{3} - \frac{1}{4}$

Adding and Subtracting Algebraic Fractions

The same method works on algebraic fractions.

EXAMPLE 2  

Subtract $\frac{1}{2x} - \frac{2y}{3x}$

Solution

Step 1 Find the LCD. Factor each denominator completely:

$$2x = 2 \cdot x$$
$$3x = 3 \cdot x$$

The LCD must include one 2, one 3, and one x:

$$\text{LCD} = 2 \cdot 3 \cdot x = 6x$$

Step 2 Build each fraction to an equivalent one with the LCD as its denominator:

$$\frac{1}{2x} = \frac{?}{6x} \quad \text{and} \quad \frac{2y}{3x} = \frac{?}{6x}$$

EXERCISE 2

Add $\frac{y}{5} + \frac{3}{y}$

The building factor for the first fraction is 3, so

$$\frac{1}{2x} = \frac{1 \cdot 3}{2x \cdot 3} = \frac{3}{6x}$$

The building factor for the second fraction is 2, so

$$\frac{2y}{3x} = \frac{2y \cdot 2}{3x \cdot 2} = \frac{4y}{6x}$$

Step 3 Subtract the resulting like fractions by combining their numerators:

$$\frac{1}{2x} - \frac{2y}{3x} = \frac{3}{6x} - \frac{4y}{6x} = \frac{3 - 4y}{6x}$$

The difference cannot be simplified further.

Now try Exercise 2.

EXAMPLE 3 

Add $\frac{1}{3a} + \frac{2}{3a^2}$

Solution

Step 1 Find the LCD. Factor each denominator completely:

$$3a = \textcircled{3} \cdot a$$
$$3a^2 = 3 \cdot \textcircled{aa}$$

The LCD must include one 3 and two *a*s.

$$\text{LCD} = 3 \cdot aa = 3a^2$$

Step 2 Build each fraction to an equivalent one with the LCD as its denominator:

$$\frac{1}{3a} = \frac{?}{3a^2} \quad \text{and} \quad \frac{2}{3a^2} = \frac{?}{3a^2}$$

The building factor for the first fraction is a, so

$$\frac{1}{3a} = \frac{1 \cdot a}{3a \cdot a} = \frac{a}{3a^2}$$

The second fraction already has the LCD for its denominator.

Step 3 Add the resulting like fractions by combining their numerators:

$$\frac{1}{3a} + \frac{2}{3a^2} = \frac{a}{3a^2} + \frac{2}{3a^2} = \frac{a + 2}{3a^2}$$

The sum cannot be simplified further.

Now try Exercise 3.

EXERCISE 3

Subtract $\frac{7}{2p^3} - \frac{9}{2p^2}$

Remember that a whole number can be written as a fraction with a denominator of 1. In the next example, we write 3 as $\frac{3}{1}$.

EXAMPLE 4 

Subtract $3 - \dfrac{2}{n}$

Solution

Step 1 Find the LCD. Since there is only one denominator, it is the LCD:

$$\text{LCD} = n$$

Step 2 Build each fraction to an equivalent one with the LCD as its denominator:

$$\frac{3}{1} = \frac{?}{n} \quad \text{and} \quad \frac{2}{n} = \frac{?}{n}$$

The building factor for the first fraction is n, so

$$\frac{3}{1} = \frac{3 \cdot n}{1 \cdot n} = \frac{3n}{n}$$

The second fraction already has the LCD for its denominator.

Step 3 Subtract the resulting like fractions by combining their numerators:

$$3 - \frac{2}{n} = \frac{3n}{n} - \frac{2}{n}$$
$$= \frac{3n - 2}{n}$$

The difference cannot be reduced. (Why?) ____________

Now try Exercise 4.

EXERCISE 4

Add $\dfrac{r}{s} + 2$

ANSWERS TO 7.4 EXERCISES

1. $\dfrac{5}{12}$
2. $\dfrac{y^2 + 15}{5y}$
3. $\dfrac{7 - 9p}{2p^3}$
4. $\dfrac{r + 2s}{s}$

HOMEWORK 7.4

1. Explain why we cannot add or subtract unlike fractions the same way we added and subtracted like fractions.

2. Explain how to add or subtract unlike fractions using three steps.

Add or subtract. Remember to follow three steps: (a) Find the LCD, (b) build each fraction, and (c) add or subtract like fractions.

3. $\dfrac{1}{2} + \dfrac{2}{3}$

4. $\dfrac{2}{3} + \dfrac{3}{4}$

5. $\dfrac{3}{4} - \dfrac{2}{3}$

6. $\frac{2}{3} - \frac{1}{4}$

7. $\frac{5}{6} + \frac{2}{9}$

8. $\frac{7}{8} + \frac{1}{2}$

9. $\frac{3}{5} - \frac{7}{10}$

10. $\frac{1}{6} - \frac{3}{4}$

11. $\frac{a}{3} + \frac{b}{5}$

12. $\frac{c}{2} + \frac{d}{7}$

13. $\frac{m}{4} - \frac{9}{10}$

14. $\frac{n}{9} - \frac{4}{15}$

15. $\frac{-3w}{8} - \frac{5z}{4}$

16. $\frac{-2h}{3} - \frac{7k}{12}$

17. $\frac{-2}{9} + \frac{5v}{6}$

18. $\frac{-7}{6} + \frac{3u}{8}$

19. $\frac{b^2}{4} + \frac{-b}{5}$

20. $\frac{a}{3} + \frac{-a^2}{8}$

21. $1 + \frac{y}{9}$

22. $1 - \frac{t}{3}$

23. $c - \frac{c}{4}$

24. $g + \frac{g}{6}$

25. $\frac{x}{10} - \frac{x}{8}$

26. $\frac{-s}{9} - \frac{s}{12}$

27. $\frac{2}{p} + \frac{q}{3}$

28. $\frac{-w}{2} - \frac{1}{v}$

29. $\frac{1}{a} - \frac{2}{b}$

30. $\frac{3}{u} + \frac{4}{z}$

31. $\frac{5}{xy} - \frac{3}{y}$

32. $\frac{2}{cd} + \frac{3}{c}$

33. $\frac{1}{st} + \frac{s}{t}$

34. $\frac{p}{r} - \frac{1}{pr}$

35. $\frac{-3}{a} - \frac{2a}{b}$

36. $\frac{-4m}{z} + \frac{5}{m}$

37. $2 + \frac{1}{v}$

38. $3 - \frac{1}{h}$

39. $\frac{2}{z} - \frac{1}{2z}$

40. $\frac{1}{3d} + \frac{3}{d}$

41. $\frac{-2}{5n} + \frac{q}{4n}$

42. $\dfrac{-b}{2k} - \dfrac{3}{7k}$

43. $\dfrac{-6}{s} + \dfrac{3}{s^2}$

44. $\dfrac{8}{g^2} - \dfrac{4}{g}$

45. $\dfrac{z}{4x^2} - \dfrac{1}{6xz}$

46. $\dfrac{2}{9bc} + \dfrac{c}{6b^2}$

Combine the fractions.

47. a. $\dfrac{1}{x} + \dfrac{x}{2}$ **b.** $\dfrac{1}{x} \div \dfrac{x}{2}$ **c.** $\dfrac{1}{x} \cdot \dfrac{x}{2}$ **d.** $\dfrac{1}{x} - \dfrac{x}{2}$

48. a. $\dfrac{2}{a} \cdot \dfrac{4}{b}$ **b.** $\dfrac{2}{a} - \dfrac{4}{b}$ **c.** $\dfrac{2}{a} \div \dfrac{4}{b}$ **d.** $\dfrac{2}{a} + \dfrac{4}{b}$

49. a. $\dfrac{3}{2a} - \dfrac{2}{a^2}$ **b.** $\dfrac{3}{2a} \cdot \dfrac{2}{a^2}$ **c.** $\dfrac{3}{2a} + \dfrac{2}{a^2}$ **d.** $\dfrac{3}{2a} \div \dfrac{2}{a^2}$

50. a. $\dfrac{2y}{x^2} \div \dfrac{6}{xy}$ **b.** $\dfrac{2y}{x^2} + \dfrac{6}{xy}$ **c.** $\dfrac{2y}{x^2} \cdot \dfrac{6}{xy}$ **d.** $\dfrac{2y}{x^2} - \dfrac{6}{xy}$

Building Number Skills

Choose the best answer. (See Appendix A.9 to review converting between fractions and percents.)

51. $\frac{1}{4}$ is
a. 4% **b.** 25% **c.** 50%

52. $\frac{3}{4}$ is
a. 34% **b.** 66% **c.** 75%

53. $\frac{1}{3}$ is about
a. 0.33% **b.** 3.3% **c.** 33%

54. $\frac{1}{6}$ is about
a. 0.16% **b.** 16% **c.** 61%

55. $\frac{2}{3}$ is about
a. 23% **b.** 66% **c.** 0.66%

56. $\frac{2}{5}$ is
a. 20% **b.** 25% **c.** 40%

57. $\frac{31}{101}$ is about
a. 30% **b.** 3% **c.** 0.3%

58. $\frac{48}{97}$ is about
a. 0.5% **b.** 5% **c.** 50%

59. $\frac{205}{793}$ is about
a. 20% **b.** 25% **c.** 30%

60. $\frac{196}{611}$ is about
a. 20% **b.** 30% **c.** 40%

7.5 Equations with Fractions

We have already solved some simple equations with fractions. (See Lesson 4.3 to review solving equations with two or more operations.)

EXAMPLE 1

Solve $\frac{2x}{3} = 5$

Solution

What operations have been performed on the variable? We want to undo those operations in reverse order.

Operations Performed on x	*Steps for Solution*
1. Multiplied by 2	**1.** Multiply by 3
2. Divided by 3	**2.** Divide by 2

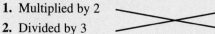

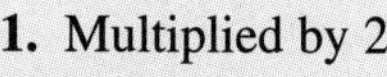

We first multiply both sides by 3 to get

$$\mathbf{3}\left(\frac{2}{3}x\right) = (5)\mathbf{3}$$
$$2x = 15$$

EXERCISE 1

Solve $\frac{-3b}{8} = -6$

and then divide both sides by 2 to get

$$\frac{2x}{\mathbf{2}} = \frac{15}{\mathbf{2}}$$

$$x = \frac{15}{2}$$

You can check that $\frac{15}{2}$ is indeed the solution; that is, $\frac{2}{3}(\frac{15}{2}) = 5$.

Now try Exercise 1.

Using Reciprocals

Now that we have studied fractions, we can solve the equation in Example 1 in one step instead of two. First, notice that

$$\frac{2x}{3} \quad \text{is the same as} \quad \frac{2}{3}x$$

because

$$\frac{2}{3}x = \frac{2}{3} \cdot \frac{x}{1} = \frac{2x}{3}$$

Thus, the original equation can be written as

$$\frac{2}{3}x = 5$$

Because x is multiplied by $\frac{2}{3}$, we will divide both sides by $\frac{2}{3}$ to solve for x.

$$\frac{\frac{2}{3}x}{\frac{\mathbf{2}}{\mathbf{3}}} = \frac{5}{\frac{\mathbf{2}}{\mathbf{3}}} \quad \text{Divide both sides by } \frac{2}{3}.$$

This is a little clumsy to write, so we take advantage of a property of fractions. Remember that dividing by a fraction is the same as multiplying by its reciprocal. Instead of dividing both sides by $\frac{2}{3}$, we can multiply both sides by $\frac{3}{2}$:

$$\frac{\mathbf{3}}{\mathbf{2}}\left(\frac{2}{3}x\right) = (5)\frac{\mathbf{3}}{\mathbf{2}}$$

Because $\frac{3}{2} \cdot \frac{2}{3} = 1$, we end up with

$$1 \cdot x = \frac{5}{1} \cdot \frac{3}{2}$$

or

$$x = \frac{15}{2}$$

the same as before.

EXAMPLE 2 

Solve $\frac{5}{2}z = -4$

Solution

Since z is multiplied by $\frac{5}{2}$, we need to divide both sides by $\frac{5}{2}$, or we can multiply both sides by $\frac{2}{5}$:

$$\frac{\mathbf{2}}{\mathbf{5}}\left(\frac{5}{2}z\right) = (-4)\frac{\mathbf{2}}{\mathbf{5}}$$

$$1 \cdot z = \frac{-4}{1} \cdot \frac{2}{5}$$

$$z = \frac{-8}{5}$$

Now try Exercise 2.

EXERCISE 2

Solve $\frac{-2}{3}p = 14$

Clearing Fractions

If there is more than one fraction in the equation, we can use their LCD to write a new equation without fractions. This is called *clearing* the fractions from the equation, and it will be useful for solving longer equations, as we'll see in Chapters 8 and 9.

EXAMPLE 3 

Solve $\frac{-3}{2}x = \frac{4}{3}$

Solution

The LCD of $\frac{-3}{2}$ and $\frac{4}{3}$ is 6. We will multiply both sides of the equation by 6:

$$\frac{\mathbf{6}}{\mathbf{1}} \cdot \frac{-3}{2}x = \frac{4}{3} \cdot \frac{\mathbf{6}}{\mathbf{1}} \quad \text{Write 6 as } \frac{6}{1}.$$

$$-9x = 8$$

When we multiply by the LCD, we get an equivalent equation without fractions. The fractions have been cleared. We can now proceed as usual. Divide both sides of the equation by -9 to obtain

$$\frac{-9x}{\mathbf{-9}} = \frac{8}{\mathbf{-9}}$$

$$x = \frac{-8}{9}$$

The solution is $x = \frac{-8}{9}$.

Now try Exercise 3.

EXERCISE 3

Solve $\frac{5}{6}n = \frac{2}{3}$

ANSWERS TO 7.5 EXERCISES

1. $b = 16$
2. $p = -21$
3. $n = \frac{4}{5}$

HOMEWORK 7.5 

1. If we want to divide by a fraction, we can get the same answer if we multiply by ______________.

2. To clear the fractions from an equation, we should multiply both sides by ______________.

Solve each equation by multiplying both sides by the reciprocal of the fraction.

3. $\frac{1}{5}x = 10$

4. $\frac{1}{6}z = 12$

5. $\frac{2}{3}y = 12$

6. $\frac{3}{4}w = 18$

7. $\frac{5}{4}a = -25$

8. $\frac{8}{5}b = -24$

9. $\frac{3}{2}p = -7$

10. $\frac{7}{4}q = -2$

11. $\frac{-2}{9}k = \frac{3}{5}$

12. $\frac{-3}{7}h = \frac{6}{35}$

13. $\frac{-4}{3}u = \frac{-8}{21}$

14. $\frac{-11}{6}v = \frac{-11}{3}$

(a) Clear the fractions from the equation and (b) solve the new equation.

15. $\frac{8}{5}q = \frac{2}{7}$

16. $\frac{2}{9}r = \frac{6}{5}$

17. $\frac{-5m}{3} = \frac{7}{2}$

18. $\frac{4s}{13} = \frac{-2}{3}$

19. $\frac{2}{b} = \frac{-3}{4}$

20. $\frac{7}{4} = \frac{-3}{p}$

21. $\frac{1}{5} = \frac{3}{10x}$

22. $\frac{3}{2x} = \frac{-5}{6}$

23. $\frac{-1}{8x} = \frac{5}{6}$

24. $\frac{7}{18} = \frac{1}{6x}$

25. $\frac{n-9}{3} = -2$

26. $-5 = \frac{a+3}{2}$

27. $1 = \frac{11}{n+1}$

28. $\frac{7}{w-4} = 1$

Solve each problem by writing and solving an algebraic equation. Don't forget to state what your variable represents. (See Lesson 4.4 to review writing equations.)

29. Darcy deposits $\frac{2}{9}$ of her paycheck in her savings account each week. If she saves \$130 a week, how much does she make?

30. A laser printer is on sale for $\frac{4}{5}$ of its regular price. If the sale price is \$688, what is the regular price?

31. Maro received only 48 votes in the election, but that was $\frac{2}{3}$ of all the votes cast. How many votes were cast in all?

32. A pool filtering system removes $\frac{3}{4}$ pound of debris from a pool in 1 hour. According to the manufacturer, the system removes $\frac{9}{10}$ of all the debris in the pool each hour. If the manufacturer's claim is accurate, how much debris was in the pool before the system was activated?

33. Some biologists catch 24 fish from a small lake, tag the fish, and return them to the lake. They later determine that $\frac{3}{20}$ of the fish in the lake are tagged. How many fish are in the lake?

34. Tsu-te has \$6000 invested in stocks. His stocks represent $\frac{3}{8}$ of all his investments. How much money does Tsu-te have invested?

Combine the fractions.

35. a. $\dfrac{2b}{c} - \dfrac{1}{2c}$ **b.** $\dfrac{2b}{c} \div \dfrac{1}{2c}$

36. a. $\dfrac{3}{tw} + \dfrac{t}{w}$ **b.** $\dfrac{3}{tw} \cdot \dfrac{t}{w}$

37. a. $\dfrac{v}{p} \cdot \dfrac{2q}{v}$ **b.** $\dfrac{v}{p} + \dfrac{2q}{v}$

38. a. $\dfrac{2x}{y} \div \dfrac{x}{3}$ **b.** $\dfrac{2x}{y} - \dfrac{x}{3}$

39. a. $\dfrac{a}{9b} \div \dfrac{a^2}{6b}$ **b.** $\dfrac{a}{9b} + \dfrac{a^2}{6b}$

40. a. $\dfrac{9w}{2z} \cdot \dfrac{8z^2}{3w}$ **b.** $\dfrac{9w}{2z} - \dfrac{8z^2}{3w}$

Evaluate each expression for $m = -4$, $n = -3$, $p = 6$, and $q = -2$. (See Lesson 4.5 to review evaluating expressions.)

41. $m(n + p)$ **42.** $mn + mp$ **43.** $m - n - q$

44. $m - (n - q)$ **45.** $m - nq$ **46.** $(m - n)q$

47. $\frac{p}{q} - m$

48. $\frac{p - m}{q}$

49. $\frac{p - q}{p + m}$

50. $\frac{pq}{p + m}$

Building Number Skills

From the following list, choose the common fraction that most closely approximates the given percent. Then use the fraction to estimate the product.

$\frac{1}{2} \quad \frac{1}{4} \quad \frac{3}{4} \quad \frac{1}{3} \quad \frac{2}{3}$

Example: 76% of 48 $\approx \frac{3}{4}(48) = 36$

51. 27% of 28

52. 65% of 27

53. 34.5% of 96

54. 49.3% of 86

55. 72% of 120

56. 30% of 150

57. 68% of 4500

58. 22% of 3200

59. 52% of 320

60. 77% of 240

7 Summary

Lesson 7.1

The **denominator** of a fraction tells us how many *equal* pieces to divide the whole into, and the **numerator** tells us how many of the pieces to take.

If the numerator is smaller than the denominator, the fraction is called a **proper fraction.** It represents a number smaller than 1.

If the numerator is larger than the denominator, the fraction is called an **improper fraction.** It represents a number larger than 1.

There are three ways to write a negative fraction:

$$\frac{-a}{b} = \frac{a}{-b} = -\frac{a}{b}$$

The positive and negative fractions and zero are called the **rational numbers.** Fractions that contain variables are called **algebraic fractions.**

Fundamental Principle of Fractions

1. We can multiply the numerator and denominator of a fraction by the same (nonzero) number and the result will be an equivalent fraction.
2. We can divide the numerator and denominator of a fraction by the same (nonzero) number and the result will be an equivalent fraction.

When we **reduce** a fraction, we replace it by an equivalent fraction whose numerator and denominator are smaller numbers or simpler expressions.

To Reduce a Fraction

1. Look for common factors in the numerator and denominator. If necessary, factor the numerator and denominator completely.
2. Write the numerator and denominator in factored form.
3. Divide the numerator and denominator by the common factors.

When we **build** a fraction we multiply numerator and denominator by the same factor.

To Build a Fraction

1. Factor the old denominator and the new denominator, if necessary.
2. Find the building factor by comparing the old denominator with the new denominator. The extra factors in the new denominator are the building factor.
3. Multiply the numerator and denominator of the old fraction by the building factor.

Lesson 7.2

To Multiply Two or More Fractions Together

1. Divide out any common factors that appear in a numerator and a denominator.
2. Multiply together the remaining factors in the numerator.
   Multiply together the remaining factors in the denominator.
3. Reduce the product, if necessary.

The **reciprocal** of a fraction is obtained by interchanging its numerator and denominator.

To Divide Two Fractions

1. Replace the second fraction by its reciprocal.
2. Multiply this reciprocal by the first fraction.
3. Reduce your answer if necessary.

Lesson 7.3

Fractions with the same denominator are called **like fractions;** fractions with different denominators are called **unlike fractions.**

To Add or Subtract Like Fractions

1. Combine the numerators over the same denominator.
2. Simplify the numerator if possible.

The **lowest common denominator** (LCD) for two or more fractions is the smallest number or the simplest expression that all the denominators divide into evenly.

To Find the LCD of Two or More Fractions

1. Factor each denominator completely, and arrange the factors in order.
2. For each factor,
   a. Which denominator has the most copies of that factor? Circle them. (If there is a tie, either denominator will do.)
   b. Include all the circled factors in the LCD.
3. Multiply together the factors of the LCD.

Lesson 7.4

To Add or Subtract Unlike Fractions

1. Find the LCD of the fractions.
2. Build the fractions so that each has the LCD for its denominator.
3. Add (or subtract) the like fractions that result.

Lesson 7.5

To clear the fractions from an equation, we multiply both sides of the equation by the LCD of the fractions.

CHAPTER 7 REVIEW 

Add the following words and phrases to your glossary, if you have not already done so.

numerator	algebraic fraction	cancel	reciprocal
rational number	building factor	lowest common denominator (LCD)	clear fractions
build	unlike fractions	improper fraction	
like fractions	proper fraction	common factor	
denominator	reduce		

True or false.

1. The fraction $\frac{4}{6}$ is twice as big as $\frac{2}{3}$.

2. To add two fractions, we add their numerators and add their denominators.

3. To multiply or divide unlike fractions, the first step is to find the lowest common denominator.

4. The fundamental principle of fractions is used both to build and to reduce fractions.

Explain the difference between each pair.

5. $-\dfrac{2}{3}$ and $\dfrac{-2}{-3}$

6. Like fractions and unlike fractions

7. Opposite and reciprocal

8. Numerator and denominator

9. $\dfrac{a}{b} + \dfrac{c}{d}$ and $\dfrac{a+b}{c+d}$

10. Integer and rational number

Choose a variable for the unknown quantity and write an algebraic fraction.

11. Five of the cats at HappyCat Boarding require special diets. What fraction of the cats is that?

12. The County Highway Commission maintains 2400 miles of roads and highways. They do repair work on a rotation schedule. What fraction of the roads are repaired each year?

13. Martin needs $\frac{1}{3}$ bag of Gummi Bears for each of the students in his second-grade class. How many bags of Gummi Bears will he use?

14. Alexis uses $\frac{1}{4}$ pint of raspberries in each Raspberry Dream dessert. How many pints will she need for her dinner party?

Graph each set of numbers on a number line.

15. $-\sqrt{3}, \frac{5}{2}, \sqrt{5}, \frac{-3}{2}, -2, \frac{0}{-2}$

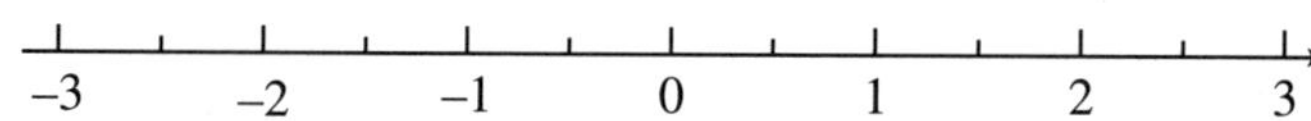

16. $\frac{-2}{3}, \frac{-3}{2}, -\sqrt{2}, \sqrt{3}, \sqrt{4}, \frac{-0}{3}$

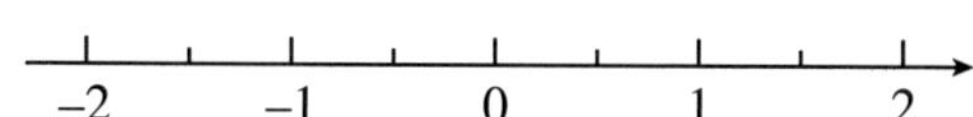

Simplify each pair of expressions.

17. a. $\dfrac{5 - 2^3}{5 - 2}$ **b.** $\dfrac{5(-2)^3}{5(-2)}$

18. a. $\dfrac{18 + \sqrt{81}}{3}$ **b.** $\dfrac{18\sqrt{81}}{3}$

19. a. $\sqrt{\dfrac{100}{25}}$ **b.** $\dfrac{\sqrt{100}}{\sqrt{25}}$

20. a. $\dfrac{3 \cdot 5 + 2}{3}$ **b.** $\dfrac{3(5 + 2)}{3}$

From each group of expressions, choose the two that are equivalent.

21. a. $\dfrac{4x}{8}$ **b.** $\dfrac{4 + x}{8}$ **c.** $\dfrac{x}{2}$ **d.** $\dfrac{x}{4}$

22. a. $\dfrac{3 - a}{a}$ **b.** $\dfrac{-3a}{a}$ **c.** -3 **d.** 2

23. a. $\dfrac{2b - c}{b + c}$ **b.** $\dfrac{-2bc}{bc}$ **c.** $\dfrac{-2bc}{b + c}$ **d.** -2

24. a. $\dfrac{x + 5y}{5x}$ **b.** $\dfrac{5xy}{5 + x}$ **c.** $\dfrac{5xy}{5x}$ **d.** y

Combine fractions. Reduce your answers to lowest terms.

25. $\dfrac{2}{3} \cdot \dfrac{-5}{4}$ **26.** $\dfrac{-1}{6} \cdot \dfrac{7}{8}$ **27.** $\dfrac{4}{5} \div \dfrac{2}{3}$

28. $\dfrac{9}{2} \div \dfrac{6}{7}$ **29.** $\dfrac{3}{7} + \dfrac{4}{7}$ **30.** $\dfrac{9}{8} + \dfrac{7}{8}$

31. $\dfrac{7}{9} - \dfrac{4}{9}$ **32.** $\dfrac{5}{6} - \dfrac{1}{6}$ **33.** $\dfrac{3x}{4} \cdot \dfrac{-6}{5x^2}$

34. $\frac{12a}{5b} \cdot \frac{25a}{9}$

35. $\frac{-6m^2}{10n} \div \frac{n}{3m}$

36. $\frac{4p}{7q} \div \frac{14q}{p^2}$

37. $\frac{3}{2r} + \frac{-1}{2r}$

38. $\frac{7t^2}{5s} + \frac{-3t^2}{5s}$

39. $\frac{4v}{7w^2} - \frac{-3v}{7w^2}$

40. $\frac{1}{6b^3} - \frac{-5}{6b^3}$

41. $\frac{1}{3} + \frac{1}{6}$

42. $\frac{1}{12} + \frac{1}{4}$

43. $\frac{1}{2} - \frac{1}{2a}$

44. $\frac{1}{y} - \frac{1}{y^2}$

45. $2 - \frac{3}{2x}$

46. $\frac{5}{7g} - 4$

47. $\frac{6}{ab} + \frac{4}{b^2}$

48. $\frac{3m}{n^2} + \frac{2}{mn}$

49. $-\left(\frac{21b}{7}\right)^4$

50. $\left(\frac{-38t}{19}\right)^4$

Solve each equation and check your solutions.

51. $\frac{1}{7}w = 2$

52. $\frac{-2}{3}v = 4$

53. $\frac{5}{2}a = \frac{-4}{3}$

54. $\frac{6}{7} = \frac{3}{4}b$

55. $\frac{9}{2d} = -5$

56. $7 = \frac{-2}{3m}$

57. $\frac{8}{3} = \frac{-16}{c}$

58. $\frac{4}{q} = \frac{-6}{7}$

59. $1 = \frac{-5}{n - 3}$

60. $\frac{4}{w + 4} = 1$

61. $5 = \frac{x + 1}{6}$

62. $\frac{y - 3}{9} = 11$

Solve each word problem by writing and solving an algebraic equation.

63. A recipe for chocolate chip cookies uses $\frac{2}{3}$ cup brown sugar per batch. How many batches can be made from 4 cups of brown sugar?

64. A motorcycle engine requires a mixture of 2 parts oil to 5 parts gasoline. How much mixture can be made with 1 quart of oil?

65. A television set is sold for $\frac{3}{4}$ of its original price. Kurt used a store credit of \$150 to bring the sale price down to \$825. What was the original price of the television set?

66. During a sale where all items were marked at $\frac{2}{3}$ of their original price, Kelly bought a computer, using a \$200 gift certificate. If Kelly paid \$466, what was the original price of the computer?

67. Only $\frac{8}{9}$ of the computer monitors coming off the assembly line actually pass inspection. How many monitors must be manufactured so that 1000 monitors pass inspection?

68. About $\frac{5}{7}$ of the drivers involved in fatal motor vehicle accidents are male. In 1999, 30,400 male drivers were involved in fatal motor vehicle accidents. How many fatal accidents were there in all?

Building Number Skills

Use rounding to choose the best answer. Do *not* use pencil and paper or a calculator.

69. $1.253 + 8.461$ is about
a. 10 **b.** 100 **c.** 1000

70. $794.46 - 305.21$ is about
a. 5 **b.** 50 **c.** 500

71. $\frac{629}{87}$ is about
a. 7 **b.** 70 **c.** 700

72. $0.92(31.4)$ is about
a. 0.3 **b.** 3 **c.** 30

73. $82.36 - 28.41$ is equal to
a. 0.5395 **b.** 5.395 **c.** 53.95

74. $123.45 + 67.89$ is equal to
a. 19.134 **b.** 191.34 **c.** 1913.4

75. $11.4(28.6)$ is equal to
a. 3.2604 **b.** 32.604 **c.** 326.04

76. $\frac{24.645}{26.5}$ is equal to
a. 0.93 **b.** 9.3 **c.** 93

77. $\frac{3}{4}$ is equal to
a. 34% **b.** 66% **c.** 75%

78. $\frac{2}{3}$ is about
a. 23% **b.** 44% **c.** 67%

79. 49% of 813 is about
a. 40 **b.** 400 **c.** 4000

80. 35% of 17.8 is about
a. 6 **b.** 60 **c.** 600

81. $\sqrt{7.1}$ is
a. between 2 and 3
b. between 3 and 4
c. greater than 4

82. $\sqrt{21.8}$ is
a. between 2 and 3
b. between 3 and 4
c. greater than 4

CHAPTER

8

Proportion and Percent

8.1 Using Percents

A. Circle Graphs

In Lesson 2.3 we used the formula

$$P = rW$$

to solve problems involving percent. In this formula W stands for the *whole* amount, r stands for the percentage *rate* (in decimal form), and P stands for the percentage, or *part,* of the whole.

EXAMPLE 1

Leslie spends 35% of her monthly income on rent. If Leslie's rent is \$840, how much is her monthly income?

Solution

Step 1 First, identify the three quantities, P, r, and W. The percentage rate is 35%, so $r = 0.35$. We are asked to find Leslie's monthly income. Leslie's rent is only a *part* of her entire income, so her rent is P and her monthly income is W. Thus, $P = 840$ and W is unknown.

Step 2 Substitute the values identified in Step 1 into the formula:

$$P = rW$$
$$840 = 0.35W$$

Step 3 Solve the equation:

$$\frac{840}{0.35} = \frac{0.35W}{0.35} \quad \text{Divide both sides by 0.35.}$$
$$2400 = W$$

Leslie's monthly income is \$2400.

Try Exercise 1 on page 530.

We can use a **circle graph** to show how a quantity is divided among several parts or categories. The graph shows the fraction of the whole amount allotted to each category.

EXAMPLE 2  

Francine lives in the campus co-op, and this weekend they are planning their monthly budget. They have a monthly revenue of $8000, and their monthly expenses are shown by the circle graph in Figure 8.1. Each budget item corresponds to a slice of the circle graph, and the size of the slice is determined by the relative size of the budget item. For instance, the co-op's mortgage payment is 30% of its monthly budget, so that slice is 30% of the whole circle.

Figure 8.1 Budget

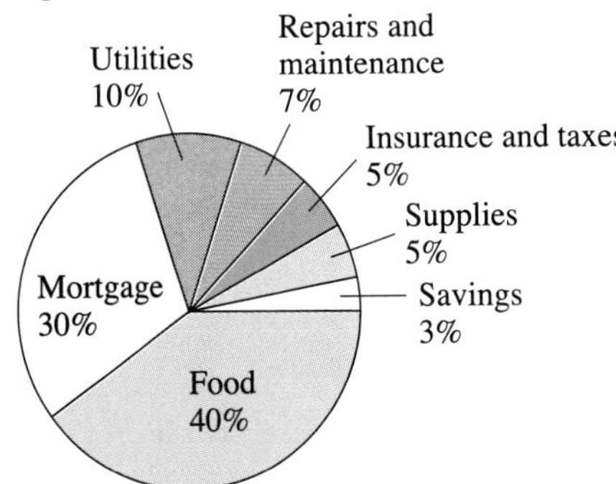

a. Which of their expenses uses the largest portion of the co-op's revenue? What percent of their budget do they spend on that item?

The co-op spends the largest portion of their budget on food: 40%.

b. Which two expenses use the same percent of the budget?

Insurance and taxes use the same percent of the budget as supplies: 5%.

c. Fill in the table to calculate the amount the co-op budgets for each of its expenses.

Mortgage payment	0.30 × 8000	=
Insurance and taxes	0.05 × 8000	=
Repairs and maintenance		=
Utilities		=
Food		=
Supplies		=
Savings		=
Total		

EXERCISE 1

Of the 15,600 students at Van Buren College, 9048 are considered full-time because they are enrolled in 12 or more units. What percent of the students are full-time?

Step 1 Identify P, r, and W.

Step 2 Substitute the values into the percent formula.

Step 3 Solve the equation and answer the question in the problem.

In Example 2, you should have found that the sum of all the co-op's expenses is equal to their total monthly budget of $8000.

Now try Exercise 2.

EXERCISE 2

Figure 8.2 shows the fraction of enlisted women who serve in each branch of the armed forces.

Figure 8.2 Women in the armed forces

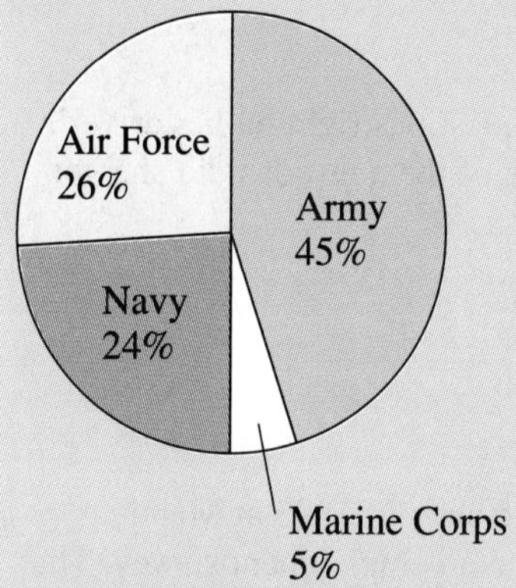

Source: U.S. Department of Defense

a. Which two branches make up half of enlisted women?

b. There are approximately 90,000 women enlisted in the armed forces. Fill in the table showing the number of women in each branch.

Branch	*Percent of women*	*Number of women*
Army		
Navy		
Air Force		
Marines		
Total		

ANSWERS TO 8.1A EXERCISES

1. 58%

2a. Air Force and Navy or Army and Marine Corps

2b.

Branch	*Percent of women*	*Number of women*
Army	45%	40,500
Navy	24%	21,600
Air Force	26%	23,400
Marines	5%	4,500
Total	100%	90,000

HOMEWORK 8.1A 

Solve each problem. Use the formula $P = rW$.

1. Orrin makes $560 a week at his job in an accounting firm. Next year he will get a 4% raise. How much will his salary increase?

2. A Sprint convertible costs $12,800, but in January the price will go up by 7%. How much will the price of a Sprint increase?

3. Ralph got 30 out of 36 questions right on a history test. What percent of the questions did he get right?

4. Wanda got 56 out of 64 questions right on her driver's license test. What percent of the questions did she get right?

5. Of the management positions at Virtual Imaging, 35% are held by women. If there are 21 women in management, how many positions are there?

6. Of the graduates from the class of 1993 at Manning College, 182 responded to an employment survey. That was 28% of the graduating class. How many people graduated from Manning College in 1993?

7. Figure 8.3 shows the ages of people who use the Internet.

Figure 8.3 Age of Internet users

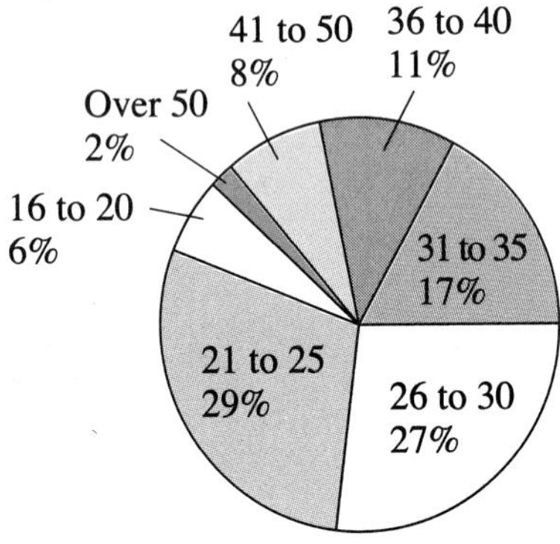

a. What percent of Internet users are over 50?

b. What percent of Internet users are between 20 and 30?

c. What percent of Internet users are over 20?

d. There are 50,000 Internet users in Delbert's home state. Use the following table to compute the number of users in each age group:

Under 20	$0.06 \times 50{,}000$	=
21 to 25		=
26 to 30		=
31 to 35		=
36 to 40		=
41 to 50		=
Over 50		=
Total		

8. A survey of 1500 high school juniors and seniors is summarized in Figure 8.4.

Figure 8.4 Where American teenagers keep their money

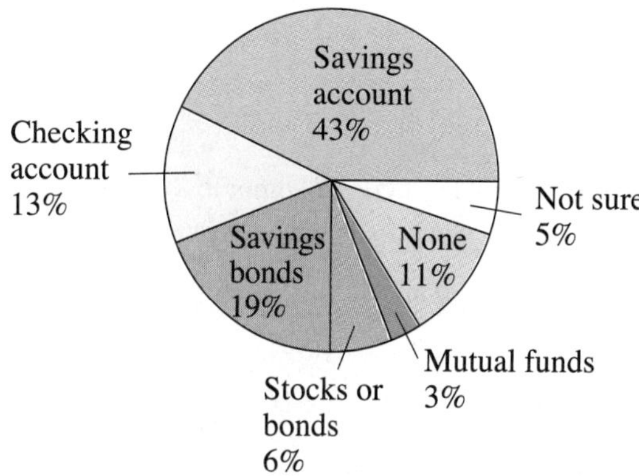

a. Where do most teenagers keep their money?

b. What percent of teenagers own savings bonds?

c. What percent of teenagers keep their money in either savings or checking accounts?

d. Use the following table to compute how many of the respondents fall into each category:

Savings account	=
Checking account	=
Savings bond	=
Stocks or bonds	=
Mutual funds	=
None	=
Not sure	=
Total	

9. In 1992 there were 71,154 American students studying abroad. Figure 8.5 shows the regions in which they studied.

Figure 8.5 Host regions

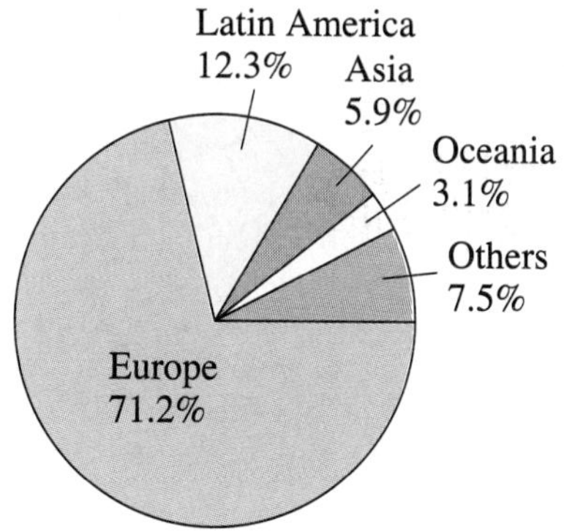

a. What percent of the students studied in Latin America?

b. In which continent did most of the American students abroad study?

c. In which continent did 5.9% of the students study?

d. Use the following table to compute the number of American students in each region:

Europe	=
Latin America	=
Asia	=
Oceania	=
Others	=
Total	

10. In 1993 there were 438,618 foreign students studying in the United States. Figure 8.6 shows their home regions.

Figure 8.6 Home regions

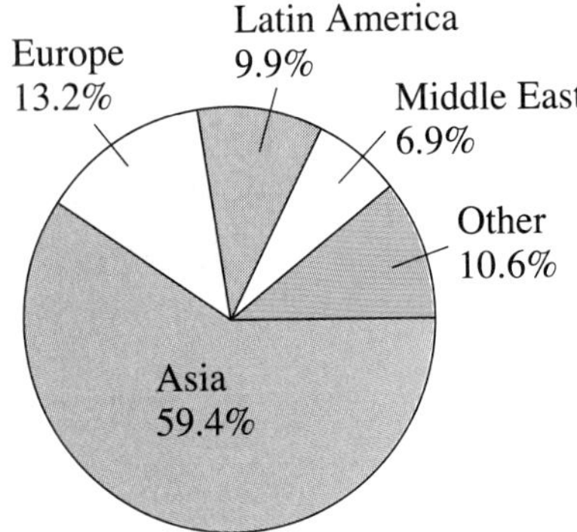

a. What percent of the students came from Europe?

b. Where did most of the students come from?

c. Where did 6.9% of the students come from?

d. Use the following table to compute the number of students who came from each region:

Asia	=
Europe	=
Latin America	=
Middle East	=
Others	=
Total	

Reduce. (See Lesson 7.1 to review reducing fractions.)

11. $\frac{34t^2}{17t}$

12. $\frac{57g^6}{-3g^3}$

13. $\frac{66bc^3}{-22bc^2}$

14. $\frac{32pr^2}{16pr}$

15. $\frac{42kl^2}{14kl^2}$

16. $\frac{52x^2y}{13x^2y}$

17. $\frac{72w^2z}{42wz^2}$

18. $\frac{21a^3b^2}{28a^2b^3}$

19. $\frac{-15kt^2}{20m^2t}$

20. $\frac{35r^2s}{-14s^2t}$

Building Number Skills

Try to do these problems without using pencil and paper or your calculator. Decide if you have enough money in each circumstance. Assume there is no tax.

21. You have $10 and want to purchase a package of shrimp ($2.99), a bag of rice ($1.69), and a bottle of wine ($5.99). Do you have enough money?

22. You have $15 and want to purchase a bag of potatoes ($1.99), a package of fish sticks ($7.99), and a bottle of cola ($2.98). Do you have enough money?

23. You have $20 and want to purchase a calendar ($6.99) and two coffee mugs ($5.99 each). Do you have enough money?

24. You have $30 and want to purchase two ties ($9.99 each) and a belt ($8.99). Do you have enough money?

25. You have $65 and want to treat your friend to visit Magic World. Parking will cost $5.00 and admission is $29.95 per person. Do you have enough money for parking and admission for two?

26. You have $15 and want to take the bus to visit the zoo. The bus fare is $2.65 each way and the admission fee at the zoo is $10.95. Do you have enough money?

27. You have $80 and want to purchase a dinner for two. The entrées are each $24.95, salads are $4.95, and desserts are $5.95. Do you have enough money for both of you to have an entrée, salad, and dessert?

28. You have $200 and want to spend a weekend at your aunt's home. The special one-way plane fares are each $79.95, and cab fare will be $10 for a trip either direction between the airport and your aunt's home. Do you have enough money for the trip and still be able to rent headphones ($6.00) for the movies on the flights?

29. You have $1000 and want to purchase 100 toy ray-guns ($2.99 each) and 100 toy spaceships ($7.99 each). Do you have enough money?

30. You have $100 and want to purchase 200 bags of jelly beans ($0.19 each) and 300 bags of licorice whips ($0.69 each). Do you have enough money?

B. Percent Increase or Decrease

Percents give us a way to compare numerical quantities. Consider the following situation. Stanley and Ernest work for Softek, a struggling software company. During a bad period, the company asks all its employees to take a \$20 per week pay cut to help them stay in business. Stanley works in the mail room and makes \$200 per week, whereas Ernest is a software designer and makes \$1000 per week. Stanley's pay cut amounts to

$$\frac{20}{200} = 0.10 = 10\%$$

of his salary, while Ernest's pay cut is

$$\frac{20}{1000} = 0.02 = 2\%$$

of his salary. Asking everyone to take the same \$20 pay cut doesn't seem exactly fair. It might be better for Softek to ask all its employees to take the same *percent* pay cut—the same fraction of their salaries.

Now suppose that business improves and Softek's income increases by \$30,000 over last year's revenue. This is good news, but how good? That depends on how much Softek made last year. For instance, if last year's revenue was \$60,000, then this year's income represents a 50% increase in revenue, quite an accomplishment. On the other hand, if Softek brought in \$6,000,000 last year, then an increase of \$30,000 represents only a 0.5% increase, not enough to make a real difference.

The *amount* of a pay cut or a revenue increase by itself is not very meaningful until we compare it with the original salary or the original revenue. By computing the **percent increase** or **percent decrease** we can see what *fraction* of the original amount the increase or decrease is.

$$\textbf{percent increase} = \frac{\textbf{amount of increase}}{\textbf{original amount}} \times \textbf{100}$$

$$\textbf{percent decrease} = \frac{\textbf{amount of decrease}}{\textbf{original amount}} \times \textbf{100}$$

EXERCISE 3

The price of an airline ticket to Toronto increased from \$280 to \$325, and the price of a train ticket increased from \$110 to \$130.

a. What was the percent increase in the airfare?

b. What was the percent increase in the price of the train ticket?

c. Which fare increased by a greater percent of the original price?

EXAMPLE 3 

Heather and Mei Li both received rent increases on their apartments this year. Heather's rent increased from \$540 to \$567, and Mei Li's rent increased from \$450 to \$477. Whose apartment had the greater percent increase in rent?

Solution

For Heather's apartment, the amount of rent increase was \$567 − \$540, or \$27. The percent increase in Heather's rent was thus

$$\text{percent increase} = \frac{27}{540} \times 100 = 5\%$$

For Mei Li's apartment, the amount of rent increase was \$477 − \$450, or \$27. The percent increase in Mei Li's rent was

$$\text{percent increase} = \frac{27}{450} \times 100 = 6\%$$

Mei Li's apartment experienced the greater percent increase in rent.

Now try Exercise 3.

Calculating the New Value

If a variable undergoes a percent increase or percent decrease from its original value, how can we calculate its new value? For instance, if gasoline prices increase by 15%, this means that the cost of a gallon of gasoline increases by 15% *of its original value.* Suppose gasoline used to cost $1.20 per gallon. The increase in price would be

$$0.15(1.20) = 0.18$$

We add the increase to the old price. The new price is then

$$\underset{\text{Old price}}{1.20} + \underset{\text{Increase}}{0.18} = \underset{\text{New price}}{1.38}$$

Now we will use a clever trick to obtain a formula for calculating percent increase or decrease. We can write the last equation like this:

$$\underset{\text{Old price}}{1(\mathbf{1.20})} + \underset{\text{Increase}}{0.15(\mathbf{1.20})} = \underset{\text{New price}}{1.38}$$

Think of the two terms on the left side as like terms; both terms are multiples of 1.20. We can combine like terms by adding (or subtracting) their coefficients. Thus, the left side is the same as

$$(1 + 0.15)(1.20) = 1.38$$

or just

$$1.15(1.20) = 1.38$$

(You can check the calculations to see that this works.)

In general, we calculate the result of a percent increase by adding the decimal form of the percent to 1, then multiplying that number times the original amount. For example, to calculate an increase of 47%, we multiply the original amount by 1.47. To calculate the result of a percent *decrease,* we *subtract* the decimal form of the percent from 1 and then multiply times the original amount.

EXAMPLE 4  

Health officials estimate that the number of reported cases of influenza declined by 36% after a vigorous immunization program. Albemarle County Hospital treated 650 cases of flu last year. How many cases can they expect this year if they carry out an immunization program?

Solution

We expect a decrease of 36% in the number of flu cases. Therefore, we subtract 0.36 from 1 and multiply the result times last year's caseload. This gives us

$$\begin{aligned}\text{expected cases} &= (1 - 0.36) \times (\text{cases last year})\\ &= 0.64(650)\\ &= 416\end{aligned}$$

The hospital should expect 416 flu cases this year.

Now try Exercise 4.

EXERCISE 4

Horatio and Nelson joined a diet plan at work. Originally, Horatio weighed 210 pounds, and Nelson weighed 190 pounds. After 6 months, Horatio had reduced his weight by 11.4%, and Nelson had reduced his weight by 12.6%.

a. What was Horatio's new weight?

b. What was Nelson's new weight?

c. Who lost more pounds?

ANSWERS TO 8.1B EXERCISES

3a. 16.1%

3b. 18.2%

3c. The train fare increased by a greater percent.

4a. 186 pounds

4b. 166 pounds

4c. Both lost 23.94 (about 24) pounds.

HOMEWORK 8.1B 

Solve. If necessary, round your answers to two decimal places.

1. Senator Fogbank has promised to cut his campaign spending by 28%. If he spent $826,000 on his last election, how much can he spend next time and still keep his promise?

2. Major Motors plans to increase the price of its vehicles by 8% next year. If a Major Motors Meteor costs $16,600 this year, how much will it cost next year?

3. Josh made $25,000 a year at his old job, but he took a 12% paycut when he moved out of town.
   a. What is Josh's new salary?
   b. After 18 months on his new job, Josh got a 12% raise. How much was he making then?

4. Gwen's sporting goods shop made a profit of $36,000 during the first quarter of the year. In the second quarter, profits rose by 8.5%.
   a. What was Gwen's profit for the second quarter?
   b. In the third quarter, profits fell by 8.5%. How much did Gwen's shop make in the third quarter?

5. Earl invested $7500 in the stock market and earned dividends of $1200. Edith invested $4800 and earned dividends of $864.
   a. What percent of his investment did Earl earn in dividends?
   b. What percent of her investment did Edith earn in dividends?
   c. Who earned the better rate of return on the investment?

6. Sulie earned $38,000 last year and paid $13,680 in taxes. Byron earned $34,500 and paid $11,040 in taxes.
   a. What percent of her income did Sulie pay in taxes?
   b. What percent of his income did Byron pay in taxes?
   c. Who paid the greater percent of annual income in taxes?

7. The population of Sterling increased from 5000 to 6200 over the last 5 years. In the same time, the enrollment of the elementary school increased from 460 to 600. Which experienced the greater percent increase in population: the school or the town as a whole?

8. The median income in Clark County increased from $18,000 to $18,600 in 2 years. In the same time, the median price of a house increased from $120,000 to $125,400. Which experienced the greater percent increase: salaries or housing prices?

9. The number of bachelor's degrees awarded in mathematics fell from 15,904 in 1988 to 15,237 in 1989; the number of degrees in the physical sciences fell from 17,806 to 17,204. Which discipline experienced the greater percent decline in degrees awarded?

10. The number of phonograph records sold in the United States declined from 865.7 million in 1990 to 801.0 million in 1991. The number of tape cassettes sold declined from 442.2 million to 360.1 million. Which medium experienced the greater percent decline in sales?

11. A personal computer lists for $1800. The store where Lupe works has the computer on sale for 15% off, and she can apply her employee discount to get 10% off the sale price. Or she can buy the computer through a catalog at 25% off the list price.

a. What will Lupe pay for the computer if she buys it through her store?

b. What will Lupe pay for the computer if she buys it through the catalog?

c. Which is the better deal?

12. Your job pays $500 a week. You have an offer to take a new job that pays 5% more than your present salary, with a 5% raise at the end of the year. Your present boss offers you a 10% raise if you stay.

a. How much would you be making at the new job at the end of the year?

b. How much would you be making at your present job at the end of the year?

c. Which option should you take, if money is the only consideration?

13. Softek sold 8000 copies of its new drafting program in 1990. In 1991 sales rose by 20%, and in 1992 sales rose another 35%.

a. How many copies of the drafting program did Softek sell in 1991? In 1992?

b. Suppose sales of the drafting program had increased by 55% over 2 years. How many copies would have been sold in 1992?

c. Is an increase of 55% the same as an increase of 20% followed by an increase of 35%?

14. Enrollment in foreign-language classes at Forrest College declined by 15% in 1992 and by 20% in 1993. In 1991 there had been 1500 foreign-language students.

a. How many students took foreign languages in 1992? In 1993?

b. Suppose enrollments had declined by 35% over 2 years. How many language students would there have been in 1993?

c. Is a decline of 35% the same as a decline of 15% followed by a decline of 20%?

15. The bar graph in Figure 8.7 shows the change in stock prices of seven companies for the week ending Friday, 19 January 2001.

Figure 8.7 Weekly change in stock price

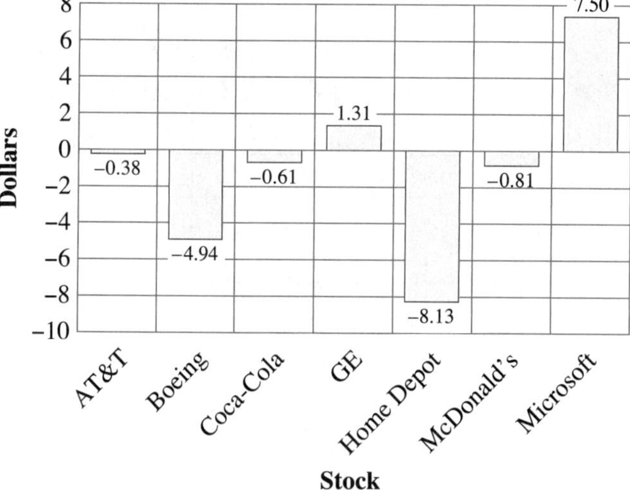

a. Did Coca-Cola stock rise or fall in value?

b. Coca-Cola stock had started the week at $56.62 per share. What was the percent change in Coca-Cola stock?

c. McDonald's had started the week at $33.62 per share. What was the percent change in McDonald's stock?

d. If you had $1000 worth of stock in both Coca-Cola and McDonald's at the start of the week, which investment was worth more at the end of the week?

16. Use Figure 8.7 to answer the following questions.

a. Did AT&T rise or fall in value?

b. AT&T stock had started the week at $24.44 per share. What was the percent change in AT&T stock?

c. If you had $1000 invested in AT&T stock at the beginning of the week, by how many dollars did your investment decrease?

d. Microsoft stock had started the week at $53.50. What was the percent change in its stock?

e. Suppose that you also had $200 invested in Microsoft at the beginning of the week. Was the increase in that investment enough to offset the decrease from AT&T?

17. The population of Silicon City was 20,000 in 1990 and has been growing at a rate of 5% a year since then.

a. Fill in the table (round answers to the nearest whole number):

Year	*Population at start of year*	*Increase in population*	*Population at end of year*
1990	20,000	1000	21,000
1991	21,000	1050	22,050
1992			
1993			
1994			
1995			
1996			
1997			
1998			
1999			
2000			

b. Let t represent the number of years since 1990 (so that $t = 0$ in 1990) and let P represent the population at the start of the year. On the grid provided, make a graph of the data in your table, using t for the independent variable and P for the dependent variable.

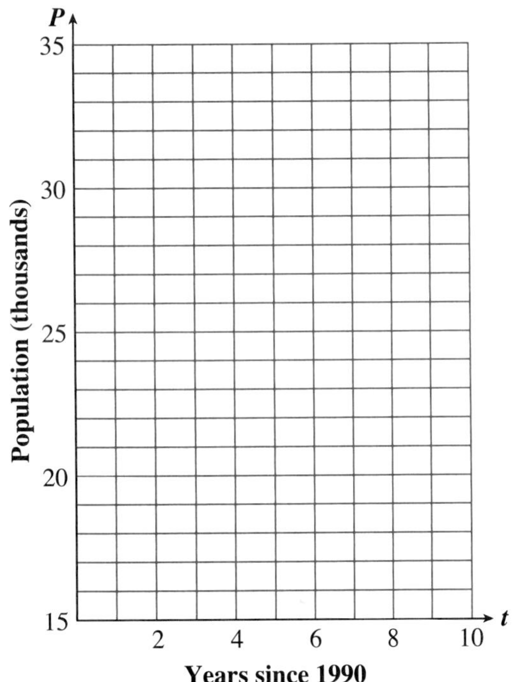

18. The town of Silver Lode had a population of 20,000 in 1890, but its population has been declining by 10% every 10 years since then.

a. Fill in the table (round answers to the nearest whole number):

Year	*Population at start of decade*	*Decrease in population*	*Population at end of decade*
1890	20,000	2000	18,000
1900	18,000	1800	
1910			
1920			
1930			
1940			
1950			
1960			
1970			
1980			
1990			

b. Let t represent the number of years since 1890 (so that $t = 0$ in 1890) and let P represent the population at the start of the year. On the grid provided, make a graph of the data in your table, using t for the independent variable and P for the dependent variable.

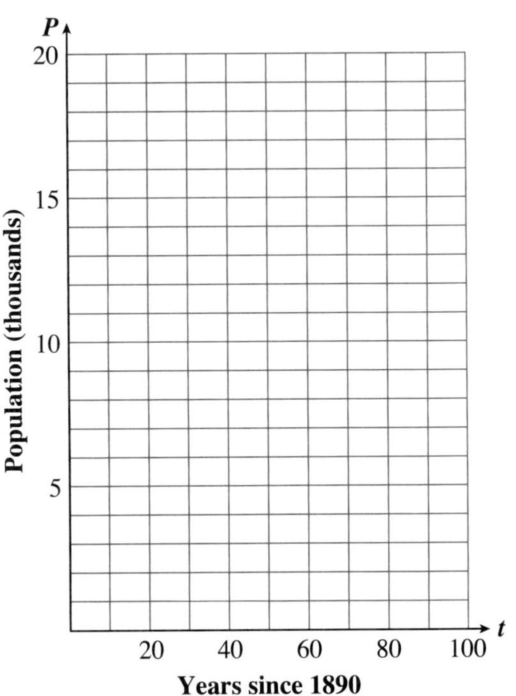

Build the given fraction to an equivalent fraction with the new denominator. (See Lesson 7.1 to review building fractions.)

19. $\frac{7uv}{9} = \frac{?}{9uv}$

20. $\frac{6p}{5} = \frac{?}{5pq^2}$

21. $\frac{c}{6} = \frac{?}{36d^2}$

22. $\frac{f^2}{9} = \frac{?}{900g^2}$

23. $1 = \frac{?}{70uv}$

24. $-1 = \frac{?}{4st}$

25. $\frac{-6}{5} = \frac{?}{15p^2q}$

26. $\frac{3}{8} = \frac{?}{800ab^2}$

27. $\frac{9x}{11y} = \frac{?}{22xy^2}$

28. $\frac{-5k}{4j} = \frac{?}{40j^3k}$

Building Number Skills

Try to answer these questions without using paper and pencil or a calculator.

29. 1 meter is equivalent to 39.37 inches. How many inches are in 10 meters?

30. 1 kilogram is equivalent to 2.205 pounds. How many pounds are in 10 kilograms?

31. Suppose a sandwich costs $4.49. How much will 10 sandwiches cost?

32. Suppose a floppy disk costs $1.19. How much will 10 floppies cost?

33. 1 quart is equivalent to 0.946 liter. How many liters are in 100 quarts?

34. 1 foot is equivalent to 0.3048 meter. How many meters are in 100 feet?

35. A box of graham crackers contains 0.454 kilogram. How many grams is that? (There are 1000 grams in 1 kilogram.)

36. A bag of rice contains 22.7 kilograms. How many grams is that?

37. 1 mile is about 1.6093 kilometers. How many meters is that? (There are 1000 meters in 1 kilometer.)

38. A road sign indicates a distance of 42.2 kilometers to Boston. How many meters is that?

8.2 Ratios and Rates

Ratios

Like percents, ratios give us a way to compare numerical quantities. For example, if we say that the ratio of men to women in the state legislature is 5 to 2, we mean that for every five men in the legislature there are two women. We often display a ratio as a fraction, because their properties are similar. For the example just mentioned, we would write

$$\frac{\text{number of men}}{\text{number of women}} = \frac{5}{2}$$

This does *not* mean that there are exactly five men and two women in the legislature! There could be 100 men and 40 women, or 185 men and 74 women (or many other combinations), because

$$\frac{100}{40} = \frac{185}{74} = \frac{5}{2}$$

EXERCISE 1

On a Saturday evening, a restaurant found that 84 of its customers asked for seating in the nonsmoking section, and 35 customers preferred the smoking section.

a. What was the ratio of smokers to nonsmokers?

b. Write the ratio in simplest form.

EXAMPLE 1

Find three different pairs of numbers whose ratio is $\frac{4}{3}$. (There are many different answers.)

Solution

This is the same as finding three different fractions equivalent to $\frac{4}{3}$. We can find equivalent fractions by reducing or building. Since $\frac{4}{3}$ is already in lowest terms, we will *build* $\frac{4}{3}$ to find equivalent fractions:

Using **2** as the building factor: $\frac{4 \cdot 2}{3 \cdot 2} = \frac{8}{6}$

Using **3** as the building factor: $\frac{4 \cdot 3}{3 \cdot 3} = \frac{12}{9}$

Using **4** as the building factor: $\frac{4 \cdot 4}{3 \cdot 4} = \frac{16}{12}$

We can use any number we like (except zero) as the building factor. Besides the numbers 4 and 3, we have found the following pairs with the desired ratio:

8 and 6, 12 and 9, 16 and 12

All the ratios $\frac{4}{3}$, $\frac{8}{6}$, and $\frac{12}{9}$ are equivalent, but we write ratios in simplest or reduced form whenever possible.

Now try Exercise 1.

In Exercise 1 we could have computed the ratio of nonsmokers to smokers. This ratio is the reciprocal of the ratio you computed in Exercise 1—namely, $\frac{84}{35}$, or $\frac{12}{5}$. Either ratio can be useful in applications; so it is important to know which quantity corresponds to the numerator and which to the denominator. When we write the ratio of a to b as a fraction, a is the numerator, and b is the denominator.

The **ratio of *a* to *b*** is written $\frac{a}{b}$.

Decimal Form for Ratios

Any ratio can be expressed as a decimal instead of a common fraction. For example, since $\frac{12}{5}$ is equal to 2.4, we might say that the ratio of nonsmokers to smokers in Exercise 1 is 2.4 to 1. We usually express a ratio as a decimal fraction if the numbers being compared are decimal numbers.

EXAMPLE 2

Major Motors budgeted \$5.6 million for research and development (R&D) next year and \$3.5 million for advertising. What is the ratio of the amount budgeted for advertising to the amount budgeted for R&D?

Solution

The ratio of amount budgeted for advertising to amount budgeted for R&D is $\frac{3.5}{5.6}$, or 0.625. (You might also recognize this fraction as $\frac{5}{8}$.)

Now try Exercise 2.

EXERCISE 2

In 1999, 938.9 million compact discs (CDs) were sold, and 123.6 million cassette tapes were sold. What is the ratio of CDs to tapes sold in 1999? Round your answer to the nearest tenth.

(Source: Recording Industry Association of America)

Rates

We use ratios to compare two quantities expressed in the same units, so the ratio itself has no units. For example, if we measure a typical dormitory room, we might find that the ratio of its length to its width is

$$\frac{15 \text{ feet}}{12 \text{ feet}}, \quad \text{or} \quad \frac{15 \text{ feet}}{12 \text{ feet}} = \frac{5}{4}$$

We would say that the ratio of length to width is 1.25 or 5 to 4, but not 5 feet to 4 feet.

We use **rates** to compare quantities that have different units. You are familiar with many rates already. You might hear that a certain car gets gas mileage of 28 miles per gallon or that bananas cost 55 cents per pound. We know that *per* indicates division, so a rate is actually a type of ratio. Thus,

$$28 \text{ miles per gallon} \quad \text{means} \quad \frac{28 \text{ miles}}{1 \text{ gallon}},$$

and

$$55 \text{ cents per pound} \quad \text{means} \quad \frac{55 \text{ cents}}{1 \text{ pound}}.$$

EXAMPLE 3

Armand made \$144 for 15 hours of work last week. Give his rate of pay first as a ratio and then as a rate in dollars per hour.

Solution

As a ratio, Armand's rate of pay is $\frac{144 \text{ dollars}}{15 \text{ hours}}$. We simplify the ratio by dividing the denominator into the numerator, to get $\frac{9.6 \text{ dollars}}{1 \text{ hour}}$, or 9.6 dollars per hour, or \$9.60 per hour.

Thus, a rate is just a ratio of quantities with different units.

Now try Exercise 3.

EXERCISE 3

Beaverhead is the largest county in Montana, with an area of 5543 square miles. The population of Beaverhead County is 8790. What is the population density, in people per square mile? Round your answer to tenths.

ANSWERS TO 8.2 EXERCISES

1a. $\frac{35}{84}$

1b. $\frac{5}{12}$

2. 7.6

3. 1.6 people per square mile

HOMEWORK 8.2 

1. For every three sailboats registered to use the lake, there are four motorboats.
   a. What is the ratio of sailboats to motorboats?
   b. What is the ratio of motorboats to sailboats?
   c. What fraction of all the boats registered are sailboats? What fraction are motorboats?
   d. If there are 36 sailboats registered, how many motorboats are registered?
   e. If there are 60 motorboats registered, how many sailboats are registered?

2. For every nine girls in a Brownie troop, there must be two adults.
   a. What is the ratio of girls to adults?
   b. What is the ratio of adults to girls?
   c. What fraction of all Brownies are adults? What fraction are girls?
   d. If there are 45 girls who want to be Brownies, how many adults are needed?
   e. If 12 adults are willing to be Brownie leaders, how many girls can be Brownies?

3. Find three pairs of numbers whose ratio is $\frac{5}{8}$.

4. Find three pairs of numbers whose ratio is $\frac{7}{2}$.

5. If there are more sheep than goats, what can you say about the ratio of sheep to goats?

6. A rate is just a ratio of quantities with different ________________.

Write each ratio as a fraction in simplest form.

7. Cary's math book has 248 pages of text and 186 pages of exercises. What is the ratio of text to exercises in the book?

8. Emmet spends \$720 a month for housing and \$500 a month for food. What is the ratio of his food expenses to his housing expenses?

9. A typical American diet includes 960 calories per day from carbohydrates, 240 calories from protein, and 840 calories from fat.

a. What is the ratio of calories from fat to calories from protein?

b. What is the ratio of calories from protein to calories from carbohydrates?

10. A recommended low-fat diet includes 1190 calories per day from carbohydrates, 170 calories from protein, and 340 calories from fat.

a. What is the ratio of calories from fat to calories from protein?

b. What is the ratio of calories from fat to calories from carbohydrates?

11. A recipe for cranberry sangria calls for, among other ingredients, 48 ounces of cranberry juice and 32 ounces of red wine. You would like to make a smaller batch of cranberry sangria, so you need to know the ratio of cranberry juice to red wine. What is that ratio?

12. A recipe in a German cookbook calls for 220 grams of sugar and 385 grams of flour. What is the ratio of flour to sugar needed in the recipe?

13. The quality-control department at Lighten Up finds that a shipment of 4500 light bulbs includes 54 defective bulbs.

a. What is the ratio of defective bulbs to the total number of bulbs?

b. What is the ratio of good bulbs to defective bulbs?

c. What percent of the bulbs are defective?

14. Village Miller distributed 12,000 boxes of Oat Toasties last month, and 15 of the boxes contain a certificate for breakfast in Paris.

a. What is the ratio of certificates to the total number of boxes?

b. What is the ratio of plain boxes to prize boxes?

c. What percent of the boxes contain certificates?

Write each ratio as a decimal fraction. If necessary, round your answers to three decimal places.

15. The average American eighth-grader spends 5.6 hours per week on homework and 21.4 hours per week watching TV. What is the ratio of time spent on homework to time spent watching TV?

16. The average American man spends 26.7 hours per week watching TV and 7.2 hours per week eating. What is the ratio of time spent eating to time spent watching TV?

17. On a scale drawing of a new public library, the length of the lobby is 14.4 centimeters, and its width is 9 centimeters. The ratio of the width to the length will be the same for the actual lobby. What is that ratio?

18. A scale model of Big Ben is 20 centimeters tall, and the clock face is 1.5 centimeters wide. What is the ratio of the width of the clock face to the height of the tower?

19. The rectangle in Figure 8.8 is called a Golden Rectangle, because the ratio of its length to its width is the Golden Ratio. This ratio occurs frequently in nature, and rectangles whose dimensions obey the Golden Ratio are supposed to be most pleasing in appearance. If the dimensions of the rectangle are 7 centimeters and 11.2 centimeters, calculate the Golden Ratio.

Figure 8.8

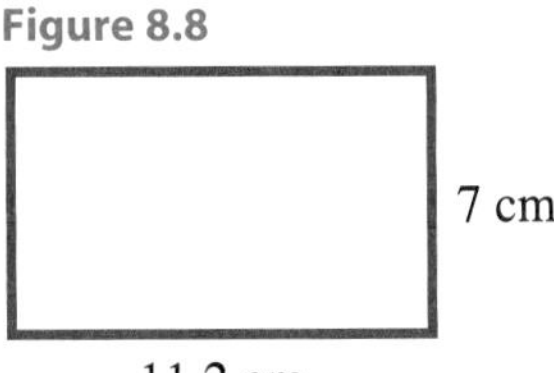

20. Measure your height and the height from the floor to your navel. For most people, the ratio of their height to their navel height is approximately equal to the Golden Ratio. Compute the ratio for your own heights. Is the result close to the value you calculated in Problem 19?

21. The Crane County commissioners have budgeted $24.9 million for education and $41.5 million for law enforcement. What is the ratio of funds budgeted for education to funds budgeted for law enforcement?

22. In 1989 Americans spent $257.8 billion on clothing and $483.5 billion on medical care. Compute the ratio, to the nearest thousandth, of amount spent on clothing to amount spent on medical care.

Write a rate to describe each situation. If necessary, round your answers to three decimal places.

23. An experimental commuter minivan can travel 432 miles on a 12-gallon tank of gas. What is its fuel efficiency, in miles per gallon?

24. Maryellen's family used 1044 kilowatt-hours of electricity during the last billing period, which was for 58 days. What was their rate of energy consumption, in kilowatt-hours per day?

25. Liesl jogs 5 miles in 40 minutes. What is her speed in minutes per mile? In miles per minute?

26. Nabil read 60 pages in 1 hour and 32 minutes. What was his reading speed in minutes per page? In pages per minute?

27. Harry tiled a 12-foot × 15-foot kitchen in $2\frac{1}{2}$ hours. How fast did he work, in square feet per hour?

28. Rita worked on a needlework chair seat for a total of 32 hours. If the finished piece measures 15 inches × 16 inches, how fast did Rita work, in square inches per hour?

29. Maurice paid $29.82, including sales tax, for a $28 sweater.

a. What was the sales tax rate, in cents per dollar?

b. Express the sales tax rate as a percentage.

30. Nicole paid $8.40 to fill her car with gasoline. Of that amount, $0.90 was for taxes.

a. What was the tax rate, in cents per dollar?

b. Express the tax rate as a percentage.

Find the product or power. Reduce your answers to lowest terms. (See Lesson 7.2 to review multiplying fractions.)

31. $\frac{10b}{5c} \cdot \frac{3c^2}{6bc}$

32. $\frac{14m^2}{2mn} \cdot \frac{3n}{21m}$

33. $\frac{2}{15a} \cdot \frac{4a}{3b} \cdot \frac{9b^2}{8}$

34. $\frac{18u}{7v^2} \cdot \frac{v}{9} \cdot \frac{21v}{u^3}$

35. $\frac{-p^3}{16q^4} \cdot \frac{28q}{3p} \cdot \frac{-6q^2}{7p}$

36. $\frac{12h}{5k^3} \cdot \frac{40k^2}{3} \cdot \frac{-3}{4k}$

37. $\left(\frac{-2r}{s}\right)^3$

38. $\left(\frac{-i}{3j}\right)^3$

Building Number Skills

Compute mentally. Do not use paper and pencil or a calculator.

39. 237 pounds of topsoil are to be spread evenly over an area of 10 square feet. How many pounds of topsoil will be on each square foot?

40. A recipe calls for 575 grams of ground beef. If the recipe makes ten servings, approximately how much ground beef is in each serving?

41. A roll of paper is 10 inches wide. How long is the roll if it has an area of 12,480 square inches?

42. The total weight of the ten hockey players on the ice is 1893 pounds. What is the average weight of the players?

43. A coin falls 490 centimeters in 1 second. How many meters did it fall? (There are 100 centimeters in 1 meter.)

44. Gerhard is 186 centimeters tall. How many meters is that?

45. 1 quart is approximately 946 milliliters. How many liters is that? (There are 1000 milliliters in 1 liter.)

46. A recipe calls for 15 milliliters of melted butter. How many liters is that?

47. How long will it take to move a distance of 938 feet when traveling at a constant speed of 100 feet per second?

48. How long will it take an iceberg to move 1.2 centimeters if it moves at a rate of 100 millimeters per year?

8.3 Proportions

A. Proportional Variables

Ratios can be useful in studying the relationship between two variables.

EXAMPLE 1

The table shows the price of gasoline. Fill in the ratios in the third column of the table.

Gallons of gasoline	*Total price*	$\frac{\textit{Price}}{\textit{Gallons}}$
4	5.20	$\frac{5.20}{4} = ?$
6	7.80	$\frac{7.80}{6} = ?$
8	10.40	?
12	15.60	?
14	18.20	?

The two variables in Example 1 are the *number of gallons* of gasoline purchased and the *total price.* The ratio $\frac{\text{total price}}{\text{number of gallons}}$, or price per gallon, is the same for each pair of values in the table. This agrees with common sense: The price per gallon of gasoline is the same no matter how many gallons you buy.

EXAMPLE 2

The table shows the population of a town. *Years* means the age of the town, or the number of years since the town was established. *Population,* of course, means the number of people who live in the town. Fill in the ratios in the third column of the table.

Years	*Population*	$\frac{\textit{People}}{\textit{Year}}$
10	432	$\frac{432}{10} = ?$
20	932	$\frac{932}{20} = ?$
30	2,013	?
40	4,345	?
50	9,380	?
60	20,251	?

The ratios in Example 2, $\dfrac{\text{number of people}}{\text{number of years}}$, give the rate of growth of the population in people per year. You can see from the table that this ratio is *not* constant; in fact, it increases as time goes on. The population of the town does not increase at a constant rate.

Two quantities are said to be **proportional** if their ratios are always equal. In Example 1, we see that the price of gasoline is proportional to the number of gallons you buy. On the other hand, in Example 2, the population of the town is *not* proportional to its age. The graphs of these two variable relationships are shown in Figure 8.9.

Figure 8.9

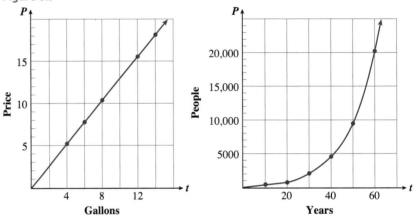

How can we decide if two variables are proportional? First, we compute several pairs of corresponding values for the variables and then we compute their ratios.

EXAMPLE 3  

Tuition at Woodrow University is $400 plus $30 per unit. Is the tuition proportional to the number of units you take?

Solution

Make a table showing several pairs of variables and compute their ratios:

Number of units	*Tuition*	$\dfrac{\textit{Tuition}}{\textit{Units}}$
6	580	96.67
8	640	80
15	850	56.67

The ratio tuition/units, or tuition per unit, is not constant, so the tuition is *not* proportional to the number of units.

Now try Exercise 1.

EXERCISE 1

Anouk is traveling by train across Alaska at 60 miles per hour. Fill in the ratios in the table. Is the distance Anouk has traveled proportional to the time she has traveled?

Hours	*Miles*	$\dfrac{\textit{Miles}}{\textit{Hours}}$
3	180	
5	300	
8	480	

In Example 1, we used the following fact to decide whether the variables were proportional.

Two variables are proportional if the ratio of related values is constant—that is, if the ratio is always the same.

We can write this in mathematical language as

y is proportional to x if $\frac{y}{x} = k$, where k is a constant.

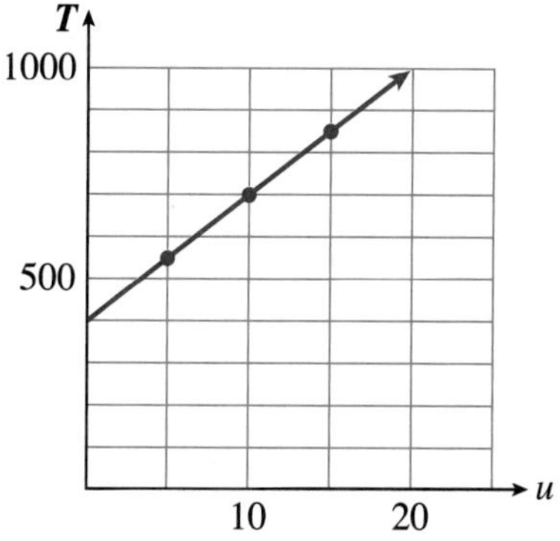

Figure 8.10

Graphs of Proportional Variables

Can we decide by looking at their graph if two variables are proportional? After studying Figure 8.9 again you may have a conjecture, or guess, about this. But before you decide, let's graph the relationship in Example 3.

EXAMPLE 4

Write an equation relating the variables in Example 3, and graph the equation.

Solution

Let u represent the number of units you enroll in and let T represent the amount of tuition you will pay. You will pay \$400 to start, plus \$30 times the number of units. Thus,

$$T = 400 + 30u$$

Use the table in Example 3, or find a few more pairs of values from the equation, and plot points. The graph is shown in Figure 8.10.

Now try Exercise 2.

From the preceding examples and exercises, what can we say about the graph of two proportional variables? It is true that the graph will be a straight line, but we must be more specific. The graph of tuition is linear, even though tuition is not proportional to the number of units. The variables in Exercise 2, distance traveled and time elapsed, *are* proportional, and their graph *passes through the origin.* We have discovered the following fact:

Two variables are proportional if their graph is a straight line that passes through the origin.

EXERCISE 2

a. Write an equation relating the variables in Exercise 1. Let t represent the number of hours Anouk has traveled and let d represent the distance she has traveled, in miles.

b. Graph the equation. Use the table in Exercise 1 and plot points on the grid in Figure 8.11.

Figure 8.11

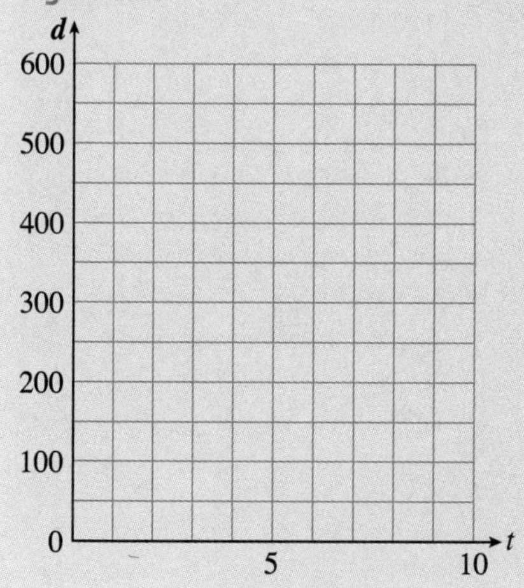

Equations Relating Proportional Variables

You may have noticed in Exercise 2 that the number of miles traveled is always 60 times the number of hours traveled. The distance is a constant multiple of the time. When two variables are related in this way their equation has the form

$$y = kx$$

where k is a constant. The graph of such an equation will always be a straight line that passes through the origin. Thus, we have another fact about proportional variables:

y is proportional to x if y is a constant multiple of x.

Stating this in another way, we see that we can always write an equation that relates two proportional variables:

If y is proportional to x then $y = kx$.

EXAMPLE 5

Do you think the following pairs of variables are proportional?

a. Your age and your height

b. The amount of a purchase and the amount of sales tax

Solution

a. Unless you are very unusual, your height is *not* proportional to your age and probably wasn't even when you were a growing child. Suppose you were 2 feet tall at the age of 1 year. If the ratio of your height to your age were a constant, we would have

$$\frac{2 \text{ feet}}{1 \text{ year}} = \frac{4 \text{ feet}}{2 \text{ years}} = \frac{6 \text{ feet}}{3 \text{ years}} \qquad \text{and so on.}$$

This is not a very realistic growth pattern.

b. Sales tax is usually expressed as a rate or as a percent. For instance, the sales tax in your state might be 5 cents on the dollar, or 5%. Thus the sales tax *is* a constant multiple of the purchase amount:

$$\text{sales tax} = 0.05(\text{purchase amount})$$

so the sales tax is proportional to the amount of the purchase.

Now try Exercise 3.

The relationship between proportional variables is a very simple and useful one. When we change one variable, we know exactly what will happen to the other variable. If you double the amount of gasoline that you buy—say, from 4 gallons to 8 gallons—the price will also double. On the other hand, when your age doubles, it is unlikely that your height will double also.

EXERCISE 3

You are making chocolate chip cookies from your grandmother's recipe.

a. Is the number of chocolate chips you need proportional to the number of cookies you bake?

b. Is the baking time for the cookies proportional to the number of cookies on the cookie sheet?

Problem Solving

We can use these observations to solve problems about proportional variables.

EXAMPLE 6 

Under normal driving conditions, the distance you drive in your car is proportional to the number of gallons of gasoline you use. Suppose Myrna used 6 gallons of gas to drive 159 miles.

a. Write an equation relating the distance Myrna drove to the amount of gas used.

b. How many gallons of gas will she need to drive 1000 miles?

Solution

a. Let x represent the number of gallons of gas used and let y represent the number of miles driven. Then y is proportional to x, and we can write

$$y = kx$$

We find the value of k by using one pair of related values for the variables. Substitute 6 for x and 159 for y, and solve for k:

$$159 = k(6) \quad \text{Divide both sides by 6.}$$

$$\frac{159}{6} = k$$

$$26.5 = k$$

Thus, $y = 26.5x$ is an equation relating the two variables.

b. If we now substitute 1000 for y, we can solve for the related value of x:

$$1000 = 26.5x \quad \text{Divide both sides by 26.5.}$$

$$\frac{1000}{26.5} = x$$

$$37.74 \approx x$$

Myrna will need approximately 38 gallons of gas in order to drive 1000 miles.

Now try Exercise 4.

EXERCISE 4

A 6-inch turkey submarine sandwich contains 282 calories.

a. Write an equation relating the number of calories, y, to the length of the sandwich, x.

b. How many calories are there in a 13-inch turkey sandwich?

ANSWERS TO 8.3A EXERCISES

1.

Hours	*Miles*	*Miles/Hours*
3	180	60
5	300	60
8	480	60

Yes

2a. $d = 60t$

2b.

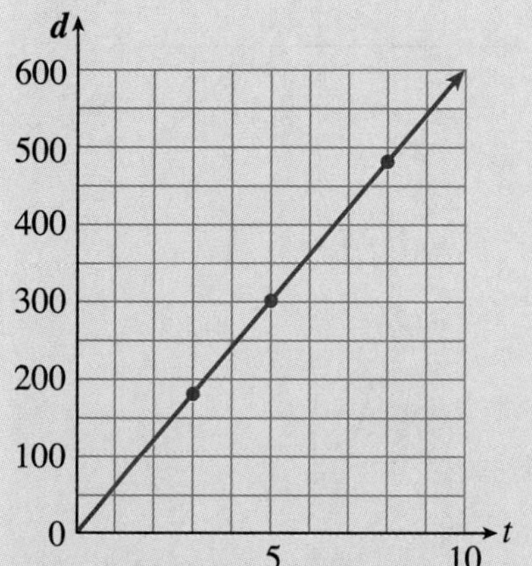

3a. Yes

3b. No

4a. $y = 47x$

4b. 611 calories

HOMEWORK 8.3A

1. If two variables are proportional, what is true about their ratio?

2. If two variables are proportional, what does their graph look like?

In problems 3–22, decide whether the two variables are proportional.

3.

h	k
3	9
6	18
8	24
12	36

4.

P	Q
8	2
20	5
28	7
52	13

5.

w	z
6	0
10	2
16	6
18	7

6.

C	R
1	15
3	35
6	65
12	125

7.

Y	V
2	4
3	9
5	25
8	64

8.

n	s
9	3
16	4
49	7
121	11

9.

H	t
2	0.5
4	0.25
5	0.2
10	0.1

10.

B	d
3	0.6
4	0.8
6	1.2
9	1.8

11.
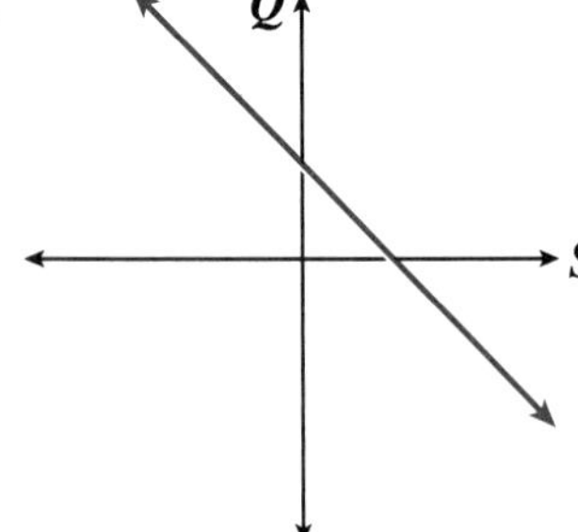

12.
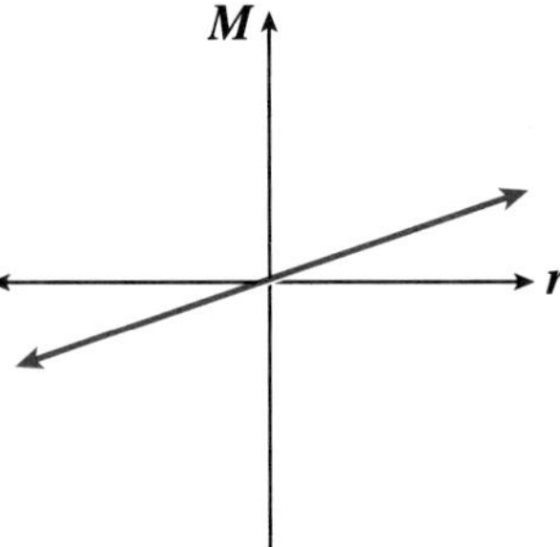

13.
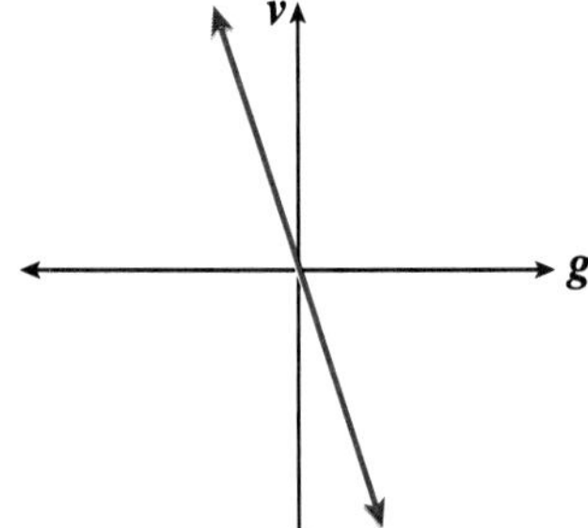

14.
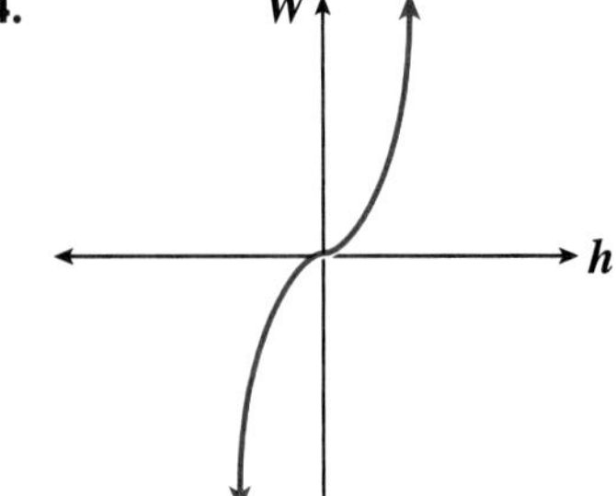

15. $f = 5.8 - d$

16. $t = 1.4q$

17. $S = \dfrac{13m}{7}$

18. $L = 12v^2$

19. To make Classic Pop, you need 3 ounces of Secret Syrup for every 32 ounces of carbonated water. Is the amount of Secret Syrup proportional to the amount of carbonated water?

20. To brew coffee, you put in one scoop of coffee for every 6 ounces of water, plus one scoop for the pot. Is the amount of coffee needed proportional to the amount of water used?

21. The thickness of a book in millimeters can be found by adding 6 millimeters (for the thickness of the covers) to 1 millimeter for every 25 pages. Is the thickness of the book proportional to the number of pages?

22. A 4-liter can of paint will cover 150 square feet of primed surface. Is the amount of paint you need proportional to the area you have to paint?

For problems 23–26:
(a) Write an equation relating the variables, (b) find several points, record them in the table, and graph your equation on the grid provided. (c) Use your equation to answer the question and verify on the graph.

23. The distance d traveled at a constant speed is proportional to the time taken t. A research submarine traveled 45 miles in 3 hours. How long will it take them to travel 300 miles?

t	d

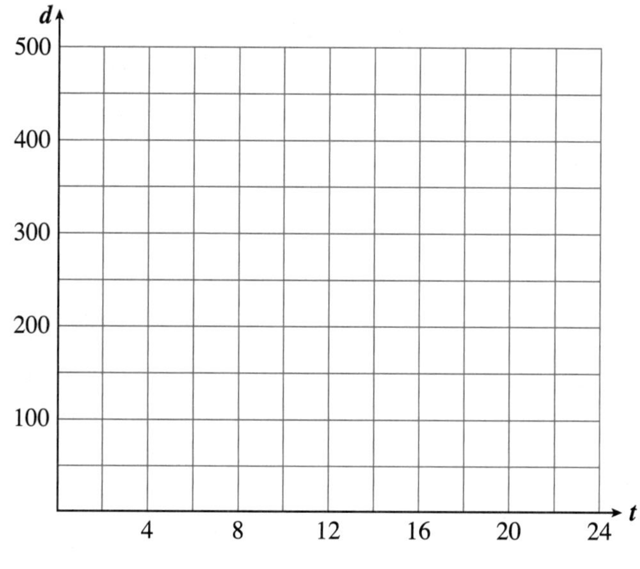

24. Megan's wages w are proportional to the time h she works. Megan made $60 for 8 hours work as a lab assistant. How long will it take her to make $500?

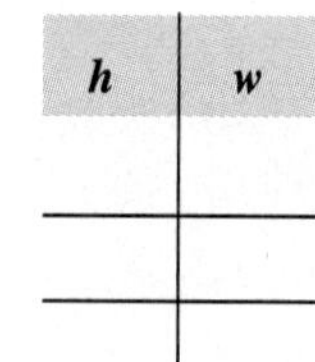

h	w

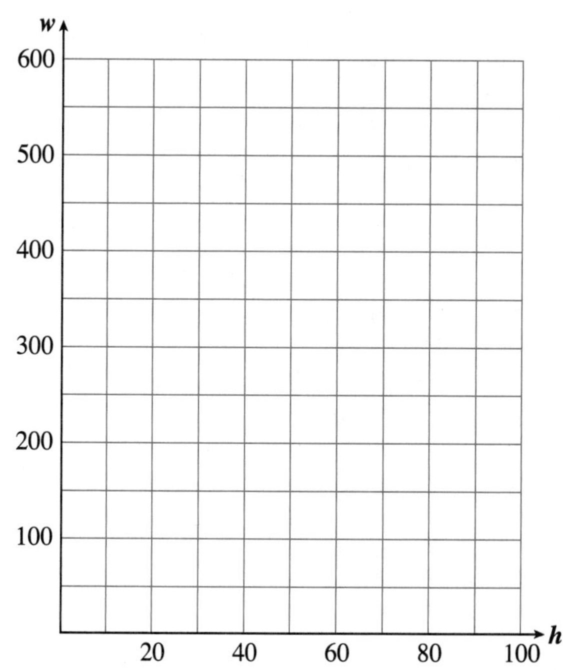

25. The cost C of a turkey is proportional to its weight w. A 16-pound turkey costs \$12.80. How much does a 22-pound turkey cost?

w	C

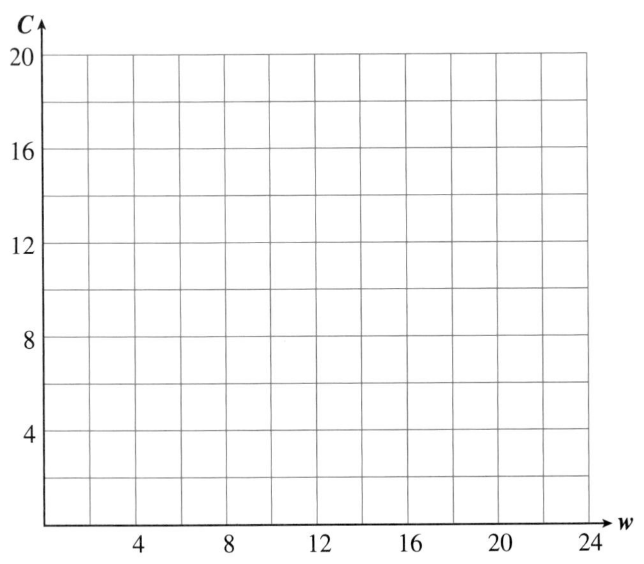

26. The amount of pure copper c produced by Copperfield Mine, is proportional to the amount of ore extracted g. A new lode produced 24 grams of copper from 800 grams of ore. How many grams of copper will 1 kilogram of ore yield?

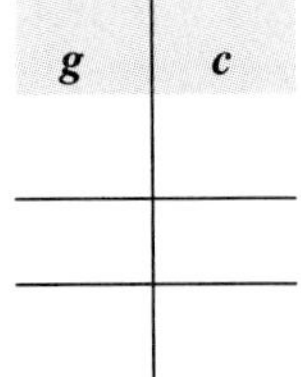

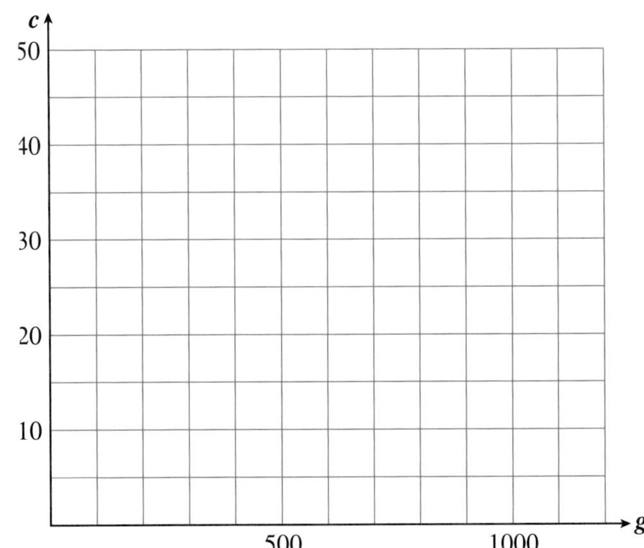

27. Trish's real estate commission is proportional to the amount of the sale. She made a \$3200 commission on a sale of \$80,000.

a. Write an equation relating Trish's sales and her commission.

b. What would be her commission on the sale of a \$200,000 property?

28. Property taxes are proportional to the value of the property. Gwen paid \$3000 in property tax on her summer home, which is valued at \$120,000.

a. Write an equation relating the property tax to the value of the property.

b. How much will Gareth pay on his cabin, which was assessed at \$50,000?

29. a. State a formula for the perimeter of a square in terms of the length of its side.

b. Complete the table showing the perimeters for squares with various sides:

s	2	5	8	11
P				

c. Graph your equation, using the values in your table.

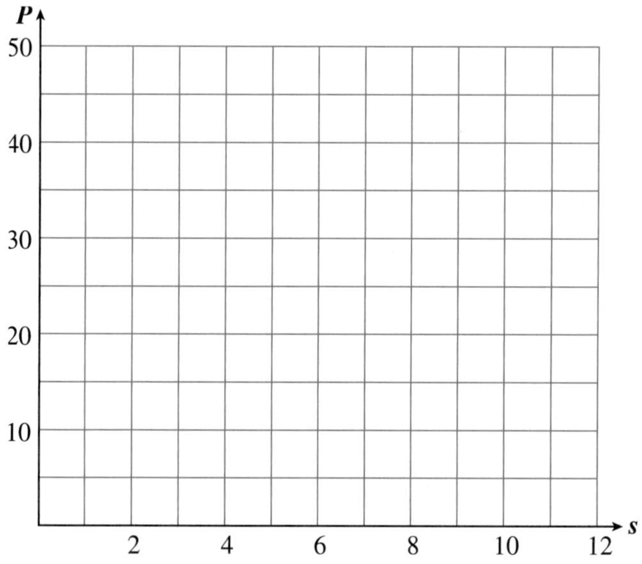

d. Is the perimeter of a square proportional to the length of its side?

30. a. State a formula for the circumference of a circle in terms of its radius.

b. Complete the table showing the circumference for circles with various radii:

r	2	5	7	10
C				

c. Graph your equation, using the values in your table.

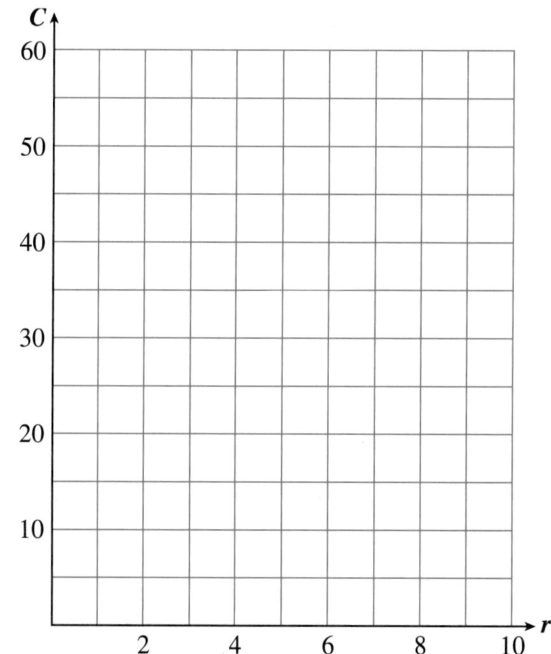

d. Is the circumference of a circle proportional to its radius?

31. a. State a formula for the area of a square in terms of the length of its side.

b. Complete the table showing the area for squares of various sides:

s	2	5	6	8
A				

c. Graph your equation, using the values in your table.

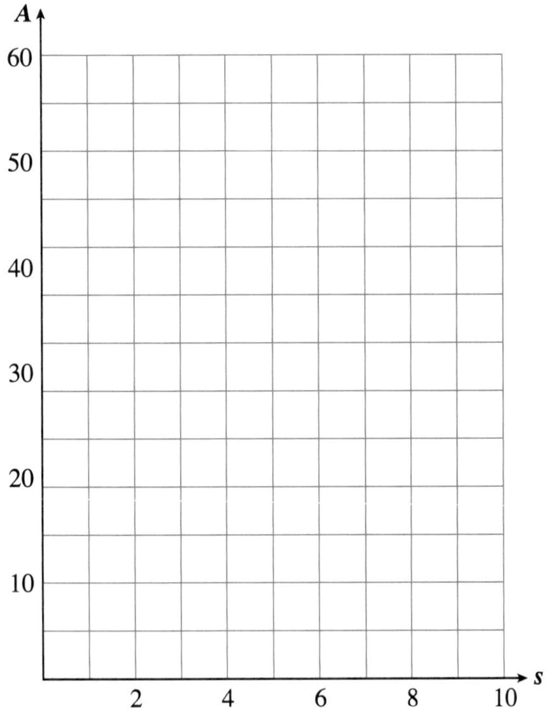

d. Is the area of a square proportional to the length of its side?

32. a. State a formula for the area of a circle in terms of its radius.

b. Complete the table showing the area for circles of various radii:

r	2	3	4	5
A				

c. Graph your equation, using the values in your table.

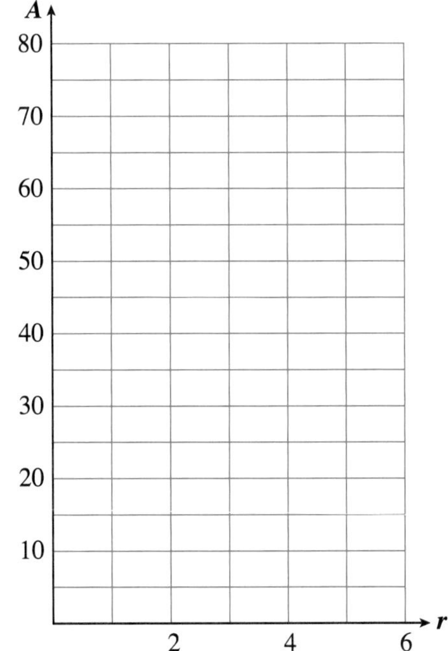

d. Is the area of a circle proportional to its radius?

Divide. Reduce your answers to lowest terms. (See Lesson 7.2B to review dividing fractions.)

33. $\frac{-s^2w}{30s} \div \frac{15sw^3}{18}$

34. $\frac{-24bn}{5b^3} \div \frac{-72n^2}{45b}$

35. $\frac{8a^3}{12h^3} \div \frac{56a^2h^4}{3a^4h^2}$

36. $\frac{6t^3}{27u^2} \div \frac{10t^2}{9ut^3}$

37. $\frac{a}{b} \div \frac{c}{d} \div \frac{e}{f}$

38. $\frac{a}{b} \div \frac{c}{d} \cdot \frac{e}{f}$

39. $\frac{a}{b} \div \left(\frac{c}{d} \div \frac{e}{f}\right)$

40. $\frac{a}{b} \div \left(\frac{c}{d} \cdot \frac{e}{f}\right)$

Building Number Skills

Some products involving decimals can be calculated quickly by converting the decimal to a common fraction.

Example: $0.25(16) = \frac{1}{4}(16) = \frac{16}{4} = 4$

Find each product, without using a calculator, by converting the decimal to a common fraction.

41. 0.5(18)

42. 0.2(15)

43. 0.125(24)

44. 0.25(32)

45. 0.6(25)

46. 1.5(6)

47. 0.75(12)

48. 0.375(16)

49. 1.25(16)

50. 1.625(40)

B. Solving Proportions

A **proportion** is a statement that two ratios are equal. In other words, *a proportion is a type of equation,* one in which both sides are ratios. For example, the equations

$$\frac{10}{15} = \frac{14}{21} \quad \text{and} \quad \frac{3.8}{5} = \frac{26.6}{35}$$

are proportions.

Like any other equation, a proportion may involve variables. Here is an example:

$$\frac{21}{x} = \frac{27}{72}$$

Can we find a value for x that makes this equation true? To clear the fractions from the equation, we first multiply both sides by the LCD, $72x$:

$$\mathbf{72x}\left(\frac{21}{x}\right) = \mathbf{72x}\left(\frac{27}{72}\right)$$
$$72(21) = 27x$$

Now we proceed as usual. First, simplify the left side:

$$1512 = 27x \quad \text{Divide both sides by 27.}$$
$$\frac{1512}{\mathbf{27}} = \frac{27x}{\mathbf{27}}$$
$$56 = x$$

We can check that 56 is in fact a solution by noticing that $\frac{21}{56}$ and $\frac{27}{72}$ both reduce to $\frac{3}{8}$.

Cross-Multiplying

We can use a special technique to simplify the solutions of proportions. In the example above, we multiplied both sides of the original proportion by $72x$. You may notice an interesting shortcut for clearing the fractions. Look at the original proportion and imagine multiplying across the equal sign, as shown by the arrows:

$$\frac{21}{x} = \frac{27}{72} \quad \begin{array}{l}\text{Multiply 21 times 72.}\\ \text{Multiply } x \text{ times 27.}\end{array}$$

Write one product on each side of the equal sign. This gives the same equation we found by clearing fractions:

$$72(21) = 27x$$

and we can then complete the solution.

The shortcut we noticed for this problem is a fundamental property of proportions.

Property of Proportions

If $\dfrac{a}{b} = \dfrac{c}{d}$, then $ad = bc$.

This cross-multiplying shortcut allows us to clear fractions from a proportion in one step. Be careful, though: You can use cross-multiplying *only* when you are solving a proportion!

EXAMPLE 7

Solve the proportion

$$\frac{\frac{4}{9}}{\frac{2}{3}} = \frac{4}{x}$$

by using the property of proportions.

Solution

By cross-multiplying we have

$$\frac{4}{9}x = \frac{2}{3}(4)$$

and solving this equation we find

$$\frac{4}{9}x = \frac{8}{3} \qquad \text{Divide both sides by } \frac{4}{9}.$$

$$x = \frac{8}{3} \div \frac{4}{9} \qquad \text{Multiply by the reciprocal of } \frac{4}{9}.$$

$$= \frac{8}{3} \cdot \frac{9}{4} = 6$$

The solution is 6.

Now try Exercise 5.

EXERCISE 5

Solve $\frac{4}{1.6} = \frac{9}{x}$

Problem Solving with Proportions

If two variables are proportional, the ratio of their corresponding values is always the same. For example, if the number of miles you drive is proportional to the gallons of gas you use, then the ratio $\frac{\text{gallons}}{\text{miles}}$ is a constant. We can compute the ratios of different pairs of values, and they will be equal. We can then write a proportion for the two ratios. This is true even if one of the values is unknown.

Here is a way to solve Example 6, using a proportion.

EXAMPLE 8

Under normal driving conditions, the distance you drive in your car is proportional to the number of gallons of gasoline you use. Suppose Myrna used 6 gallons of gas to drive 159 miles. How many gallons of gas will she need to drive 1000 miles?

Solution

Step 1a What are we asked to find?

The amount of gas needed to drive 1000 miles

Step 1b Choose a variable to represent the unknown quantity:

Number of gallons needed: *g*

EXERCISE 6

Use a proportion to solve Exercise 4: If a 6-inch turkey submarine sandwich contains 282 calories, how many calories are in a 13-inch turkey sandwich?

Step 1 Choose a variable for the unknown quantity.

Step 2 Write a proportion for the ratio $\frac{\text{calories}}{\text{inches}}$.

Step 3 Solve the proportion.

ANSWERS TO 8.3B EXERCISES

5. 3.6

6. $\frac{282}{6} = \frac{x}{13}$, 611 calories

Step 2a Write two ratios for the quantities in the problem, one using known values and the second using the unknown value:

Use the ratio of gallons of gas used to miles driven.

First ratio: $\frac{6 \text{ gallons}}{159 \text{ miles}}$ Second ratio: $\frac{g \text{ gallons}}{1000 \text{ miles}}$

Step 2b Write an equation using the ratios from Step 2a. You should omit the units from the equation itself, but it is a good idea to write down the ratio of units as a guide:

$$\frac{\textbf{gallons}}{\textbf{miles}}: \frac{\mathbf{6}}{\mathbf{159}} = \frac{g}{\mathbf{1000}}$$

Step 3a Solve the equation:

$$\frac{6}{159} = \frac{g}{1000} \quad \text{Multiply both sides by 1000.}$$

$$(\mathbf{1000})\frac{6}{159} = \frac{g}{1000}(\mathbf{1000}) \quad \text{Simplify.}$$

$$37.74 \approx g$$

Step 3b Answer the question in the problem:

Myrna will need approximately 38 gallons of gas to drive 1000 miles.

CAUTION!

When using a proportion, make sure that both ratios have the same units in numerator and denominator. In Example 8, it would be incorrect to write

$$\frac{159 \text{ miles}}{6 \text{ gallons}} = \frac{g \text{ gallons}}{1000 \text{ miles}} \quad \leftarrow \textit{Incorrect!}$$

Now try Exercise 6.

HOMEWORK 8.3B

1. What is a proportion?

2. State the property of proportions.

Which of the expressions in Problems 3–10 are proportions?

3. $\frac{3}{x} = \frac{8}{15}$

4. $\frac{7}{3} = \frac{20}{z}$

5. $\frac{v}{9} - \frac{12}{5}$

6. $\frac{1}{8} + \frac{t}{3}$

7. $\frac{w}{6} + \frac{2}{5} = \frac{3}{2}$

8. $\frac{p}{4} = \frac{9}{4} - \frac{4}{5}$

9. $\frac{2.6}{m} = \frac{m}{1.6}$

10. $\frac{g}{5.4} = \frac{9.3}{g}$

Use the property of proportions to decide whether each pair of fractions is equal.

11. $\frac{85}{51} = \frac{97}{55}$

12. $\frac{234}{73} = \frac{223}{68}$

13. $\frac{10.4}{6.5} = \frac{31.2}{19.5}$

14. $\frac{1.52}{5.32} = \frac{2.02}{7.07}$

Solve each proportion.

15. $\frac{12}{21} = \frac{28}{n}$

16. $\frac{104}{39} = \frac{m}{15}$

17. $\frac{b}{7} = \frac{9}{5}$

18. $\frac{3}{r} = \frac{4}{11}$

19. $\frac{3.6}{8} = \frac{y}{10}$

20. $\frac{h}{6} = \frac{5.3}{15}$

21. $\frac{\frac{7}{3}}{3} = \frac{z}{\frac{9}{2}}$

22. $\frac{q}{\frac{16}{5}} = \frac{\frac{5}{4}}{4}$

Use a proportion to solve each problem.

23. Ben has a 4-inch wide × 6-inch long photo of himself scaling Half Dome in Yosemite. If he has it enlarged to a poster 4 feet long, how wide will the poster be?

24. A general-purpose birdseed contains 2 cups of sunflower seed for every 5 cups of millet. How much millet should be mixed with 5 quarts of sunflower seed?

25. The road to the camp grounds at Pine Lake State Park is 1.5 miles long. On the map, the road measures $\frac{3}{4}$ inch. If Pine Lake is $3\frac{1}{4}$ inches long on the map, how long is the lake actually?

26. A cinnamon bread recipe calls for $1\frac{1}{4}$ tablespoons of cinnamon and 5 cups of flour. How much cinnamon would be needed if 8 cups of flour are used?

27. Gary is 6 feet tall. At 4 P.M. his shadow is 10 feet long. How tall is the spruce tree on his front lawn if its shadow is 38 feet long at 4 P.M.?

28. In a survey of 800 voters, 625 favored a gun control ballot measure. If there are 15,000 voters in Senator Fogbank's district, how many can be expected to favor gun control?

29. To estimate the number of deer in a certain wooded area, the Forest Service tagged and released 50 deer. A month later they captured 60 deer and found that 9 of them were tagged. Approximately how many deer are there in the woods?

30. The quality-control department at Major Motors found that in a sample of 500 electronic toggle units, 12 were defective. How many defective units should they expect in a shipment of 3600 units?

Find the area and perimeter of each figure. (See Lessons 2.4 and 5.2B to review area and perimeter.) Round your answers to two decimal places.

31.
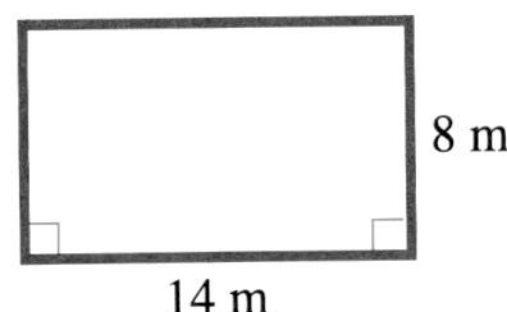

32.
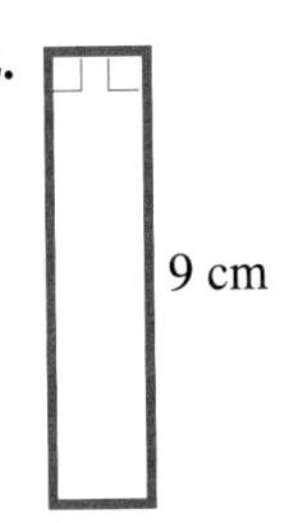

33.
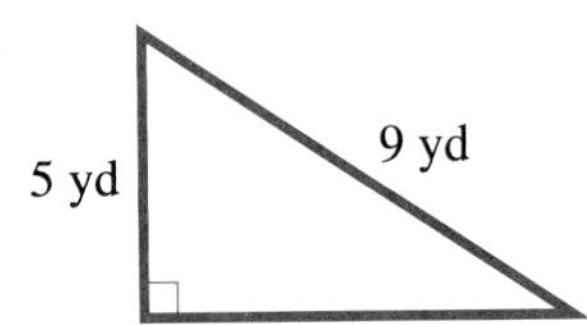

34.
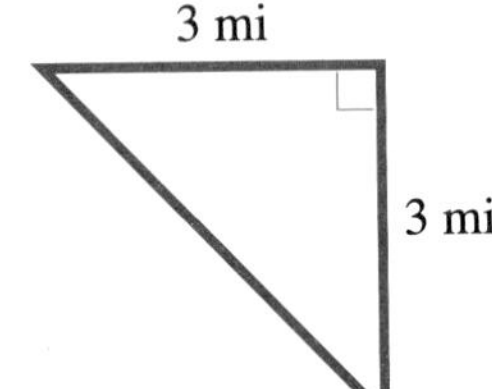

35.
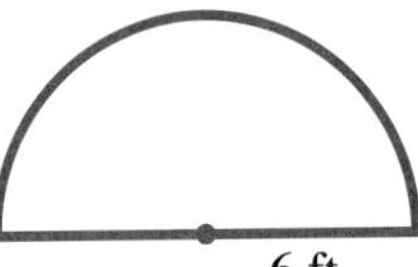

36.
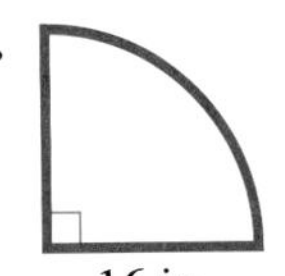

37.

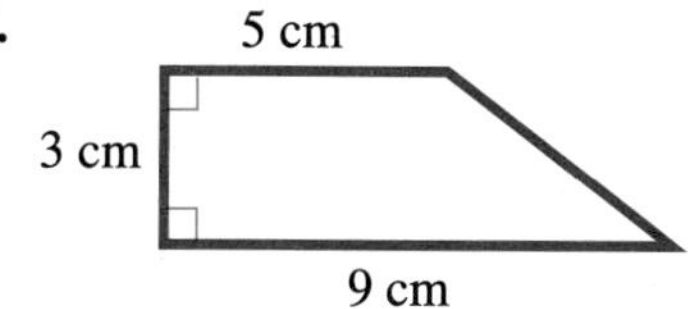

38.

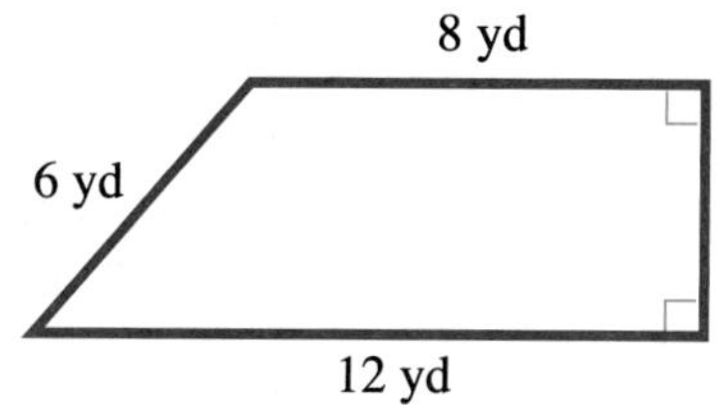

39.

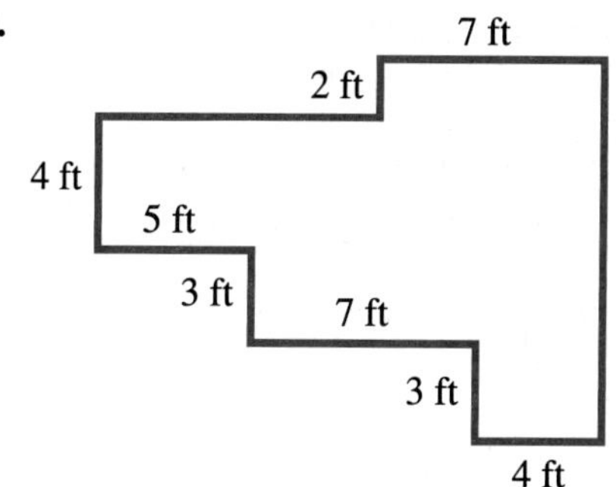

40.

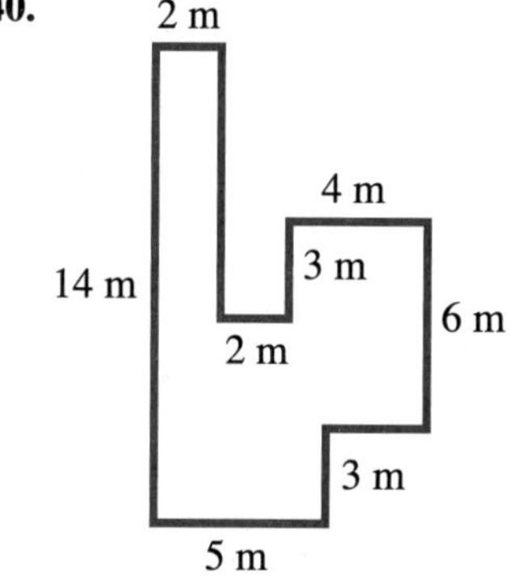

Building Number Skills

Here is a way to compute 15% of a bill mentally.

Step 1 Compute 10% of a bill by shifting the decimal one place to the left.
Step 2 Divide this new amount by 2 to get 5% of the bill.
Step 3 Add the two amounts you computed to get 15% of the bill.

Example Compute 15% of $22.00.

Solution **Step 1** 10% of $22 is $2.20.
Step 2 5% of $22.00 is $2.20 ÷ 2 = $1.10.
Step 3 15% of $22 is $2.20 + $1.10 = $3.30.

Compute 15% of each amount mentally.

41. 15% of $40

42. 15% of $50

43. 15% of $32

44. 15% of $28

45. 15% of \$19.93 is about
a. \$0.30 **b.** \$3.00 **c.** \$30.00

46. 15% of \$38.46 is about
a. \$0.60 **b.** \$6.00 **c.** \$60.00

47. 15% of \$59.96 is about
a. \$8.00 **b.** \$9.00 **c.** \$10.00

48. 15% of \$17.83 is about
a. \$2.70 **b.** \$2.80 **c.** \$2.90

49. 15% of \$8.99 is about
a. \$1.25 **b.** \$1.35 **c.** \$1.45

50. 15% of \$6.99 is about
a. \$0.95 **b.** \$1.05 **c.** \$1.15

8.4 Area and Volume

Areas of Proportional Figures

The Pizza Company sells its 8-inch tuna pizza for \$5.00 and its 16-inch tuna pizza for \$15.00. Which is the better deal: three 8-inch pizzas for \$15.00 or one 16-inch pizza for \$15.00? (See Figure 8.12.)

Figure 8.12

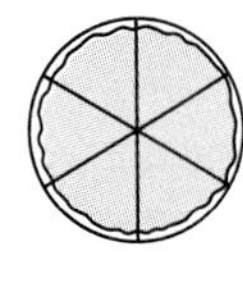

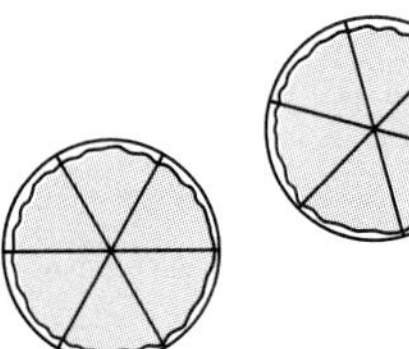

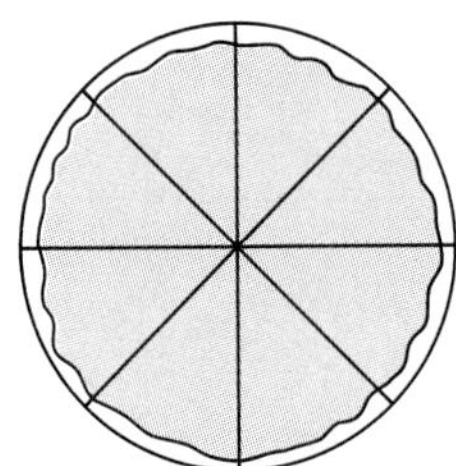

At first it might seem that the smaller pizzas are a better deal, since *doubling* the diameter of the pizza *tripled* its price. But with pizza, the better deal is the one that gives you the most pizza for your money, and since pizzas are flat, that means the most *area.* Does doubling the diameter (or radius) of a pizza double its area as well?

What about square pizzas? Does a 20-inch square pizza have twice the area of a 10-inch square pizza? In general, if we multiply the dimensions of a plane figure by some number k, will its area also increase by a factor of k? Multiplying the dimensions by k is called **scaling** the dimensions by k, and k is called the **scale factor.**

We will test this hypothesis by computing some areas.

Activity 1 Compute the area of each square.

a. Original square: $s = 3$ inches

area = ____________________

b. Twice the original dimensions:

$s = 6$ inches

area = ____________________

To find the *scale factor,* compute the following ratio:

scale factor = ratio of new side to original side:

$$\frac{\text{new } s}{\text{original } s} = \underline{\qquad\qquad}$$

Thus, the original dimensions are scaled by a factor of $k =$ __________

Ratio of new area to original area: $\dfrac{\text{new area}}{\text{original area}} =$ __________

(How many of the original squares fit inside this new square? See Figure 8.13a.)

c. Three times the original dimensions:

$s = 9$ inches

area = ____________________

Scale factor = ratio of new side to original side = __________

Ratio of new area to original area = __________

(How many of the original squares fit inside this new square? See Figure 8.13b.)

d. Can you see a pattern in the results of your calculations?

Suppose the sides of the new square are 5 times the sides of the original square. Then the area of the new square will be _____ times the area of the original square.

Check your answer by performing the calculations:

side of new square: $s =$ ____________________

area of new square = ____________________

ratio of new area to original area = ____________________

Figure 8.13a

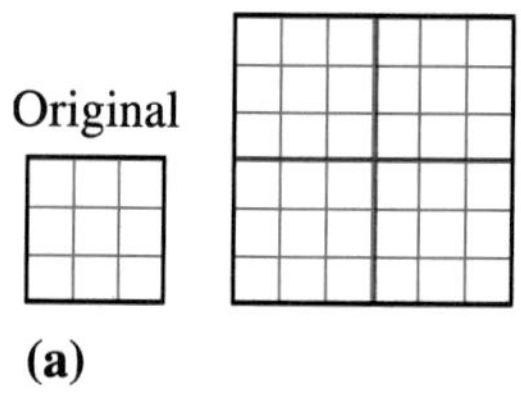

Figure 8.13b

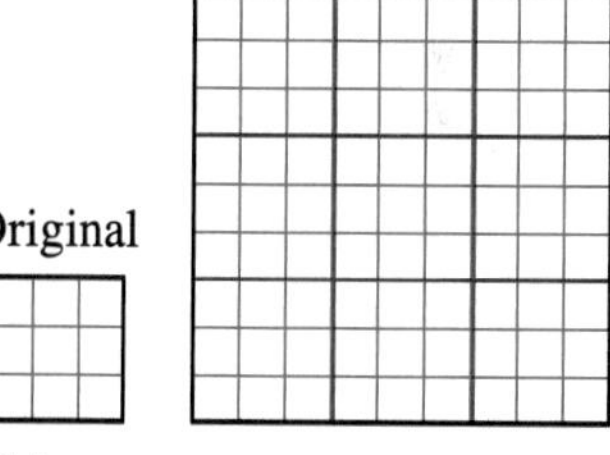

You should now have a conjecture about the areas of squares. Fill in the blank:

> **If we scale (multiply) the sides of a square by k, then the area of the new square is _____ times the area of the original square.**

Let's see if the same pattern holds for triangles. First notice that, if we want the new triangle to have the same *shape* as the original triangle, we must scale (multiply) *both* dimensions by the same amount. (See Figure 8.14.)

Figure 8.14 Triangles A and D have the same shape.

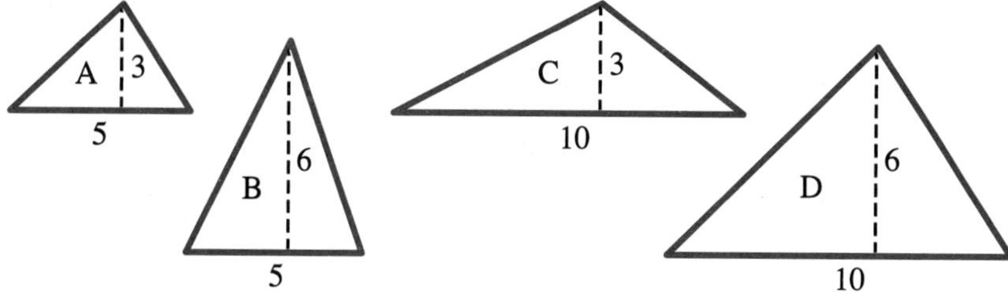

Activity 2 Compute the area of each triangle.

a. Original triangle:

$b = 4$ centimeters,

$h = 3$ centimeters

area = ____________________

b. Twice the original dimensions:

$b = 8$ centimeters

$h = 6$ centimeters

area = ______________________

Scale factor = ratio of new dimensions to original dimensions = ___________

Ratio of new area to original area = ___________

c. Three times the original dimensions:

$b = 12$ centimeters

$h = 9$ centimeters

area = ______________________

Scale factor = ratio of new dimensions to original dimensions = ___________

Ratio of new area to original area = ___________

d. Can you see a pattern in the results of your calculations?

Suppose the dimensions of the new triangle are 6 times the dimensions of the original triangle. Then the area of the new triangle will be ____ times the area of the original triangle.

Check your answer by performing the calculations:

Dimensions of new triangle:

$b =$ ___________, $h =$ ___________

area of new triangle = ______________________

ratio of new area to original area = ______________________

Finish the following conjecture about the areas of triangles:

If we scale (multiply) the dimensions of a triangle by *k*, then the area of the new triangle is ____ times the area of the original triangle.

Finally, let's consider the areas of circles.

Activity 3 Compute the area of each circle.

a. Original circle: $r = 2$ inches

area = ______________________

b. Twice the original dimension:

$r = 4$ inches

area = ______________________

Scale factor = ratio of new radius to original radius: ___________

Ratio of new area to original area: ___________

c. Three times the original dimension:

$r = 6$ inches

area = ______________________

Scale factor = ratio of new radius to original radius: ___________

Ratio of new area to original area: ___________

d. Can you see a pattern in the results of your calculations?

Suppose the radius of the new circle is 10 times the radius of the original circle. Then the area of the new circle will be _____ times the area of the original circle.

Check your answer by performing the calculations:

radius of new circle: $r =$ ______________________

area of new circle = ______________________

ratio of new area to original area = ______________________

Finish the following conjecture about the areas of circles:

If we scale (multiply) the radius of a circle by k, then the area of the new circle is _____ times the area of the original circle.

If two plane figures have the same shape but different sizes they are called **proportional** or **similar figures.** The discovery you made in Activities 1 through 3 about squares, triangles, and circles is true for any similar figures.

Area Principle for Similar Figures

If we multiply the dimensions of a plane figure by k, then the new figure is similar to the original figure, and its area is times k^2 the area of the original figure.

We can use this principle to solve problems about area.

EXAMPLE 1  

An architect is drawing up floor plans for a new housing development. The dining room in her plan has an area of only 120 square feet. If she can increase the dimensions of the room by 25%, what will be the new area?

Solution

In Lesson 8.1 we learned that to calculate an increase of 25% we multiply the amount by 1.25. Thus, the architect is planning to scale the dimensions of the dining room by $k = 1.25$. According to the area principle, the area of the new dining room will be $(1.25)^2$, or 1.5625, times the old area. The new area will then be

$$k^2(\text{old area}) = \text{new area}$$
$$1.5625(120) = 187.5 \text{ square feet}$$

Now try Exercise 1.

EXERCISE 1

Alys wants to replace her old hot tub with a new circular spa. If the radius of the spa is 15% greater than the radius of the hot tub, how much greater will the area of the spa be?

Volumes of Proportional Solids

First, let's recall some formulas for the volumes of solid objects (Figure 8.15).

Figure 8.15

Box

h w l

$V = lwh$

Sphere

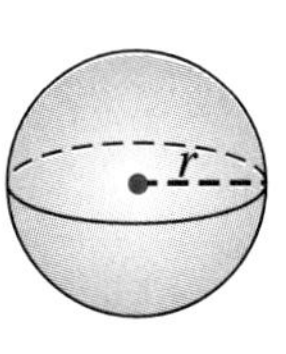

$V = \frac{4}{3}\pi r^3$

Cylinder

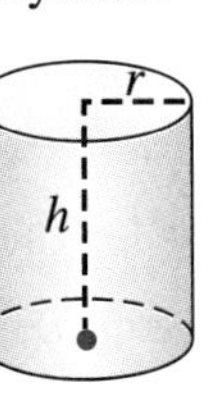

$V = \pi r^2 h$

Cone

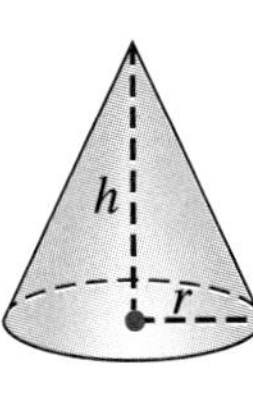

$V = \frac{1}{3}\pi r^2 h$

Two solid objects are called **proportional** or **similar** if they have the same shape but different sizes.

If we scale all the dimensions of a particular solid object—say, a cylinder—by the same number, then the new object will be similar to the old one. In Figure 8.16, we multiplied both the radius and the height of the first cylinder by $\frac{3}{4}$. The new cylinder has the same shape as the first one, but of course it is smaller.

Figure 8.16

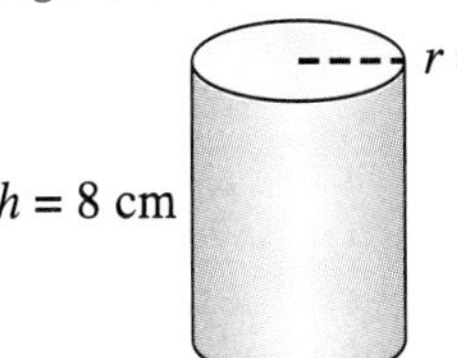

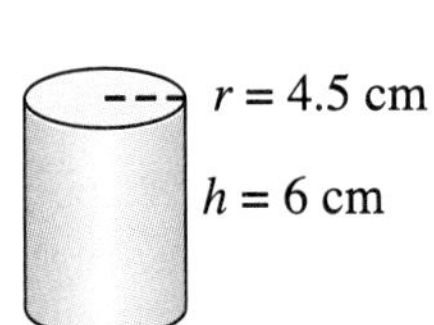

Is there a relationship between the *volumes* of similar objects? Let's look at some examples to find out.

Activity 4 Compute the volume of each box.

a. Original box:

l = 3 inches
w = 2 inches
h = 5 inches
volume = __________

b. Twice the original sides:

l = __________, w = __________, h = __________

volume = __________

Scale factor = ratio of new sides to original sides:

$$\frac{\text{new } l}{\text{original } l} = \frac{\text{new } w}{\text{original } w} = \frac{\text{new } h}{\text{original } h} = \_\_\_\_\_\_\_\_\_\_$$

The dimensions are scaled by a factor of k = __________.

Ratio of new volume to original volume:

$$\frac{\text{new volume}}{\text{original volume}} = \_\_\_\_\_\_\_\_\_\_$$

(How many of the original boxes fit inside this new box? See Figure 8.17a.)

Figure 8.17a

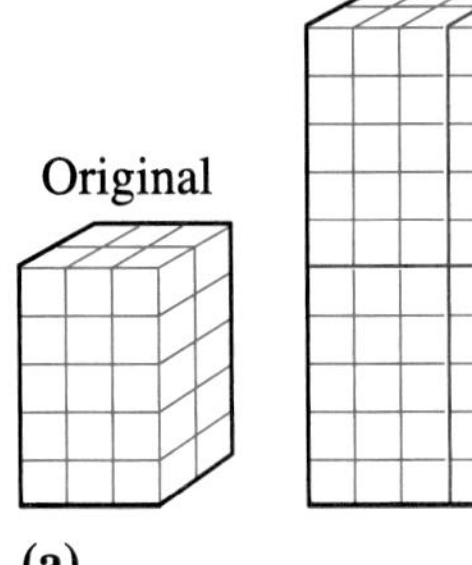

(a)

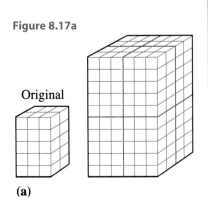

c. Three times the original sides:

$l =$ __________, $w =$ __________, $h =$ __________

volume = __________

Scale factor = ratio of new sides to original sides = __________

Ratio of new volume to original volume = __________

(How many of the original boxes fit inside this new box? See Figure 8.17b.)

d. Can you see a pattern in the results of your calculations?

Suppose the sides of the new box are $\frac{1}{2}$ times the dimensions of the original box. Then the volume of the new box will be _____ times the volume of the original box.

Check your answer by performing the calculations:

length of new box: $l =$ __________

width of new box: $w =$ __________

height of new box: $h =$ __________

volume of new box = __________

ratio of new volume to original volume = __________

Figure 8.17b

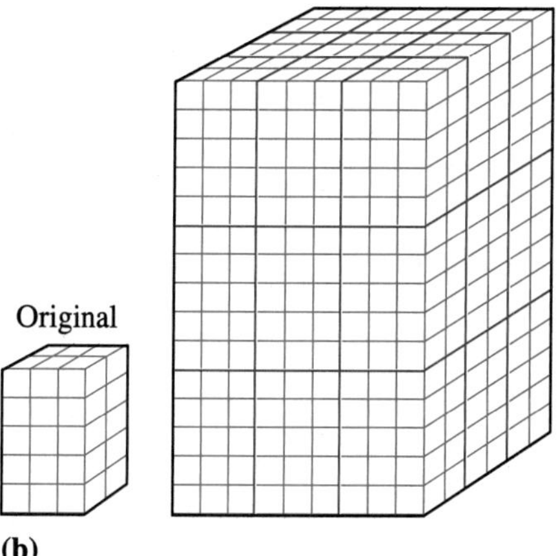

(b)

You should now have a conjecture about the volumes of boxes. Fill in the blank:

If we scale the sides of a box by k, then the volume of the new box is _____ times the volume of the original box.

Maybe you are ready to believe that the same relationship holds for any kind of similar solid objects. Nevertheless, it might be a good idea to test our hypothesis on some other type of figures—say, spheres.

Activity 5 Compute the volume of each sphere.

a. Original sphere: $r = 10$ feet

volume = __________

b. Twice the original radius: $r =$ __________

volume = __________

Scale factor = ratio of new radius to original radius:

$$\frac{\text{new } r}{\text{original } r} = \_\_\_\_\_$$

The radius is scaled by a factor of $k =$ __________

Ratio of new volume to original volume:

$$\frac{\text{new volume}}{\text{original volume}} = \_\_\_\_\_$$

c. Three times the original radius: $r =$ __________

volume = __________

Scale factor = ratio of new radius to original radius =

____

Ratio of new volume to original volume = __________

d. Can you see a pattern in the results of your calculations?

Suppose the radius of the new sphere is 1.5 times the radius of the original sphere. Then the volume of the new sphere will be ________ times the volume of the original sphere.

Check your answer by performing the calculations:

radius of new sphere = ________________

volume of new sphere = ________________

ratio of new volume to original volume = ________________

Finish the following conjecture about the volumes of spheres:

If we scale the radius of a sphere by *k*, then the volume of the new sphere is ____ times the volume of the original sphere.

In fact, this relationship holds for the volumes of any similar solid objects, and we can state the following volume principle.

Volume Principle for Similar Solids

If we multiply the dimensions of a solid object by k, then the new object is similar to the original object, and its volume is k^3 times the volume of the original object.

EXERCISE 2

The radius of Jupiter is about 10.7 times the radius of Earth. How does the volume of Jupiter compare to the volume of Earth?

EXAMPLE 2 

In *Gulliver's Travels,* the Lilliputians are tiny people about 6 inches tall. How would the volume of a Lilliputian compare with that of a human who is 6 feet tall?

Solution

The scale factor is the ratio of the Lilliputian's height to the human's height:

$$\frac{6 \text{ inches}}{6 \text{ feet}} = \frac{6 \text{ inches}}{72 \text{ inches}} = \frac{1}{12}$$

Thus, the volume is scaled by $\left(\frac{1}{12}\right)^3 = \frac{1}{1728}$. The volume of the Lilliputian would be $\frac{1}{1728}$ times the volume of the 6-foot tall human.

Now try Exercise 2.

ANSWERS TO 8.4 EXERCISES

1. 32.25% greater
2. About 1225 times greater

HOMEWORK 8.4 

In each group of objects, find two that are similar.

1. 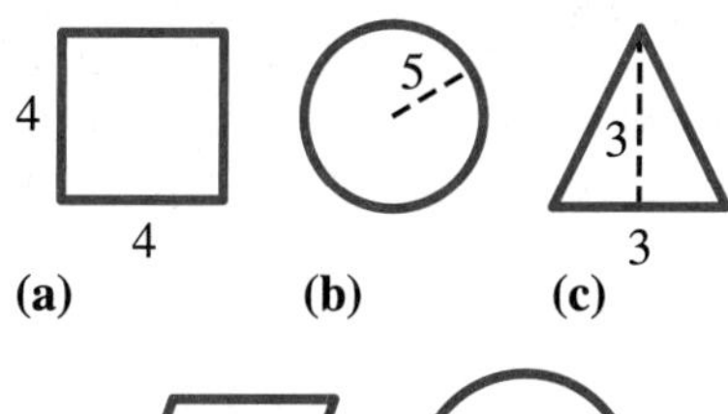

2. 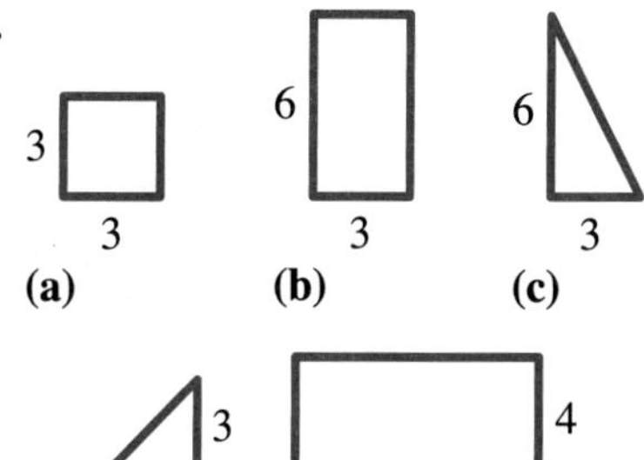

3. 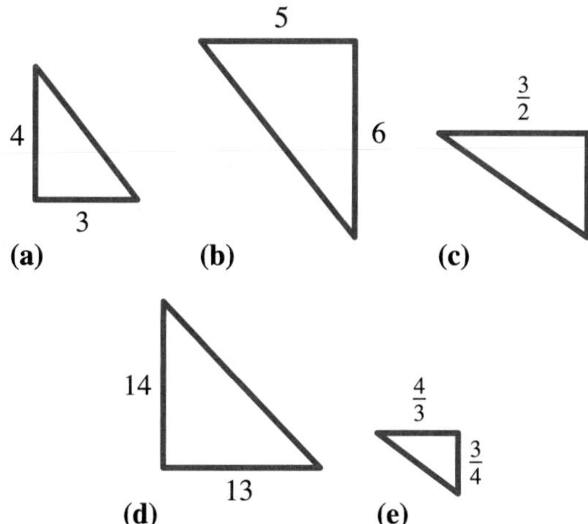

4. 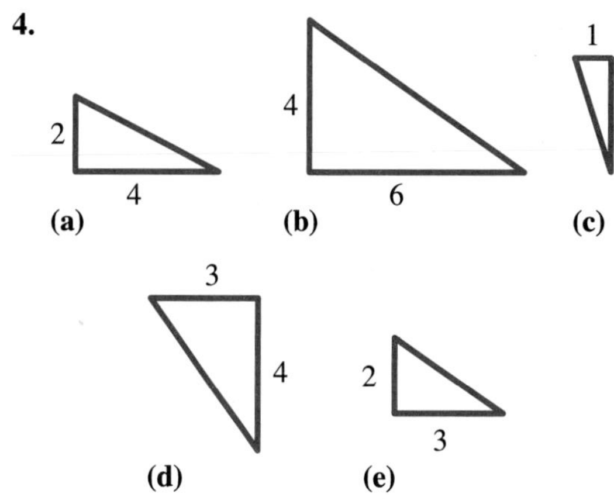

5. 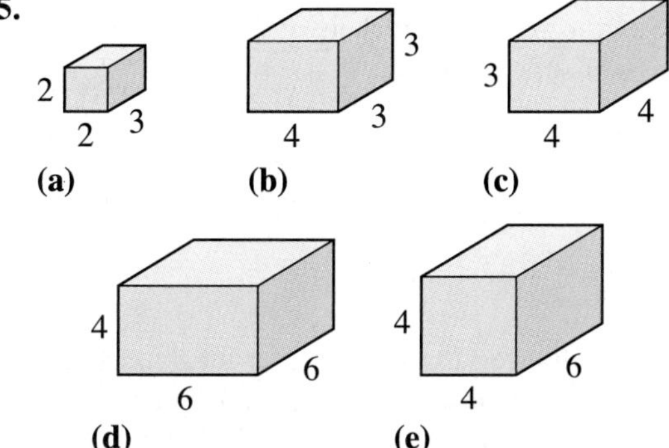

6. 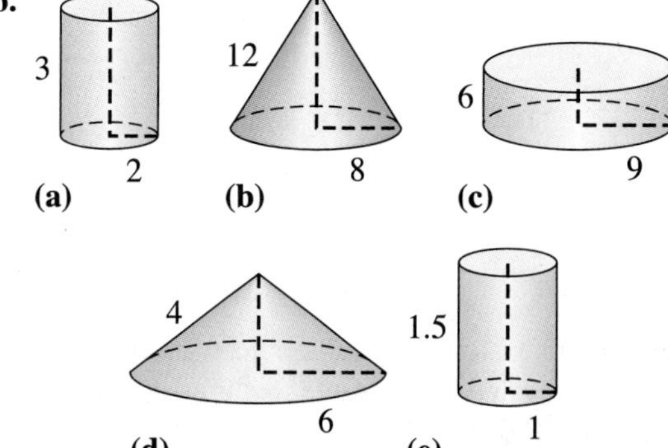

Solve each problem.

7. Elina has a 3-inch × 5-inch photograph of her cat, Sinbad. She wants to enlarge the photo by 50% in each dimension. By what factor will the area of the photo increase? What will be the area of the enlarged photograph?

8. For their publicity campaign, the Magazine Sweepstakes staff is making a poster-size version of a check for $10 million. A normal check is $2\frac{1}{2}$ inches wide × 6 inches long, and the poster-size version will be 12 times as big in both dimensions. By what factor will the area increase? What will be the area of the poster-size check?

9. Emily made a scale model of her sailboat, in which each dimension of her model is one-tenth that of the real sailboat. If the real sail has an area of 75 square feet, what is the area of the model's sail?

10. Nora wants to make a copy of her favorite artist's masterpiece, a mural covering an area of 10 square meters. Since she doesn't have a wall large enough for a full-size copy, her version will be only one-half the size in each dimension. What will be the area of the copy?

11. War Games produces a scale-model replica of the B2 bomber. If the scale is 1 to 500, how does the volume of the model compare with that of the real bomber?

12. Ariff is making a 1-foot tall model of the Eiffel Tower. The original Eiffel Tower is 984 feet tall. How will the volume of the model compare with the volume of the original?

13. Solve the problem about pizzas that opened this lesson. Which is the better deal: $15 for three 8-inch pizzas or $15 for one 16-inch pizza?

14. At the Chicken Coop, each dimension of the superbucket of fried chicken is 1.5 times the corresponding dimension of the minibucket. Which is the better deal: $24 for three minibuckets of fried chicken or $24 for one superbucket?

15. When the dimensions of a solid object are all scaled up, its weight is scaled by the same factor as its volume. If a 6-foot tall, 180-pound human being is scaled up to a height of 60 feet, what will be the giant's weight?

16. A typical boll weevil is 1 centimeter long and weighs 0.0003 gram. If gamma rays produce a mutant boll weevil 5 meters long, how much will it weigh? (See Problem 15.)

Evaluate. Round to three decimal places. (See Lessons 5.2 and 5.4 to review order of operations.)

17. $-2w(3w - 2)$, for $w = -3$

18. $-6t(4 - 2t)$, for $t = -5$

19. $3x - 5x^2 - x^3$, for $x = -2$

20. $2x^4 - x^2 + 3x$, for $x = -3$

21. $\dfrac{1 - m^2}{1 + m^2}$, for $m = -1.5$

22. $\dfrac{n(1 - n)}{1 + n}$, for $n = -2.4$

23. $2h^2\sqrt{6h}$, for $h = 12.8$

24. $4k^3\sqrt{5k}$, for $k = 8.5$

25. $\dfrac{-b + \sqrt{b^2 + 8a}}{2a}$, for $a = 2$ and $b = -3$

26. $\dfrac{-b - \sqrt{b^2 + 8a}}{2a}$, for $a = 3$ and $b = 4$

Building Number Skills

We sometimes can approximate percentages in three steps.

Step 1 Approximate the percentage rate with an ordinary fraction.
Step 2 Round the whole to a convenient value.
Step 3 Multiply the two approximations.

Example Approximate 23% of 769.

Solution **Step 1** Approximate 23% by $\frac{1}{4}$.
Step 2 Round 769 to 800.
Step 3 Multiply $\frac{1}{4}(800) = 200$.

Choose the best answer.

27. 52% of 78 is about
a. 40 **b.** 50 **c.** 60

28. 47% of 822 is about
a. 300 **b.** 400 **c.** 500

29. 32% of 291 is about
a. 10 **b.** 100 **c.** 1000

30. 35% of 88 is about
a. 10 **b.** 20 **c.** 30

31. 27% of 7.6 is about
a. 1 **b.** 2 **c.** 3

32. 24% of 12.3 is about
a. 3 **b.** 4 **c.** 5

33. 65% of 302 is about
a. 100 **b.** 200 **c.** 300

34. 67% of 587 is about
a. 400 **b.** 500 **c.** 600

35. 76.1% of 39 is about
a. 20 **b.** 30 **c.** 40

36. 74.7% of 810 is about
a. 500 **b.** 600 **c.** 700

Challenge Problems

37. If you want to increase the area of a photograph by a factor of 4, by what factor should you scale its dimensions?

38. If you want to increase the area of a photograph by a factor of 2, by what factor should you scale its dimensions?

39. If you want to increase the volume of a box by a factor of 8, by what factor should you scale its dimensions?

40. If you want to increase the volume of a box by 2, by what factor should you scale its dimensions?

8.5 Slope

Unit Pricing

Delbert and Francine are classmates at City College. Before handing in big assignments such as term papers or lab reports, they make copies of their work. Delbert uses QuikCopy and keeps track of the copying fee for each assignment:

Number of pages	14	27	35
Cost (¢)	112	216	280

Francine copies her assignments at Copy World:

Number of pages	12	26	32
Cost (¢)	72	156	192

Who is paying the higher price for copying? In Lesson 2.3 we learned a formula for unit cost, $u = \frac{p}{n}$, where n is the number of items and p is their total cost. Delbert and Francine compute the cost per page for each of their assignments:

Delbert	*Francine*
$\frac{112 \text{ cents}}{14 \text{ pages}} = 8 \text{ cents/page}$	$\frac{72 \text{ cents}}{12 \text{ pages}} = 6 \text{ cents/page}$
$\frac{216 \text{ cents}}{27 \text{ pages}} = 8 \text{ cents/page}$	$\frac{156}{26 \text{ pages}} = 6 \text{ cents/page}$
$\frac{280 \text{ pages}}{35 \text{ cents}} = 8 \text{ cents/page}$	$\frac{192 \text{ cents}}{32 \text{ pages}} = 6 \text{ cents/page}$

Delbert is paying the higher price for copying.

The results can also be seen from a graph. (See Figure 8.18.) When we plot the number of pages on the horizontal axis and the cost on the vertical axis, we see that, as the number of pages increases, Delbert's cost is increasing more rapidly than Francine's.

p	D
14	112
27	216
35	280

p	F
12	72
26	156
32	192

Figure 8.18

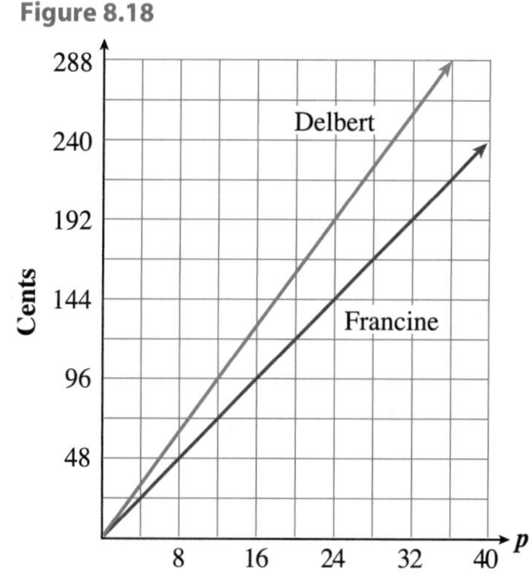

Now try Exercise 1.

Steepness of the Graph

Because his cost is increasing more rapidly, the graph of Delbert's cost is steeper than the graph of Francine's cost. When Delbert's assignment increased by 13 pages (from 14 to 27 pages), his cost increased by 104 cents (from 112 to 216 cents). The rate of increase is

$$\frac{104 \text{ cents}}{13 \text{ pages}} = 8 \text{ cents per page}$$

the same unit cost we calculated above. When Francine's assignment increased by 14 pages (from 12 to 26 pages), her cost increased by 84 cents (from 71 to 156 cents). So her unit cost was

$$\frac{84 \text{ cents}}{14 \text{ pages}} = 6 \text{ cents per page}$$

as we saw earlier.

The unit cost in each case is a measure of the steepness of the graph. It tells us how much to increase the vertical coordinate *per unit* of increase in the hori-

EXERCISE 1

Nazli and Marya both work part-time as data-entry technicians at different companies. Figure 8.19 shows their earnings in terms of the number of hours they work.

Figure 8.19

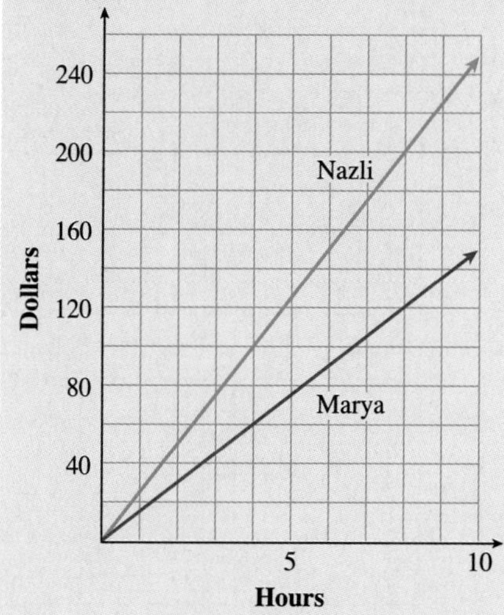

a. Who earns more per hour? Explain how you know.

b. Use the graph to fill in the table:

Hours	4	8	10
Nazli's earnings			
Marya's earnings			

c. Compute each woman's wages in dollars per hour.

EXERCISE 2

Kira buys granola in bulk at the health food store. There are three standard containers customers can use, and the price of each container is shown in Figure 8.20.

Figure 8.20

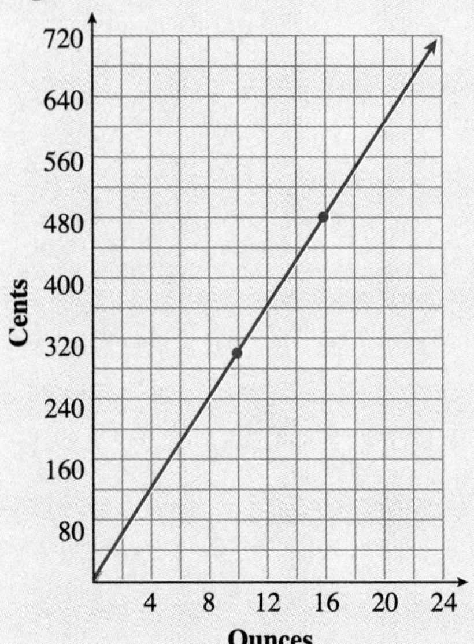

a. What is the unit cost of the granola, in cents per ounce?

b. If we extend the graph to include 25 ounces of granola, how much taller should we make the vertical axis?

zontal coordinate. For example, if the length of Delbert's assignment increases by one page, his copying cost will increase by 8 cents, and the graph will climb 8 units for each unit we move to the right.

Now try Exercise 2.

Rate of Growth

We can also use ratios to compare rates of growth. For example, Sara bought two kinds of tomato plants at the nursery. The Spring Giant seedlings were 2 inches tall, and the Jubilee seedlings were 16 inches tall. Sara monitored the growth of both varieties of tomato by recording the height h of the vines every 2 weeks. She also plotted their heights on a graph, shown in Figure 8.21.

From the graph, we can see that the Spring Giant tomato vines are growing faster than the Jubilee vines, because their graph increases more steeply. The steepness of the graph is a measure of how fast the vines are growing.

Spring Giant

t (weeks)	*h (inches)*
0	2
2	10
4	18
6	26
8	34

Jubilee

t (weeks)	*h (inches)*
0	16
2	21
4	26
6	31
8	36

Figure 8.21

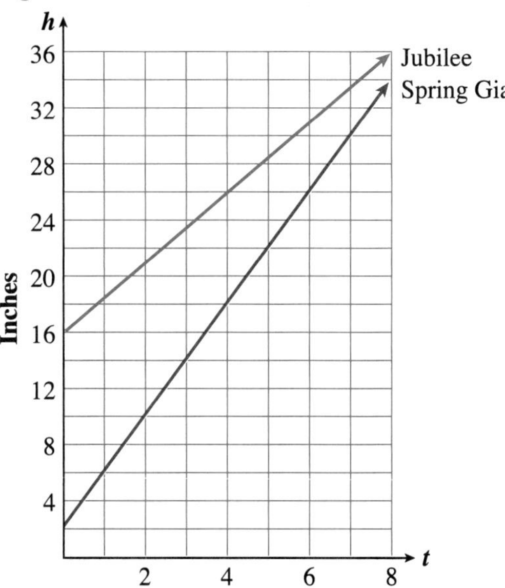

To see exactly how fast the vines grow, we compute their rate of growth. During the first 2 weeks, the Spring Giant vines grew 8 inches (from 2 inches to 10 inches), so their growth rate was

$$\frac{8 \text{ inches}}{2 \text{ weeks}} = 4 \text{ inches/week}$$

The Spring Giant tomato vines continued to grow at the same rate. For example, from the second to the fourth week we find

$$\frac{\text{increase in height}}{\text{time interval}} = \frac{(18 - 10) \text{ inches}}{(4 - 2) \text{ weeks}}$$
$$= \frac{8 \text{ inches}}{2 \text{ weeks}} = 4 \text{ inches/week}$$

Now try Exercise 3.

Visualizing Growth Rate on a Graph

We can illustrate the growth rate of each tomato vine on its graph. In Figure 8.22, the vertical line segment labeled Δh represents the increase in the vine's height from the second week to the sixth week. (The symbol Δ is the Greek letter *delta.* It stands for "change in.") The horizontal line segment labeled Δt represents the time interval, 4 weeks. The vine grew at a rate of

$$\frac{\Delta h}{\Delta t} = \frac{(26 - 10) \text{ inches}}{(6 - 2) \text{ weeks}}$$
$$= \frac{16 \text{ inches}}{4 \text{ weeks}} = 4 \text{ inches/week}$$

Figure 8.22

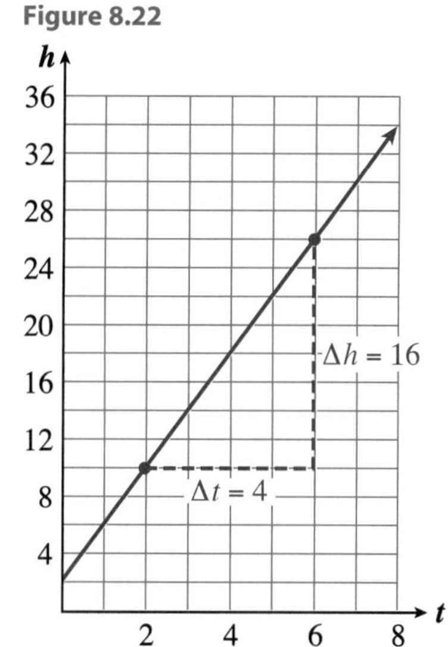

What does this ratio tell us about the graph? Place your pencil tip at any point on the graph and then move 1 unit to the right along the grid lines. To get back to the graph, you must now move up by 4 units (two grid lines.) The graph increases by climbing 4 units vertically for each unit we move horizontally.

Now try Exercise 4.

Slope of a Line

The ratio $\dfrac{\Delta h}{\Delta t}$ is called the **slope** of the line in Figure 8.22. The slope measures how steeply the line increases (or decreases) as we move from left to right across the graph. In general, we define the slope for any line as follows:

> The **slope,** m, of a line is the ratio
>
> $$m = \frac{\Delta y}{\Delta x} = \frac{\textbf{change in the } y\textbf{-coordinate}}{\textbf{change in the } x\textbf{-coordinate}}$$
>
> as we move from any point on the line to another point on the line.

EXERCISE 3

a. Compute the ratio $\dfrac{\text{increase in height}}{\text{time interval}}$ to find the growth rate of the Spring Giant tomato vines from the fourth week to the eighth week.

b. Find the growth rate of the Jubilee tomato vines over the first 4 weeks and from the fourth week to the eighth week.

EXERCISE 4

a. On the graph in Figure 8.23, illustrate the growth rate of the Jubilee tomato vines from the fourth week to the eighth week.

$$\frac{\Delta h}{\Delta t} =$$

b. Start at any point on the graph. If you move 1 unit horizontally, how many units does the graph climb vertically? If you move 2 units horizontally, how many units does the graph climb vertically?

Figure 8.23

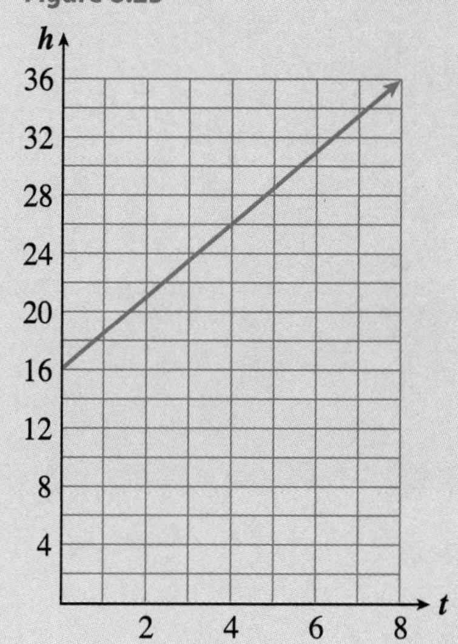

Figure 8.24

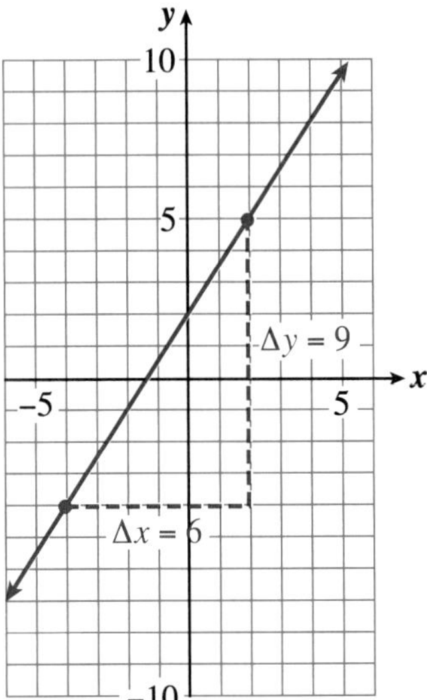

EXAMPLE 1  

Find the slope of the line in Figure 8.24.

Solution

Choose two points on the line—say and (−4, −4) and (2, 5). Compute the change in the vertical coordinate

$$\Delta y = 5 - (-4) = 9$$

and the change in the horizontal coordinate

$$\Delta x = 2 - (-4) = 6$$

as shown in the figure. Then calculate the slope:

$$m = \frac{\Delta y}{\Delta x} = \frac{9}{6} = \frac{3}{2}$$

The slope of the line is $\frac{3}{2}$.

You can check that you will get the same value for the slope, $\frac{3}{2}$, if you use any other points on the line for the calculations.

CAUTION!

It is important that the numerator (top) of the slope ratio give the change in the *vertical* coordinate of points on the graph. In Example 1, it would be *incorrect* to compute $\frac{\Delta x}{\Delta y}$ for the slope.

Now try Exercise 5.

EXERCISE 5

a. Calculate the slope of the line shown in Figure 8.25.

$$m = \frac{\Delta y}{\Delta x} =$$

b. Illustrate your calculation on the graph.

Figure 8.25

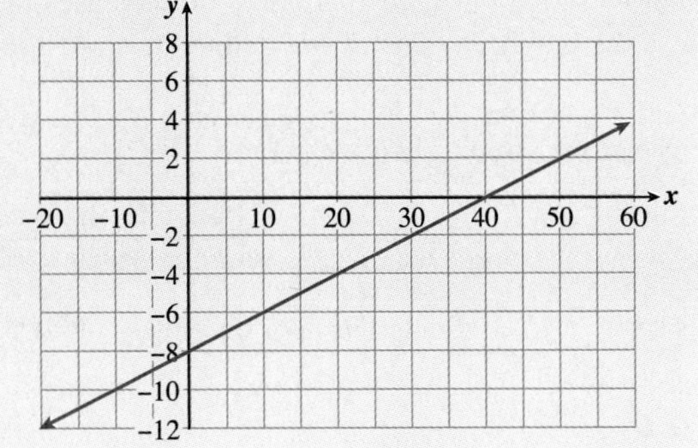

Decreasing Graphs

If the y-values of points on a line decrease as we move from left to right, the line has a negative slope.

EXAMPLE 2 

At noon the temperature was 18°F, and it has been getting colder by 2° every hour.

a. Graph the temperature from noon to midnight.

b. Find the slope of the graph. What does the slope tell us about the problem situation?

Solution

a. Make a table showing the temperature T at h hours after noon:

h	0	2	5	9	12
T	18	14	8	0	−6

Plot the points from the table on a grid, as shown in Figure 8.26. Connect the points with a straight line.

b. To compute the slope of the line, choose two points on the graph—say, (5, 8) and (12, −6). Compute the change in the vertical coordinate

$$\Delta T = -6 - 8 = -14 \text{ degrees}$$

and the change in the horizontal coordinate

$$\Delta h = 12 - 5 = 7 \text{ hours}$$

Calculate the slope ratio:

$$\frac{\Delta T}{\Delta h} = \frac{-14 \text{ degrees}}{7 \text{ hours}} = -2 \text{ degrees/hour}$$

The slope is −2. It represents the rate at which the temperature is decreasing, degrees per hour.

Figure 8.26

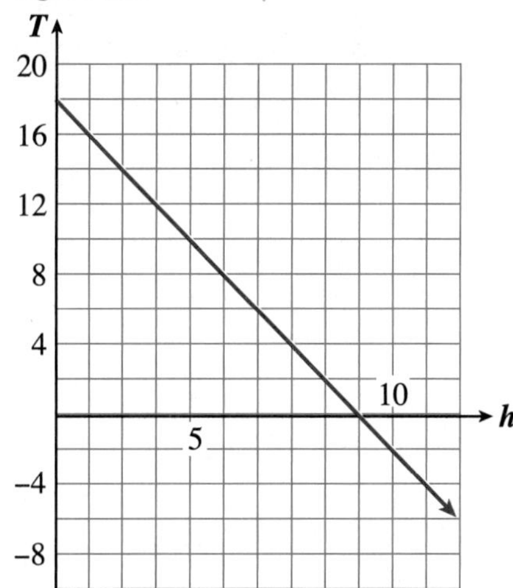

CAUTION!

In Example 2, note that ΔT, the change in temperature, is negative. To compute the change in a coordinate, always subtract the initial value from the final value:

change in coordinate = final value − initial value

For this example,

$$\Delta T = \text{final temperature} - \text{initial temperature}$$

Now try Exercise 6.

EXERCISE 6

Find the slope of the graph in Figure 8.27. Use the points (2, −10) and (4, −22). Illustrate your calculations on the graph.

$$m = \frac{\Delta y}{\Delta x} =$$

Figure 8.27

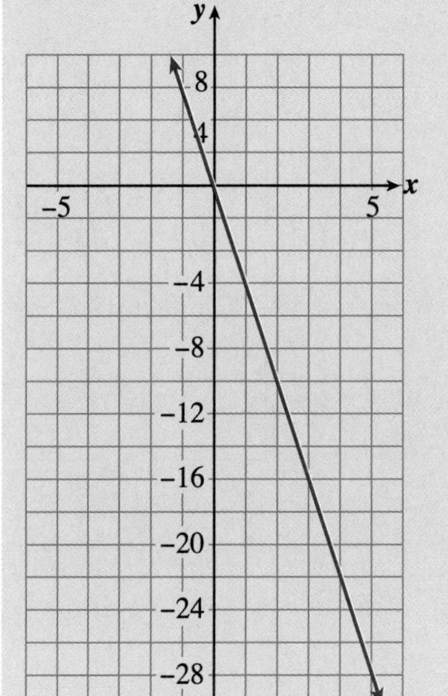

ANSWERS TO 8.5 EXERCISES

1a. Nazli earns more. The graph of her earnings is steeper.

1b.

Hours	4	8	10
Nazli's earnings	100	200	250
Marya's earnings	60	120	150

Nazli earns $25 per hour; Marya earns $15 per hour.

2a. 30 cents per ounce **2b.** 30 units taller

3a. 4 inches per week **3b.** 2.5 inches per week; 2.5 inches per week

4a.

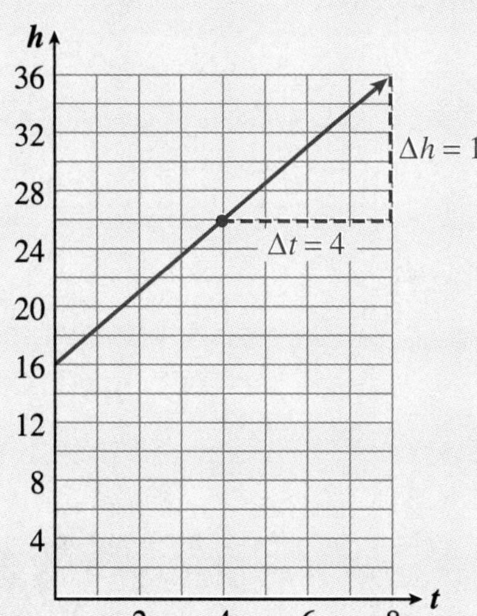

$\dfrac{\Delta h}{\Delta t} = 2.5$ inches per week

4b. 2.5 units; 5 units

5a. $m = \dfrac{1}{5}$

5b.

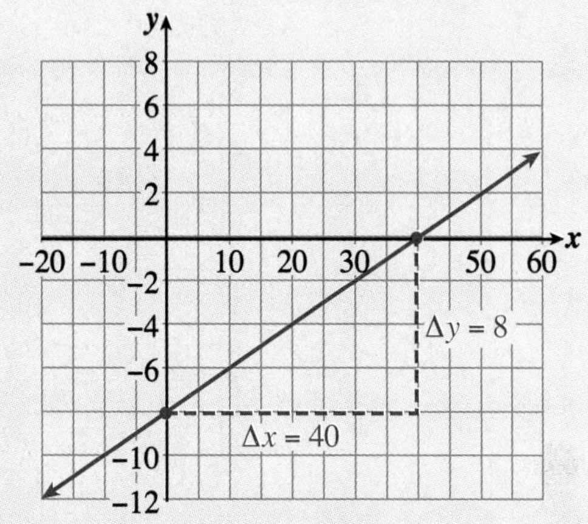

6. $m = -6$

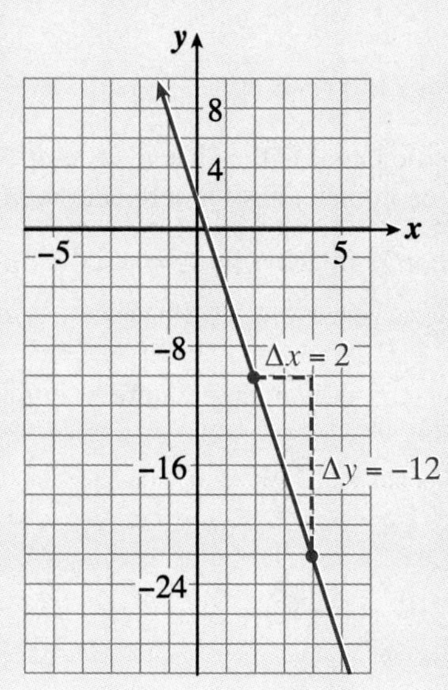

HOMEWORK 8.5 

1. Lynette is saving money for the down payment on a new car. Figure 8.28 shows the amount A she has saved, in dollars, w weeks after the first of the year.

Figure 8.28

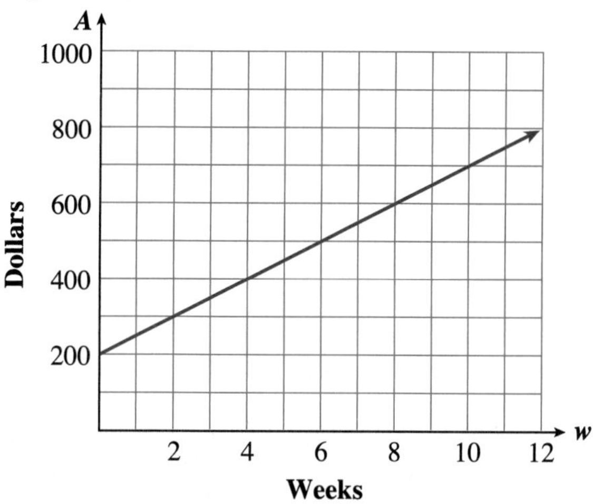

a. How much does Lynette save each week?

b. Give the coordinates (w, A) of any two points on the graph. Use those coordinates to compute the slope of the graph, $\dfrac{\Delta A}{\Delta w}$.

c. What are the units of the slope? What does the slope tell you about the problem?

2. Tyrel is saving newspapers for the paper drive at his son's school. Figure 8.29 shows the amount of newspaper P he has, in pounds, t weeks after the paper drive began.

Figure 8.29

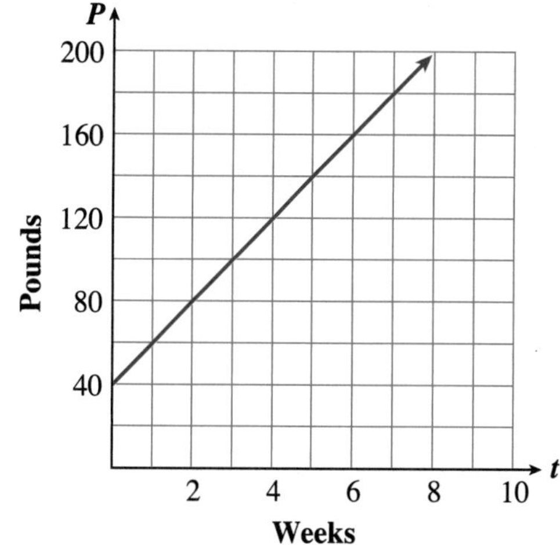

a. How many pounds of newspaper does Tyrel save each week?

b. Give the coordinates (t, P) of any two points on the graph. Use those coordinates to compute the slope of the graph, $\dfrac{\Delta P}{\Delta t}$.

c. What are the units of the slope? What does the slope tell you about the problem?

3. Britney is raising a goat to show at the state fair. Figure 8.30 shows the goat's weight W when it was t months old.

Figure 8.30

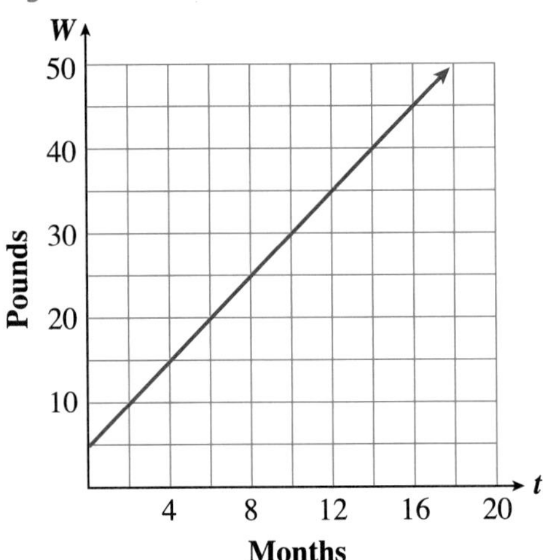

a. How much did the goat's weight increase from the second month to the fourth month? From the fourth month to the eighth month?

b. Compute the goat's rate of growth, including units.

c. Illustrate the rate of growth, $\frac{\Delta W}{\Delta t}$, on the graph.

4. Jason is raising a rabbit for the county fair. Figure 8.31 shows the rabbit's weight W when it was t weeks old.

Figure 8.31

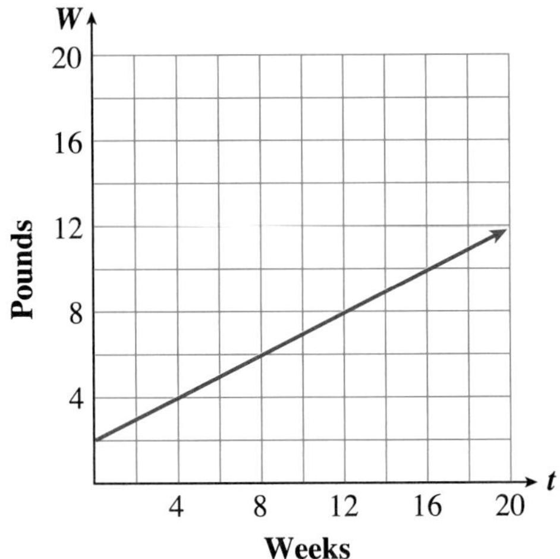

a. How much did the rabbit's weight increase from the 4th week to the 12th week? From the 2nd week to the 8th week?

b. Compute the rabbit's rate of growth, including units.

c. Illustrate the rate of growth, $\frac{\Delta W}{\Delta t}$, on the graph.

5. Figure 8.32 shows the wholesale cost C in dollars of p pounds of dry-roasted peanuts.

Figure 8.32

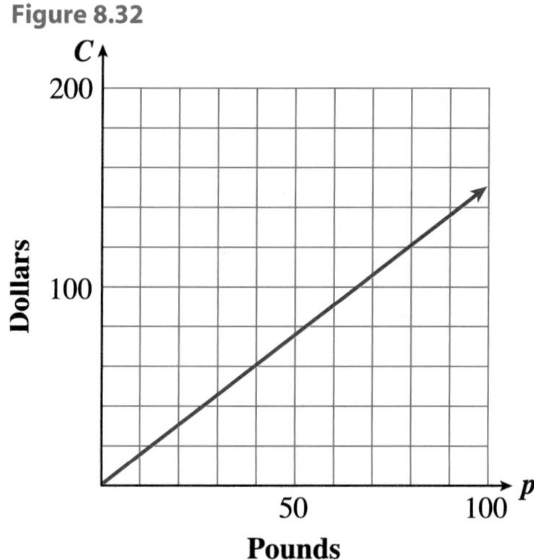

a. What does the slope of this graph tell you about dry-roasted peanuts?

b. Compute the slope, including units.

c. If the Lone Star Barbecue increases its weekly order of peanuts from 100 to 120 pounds, how much will the cost increase?

6. Figure 8.33 shows the cost C in thousands of dollars for s thousand square feet of grass sod for a lawn.

Figure 8.33

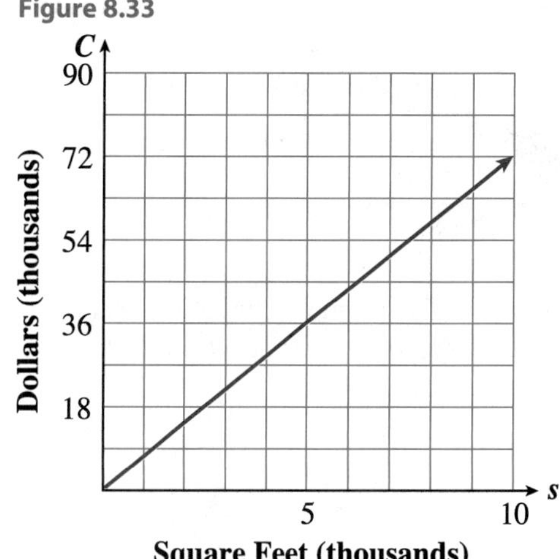

a. What does the slope of this graph tell you about sod?

b. Compute the slope, including units.

c. The Huron Breeze golf course is considering expanding its proposed putting green from 12,000 square feet to 15,000 square feet. How much will the cost of the sod increase?

7. The steepness of an incline, like a ramp or a staircase, can be measured by its slope (see Figure 8.34).

Figure 8.34

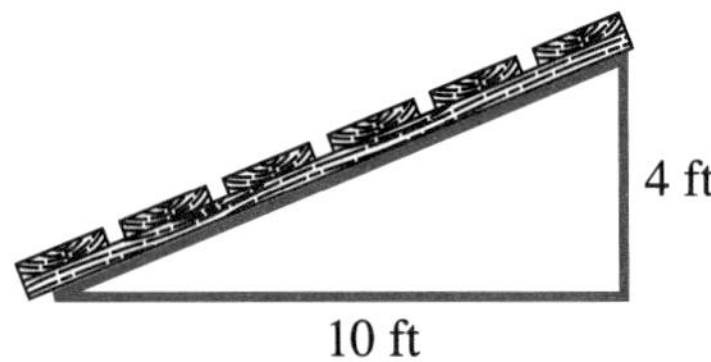

a. Find the slope of the ramp pictured if its height is 4 feet and the horizontal distance from its base to its top is 10 feet.

b. Would a ramp of slope $\frac{3}{5}$ be steeper than the one pictured, or not as steep?

8. a. Find the slope of the staircase in Figure 8.35 if its height is 12 feet and the horizontal distance between the foot and the head of the staircase is 8 feet.

Figure 8.35

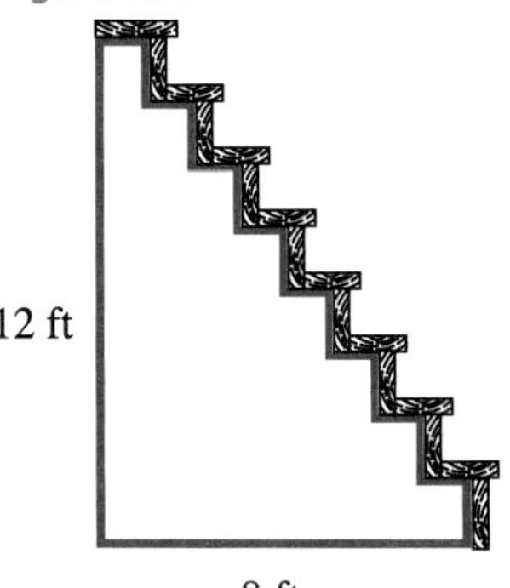

b. Would a staircase of slope $\frac{5}{4}$ be steeper than the one pictured, or not as steep?

9. Roy is traveling home by train. Figure 8.36 shows his distance d from home in miles h hours after the train started.

Figure 8.36

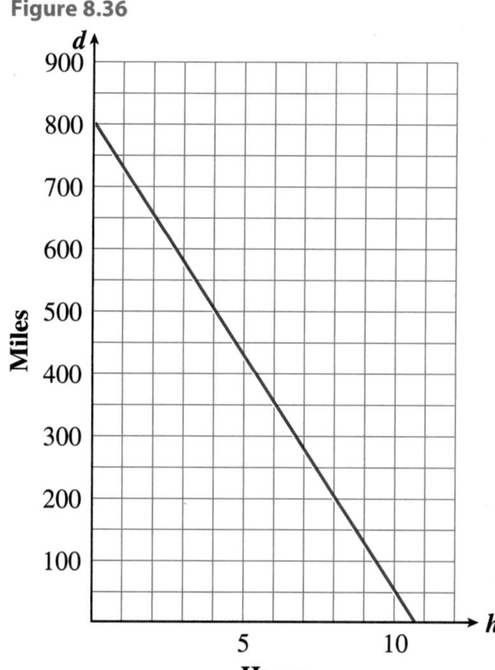

a. Compute the slope of the graph, including units.

b. What does the slope tell you about Roy's journey?

c. Find the vertical intercept of the graph. What is its meaning for the problem?

d. Compute the horizontal intercept of the graph. What is its meaning for the problem?

10. Antoine is living off his savings while he goes to college this year. Figure 8.37 shows the amount of Antoine's savings, S dollars, left after t months.

Figure 8.37

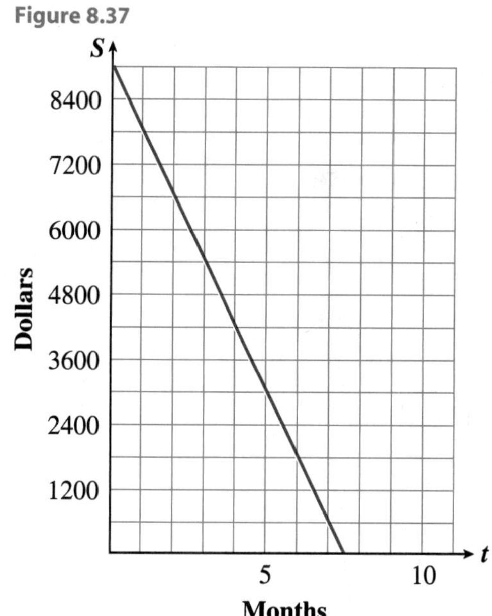

a. Compute the slope of the graph, including units.

b. What does the slope tell you about Antoine's savings?

c. Find the vertical intercept of the graph. What is its meaning for the problem?

d. Compute the horizontal intercept of the graph. What is its meaning for the problem?

11. During a chemistry experiment, Ysabel lowers the temperature of an unknown compound to test its properties. Figure 8.38 shows the temperature T of the compound, in degrees Celsius, m minutes after the experiment starts.

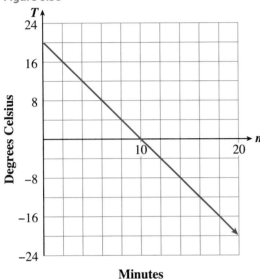

Figure 8.38

a. Compute the slope of the graph, including units.

b. What does the slope tell you about the temperature?

c. Find the vertical intercept of the graph. What is its meaning for the problem?

d. Compute the horizontal intercept of the graph. What is its meaning for the problem?

12. On a tour of a copper mine, Cara rides the elevator down to the lowest level. Figure 8.39 shows her elevation H in feet relative to ground level t minutes after the elevator starts.

Figure 8.39

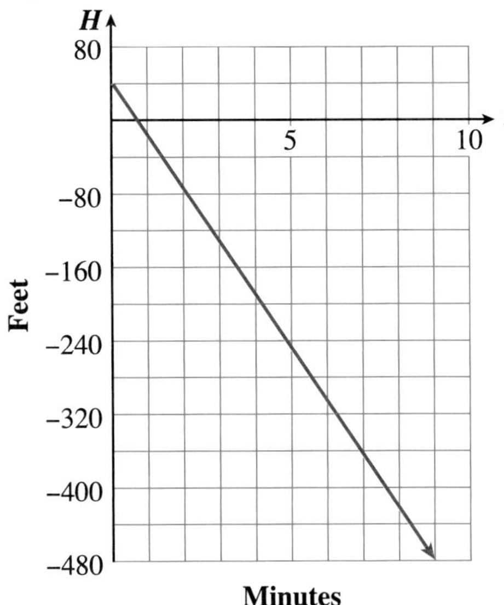

a. Compute the slope of the graph, including units.

b. What does the slope tell you about Cara's elevation?

c. Find the vertical intercept of the graph. What is its meaning for the problem?

d. Compute the horizontal intercept of the graph. What is its meaning for the problem?

(a) Use the given points to compute the slope of the line and (b) illustrate the slope on the graph.

13.

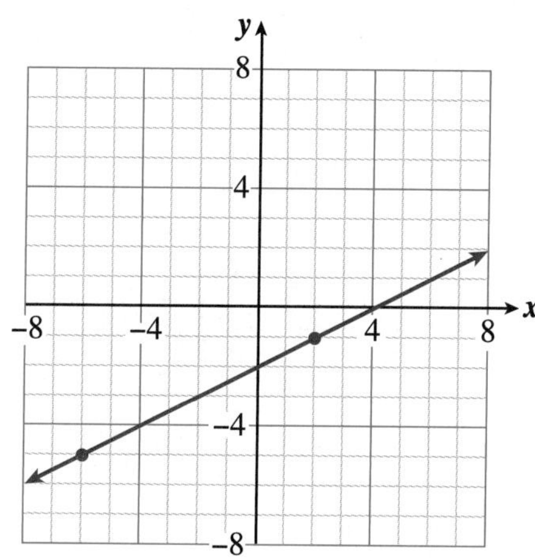

14.

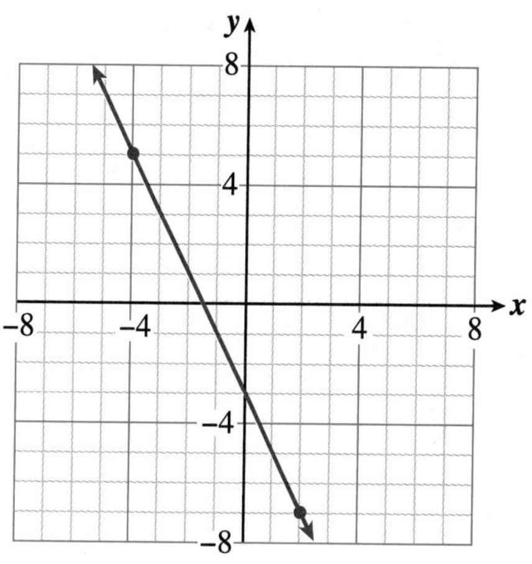

15.

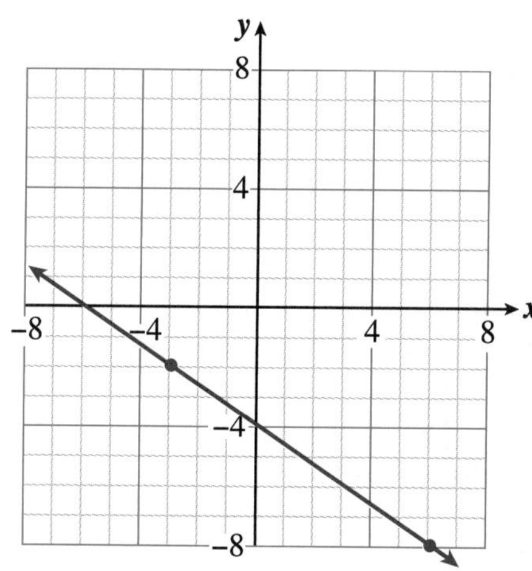

16.

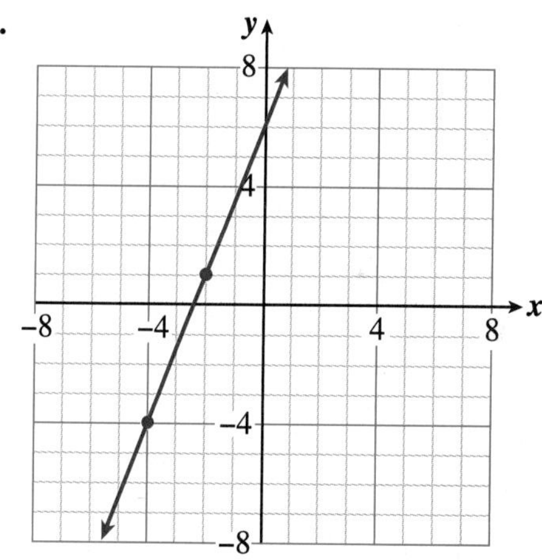

(a) Make a table of values and graph the line, (b) compute the slope of the line, and (c) find the *y*-intercept of the line.

17. $y = 2x + 3$

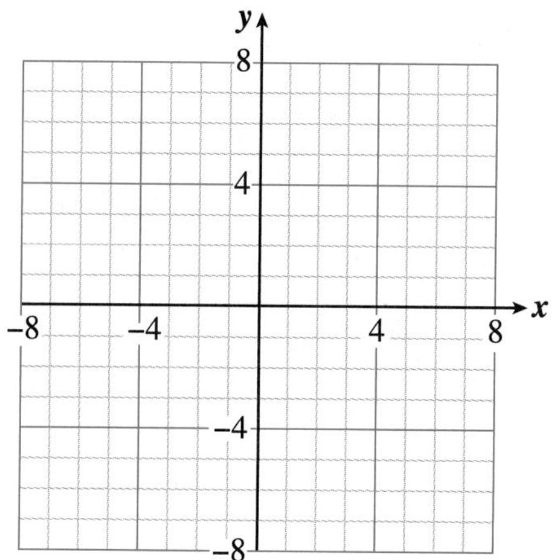

18. 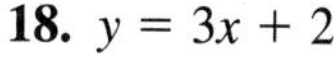 $y = 3x + 2$

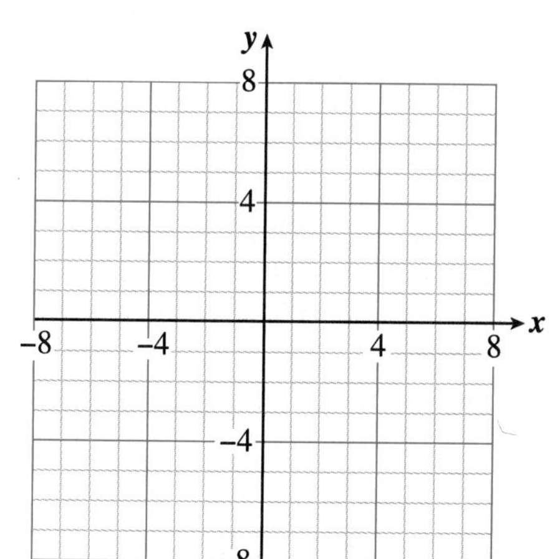

19. $y = \frac{3}{4}x - 6$

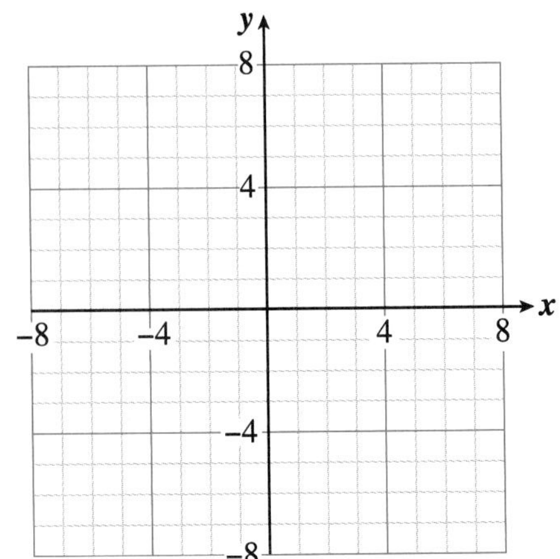

20. $y = \frac{3}{2}x - 3$

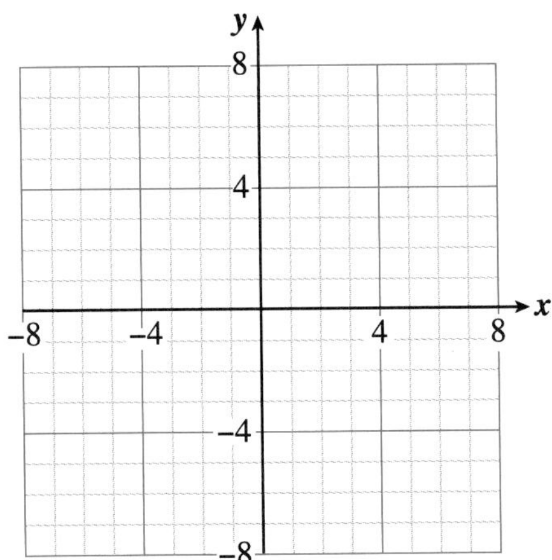

21. $y = \frac{-4}{3}x + 2$

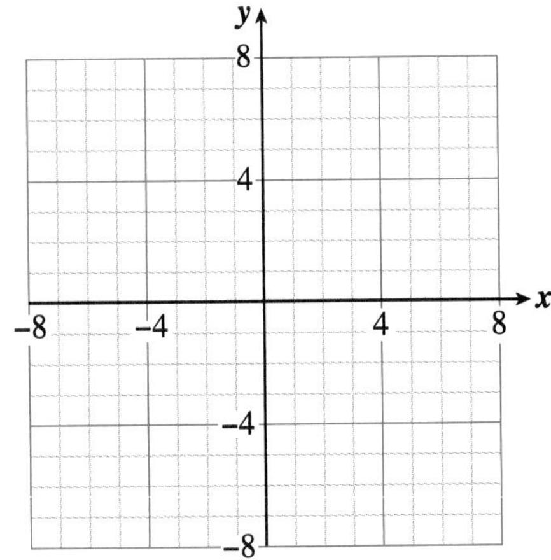

22. $y = \frac{-2}{3}x + 4$

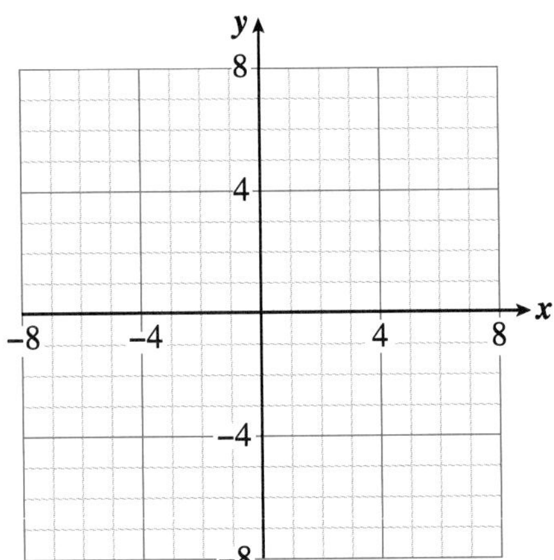

Add or subtract. (See Lesson 7.4 to review adding and subtracting fractions.) Remember to follow three steps: (1) Find the LCD, (2) build each fraction, and (3) add or subtract like fractions.

23. $\frac{3}{2p^2q} + \frac{3}{8pq^2}$

24. $\frac{9}{2vw} - \frac{2}{vw^2}$

25. $\frac{g^2}{3h^2} - \frac{3}{g}$

26. $\frac{5}{6ac^2} + \frac{2c}{3a^2}$

27. $b - \frac{2}{b}$

28. $\frac{1}{ak} + 2k$

29. $1 - \frac{1}{x+1}$

30. $1 + \frac{x+1}{x}$

Building Number Skills

Choose the best answer. Do *not* use pencil and paper or your calculator.

31. A store offers 50% off the price on certain clearance items. The discount for an item that normally sells for $59.98 is about
a. $20 **b.** $30 **c.** $40

32. In an attempt to increase its revenue, a restaurant raises the prices of all its items by 10%. For a $13.95 menu item, this increase is about
a. $1.40 **b.** $1.50 **c.** $140

33. 78% of the nursing students graduate with honors. If there are 793 nursing students, the number graduating with honors is about
a. 60 **b.** 500 **c.** 600

34. Approximately 71% of American females are 18–44 years old. In a group of 700 American females, the number aged 18–44 is about
a. 10 **b.** 50 **c.** 500

35. All employees received a 3.9% raise this year. For an annual salary of $25,000, this increase is about
a. $10 **b.** $100 **c.** $1000

36. Last year a company had $198,000 in sales. An increase of 6% is expected. This increase is about
a. $1200 **b.** $12,000 **c.** $120,000

37. Of the 35 students in the class, 17 are men. The percent of students in the class who are men is about
a. 17% **b.** 35% **c.** 50%

38. In a package of 400 computer chips, 98 of them are defective. The percent of chips that are defective is about
a. 25% **b.** 50% **c.** 98%

39. The sales tax on a $99 radio was $6.68. The tax rate is about
a. 6% **b.** 7% **c.** 60%

40. The discount on a $795 printer was $198.75. The discount rate is about
a. 25% **b.** 30% **c.** 40%

8 Summary

Lesson 8.1

Percents give us a way to compare numerical quantities.

We use a **circle graph** to show how a quantity is divided among several parts or categories.

$$\textbf{Percent increase} = \frac{\textbf{amount of increase}}{\textbf{original amount}} \times 100$$

$$\textbf{Percent decrease} = \frac{\textbf{amount of decrease}}{\textbf{original amount}} \times 100$$

To calculate the result of a percent increase, add the decimal form of the percent to 1 and then multiply that number times the original amount. To calculate the result of a percent decrease, subtract the decimal form of the percent from 1 and then multiply times the original amount.

Lesson 8.2

Like percents, ratios give us a way to compare numerical quantities.

The **ratio** of a to b is written $\frac{a}{b}$.

A ratio can be expressed as a decimal fraction or as a common fraction.

A **rate** is a ratio of quantities that have different units.

Lesson 8.3

Two quantities are said to be **proportional** if their ratios are always equal.

y is proportional to x if $\frac{y}{x} = k$, where k is a constant.

Two variables are proportional if their graph is a straight line that passes through the origin.

y is proportional to x if y is a constant multiple of x.

If y is proportional to x, then $y = kx$.

A **proportion** is a statement that two ratios are equal.

Property of Proportions

If $\frac{a}{b} = \frac{c}{d}$, then $\boldsymbol{ad = bc}$.

When using a proportion to solve a problem, make sure that both ratios have the same units in numerator and denominator.

Lesson 8.4

Multiplying the dimensions of a figure by k is called **scaling** the dimensions, and k is called the **scale factor.**

If two plane figures have the same shape but different sizes, they are called **proportional** or **similar** figures.

Area Principle for Similar Figures

If we multiply the dimensions of a plane figure by k, then the new figure is similar to the original figure, and its area is k^2 times the area of the original figure.

Formulas for the volumes of solid objects:

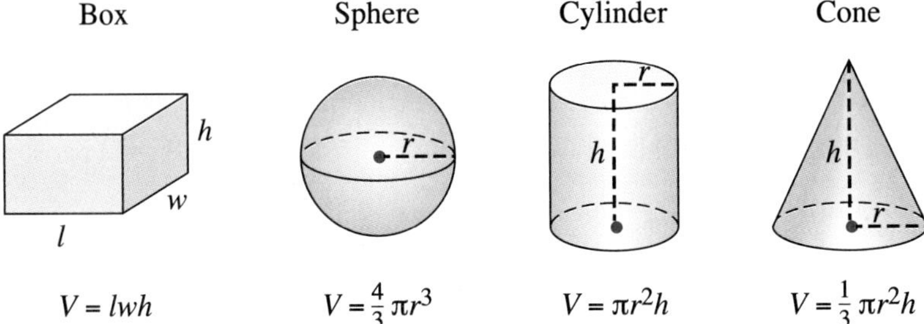

Two solid objects are called **proportional** or **similar** if they have the same shape but different sizes.

Volume Principle for Similar Solids

If we multiply the dimensions of a solid object by k, then the new object is similar to the original object, and its volume is k^3 times the volume of the original object.

Lesson 8.5

The **slope,** m, of a line is the ratio

$$m = \frac{\Delta y}{\Delta x} = \frac{\textbf{change in the } y\textbf{-coordinate}}{\textbf{change in the } x\textbf{-coordinate}}$$

as we move from any point on the line to another point on the line.

The slope of a line measures how steeply the line increases (or decreases) as we move from left to right across the graph.

If the y-values of points on a line decrease as we move from left to right, the line has a negative slope.

The slope is a ratio that tells us the rate of change of the second variable with respect to the first variable.

CHAPTER 8 REVIEW 

Add the following words and phrases to your glossary, if you have not already done so.

ratio	proportional variables	scale factor
rate	proportion	similar figures
slope	percent increase	circle graph

1. A percent is just the numerator of a fraction whose denominator is ________.

2. Explain how to convert a decimal fraction to a percent.

3. What do the variables stand for in the formula $P = rW$?

4. Explain the difference between a *ratio* and a *rate*.

5. To calculate a percent increase, we divide the ______________ by the ______________.

6. If two variables are proportional, describe their graph.

7. How can you tell if two variables are proportional by looking at a table of values?

8. A proportion is a type of ______________ in which both sides are ______________.

9. What is cross-multiplying? When can you use it?

10. Explain the statement: A slope is a kind of ratio.

For Problems 11–16, fill in the table with equivalent values.

	Fraction	*Decimal*	*Percent*
11.	$\frac{1}{8}$		
12.		0.80	
13.			8%
14.			3.5%
15.		0.35	
16.	$\frac{3}{5}$		

Write an equation and solve it to answer the question.

17. Cmart is taking a 15% discount off the regular price of all its items. A metric frimble normally sells for $13.00. How much is the discount, and what is the sales price?

18. All employees of Big Bob's Restaurant will get a 7% raise. The manager currently has a salary of $18,000 a year. How large will the manager's raise be, and what will be her new salary?

19. The sales tax alone on a new computer comes to \$424.15. If the tax rate is 8.5%, what was the price of the computer? How much is the total bill for the computer when tax is included?

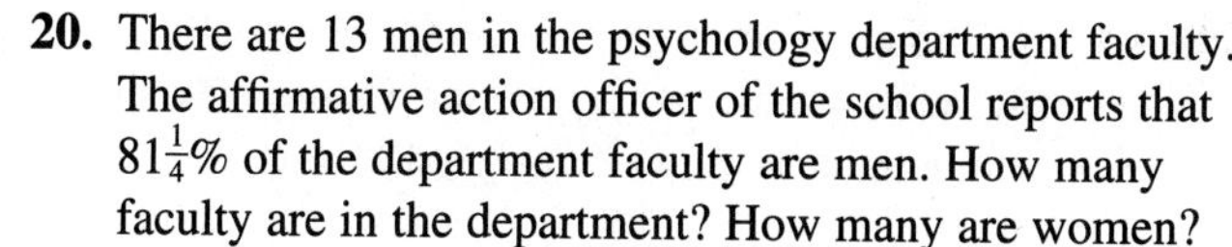

20. There are 13 men in the psychology department faculty. The affirmative action officer of the school reports that $81\frac{1}{4}\%$ of the department faculty are men. How many faculty are in the department? How many are women?

21. Figure 8.40 shows the major types of materials found in municipal landfills around the country.

Figure 8.40 Municipal sold-waste generation

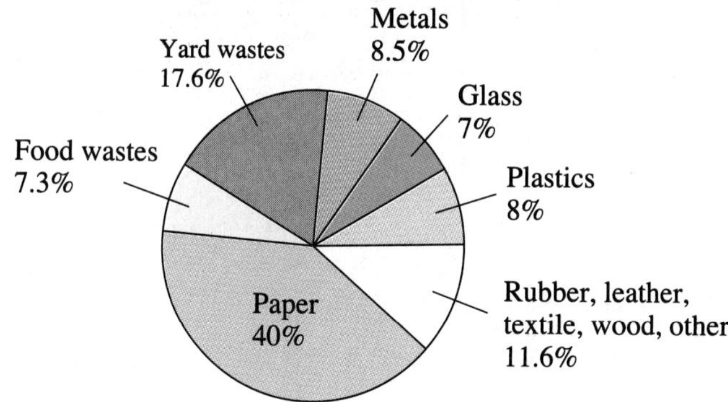

a. What percent of municipal waste comes from yard waste?

b. What material makes up 8% of municipal waste?

c. What material makes up the largest fraction of municipal waste?

d. The Environmental Protection Agency estimates that the United States will generate 179.6 million tons of solid waste between 1960 and 2010. Complete the table showing the amount of each type of waste.

Material	*Waste (%)*	*Waste (millions of tons)*
Paper		
Food waste		
Yard waste		
Metal		
Glass		
Plastic		
Other		

22. A typical home in southern California uses 7000 kilowatt-hours of electricity per year. Figure 8.41 shows how the power is used. (Source: *Los Angeles Times*)

Figure 8.41 Typical power usage

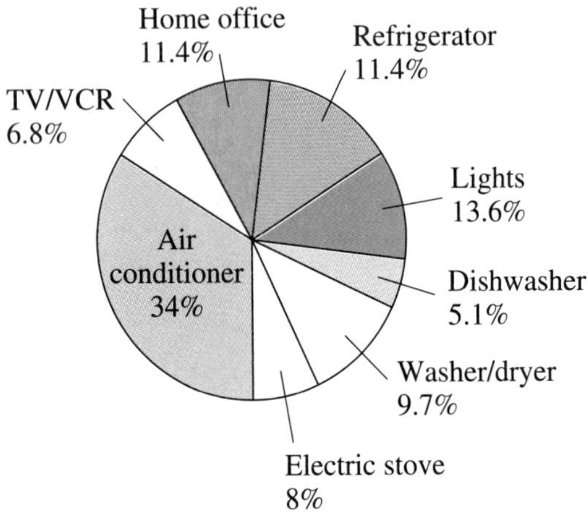

a. Which two appliances use about the same amount of electricity?

b. Which appliance uses about half as much power as electric lights?

c. Complete the table showing the number of kilowatt-hours used per year by each appliance.

Appliance	*Power (%)*	*Kilowatt-hours/Year*
Refrigerator		
Lights		
Dishwasher		
Washer and dryer		
Stove		
Air Conditioner		
TV/VCR		
Home Office		
Total		

Solve each problem.

23. The labor union negotiates a deal with management for salaries over the next 2 years. During the next year, all employees will receive a 10% salary cut from their current salary, but the following year they will receive an 11% salary increase. If Lana's current salary is \$30,000, what will be her salary next year and the following year?

24. Sherif now earns \$460 per week. When he gets his degree, his company will give him a 5% raise. By how much will his salary increase, and how much will his new salary be?

25. In 1991 the median income for a full-time worker aged 25–34 years was $26,100, and the income for a worker aged 65 or older was $34,473. In 1992, the median income for the 25–34 year-old worker was $26,533, and for the worker 65 years or older it was $35,256. Which age group had the larger percent increase in income?

26. In 1991 the national park system had expenditures of $1,104,400,000 and hosted 267,800,000 visitors. In 1992 the system spent $1,268,700,000 with 274,700,000 visitors. Which was greater, the percent increase in expenditures or the percent increase in visitors?

27. The bar graph in Figure 8.42 shows the change in stock prices of 3 Dow-Jones 30 companies for the week that ended Friday, 19 January 2001.

a. The ending price of Dupont was $42.63 per share. What was the starting price?

b. What was the percent change in the price of Dupont stock?

Figure 8.42 Change in stock prices

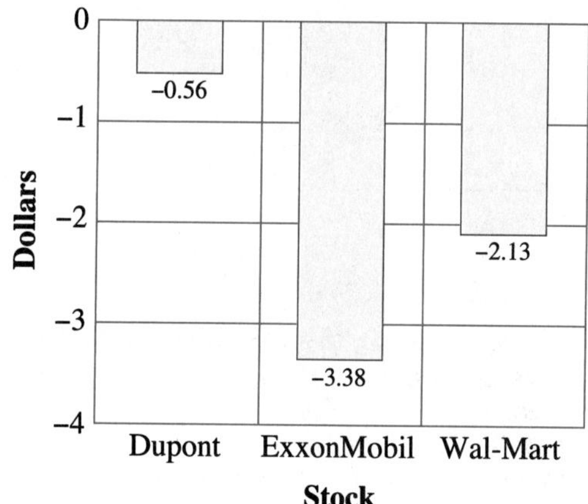

28. Use Figure 8.42 to answer the following.

a. The ending price of ExxonMobil was $79.44 per share. What was the starting price?

b. What was the percent change in the price of ExxonMobil stock?

Write ratios for each problem.

29. A serving of Healthy Grain cereal contains 65 grams of fat and 300 grams of carbohydrates. What is the ratio of grams of fat to grams of carbohydrates? Give your answer first as a common fraction and then as a decimal fraction.

30. The Chess Club has 48 men and 42 women. What is the ratio of men to women? Give your answer first as a common fraction and then as a decimal fraction.

31. Christine sewed 26 Batman costumes in 4 days. At what rate did she sew, in costumes per day?

32. Ron and Toni cross-country ski 9 miles in $1\frac{1}{2}$ hours. At what rate did they ski, in miles per hour?

For Problems 33–38, decide whether the variables are proportional.

33.

t	w
2	3.6
5	9
6	10.8
15	27

34.

b	h
3	4.5
5	12.5
6	18
8	32

35. $z = 6.5u - 2$

36. $d = \dfrac{4s}{3}$

37.

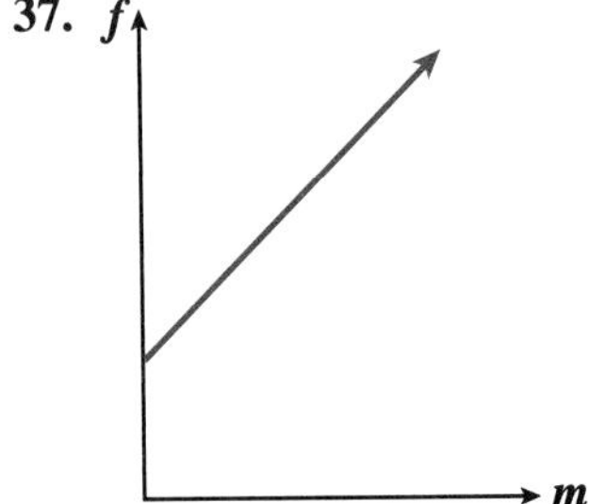

38.

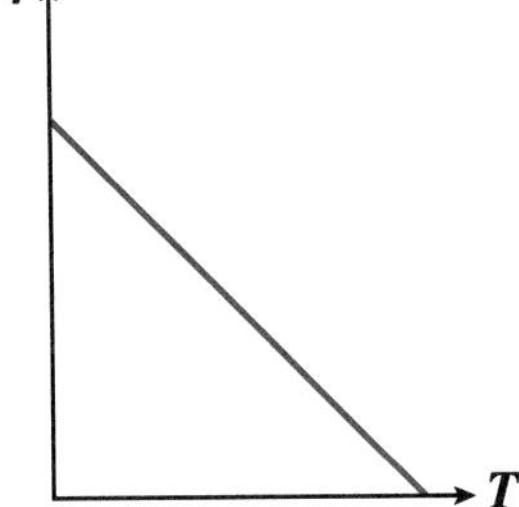

(a) Write an equation relating the variables; (b) find several points and graph your equation on the grid provided; (c) use your equation to answer the question and verify on the graph.

39. One way to assess the nutritive value of different foods is to compare the "calorie cost" of the usable protein in the food. For example, the number of calories, c, in a serving of salmon is proportional to the number of grams of protein, p, in the serving. A serving that provides 4 grams of protein costs 32 calories. How many calories do 10 grams of protein cost?

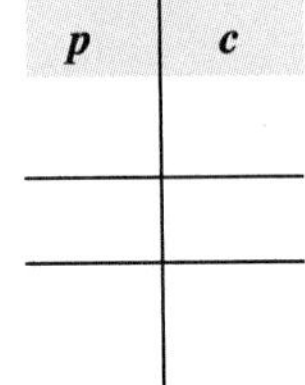

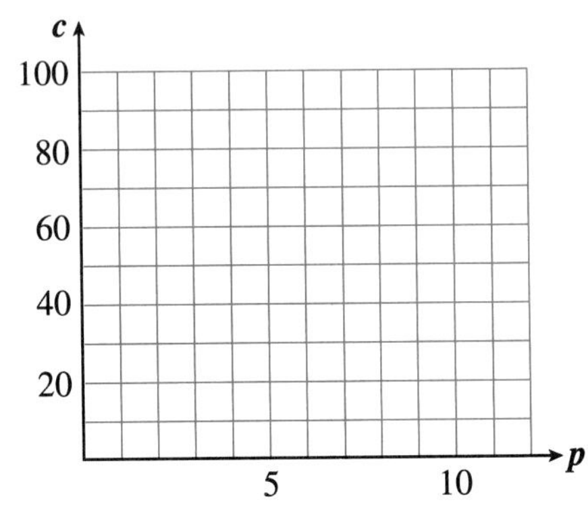

40. As you might expect, the calorie cost of hamburger is higher than the cost of salmon. (See Problem 39.) Ten grams of protein from hamburger costs 150 calories. How many calories do 8 grams of protein cost?

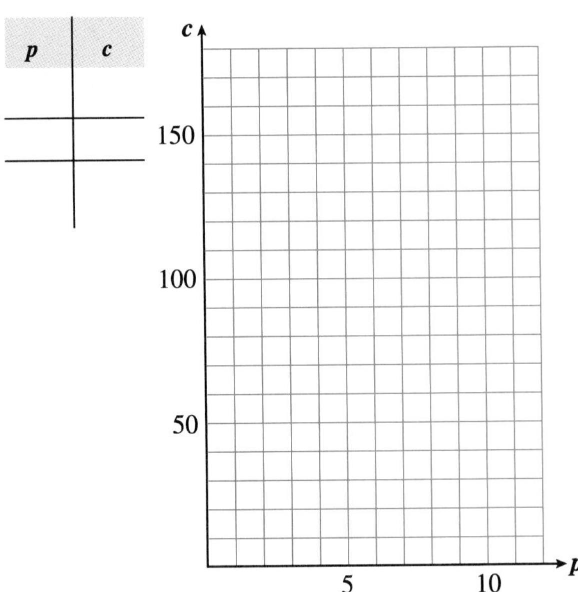

Solve each proportion.

41. $\frac{18}{n} = \frac{15}{17.5}$

42. $\frac{\frac{3}{8}}{\frac{9}{4}} = \frac{b}{2}$

43. $\frac{4}{x} = \frac{x}{9}$

44. $\frac{w}{12} = \frac{5}{2w}$

Write proportions to solve each problem.

45. For "Kids' Night" at the museum, three children are admitted free of charge if accompanied by two adults. The Isaac Newton Elementary School Boosters' Club decides to gather enough adults to go to the museum so that an entire class of children can go free.
   a. What is the ratio of children to adults if there are exactly enough of each for the children to go in free?
   b. If there are 18 children in the class, how many adults are needed?
   c. If there are 16 adults who are willing to participate, how many children can go in free?

46. Regina finds that her rental car used 24 gallons of gasoline for a trip of 768 miles.
   a. What is the ratio of miles driven to gallons used? Express the answer as a rate.
   b. If the car maintains the same gas mileage, how much gasoline will Regina use for a trip of 1552 miles?
   c. Ayman's car needed 8.5 gallons of gasoline for a trip of 306 miles. How does his car's gas mileage compare with that of Regina's rental car?

47. Trudy is walking up a hill with a long, constant incline. After walking 300 meters, she has gained 45 meters in elevation.

a. What is the ratio of elevation gained to distance walked? Express the answer as a decimal.

b. How far will Trudy have to walk on the hill before she has gained 105 meters in elevation?

c. Another hill has an increase in altitude of 1 meter over a distance of 6 meters. Which hill is steeper?

48. The Crunchy Potato Chip Company inspects a sample from each batch of potatoes delivered before the company accepts delivery. In a shipment of 600,000 potatoes, a sample of 50 is inspected. Three of the inspected potatoes have unacceptable quality.

a. What is the ratio of bad potatoes to all the potatoes in the sample? Write your answer as a decimal.

b. If the ratio of all the bad potatoes to the total 600,000 potatoes delivered is the same as the ratio from part (a), how many bad potatoes are in the shipment?

c. The company rejects any shipment if more than 5% of their sample is unacceptable. Will the company reject this shipment?

49. A half-cup serving of Nutty Nuggets contains 6 grams of sugar. How much sugar is in 1.2 cups of Nutty Nuggets?

50. On a map of the California coastline, the scale shows that 5 inches represents 3 miles. The Sycamore Canyon Trail measures 14.5 inches long on the map. How long is the actual trail?

Determine which two of the figures are similar. Then answer the questions that follow.

51.

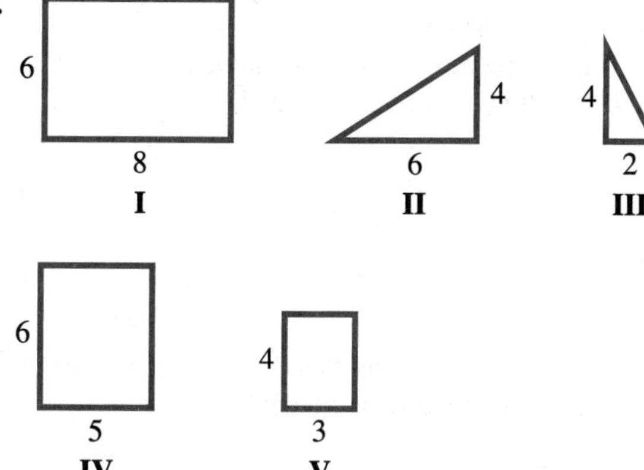

a. Find the scale factor between the similar figures.

b. What is the ratio of the areas of the two similar figures?

52.

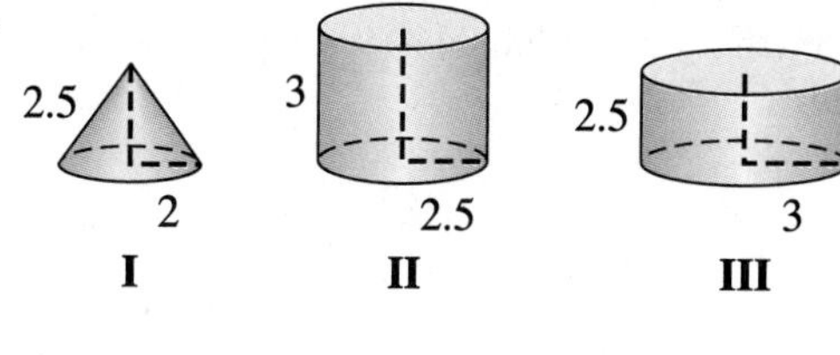

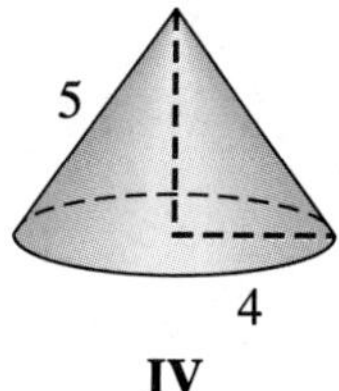

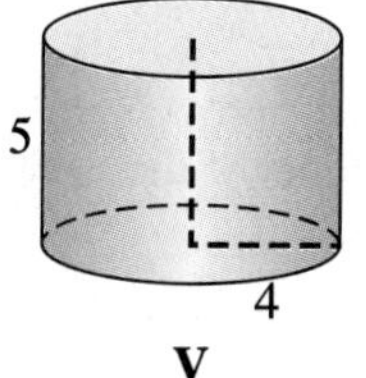

a. Find the scale factor between the similar figures.

b. What is the ratio of the volumes of the two similar figures?

53. Ida is making a scale model of her farm. The actual farm is a rectangle 1200 meters × 900 meters.

a. If the length of the model is 4 meters, what is its width?

b. If the water tower is 30 meters tall, how tall should the scale model of the tower be?

c. If the plot used for growing soybeans has an area of 60,000 square meters, what is the corresponding area in the model?

d. If the actual barn has a volume of 27,000 cubic meters, what should be the volume of the model barn?

54. Jeff created a 60-foot tall statue of a rodent to advertise a pizza restaurant. He first created a 12-inch scale model to help with the design.

a. If the scale model has a thumb that is $\frac{1}{2}$-inch tall, how tall is the statue's thumb?

b. If the statue has a tail that is 40 feet long, how long is the tail on the scale model?

c. If the statue's foot covers an area of 288 square feet, how much area does the model's foot cover?

d. If the scale model has a volume of 216 cubic inches, what is the volume of the statue?

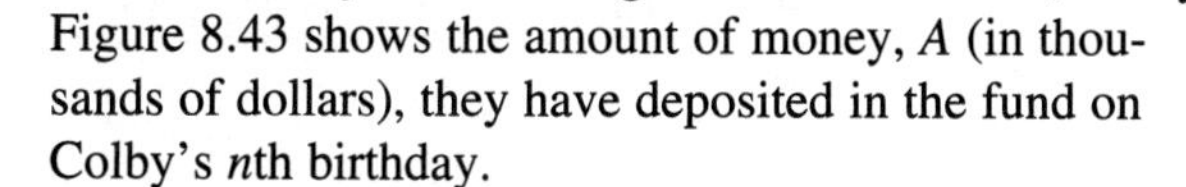

55. Rani and Larry start a college fund for their son, Colby. Figure 8.43 shows the amount of money, A (in thousands of dollars), they have deposited in the fund on Colby's nth birthday.

Figure 8.43

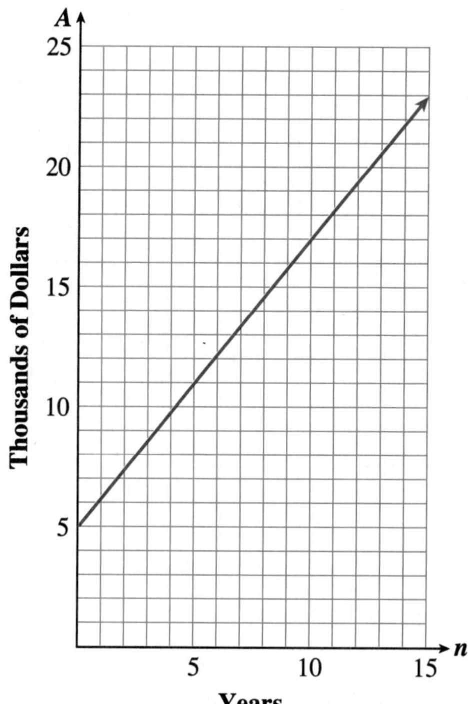

a. Read values from the graph to complete the table:

n	0	5	10
A			

b. What does the vertical intercept tell you about the college fund?

c. Use values from the table to compute the slope of the graph, including units.

d. What does the slope tell you about the college fund?

56. Audra is flying to Paris for her vacation. Figure 8.44 shows her distance d from Paris, in miles, t hours into the flight.

Figure 8.44

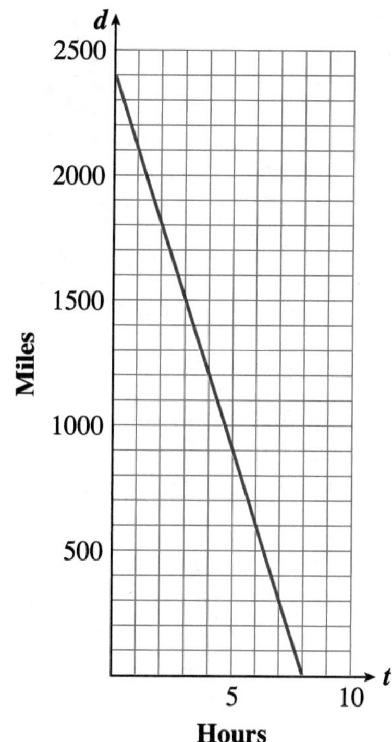

a. Read values from the graph to complete the table:

t	0	3	6
d			

b. Use values from the table to compute the slope of the graph, including units.

c. What does the slope tell you about Audra's flight?

d. What does the horizontal intercept tell you about the trip?

(a) Make a table of values and graph the line and (b) compute the slope of the line.

57. $y = \frac{5}{4}x - 2$

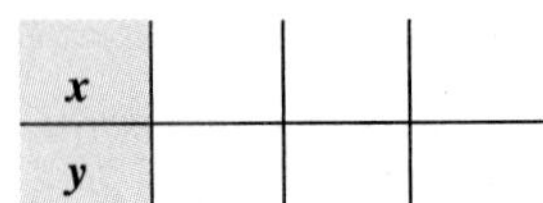

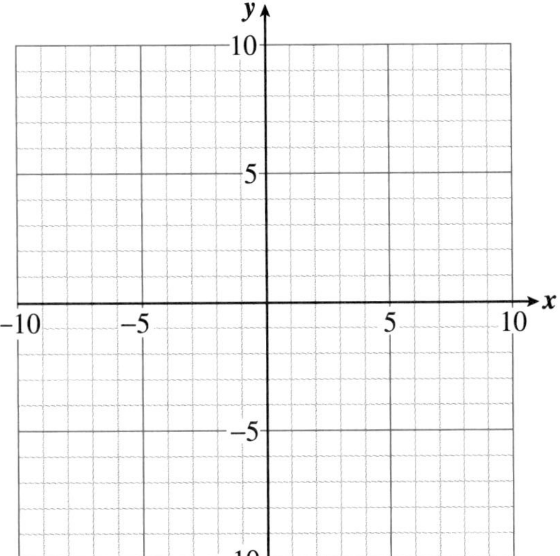

58. $y = -2x + 6$

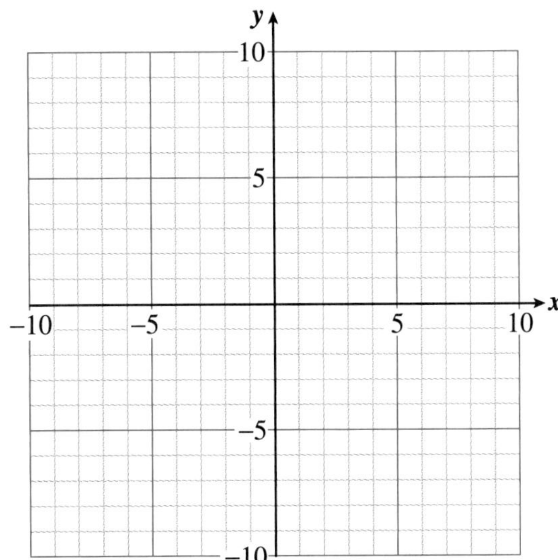

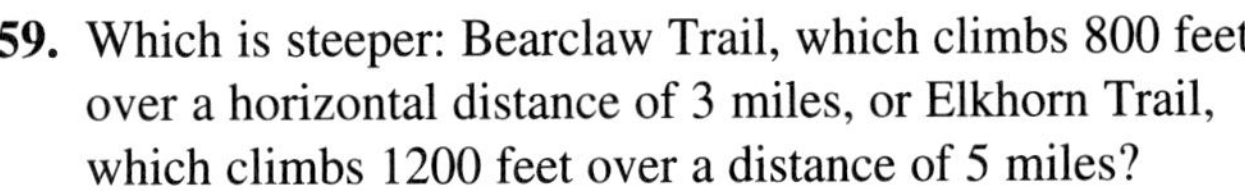

59. Which is steeper: Bearclaw Trail, which climbs 800 feet over a horizontal distance of 3 miles, or Elkhorn Trail, which climbs 1200 feet over a distance of 5 miles?

60. The tallest water slide at WaterWorld is 170 feet long, and the ladder up to the top of the slide is 50 feet tall (see Figure 8.45).

Figure 8.45

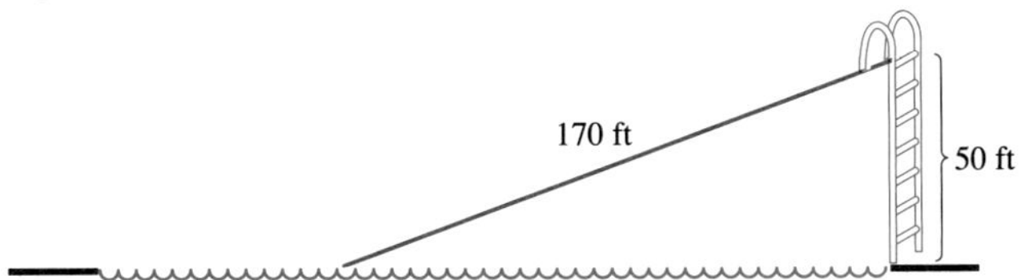

a. Find the horizontal distance from the base of the ladder to the bottom of the slide.

b. What is the slope of the slide?

Building Number Skills

Try to answer each question without using pencil and paper or a calculator. (For Problems 61 and 62, assume prices include tax.)

61. You have \$12 and you want to buy a package of pencils at \$2.19, a package of floppy disks at \$5.98, and three notebooks at \$1.47 each. Do you have enough money?

62. You have \$100 and you want to buy 10 bags of chips at \$2.39 each, 10 bottles of salsa at \$2.98 each, and 10 six-packs of soda at \$4.19 each. Do you have enough money?

63. 100 miles is about 160.9 kilometers. How many kilometers are in 1 mile?

64. 1 yard is about 91.5 centimeters. How many centimeters are there in 1 foot?

65. A chocolate dessert that serves 6 calls for 250 grams of chocolate. How much chocolate do you need to serve this dessert at a banquet of 60 people?

66. How long will it take a hollyhock to grow 2 meters if it grows 10 centimeters per week?

For Problems 67 and 68, compute mentally by converting the decimal to a common fraction.

67. 0.125(480)

68. 0.6(45)

69. 15% of \$800 is __________.

70. 15% of \$58.43 is about __________.

71. 27% of \$798 is about __________.

72. 65% of \$359 is about __________.

73. A summer dress that sold for \$68.95 is marked down 20%. The discount is about
a. \$2 **b.** \$14 **c.** \$20 **d.** \$50

74. Michiko's current salary is \$1300 per week, but she will get a 4% raise in the spring. Her new salary will be about
a. \$1304 **b.** \$1340 **c.** \$1350 **d.** \$1800

CHAPTER

9

Applications to Problem Solving

9.1 More Equations

A. Equations with Like Terms

Before we begin solving an equation, we should simplify both sides of the equation as much as possible. In particular, we should combine like terms when they occur on one side of the equation. (See Lesson 5.3 to review like terms.) Here is an example of such an equation.

EXAMPLE 1

Kyle worked 18 hours last week and 16 hours this week. His paycheck for these 2 weeks, before deductions, was \$292.40. What is Kyle's hourly wage?

Solution

Step 1 Let w represent Kyle's hourly wage. Write algebraic expressions for the amount of money Kyle made last week, and the amount he made this week:

Kyle's wages last week: 18*w*
Kyle's wages this week: 16*w*

Step 2 Write an equation about Kyle's paycheck for 2 weeks:

Kyle's paycheck is equal to the amount he made last week plus the amount he made this week.

$$\mathbf{18w + 16w = 292.40}$$

Step 3 Solve the equation:

$$18w + 16w = 292.40 \quad \text{Combine like terms.}$$
$$34w = 292.40 \quad \text{Divide both sides by 34.}$$
$$\frac{34w}{\mathbf{34}} = \frac{292.40}{\mathbf{34}}$$
$$w = 8.60$$

Kyle's hourly wage is \$8.60.

EXERCISE 1

Arancia worked 10 hours last week and 15 hours this week.

a. Let h represent Arancia's hourly wage. Write algebraic expressions for her wages last week and her wages this week.

b. Arancia's paycheck for the 2 weeks was \$462.50. Write an equation about her paycheck.

c. Solve your equation to find Arancia's hourly wage.

EXERCISE 2

Solve $5w - 8 - 3(4w) = 48$

ANSWERS TO 9.1A EXERCISES

1a. $10h$, $15h$

1b. $10h + 15h = 462.50$

1c. \$18.50

2. -8

When we check the solution to the equation in Example 1, we substitute 8.60 for w each time it occurs in the equation. Thus,

$$18(\mathbf{8.60}) + 16(\mathbf{8.60}) = 292.40?$$
$$154.80 + 137.60 = 292.40 \quad \text{True}$$

Now try Exercise 1.

For more complicated equations, we follow the order of operations.

EXAMPLE 2  

Solve $18 - 2t + 12 - 6(3t) = 80$

Solution

Simplify the left side of the equation as much as possible. Following the order of operations, we do the multiplication first:

$$18 - 2t + 12 - \mathbf{6(3t)} = 80 \quad \text{Multiply.}$$
$$18 - 2t + 12 - 18t = 80 \quad \text{Combine like terms.}$$
$$30 - 20t = 80$$

Once we have simplified each side of the equation, we proceed as usual to solve. First, isolate the variable term, $-20t$:

$$30 - 20t = 80 \quad \text{Subtract 30 from both sides.}$$
$$\mathbf{-30} \qquad \mathbf{-30}$$
$$-20t = 50 \quad \text{Divide both sides by } -20.$$
$$\frac{-20t}{\mathbf{-20}} = \frac{50}{\mathbf{-20}}$$
$$t = -2.5$$

The solution is $t = -2.5$.

Check: $18 - 2(\mathbf{-2.5}) + 12 - 6[3(\mathbf{-2.5})] = 80?$

$$18 + 5 + 12 + 45 = 80 \quad \text{True.}$$

Now try Exercise 2.

HOMEWORK 9.1A

Solve. Simplify each side of the equation first.

1. $8x - 3x = 20$

2. $2y + 4y = 42$

3. $-36 = 2z + 8 - 6z$

4. $29 = -3w + 14 - 2w$

5. $-7a - 5a - 12 = 0$

6. $-4b + 10b + 18 = 18$

7. $16 - 4(9c) = 34$

8. $5 + 3(6d) = -4$

9. $9 + 4f - 6f = 12$

10. $-8 - 6g + 9g = -16$

11. $-7m + 17 + 3(2m) = 3$

12. $15n - 4 - 4(3n) = -7$

13. $-24 = 26p + 2(15p) - 24$

14. $0 = 32q - 4(9q) + 4$

15. $-6s - s - 10 = 0$

16. $-6s - 1 - 10s = 0$

17. $-6s - s - 10s = 0$

18. $-6s - 1 - 10 = 0$

19. $-3(-2v) - 3(-2v) - 60 = 36$

20. $4(-3u) - u - 3u - 29 = 39$

(a) Choose the correct equation for each situation and (b) solve the equation.

$3y + 6y = 72$	$3y + 6 = 72$
$3(6y) = 72$	$3(y + 6) = 72$

21. Harlan paid \$72 for three packages of videocassette tapes. Each package contains six tapes. How much did each tape cost?

22. Thu worked a total of 72 hours in 3 weeks. If she works 6 hours each Saturday, how many hours does she work during the week?

23. Three troops of cub scouts and their six leaders attended a weekend jamboree. There were 72 people altogether. How many cub scouts are in each troop?

24. Armand bought three jazz compact discs and six rock compact discs at a special sale price for a total of \$72. What was the sale price?

Building Number Skills

Because $50 = \frac{1}{2} \times 100$, multiplying by 50 is the same as first dividing by 2 and then multiplying the result by 100. Similarly, multiplying by 25 is the same as first dividing by 4 and then multiplying by 100.

Examples: **a.** $26 \times 50 = \dfrac{26}{2} \times 100 = 13 \times 100 = 1300$

b. $16 \times 25 = \dfrac{16}{4} \times 100 = 4 \times 100 = 400$

Perform each computation mentally, without using a calculator or a pencil.

25. 42×50

26. 88×50

27. 28×25

28. 32×25

29. 50×72

30. 50×96

31. 25×48

32. 25×64

33. 56×25

34. 98×50

B. Variables on Both Sides of an Equation

When we solve an equation, we try to isolate the variable on one side of the equation. Consider the equation

$$5x - 3 = 2x + 9$$

This equation has terms containing the variable on *both* sides of the equal sign. We must first collect all terms containing the variable on the same side of the equation.

Let's collect all variable terms on the left side of the equation. We need to get rid of the term $2x$ on the right side. We do this in the same way that we get rid of constant terms from one side of an equation: Because $2x$ is *added* to 9, we *subtract* $2x$ from both sides of the equation:

$$\begin{array}{rcr} 5x - 3 & = & 2x + 9 \\ \mathbf{-2x} & & \mathbf{-2x} \\ \hline 3x - 3 & = & 9 \end{array}$$

Subtract $2x$ from both sides.

Now that we have all the variable terms on one side of the equation, we can proceed with the solution as usual:

$$\begin{aligned} 3x - 3 &= 9 \quad \text{Add 3 to both sides.} \\ \mathbf{+3} \quad & \quad \mathbf{+3} \\ 3x &= 12 \quad \text{Divide both sides by 3.} \\ \frac{3x}{\mathbf{3}} &= \frac{12}{\mathbf{3}} \\ x &= 4 \end{aligned}$$

The solution is 4.

We can also solve the equation above by getting the variable terms on the *right* side of the equation and end up with the same solution. Here is the same equation, but this time we solve it by getting all variable terms on the right side of the equal sign.

EXERCISE 3

Solve $-3b - 6 = 2b + 19$

EXAMPLE 3  

Solve $5x - 3 = 2x + 9$

Solution

Collect all variable terms on the right side of the equation:

$$\begin{aligned} 5x - 3 &= 2x + 9 \\ \mathbf{-5x} \quad & \quad \mathbf{-5x} \quad \text{Subtract } 5x \text{ from both sides.} \\ -3 &= -3x + 9 \\ \mathbf{-9} \quad & \quad \mathbf{-9} \quad \text{Subtract 9 from both sides.} \\ -12 &= -3x \\ \frac{-12}{\mathbf{-3}} &= \frac{-3x}{\mathbf{-3}} \quad \text{Divide both sides by } -3. \\ 4 &= x \end{aligned}$$

The solution is 4, the same as before.

Now try Exercise 3.

Problem Solving

Equations that arise in problem solving may have variables on both sides of the equal sign.

EXAMPLE 4

To get the score he wants on his take-home exam, Barry can either work 30 problems and write an essay worth 36 points, or he can work 24 problems and complete a project worth 60 points. How much is each problem worth?

Solution

Step 1a What are we asked to find in the problem?

The point value of each problem

Step 1b Choose a variable to represent the unknown quantity:

Point value of each problem: p

Step 2a Find something in the problem that we can express in two different ways:

The score Barry wants.

First way: 30 problems plus the essay: $\mathbf{30p + 36}$
Second way: 24 problems plus the project: $\mathbf{24p + 60}$

Step 2b Write an equation using the expressions from Step 2a:

$$\mathbf{30p + 36 = 24p + 60}$$

Step 3a Solve the equation:

$$
\begin{aligned}
30p + 36 &= 24p + 60 \\
-24p \quad &\quad -24p && \text{Subtract } 24p \text{ from both sides.} \\
6p + 36 &= 60 \\
-36 &\quad -36 && \text{Subtract 36 from both sides.} \\
6p &= 24 \\
\frac{6p}{6} &= \frac{24}{6} && \text{Divide both sides by 6.} \\
p &= 4
\end{aligned}
$$

Step 3b Answer the question in the problem:

Each problem is worth 4 points.

Now try Exercise 4.

EXERCISE 4

On her last history test, Parisa answered nine short-answer questions correctly and earned 15 points on the essay. Seana answered ten short-answer questions correctly and got 23 points on the essay. Seana's score on the test was 14 points higher than Parisa's. How many points was each short-answer question worth?

a. Choose a variable and write an equation for the problem.

b. Solve your equation and answer the question in the problem.

ANSWERS TO 9.1B EXERCISES

3. -5

4a. $(9p + 15) + 14 = 10p + 23$

4b. 6 points

HOMEWORK 9.1B 

Solve each equation.

1. $5z + 23 = 2z + 14$

2. $6a - 13 = 2a + 19$

3. $13 - 2b = 3b - 32$

4. $8 - 4y = -7 - 7y$

5. $-2 - 3w + 5w = -4w - 44$

6. $2c - 5 + 4c = 18 - 3c + 4$

7. $2x - 7 + 3x = 1 + 8x + 4$

8. $10 - 6d - 3d = 2 - 5d$

9. $-12n - 20 = 15n + 4 - 3n$

10. $-5p - 4 + 3p = 2 + 7p + 3$

Solve each problem using three steps: (a) Identify the unknown quantity and choose a variable to represent it. (b) Find some quantity that can be expressed in two different ways and write an equation. (c) Solve the equation and answer the question in the problem. Write your answer in a sentence.

11. Last summer Kevin made eight batches of barbecue sauce using one bottle of catsup. This summer he made only five batches and had 12 ounces of catsup left over. How much catsup does Kevin use in a batch of barbecue sauce?

12. Barbara used six skeins of yarn to make a baby blanket for her granddaughter. She made another identical blanket for her grandniece, using two skeins of yarn plus a 12-ounce ball of yarn. How many ounces are in each skein of yarn that Barbara uses?

13. Arch has two wooden beams of equal length. He makes three posts for mailboxes with one beam and has 33 inches left. He makes another mailbox post with the other beam and has 131 inches left. How long is each mailbox post?

14. George and Dan work as computer programmers for the same hourly wage. Last week George worked 35 hours and got a $60 bonus. Dan worked 28 hours and received a $144 stipend for attending a course in the language C. If their paychecks were for the same amount, what is their hourly wage?

15. Quentin bought 12 cans of cat food and a half-gallon of milk that cost $1.80. Tilda bought 8 cans of cat food and a $5 bag of kibble. They paid the same price at the checkout stand. How much does one can of cat food cost?

16. Laurie went to the Iosco County fair in Tawas City. She calculates that with the money she has she can go on six rides and spend $5.50 on food, or she can go on nine rides and spend $1.75 in the penny arcade. How much does one ride cost?

17. If Sarah drinks three glasses of milk a day, she will exceed her minimum daily requirement for calcium by 70 milligrams. If she drinks only two glasses of milk, she will still need another 220 milligrams of calcium. How many milligrams of calcium does a glass of milk contain?

18. Mr. Owsley starts his morning chores when the first ferry leaves for the Charity Islands. When the ferry returns from its third trip, he still has 30 minutes of work to do. When the ferry returns from its fifth trip, Mr. Owsley has been done for 50 minutes. How long does the ferry take for one round-trip?

Building Number Skills

Because 18 is exactly 2 less than 20, adding 18 is the same as adding 20 and subtracting 2 from the result. Adding other numbers that are nearly multiples of 10 can be done similarly.

Examples: **a.** $34 + 18 = (34 + 20) - 2 = (54) - 2 = 52$

b. $53 + 29 = (53 + 30) - 1 = (83) - 1 = 82$

Perform each computation mentally, without using a calculator or pencil.

19. $15 + 39$

20. $46 + 39$

21. $45 + 28$

22. $53 + 38$

23. $59 + 34$

24. $19 + 72$

25. $48 + 27$

26. $78 + 14$

27. $37 + 26$

28. $15 + 47$

9.2 The Distributive Law

A. Simplifying Expressions

To simplify the expression

$$3(5 + 7)$$

we follow the order of operations: First add the terms within parentheses and then multiply to get

$$3(5 + 7) = 3(12) = 36$$

But how can we simplify an expression like the following?

$$3(5 + 7x)$$

Figure 9.1

15 ft

18 ft 12 ft

We cannot add the terms inside the parentheses because they are not like terms. We will soon encounter equations in which such expressions occur. A property called the **distributive law** allows us to simplify these expressions.

Even though the distributive law is most useful for expressions containing variables, we'll start with an example involving numbers only. Suppose you plan to convert your garage into a small apartment with a bedroom and a sitting room, with the dimensions shown in Figure 9.1.

What will be the total living space (the area) of the apartment? We can calculate the area in two different ways:

Method 1 Think of the apartment as one large space.

Method 2 Find the area of each room and add them together.

We should get the same value for the area with either method. Using the formula $A = lw$ for the area of a rectangle, we have the following equation:

$$\underbrace{15(18 + 12)}_{\text{Method 1}} = \underbrace{15(18) + 15(12)}_{\text{Method 2}}$$

You can check that both of these methods give a total area of 450 square feet. However, the equation itself is more important than the answer here because it is an example of the distributive law. Think of distributing the multiplication by 15 over each term inside the parentheses:

$$15(18 + 12) = 15(18) + 15(12)$$

This gives us an alternate way of simplifying the expression $15(18 + 12)$. Although we don't really need another method for this example, we can apply the method to expressions involving variables as well.

In general, the distributive law looks like this.

The Distributive Law

$$a(b + c) = ab + ac$$

Using the Distributive Law with Variables

The distributive law also applies to variable expressions. Before continuing, you may want to review Lesson 5.3 on multiplying variable terms.

EXAMPLE 1 

Use the distributive law to simplify.

a. $3(5 + 7x)$ **b.** $-4(8 - 5w)$

Solution

a. Distribute the multiplication by 3 to each term inside the parentheses:

$$3(5 + 7x) = 3(5) + 3(7x)$$
$$= 15 + 21x$$

b. Distribute the multiplication by -4 to each term inside the parentheses:

$$-4(8 - 5w) = -4(8) - 4(-5w)$$
$$= -32 + 20w$$

EXERCISE 1
Simplify $-6(4b - 9)$

Now try Exercise 1.

Recall that $-x$ means $-1 \cdot x$. Similarly, $-(a + b)$ means $-1 \cdot (a + b)$. Thus, if a negative sign appears outside the parentheses, we can use the distributive law to simplify the expression.

EXAMPLE 2  

Simplify $-(6z - 5)$

Solution

$$-(6z - 5) = -1 \cdot (6z - 5) \quad \text{Apply the distributive law.}$$
$$= -6z + 5$$

EXERCISE 2
Simplify $-(1 - x)$

Now try Exercise 2.

A negative sign outside the parentheses changes the sign of *each term inside the parentheses.*

We can also use the distributive law to multiply by a number on the right of the parentheses:

$$(b + c)a = ba + ca$$

EXAMPLE 3 

$$(-3p + 6q)(-8) = -3p(-8)6q(-8)$$
$$= 24p - 48q$$

EXERCISE 3
Simplify $(4g - 8h)(-3)$

Now try Exercise 3.

Applying the distributive law is a form of multiplication: We multiply the outside factor by each term inside the parentheses. On the other hand, combining like terms involves addition and subtraction. Thus, the order of operations tells us to apply the distributive law before combining like terms.

EXERCISE 4
Simplify
$8x - 4(3x + 5) - (1 - 2x)$

ANSWERS TO 9.2A EXERCISES

1. $-24b + 54$
2. $-1 + x$
3. $-12g + 24h$
4. $-2x - 21$

EXAMPLE 4 

Simplify and combine like terms: $9(a + 2) - 3(2a - 5) + 4a$

Solution

Since multiplication comes before addition in the order of operations, we apply the distributive law to remove the parentheses before we combine the like terms:

$$
\begin{aligned}
&9(a + 2) - 3(2a - 5) + 4a && \text{Apply the distributive law.}\\
&\quad = 9a + 18 - 6a + 15 + 4a\\
&\quad = 7a + 33 && \text{Combine like terms}
\end{aligned}
$$

Now try Exercise 4.

HOMEWORK 9.2A 

Use the distributive law to write each expression without parentheses.

1. $3(x - 4)$
2. $4(x + 2)$
3. $5(2y - 3)$
4. $8(3y - 6)$
5. $-2(4x + 8)$
6. $-3(5x - 7)$
7. $-5(4 - 5a)$
8. $-9(6 - 2a)$
9. $-(5b - 3)$
10. $-(-8b - 5)$
11. $(-6 + 2t)(-6)$
12. $(3y + 3)(-7)$

For Problems 13–18, choose a value for the variable and evaluate the two expressions. Are they equal?

13. $2(b + 3)$; $2b + 6$

14. $3(a - 5)$; $3a - 15$

15. $-5(2z - 1)$; $-10z + 5$

16. $-2(4x + 3)$; $-8x - 6$

17. $(6 - 4k)(3)$; $18 - 12k$

18. $(-2 + 7t)(-1)$; $2 - 7t$

19. Jim and Nora collect aluminum cans to recycle for the school where they teach. Jim collected 4 pounds of cans, and Nora collected 6 pounds. The recycling center pays 55 cents per pound. Compute Jim and Nora's income in two ways:
    a. How much did Jim make? How much did Nora make? What was their total income?

    b. How many pounds of cans did Jim and Nora collect? What was their income?

20. Ralph sold 16 tickets to the Light Opera production of *H.M.S. Pinafore,* and Josephine sold 12 tickets. The tickets sell for \$8 each. Compute the amount of money Ralph and Josephine raised in two ways:
    a. How much did Ralph raise? How much did Josephine raise? How much did they raise together?

    b. How many tickets did Ralph and Josephine sell? How much money did they raise?

21. **a.** Use the distributive law to simplify $3(2x + 5x)$.

    b. Use the order of operations to simplify $3(2x + 5x)$.

22. **a.** Use the distributive law to simplify $5y(23 - 18)$.

    b. Use the order of operations to simplify $5y(23 - 18)$.

For Problems 23–32, simplify and combine like terms.

23. $-6(x + 1) + 2x$

24. $5(1 - x) - 3x$

25. $4x - 9(2 - 3x)$

26. $-7x - 3(6 + 4x)$

27. $5 - 2(4x - 9) + 9x$

28. $8 - 3(12x - 4) - 2x$

29. $5(y + 3) - 2(y - 3)$

30. $-7(3z - 5) - 2(6 + 2z)$

31. $-3(2t - 5) + 2t - 5(t + 2)$

32. $8(1 - 2x) + 3x - 4(2x - 2)$

33. **a.** Explain in words the difference between the expressions $5b$ and $5 + b$.

b. Explain in words the difference between the expressions $2(5b)$ and $2(5 + b)$.

c. Simplify each expression in part (b).

34. **a.** Explain in words the difference between the expressions $4 - z$ and $4(-z)$.

b. Explain in words the difference between the expressions $-2(4 - z)$ and $-2(4)(-z)$.

c. Simplify each expression in part (b).

35. True or false:

a. The distributive law does not apply to the expression $7(5t)$.

b. When using the distributive law, we must multiply the factor outside the parentheses by each term inside the parentheses.

c. The distributive law cannot be used to simplify expressions that involve multiplication only.

d. We can only use the distributive law if the terms inside the parentheses are like terms.

36. Which of the following is a correct application of the distributive law?

a. $3(2x) = 6 + 3x$

b. $3(2 + x) = 6 + x$

c. $3(2x) = 6x$

d. $3(2 + x) = 6 + 3x$

Building Number Skills: Divisibility by 2 or 4

All the even numbers are divisible by 2. If the last two digits of a number are divisible by 4, the original number is divisible by 4.

Examples: **a.** 316 is even, so it is divisible by 2. (You can check that $316 \div 2 = 158$.)

b. 16 is divisible by 4, so 316 is also divisible by 4.

c. 841 is odd, so it is not divisible by 2 nor by 4.

Determine if the number is divisible by 2 and 4.

37. 548 **38.** 916 **39.** 2357 **40.** 922

41. 4528 **42.** 1234 **43.** 8762 **44.** 3701

45. 1004 **46.** 8309

B. Using the Distributive Law in Equations

In Lesson 9.1, we saw that we must combine any like terms before we can solve an equation. Sometimes we must apply the distributive law to remove parentheses as well.

EXAMPLE 5  

Solve $2(4x - 7) - 5x = 43$

Solution

We begin by simplifying the left side of the equation. Since multiplication comes before addition and subtraction in the order of operations, we first perform the multiplication by applying the distributive law:

$$2(4x - 7) - 5x = 43 \quad \text{Apply the distributive law.}$$
$$8x - 14 - 5x = 43$$
$$3x - 14 = 43 \quad \text{Combine like terms.}$$
$$\mathbf{+14} \quad \mathbf{+14} \quad \text{Add 14 to both sides.}$$
$$3x = 57$$
$$\frac{3x}{\mathbf{3}} = \frac{57}{\mathbf{3}} \quad \text{Divide both sides by 3.}$$
$$x = 19$$

The solution is 19. (You should check the solution.)

EXERCISE 5
Solve $-13 = 18 - (4t - 9) + 2t$

Now that we have considered some rather complicated equations, it might be helpful to organize our equation-solving techniques into a strategy. The following guidelines should be sufficient for all the equations we have studied so far.

Steps for Solving Equations

1. Simplify each side of the equation if necessary:
   a. Apply the distributive law to remove parentheses.
   b. Combine like terms.
2. Get all variable terms on one side of the equation by adding or subtracting terms.
3. Isolate the term containing the variable on one side of the equation by adding or subtracting to cancel the constant term from that side.
4. Divide both sides by the coefficient of the variable.

ANSWERS TO 9.2B EXERCISE
5. 20

Now try Exercise 5.

HOMEWORK 9.2B

1. Solve the equation $2(x - 5) = 6$ in two ways:
   a. By undoing the operations in reverse order

   b. By applying the distributive law to the left side first

2. Solve the equation $-3(y + 4) = 9$ in two ways:
   a. By undoing the operations in the reverse order

   b. By applying the distributive law to the left side first

Solve each equation.

3. $-4(x - 6) = 8$

4. $5(x - 2) = -15$

5. $5 - (2b + 9) = 16$

6. $3 - (3b - 3) = -9$

7. $-15 = 3(2 + 7x)$

8. $22 = -2(6x - 1)$

9. $14 + 8(3z - 5) = 34$

10. $-9 + 2(9 + 4z) = -23$

11. $-64 = 2(1 - 3w) - 5w$

12. $35 = 4(2w + 5) - 3w$

13. $9 = 6p - (p - 4) - 20$

14. $-20 = -4p - (7 - p) + 5$

15. $-30c - 30 + 5(3c + 4) = -10$

16. $18t - 6 - 2(6t - 4) = 26$

17. $0.6b + 0.3(b - 6) = 9$

18. $0.5(b - 30) + 0.2b = 55$

19. $0.7(d + 20) - 0.5(d + 30) = 13$

20. $0.4(d - 30) - 0.2(d - 40) = 6$

Write an algebraic expression for each situation.

21. At 8 A.M. the temperature was 72°, and it has been rising by 6° every hour. At this rate, what will be the temperature after h hours?

22. Avram has typed 480 words of his term paper and is still typing at a rate of 30 words per minute. How many words will Avram have typed after m minutes?

23. Gil's math notebook has 300 pages, and he uses on average 6 pages per day for his notes and homework. How many pages will Gil have left after d days?

24. Francine saved \$5000 to go to school full-time. She spends \$150 a week on food and rent. How much of her savings will be left after w weeks?

25. For her mother's retirement party, Daniella spent \$50 on a gift plus her share of the cost of the party. If the party costs P dollars and 12 people are contributing (including Daniella), how much did Daniella spend?

26. Delbert shares a house with four roommates. He pays \$200 rent per month, plus an equal share of the utilities. If the utilities cost U dollars this month, how much does Delbert owe?

27. Brian reserved \$60 from the office Sunshine Fund for a party and used the rest to buy new desk sets for the 15 people in the office. If the Sunshine Fund originally had S dollars, how much can Brian spend on each desk set?

28. Vincent's school provided him with C colored pencils to distribute to the students in his class, and he bought 8 more pencils so that each student would get the same number. If there are 23 students in Vincent's class, how many pencils does each get?

29. Raylyn's auto registration fee is \$20 plus 2% of the value of her car. If Raylyn's car is worth B dollars, how much will her registration fee be?

30. Maryam has to carry a dummy weighing 50 pounds plus 60% of her own weight for 20 yards as part of the test to become a lifeguard. If Maryam weighs W pounds, how much must she carry?

Building Number Skills: Divisibility by 3 or 9

If the digits of a number add up to a number divisible by 3, then the original number is divisible by 3. If the digits of a number add up to a number divisible by 9, then the original number is divisible by 9.

Example: Is 345 divisible by 3? $3 + 4 + 5 = 12$, which is divisible by 3, so 345 is divisible by 3. But since 12 is not divisible by 9, 345 is not divisible by 9.

Determine if the number is divisible by 3 and 9.

31. 458

32. 691

33. 714

34. 243

35. 1116

36. 9996

37. 8197

38. 4074

39. 9018

40. 8001

9.3 Problem Solving

A. Algebraic Expressions

Often we need to express two or more unknown quantities in terms of just *one* variable.

EXAMPLE 1  

Patrick completed four more units this semester than he did last semester.

a. Choose a variable for the number of units Patrick took last semester:

Number of units last semester: u

b. Write an expression in terms of u for the number of units he completed this semester:

Number of units this semester: $u + 4$

c. Write an expression for the total number of units Patrick completed this year:

Total units completed: $u + (u + 4)$, **or** $2u + 4$

EXAMPLE 2

Paul has scored 15 fewer points than his bowling partner, Betty.

a. Chose a variable for the number of points Betty has scored:

Betty's points: p

b. Write an expression in terms of p for the number of points Paul has scored:

Paul's points: $p - 15$

c. Write an expression for the total number of points Betty and Paul have scored:

Total points: $p + (p - 15)$, **or** $2p - 15$

Now try Exercise 1.

EXERCISE 1

A hamburger and large order of fries together contain 900 calories.

a. Let h represent the number of calories in a hamburger. Write an expression in terms of h for the number of calories in a large order of fries.

b. Write an expression for the number of calories in two hamburgers and three orders of fries.

c. Simplify your expression in part (c).

ANSWERS TO 9.3A EXERCISES

1a. $900 - h$

1b. $2h + 3(900 - h)$

1c. $2700 - h$

HOMEWORK 9.3A 

1. Edith's reading assignment in history is 12 pages longer than her assignment in English.
   a. Choose a variable for the number of pages in Edith's English assignment.
   b. Write an expression in terms of your variable for the number of pages in Edith's history assignment.
   c. Write an expression for the total number of pages in Edith's two assignments.

2. Read Problem 1 again.
   a. Choose a variable for the number of pages in Edith's history assignment.
   b. Write an expression in terms of your variable for the number of pages in Edith's English assignment.
   c. Write an expression for the total number of pages in Edith's two assignments.

3. The price of a warm-up suit is $20 less than a pair of jogging shoes.
   a. Choose a variable for the price of the jogging shoes.
   b. Write an expression in terms of your variable for the price of the warm-up suit.
   c. Write an expression for the total cost of the jogging shoes and the warm-up suit.

4. Read Problem 3 again.
   a. Choose a variable for the price of the warm-up suit.
   b. Write an expression in terms of your variable for the price of the jogging shoes.
   c. Write an expression for the total cost of the jogging shoes and the warm-up suit.

5. Patel's Pet Clinic treats twice as many cats as dogs.
   a. Choose a variable for the number of dogs treated last week.
   b. Write an expression in terms of your variable for the number of cats treated last week.
   c. Last week, Patel's Clinic treated five fewer pygmy goats than the total number of cats and dogs treated. In terms of your variable, how many goats is that?

6. Bruce jogs three times as far on Saturdays as he does on Fridays.
   a. Choose a variable for the number of miles Bruce jogged last Friday.
   b. Write an expression in terms of your variable for the number of miles Bruce jogged last Saturday.
   c. Last Sunday, Bruce competed in a race that was 1 mile longer than the total distance he ran on Friday and Saturday. In terms of your variable, how many miles is that?

7. Gabriel can take 24 people, children and adults, on his boat for a whale-watching excursion.

a. Choose a variable for the number of children on the excursion.

b. Write an expression in terms of your variable for the number of adults Gabriel can take on the excursion.

c. If double the number of children are signed up for next week's trip, how many adults can go next week (in terms of your variable)?

8. Delbert and Francine together made $50,000 last year.

a. Choose a variable for the amount Delbert made last year.

b. Write an expression in terms of your variable for the amount Francine made.

c. This year, Francine doubled her salary. Write an expression for Delbert and Francine's total income this year.

9. Premium ice cream is 12% butterfat, and chocolate syrup is 55% butterfat. A chocolate sundae, without whipped cream and a cherry, weighs 10 ounces.

a. Choose a variable for the number of ounces of ice cream in the sundae.

b. Write an expression in terms of your variable for the number of ounces of chocolate syrup in the sundae.

c. Write an expression for the amount of fat in the ice cream.

d. Write an expression for the amount of fat in the chocolate syrup.

e. Write an expression for the total amount of fat in the sundae.

10. Polls show that 70% of Democrats and 40% of Republicans support a school bond measure. There are 30,000 registered voters in Kent County.

a. Choose a variable for the number of registered Democrats in Kent County.

b. Write an expression in terms of your variable for the number of registered Republicans in Kent County. (Assume there are no voters registered with a third party.)

c. Write an expression for the number of Democratic voters who support the bond measure.

d. Write an expression for the number of Republican voters who support the bond measure.

e. Write an expression for the total number of voters who support the bond measure.

Solve.

11. $4u - 2 - 7u = 10$

12. $-6y + 8 - 3y = -10$

13. $-2(-5h) - 8 = -20$

14. $3(-6k) + 7 = -20$

15. $-15 = 5(3x - 4)$

16. $-24 = 4(6x - 3)$

17. $5 + 3(-2b + 4) = 17$

18. $11 - 4(-4z + 3) = -1$

19. $-5 = 12 - (3M - 1)$

20. $-7 = 16 - (2N + 5)$

Building Number Skills: Divisibility by 5 or 10

When the last digit of a whole number is 0, then the number is divisible by 10. When the last digit of a number is 0 or 5, then the number is divisible by 5.

Example: Because the last digit of 1115 is 5, the number 1115 is evenly divisible by 5. But because the last digit is not 0, 1115 is not divisible by 10.

Determine if the number is divisible by 5 and 10.

21. 170

22. 235

23. 4355

24. 2225

25. 5050

26. 1010

27. 1001

28. 5551

29. 9015

30. 8000

B. Modeling Equations

We can now solve problems that involve more than one unknown quantity. For such problems it is often helpful to write algebraic expressions for each unknown before trying to form an equation. Here is a revised outline of our method for solving applied problems.

Steps for Solving Applied Problems

1. *Writing algebraic expressions*

- **a.** Describe each unknown quantity in words.
- **b.** Choose a variable to represent *one* of the unknown quantities.
- **c.** Write an algebraic expression for any other unknown quantities in terms of your variable.

2. *Writing an equation*

- **a.** Find some quantity that can be expressed in two different ways.
- **b.** Write an equation, using your expressions from step 1.

3. *Solving*

- **a.** Solve the equation.
- **b.** Answer the question in the problem.

EXAMPLE 3  

The winning candidate beat her opponent for a seat on the city council by 790 votes. A total of 8456 votes were cast for two candidates. How many votes did each candidate receive?

Solution

Step 1 First, identify the unknown quantities in the problem:

Number of votes for loser

Number of votes for winner

Next, assign a variable to *one* of these quantities. We will let n stand for the number of votes for the loser. We must then write an algebraic expression for the number of votes for the winner in terms of n:

Number of votes for loser: n

Number of votes for winner: $n + 790$

Step 2 Look for something that can be expressed in two different ways:

The total number of votes cast

First way: We know that the total number of votes was **8456.**

Second way: We can add the number of votes for the two candidates.

Total number of votes: $\underset{\text{Votes for loser}}{n} + \underset{\text{Votes for winner}}{n + 790}$

EXERCISE 2

a. The length of a rectangle is 6 inches more than its width. Write an algebraic expression for the perimeter of the rectangle.

b. One side of a triangle is 20 inches long. The other two sides are the same as the width of the rectangle in part (a). Write an expression for the perimeter of the triangle.

c. The perimeter of the rectangle is 16 inches greater than the perimeter of the triangle. Write an equation and find the dimensions of each figure.

Now write an equation using your expressions above:

$$\mathbf{n + n + 790 = 8456}$$

Step 3 Solve the equation. Begin by simplifying the left side:

$$n + n + 790 = 8456 \quad \text{Combine like terms.}$$
$$2n + 790 = 8456 \quad \text{Subtract 790 from both sides.}$$
$$\underline{-790 \quad -790}$$
$$2n = 7666 \quad \text{Divide both side by 2.}$$
$$\frac{2n}{2} = \frac{7666}{2}$$
$$n = 3833$$

Thus, the loser received 3833 votes. To find the number of votes for the winner, go back to Step 1, where we have the expression

Number of votes for winner: $n + 790$

Substitute 3833 for n to find

$$\mathbf{n + 790 = 3833 + 790 = 4623}$$

The winner received 4623 votes.

Now try Exercise 2.

ANSWERS TO 9.3B EXERCISES

2a. $2w + 2(w + 6)$

2b. $2w + 20$

2c. $2w + 2(w + 6) = (2w + 20) + 16$; rectangle: 12×18 inches, triangle: 12 inches $\times$ 12 inches $\times$ 20 inches

HOMEWORK 9.3B 

Solve each problem by following the three steps on page 633.

1. Ralph and Wanda spent the afternoon stuffing envelopes for their community center. Wanda stuffed 57 more envelopes than Ralph, and together they stuffed 349 envelopes. How many envelopes did each stuff?

2. A new bicycle helmet costs \$53 less than new saddle bags. Together the helmet and saddle bags cost \$222. How much does each cost?

3. A restaurant has twice as many tables in its nonsmoking section as in its smoking section. If there are 54 tables all together, how many are designated for smokers?

4. Eight times as many students as faculty members attended the Honor Society's tea. If 144 people attended the tea, how many were faculty members?

5. Kiri is driving a bus full of children to summer camp. The bus holds 64 children, and there are six fewer girls than boys on the bus. How many boys are on the bus?

6. There are 18 fewer men than women enrolled in the Physical Therapy program. If the total enrollment is 86 students, how many are women?

7. One child and one adult together can buy tickets to Mystery Ridge for $32. Tickets for four children and two adults cost $82. How much is a child's ticket?

8. A cheeseburger and a pineapple shake together contain 1030 calories. Two shakes and three cheeseburgers contain 2710 calories. How many calories are there in a pineapple shake?

9. The central table in a U-shaped display booth has the shape of a trapezoid. (See the figure.) The area of the table should be 28 square feet, and it should be 4 feet wide. If one base of the trapezoid is 6 feet longer than the shorter base, how long is each base? (See Homework 4.2B for an appropriate formula.)

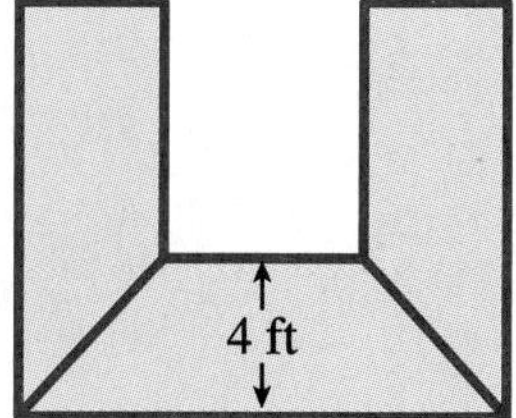

10. Allen's latest mural is a rectangular canvas whose length is 3 feet less than twice its width. Allen calculates that he will need 42 feet of 2-inch-wide fir stripping to make a frame for the mural. What are the dimensions of the mural? (*Hint:* What are "the dimensions" of a rectangle?) (See Homework 4.2B for an appropriate formula.)

11. Jetstream Airlines sold 140 tickets to Phoenix and took in $22,600. First-class tickets cost $230, and coach tickets cost $150.

a. Write algebraic expressions for the number of coach tickets sold and the number of first-class tickets sold.

b. Write algebraic expressions for the receipts from the coach tickets and the receipts from the first-class tickets.

c. Write an equation about the ticket receipts.

d. How many coach tickets were sold, and how many first-class tickets?

12. Axel's truck gets 24 miles to the gallon in town and 35 miles to the gallon on the highway. Last month Axel purchased 40 gallons of gasoline and traveled 1235 miles.

a. Write algebraic expressions for the number of gallons Axel used in town and the number of gallons he used on the highway.

b. Write algebraic expressions for the number of miles Axel drove in town and the number of miles he drove on the highway.

c. Write an equation about the number of miles Axel drove.

d. How many gallons of gas did Axel use in town and how many on the highway?

Simplify.

13. $7(y - 5) + 3(5 - y)$

14. $-4(3 + 2x) - 3(2x + 1)$

15. $6 - 2(a - 3) - 4(1 + 3a) + 12 - 2$

16. $-5a + 7 - 2(4a - 3) - 6a + 4(2 - 3a)$

Solve.

17. $24 - 8n + 4 = -6 - 10n$

18. $10 - (x - 6) = 3(x - 8) + x$

19. $0.30(h + 200) + 0.50h = 0.60h + 220$

20. $0.80t - 180 = 0.40t + 0.20(t - 300)$

Building Number Skills

We can perform many subtraction problems mentally by regrouping the first number.

Examples: **a.** $53 - 6 = (40 + 13) - 6 = 40 + (13 - 6) = 40 + 7 = 47$

b. $37 - 9 = (20 + 17) - 9 = 20 + (17 - 9) = 20 + 8 = 28$

Subtract mentally by regrouping the first number as a multiple of 10 plus a number between 10 and 20.

21. $44 - 8$

22. $81 - 5$

23. $73 - 6$

24. $62 - 9$

25. $96 - 7$

26. $36 - 9$

27. $54 - 5$

28. $93 - 4$

29. $68 - 9$

30. $44 - 6$

9.4 Equations with Fractions

Clearing Fractions from an Equation

In Lesson 7.5 we solved equations by first clearing the fractions. Now let's solve a more complicated equation:

$$\frac{3}{4}x - \frac{1}{2} = \frac{5}{4}$$

Here is a possible solution. First, analyze what operations have been performed on the variable and then undo those operations in reverse order:

Operations Performed on x	*Steps for Solution*
1. Multiplied by $\frac{3}{4}$	**1.** Add $\frac{1}{2}$
2. Subtracted $\frac{1}{2}$	**2.** Divide by $\frac{3}{4}$

Now carry out the steps of the solution:

$$\frac{3}{4}x - \frac{1}{2} + \mathbf{\frac{1}{2}} = \frac{5}{4} + \mathbf{\frac{1}{2}} \qquad \text{Add } \frac{1}{2} \text{ to both sides:}$$

$$\frac{3}{4}x = \frac{7}{4} \qquad \frac{5}{4} + \frac{1}{2} = \frac{5}{4} + \frac{2}{4} = \frac{7}{4}$$

$$\mathbf{\frac{4}{3}}\left(\frac{3}{4}x\right) = \left(\frac{7}{4}\right)\left(\mathbf{\frac{4}{3}}\right) \qquad \text{Divide both sides by } \frac{3}{4}\text{, or multiply by } \frac{4}{3}.$$

$$x = \frac{7}{3}$$

You can see that a fair amount of calculation with fractions was necessary in this solution. Is there a better way to solve the equation? Yes! We can avoid most of the calculation by clearing *all* the fractions before we begin. Look at the original equation again:

$$\frac{3}{4}x - \frac{1}{2} = \frac{5}{4}$$

We would like to clear all the fractions in one step, so what number should we multiply by? If you said the LCD of the fractions, you are right. For this equation, the LCD is 4.

We are going to multiply both sides of the equation by 4. As long as we perform exactly the same operation on both sides of an equation, we don't alter its solution. We must multiply *every* term on *both* sides of the equal sign by 4:

$$\mathbf{4}\left(\frac{3}{4}x - \frac{1}{2}\right) = \left(\frac{5}{4}\right)\mathbf{4} \qquad \text{Apply the distributive law.}$$

$$4\left(\frac{3}{4}x\right) - 4\left(\frac{1}{2}\right) = \left(\frac{5}{4}\right)4$$

Simplifying each term gives

$$3x - 2 = 5$$

We have cleared the fractions from the equation, and can now complete the solution without worrying about fractions any more:

$$3x - 2 = 5 \quad \text{Add 2 to both sides.}$$
$$3x = 7 \quad \text{Divide both sides by 3.}$$
$$x = \frac{7}{3}$$

We will use this technique of clearing fractions whenever it will simplify our work.

EXAMPLE 1 

Solve $\frac{2}{3}x + 2 = \frac{4}{5} - \frac{2}{5}x$

Solution

First find the LCD for all the fractions in the equation. The LCD for $\frac{2}{3}$, $\frac{4}{5}$, and $\frac{2}{5}$ is 15. Multiply both sides of the equation by 15 to clear the fractions.

$$\mathbf{15}\left(\frac{2}{3}x + 2\right) = \left(\frac{4}{5} - \frac{2}{5}x\right)\mathbf{15}$$

Apply the distributive law. This means that we must multiply *every* term by 15, including any terms that do not involve fractions. It may also be helpful to write 15 as $\frac{15}{1}$:

$$\frac{\mathbf{15}}{\mathbf{1}}\left(\frac{2}{3}x + 2\right) = \left(\frac{4}{5} - \frac{2}{5}x\right)\frac{\mathbf{15}}{\mathbf{1}}$$
$$\frac{\mathbf{15}}{\mathbf{1}} \cdot \frac{2}{3}x + \frac{\mathbf{15}}{\mathbf{1}} \cdot 2 = \frac{4}{5} \cdot \frac{\mathbf{15}}{\mathbf{1}} - \frac{2}{5}x \cdot \frac{\mathbf{15}}{\mathbf{1}}$$

Simplify each term. Then continue the solution as usual:

$$10x + 30 = 12 - 6x$$
$$16x + 30 = 12 \quad \text{Add } 6x \text{ to both sides.}$$
$$16x = -18 \quad \text{Subtract 30 from both sides.}$$
$$x = \frac{-18}{16} \quad \text{Divide both sides by 16.}$$
$$x = \frac{-9}{8} \quad \text{Reduce the answer.}$$

The solution is $\frac{-9}{8}$.

CAUTION!

We must multiply each term in the equation by the LCD, even terms that are not fractions. On the left side of the equation in Example 1 we multiplied both terms of $\frac{2}{3}x + 2$ by 15, even though 2 is not a fraction.

Now try Exercise 1.

EXERCISE 1

Solve

$$\frac{3}{4}x - \frac{5}{2} = 1 + \frac{1}{6}x$$

by first clearing the fractions.

At this point it might be helpful to review all our techniques for solving equations.

To Solve an Equation

First Simplify the Equation as Much as Possible

1. Clear any fractions by multiplying both sides by the LCD.
2. Remove any parentheses by applying the distributive law.
3. Combine any like terms appearing on the same side of the equation.

Then Isolate the Variable

4. Add or subtract the same quantity on both sides to get all the variable terms on one side of the equation.
5. Add or subtract the same quantity on both sides to get all the constant terms on the other side.
6. Divide both sides by the coefficient of the variable.

You may not have to use all these steps to solve a particular equation. (For instance, the equation may not include any fractions.) However, these steps should be sufficient to solve any **linear equation.** Linear equations have no exponents on their variables, and no variables appear in the denominators of fractions.

Problem Solving

Applied problems may lead to equations involving fractions.

EXAMPLE 2  

Millicent Welloff plans to leave $\frac{3}{5}$ of her fortune to her daughter and another $\frac{1}{4}$ to her sister. The rest of the money, \$375,000, will go to her secretary and companion, Ivy. What is the total value of Millicent's fortune?

Solution

Step 1 What are we asked to find?

The total value of Millicent's fortune

Choose a variable to represent the unknown quantity:

Total value of the fortune: ***m***

Step 2 Find something in the problem that we can express in two different ways:

The total value of the fortune

First way: We have called the total value m.

Second way: We can add up the amounts left to each beneficiary:

$$\underbrace{\frac{3}{5}m}_{\text{Daughter's share}} + \underbrace{\frac{1}{4}m}_{\text{Sister's share}} + \underbrace{375{,}000}_{\text{Ivy's share}}$$

Write an equation using these two expressions:

$$\frac{3}{5}m + \frac{1}{4}m + 375{,}000 = m$$

Step 3 Solve the equation:

$$20\left(\frac{3}{5}m + \frac{1}{4}m + 375{,}000\right) = 20(m) \quad \text{Multiply both sides by the LCD, 20.}$$

$$20\left(\frac{3}{5}m\right) + 20\left(\frac{1}{4}m\right) + 20(375{,}000) = 20m \quad \text{Apply the distributive law.}$$

$$12m + 5m + 7{,}500{,}000 = 20m \quad \text{Simplify each term.}$$

$$17m + 7{,}500{,}000 = 20m \quad \text{Combine like terms on the left side.}$$

$$7{,}500{,}000 = 3m \quad \text{Subtract } 17m \text{ from both sides.}$$

$$2{,}500{,}000 = m \quad \text{Divide both sides by 3.}$$

Answer the question in the problem.

The total value of Millicent's fortune is $2,500,000.

Now try Exercise 2.

EXERCISE 2

Delbert will need to use $\frac{3}{10}$ of his brown sugar for a batch of chocolate chip cookies and another $\frac{2}{5}$ of his brown sugar for a batch of granola cookies, leaving $\frac{1}{2}$ cup of brown sugar. How much brown sugar does Delbert have?

a. Choose a variable and write an equation about the problem.

b. Solve your equation and answer the question in the problem.

ANSWERS TO 9.4 EXERCISES

1. 6

2a. $x - \frac{3}{10}x - \frac{2}{5}x = \frac{1}{2}$

2b. $\frac{5}{3}$ cup

HOMEWORK 9.4 

1. If we want to clear all the fractions from an equation, we can multiply both sides by ______________.

2. What is wrong with this solution?

$$\frac{2}{3}x + 3 = \frac{1}{3}$$
$$2x + 3 = 1$$
$$2x = -2$$
$$x = -1$$

Solve each equation by first clearing the fractions.

3. $\frac{1}{6}m + 3 = \frac{-5}{6}$

4. $\frac{1}{3}n - 2 = \frac{5}{3}$

5. $\frac{2}{3}c + \frac{4}{3} = 2c + 4$

6. $3d - 6 = \frac{2}{5} - \frac{1}{5}d$

7. $\frac{2}{3}s - 6 = \frac{8}{3}s - 3$

8. $t + \frac{1}{2} = \frac{5}{2}t - 1$

9. $z + \frac{1}{7} = \frac{2}{3}z$

10. $\frac{1}{5}w - 1 = \frac{2}{3}$

11. $\frac{1}{6}p - \frac{7}{3} = \frac{2}{9}p - \frac{1}{4}p$

12. $\frac{2}{3}q - \frac{1}{4} = \frac{25}{12} + \frac{1}{3}q$

Solve each problem by writing and solving an equation.

13. Jenny's plot in a communal garden is a rectangular strip with an area of $7\frac{1}{3}$ square yards. If the strip is $1\frac{3}{8}$ yards wide, how long is it?

14. A rectangular circuit board for an appliance should have an area of $9\frac{3}{4}$ square inches. If the board can be no wider than $2\frac{1}{4}$ inches, how long must it be?

15. Julia is reading *Don Quixote* in her Spanish literature class. She read $\frac{2}{5}$ of the novel last week, read another $\frac{1}{4}$ this week, and has 308 pages left to read. How many pages are in Julia's edition of *Don Quixote*?

16. Koichi was offered a job in Santa Fe after graduation. He planned to drive out to New Mexico over 3 days. On the first day, he drove $\frac{1}{3}$ of the total distance, and he drove $\frac{5}{12}$ of the total distance the next day. This left 428 miles to drive the third day. What is the total distance Koichi drove to Santa Fe?

Use only one variable for each problem. (See Lesson 9.3 to review writing variable expressions.)

17. Carol's Cookie Jar employs 14 people who work as bakers or as cashiers.
a. Choose a variable for the number of bakers.

b. Write an expression for the number of cashiers.

18. Technicraft employs 38 people who work as engineers or as technicians.
a. Choose a variable for the number of engineers.

b. Write an expression for the number of technicians.

19. (Refer to Problem 17.) After an especially good year, each baker gets a bonus of \$150, and each cashier gets a bonus of \$100. Use your variable from Problem 17 to write expressions for
a. The amount of money Carol's Cookie Jar distributed in baker's bonuses.

b. The amount of money Carol's Cookie Jar distributed in cashier's bonuses.

c. The total amount of money Carol's Cookie Jar distributed in bonuses.

20. (Refer to Problem 18.) The employees at Technicraft agree that each engineer will contribute \$25 to support a Little League team, and each technician will contribute \$15. Use your variable from Problem 18 to write algebraic expressions for
a. The amount of money the engineers contribute.

b. The amount of money the technicians contribute.

c. The total amount of money the employees contribute.

21. When they reseed a tract of forest, EarthCare plants three times as many pine trees as oak trees.

a. Choose a variable for the number of oak trees planted in each tract.

b. Write an expression for the number of pine trees planted in each tract.

22. To meet market demand, each member of a farming co-op agrees to plant half as many acres of corn as of soybeans.

a. Choose a variable for the number of acres of soybeans each farmer plants.

b. Write an expression for the number of acres of corn each farmer plants.

23. (Refer to Problem 21.) Each pine tree needs 9 square yards of ground space, and each oak tree needs 20 square yards. Use your variable from Problem 21 to write algebraic expressions for

a. The area needed for the oak trees in each tract of forest.

b. The area needed for the pine trees in each tract of forest.

c. The total area needed for the trees in each tract.

24. (Refer to Problem 22.) Each acre of soybeans needs 40 pounds of fertilizer, and each acre of corn needs 100 pounds of fertilizer. Use your variable from Problem 22 to write algebraic expressions for

a. The amount of fertilizer each farmer needs for soybeans.

b. The amount of fertilizer each farmer needs for corn.

c. The total amount of fertilizer each farmer needs.

25. Students can fulfill their algebra requirement in two ways: They can pass the algebra course, or they can pass a competency exam. Experience has shown that 40% of the students enrolled in the algebra course pass, and 10% of those taking the competency exam pass. This year 840 students will attempt to fulfill their algebra requirement.

a. Choose a variable for the number of students enrolled in the algebra course.

b. Write an expression for the number of students who took the competency test.

c. Write an expression for the number of students expected to pass the algebra course.

d. Write an expression for the number of students expected to pass the competency exam.

e. Write an expression for the number of students expected to fulfill the algebra requirement.

26. The Alumni Association sold 460 tickets to the Pierce College Rodeo. Student tickets cost $5, and regular admission cost $8.

a. Choose a variable for the number of student tickets sold.

b. Write an expression for the number of general admission tickets sold.

c. Write an expression for the revenue generated by the student tickets.

d. Write an expression for the revenue generated by the general admission tickets.

e. Write an expression for the total revenue.

Challenge Problems

The following problems appeared in the *Greek Anthology*, which was compiled about 500 B.C. Algebra as we know it had not been invented yet, so you have a great advantage over the Greek students who tackled these problems!

27. Six people are going to share a basket of apples. The first four people receive $\frac{1}{3}$, $\frac{1}{8}$, $\frac{1}{4}$, and $\frac{1}{5}$ of the apples, respectively. The fifth person receives ten apples, and one apple remains for the sixth person. How many apples were in the basket originally?

28. Demochares lived $\frac{1}{4}$ of his life as a boy, $\frac{1}{5}$ as a youth, and $\frac{1}{3}$ in middle age and has lived 13 years in his old age. How old is Demochares?

Building Number Skills

Here is another way to subtract mentally: Hold aside the units digit of the first number and then add it at the end.

Examples: **a.** $53 - 6 = (50 + 3) - 6 = (50 - 6) + 3 = 44 + 3 = 47$

b. $41 - 24 = (40 + 1) - 24 = (40 - 24) + 1 = 16 + 1 = 17$

Subtract mentally by rewriting the first number as a sum of a multiple of 10 plus a single digit number.

29. $81 - 5$ **30.** $44 - 8$ **31.** $62 - 9$ **32.** $73 - 6$

33. $36 - 9$ **34.** $96 - 7$ **35.** $93 - 4$ **36.** $54 - 5$

37. $44 - 6$ **38.** $68 - 9$

9.5 Applications

A. Problems Involving Interest

Review of the Interest Formula

In Chapter 2 we studied problems involving the formula for simple interest,

$$I = Prt$$

where I stands for the interest earned on a principal of P dollars invested at interest rate r for a period of t years. For example, if you invest \$1000 at 8% annual interest rate for 3 years, your interest will be

$$\begin{aligned} I &= Prt \\ &= (1000)(0.08)(3) \\ &= \$240 \end{aligned}$$

EXAMPLE 1

How much should Alice invest in her savings account, which pays 5.5% annual interest, if she would like to earn \$300 interest in 2 years?

Solution

Step 1 Amount Alice should invest: P

Step 2 Substitute the given values into the interest formula. Remember to convert the interest rate, 5.5%, to a decimal fraction:

$$I = Prt$$
$$300 = P(0.055)(2)$$

Step 3 Solve the equation:

$$300 = P(0.055)(2) \quad \text{Simplify the right side.}$$
$$300 = P(0.11) \quad \text{Divide both sides by 0.11.}$$
$$\frac{300}{\mathbf{0.11}} = \frac{P(0.11)}{\mathbf{0.11}}$$
$$2727.\overline{27} = P$$

In order to make $300 in interest, Alice should round up the principal to $2727.28.

Now try Exercise 1.

EXERCISE 1

Shanil has $2000 to invest and would like to earn $600 interest in 4 years time. What interest rate does she need to find?

Two Investments

In this lesson we consider some interest problems that involve more than one investment. For example, suppose you invest $1000 in an account that pays 6% annual interest and $3000 in another account that pays 9% annual interest. To find the total amount of interest you will earn in 1 year, we find the interest earned on each account and add them together:

Interest on first account: $I = Prt$
$= (1000)(0.06)(1) = \$60$

Interest on second account: $I = Prt$
$= (3000)(0.09)(1) = \$270$

The total interest earned after 1 year is thus $60 + $270, or $330.

We can apply the same ideas to problems involving variables. In the example and exercise that follow, all investment periods are 1 year, unless stated otherwise.

EXAMPLE 2 

Clarissa has $6000 to invest. She decides to put part of the money in her savings account, which pays 3% annual interest, and part in a certificate of deposit (CD) that pays 7% interest.

a. Write variable expressions for the principal Clarissa invests in each account:

Amount invested in savings: x
Amount invested in a CD: $6000 - x$

b. Write variable expressions for the interest Clarissa earns on each account in 1 year:

Interest earned on savings: $I = Prt$
$= \mathbf{x(0.03)(1) = 0.03x}$

Interest earned on the CD: $I = Prt$
$= \mathbf{(6000 - x)(0.07)(1) = 0.07(6000 - x)}$

EXERCISE 2

a. Ornette wants to invest \$1000 in two stocks. If he invests x dollars in Pacificon stocks, how much can he invest in Nextel stocks?

b. If Pacificon earns 6% dividends and Nextel earns 8%, write an algebraic expression for Ornette's total earnings.

c. It turned out that Ornette earned \$64 dividends on his stocks. Write an equation and solve it to discover how much he invested in each company.

c. Write an expression for the total interest Clarissa earns in 1 year:

$$\textbf{total interest} = \textbf{interest from savings} + \textbf{interest from CD}$$
$$= \mathbf{0.03}x + \mathbf{0.07(6000} - x\mathbf{)}$$

d. Suppose Clarissa earned \$340 interest on her investments. Using your work above, write an equation about this:

$$\mathbf{0.03}x + \mathbf{0.07(6000} - x\mathbf{)} = \mathbf{340}$$

e. Solve your equation to find out how much Clarissa invested in each account. Begin by simplifying the left side:

$0.03x + 0.07(6000 - x) = 340$ Apply the distributive law.

$0.03x + 420 - 0.07x = 340$ Combine like terms.

$-0.04x + 420 = 340$ Subtract 420 from both sides.

$\mathbf{-420} \quad \mathbf{-420}$

$-0.04x = -80$ Divide both sides by -0.04.

$$\frac{-0.04x}{\mathbf{-0.04}} = \frac{-80}{\mathbf{-0.04}}$$

$$x = 2000$$

Looking back at part (a), we see that x stands for the amount Clarissa invested in her savings account. Thus, Clarissa invested \$2000 in savings. The amount Clarissa invested in the CD is $6000 - x$, or $6000 - 2000$, or \$4000.

Now try Exercise 2.

ANSWERS TO 9.5A EXERCISES

1. 7.5%

2a. $1000 - x$

2b. $0.06x + 0.08(1000 - x)$

2c. Ornette invested \$800 in Pacificon and \$200 in Nextel.

HOMEWORK 9.5A

1. a. How much interest will you earn if you invest x dollars at 5% for 1 year? (Your answer will be in terms of x.)

b. If you earned \$140 interest, how much did you invest? (Set up an equation and solve.)

2. a. How much interest will you earn if you invest d dollars at 9% for 1 year? (Your answer will be in terms of d.)

b. If you earned \$315 interest, how much did you invest? (Set up an equation and solve.)

3. Reginald asks his wealthy uncle, Dupont Tightfist, to loan him \$50,000 for a new sports car. Reginald promises to repay the loan in 4 years when he receives his inheritance. What annual interest rate should Uncle Dupont charge if he wants to earn \$25,000 interest from the loan?

4. Aunt Ethel's investment brokers, Connive and Finagle, promise her earnings of \$180,000 in 5 years if she will invest her life savings of \$150,000 in their plan. What annual interest rate must the plan earn in order to make good their promise?

5. Mariah invested \$1200 in her credit union that pays 6.5% annual interest and \$600 in a city bond that pays 8% annual interest. How much interest will Mariah earn after 5 years?

6. Farnsworth invested \$20,000 in Softek stock that pays annual dividends of 9% and another \$15,000 in Green Products, for an annual return of 7.2%. How much income will Farnsworth's investments earn after 3 years?

7. Wynton plans to borrow \$25,000 to open a print shop. He will borrow part of the money from a city fund for small businesses that charges 5% annual interest and the rest from a savings and loan that charges 8.4% annual interest.

a. Write variable expressions for the amount of money Wynton borrows from the city fund and the amount he borrows from the savings and loan.

b. Write algebraic expressions for the amount of interest Wynton will owe 1 year later on his loan from the city and the amount of interest he will owe the savings and loan.

c. Write an algebraic expression for the total amount of interest Wynton will owe after 1 year. Simplify your expression.

8. Camilla borrowed a total of \$30,000 for her last year in medical school. Some of the money came from a student loan that charges 4% annual interest, and the rest came from her father, who is charging Camilla 5.25% annual interest.

a. Write variable expressions for the amount of money Camilla borrowed as a student loan and the amount she borrowed from her father.

b. Write algebraic expressions for the amount of interest that Camilla will owe after 1 year on the student loan and the amount of interest she will owe her father.

c. Write an algebraic expression for the total amount of interest Camilla will owe after 1 year. Simplify your expression.

9. Risa has $1200 to invest in two different accounts. Her savings account pays 5% annual interest, and a new T-bill (treasury bill) pays 7.5% annual interest, but she can't make withdrawals from the T-bill. To help her decide how much to invest in each account, Risa makes a table as shown below.

a. Fill in the table:

Amount invested in savings ($)	*Amount invested in T-bill ($)*	*Interest from savings ($)*	*Interest from T-bill ($)*	*Total interest ($)*
100				
200				
300				
400				
500				
600				
700				
800				
900				
1000				
1100				

b. Can you write an algebraic expression that can be used to calculate Risa's total interest for different combinations of investments? (*Hint:* Let x stand for the amount Risa invests in her savings account.)

10. Suraya plans to borrow $12,000 from two sources to finance an archaeological expedition. If she borrows money from her retirement account, she will pay 3% annual interest, but she must repay the entire loan after 1 year. The Archaeological Society will loan her money at an annual interest rate of 8.5%. Suraya makes a table to see what her interest payment will be for various combinations of loans.

a. Fill in the table:

Amount borrowed from retirement ($)	*Amount borrowed from society ($)*	*Interest on retirement loan ($)*	*Interest on society's loan ($)*	*Total interest ($)*
1,000				
2,000				
3,000				
4,000				
5,000				
6,000				
7,000				
8,000				
9,000				
10,000				
11,000				

b. Can you write an algebraic expression that can be used to calculate Suraya's interest payment for various combinations of loans? (*Hint:* Let x stand for the amount Suraya borrows from her retirement account.)

11. You have \$5000 to invest. You put part of the money in stocks that pay 12% annual interest and the rest in bonds that pay 7% interest.

a. If you put x dollars in stocks, how much is left for you to put into bonds? (Your answer will be in terms of x.)

b. How much interest will you earn on the stocks in 1 year? How much interest will you earn on the bonds? How much will you earn total? (All answers in terms of x.)

c. Suppose you earned \$440 in interest on your investments. How much did you invest in stocks, and how much in bonds? (Set up an equation and solve.)

12. Delbert's life savings is \$900. He keeps part of his money in the credit union that pays 6% annual interest and part in a savings account that pays 6.5% interest.

a. If Delbert has x dollars in the credit union, how much does he have in the savings account? (In terms of x.)

b. In 1 year how much interest does Delbert earn from the credit union? How much does he earn from the savings account? How much total interest will Delbert earn? (All answers in terms of x.)

c. Delbert earned \$55 interest last year. How much of his savings did he invest in the credit union, and how much in the savings account? (Set up an equation and solve.)

13. Francine invested some money in Movies with Pepperoni, a combination video-rental and take-out pizza parlor venture, but unfortunately she *lost* 9.6% on her investment. Luckily, she kept \$6000 more than her ill-fated investment in a T-bill that earned 7.3% interest. Her net earnings for the year were \$334.50. How much did she invest in the video–pizza venture?

14. Maxwell bought some shares of Octopus Conglomerate stock that *lost* 16% last year. He also invested \$2500 less in Co-op Earth, and those shares earned $12\frac{1}{2}\%$. Nevertheless, Maxwell *lost* \$522.50 last year. How much did he invest in Octopus Conglomerate stock?

Building Number Skills

Another way of subtracting involves doing two simpler subtractions.

Examples: **a.** $95 - 7 = 95 - (5 + 2) = (95 - 5) - 2 = 90 - 2 = 88$

b. $41 - 24 = 41 - (21 + 3) = (41 - 21) - 3 = 20 - 3 = 17$

Subtract mentally by regrouping the second number to match the units digit of the first number.

15. $44 - 8$

16. $81 - 5$

17. $73 - 26$

18. $62 - 39$

19. $96 - 57$

20. $36 - 19$

21. $54 - 35$

22. $93 - 24$

23. $68 - 19$

24. $44 - 26$

B. Problems Involving Mixtures

Review of the Percent Formula

In Lesson 8.1 we reviewed the formula

$$P = rW$$

for solving problems involving percent. In this lesson we study mixtures and combinations involving percent.

First, make sure that you can use the percent formula to solve for any one of its three variables.

EXAMPLE 3  

a. A jar of marbles contains 14 red marbles and the rest blue marbles. Of the marbles in the jar, 35% are red. How many marbles does the jar contain?

b. A jar contains 60 marbles, of which 30% are red. How many red marbles are in the jar?

Solution

a. In this problem we are given the percent *rate* (35%) and the *part* of the marbles that is red (14 marbles). We are asked to find the *whole* number of marbles in the jar:

$$P = rW$$

$$14 = 0.35W \quad \text{Solve for } W.$$

$$\frac{14}{\mathbf{0.35}} = \frac{0.35W}{\mathbf{0.35}}$$

$$40 = W$$

There are 40 marbles in the jar.

b. We are given the *whole* number of marbles (60) and the percent *rate* (30%), and we are asked to find the *part* that is red:

$$P = rW$$

$$P = 0.30(60)$$

$$P = 18$$

There are 18 red marbles in the jar.

EXERCISE 3

A jar contains 80 marbles, of which 16 are red. What percent of the marbles in the jar are red?

Now try Exercise 3.

We can write the percent formula in three equivalent forms, depending upon which of its three variables is unknown:

If P is unknown: $P = rW$

If r is unknown: $r = \frac{P}{W}$

If W is unknown: $W = \frac{P}{r}$

These forms will be handy for the mixture problems we wish to study.

Percents Don't Add

Now consider mixing together two jars of marbles. The first jar contains 40 marbles, of which 30 are red, and the second jar contains 60 marbles, of which 30 are red (see Figure 9.2).

Figure 9.2

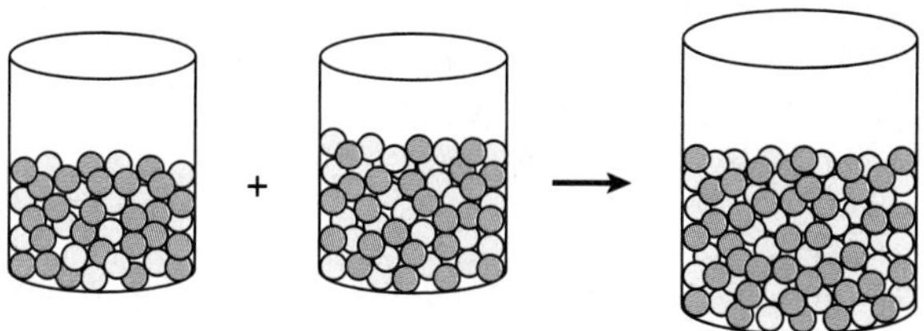

Let's compute the percent of red marbles in each of the two jars:

$$\text{First jar: percent red} = \frac{\text{number of red marbles}}{\text{total number of marbles}} = \frac{P}{W} = \frac{30}{40} = 0.75$$

The marbles in the first jar are 75% red.

$$\text{Second jar: percent red} = \frac{\text{number of red marbles}}{\text{total number of marbles}} = \frac{P}{W} = \frac{30}{60} = 0.50$$

The marbles in the second jar are 50% red.

Next, we'll pour both jars of marbles into a large box and mix the contents. Let's answer three questions about the mixture:

1. How many marbles are in the mixture?

   Add the total marbles in the two jars: 40 + 60 = 100.

2. How many red marbles are in the mixture?

   Add the red marbles from each jar: 30 + 30 = 60.

3. What is the percent of red marbles in the mixture?

$$\textbf{percent red} = \frac{\textbf{number of red marbles}}{\textbf{total number of marbles}} = \frac{\mathbf{60}}{\mathbf{100}} = \mathbf{0.60}$$

   The mixture is 60% red.

Finally, here is a very important point about mixtures: The percents of red marbles in the two jars do *not* add up to the percent of red marbles in the mixture.

$$75\% + 50\% \neq 60\%$$

This makes sense if you think about it. Percents are fractions, and in this example they are fractions of *different* whole amounts. We cannot add fractions of different quantities. We cannot find the percent of red marbles in the mixture by adding the percents of the two ingredients.

Percents don't add.

We can, however, compute the *number* of red marbles in each ingredient and add those. Then we can find the percent red of the mixture as we did above.

Solving Mixture Problems

It is helpful to organize all the information in a mixture problem into a table. For the marbles problem, the table looks like this:

	Number of marbles (W)	*Percent red (r)*	*Number of red marbles (P)*
First jar	40		30
Second jar	60		30
Mixture	**100**		**60**

The table has three rows and three columns. We label the three horizontal rows with the two components of the mixture (the two jars of marbles) and the mixture itself. We label the three columns with the number of marbles in each jar, the *percent* of the important ingredient (red marbles in this case), and the *amount* of important ingredient (number of red marbles) in each jar.

After we have entered the numerical values from the problem into the table, we can obtain the bold-faced values in the first and third columns by *adding down.* This makes sense because the marbles in the mixture come from combining the marbles in the two jars. However, if you fill in the percent column with the values we calculated above, you will see that *we cannot add the values in the percent column.* Percents don't add!

Jars of marbles are easy to visualize, but the same algebraic principles apply to a great variety of problems involving percentages and mixtures of ingredients. Here is a general form for a mixture table:

	Total amount (W)	*Percent of important ingredient (r)*	*Amount of important ingredient (P)*
First component			
Second component			
Mixture			

EXAMPLE 4

Polls show that 70% of the urban voters in Elysian County are liberal and 30% are conservative, whereas 20% of the rural voters are liberal and 80% are conservative. If there are 600,000 registered voters in urban districts and 400,000 registered voters in rural districts, what percent of the electorate of Elysian County is liberal?

Solution

We cannot just add the percents of liberals in urban and rural districts. We must first find out how *many* urban and rural liberals there are. We will do this using a table. For this problem, the two components of the mixture are the rural and urban voters. The important ingredient is the liberal vote.

	Number of voters (W)	*Percent liberal (r)*	*Number of liberals (P)*
Urban voters	600,000	0.70	**420,000**
Rural voters	400,000	0.20	**80,000**
All voters	**1,000,000**		**500,0000**

Fill in the table with the given information. Then find the bold-faced values as follows:

1. Multiply across each row.
2. Add down the first and third columns.

(Can you explain why this works?) Remember that we *cannot* add the percent values in the second column. To fill in the last block of the table and answer the question in the problem, we use the percent formula:

$$\text{percent liberals} = \frac{P}{W} = \frac{500{,}000}{1{,}000{,}000} = 0.50$$

Among all voters in Elysian County, liberals make up 50%.

Now try Exercise 4.

EXERCISE 4

The city of Hundred Pines has two hospitals, Adam General Hospital and Brian Medical Hospital. Of the patients admitted into Adam General's emergency room, 80% survive and 20% do not. Of the patients admitted to Brian Medical's emergency room, 90% survive and 10% do not. Last year Adam General admitted 1200 patients into its emergency room, and Brian Medical admitted 800. What percent of the patients admitted to the emergency rooms of the two hospitals survived last year? Use the table to organize your calculations.

	Number admitted	*Percent survivors*	*Number of survivors*
Adam General			
Brian Medical			
Total			

ANSWERS TO 9.5B EXERCISES

3. 20%

4. 84%;

	Number admitted	*Percent survivors*	*Number of survivors*
Adam General	1200	0.80	960
Brian Medical	800	0.90	720
Total	2000	0.84	1680

HOMEWORK 9.5B 

1. A fruit juice cocktail is 30% grape juice.
   a. How much grape juice is in 8 gallons of the cocktail?

   b. How much grape juice is in x gallons of the cocktail?

2. Eggnog is 60% milk.
   a. How much milk is in 2 gallons of eggnog?

   b. How much milk is in $12 - x$ gallons of eggnog?

3. Delbert has two bags of M & M's. The first bag contains 50 M & M's, of which 30% are red. The second bag contains 30 M & M's, of which 70% are red. Delbert pours both bags into a jar and mixes the contents. Use the table to help you answer the following questions.

	Number of M & M's	*Percent red*	*Number of red M & M's*
First bag			
Second bag			
Mixture			

   a. How many red M & M's are there in the first bag? In the second bag?

   b. How many M & M's total are there in the mixture? How many red M & M's are in the mixture?

   c. What percent of the M & M's in the mixture are red?

   d. Check your answer to part (c) by multiplying across the last row to verify the number of red M & M's in the mixture.

4. Francine has two bottles filled with a punch made of club soda and prune juice. The first bottle holds 2 quarts of punch and is 60% prune juice. The second bottle holds 6 quarts of punch and is 20% prune juice. Francine pours both bottles into a large punch bowl. Use the table to help you answer the following questions.

	Quarts of punch	*Strength (% prune juice)*	*Quarts prune juice*
60% punch			
20% punch			
Mixture			

   a. How much prune juice is in the first bottle? In the second bottle?

   b. How much punch is in the punch bowl? How much prune juice is in the punch bowl?

   c. What percent prune juice is the mixture?

   d. Check your answer to part (c) by multiplying across the last row to verify the amount of prune juice in the mixture.

5. A chemist keeps a supply of acid in two strengths. The first solution is 20% acid (mixed with water), and the second solution is 30% acid.

a. How much acid is in 5 quarts of the 20% solution? How much acid is in 10 quarts of the 30% solution?

b. If the chemist mixes 5 quarts of the 20% solution with 10 quarts of the 30% solution, how much acid will be in the mixture? How many total quarts of solution will she have?

c. What will be the strength (percent acid) of the mixture?

d. Fill in the table:

	Quarts of solution	*Strength (% acid)*	*Quarts of acid*
20% solution			
30% solution			
Mixture			

6. An auto mechanic keeps a supply of antifreeze in two strengths. The first solution is 20% antifreeze (mixed with water), and the second solution is 50% antifreeze.

a. How much antifreeze is in 15 quarts of the 20% solution? How much antifreeze is in 30 quarts of the 50% solution?

b. If the mechanic mixes 15 quarts of the 20% solution with 30 quarts of the 50% solution, how much antifreeze will be in the mixture? How many total quarts of antifreeze mixture will he have?

c. What will be the strength (percent antifreeze) of the mixture?

d. Fill in the table:

	Quarts of solution	*Strength (% antifreeze)*	*Quarts of antifreeze*
20% Solution			
50% Solution			
Mixture			

7. Suppose the chemist in Problem 5 wants to make 15 quarts of a 28% solution of acid by mixing the two solutions she has on hand. How many quarts of each should she use?

a. Let x represent the number of quarts of the 20% solution she will need. How many quarts of the 30% solution will she need?

b. Make a new table like the one in Problem 5 and fill it in with the information for this problem.

8. Suppose the auto mechanic from Problem 6 wants to make 45 quarts of a 30% solution of antifreeze by mixing the two solutions he has on hand. How many quarts of each should he use?

a. Let x represent the number of quarts of the 20% solution he will need. How many quarts of the 50% solution will he need?

b. Make a new table like the one in Problem 6 and fill it in with the information for this problem.

9. Major Motors manufactures two kinds of vehicles, pickup trucks and compact cars. Next year they must make 60% of their vehicles comply with emission standards. Their engineers figure that by that time they can modify their production lines so that 80% of their trucks will comply with standards, but only 30% of their cars. If Major Motors would like to produce 50,000 cars next year, how many trucks should they produce in order to meet the standards? Use the table and the questions that follow, to help you solve this problem.

	Number produced	***Percent meeting standards***	***Number meeting standards***
Cars			
Trucks			
All vehicles			

a. What are we asked to find in this problem? Let x represent that quantity.

b. Fill in the first column of the table with numbers or algebraic expressions.

c. Fill in the second column of the table with the percents given in the problem.

d. Find expressions for the number of cars meeting standards and the number of trucks meeting standards.

e. Use the table to write *two different* expressions for the total number of vehicles meeting standards. Write an equation.

f. Solve your equation to answer the question in the problem.

10. A survey of 2000 students reported that 58% favored shortening the semester by 1 week. The survey also stated that 70% of seniors favored the shorter semester, whereas only 50% of other students were in favor of it. How many seniors were surveyed? Use the table and the questions that follow, to help you solve this problem.

	Number surveyed	***Percent in favor***	***Number in favor***
Seniors			
Others			
All students			

a. What are we asked to find in this problem? Let x represent that quantity.

b. Fill in the first column of the table with numbers or algebraic expressions.

c. Fill in the second column of the table with the percents given in the problem.

d. Find expressions for the number of seniors in favor of the shortened semester and the number of other students in favor.

e. Use the table to write *two different* expressions for the total number of students who favor the shorter semester. Write an equation.

f. Solve your equation and answer the question in the problem.

Building Number Skills

Sometimes we can add mentally by grouping the terms to make multiples of 10.

Example: $13 + 6 + 7 + 24 = (13 + 7) + (6 + 24) = 20 + 30 = 50$

Compute each sum without calculator or paper.

11. $8 + 29 + 32$

12. $16 + 47 + 4$

13. $14 + 3 + 27 + 6$

14. $9 + 23 + 7 + 31$

15. $26 + 32 + 4 + 8$

16. $35 + 18 + 5 + 2$

17. $25 + 6 + 25 + 34$

18. $8 + 29 + 21 + 32$

19. $17 + 36 + 24 + 33$

20. $22 + 39 + 11 + 18$

C. Problems Involving Motion

Review of the Distance Formula

In previous lessons we have used the distance formula

$$d = rt$$

to solve problems involving motion at a constant speed. You should be able to solve the distance formula for any one of its three variables. Keep in mind that the units on the three variables must be compatible.

EXAMPLE 5

a. If you travel at 40 miles per hour for 3 hours, how far will you travel?

b. If you jog 600 yards at 5 yards per second, how long will it take you?

EXERCISE 5
If you want to bicycle 270 kilometers in 6 days, how fast should you go?

Solution

For each problem, substitute the known quantities into the distance formula and solve for the unknown quantity.

a. For this problem, the distance is unknown:

$$\begin{aligned} d &= rt \\ &= \left(40\ \frac{\text{miles}}{\text{hour}}\right)(3\ \text{hours}) = 120\ \text{miles} \end{aligned}$$

You will travel 120 miles.

b. For this problem, the time is unknown:

$$\begin{aligned} d &= rt \\ 600 &= 5t \\ \frac{600}{\mathbf{5}} &= \frac{\mathbf{5}t}{\mathbf{5}} \quad \text{Divide both sides by 5.} \\ 120 &= t \end{aligned}$$

It will take you 120 seconds, or 2 minutes.

Now try Exercise 5.

Solving Motion Problems

If a problem involves two or more moving objects, it is helpful to draw a sketch of their relative positions.

EXAMPLE 6

Lisa and Carin start at the same place and canoe down the Lazy River. If Lisa paddles at 6 miles per hour and Carin paddles at 10 miles per hour, how far apart will they be after $\frac{3}{4}$ of an hour?

Solution

First decide whether we are looking for a time, a rate, or a distance. "How far" indicates that we are looking for a distance.

Step 1 Distance between Lisa and Carin after $\frac{3}{4}$ hour: m

Next, draw a sketch of the situation (see Figure 9.3).

Figure 9.3

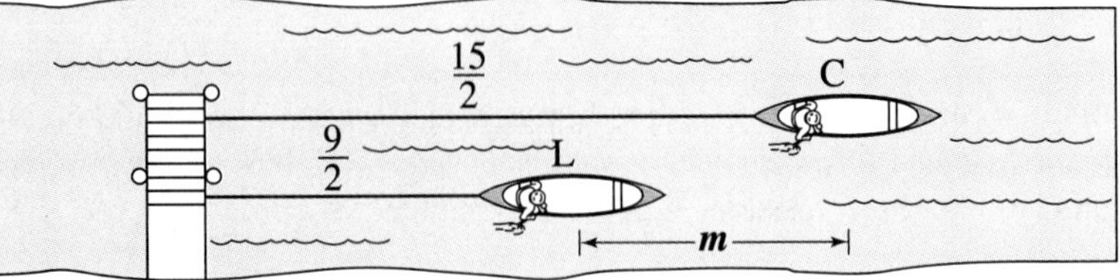

Label the sketch with the *distances* involved or with algebraic expressions for the distances. (Do *not* label your sketch with times or rates!) For this sketch, we can calculate the distance Lisa paddled and the distance Carin paddled in $\frac{3}{4}$ hour:

$$\text{Distance Lisa paddled:} \quad d = rt$$
$$= 6\left(\frac{3}{4}\right) = \frac{9}{2} \text{ miles}$$
$$\text{Distance Carin paddled:} \quad d = rt$$
$$= 10\left(\frac{3}{4}\right) = \frac{15}{2} \text{ miles}$$

The distance between Lisa and Carin is m miles, from Step 1.

Step 2 Now try to write an equation. Use your sketch to find some distance that can be expressed in two different ways. Notice that the distance Carin paddled is $\frac{15}{2}$ miles, but it is also $\frac{9}{2} + m$ miles. Thus,

$$\frac{9}{2} + m = \frac{15}{2}$$

Step 3 Solve the equation:

$$\frac{9}{2} + m = \frac{15}{2} \qquad \text{Subtract } \frac{9}{2} \text{ from both sides.}$$
$$m = \frac{15}{2} - \frac{9}{2} = \frac{6}{2} = 3$$

After $\frac{3}{4}$ hour, Lisa and Carin will be 3 miles apart.

Using a Table

For some motion problems, it may be helpful to make a table. The table helps you write algebraic expressions for the distances in the problem. Try Exercise 6.

ANSWERS TO 9.5C EXERCISES

5. 45 kilometers per day

6a. $5x$ miles

6b. $x + 15$ miles per hour; $5(x + 15)$ miles

6c.

	Rate	*Time*	*Distance*
Jerry	x	5	$5x$
Lou	$x + 15$	5	$5(x + 15)$

6d. $5x + 5(x + 15)$

6e. $5x + 5(x + 15) = 525$; Jerry's speed is 45 miles per hour and Lou's speed is $45 + 15$, or 60 miles per hour.

EXERCISE 6

Jerry and Lou start in Needles, California, and drive in opposite directions through the desert.

a. Jerry drives at x miles per hour. How far will he travel in 5 hours?

Jerry's distance:

b. Lou drives 15 miles per hour faster than Jerry. What is Lou's speed, and how far will he travel in 5 hours?

Lou's speed:
Lou's distance:

c. Summarize the work done above in a table:

	Rate	*Time*	*Distance*
Jerry			
Lou			

d. How far apart are Jerry and Lou after 5 hours? (Draw a sketch of the situation, and label with the distances involved. See Figure 9.4.)

Figure 9.4

e. Jerry and Lou are 525 miles apart after 5 hours. Write an equation and solve it to find out how fast each is driving.

HOMEWORK 9.5C 

1. Carl and Wendy leave Magic Mountain and drive in opposite directions for 2 hours. Wendy drives at 60 miles per hour, and Carl drives at 45 miles per hour.
   a. How far apart are Carl and Wendy after 2 hours?

   b. Draw a sketch illustrating the problem.

2. Ivan and Raissa leave the harbor at the same time and sail in the same direction. Ivan sails at 18 miles per hour, and Raissa sails at 12 miles per hour.
   a. How far apart are Ivan and Raissa after 3 hours?

   b. Draw a sketch illustrating the problem.

3. The distance from Hampton to Fell's Crossing is 12 miles along the Rifle River. The commuter ferry on the river has a cruising speed of 21 miles per hour, and the current in the Rifle River is usually 3 miles per hour.
   a. How long is the ferry trip upstream from Hampton to Fell's Crossing?

   b. How long is the return trip?

4. Emilio is training for a 5-day bicycle race. He rode 60 miles in 3 hours against a headwind of 5 miles per hour.
   a. What would Emilio's speed have been without the wind?

   b. How long will the return trip take Emilio, with the wind at his back?

5. Isabelle leaves the city on a Friday afternoon for a long weekend in the mountains, 120 miles away. She drives the first 60 miles through the suburbs at an average speed of 30 miles per hour, but for the last 60 miles she averages 50 miles per hour. What is her average speed for the whole trip? (*Hint:* The answer is not 40 miles per hour! Remember that

$$\textit{average speed} = \frac{\textit{total distance}}{\textit{total time}}.$$

   Find the time it took for Isabelle to complete each part of her trip.)

6. Rosa swam the first four laps of an 800-meter workout at an average speed of 60 meters per minute, and the second four laps at an average speed of 48 meters per minute. What was her average speed for the whole workout? (*Hint:* The answer is not 54 meters per minute! Remember that

$$\textit{average speed} = \frac{\textit{total distance}}{\textit{total time}}.$$

   Find the time it took for Rosa to complete each four laps of her workout.)

7. Brooke and Claire start at opposite ends of a hiking trail and walk toward each other. Brooke walks 2 miles per hour faster than Claire. They meet after 4 hours.

a. If Claire walks at w miles per hour, how fast does Brooke walk?

b. How far does Claire walk in 4 hours? How far does Brooke walk in 4 hours? (Your answers will be in terms of w.)

c. Draw a sketch illustrating the problem.

8. Two planes, a jet airliner and a small private plane, leave Chicago at the same time and fly in opposite directions. The jet airliner flies four times as fast as the small plane.

a. If the speed of the small plane is s, what is the speed of the jet airliner?

b. How far does the small plane fly in 3 hours? How far does the jet fly? (Your answers will be in terms of s.)

c. Draw a sketch illustrating this problem.

9. Oscar and Felix travel from Eastbank to Westbend on the same day. Oscar drives at an average speed of 45 miles per hour. Felix takes the train at an average speed of 70 miles per hour. Felix leaves 1 hour after Oscar and arrives 2 hours ahead of him.

a. If Oscar traveled for h hours, how long did Felix travel?

b. How far did Oscar travel, in terms of h? How far did Felix travel?

c. Draw a sketch illustrating the problem.

10. Sabra's bike club rides at an average speed of 15 miles per hour on the first day of their annual tour. That evening, Sabra discovers that she forgot her rain gear and calls her brother at home. He drives to the bike club's camp ground at an average speed of 60 miles per hour, completing the trip in 6 hours less time than Sabra did.

a. If Sabra rode for h hours, how long did her brother drive?

b. How far did Sabra travel, in terms of h? How far did her brother travel?

c. Draw a sketch illustrating this problem.

11. Delbert and Francine leave Fresno at the same time and drive in opposite directions. Francine drives 5 miles per hour slower than Delbert. After 6 hours they are 570 miles apart.

a. Make a table for this problem. Fill it in with your answers to parts (b) and (c).

b. If Delbert's speed is s miles per hour, what is Francine's speed?

c. How far did Delbert travel in 6 hours? How far did Francine travel in 6 hours?

d. Draw a sketch illustrating the problem.

e. How far apart are Delbert and Francine after 6 hours? Write an equation for the distance between Delbert and Francine after 6 hours.

f. Solve the equation. What was Delbert's driving speed? What was Francine's driving speed?

12. Bonnie left Dallas and drove north at 40 miles per hour. Three hours later, Clyde headed north from Dallas on the same road at 70 miles per hour. He kept driving until he caught up with Bonnie.

a. Make a table for this problem. Fill it in with your answers to parts (b) and (c).

b. If Clyde drove for h hours, how long did Bonnie drive?

c. How far did Bonnie drive? How far did Clyde drive?

d. Draw a sketch illustrating the problem.

e. What can you say about the two distances in part (b)? Write an equation about the distances.

f. Solve the equation. How long did Clyde drive before catching Bonnie?

13. Amelia flew her light plane against a steady headwind, and the trip took 6 hours and 40 minutes. On the return trip, the same wind was at her back, and the journey took only 5 hours. Amelia's plane flies at 224 miles per hour if there is no wind. What was the speed of the wind?

a. What are you asked to find in this problem? Assign a variable.

b. Make a table for this problem with two rows: initial trip and return trip. Fill in the table.

c. Draw a sketch describing the problem.

d. Find some quantity that can be expressed in two different ways and write an equation.

e. Solve your equation and answer the question in the problem.

14. Jeremy can paddle his canoe $1\frac{1}{3}$ miles per hour in still water. He paddles for $1\frac{1}{2}$ hours downstream in the Au Sable River, and to his surprise it takes him $4\frac{1}{2}$ hours to return upstream. What is the speed of the current in the river?

a. What are you asked to find in this problem? Assign a variable.

b. Make a table for this problem with two rows: downstream and upstream. Fill in the table.

c. Draw a sketch describing the problem.

d. Find some quantity that can be expressed in two different ways and write an equation.

e. Solve your equation and answer the question in the problem.

Building Number Skills

Some products can be computed mentally by rewriting one of the factors.

Examples: **a.** $45 \times 102 = 45 \times (100 + 2) = 45 \times 100 + 45 \times 2$

$= 4500 + 90 = 4590$

b. $6 \times 99 = 6 \times (100 - 1) = 6 \times 100 - 6 \times 1 = 600 - 6 = 594$

Compute each product without a calculator or a pencil.

15. 7×98 **16.** 8×198 **17.** 80×99 **18.** 25×102

19. 101×50 **20.** 98×45 **21.** 202×45 **22.** 198×15

23. 201×36 **24.** 302×25

9 Summary

Lesson 9.1

Before we begin solving an equation, we combine any like terms on each side.

For more complicated equations, we follow the order of operations.

If an equation has the variable on *both* sides of the equal sign, we must collect all terms containing the variable on the same side of the equation.

Lesson 9.2

Distributive Law

$$a(b + c) = ab + ac \qquad \text{and} \qquad (b + c)a = ba + ca$$

A negative sign outside the parentheses changes the sign of *each term inside the parentheses.*

The order of operations tells us to apply the distributive law before combining like terms.

Steps for Solving Equations

1. Simplify each side of the equation if necessary:
   - **a.** Apply the distributive law to remove parentheses.
   - **b.** Combine like terms.

2. Isolate the term containing the variable on one side of the equation by adding or subtracting to cancel the constant term from that side.
3. Divide both sides by the coefficient of the variable.

Lesson 9.3

Often we need to express two or more unknown quantities in terms of just *one* variable.

Steps for Solving Applied Problems

1. Writing algebraic expressions
   a. Describe each unknown quantity in words.
   b. Choose a variable to represent one of the unknown quantities.
   c. Write an algebraic expression for any other unknown quantities in terms of your variable.
2. Writing an equation
   a. Find some quantity that can be expressed in two different ways.
   b. Write an equation, using your expressions from step 1.
3. Solving
   a. Solve the equation.
   b. Answer the question in the problem.

Lesson 9.4

To clear the fractions from an equation, we multiply *each* term on *both* sides of the equation by the LCD of the fractions.

To Solve a Linear Equation:
First simplify the equation as much as possible.

1. Clear any fractions by multiplying both sides by the LCD.
2. Remove any parentheses by applying the distributive law.
3. Combine any like terms appearing on the same side of the equation.

Then isolate the variable.

4. Add or subtract the same quantity on both sides to get all the variable terms on one side of the equation.
5. Add or subtract the same quantity on both sides to get all the constant terms on the other side.
6. Divide both sides by the coefficient of the variable.

Lesson 9.5

We can use the interest formula, $I = Prt$, to solve problems involving two or more investments.

We can use the percent formula, $P = rW$, to solve problems about mixtures.

We cannot add percents if they are fractions of different whole amounts.

We use a table to help us solve mixture problems:

	Total amount (W)	Percent of important ingredient (r)	Amount of important ingredient (P)
First component			
Second component			
Mixture			

We can use the distance formula, $d = rt$, to solve problems involving motion at a constant speed.

Diagrams and tables can be helpful for solving problems involving motion.

CHAPTER 9 REVIEW 

True or False.

1. The distributive law is used to multiply two like terms.

2. To clear fractions from an equation, multiply each fraction by the LCD.

3. If there are variable terms on both sides of an equation, we must collect all the variable terms on one side.

4. To simplify each side of an equation, we should apply the distributive law to remove the parentheses and then collect like terms.

5. If you mix some grape juice that is 40% concentrate with a container of grape juice that is 20% concentrate, the mixture will be 30% concentrate.

6. When solving an applied problem, we try to express all the unknown quantities in terms of a single variable.

7. $a(b + c) = (a + b)c$

8. $-a(b + c) = -ab + ac$

9. $-a(b - c) = -ab + ac$

10. $2 + 3x = 5x$

Simplify by applying the distributive law and combining like terms.

11. $3x - 5x - (-7x)$

12. $-6b + 9b - (-b)$

13. $10 - 8(2s)$

14. $-12 + 6(-3d)$

15. $-6(7 - 5w)$

16. $4(-3 + 8t)$

17. $5 + (-3c - 5c)$

18. $-3 - (4v - 8v)$

19. $-6a - 4(-6a - 4)$

20. $3y - 7(3y - 7)$

21. $-3(2 - 2N) + 4(6N - 5)$

22. $5(3B - 4) - 2(7 - 6B)$

Solve.

23. $3 - 2d - d = 0$

24. $6r - 3 - 2r = 9$

25. $5r + 7(8 - r) = 12(4)$

26. $17t + 4(2t - 1) = 96$

27. $4(d - 3) = -2(3d - 6)$

28. $-3(3v + 1) = 4(2v + 12)$

29. $6u - 9(2u - 5) = 6 - 3(4 + 3u)$

30. $5(4 - 2g) + 7g = 9g - 2(g - 5)$

Write an equation and solve it to answer the question.

31. There are seven times as many women as there are men in the nursing program. If there are a total of 56 people in the program, how many are men?

32. Toni cross-country skis three times as far as she walks. If the total distance she covers is 12 kilometers, how far has she walked?

33. 120 meters of fencing are needed to enclose a rectangular yard. If the length of the yard is 38 meters, what is the width?

34. Clark is driving to Smallville, a distance of 278 miles. After driving for 5 hours, he passes a sign indicating that Smallville is another 3 miles away. What was Clark's average speed for the first 5 hours?

Solve by clearing the fractions.

35. $\frac{-5}{6}t = \frac{-10}{3}$

36. $\frac{3}{10}m = \frac{2}{5}$

37. $\frac{2}{3}b + 1 = \frac{7}{3}$

38. $\frac{1}{4}c - 2 = \frac{-5}{4}$

39. $\frac{1}{2}d + 5 = \frac{23}{4} - d$

40. $\frac{7}{5}h + 3 = \frac{31}{10} + h$

(a) Write an algebraic expression in terms of the variable, (b) evaluate your expression for the given value of the variable, and (c) write an equation and solve it.

41. a. In a certain state, the income tax is 8% of your adjusted gross income, which is your gross income minus your deductions, *d*. A new teacher makes $20,000 a year. How much will a new teacher pay in state income tax?

b. Aaron is a new teacher, and his deductions this year total $3400. How much income tax will he pay?

c. Masoud is also a new teacher. How much must Masoud have in deductions if he wants to pay no more than $1200 in income tax?

42. a. A home buyer will pay a discount fee of 2% of the mortgage loan, which will be the cost of the house minus the buyer's down payment. A house in a new tract costs $300,000. How much is the discount fee for a down payment of *D* dollars?

b. How much will be Francine's discount fee if she purchases a new house in this tract with a down payment of $60,000?

c. How large a down payment must Delbert make for one of these houses if he wants his discount fee to be $3000?

43. a. Julie will receive $\frac{2}{3}$ of the profit from holding a concert, where the profit is the difference between the revenue from ticket sales and the $45,000 costs for the performers, renting the venue, and so on. How much will Julie receive if the revenue is *R* dollars?

b. How much will Julie receive if the revenue is $120,000?

c. How much revenue is required for Julie to receive $75,000?

44. a. Lynette will receive $\frac{3}{4}$ of a shipment of coal, but Jaime receives $\frac{1}{3}$ of a shipment, which he takes from Lynette's share. If the coal shipment is *c* tons, how much coal will Lynette have after Jaime has taken his share?

b. How much will Lynette have if the shipment is 6 tons?

c. How large must the shipment be if Lynette ends up with 2 tons?

45. Suppose you split $2000 into two investments. Part of the money goes into a savings account and the rest into a money market account. At the end of the year, the savings account has earned 4% interest, and the money market has earned 6% interest.

Investment ($) in savings	*Investment ($) in money market*	*Earnings ($) from savings*	*Earnings ($) from money market*	*Total ($) earnings*
100				
500				
1000				
1200				
x				

a. Fill in the table.

b. If your total interest earnings are $100.80, how much did you invest in each account?

46. Suppose you decide to invest money in two new companies. You invest some money into your friend's new company Risky Ventures. You invest twice as much in Roc Solid Aeronautics. At the end of the year, the Risky Ventures investors had earned 14% interest, and Roc Solid investors had lost 3% of their investment.

Investment ($) in Risky Ventures	*Investment ($) in Roc Solid*	*Earnings ($) from Risky Ventures*	*Earnings ($) from Roc Solid*	*Total ($) earnings*
2,000				
8,000				
12,500				
17,000				
x				

a. Fill in the table.

b. If your total earnings are $12,000, how much did you invest in each company?

47. Woodland Hills University has only two graduate programs, English and mathematics. For the English program, 120 women applied and 35% were admitted, whereas 80 men applied and 30% were admitted. For the math program, 60 women applied and 80% were admitted; 160 men applied and 75% were admitted.

a. Did English admit a higher percentage of the women candidates or of the men? What about math?

b. How many women were admitted in English? How many in math?

c. What was the total number of women candidates to the university? What percent of those women were admitted?

d. How many men were admitted in English? How many in math?

e. What was the total number of men candidates to the university? What percent of those men were admitted? Is this percent higher or lower than that for the women?

48. George's old motor scooter required a fuel that is 5% oil and the rest gasoline. He has a 6-gallon supply of fuel for his old scooter. But he bought a new scooter requiring a fuel with only 4% oil, so he plans to add an amount of pure gasoline to his old fuel.

	Gallons of fuel	*Percent oil*	*Gallons of oil*
Old fuel			
Pure gasoline			
Mixture			

a. Fill in the table, assuming that George adds x gallons of pure gasoline.

b. How much gasoline should George add? How much fuel with 4% oil will he have in all?

49. The Riddler sends out a missile from Gotham City to blow up Metropolis. A quarter of an hour later, Batman sets off in his Bat-jet to intercept the missile. The missile travels at 600 miles per hour, and the Bat-jet flies at 720 miles per hour.

	Rate (mph)	*Time (h)*	*Distance (mi)*
Missile			
Bat-jet			

a. Fill in the table, assuming that the missile flies for t hours.

b. How long will it take the Bat-jet to overtake the missile?

c. Suppose that Metropolis is 950 miles from Gotham City. Will Batman catch the missile before it destroys Metropolis?

50. Archie's boat travels 35 miles per hour in still water. He travels at top speed for $\frac{1}{2}$ hour with a current. His return trip against the current takes $\frac{3}{4}$ hour.

	Rate (mph)	*Time (h)*	*Distance (mi)*
Trip with current			
Trip against current			

a. Fill in the table above, assuming that the current's speed is v miles per hour.

b. How fast is the current?

c. How far was Archie's round trip?

Building Number Skills

Perform each computation mentally, without using a calculator or a pencil.

51. 62×50

52. 44×25

53. $27 + 48$

54. $17 + 59$

55. $47 - 18$

56. $62 - 29$

57. $14 + 38 + 16 + 12$

58. $35 + 17 + 13 + 35$

59. 6×299

60. 23×102

Determine if the number is divisible by 2, 3, 4, 5, 9, and 10.

61. 45, 678

62. 98, 765

APPENDIX

Review of Arithmetic Skills

A.1 Reducing and Building Fractions

A **fraction** is a quotient of two whole numbers. For example, $\frac{2}{3}$ and $\frac{7}{4}$ are fractions. The expression $\frac{2}{3}$ means 2 divided by 3, and $\frac{7}{4}$ means 7 divided by 4. The number above the fraction bar is called the **numerator,** and the number on the bottom is called the **denominator.** The denominator of a fraction can never be zero because we cannot divide by zero.

If the numerator is less than the denominator, then the fraction is called a **proper fraction.** A proper fraction is a number less than one whole. Thus,

$$\frac{2}{3} \quad \text{and} \quad \frac{5}{7}$$

are proper fractions. If the numerator is greater than or equal to the denominator, then the fraction is called an **improper fraction.** An improper fraction represents a number greater than or equal to one whole. For example,

$$\frac{7}{4} \quad \text{and} \quad \frac{5}{5}$$

are examples of improper fractions.

Fractions are used to describe a portion or part of some whole amount. For example, the fraction $\frac{2}{3}$ tells us to separate the whole amount into 3 pieces and take 2 of them. Thus, to take $\frac{2}{3}$ of 12 items we separate the 12 items into 3 groups of 4 items each, and then take 2 groups, to get 8 items. We see that $\frac{2}{3}$ of 12 is equal to 8.

Equivalent Fractions

Any fraction can be rewritten in a different but equivalent form by multiplying both its numerator and denominator by the same number (except zero). For example,

$$\frac{2}{3} = \frac{2 \cdot \mathbf{6}}{3 \cdot \mathbf{6}} = \frac{12}{18}$$

The two fractions $\frac{2}{3}$ and $\frac{12}{18}$ are called **equivalent fractions.** Equivalent fractions are equal, that is, they represent the same portion of a whole.

This rule for forming equivalent fractions is called the **fundamental principle of fractions.** In words, we have the following.

Fundamental Principle of Fractions

If we multiply the numerator and denominator of a fraction by the same (nonzero) number, the new fraction is equivalent to the old one.

Reducing Fractions

We can also form an equivalent fraction by *dividing* the numerator and the denominator by the same (nonzero) number. This is another version of the fundamental principle of fractions. It enables us to reduce a fraction: A fraction is **reduced,** or in **lowest terms,** if no number greater than 1 divides evenly into both the numerator and the denominator. For example, we can reduce $\frac{10}{15}$ to the equivalent fraction $\frac{2}{3}$.

To Reduce a Fraction

1. Factor numerator and denominator completely.
2. Divide numerator and denominator by any common factors.
3. Simplify the result.

EXAMPLE 1

Reduce the fraction $\dfrac{21}{30}$ to lowest terms.

Solution

Follow the three steps given above.

$$\frac{21}{30} = \frac{3 \cdot 7}{2 \cdot 3 \cdot 5} \quad \text{Factor numerator and denominator.}$$

$$= \frac{\cancel{3} \cdot 7}{2 \cdot \cancel{3} \cdot 5} \quad \text{Divide numerator and denominator by any common factors.}$$

$$= \frac{7}{2 \cdot 5} \quad \text{Multiply any remaining factors in numerator and denominator.}$$

$$= \frac{7}{10}$$

Since no number bigger than 1 divides evenly into both 7 and 10, the fraction is reduced to lowest terms.

Remember that any number divided by itself is 1.

EXAMPLE 2

Reduce to lowest terms.

a. $\dfrac{2}{6}$ **b.** $\dfrac{18}{3}$

Solution

a. $\dfrac{2}{6} = \dfrac{2}{2 \cdot 3} = \dfrac{\not{2}}{\not{2} \cdot 3} = \dfrac{1}{3}$

b. $\dfrac{18}{3} = \dfrac{3 \cdot 6}{3} = \dfrac{\not{3} \cdot 6}{\not{3}} = \dfrac{6}{1} = 6$

Building Fractions

We should always reduce a fraction if it is the final answer to a problem. However, when we are adding or subtracting fractions we need to **build** the fractions to equivalent forms with larger numerators and denominators before we can proceed with the computation. For example, we might want to build the fraction $\frac{3}{5}$ to an equivalent form with denominator 100; that is,

$$\frac{3}{5} = \frac{?}{100}$$

Instead of *dividing* numerator and denominator as we do when reducing a fraction, when building a fraction we *multiply* numerator and denominator by the **building factor.** Here are the steps for building fractions.

To Build a Fraction

1. Divide the new denominator by the old denominator. The quotient is called the building factor.
2. Multiply the numerator and denominator of the old fraction by the building factor.

EXAMPLE 3  

Find the building factor (BF) and build the fraction:

$$\frac{3}{5} = \frac{3 \cdot \text{BF}}{5 \cdot \text{BF}} = \frac{?}{100}$$

Solution

Divide the new denominator by the old denominator:

$$100 \div 5 = 20$$

The building factor is 20. Multiply the numerator and denominator of the old fraction by 20:

$$\frac{3}{5} = \frac{3 \cdot 20}{5 \cdot 20} = \frac{60}{100}$$

The new fraction is $\dfrac{60}{100}$.

EXERCISE A.1

Reduce to lowest terms. See Examples 1 and 2.

1. $\frac{6}{10}$
2. $\frac{14}{21}$
3. $\frac{12}{75}$
4. $\frac{45}{70}$
5. $\frac{24}{72}$
6. $\frac{30}{120}$
7. $\frac{36}{48}$
8. $\frac{100}{120}$
9. $\frac{63}{105}$
10. $\frac{75}{180}$
11. $\frac{216}{18}$
12. $\frac{175}{35}$

Find the building factor (BF) and build each fraction. See Example 3.

13. $\frac{2}{3} = \frac{2 \cdot \text{BF}}{3 \cdot \text{BF}} = \frac{?}{9}$
14. $\frac{5}{7} = \frac{5 \cdot \text{BF}}{7 \cdot \text{BF}} = \frac{?}{14}$
15. $\frac{3}{2} = \frac{3 \cdot \text{BF}}{2 \cdot \text{BF}} = \frac{?}{8}$
16. $\frac{4}{3} = \frac{4 \cdot \text{BF}}{3 \cdot \text{BF}} = \frac{?}{12}$
17. $\frac{2}{5} = \frac{2 \cdot \text{BF}}{5 \cdot \text{BF}} = \frac{?}{100}$
18. $\frac{1}{4} = \frac{1 \cdot \text{BF}}{4 \cdot \text{BF}} = \frac{?}{56}$
19. $\frac{1}{3} = \frac{1 \cdot \text{BF}}{3 \cdot \text{BF}} = \frac{?}{72}$
20. $\frac{4}{5} = \frac{4 \cdot \text{BF}}{5 \cdot \text{BF}} = \frac{?}{60}$
21. $\frac{5}{8} = \frac{5 \cdot \text{BF}}{8 \cdot \text{BF}} = \frac{?}{144}$
22. $\frac{5}{16} = \frac{5 \cdot \text{BF}}{16 \cdot \text{BF}} = \frac{?}{112}$
23. $\frac{0}{3} = \frac{0 \cdot \text{BF}}{3 \cdot \text{BF}} = \frac{?}{6}$
24. $\frac{0}{11} = \frac{0 \cdot \text{BF}}{11 \cdot \text{BF}} = \frac{?}{88}$

A.2 Multiplying and Dividing Fractions

Multiplying Fractions

To multiply two fractions together, we multiply their numerators together, and multiply their denominators together.

EXAMPLE 1

Multiply.

a. $\frac{2}{5} \cdot \frac{3}{7}$ **b.** $\frac{1}{3} \cdot \frac{3}{8}$

Solution

a. $\frac{2}{5} \cdot \frac{3}{7} = \frac{2 \cdot 3}{5 \cdot 7}$ Multiply numerators; multiply denominators.

$= \frac{6}{35}$

b. $\frac{1}{3} \cdot \frac{3}{8} = \frac{1 \cdot 3}{3 \cdot 8}$ Multiply numerators; multiply denominators.

$= \frac{3}{24} = \frac{\not{3}}{\not{3} \cdot 2 \cdot 2 \cdot 2}$ Reduce.

$= \frac{1}{8}$

We can save steps in Example 1b by dividing out any common factors from numerator and denominator of the product before we multiply, like this:

$$\frac{1}{3} \cdot \frac{3}{8} = \frac{1}{\not{3}} \cdot \frac{\not{3}}{8} = \frac{1}{8}$$

The common factors do not have to be in the numerator and denominator of the same fraction. In general, we have the following steps for multiplying fractions.

To Multiply Fractions

1. Divide numerators and denominators by any common factors.
2. Multiply together the remaining factors in the numerators; multiply together the remaining factors in the denominators.
3. Reduce the product if necessary.

EXAMPLE 2

Multiply $\frac{9}{4} \cdot \frac{2}{5}$.

Solution

Divide numerator and denominator of the product by 2:

$$\frac{9}{4}\cdot\frac{2}{5}=\frac{9}{\not{4}}\cdot\frac{\not{2}}{5}$$ Divide common factors from numerator and denominator.

$$=\frac{9\cdot 1}{2\cdot 5}$$ Multiply numerators; multiply denominators.

$$=\frac{9}{10}$$

Dividing Fractions

The **reciprocal** of a fraction is found by interchanging the numerator and denominator. For example,

The reciprocal of $\frac{5}{6}$ is $\frac{6}{5}$,

and the reciprocal of $\frac{1}{8}$ is $\frac{8}{1}$, or 8.

We can change any division problem into a multiplication problem with the same answer by using reciprocals. This gives us a method for dividing fractions.

To Divide Fractions

1. Replace the second fraction by its reciprocal and change the division to multiplication.
2. Follow the rules for multiplication.

EXAMPLE 3  

Divide.

a. $\frac{5}{9}\div\frac{2}{3}$

b. $\frac{16}{3}\div 4$

Solution

a. Replace $\frac{2}{3}$ by its reciprocal, $\frac{3}{2}$. Change to multiplication:

$$\frac{5}{9}\div\frac{2}{3}=\frac{5}{\not{9}}\cdot\frac{\not{3}}{2}$$ Divide common factors from numerator and denominator.

$$=\frac{5\cdot 1}{3\cdot 2}$$ Multiply numerators; multiply denominators.

$$=\frac{5}{6}$$

b. Replace 4 by its reciprocal, $\frac{1}{4}$. Change to multiplication:

$$\frac{16}{3}\div 4=\frac{\overset{4}{\not{16}}}{3}\cdot\frac{1}{\not{4}}$$ Divide common factors from numerator and denominator.

$$=\frac{4\cdot 1}{3\cdot 1}$$ Multiply numerators; multiply denominators.

$$=\frac{4}{3}$$

EXERCISE A.2 

Multiply. See Examples 1 and 2.

1. $\frac{2}{3} \cdot \frac{5}{7}$

2. $\frac{4}{3} \cdot \frac{11}{5}$

3. $\frac{6}{7} \cdot \frac{14}{15}$

4. $\frac{15}{11} \cdot \frac{22}{35}$

5. $\frac{12}{16} \cdot \frac{18}{27}$

6. $\frac{8}{75} \cdot \frac{15}{20}$

7. $\frac{28}{56} \cdot \frac{10}{15}$

8. $\frac{16}{38} \cdot \frac{19}{12}$

9. $\frac{21}{48} \cdot \frac{88}{77}$

10. $\frac{18}{121} \cdot \frac{99}{90}$

11. $\frac{24}{20} \cdot \frac{24}{36} \cdot \frac{3}{4}$

12. $\frac{18}{30} \cdot \frac{6}{8} \cdot \frac{4}{20}$

Divide. See Example 3.

13. $\frac{3}{4} \div \frac{5}{8}$

14. $\frac{15}{32} \div \frac{25}{48}$

15. $\frac{7}{3} \div \frac{28}{5}$

16. $\frac{9}{2} \div \frac{18}{7}$

17. $\frac{4}{5} \div 6$

18. $\frac{5}{6} \div 10$

19. $4 \div \frac{2}{9}$

20. $5 \div \frac{15}{32}$

21. $\frac{11}{2} \div \frac{3}{4}$

22. $\frac{10}{3} \div \frac{3}{5}$

23. $\frac{30}{24} \div \frac{18}{72}$

24. $\frac{36}{42} \div \frac{48}{63}$

A.3 Adding and Subtracting Fractions

Like Fractions

Fractions that have the same denominator are called **like fractions.** For example,

$$\frac{7}{8} \text{ and } \frac{3}{8} \text{ are like fractions,}$$

$$\text{but } \frac{2}{5} \text{ and } \frac{4}{9} \text{ are not like fractions.}$$

It is easy to add or subtract like fractions.

To Add or Subtract Like Fractions

1. Add or subtract the numerators.
2. Keep the same denominator.
3. Reduce if necessary.

EXAMPLE 1

Add the fractions.

a. $\frac{1}{5} + \frac{3}{5}$ **b.** $\frac{1}{12} + \frac{7}{12}$

Solution

a. $\frac{1}{5} + \frac{3}{5} = \frac{1+3}{5} = \frac{4}{5}$ Add the numerators; keep the same denominator.

b. $\frac{1}{12} + \frac{7}{12} = \frac{1+7}{12}$ Add the numerators; keep the same denominator.

$= \frac{8}{12} = \frac{2}{3}$ Reduce.

EXAMPLE 2

Subtract the fractions.

a. $\frac{11}{5} - \frac{4}{5}$ **b.** $\frac{13}{4} - \frac{7}{4}$

Solution

a. $\frac{11}{5} - \frac{4}{5} = \frac{11-4}{5} = \frac{7}{5}$ Subtract the numerators; keep the same denominator.

b. $\frac{13}{4} - \frac{7}{4} = \frac{13-7}{4}$ Subtract the numerators; keep the same denominator.

$= \frac{6}{4} = \frac{3}{2}$ Reduce.

Lowest Common Denominator

Before we can add two fractions with *different* denominators, we must rewrite both fractions as equivalent ones with the same denominator. This new denominator is called a **common denominator.** The first step in an addition problem is to discover a suitable common denominator. We will look for the smallest, or lowest common denominator.

The **lowest common denominator** (LCD) for two or more fractions is the smallest number that each of the denominators will divide into evenly. For instance, the LCD for $\frac{1}{4}$ and $\frac{5}{6}$ is 12, because 12 is the smallest number that 4 and 6 both divide into evenly. Sometimes it is easy to find the LCD, but if not we can use the following steps.

To Find the Lowest Common Denominator (LCD)

1. Factor each denominator completely. List all the different factors that appear.
2. Choose one of the factors. Which denominator has the most copies of that factor? Write down that many copies of the factor for the LCD.
3. Repeat step 2 for each of the different factors in your list.
4. Multiply all the factors from steps 2 and 3 to get the LCD.

EXAMPLE 3

Find the LCD for the following fractions.

a. $\frac{7}{12}, \frac{5}{18}$ **b.** $\frac{2}{15}, \frac{13}{12}, \frac{23}{30}$

Solution

a. Factor each denominator completely:

$$12 = \boxed{2 \cdot 2} \cdot 3$$
$$18 = 2 \cdot \boxed{3 \cdot 3}$$

The factors are 2 and 3.

The factor 2 occurs twice in the factorization of 12, so we use two 2s in the LCD.

The factor 3 occurs twice in the factorization of 18, so we use two 3s in the LCD.

Thus, the LCD is

$$\text{LCD} = 2 \cdot 2 \cdot 3 \cdot 3, \text{ or } 36$$

b. Factor each denominator completely:

$$15 = \boxed{3} \cdot \boxed{5}$$
$$12 = \boxed{2 \cdot 2} \cdot 3$$
$$30 = 2 \cdot 3 \cdot 5$$

The factors are 2, 3, and 5.

The factor 2 occurs twice in the factorization of 12, so we use two 2s in the LCD.

The factors 3 and 5 both occur at most once in any *one* denominator, so we use one 3 and one 5 in the LCD. Thus, the LCD is

$$\text{LCD} = 2 \cdot 2 \cdot 3 \cdot 5, \text{ or } 60$$

Unlike Fractions

Fractions with different denominators are called **unlike fractions.** To add unlike fractions, we must first build the fractions so that they have the same denominator.

To Add or Subtract Unlike Fractions

1. Find the LCD for the fractions.
2. Build each fraction to an equivalent one with the LCD as its denominator.
3. Add or subtract the numerators. Keep the same denominator.
4. Reduce if necessary.

EXAMPLE 4  

Add $\frac{7}{10} + \frac{5}{6}$.

Solution

Find the LCD for the fractions. Factor each denominator:

$$10 = \boxed{2} \cdot \boxed{5}$$

$$6 = 2 \cdot \boxed{3}$$

The LCD is $2 \cdot 3 \cdot 5$, or 30. Build each fraction so that its denominator is 30. For the first fraction, the building factor is $30 \div 10$ or 3, so

$$\frac{7}{10} \cdot \frac{3}{3} = \frac{21}{30}$$

For the second fraction, the building factor is $30 \div 6$ or 5, so

$$\frac{5}{6} \cdot \frac{5}{5} = \frac{25}{30}$$

Finally, add the two like fractions $\frac{21}{30}$ and $\frac{25}{30}$. Thus,

$$\frac{7}{10} + \frac{5}{6} = \frac{21}{30} + \frac{25}{30} = \frac{46}{30}$$

Reduce the sum to obtain

$$\frac{46}{30} = \frac{\not{2} \cdot 23}{\not{2} \cdot 15} = \frac{23}{15}$$

EXAMPLE 5

Subtract $\frac{7}{6} - \frac{3}{4}$.

Solution

Find the LCD. Factor each denominator:

$$6 = 2 \cdot \boxed{3}$$
$$4 = \boxed{2 \cdot 2}$$

The LCD is $2 \cdot 2 \cdot 3$, or 12. Build each fraction to an equivalent one with denominator 12:

$$\frac{7}{6} \cdot \frac{2}{2} = \frac{14}{12} \quad \text{and} \quad \frac{3}{4} \cdot \frac{3}{3} = \frac{9}{12}$$

Subtract the new fractions to obtain

$$\frac{7}{6} - \frac{3}{4} = \frac{14}{12} - \frac{9}{12} = \frac{5}{12}$$

EXERCISE A.3

Add or subtract. See Examples 1 and 2.

1. $\frac{1}{11} + \frac{3}{11}$
2. $\frac{1}{9} + \frac{7}{9}$
3. $\frac{14}{15} - \frac{7}{15}$
4. $\frac{17}{20} - \frac{14}{20}$
5. $\frac{1}{6} + \frac{3}{6}$
6. $\frac{3}{8} + \frac{1}{8}$
7. $\frac{9}{10} - \frac{4}{10}$
8. $\frac{3}{4} - \frac{1}{4}$
9. $\frac{1}{6} + \frac{2}{6} + \frac{3}{6}$
10. $\frac{2}{10} + \frac{3}{10} + \frac{5}{10}$
11. $\frac{1}{5} + \frac{3}{5} - \frac{2}{5}$
12. $\frac{2}{3} + \frac{4}{3} - \frac{5}{3}$
13. $\frac{7}{8} - \frac{3}{8} - \frac{2}{8}$
14. $\frac{8}{9} - \frac{2}{9} - \frac{3}{9}$
15. $\frac{19}{25} - \frac{8}{25} + \frac{4}{25}$
16. $\frac{16}{21} - \frac{4}{21} + \frac{2}{21}$

Find the lowest common denominator (LCD) of the fractions. See Example 3.

17. $\frac{3}{6}, \frac{2}{3}$

18. $\frac{1}{2}, \frac{3}{4}$

19. $\frac{5}{4}, \frac{5}{6}$

20. $\frac{3}{10}, \frac{2}{15}$

21. $\frac{4}{3}, \frac{2}{7}$

22. $\frac{4}{5}, \frac{8}{11}$

23. $\frac{11}{12}, \frac{7}{30}$

24. $\frac{29}{30}, \frac{17}{45}$

25. $\frac{1}{2}, \frac{2}{3}, \frac{3}{5}$

26. $\frac{1}{4}, \frac{4}{5}, \frac{8}{7}$

27. $\frac{19}{6}, \frac{5}{9}, \frac{8}{15}$

28. $\frac{5}{8}, \frac{11}{12}, \frac{1}{20}$

29. $\frac{21}{9}, \frac{11}{4}, \frac{1}{12}$

30. $\frac{18}{25}, \frac{1}{50}, \frac{3}{4}$

31. $\frac{23}{24}, \frac{19}{36}, \frac{7}{12}$

32. $\frac{13}{60}, \frac{3}{40}, \frac{11}{30}$

Add or subtract. See Examples 4 and 5.

33. $\frac{1}{2} + \frac{1}{3}$

34. $\frac{1}{3} + \frac{1}{4}$

35. $\frac{3}{4} - \frac{2}{3}$

36. $\frac{2}{3} - \frac{1}{2}$

37. $\frac{5}{8} + \frac{1}{12}$

38. $\frac{11}{18} + \frac{5}{24}$

39. $\frac{14}{15} - \frac{7}{10}$

40. $\frac{9}{14} - \frac{1}{3}$

41. $\frac{1}{2} + \frac{1}{3} + \frac{2}{5}$

42. $\frac{1}{7} + \frac{5}{12} + \frac{1}{4}$

43. $\frac{13}{20} + \frac{3}{8} - \frac{5}{12}$

44. $\frac{5}{6} + \frac{4}{9} - \frac{7}{15}$

45. $\frac{49}{50} - \frac{8}{25} - \frac{1}{4}$

46. $\frac{9}{4} - \frac{10}{9} - \frac{5}{12}$

47. $\frac{29}{30} - \frac{21}{40} + \frac{9}{70}$

48. $\frac{11}{36} - \frac{7}{72} + \frac{5}{24}$

A.4 Mixed Numbers and Improper Fractions

Writing a Mixed Number as an Improper Fraction

A **mixed number** is the sum of a whole number and a fraction, such as $5 + \frac{2}{3}$. A mixed number is usually written without the addition sign, like this: $5\frac{2}{3}$. (We have to remember that $5\frac{2}{3}$ means 5 *plus* $\frac{2}{3}$.) Since a mixed number is greater than 1, we can also write it as an improper fraction. We do this by adding the whole number to the fraction.

Recall that any whole number can be treated as a fraction by writing the whole number as a numerator over a denominator of 1. For our example, $5 = \frac{5}{1}$; thus,

$$5\frac{2}{3} = \frac{5}{1} + \frac{2}{3}$$

We can now add the fractions by building $\frac{5}{1}$ to a fraction with denominator 3 as follows:

$$\frac{5}{1} + \frac{2}{3} = \frac{5 \cdot 3}{1 \cdot 3} + \frac{2}{3} \qquad \text{Building factor is 3.}$$

$$= \frac{15}{3} + \frac{2}{3} = \frac{17}{3}$$

Thus, the mixed number $5\frac{2}{3}$ is equal to the improper fraction $\frac{17}{3}$.

We can state the following rule for converting a mixed number to an improper fraction.

To Convert a Mixed Number to an Improper Fraction

1. Write the whole number over a denominator of 1.
2. Multiply numerator and denominator of the whole number by the denominator of the fraction part.
3. Add the two fractions.

EXAMPLE 1

Convert the mixed number $2\frac{5}{7}$ to an improper fraction.

Solution

Write the mixed number as the sum of a whole number and a fraction; then add.

$$2\frac{5}{7} = \frac{2}{1} + \frac{5}{7} \qquad \text{Build the first fraction; the LCD is 7.}$$

$$= \frac{2 \cdot 7}{1 \cdot 7} + \frac{5}{7} \qquad \text{Simplify; add the like fractions.}$$

$$= \frac{14}{7} + \frac{5}{7} = \frac{19}{7}$$

Thus, $2\frac{5}{7} = \frac{19}{7}$.

Operations with Mixed Numbers

It is usually easier to convert all mixed numbers to improper fractions before performing calculations.

EXAMPLE 2  

Divide $3\frac{3}{4} \div 2\frac{1}{2}$.

Solution

Convert each mixed number to an improper fraction:

$$3\frac{3}{4} = \frac{3}{1} + \frac{3}{4} = \frac{3 \cdot 4}{1 \cdot 4} + \frac{3}{4}$$
$$= \frac{12}{4} + \frac{3}{4} = \frac{15}{4}$$
$$2\frac{1}{2} = \frac{2}{1} + \frac{1}{2} = \frac{2 \cdot 2}{1 \cdot 2} + \frac{1}{2}$$
$$= \frac{4}{2} + \frac{1}{2} = \frac{5}{2}$$

Divide the improper fractions: Take the reciprocal of the divisor and multiply:

$$3\frac{3}{4} \div 2\frac{1}{2} = \frac{15}{4} \div \frac{5}{2}$$
$$= \frac{15}{4} \cdot \frac{2}{5} = \frac{\overset{3}{\cancel{15}}}{\underset{2}{\cancel{4}}} \cdot \frac{\cancel{2}}{\cancel{5}}$$
$$= \frac{3}{2}$$

EXAMPLE 3

Add $1\frac{2}{5} + 2\frac{3}{8}$.

Solution

Convert each mixed number to an improper fraction:

$$1\frac{2}{5} = \frac{1}{1} + \frac{2}{5} = \frac{1 \cdot 5}{1 \cdot 5} + \frac{2}{5}$$
$$= \frac{5}{5} + \frac{2}{5} = \frac{7}{5}$$
$$2\frac{3}{8} = \frac{2}{1} + \frac{3}{8} = \frac{2 \cdot 8}{1 \cdot 8} + \frac{3}{8}$$
$$= \frac{16}{8} + \frac{3}{8} = \frac{19}{8}$$

Add the improper fractions. Since the denominators are 5 and 8, the LCD is $5 \cdot 8$, or 40. Build each fraction to an equivalent one with denominator 40:

$$\frac{7}{5} = \frac{7 \cdot 8}{5 \cdot 8} = \frac{56}{40} \quad \text{and} \quad \frac{19}{8} = \frac{19 \cdot 5}{8 \cdot 5} = \frac{95}{40}$$

Add the new fractions:

$$\frac{7}{5} + \frac{19}{8} = \frac{56}{40} + \frac{95}{40} = \frac{56 + 95}{40} = \frac{151}{40}$$

Thus, $1\frac{2}{5} + 2\frac{3}{8} = \frac{151}{40}$.

Writing an Improper Fraction as a Mixed Number

Recall that the fraction bar is a division symbol; for example, $\frac{3}{4}$ means 3 divided by 4. If the numerator and denominator of a fraction are equal, then the fraction is equal to 1. For example, $\frac{3}{3} = 1$ and $\frac{12}{12} = 1$. If the numerator is larger than the denominator, then the fraction is a number greater than 1. By dividing the denominator into the numerator, we can write an improper fraction as a whole number or as a mixed number.

EXAMPLE 4

Write each improper fraction as a whole number or as a mixed number.

a. $\frac{8}{4}$ **b.** $\frac{17}{5}$

Solution

a. Divide the denominator into the numerator: $8 \div 4 = 2$. Thus, $\frac{8}{4} = 2$, a whole number.

b. When we divide the denominator into the numerator, there is a remainder:

$$\begin{array}{r} 3 \\ 5\overline{)17} \\ -15 \\ \hline 2 \end{array} \quad \begin{array}{l} \leftarrow \text{Quotient} \\ \\ \\ \leftarrow \text{Remainder} \end{array}$$

The quotient 3 is the whole number, and the remainder 2 is the numerator of the proper fraction. Thus,

$$\frac{17}{5} = 3 + \frac{2}{5} = 3\frac{2}{5}$$

The following rule summarizes the method shown in the example.

To Convert an Improper Fraction to a Mixed Number

1. Divide the denominator into the numerator.
2. The quotient is the whole number part of the mixed number.

   The remainder is the numerator of the fraction part, with the original denominator.
3. If there is no remainder, then the improper fraction is equivalent to a whole number.

EXERCISE A.4

Convert each mixed number to an improper fraction. See Example 1.

1. $3\frac{2}{3}$ **2.** $1\frac{2}{13}$ **3.** $12\frac{1}{2}$ **4.** $11\frac{2}{7}$

5. $20\frac{4}{5}$ **6.** $7\frac{11}{20}$ **7.** $4\frac{21}{50}$ **8.** $5\frac{5}{6}$

Convert each mixed number to an improper fraction; then perform the indicated operation. See Examples 2 and 3.

9. $3\frac{1}{4} + 1\frac{5}{8}$ **10.** $5\frac{3}{4} + 2\frac{7}{8}$ **11.** $5\frac{2}{3} + 6\frac{3}{4}$ **12.** $4\frac{1}{3} + 2\frac{1}{4}$

13. $9\frac{3}{8} - 2\frac{1}{2}$ **14.** $5\frac{1}{5} - 4\frac{7}{10}$ **15.** $7\frac{3}{8} - 1\frac{7}{12}$ **16.** $6\frac{5}{8} - 2\frac{5}{6}$

17. $7\frac{1}{3} \cdot 2\frac{1}{4}$ **18.** $5\frac{2}{5} \cdot 3\frac{1}{3}$ **19.** $3\frac{3}{7} \cdot 2\frac{1}{12}$ **20.** $2\frac{2}{9} \cdot 3\frac{3}{5}$

21. $2\frac{1}{3} \div 5\frac{3}{5}$ **22.** $4\frac{1}{2} \div 2\frac{4}{7}$ **23.** $3\frac{3}{4} \div 1\frac{7}{8}$ **24.** $5\frac{5}{6} \div 2\frac{5}{8}$

Convert each improper fraction to a whole number or a mixed number. See Example 4.

25. $\frac{11}{3}$ **26.** $\frac{15}{4}$ **27.** $\frac{43}{8}$ **28.** $\frac{37}{8}$

29. $\frac{107}{16}$ **30.** $\frac{123}{16}$ **31.** $\frac{317}{32}$ **32.** $\frac{361}{32}$

A.5 Decimal Fractions

A **decimal fraction** is a fraction whose denominator is a power of 10, such as 10, 100, 1000. A decimal fraction is not written with a fraction bar, and the denominator is not written explicitly. Instead, we show the denominator using place value notation.

Place Value Notation

In a whole number, the value of each digit is determined by its position. For example, the 6 in 68 stands for 6 tens, whereas the 6 in 642 stands for 6 hundreds. The decimal point comes after the ones digit, but in a whole number we usually do not write the decimal point.

Positions to the right of the decimal are called **decimal places,** and digits in those positions represent decimal fractions. For example, the number 0.3 represents three-tenths, or $\frac{3}{10}$, and the number 0.03 represents three-hundredths, or $\frac{3}{100}$. The place values of the first few positions on either side of the decimal point are shown in Figure A.1.

Figure A.1

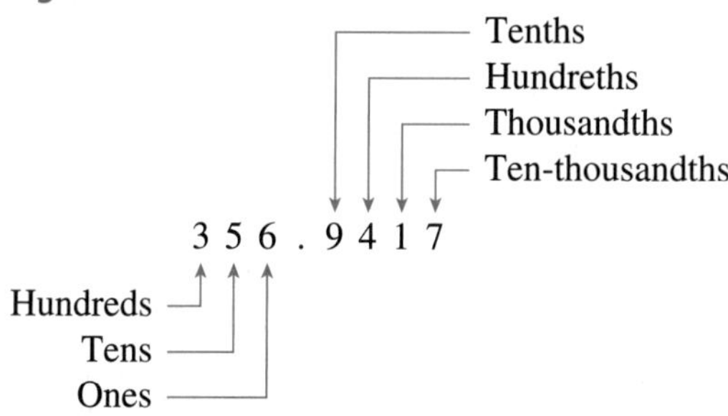

Converting a Decimal Fraction to a Common Fraction

The first place to the right of the decimal is the tenths place, and the second is the hundredths place. For example, the digits of 0.57 stand for 5 tenths and 7 hundredths. That is,

$$0.57 = \frac{5}{10} + \frac{7}{100}$$

If we add these fractions, we find that

$$\frac{5}{10} + \frac{7}{100} = \frac{5 \cdot 10}{10 \cdot 10} + \frac{7}{100} = \frac{57}{100}$$

Thus, $0.57 = \frac{57}{100}$. We see that we can read a decimal fraction just as we do a whole number and then use the place value of the *last* digit for the denominator. This gives us a rule for writing a decimal fraction as a common fraction.

To Convert a Decimal Fraction to a Common Fraction

1. The digits after the decimal are the numerator of the fraction.
2. The place value of the last digit is the denominator of the fraction.
3. Reduce the fraction if possible.

EXAMPLE 1  

Convert each decimal fraction to a common fraction.

a. 0.007 **b.** 0.35 **c.** 4.9

Solution

a. The numerator of the fraction is 007, or 7. The last digit, 7, is in the thousandths place, so the denominator of the fraction is 1000. Thus, $0.007 = \frac{7}{1000}$.

b. The numerator of the fraction is 035, or 35. The last digit, 5, is in the hundredths place, so the denominator of the fraction is 100. Thus, $0.35 = \frac{35}{100}$. This fraction can be reduced to obtain

$$0.35 = \frac{35}{100} = \frac{\not{5} \cdot 7}{2 \cdot 2 \cdot \not{5} \cdot 5} = \frac{7}{20}$$

c. The decimal number 4.9 represents a mixed number with whole number part 4 and fraction part 0.9, or 9 tenths. Thus,

$$4.9 = 4\frac{9}{10} = \frac{4 \cdot 10}{1 \cdot 10} + \frac{9}{10} = \frac{49}{10}$$

Converting a Common Fraction to a Decimal Fraction

To convert a common fraction to a decimal fraction, we divide the denominator into the numerator. (Recall that the fraction bar is a division symbol.) This is easily done with a calculator. If you must perform the division by hand, however, you will need to write a decimal point and one or more zeros after the numerator.

EXAMPLE 2

Convert the fraction $\frac{4}{5}$ to decimal form.

Solution

First, write a decimal point and a zero after the numerator and then divide the denominator into the numerator. To begin the division, place another decimal point in the quotient directly above the decimal in the dividend:

$$\begin{array}{r} .8 \\ 5\overline{)4.0} \\ -2\,0 \\ \hline \end{array}$$

After placing the decimal points, you may ignore them and proceed as if the numbers were whole numbers. We find that the decimal equivalent of $\frac{4}{5}$ is 0.8.

On a calculator, we enter the division in Example 2 as 4 [÷] 5 [=], and the calculator returns the answer, 0.8.

If the division does not terminate after the first step, we must add more zeros after the numerator and continue dividing.

EXAMPLE 3

Convert each fraction to decimal form.

a. $\frac{3}{8}$ **b.** $\frac{6}{11}$

Solution

a. Divide the denominator into the numerator. Continue to add zeros after the numerator until the division terminates:

$$\begin{array}{r} .375 \\ 8\overline{)3.000} \\ -2\,4 \\ \hline 60 \\ -56 \\ \hline 40 \\ -40 \\ \hline \end{array}$$

Place decimal in quotient above decimal in numerator

Thus, $\frac{3}{8} = 0.375.$

b. Divide the denominator into the numerator:

$$\begin{array}{r} .5454\ldots \\ 11\overline{)6.0000\ldots} \\ -5\,5 \\ \hline 50 \\ -44 \\ \hline 60 \\ -55 \\ \hline 50\ldots \end{array}$$

Place decimal in quotient above decimal in numerator

After a few steps, we see that the division is in an infinite loop—it repeats the same digits over and over without ending. The quotient, 0.545454 . . . , is called a **repeating decimal.** We indicate a repeating decimal by writing a bar over the block of repeated digits, thus: $0.\overline{54}$. So

$$\frac{6}{11} = 0.545454\ldots, \quad \text{or} \quad \frac{6}{11} = 0.\overline{54}$$

You should be aware that your calculator may round off a repeating decimal to the last digit of its display. For example, the decimal equivalent of $\frac{8}{9}$ might be displayed as 0.8888889. This is an approximation to the actual decimal fraction

$$\frac{8}{9} = 0.\overline{8}$$

EXERCISE A.5 

Convert each decimal fraction to a common fraction. See Example 1.

1. 0.17 **2.** 0.81 **3.** 0.07 **4.** 0.03

5. 0.023 **6.** 0.049 **7.** 0.6 **8.** 0.2

9. 0.26 **10.** 0.15 **11.** 0.375 **12.** 0.864

13. 2.25 **14.** 1.75 **15.** 3.60 **16.** 4.80

Convert each common fraction to a decimal fraction. See Examples 2 and 3.

17. $\frac{21}{25}$ **18.** $\frac{7}{20}$ **19.** $\frac{23}{50}$ **20.** $\frac{19}{20}$

21. $\frac{31}{100}$ **22.** $\frac{73}{100}$ **23.** $\frac{31}{1000}$ **24.** $\frac{73}{1000}$

25. $\frac{5}{16}$ **26.** $\frac{13}{8}$ **27.** $\frac{3}{8}$ **28.** $\frac{19}{16}$

29. $\frac{5}{6}$ **30.** $\frac{2}{3}$ **31.** $\frac{3}{11}$ **32.** $\frac{9}{11}$

A.6 Rounding Decimal Numbers

Sometimes it is more practical to use an approximate value rather than the exact result of a calculation. A common method of approximating numbers is called **rounding.** For example, the sales tax on a purchase is usually rounded *up* to the nearest cent.

To round a number we consider two of its digits, which we call the *target digit* and the *test digit.* Consider the number 12,469. If we want to round to thousands, then we are looking for the multiple of 1000 that is closest to our number. The thousands digit, 2, is the target digit in this case. Is 12,469 closer to 1**2**,000 or to 1**3**,000? Since **4**69 is less than halfway (500), our number is closer to 12,000. We used the number 4 to test which way to round, so 4 is the test digit.

Target digit — Test digit

1 2, 4 6 9 → 12,000

Here are the rules for rounding decimal numbers. We consider two cases: rounding a number to a decimal place and rounding a number to a nondecimal place.

To Round a Number to a Decimal Place

1. Circle the digit in the round-off place; this is the target digit.
2. Underline the digit to the right of the target digit; this is the test digit.
3. If the test digit is 5 or greater (5, 6, 7, 8, or 9), increase the target digit by one.

   If the test digit is less than 5 (0, 1, 2, 3, or 4), keep the target digit.
4. Discard the test digit and any following digits.

To Round a Number to a Nondecimal Place

1 Circle the digit in the round-off place; this is the target digit.
2. Underline the digit to the right of the target digit; this is the test digit.
3. If the test digit is 5 or greater (5, 6, 7, 8, or 9), increase the target digit by 1. If the test digit is less than 5 (0, 1, 2, 3, or 4), keep the target digit.
4. Replace the test digit and any nondecimal digits that follow by zeros.
5. Discard any decimal digits.

EXAMPLE 1 

Round 349.0258 to the indicated place.

a. To the nearest tenth **b.** To the nearest thousandth

Solution

a. The target digit is the tenths digit, 0, and the test digit is 2. Because the test digit is less than 5, we keep the target digit. We then discard the test digit and the digits following. The result is 349.0.

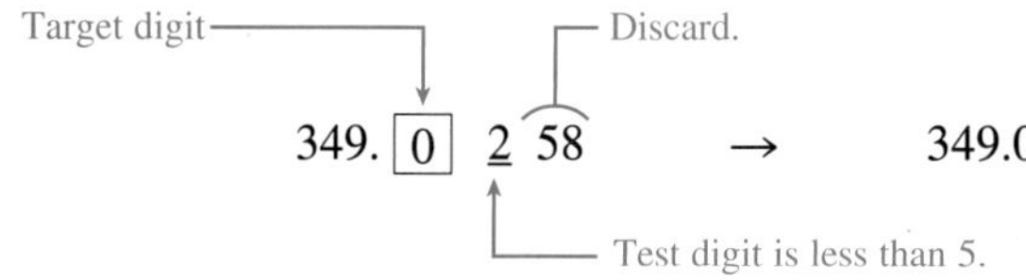

b. The target digit is the thousandths digit, 5, and the test digit is 8. Because the test digit is greater than 5, we increase the target digit by 1. We then discard the test digit. The result is 349.026.

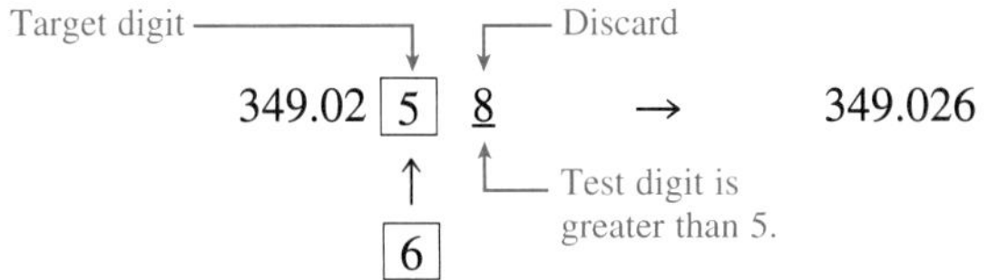

EXAMPLE 2 

Round 2479.83 to the indicated place.

a. To the nearest thousand **b.** To the nearest whole number

Solution

a. The target digit is the thousands digit, 2, and the test digit is 4. Because the test digit is less than 5, we keep the target digit. We then replace the test digit and the remaining digits up to the decimal point (4, 7, and 9) by zeros, and discard the digits after the decimal point. The result is 2000.

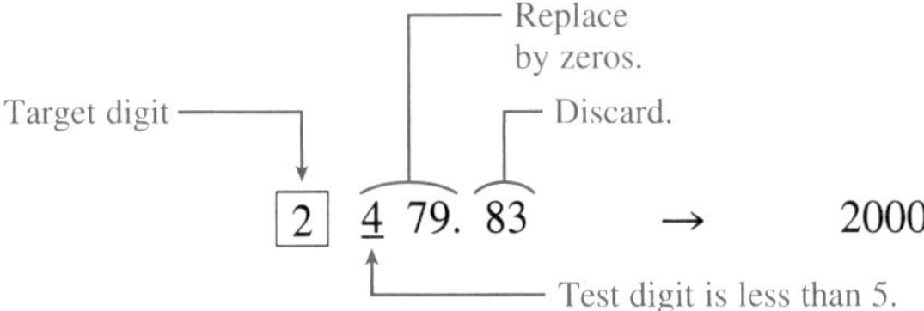

b. The target digit is the ones digit, 9, and the test digit is 8. Because the test digit is greater than 5, we add 1 to the target digit. (Note that this also changes the tens digit.) We then discard the digits after the decimal point. The result is 2480.

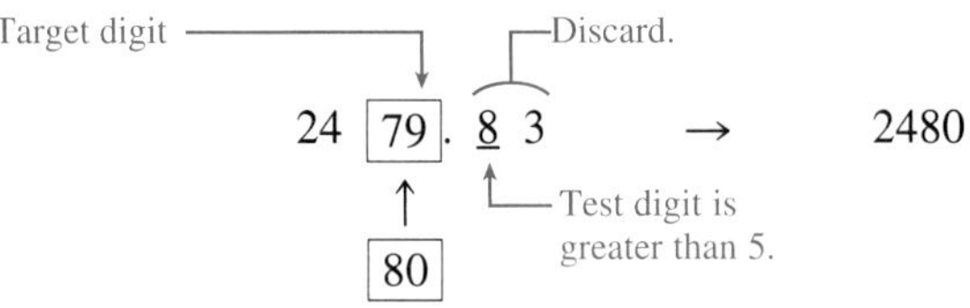

EXERCISE A.6 

Round each number to the nearest (a) ten, (b) tenth, (c) hundredth, and (d) thousandth.

1. 14.7742

2. 21.6344

3. 76.28256

4. 54.60791

5. 169.8991

6. 832.8196

7. 5545.9098

8. 9989.8982

9. 700.9597

10. 508.9595

11. 19.95059

12. 97.93965

Round each number to (a) one, (b) two, and (c) three decimal places.

13. 1.9069

14. 2.2591

15. 0.91994

16. 0.65232

17. 0.09857

18. 0.07579

19. 6.1695

20. 4.2945

A.7 Adding and Subtracting Decimal Fractions

Recall that when we add or subtract whole numbers we align the numbers vertically so that digits with the same place value are in the same column. The same is true when we add decimal numbers. The easiest way to make sure that the digits are aligned correctly is to align the decimal points vertically.

EXAMPLE 1

Add 15.263 + 6.74.

Solution

Align the decimal points vertically and add just as you would add whole numbers. Place the decimal point for the sum directly below the decimals in the two numbers:

$$\begin{array}{r} 15.263 \\ +\ 6.740 \\ \hline 22.003 \end{array}$$

We may add a zero at the end of 6.74 as shown so that each number has the same number of decimal places. The sum is 22.003.

EXAMPLE 2

Subtract 10 − 0.06.

Solution

The decimal point in the whole number 10 is on the right, after the units place. Include the decimal point and two zeros to get 10.00 so that both numbers have the same number of decimal places. Align the decimal points and subtract.

$$\begin{array}{r} 10.00 \\ -\ 0.06 \\ \hline 9.94 \end{array}$$

The difference is 9.94.

Here is the procedure for adding and subtracting decimal numbers.

To Add or Subtract Decimal Numbers

1. Write the numbers with the decimal points aligned vertically.

   If necessary, include zeros at the far right so that both numbers have the same number of decimal places.
2. Add or subtract just as you would for whole numbers.
3. Place the decimal point in the sum or difference directly below the decimal points in the numbers.

Of course, if a calculator is available, we can let it do the work.

EXAMPLE 3

Use a calculator to add or subtract.

a. 15.263 + 6.74 **b.** 10 − 0.06

Solution

Key the calculations as shown.

a. 15.263 [+] 6.74 [=]. The calculator returns the sum, 22.003.

b. 10 [−] 0.06 [=]. The calculator returns the difference 9.94.

EXERCISE A.7

Add. See Examples 1 and 3.

1. 1.46 + 3.27 **2.** 8.01 + 1.90 **3.** 3.64 + 0.36 **4.** 11.22 + 1.08

5. 0.26 + 1.4 **6.** 0.91 + 7.3 **7.** 13 + 0.26 **8.** 5 + 0.55

9. 6.2 + 0.027 **10.** 7.1 + 0.109 **11.** 31.2 + 3.25 **12.** 42.3 + 2.48

Subtract. See Examples 2 and 3.

13. 12.63 − 9.16 **14.** 6.31 − 2.26 **15.** 7 − 1.26 **16.** 10 − 7.11

17. 6.02 − 0.95 **18.** 5.03 − 0.87 **19.** 12.1 − 2.36 **20.** 14.4 − 8.43

21. 438.4 − 76.25 **22.** 587.3 − 32.91 **23.** 13 − 0.0006 **24.** 18 − 0.0002

Compute.

25. $2.49 + 6.3 - 5.08$

26. $7.92 - 1.48 + 3.6$

27. $573.1 - 28.96 - 2.094$

28. $348.8 - 59.29 - 6.123$

29. $465.5 - 127 + 238.58$

30. $651.3 - 249 + 367.42$

A.8 Multiplying and Dividing Decimal Fractions

Products

We multiply and divide decimal numbers the same way we multiply and divide whole numbers, except that we must keep track of the decimal point. Recall that a decimal place is the position for a digit *after,* or to the right of, the decimal point. To find the number of decimal places in a product, we add the number of decimal places in the two numbers being multiplied.

EXAMPLE 1

Multiply.

a. (3.4)(0.68) **b.** (0.4)(0.2)

Solution

a. Multiply the numbers, ignoring their decimal points. (The decimal points do not have to be aligned as they do in addition.)

$$\begin{array}{r} 0.68 \\ \times\ \underline{3.4} \\ 272 \\ \underline{204} \\ 2312 \end{array}$$

Next, locate the decimal point in the product. We see that 3.4 has one decimal place and 0.28 has two decimal places, so their product has three decimal places. Count three places from the *right* side of 2312:

2 . 3 1 2

The product is 2.312.

b. Multiply the numbers, ignoring their decimal points:

$$\begin{array}{r} 0.4 \\ \times\ \underline{0.2} \\ 8 \end{array}$$

Locate the decimal point. We see that 0.4 has one decimal place and 0.2 has one decimal place, so their product has two decimal places. Count two places from the right side of 8, inserting zeros as necessary.

$$.\,0\,8$$

The product is 0.08.

Here is the rule for multiplying decimal numbers.

To Multiply Two Decimal Numbers

1. Multiply the numbers, ignoring their decimal points.
2. Add the number of decimal places in the two numbers being multiplied.
3. Count this many decimal places from the right of the product and place the decimal point there.

Quotients

We have just seen that we locate the decimal point in a product *after* we perform the computation. To find a quotient of decimal numbers, we locate the decimal point *before* we divide. If the divisor is not already a whole number, we must move the decimal to the right side of the number, after the last digit. Then we *also* move the decimal in the dividend the *same number of places.* Once this is done, we write the decimal point for the quotient directly above the decimal of the dividend. Finally, we divide the same way we divide whole numbers.

EXAMPLE 2

Divide.

a. $2.12 \div 0.25$ **b.** $3 \div 0.006$

Solution

a. Move the decimals in both divisor and dividend two places to the right, to get

$$25.\overline{)212}$$

Place a decimal on the quotient bar directly above the decimal in 212 and begin dividing. Add as many zeros to the right of the decimal in the dividend as necessary:

$$\begin{array}{r} 8.48 \\ 25\overline{)212.00} \\ -200 \\ \hline 12\,0 \\ -10\,0 \\ \hline 2\,00 \\ -2\,00 \\ \hline 0 \end{array}$$

The quotient is 8.48.

b. Move the decimals in both divisor and dividend three places to the right, adding zeros to the dividend, to get

$$006.\overline{)3000}$$

Place a decimal for the quotient directly above the decimal in the dividend and begin dividing:

$$\begin{array}{r} 500 \\ 6\overline{)3000} \\ -\underline{30} \\ 000 \end{array}$$

The quotient is 500.

Here is the rule for dividing decimal numbers.

To Divide One Decimal Number by Another

1. If the divisor is not a whole number, move the decimal to the right side of the number, after the last digit.
2. Move the decimal of the divisor the same number of places to the right, adding zeros after the last digit as necessary.
3. Place the decimal for the quotient directly above the decimal of the dividend.
4. Divide as usual.

We can find products and quotients of decimal numbers quite easily with a calculator. In Example 3 we give the keying sequences for the products and quotients in Examples 1 and 2.

EXAMPLE 3

Use a calculator to find the following products and quotients.

a. (3.4)(0.68) **b.** (0.4)(0.2)

c. 2.12 ÷ 0.25 **d.** 3 ÷ 0.006

Solution

Use the keying sequences shown.

a. 3.4 [×] 0.68 [=]. The calculator displays the product 2.312.

b. 0.4 [×] 0.2 [=]. The calculator displays the product 0.08.

c. 2.12 [÷] 0.25 [=]. The calculator displays the quotient 8.48.

d. 3 [÷] 0.006 [=]. The calculator displays the quotient 500.

Remember that the calculator will round off any result to the number of digits it can show on its display. For example, the calculator displays the quotient

5 [÷] 3 [=]

as 1.6666667, instead of 1.66 . . . or $1.\overline{6}$.

EXERCISE A.8 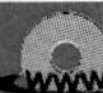

Multiply. See Examples 1 and 3.

1. (6.8)(0.6)
2. (4.7)(0.5)
3. (0.32)(0.4)
4. (0.47)(0.3)

5. (2.04)(0.02)
6. (3.07)(0.04)
7. (4.012)(0.03)
8. (5.007)(0.05)

9. (2.5)(1.3)
10. (4.7)(3.2)
11. (4.32)(2.4)
12. (5.07)(8.4)

13. (4.12)(0.42)
14. (6.28)(0.31)
15. (0.032)(0.12)
16. (0.041)(0.32)

Divide. See Examples 2 and 3.

17. $31.57 \div 7$
18. $50.58 \div 9$
19. $67.5 \div 16$
20. $97.2 \div 17$

21. $6.85 \div 0.5$
22. $8.64 \div 0.4$
23. $71.91 \div 4.7$
24. $593.14 \div 9.4$

25. $229.9 \div 0.38$
26. $374.4 \div 0.78$
27. $4.3776 \div 0.019$
28. $10.9752 \div 0.024$

29. $7 \div 1.54$
30. $8 \div 2.45$
31. $64 \div 0.08$
32. $45 \div 0.09$

A.9 Percents

A **percent** is a fraction whose denominator is 100. Percent means per 100 or divided by 100. Thus, 75% means $\frac{75}{100}$, or $\frac{3}{4}$.

Before using a percent in a calculation, we must rewrite it as a decimal fraction or as a common fraction.

Converting Percents to Decimal Fractions

To convert a percent to a decimal fraction, we divide by 100 and then discard the percent symbol. For example, 75% means $\frac{75}{100}$; so to write the decimal equivalent for 75%, we divide 75 by 100. Notice that we do not have to perform the division longhand, since dividing any number by 100 merely moves the decimal point two places to the left. That is,

$$75 \div 100 = 0.75$$

Thus, 75% is equal to 0.75.

Convert each percent to a decimal fraction.

a. 5% **b.** 135%
c. 0.8% **d.** $37\frac{1}{2}\%$

Solution

For parts (a)–(c) move the decimal point two places to the left. Don't forget to remove the % symbol.

a. 5% = 0.05 **b.** 135% = 1.35 **c.** 0.8% = 0.008

For part (d), first write the percent in decimal form and proceed as above.

d. $37\frac{1}{2}\% = 37.5\% = 0.375$

Here is the rule for changing a percent to a decimal fraction.

To Change a Percent to a Decimal Fraction

1. If the percent involves a common fraction, write it in decimal form.
2. Discard the % symbol and move the decimal point two places to the *left.*

Converting Decimal Fractions to Percents

To convert a decimal fraction to percent form, we reverse the process described above.

To Change a Decimal Fraction to Percent Form

1. Move the decimal point two places to the *right.*
2. Write the % symbol after the number.

EXAMPLE 2

Convert each decimal to a percent.

a. 0.3 **b.** 0.258 **c.** 3.4

Solution

Move each decimal point two places to the right. Write the percent symbol at the end.

a. $0.3 = 30\%$ **b.** $0.258 = 25.8\%$ **c.** $3.4 = 340\%$

Converting Percents to Common Fractions

There are several ways to change a percent to a common fraction. The most direct method is to write the percent as the numerator of a fraction with denominator 100, and then reduce the fraction. For example,

$$60\% = \frac{60}{100} = \frac{3}{5}$$

If the percent itself involves decimals, it is easier to convert the percent to a decimal fraction first. For example,

$$8.4\% = 0.084$$

Now we can convert the decimal to a common fraction to obtain

$$0.084 = \frac{84}{1000} = \frac{4 \cdot 21}{4 \cdot 250} = \frac{21}{250}$$

If the percent involves a mixed number, we can change the mixed number to an improper fraction and then divide by 100. For example, to convert $12\frac{2}{3}\%$ to a fraction, first write

$$12\tfrac{2}{3} = \frac{12}{1} + \frac{2}{3} = \frac{12 \cdot 3}{1 \cdot 3} + \frac{2}{3} = \frac{36 + 2}{3} = \frac{38}{3}$$

Now divide by 100 to find

$$\begin{aligned}\frac{38}{3} \div 100 &= \frac{38}{3} \cdot \frac{1}{100} = \frac{38}{300} \\ &= \frac{2 \cdot 19}{2 \cdot 150} = \frac{19}{150}\end{aligned}$$

EXAMPLE 3

Convert each percent to a common fraction.

a. 71.2% **b.** $14\frac{1}{6}\%$

Solution

a. First change the percent to a decimal by moving the decimal point two places to the left:

$$71.2\% = 0.712$$

Then convert the decimal fraction to a common fraction and reduce:

$$0.712 = \frac{712}{1000} = \frac{8 \cdot 89}{8 \cdot 125} = \frac{89}{125}$$

b. First write $14\frac{1}{6}$ as an improper fraction:

$$\frac{14}{1} + \frac{1}{6} = \frac{14 \cdot 6}{1 \cdot 6} + \frac{1}{6} = \frac{84 + 1}{6} = \frac{85}{6}$$

Thus, $14\frac{1}{6}\% = \frac{85}{6}\%$. Now divide by 100:

$$\frac{85}{6} \div 100 = \frac{85}{6} \cdot \frac{1}{100} = \frac{17}{120}$$

To Change a Percent to a Common Fraction

1. Write the percent as a fraction with denominator 100.
2. If the percent involves decimals, change the percent to a decimal fraction first. Then convert the decimal fraction to a common fraction.
3. If the percent involves a mixed number, write the mixed number as an improper fraction first. Then divide by 100.

Converting Fractions to Percents

The easiest way to convert a common fraction to percent form is to write the fraction as a decimal fraction first.

To Change a Common Fraction to a Percent

1. First convert the fraction to a decimal fraction.
2. Then move the decimal point two places to the right and add the % symbol.

EXAMPLE 4  

Convert $\frac{37}{80}$ to a percent.

Solution

First change $\frac{37}{80}$ to a decimal fraction; divide the numerator by the denominator:

$$\frac{37}{80} = 37 \div 80 = 0.4625$$

To change 0.4625 to a percent, move the decimal point two places to the right:

$$0.4625 = 46.25\%$$

EXERCISE A.9

Write each percent as a decimal. See Example 1.

1. 15% **2.** 37% **3.** 0.4% **4.** 0.7% **5.** 6.8% **6.** 1.1%

7. 119% **8.** 652% **9.** $3\frac{1}{4}\%$ **10.** $6\frac{3}{8}\%$ **11.** $\frac{2}{5}\%$ **12.** $\frac{3}{2}\%$

Write each decimal as a percent. See Example 2.

13. 0.33 **14.** 0.54 **15.** 0.504 **16.** 0.686 **17.** 0.787 **18.** 0.074

19. 0.0201 **20.** 0.005 **21.** 0.008 **22.** 0.0008 **23.** 5.5 **24.** 6

Write each percent as a fraction. See Example 3.

25. 35% **26.** 10% **27.** 125% **28.** 150% **29.** 60% **30.** 48%

31. 0.90% **32.** 0.11% **33.** $37\frac{1}{2}\%$ **34.** $87\frac{1}{2}\%$ **35.** $33\frac{1}{3}\%$ **36.** $66\frac{2}{3}\%$

Write each fraction as a percent. See Example 4.

37. $\frac{3}{4}$ **38.** $\frac{1}{5}$ **39.** $\frac{3}{8}$ **40.** $\frac{5}{5}$ **41.** $\frac{9}{9}$ **42.** $\frac{3}{20}$

43. $\frac{9}{4}$ **44.** $\frac{8}{5}$ **45.** $\frac{12}{5}$ **46.** $\frac{11}{4}$ **47.** $\frac{1}{250}$ **48.** $\frac{1}{500}$

A.10 Laws of Arithmetic

When we add two numbers together, we can compute the sum in either order. For example,

$$3 + 7 = 10 \qquad \text{and} \qquad 7 + 3 = 10$$

This property is called the **commutative law of addition.** In symbols, we write

Commutative Law of Addition

If a and b are numbers, then

$$a + b = b + a$$

Note that *subtraction is not commutative,* since, for example, $9 - 5$ is not the same as $5 - 9$.

When we multiply two numbers together, we can compute the product in either order. For example,

$$4 \cdot 5 = 20 \qquad \text{and} \qquad 5 \cdot 4 = 20$$

This property is called the **commutative law of multiplication.** In symbols, we write

Commutative Law of Multiplication

If a and b are numbers, then

$$a \cdot b = b \cdot a$$

Note that *division is not commutative,* since, for example, $12 \div 3$ is not the same as $3 \div 12$.

When we add three or more numbers together, we can group the numbers in any way we like. That is, we can add any two of the numbers first, and then add the third number to the sum. For example,

$$9 + (6 + 4) = 9 + 10 = 19$$

and

$$(9 + 6) + 4 = 15 + 4 = 19$$

This property is called the **associative law of addition.** In symbols, we write

Associative Law of Addition

If a, b, and c are numbers, then

$$a + (b + c) = (a + b) + c$$

Note that *subtraction is not associative,* since, for example,

$$12 - (6 - 2) = 12 - 4 = 8$$

whereas

$$(12 - 6) - 2 = 6 - 2 = 4$$

When we multiply three or more numbers together, we may group them in any order. Thus,

$$2 \cdot (3 \cdot 5) = 2 \cdot 15 = 30$$

and

$$(2 \cdot 3) \cdot 5 = 6 \cdot 5 = 30$$

This property is called the **associative law of multiplication.** In symbols, we write

Associative Law of Multiplication

If a, b, and c are numbers, then

$$a \cdot (b \cdot c) = (a \cdot b) \cdot c$$

Note that *division is not associative,* since, for example,

$$12 \div (6 \div 2) = 12 \div 3 = 4$$

whereas

$$(12 \div 6) \div 2 = 2 \div 2 = 1$$

EXAMPLE 1  

Fill in the blank according to the indicated law.

a. Commutative law:

$3 \cdot 15 = 15 \cdot$ __________

b. Associative law:

$2 + (4 + 6)$
$= (2 +$ __________ $) + 6$

Solution

a. $3 \cdot 15 = 15 \cdot \mathbf{3}$

b. $2 + (4 + 6) = (2 + \mathbf{4}) + 6$

EXAMPLE 2  

Use the commutative and associative laws to simplify the computations.

a. $24 + 18 + 6$

b. $4 \cdot 27 \cdot 25$

Solution

a. Apply the commutative law of addition:

$$\begin{aligned} 24 + 18 + 6 &= (24 + 6) + 18 \\ &= 30 + 18 = 48 \end{aligned}$$

b. Apply the commutative law of multiplication:

$$\begin{aligned} 4 \cdot 27 \cdot 25 &= (4 \cdot 25) \cdot 27 \\ &= (100) \cdot 27 = 2700 \end{aligned}$$

EXERCISE A.10 

Fill in the blank according to the indicated law. See Example 1.

1. Commutative law

$7 + 10 = 10 + \_\_\_\_$

2. Associative law

$(6 \cdot 4) \cdot 3 = 6 \cdot (4 \cdot \_\_\_\_)$

3. Associative law

$(3 + 6) + 9 = \_\_\_\_ + (6 + 9)$

4. Commutative law

$8 \cdot 12 = \_\_\_\_ \cdot 8$

5. Commutative law

$36 \cdot 147 = \_\_\_\_ \cdot 36$

6. Commutative law

$13 + 87 = 87 + \_\_\_\_$

7. Associative law

$(17 \cdot 2) \cdot 5$
$= 17 \cdot (\_\_\_\_ \cdot \_\_\_\_)$

8. Associative law

$(44 + 12) + 8$
$= 44 + (\_\_\_\_ + \_\_\_\_)$

9. Commutative law

$(5 + 9) + 4 = (9 + \_\_\_\_) + 4$

10. Commutative law

$(8 \cdot 9) \cdot 3 = (9 \cdot \_\_\_\_) \cdot 3$

Use both the commutative and associative laws to simplify the computations. See Example 2.

11. $47 + 28 + 3$

12. $12 + 147 + 8$

13. $26 + 37 + 3 + 4$

14. $55 + 32 + 5 + 8$

15. $2 \cdot 7 \cdot 5$

16. $15 \cdot 6 \cdot 2$

17. $50 \cdot 13 \cdot 2$

18. $4 \cdot 26 \cdot 25$

19. $4 \cdot 6 \cdot 5 \cdot 5$

20. $8 \cdot 8 \cdot 5 \cdot 5$

Answers to Odd-Numbered Problems

Homework 1.1A

1. Janet's laps −8 **3.** number of packages × 6
5. 200 − number of pages read **7.** hours worked × 7
9. 18.3 **11.** 8.7 **13.** 45 **15.** 0.75 **17.** 8

Homework 1.1B

1a. Percent of children born outside marriage **b.** 15%
c. Ireland **d.** Denmark **e.** Greece
3a. Temperture in Chicago **b.** 54° **c.** May
d. April **e.** November
5a.

Falls	60
Choking	125
Suffocation	350
Drowning	450
Fires	750
Motor vehicles	1000

b. Motor vehicles **c.** 0.132 **d.** Approximately 665
7a. United States; Mexico **b.** Japan; Mexico
c. Mexico; 41 yr **d.** 8 yr **9.** variable, constant, bar graph
11. 0.25 **13.** 0.62 **15.** 0.06 **17.** 1.5 **19.** 0.046
21. 0.6 **23.** 0.09 **25.** 0.085 **27.** 2.5 **29.** 0.004

Homework 1.2A

1a.

Regular price	18	25	54
(Calculation)	18 − 12	25 − 12	54 − 12
Delbert's price	6	13	42

Regular price	76	115	130
(Calculation)	76 − 12	115 − 12	130 − 12
Delbert's price	64	103	118

b. Subtract $12 from the regular price **c.** Regular price −12
d. $p - 12$
e.

p	18	25	54	76	115	130
$p - 12$	6	13	42	64	103	118

3a.

Total calories	400	550	700
(Calculation)	400 ÷ 5	550 ÷ 5	700 ÷ 5
Calories from fat	80	110	140

Total calories	925	1200	1500
(Calculation)	925 ÷ 5	1200 ÷ 5	1500 ÷ 5
Calories from fat	185	240	300

b. Divide the total calories by 5. **c.** Total calories ÷ 5
d. $c \div 5$
e.

c	400	550	700	925	1200	1500
$c \div 5$	80	110	140	185	240	300

5a.

Total bill	20	26	32
(Calculation)	0.15 × 20	0.15 × 26	0.15 × 32
Tip	3.00	3.90	4.80

Total bill	48	52	60
(Calculation)	0.15 × 48	0.15 × 52	0.15 × 60
Tip	7.20	7.80	9.00

b. Multiply the total bill by 0.15. **c.** 0.15 × total bill
d. $0.15 \times B$
e.

B	20	26	32	48	52	60
$0.15 \times B$	3.00	3.90	4.80	7.20	7.80	9.00

7a. 62% **b.** 50% **c.** S.E. conference **d.** S.E. conference **e.** U of Michigan

9a.

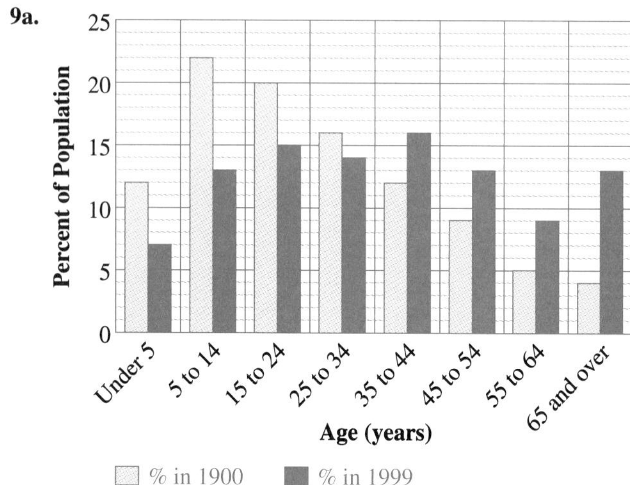

b. 5–14; 35–44 **c.** 54%; 35% **d.** 18%; 35%
11. 12 **13.** 9 **15.** 5 **17.** 7 **19.** 6 **21.** 78
23. 137 **25.** 27 **27.** 16 **29.** 13

Homework 1.2B

1a. multiplication **b.** subtraction **c.** addition
d. division **e.** multiplication **f.** addition **g.** division
3. With a raised dot $(4 \cdot 5)$ or by juxtaposition $(3x)$

5. terms sum **7.** $8 + z$ **9.** $h - 5$ **11.** $\frac{t}{20}$

13. $\frac{16}{v}$ **15.** a subtracted from 7.1 **17.** 15 times j
19. A divided by c **21.** The sum of y and h
23. 2.5 subtracted from H **25.** Three-fifths of b

27a. $\frac{m}{5}$

b.

Spending money ($)	m	20	25	60	80
Amount saved ($)	$\frac{m}{5}$	4	5	12	16

29a. $\frac{W}{4}$

b

Width of photo (in.)	W	6	10	24	30
Height of photo (in.)	$\frac{W}{4}$	$1\frac{1}{2}$	$2\frac{1}{2}$	6	$7\frac{1}{2}$

31. 12 **33.** 70 **35.** 9 **37.** 49 **39.** 28 **41.** 20
43. 30 **45.** 12.8 **47.** 13.5 **49.** 29.25

Homework 1.3A

1. $v - 8$ **3.** $H + 14$ **5.** $\frac{t}{5}$ **7.** $\frac{2}{3}w$ **9.** $\frac{15}{b}$

11. $R - 2.6$ **13.** $\frac{P}{3}$ **15.** $0.18N$ **17.** $r - d$
19. $P + I$ **21.** 3 subtracted from x, the difference of x and 3, 3 less than x, x decreased by 3 **23.** The product of 9 and h; 9 times h; 9 multiplied by h; h multiplied by 9

25a. 24 divided by 6, 4 **b.** 6 divided by 24, $\frac{1}{4}$

c. The ratio of 6 to 24, $\frac{1}{4}$ **d.** The ratio of 24 to 6, 4

(a) and (d) are the same; (b) and (c) are the same **27a.** $\frac{2}{3}p$

b.

Members present	p	90	96	120	129
Votes needed	$\frac{2}{3}p$	60	64	80	86

29. $\frac{3}{4}p$ **31.** 21 **33.** 9 **35.** 18.56 **37.** 5.81
39. 0.36

Homework 1.3B

1a. 500 **b.** Subtract c from 1200.

c.

c	200	350	425	515	640	870
$1200 - c$	1000	850	775	685	560	330

3a. 10 hours **b.** Divide 600 by r.

c.

r	30	40	45	50	80
$\frac{600}{r}$	20	15	$13.\overline{3}$	12	7.5

5a. 1.5 quarts **b.** Multiply q by 0.10.

c.

q	10	20	45	58	260	1250
$0.10q$	1	2	4.5	5.8	26	125

7a. $140 **b.** Multiply 2000 by r.

c.

r	6%	6.5%	8%	9.25%	10%
$2000r$	120	130	160	185	200

9. 5 **11.** 15.4 **13.** 30 **15.** 48 **17.** 1.2 **19.** 2
21. $\frac{1}{3}$ **23.** 0.4 **25.** 7.2
27.

Length of regular skirt	24	26.5	31	34	L
Length of petite skirt	21.5	24	28.5	31.5	$L - 2.5$

29.

Area of room	100	150	200	360	A
Cost of carpet	800	1200	1600	2880	$8A$

31. 15 **33.** 6 **35.** 18 **37.** 48 **39.** 59.2
41. 9.75 **43.** 7.31 **45.** 35.625

Homework 1.4A

1. $m - 15$ **3.** $m + 15$ **5.** $15 - m$ **7.** $12p$ **9.** $\frac{p}{12}$

11. $\frac{12}{p}$ **13.** $\frac{36}{c}$ **15.** $c + 36$ **17.** $36 - c$ **19.** $36c$

21. $3w$ **23.** $r - 6$ **25.** $\frac{2}{5}B$ **27.** $\frac{p}{4}$ **29.** $p - 15$
31. 0.25 **33.** 0.625 **35.** 0.4 **37.** 0.35 **39.** 0.3125
41. 6.25 **43.** 4.125 **45.** 7.4 **47.** 15.4375 **49.** 2.45

Homework 1.4B

1. $150t$ **b.** 450 mi; 1200 mi; 675 mi **3a.** $1000/r$
b. 20 h; 2.5 h; 125 h **5a.** $1600(0.04)t$ **b.** \$64; \$128; \$320
7a. $6000(3)r$ **b.** \$900; \$1485; \$2160 **9a.** $0.20p$
b. 0.6 qt; 2.8 qt; 12 qt **11a.** $S/12$ **b.** 8; 6.5; 9.25
13a. $A + h$ **b.** 11 cups **15a.** $T/600$ **b.** 0.80
17a. **b.**

City	Rush hour traffic flow (mph)	Commute time (h)
Singapore	37	0.270
Sydney	11	0.909
Hong Kong	12	0.833
Bangkok	13	0.769
Shanghai	15	0.667
Manila	7	1.429
Jakarta	16	0.625
Los Angeles	19	0.526
Seoul	14	0.714
Tokyo-Yokohama	28	0.357

c. Manila; Seoul **d.** Tokyo-Yokohama
19. $0.\overline{6}$ **21.** $0.8\overline{3}$ **23.** $0.\overline{7}$ **25.** $0.\overline{72}$ **27.** $0.08\overline{3}$

Homework 1.5

1.

Simone's age	6	10	15	18	30	24	s
Rachel's age	14	18	23	26	38	32	$s + 8$

a. Add 8 to Simone's age. **b.** Subtract 8 from Rachel's age.
c. $r = s + 8$ **d.** Add 8 to s. **e.** Subtract 8 from r.
3.

Bank statement	100	138	188	276	332	352	b
Actual balance	24	62	112	200	256	276	$b - 76$

a. Subtract 76 from the statement. **b.** Add 76 to actual balance.
c. $b - 76$ **d.** Subtract 76 from b. **e.** Add 76 to a.
5. (a) q is 8 less than r. (b) $q = r - 8$ **7.** (a) m is $\frac{1}{3}$ of h.
(b) $m = \frac{h}{3}$ **9.** (a) k is 0.6 more than c. (b) $k = 0.6 + c$
11. (a) d is $\frac{3}{4}$ of a. (b) $d = \frac{3}{4}a$ **13.** $8p = 40$
15. $p - 8 = 40$ **17.** $p + 8 = 40$ **19.** $8p = 40$
21. $b - 18 = 10$ **23.** $18 - b = 10$ **25.** $18 - b = 10$
27. $16t = 80$ **29.** $16t = 80$ **31.** $n - 5 = 30$
33. $5n = 30$ **35.** $n + 5 = 30$ **37.** $5n = 30$

Fraction	Decimal	Percent (%)	Fraction	Decimal	Percent (%)
$\frac{1}{2}$	0.5	50	$\frac{3}{5}$	0.6	60
$\frac{1}{3}$	$0.\overline{3}$	$33\frac{1}{3}$	$\frac{4}{5}$	0.8	80
$\frac{2}{3}$	$0.\overline{6}$	$66\frac{2}{3}$	$\frac{1}{8}$	0.125	12.5
$\frac{1}{4}$	0.25	25	$\frac{3}{8}$	0.375	37.5
$\frac{3}{4}$	0.75	75	$\frac{5}{8}$	0.625	62.5
$\frac{1}{5}$	0.2	20	$\frac{7}{8}$	0.875	87.5
$\frac{2}{5}$	0.4	40	1	1.0	100

39. $\frac{1}{8}$ **41.** 0.375 **43.** $\frac{2}{3}$ **45.** 0.4 **47.** $0.\overline{3}$ **49.** $\frac{5}{8}$

Chapter 1 Review

1. product **3.** factors **5.** ratio **7.** variable
9. evaluate **11.** terms
13a. Add 3 to the years Brianna has studied dance.
b. $B + 3$ **15a.** Subtract 500 from her paycheck.
b. $p - 500$ **17a.** July 4; 95° **b.** July 9; 71°
c. July 7; 5° **d.** July 8; 36° **19.** 4.2 **21.** 4 **23.** 84
25. rt rate · time **27.** Prt Principal · interest rate · time
29. 2,068,000 mi^2 **31.** 1.5 h **33.** (a) Fraction of a bottle;
(b) $50n = 10$ **35.** (a) Tax increase; (b) $\frac{n}{50} = 10$
37. $v = d + 0.4$ **39.** $y = x + 2.5$ **41.** 1.3 **43.** 0.29
45. $\frac{36}{3} = 12$ **47.** $\frac{48}{8} = 6$ **49.** 10 **51.** 35 **53.** 75
55. 2.34 **57.** 5.75 **59.** 7.6 **61.** $0.\overline{18}$ **63.** $4.1\overline{6}$

Homework 2.1A

1. A statement that two expressions are equal. **3.** Yes
5. Yes **7.** Yes **9.** Yes **11.** No: no "=" **13.** Yes
15. Yes **17.** Yes **19.** No **21.** Yes **23.** No
25. No **27.** No
29a.

x	0.6	0.8	1.0	1.2	1.4	1.6
$1.6x$	0.96	1.28	1.6	1.92	2.24	2.56

b. $x = 1.2$
31a.

z	14	21	28	35	42	49
$87 - z$	73	66	59	52	45	38

b. $z = 49$ **33a.** $w - 4.5$ **b.** $n = w - 4.5$

c.

w	20	18	15
n	15.5	13.5	10.5

35a. $1.60p$ **b.** $d = 1.60p$

c.

p	5	10	20
d	8	16	32

37. 0.0825 **39.** 1.086 **41.** 0.00625 **43.** $0.\overline{3}$
45. 0.10375

Homework 2.1B

1. To find the solution(s) of the equation **3.** See page 64.
5. $b = 7$ **7.** $t = 2$ **9.** $x = 3$ **11.** $h = 2.9$
13. $w = 12$ **15.** $d = 24$ **17.** $p = 30$ **19.** $x = 0.85$
21. $g = 6$ **23.** $h = 13$ **25.** $p = 5$ **27.** $z = 4.8$
29. $m = 28$ **31.** $x = 42$ **33.** $s = 208$ **35.** $v = 7.8$
37. $d = 32$ **39.** $f = 37$ **41.** $x = 0$ **43.** $v = 22.5$
45. $n = 8.3$ **47.** $w = 41.8$ **49.** 8.3 **51.** 4.0800
53. 0.20 **55.** 606.060 **57.** 0.0110

Homework 2.2A

1. Distance around the border; centimeters, feet
3. 20 cm; 25 sq. cm **5.** 26 cm; 36 sq. cm
7. 12 cm; 5 sq. cm **9.** 16 cm; 10 sq. cm **11.** $m = 27$
13. $a = 21.4$ **15.** $d = 8$ **17.** $t = 0.18$ **19a.** $\frac{2}{3}l$
b. 52 pages **21.** $\frac{1}{8}$ **23.** 0.38 **25.** $\frac{1}{3}$ **27.** 0.23
29. $\frac{3}{11}$

Homework 2.2B

1. $\left(3\frac{1}{2} \text{ inch} \approx 8.9 \text{ cm}\right)$ **3.** $\left(2\frac{1}{2} \text{ inch} \approx 6.4 \text{ cm}\right)$
7a. 30 **b.** 18 **c.** 6 **d.** 2.5 **9a.** 4 **b.** 12
c. 10 **d.** 10 **11.** 18 sq. cm; 20.5 cm **13.** 22.5 sq. cm; 23 cm **15.** 41 sq. cm; 23.3 cm **17. a.** $p + 28.35 = 2652$
19. b. $r - 3.8 = 12.7$ **21.** 2.4 **23.** 21.1 **25.** 8.2
27. 48.8 **29.** 0.91

Homework 2.3A

1. An equation relating two or more variables
3. $A = \frac{S}{n}$: A = average; S = sum of values; n = number of values
5. $e = \frac{m}{g}$: e = efficiency; m = miles; g = gallons
7. $P = rW$: P = part; r = percentage rate; w = whole
9. $s = p - d$: s = sale price; p = regular price; d = discount
11. 84 mi **13.** \$1365 **15.** \$316.80 **17.** 7.5 mi/gal
19. \$45 **21.** \$525 **23.** Yes; yes **25.** No
27. Yes; no **29.** Yes; yes **31.** No

33.

t	w
40	25
55	40
60	45
80	65
90	75
110	95

$w = t - 15$

35.

D	H
3	12
5	20
6	24
9	36
11	12
42	64

$H = 4D$

37. 28.467; 28.5 **39.** 5.667; 5.7 **41.** 0.006; 0.0
43. 5.000; 5.0 **45.** 1.000; 1.0

Homework 2.3B

1. $f = 0.34$ **3.** $y = 56$ **5.** $u = 18$ **7.** $p = 54$
9. $e = \frac{m}{g}$, $26 = \frac{m}{14}$; $m = 364$ mi
11. $P = R - C$, $1500 = R - 215$; $R = \$1715$
13. $P = rW$, $162 = 0.36w$; $w = 450$ freshmen
15. $I = Prt$, $98.80 = 1300r(1)$; $r = 0.076 = 7.6\%$
17. $P = rW$, $540 = 0.45W$; $W = 1200$ people voted
19. $I = Prt$, $171 = P(0.095)1$; $P = \$1800$
21. 13 cm; 9 sq. cm **23.** 28 cm; 28 sq. cm **25.** 0.83
27. 0.29 **29.** 0.38 **31.** 0.87 **33.** 0.19

Activity 1: Rectangles

1. 7; 6; 42 **3.** 4; 2; 8 **5a.** Yes; $A = lw$ **b.** Multiply the number of boxes in each row by the number of rows.

Activity 2: Triangles

7. 8; 4; 16 **9.** 4; 6; 12

Activity 3: Parallelograms

11. 2; 3; 6 **13.** 3; 7; 21 **15a.** Yes; $A = bh$
b. The parallelogram can be cut and the pieces rearranged to form a rectangle, whose dimensions are the base and height of the parallelogram.

Homework 2.4A

1. 20 sq. units **3.** 10 sq. units **5.** 14 sq. units
7. 144 sq. ft **9.** 36 sq. m. **11.** 12 sq. cm **13.** \$3000
15. 40 words per minute **17.** $y = \frac{2}{3}x$ **19.** $R = \frac{3}{2}N$
21. 1.83 **23.** 6.4 **25.** 3.57 **27.** 2.92 **29.** 2.13

Homework 2.4B

1. 20 sq. cm **3.** 24 sq. cm **5.** 15 sq. cm **7.** 63 sq. m
9. 64 sq. yd **11.** Area: 162 sq. m; perimeter: 66 m
13. 1.2 sq. in. **15.** 5.85 sq. in. **17.** 13 sq. in.

19. 102 sq. in. **21.** Red: 8 sq. cm; Blue: 4 sq. cm; Yellow: 24 sq. cm **23.** Yellow: 4 sq. cm; Blue: 6 sq. cm; Red: 6 sq. cm. **25.** $d = rt$, $364 = r(7)$; $r = 52$ miles per hour **27.** $d = rt$, $2800 = 560(t)$; $t = 5$ hours **29.** $A = \frac{S}{n}$, $13.25 = \frac{S}{16}$; $S = \$212$ **31.** 5.714 **33.** 29.167 **35.** 16.889 **37.** 13.333 **39.** 33.917

Homework 2.5A

1. (a) $6n = 162$; (b) 27 **3.** (a) $n - 24 = 38$; (b) 62 **5.** (a) $n/2.5 = 6.6$; (b) 16.5 **7.** (a) $20 = n + 15.3$; (b) 4.7

9a.

Country	*Family size*	*Women's average age at marriage*
Iraq	7.1	22.3
Bangladesh	5.7	16.7
Sudan	5.6	20.7
Venezuela	5.1	21.3
Ethiopia	4.5	18.9
Spain	3.5	23.1
Australia	3	22.0
Japan	3	25.4
Canada	2.7	24.3
Denmark	2.2	25.6

b.

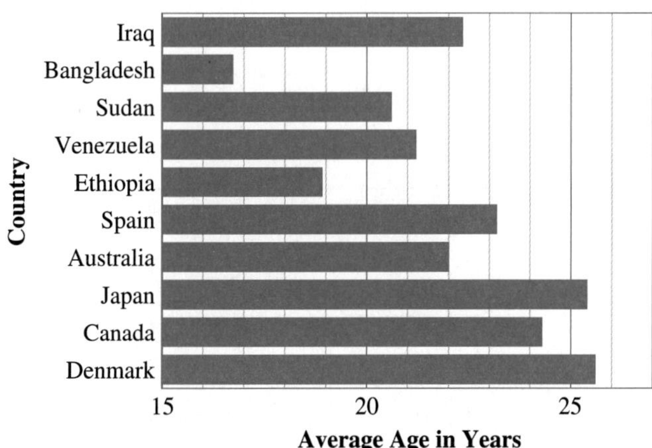

c. Bangladesh; Denmark **d.** Women many at a younger age in countries with large average family size. Other factors must be involved, because Bangladesh has the youngest brides but only the second largest family size. **11.** Blue: 4 sq. in.; Red: 4 sq. in.; Yellow: 8 sq. in. **13.** Red: 8 sq. in.; Gray: 12 sq. in.; Yellow: 4 sq. in.; Light blue: 4 sq. in.; Dark blue: 8 sq. in. **15.** 1.175 **17.** 0.75 **19.** $5.\overline{6}$ **21.** 1.5 **23.** 10.025

Homework 2.5B

1a. Amount Lupé had before the Craft Fair: A **b.** $A - 24 = 39$ **c.** \$63 **3a.** Brenda's weight: B **b.** $B + 32 = 157$ **c.** 125 lb **5a.** Miranda's hourly wage: w **b.** $20w = 136$ **c.** \$6.80 an hour **7a.** Total profit: P **b.** $\frac{P}{8} = 64$ **c.** \$512 **9.** 13 h **11.** 7 yr **13.** 46 **15.** \$1,400,000 **17.** \$4356

19a.

City	*People per room*	*Deaths per 1000 births*
Singapore	1.2	7
Sydney	0.5	10
Hong Kong	1.6	7
Bangkok	3.2	27
Shanghai	2	14
Manila	3	36
Jakarta	3.4	45
Los Angeles	0.5	9
Seoul	2	12
Tokyo-Yokohama	0.9	5

b. Jakarta; Jakarta **c.** Tokyo-Yokohama; Sydney and Los Angeles **d.** High infant mortality rates are associated with crowded living conditions. Poverty, cultural norms, and a harsh environment might contribute to both crowding and infant mortality. **21.** 0.7 **23.** 35 **25.** 1.2 **27.** 1.6 **29.** 2.5

Chapter 2 Review

1. solution **3.** parallelogram **5.** formula **7.** area **9.** A variable can take on different values; a constant does not change. **11.** Terms are expressions that are added or subtracted; factors are multiplied. **13.** Perimeter is the distance around the border of a figure; area is a measure of the amount of space enclosed by the figure. **15.** Revenue is the amount of money taken in on a business venture; profit is the revenue minus the costs. **17.** The base of a triangle is the length of one side. The base of a parallelogram or a trapezoid is the length of one of the parallel sides. The height is the distance from the base to the opposite vertex or side. **19.** 28 **21.** 78 **23.** 16 **25.** 8.1 **27.** 24 **29.** Height of normal doorway: h, $5h$ **31.** Price of book: p, $p - 8$ **33.** Number of sheep: s, $\frac{156}{s}$ **35.** Regular fare: r, $r - 250$ **37.** Length of play: l, $l - 20$ **39.** P = Part, r = percentage rate, W = Whole **41.** d = distance, r = rate, t = time **43.** e = efficiency, m = miles, g = gallon **45.** I = interest, P = principal, r = rate, t = time **47.** A = area; b = base; h = height

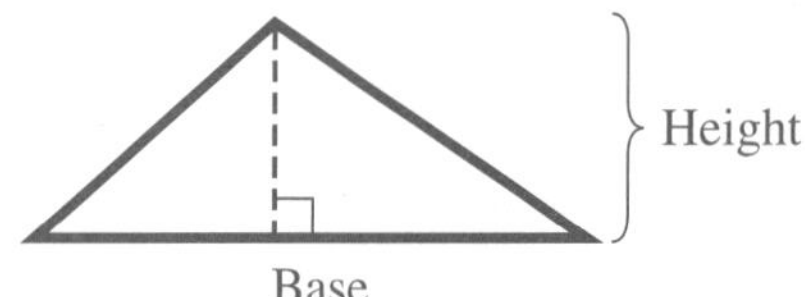

49. $y = 2x$ **51.** $j = i - 9$ **53.** $q = \frac{3}{2}p$ **55.** $v = \frac{60}{u}$
57. $a = 13$ **59.** $c = 33$ **61.** $h = 4$ **63.** $m = 32$
65. $p = 6$ **67.** 26 ft **69.** \$5.27 **71.** \$645 **73.** 8 m
75. 2.2 h **77.** (a) Age of older church: a; (b) $173 = a - 238$; (c) 411 years old **79.** (a) Weight of son: w; (b) $650 = 483 + s$; (c) 167 lb **81.** (a) Size of jackpot: J; (b) $\frac{J}{6} = 2{,}535{,}000$; (c) \$15,210,000 **83.** (a) Time (seconds) to swing back and forth once: t; (b) $4 = 5t$; (c) 0.8 sec
85. (a) Number polled: N; (b) $0.65n = 663$; (c) 1020 voters
87. 20 cm; 18 sq. cm **89.** 16 cm; 14 sq. cm
91. $A = 8$ sq. cm **93.** $A = 15.5$ sq. cm **95.** 3.11
97. $\frac{5}{6}$ **99.** 4.79 **101.** 0.08125 **103.** 2.09 **105.** 1.00
107. 0.846 **109.** 2.167 **111.** $9.1\overline{6}$ **113.** $8.\overline{3}$

Activity

3. Lose 15 yd **5.** 6 lb lighter **7.** Lose \$5
9. Descend 400 ft

Homework 3.1A

1. −\$100 **3.** 2 in. **5.** −450 ft **7.** −\$3 million
9. −384

11.
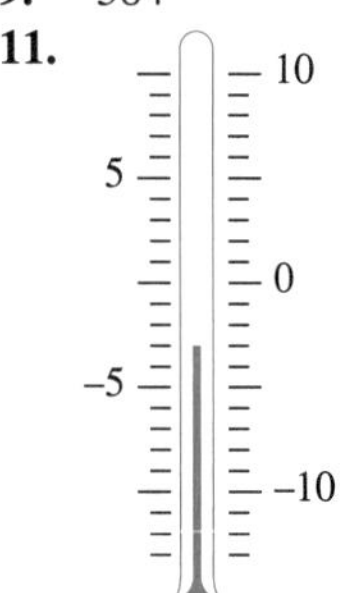

13.
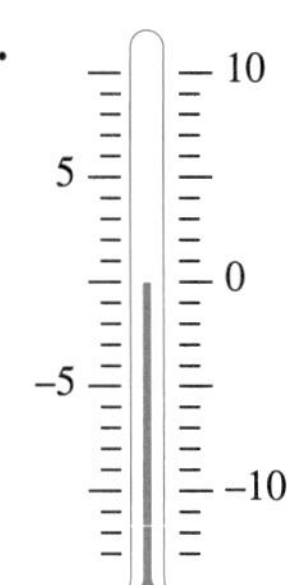

15.
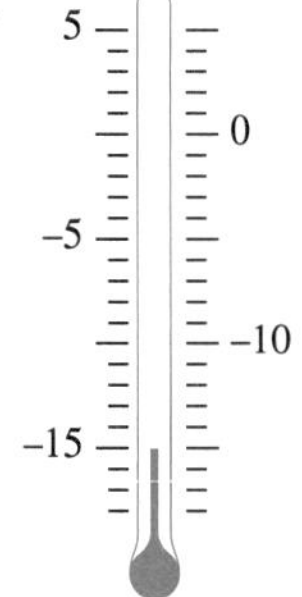

17.
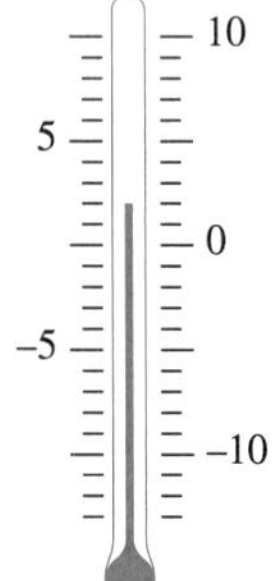

19. 3°
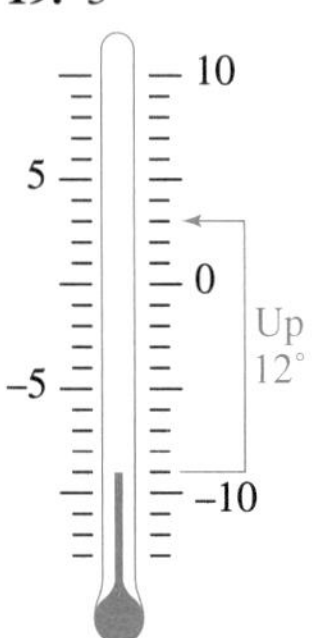

21. −8°
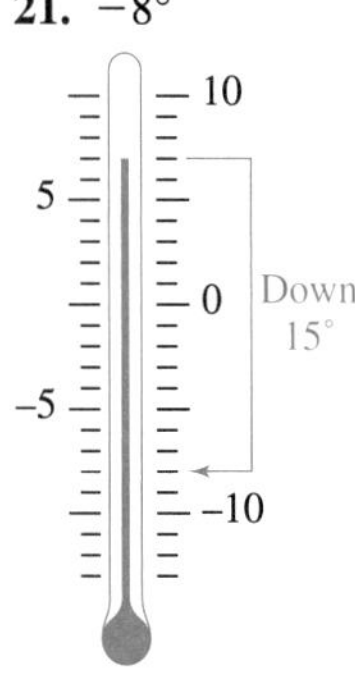

23. −13°
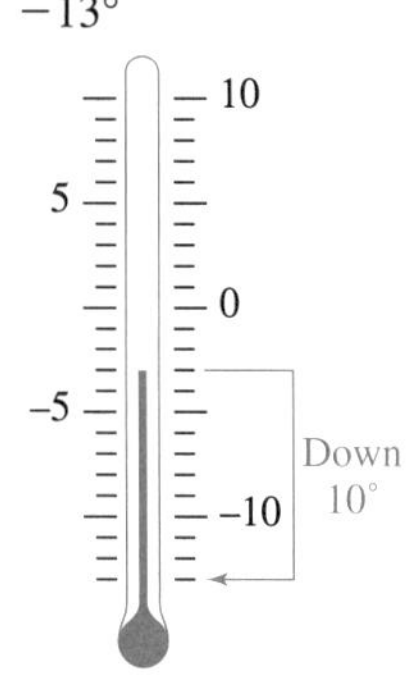

25a. −3°C **b.** Coldest: January at −11°C; warmest: July at 17°C **c.** 22°C

27.
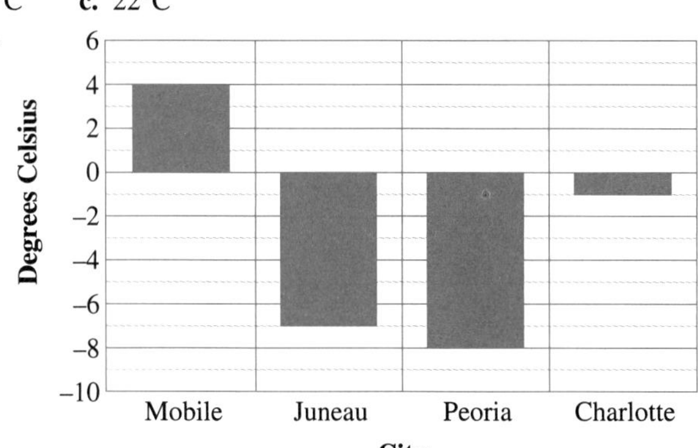

29. Perimeter **31.** Perimeter **33.** Area **35.** Area
37. Perimeter

39.

41.
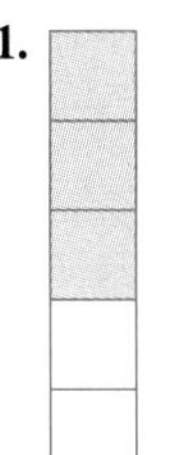

43.

45.

47.
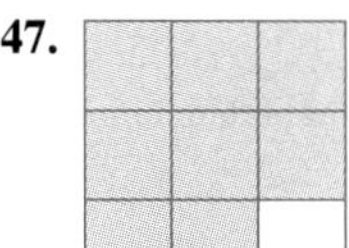

Homework 3.1B

1.
−3, $-2\frac{3}{4}$, $1\frac{1}{2}$, 3.5, 5

3.
$-3\frac{1}{2}$, −2, $-\frac{2}{3}$, 2, 3.5

5. $6.5 > 6.07$ **7.** $-12 < -2$ **9.** $-\frac{3}{4} < -\frac{1}{2}$ **11.** True
13. False **15.** True **17.** False **19.** False
21. 3, 4, 2.01 **23.** −10, −4, −3.1 **25.** −10, −5, −4.1
27. −4 **29.** 15 **31.** 2 **33.** −7 **35.** $E > 65$
37. $T > 1083$ **39.** $s < 85$ **41.** $d > -50$

43.

Country	Railroads in 1900 (thousands of miles)	Railroads in 2000 (thousands of miles)	Change in miles
U.S.	193	149	−44
Russia	33	54	21
India	26	38	12
France	24	21	−3
Britain	19	11	−8

45. 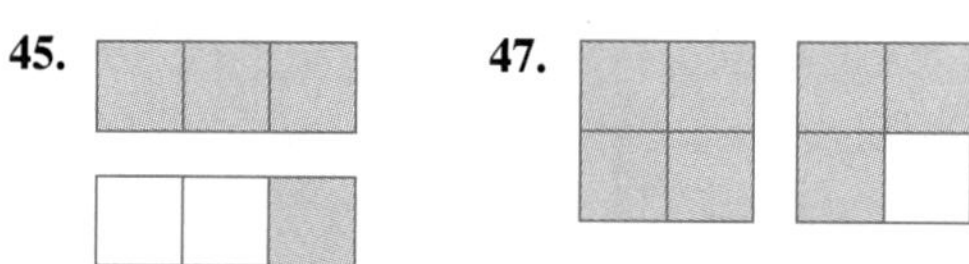

47. 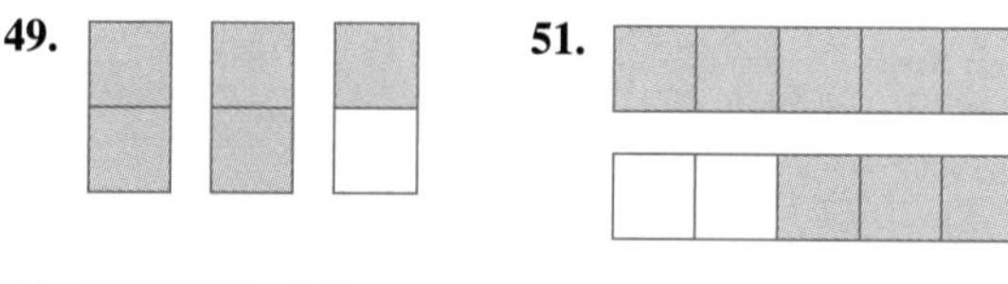

49. 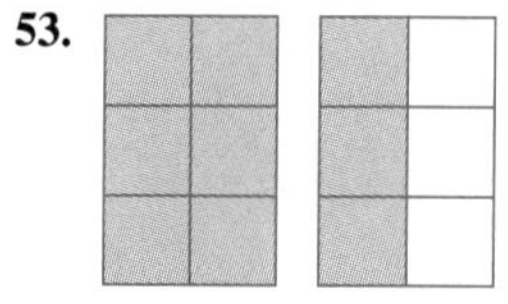

51.

53.

Homework 3.2A

1.

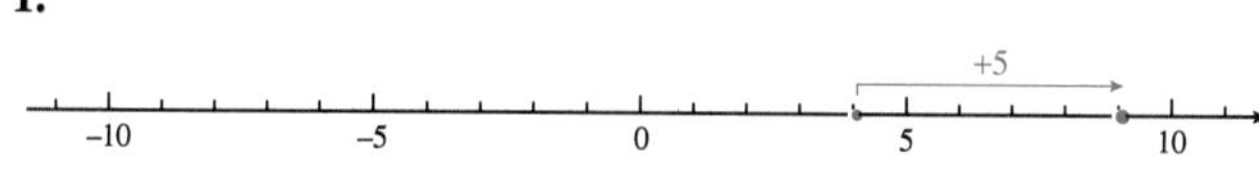

3.

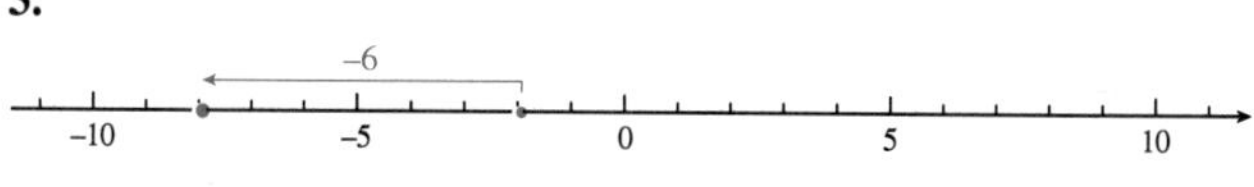

5.

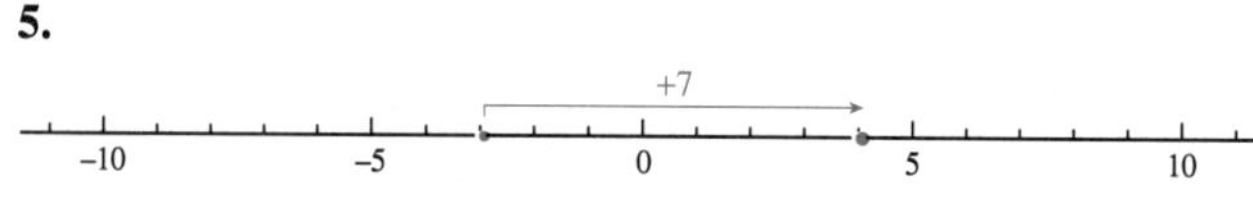

7.

9.

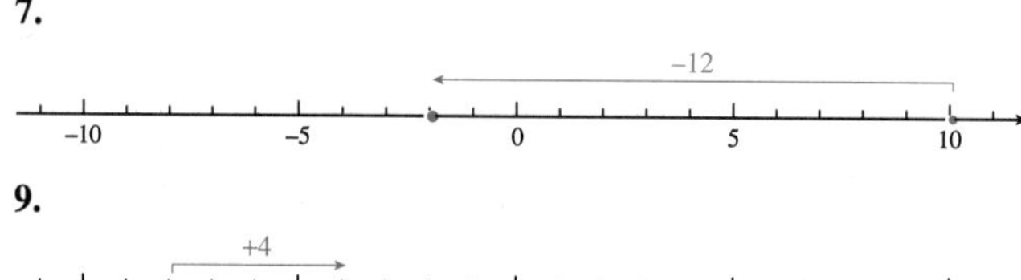

11. Two decreases combine to be a decrease. **13.** If last year the profit was −\$1000 (that is, a loss of \$1000) and this year the profit is \$1500, the total profit is −\$1000 + \$1500 = \$500.

15. 16 **17.** −11 **19.** −29 **21.** −16 **23.** −75 **25.** 7 **27.** 13 **29.** 8 **31.** 7 **33.** 0 **35.** −6 **37.** −13 **39.** −9 **41.** 6 **43.** −34 **45.** −22 **47.** −114 **49.** −0.8 **51.** −381 **53.** −4.5 **55.** 4.49 **57.** −21 lb **59.** −200 ft **61.** −6800 ft **63.** 0 **65.** 0 **67.** 0 **69.** Combining a gain and loss of equal size causes a net change of zero. **71.** $\frac{2}{3}$ **73.** $\frac{3}{5}$ **75.** $\frac{4}{9}$ **77.** $\frac{5}{12}$ **79.** $\frac{1}{4}$

Homework 3.2B

1. −450 ft **3.** −\$9050 **5.** \$280.85 **7.** 2 **9.** −2 **11.** −50 **13a.** \$1.3 million; −\$0.8 million; −\$0.2 million; \$2.1 million **b.** \$2.4 million

15a.

Month	Jan	Feb	Mar	Apr	May	June
Relative water level (ft)	3	6	8	7	6	4
Water depth (ft)	37	40	42	41	40	38

Month	July	Aug	Sep	Oct	Nov	Dec
Relative water level (ft)	1	−3	−5	−7	−6	−4
Water depth (ft)	35	31	29	27	28	30

b.

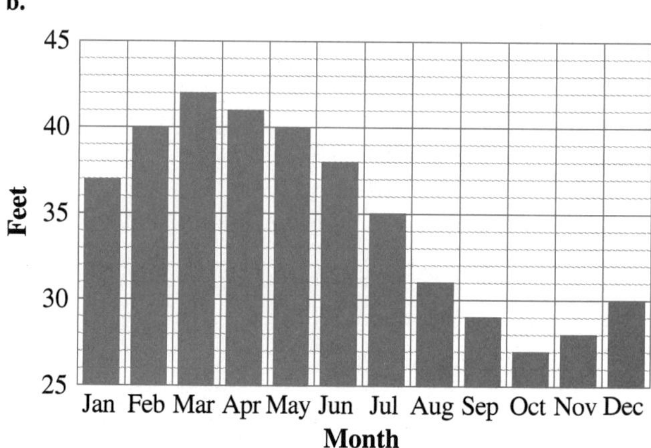

17. −12 **19.** −2 **21.** −6 **23.** 8 **25.** $x = 7$ **27.** $a = -14$ **29.** $v = -9$ **31.** $g = 0$ **33.** $n = -2.7$ **35.** $s = 70.6$ **37.** $j = -0.56$ **39.** −4.7 **41.** −2.3 **43.** −14.4 **45.** −2.2 **47.** $6p$ **49.** $\frac{1200}{v}$ **51.** $1800 - c$ **53.** $m - 28$ **55.** $\frac{5000}{w}$ **57.** $1\frac{4}{5}$ **59.** $1\frac{5}{6}$ **61.** $1\frac{3}{7}$ **63.** $1\frac{1}{4}$ **65.** $1\frac{5}{8}$

Homework 3.3A

1. $8 + 4 = 12$

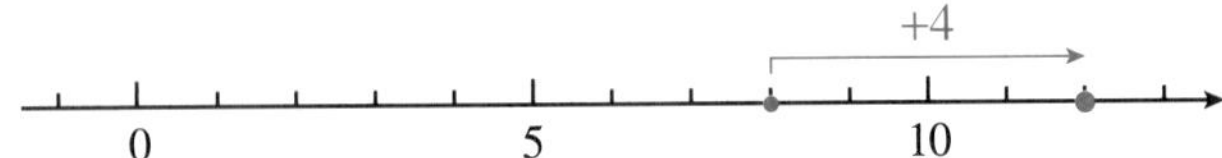

3. $-3 + (-8) = -11$

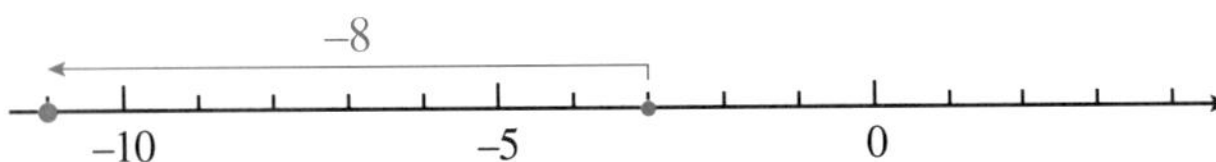

5. $11 - (+5) = 6$

7. $-9 + 2 = -7$

9. Taking away a debt adds to the net worth. **11.** 9 **13.** -6 **15.** -15 **17.** -12 **19.** -10 **21.** 0 **23.** 17 **25.** 37 **27.** -22 **29.** -31 **31.** -34 **33.** -11 **35.** 6 **37.** 0 **39a.** Round 1: -3; round 2: -2; round 3: 3 **b.** -2 **41.** 106.7 **43.** -97 **45.** 0.402 **47.** -0.9 **49a.** 12; 12 **b.** Both interpretations give 12. **51a.** -12; -12 **b.** Both interpretations give -12. **53a.** $8 + (-15)$; $8 - (+15)$ **55a.** -12 **b.** -7 **57.** 8 **59.** -40 **61.** 24 **63.** -70 **65.** 35 **67.** 2 **69.** -2

71.

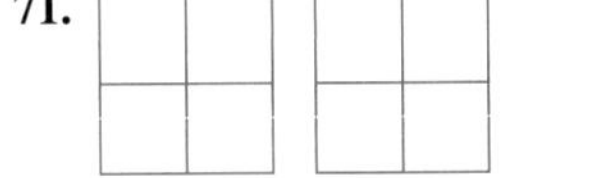

8

73. 9

75.

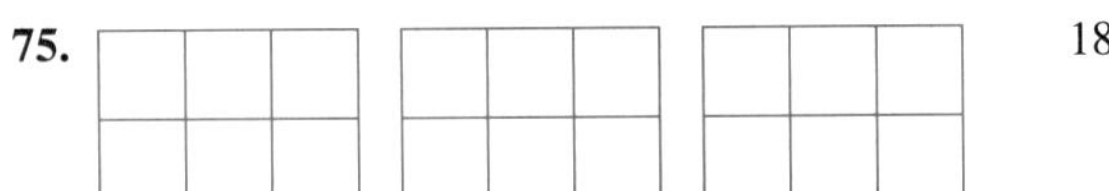

18

77.

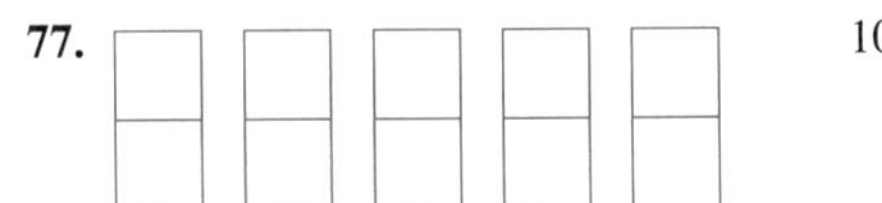

10

79.

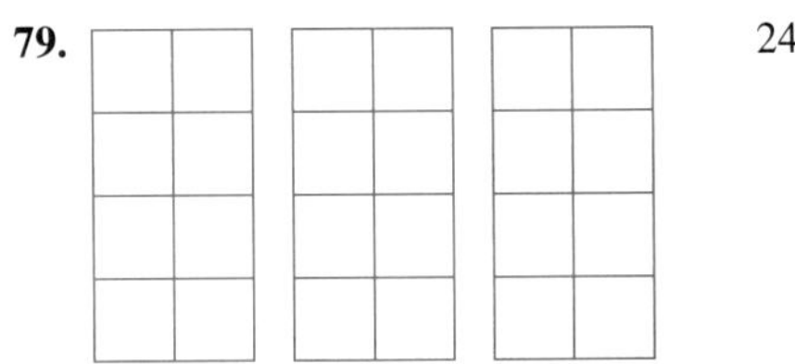

24

Homework 3.3B

1. -8 **3.** 6 **5.** 6 **7.** 0 **9.** -36 **11.** 3 **13.** -7 **15.** -5 **17.** 18 **19.** -18

21a.

Year	1995	1996	1997	1998	1999
Change in exports (billion \$)	0.6	0.1	1.9	0.1	−1.1

\$0.6 billion increase; \$0.1 billion

b.

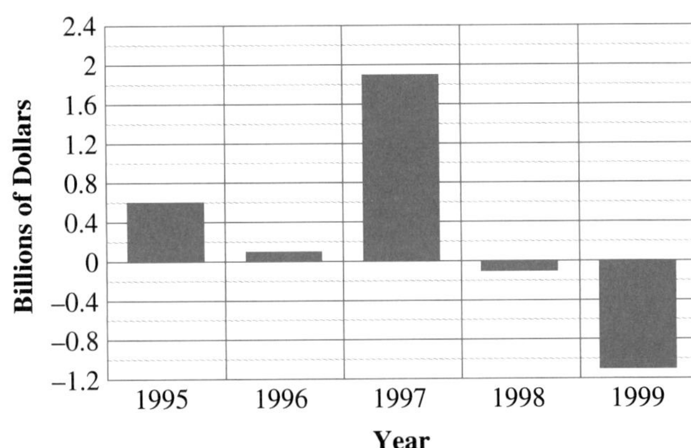

23a.

Year	1996	1997	1998	1999	2000
Aircraft	754	568	656	391	585

b. 1996 **25.** $a = -4$ **27.** $p = -22$ **29.** $h = 3$ **31.** $m = -5$ **33.** $t = 14$ **35.** $u = -60$ **37.** $c = 2.9$ **39.** $f = -7.1$ **41.** -3.5 **43.** -3.7 **45.** 0.9

47.

x	$x + 4$
−7	−3
−5	−1
−3	1
−1	3

49.

h	$h - 5$
−8	−13
−5	−10
−2	−7
2	−3

51. $3k$ **53.** $2000t$ **55.** $4g$ **57.** $\frac{z}{16}$ **59.** $\frac{n}{12}$

61. 2

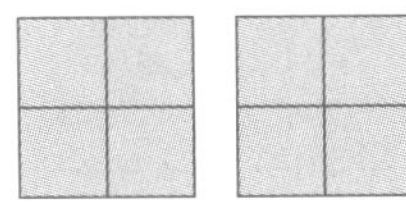

63. 3

65. 4

67. 1

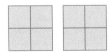

69. 4

Divide the denominator into the numerator.

For example, $\frac{10}{5} = 10 \div 5 = 2$

Homework 3.4A

1. −18 **3.** 18 **5.** −28 **7.** −9 **9.** 7 **11.** −3 **13.** 0 **15.** Undefined **17.** −1 **19.** 14 **21.** −16 **23.** 0 **25.** We think of the positive number as telling how many copies of the negative number are being added. For example, $2(-3) = (-3) + (-3) = -6$. **27a.** 48 **b.** −14 **29a.** 16 **b.** −3 **31a.** −32 **b.** 5 **33a.** −11 **b.** 24 **35a.** −63 **b.** −2 **37.** $8(-5) = -40$ **39.** $-4(-14) = 56$ **41.** $-7(28) = a$ **43.** 414 **45.** 31 **47.** −18 **49.** −7.83 **51.** (a) **53.** (b) **55.** (a)

57. $\frac{7}{4}$

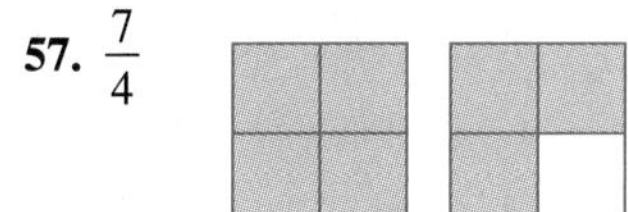

59. $\frac{10}{3}$

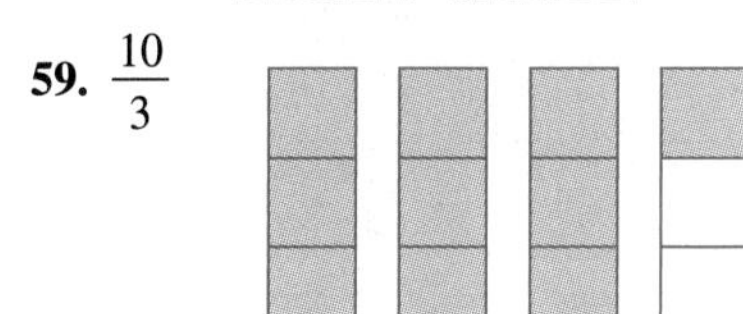

61. $\frac{19}{8}$

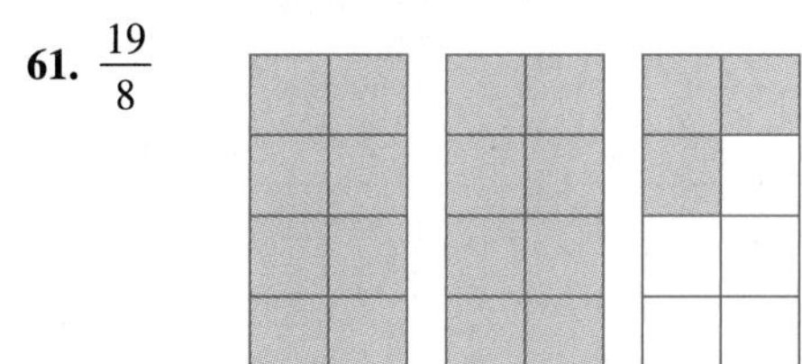

63. $\frac{9}{2}$

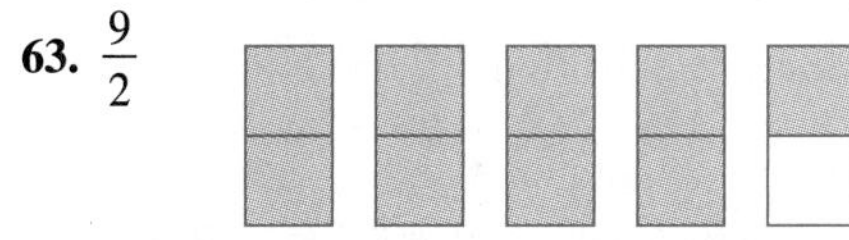

65. $\frac{14}{9}$

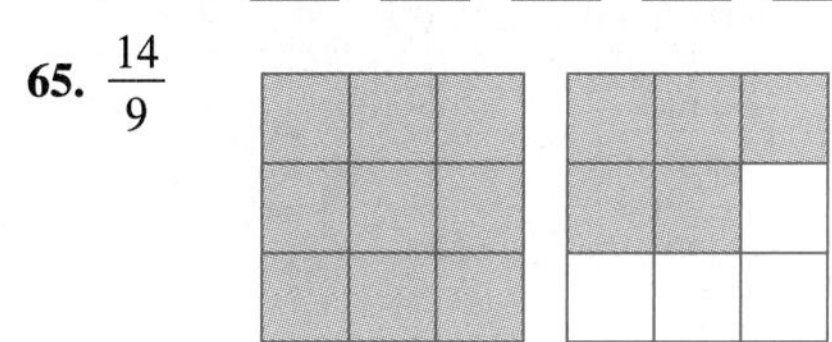

We use the denominator of the mixed number as the denominator of the improper fraction. For the numerator of the improper fraction, multiply the whole number times the denominator and add the numerator of the mixed number.

For example, $2\frac{2}{3} = \frac{(2 \cdot 3) + 2}{3} = \frac{6 + 2}{3} = \frac{8}{3}$.

Homework 3.4B

1. −72 **3.** 625 **5.** 5 **7.** −96 **9.** −1 **11.** −24 **13.** −13 **15.** $-\frac{5}{2}$ **17.** −64 **19.** −12 **21.** −0.1 **23.** −0.75 **25.** 3.84 **27.** 0 **29a.** −\$1 **b.** −\$50 **c.** AmExp, \$12.50; Disney, −\$12.50; ExxonMob, −\$75; GnMotr, −\$37.50 **d.** −\$162.50 **31.** $q = -3$ **33.** $r = 9$ **35.** $y = -7$ **37.** $p = -81$ **39.** $s = 75$ **41.** $f = -72$ **43.** $h = -28$ **45.** $z = -0.059$ **47.** $c = -60$ **49.** $b = 0.4$ **51.** $x = -11$ **53.** $x = 8$ **55.** $x = 10$ **57.** $x = 16$ **59.** 45 **61.** 2.7 mi **63.** 800

65.

b	$-4b$
−6	24
−4	16
4	−16
6	−24

67.

n	$\frac{-12}{n}$
−18	$\frac{2}{3}$
−8	$\frac{3}{2}$
8	$-\frac{3}{2}$
18	$-\frac{2}{3}$

69. $1\frac{2}{3}$

71. $2\frac{1}{4}$

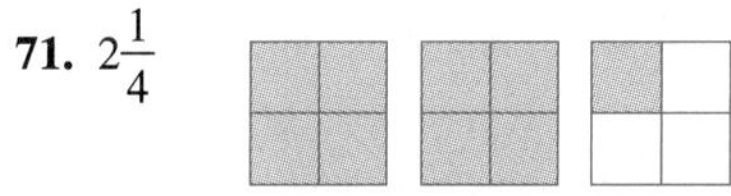

73. $2\frac{3}{8}$

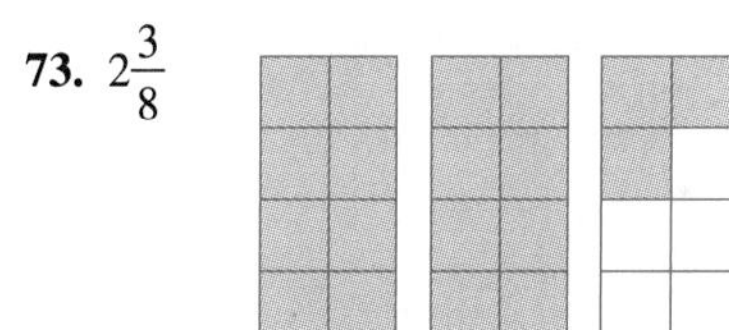

75. $5\frac{1}{3}$

77. $2\frac{1}{2}$

Divide the denominator into the numerator and use the quotient as the whole-number part and the remainder as the numerator of the fraction part of the mixed number.

For example, $\frac{15}{6} = 2\frac{3}{6} = 2\frac{1}{2}$.

Homework 3.5

1. $n + 12 = 5$ **3.** $-12 + n = -5$ **5.** $n - 3 = -30$ **7.** $\frac{n}{3} = -30$ **9.** (a)Temperature last night: T (b) $T + 17 = -6$ (c) −23° **11.** (a) Yesterday's balance: B (b) $B + 132 = 74$ (c) −\$58 **13.** (a) Elevation before lunch: E (b) $E + 248 = 109$ (c) −139 ft **15.** (a) Socrates' birth year: y (b) $y + 71 = -399$ (c) 470 B.C. **17.** (a) Game yardage: y (b) $-18 + y = -26$ (c) −8 yd **19.** (a) Change in elevation: c (b) $-286 + c = -159$ (c) 127 ft **21.** −30 **23.** −12 **25.** −4 **27.** 64 **29.** Undefined **31a.** −\$1.50; each share lost \$1.50 in value. **b.** \$82.50 **33.** 16 h **35.** 20 h **37.** 15 h **39.** 6 h **41.** 3 h

Chapter 3 Review

1. Profit; change in weight; change in altitude **3.** An equation has = between two algebraic expressions, an inequality has instead an inequality symbol like > or <. **5.** If both numbers are negative, ignore the signs and add, and then make the answer negative. When the numbers are opposite in sign, the number farther from 0 on the number line determines the sign of the answer. Find the numerical part of the answer by computing the difference between the unsigned values of the two given signed numbers. **7.** Ignore the signs and multiply or divide the numbers. If they have opposite signs, the answer is negative. If they have the same sign the answer is positive. **9.** undefined
11. > **13.** < **15.** = **17.** (a) Rent: R (b) $R < 650$
19. (a) Last night's temperature: T (b) $T < -15$
21. -6

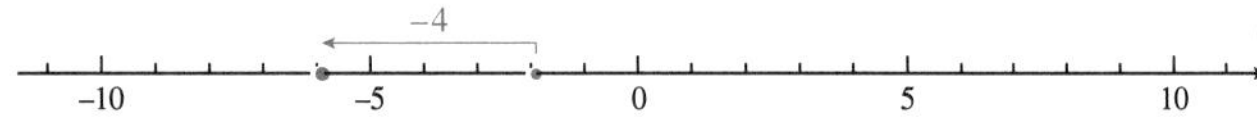

23. $-5 + 7$ **25.** negative **27.** zero **29.** 32
31. -9 **33.** -34 **35.** -7 **37.** -2 **39.** 0
41. -25 **43.** -18 **45.** -54 **47.** 4 **49.** $-\frac{3}{2}$
51. 160 **53a.** Imports exceed exports. **b.** $-\$1048$ billion
c.

Year	1991	1992	1993	1994	1995
Increase in trade balance (billion $)	50.4	−5	−33.2	−28.1	1.1

Year	1996	1997	1998	1999
Increase in trade balance (billion $)	−6.2	−2.6	−62.2	−98.1

55. $m = -14$ **57.** $n = 9$ **59.** $b = 36$ **61.** $h = -72$
63. $d = -7.25$ **65.** $y = 11.5$ **67.** (a) $D + 2500 = -4800$ (b) $-\$7300$ **69.** (a) $-0.45n = -542.70$ (b) 1206
71a. \$10 **b.** \$750
73.

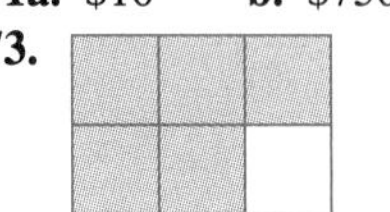

75. 20

77. $\frac{9}{4}$

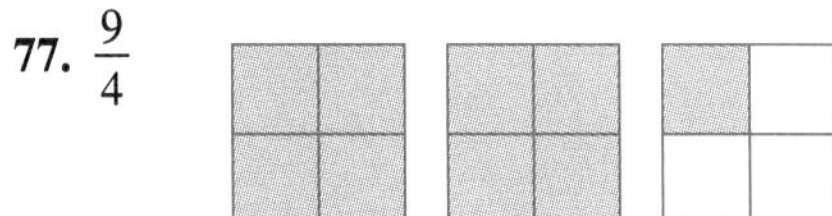

79. $4\frac{1}{2}$

81. $\frac{7}{9}$ **83.** 12

Homework 4.1A

1. 5 **3.** 14 **5.** 5.8 **7.** 6 **9.** 73 **11.** 24
13. $z = 9$ **15.** $x = -5$ **17.** $a = -7$ **19.** $-\$38$
21. $-\$0.54$ **23.** 27 **25.** 24 **27.** 54 **29.** 12
31. 23 **33.** 8 **35.** -16 **37.** 13 **39.** 9 **41.** 0
43. -8 **45a.** 4 sq. cm; 8 sq. cm; 12 sq. cm **b.** 2 sq. cm
c. 22 sq. cm **47.** 3700 **49.** 7430 **51.** 0.84
53. 0.00036 **55.** 0.286

Homework 4.1B

1a. 18 **b.** 50 **3a.** 3 **b.** 18 **5a.** 2 **b.** 8
7a. 42 **b.** 12 **9a.** 2 **b.** 18 **11a.** $3(5 + 8)$
b. $3 \cdot 5 + 8$ **13a.** $25 - (12 - 8)$ **b.** $25 - 12 - 8$
15. 32 **17.** 10 **19.** 2 **21.** 16 **23.** 13 **25.** 4
27. 24 **29.** 52 **31.** 6 **33.** 2 **35.** 7 **37.** 10.69
39. 0.69 **41.** 5.76 **43.** $x = -8$ **45.** $x = 48$
47. $w = 27$ **49.** $w = -2$ **51.** 5.96 **53.** 0.01765
55. 0.5637 **57.** 12.48 **59.** 0.0008

Homework 4.2A

1.

Minutes elapsed	1	3	5
(Calculation)	$75 + 30$	$75 + 30 \cdot 3$	$75 + 30 \cdot 5$
Oven temperature	105	165	225

Minutes elapsed	6	8	10
(Calculation)	$75 + 30 \cdot 6$	$75 + 30 \cdot 8$	$75 + 30 \cdot 10$
Oven temperature	255	315	375

1a. Add 30 times the number of minutes to 75. **b.** $75 + 30m$
3.

Living expenses	2400	3000	3600
(Calculation)	$800 + \frac{2400}{2}$	$800 + \frac{3000}{2}$	$800 + \frac{3600}{2}$
Parents will pay	2000	2300	2600

Living expenses	4000	4500	5000
(Calculation)	$800 + \frac{4000}{2}$	$800 + \frac{4500}{2}$	$800 + \frac{5000}{2}$
Parents will pay	2800	3050	3300

3a. Add half of Luisa's living expenses to 800. **b.** $800 + \frac{a}{2}$

5.

Pints kept	4	8	12
(Calculation)	$\frac{80-4}{4}$	$\frac{80-8}{4}$	$\frac{80-12}{4}$
Pints for each daughter	19	18	17

Pints kept	20	28
(Calculation)	$\frac{80-20}{4}$	$\frac{80-28}{4}$
Pints for each daughter	15	13

5a. Subtract the number of pints Mildred kept from 80 and then divide the result by 4. **b.** $\frac{80-M}{4}$ **7.** $12 - 2t$ **9.** $2t - 12$ **11.** $\frac{12}{m-3}$ **13.** $\frac{m-3}{12}$ **15.** $400 + 30u$ **17.** $\frac{R-560}{3}$ **19.** $2000 - 60p$ **21.** $l + w$ **23.** bh **25.** $\frac{A}{V}$ **27.** $7t + 5$ **29.** $r + u - 60$ **31.** $\frac{1}{2}(r + 3)$ **33.** 3544 **35.** 3546 **37.** 8.48 **39.** 5 **41.** 31 **43.** $<$ **45.** $>$ **47.** $>$ **49.** $<$ **51.** 1200 **53.** 1100 **55.** 150 **57.** 130 **59.** 3200

Homework 4.2B

1.

z	0	3	10
$5z + 4$	4	19	54

3.

b	4	6	9
$26 - 2b$	18	14	8

5.

h	13	7	24
$\frac{h-5}{4}$	2	$\frac{1}{2}$	$\frac{19}{4}$

7.

d	3	6.5	0.2
$7(d + 1)$	28	52.5	8.4

9.

m	3	12	0.5
$\frac{m}{3+m}$	0.5	0.8	0.14

11. 5 **13.** 280 **15.** 26 **17.** 5 **19.** 1.2 **21.** 24.4 ft **23.** 525 sq. ft **25.** 37°C **27.** $65,000 **29.** −24 **31.** −5 **33.** −4 **35.** 100 **37.** 25

39a.

Month	Jan	Feb	Mar	Apr	May	June
$20 - T(°C)$	31	29	23	18	13	7

Month	July	Aug	Sep	Oct	Nov	Dec
$20 - T(°C)$	3	4	8	15	21	28

b.

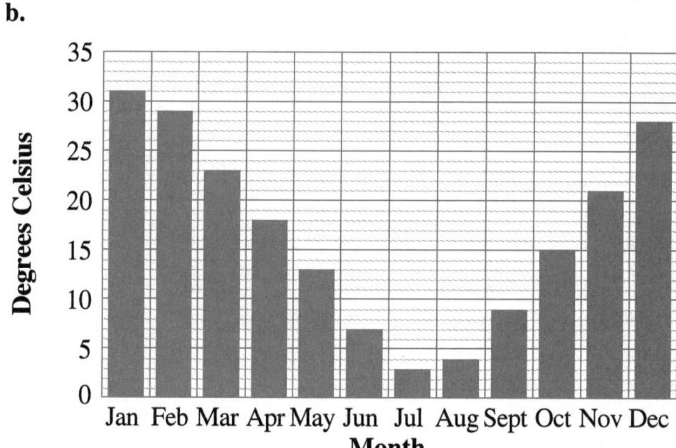

41. 3200 **43.** 3,500,000 **45.** 60 **47.** 50 **49.** 640,000

Homework 4.3A

1. $x = 11$ **3.** $a = 5$ **5.** $t = 3$ **7.** $h = 2$ **9.** $m = 28$ **11.** $v = 6$ **13.** $p = 10$ **15.** $z = 3$ **17.** $f = 11$ **19.** $m = 0.55$ **21.** $d = 600$ **23.** $w = 13.608$ **25.** $S - R$ **27.** $\frac{g}{A}$ **29.** $F + 0.015P$ **31.** $0.62(T + H)$

33a.

x	2	4	6	8	10	12
$3x - 6$	0	6	12	18	24	30

b. $x = 6$

35a.

x	4	6	8	10	12	14	16
$\frac{4+2x}{3}$	4	$\frac{16}{3}$	$\frac{20}{3}$	8	$\frac{28}{3}$	$\frac{32}{3}$	12

b. $x = 16$ **37.** $h = 28$ **39.** $t = 20$ **41.** $w = 29.36$ **43.** $g = 14$ **45.** $z = 8$ **47.** $y = 8$ **49.** $b = 3$ **51.** 5700 **53.** 4250 **55.** 8200 **57.** 0.78 **59.** 5.4562

Homework 4.4A

1. 15 ft **3.** 2628 lb **5.** 8.5% **7.** 25 m **9.** 300 mi **11.** 6% **13.** 10 in.

15a.

m	0.5	0.75	1	1.25	2
$12m + 9$	15	18	21	24	33

b. $12m + 9 = 24$; 1.25 in.

17a.

s	82	84	86	88	90
$\frac{s+156}{3}$	$79.\overline{3}$	80	$80.\overline{6}$	$81.\overline{3}$	82

b. $\frac{s+156}{3} = 80$; 84 points **19.** 51.75 sq. in. **21.** $x = 23$
23. $x = 7$ **25.** $x = 10$ **27.** 6 **29.** 28 **31.** 28
33. 12 **35.** 2 **37.** −5 **39.** −21 **41.** −99
43. 17 **45.** 25 **47.** 300 **49.** 3600 **51.** 300
53. 3000 **55.** 240,000

Homework 4.4B

1. $3y - 12 = 30$; 14 **3.** $\frac{y}{3} - 12 = 30$; 126 mi
5. $4x + 20 = 80$; \$15 **7.** $\frac{x}{4} + 20 = 80$; \$240
9. (a) Number of tapes bought: n (b) $7n + 2 = 30$ (c) Irwin bought four tapes. **11.** (a) Hours Fran must work: h (b) $58 = 4h + 30$ (c) Fran must work 7 h. **13.** (a) Cost of a camera: c (b) $269 = 4c + 9$ (c) The camera costs \$65.
15. (a) Georgia's adjusted income: A (b) $2840 = 600 + 0.07A$ (c) Georgia's adjusted income was \$32,000. **17.** −14
19. 40 **21.** 24 **23.** 1 **25.** 81 **27a.** 5 cm, 10 cm, $4\frac{1}{3}$ cm **b.** 390 sq. cm **29.** 1400 **31.** 3700 **33.** 400
35. 70 **37.** 120

Homework 4.5A

1. $6 + 2 + (-12) + 3 = -1$ **3.** $-3 + (-4) + 2 = -5$
5. $6 + (-8) + 9 + (-1) = 6$ **7.** $-28 + 35 + (-63) = -56$
9. $-12.8 + 21.7 + (-19.2) + 17.5 = 7.2$ **11a.** −8
b. −240 **c.** 32 **13a.** −10 **b.** 192 **15a.** −149
b. −61 **17a.** 0 **b.** 30 **19.** (a) How long Alida must wait: t (b) $15 + 6t = 75$ (c) Alida must wait 10 min.
21. (a) Pak's old salary: s (b) $\frac{s}{3} + 500 = 8500$ (c) His old salary was \$24,000. **23.** Two terms: 21
25. One term: −24 **27.** Two terms: −10 **29.** Four terms: 33
31. Three terms: −280 **33.** Four terms: 72 **35.** 40
37. 10 **39.** −90 **41.** −120 **43.** 20 **45.** −40
47. −1200 **49.** −400

Homework 4.5B

1a. −3 **b.** −28 **c.** 11 **d.** 3 **3.** −2 **5.** −48
7. 40 **9.** −13 **11.** −3 **13.** 13 **15.** 20
17. −24 **19.** −64 **21.** −33 **23.** −17 **25.** −9
27. 3 **29.** 2 **31.** −2

33.

z	−2	0	−5
$15 - 5z$	25	15	40

35.

a	−4	4	2
$-2(3a - 8)$	40	−8	4

37.

c	−7	−2	2
$(c + 3)(c - 4)$	44	−6	−10

39.

u	−4	3	5
$\frac{3u}{u-5}$	$\frac{4}{3}$	$-\frac{9}{2}$	Undefined

41.

r	−6	0	2
$\frac{r-2}{r+2}$	2	−1	0

43. −63 **45.** 136 **47.** 36 cm **49.** $13\frac{1}{3}$°C
51. −2800 **53.** 2000 **55.** −1300 **57.** 600
59. −2000

Chapter 4 Review

1. False **3.** True **5.** False **7a.** $2 + 5 \cdot (12 \div 3)$
b. $(2 + 5) \cdot 12 \div 3$ **9a.** Multiplication; 4
b. Perform 3 − 4 first; −7 **11.** 5 **13.** −19 **15.** −75
17. 4 **19.** −2 **21.** 6 **23.** 276 **25.** $\frac{9}{2}$ **27.** 0.84
29. −2.5 **31.** \$3437.50 **33.** 157.5 sq. cm
35a. $400 + 80m$ **b.** \$1120 **c.** 13 mo **37a.** $5(d + 0.5)$
b. 32.5 mi **c.** 4 mi **39a.** $\frac{C - 50}{10}$ **b.** \$62 **c.** \$840
41. (1) List the operations perfomed on the variable *in order.* (2) Undo those operations *in reverse order.* **43.** $x = \frac{26}{5}$
45. $z = 15$ **47.** $a = 9$ **49.** $h = 8$ **51.** $v = 18$
53. 250 seeds **55.** \$164 **57.** 3570 **59.** 5.64
61. 120 **63.** 180,000 **65.** 2.35 **67.** 800
69. 16,000 **71.** −400 **73.** −1500

Homework 5.1A

1a. 64 sq. in. **b.** 512 cu. in. **3.** 2560 **5a.** 400 sq. cm
b. 6; 2400 sq. cm **c.** 8000 cu. cm **9.** In $(-5)^2$, −5 is squared: $(-5)^2 = 25$. In -5^2, only 5 is squared: $-5^2 = -25$.
11. 4^8 **13.** $(0.5)^3$ **15.** $(-1)^2$ **17.** a^6 **19.** 2^4q^3

21a.

1^2	2^2	3^2	4^2	5^2
1	4	9	16	25

6^2	7^2	8^2	9^2	10^2
36	49	64	81	100

b.

1^3	2^3	3^3	4^3	5^3
1	8	27	64	125

6^3	7^3	8^3	9^3	10^3
216	343	512	729	1000

23a. 10, 100, 1000, 10,000, 100,000 **b.** One followed by 24 zeros **25a.** 8 **b.** 25 **c.** 49 **d.** 6 **27a.** 16 **b.** 2 **c.** 64 **d.** 1 **29a.** 81 **b.** 216 **c.** 10 **d.** 144 **31a.** 9 **b.** −8 **33a.** −125 **b.** 81 **35a.** −9 **b.** −8 **37a.** −16 **b.** −64 **39a.** 6.76 **b.** 10.24 **c.** 117.649 **d.** 0.09 **41.** −24 **43.** 19 **45.** 2 **47a.** 3500 **b.** 35,000 **c.** 350,000 **49a.** 7.4 **b.** 74 **c.** 740 **51.** $\frac{1}{2}$ **53.** $\frac{2}{3}$ **55.** $\frac{1}{4}$ **57.** $\frac{5}{8}$ **59.** $\frac{5}{6}$
Find a whole number that divides evenly into both numerator and denominator. Divide numerator and denominator by that number. For example, $\frac{3}{6} = \frac{3 \div 3}{6 \div 3} = \frac{1}{2}$.

Homework 5.1B

1. 60,000 **3.** 275,000 **5.** 4862.3 **7.** 7,280,000,000,000 **9.** 74,000,000 **11.** 0.6 **13.** 10^3 **15.** 10^5 **17.** 10^3 **19.** 10^{17} **21.** 3.5×10^6 **23.** 2.76×10^1 **25.** 6.752×10^4 **27.** 7.92×10^{21} **29.** 4.3945×10^{18} **31.** 245,500 **33.** 12,000 **35.** 125 **37.** 5.12×10^{38} **39.** $133{,}333{,}333.\overline{3}$ sec = 1543.2 days **41.** 345.1 lb/cu. ft **43a.** $3.2^4 = 3.2(3.2)(3.2)(3.2)$; 3.2×10^4 is scientific notation. **b.** $3.2^4 = 104.8576$; $3.2 \times 10^4 = 32{,}000$ **45a.** 5.28×10^3 **b.** 3.65×10^2 **c.** 6.048×10^5 **d.** 2.788×10^7 **47.** 36 **49.** 13 **51.** −15 **53.** 7 **55.** −60 **57.** $\frac{4}{5}$ **59.** $\frac{9}{20}$ **61.** $\frac{6}{25}$ **63.** $\frac{19}{50}$ **65.** $\frac{24}{25}$

Homework 5.2A

1a. −16 **b.** 16 **3a.** 96 **b.** 4 **5a.** −4 **b.** 192 **7.** 32 **9.** 160 **11.** 144 **13.** 100 **15.** −27 **17.** 28 **19a.** $x + x + x$ **b.** $x \cdot x \cdot x$ **21.** −28 **23.** 24 **25.** −10 **27.** 5 **29.** −31 **31.** −32 **33.** 16 **35.** 162 **37.** 4 **39.** −6 **41.** 18 **43.** $2x = x + x$; $x^2 = x(x)$ **45.** $16w = 400$; $w = 25$ ft **47.** $576 = r \cdot 1800$; $r = 0.32$, or 32% **49.** $222{,}960 = S/50$; $S = 11{,}148{,}000$ acres **51.** $1\frac{1}{5}$ **53.** $1\frac{1}{2}$ **55.** $1\frac{2}{3}$ **57.** $1\frac{2}{3}$ **59.** $1\frac{1}{4}$

Homework 5.2B

1. $C = \pi d$ **3.** $A = \pi r^2$ **5.** 50.27 in. **7.** 37.70 ft **9.** 3.77 m **11.** 78.54 cm **13.** 9.55 yd **15.** 6.76 in. **17.** 50.27 sq. ft **19.** 0.13 sq. cm **21.** 490.87 sq. yd **23.** 3.35×10^1 m^3 **25.** 4.19×10^{-6} cu. cm **27.** 5.89×10^8 cu. ft **29.** 153.94 sq. in. **31.** 23 in. **33.** 105,904.14 cu. ft **35.** 452.39 sq. ft **37a.** 1.3914×10^6; 6.957×10^5 km **b.** 1.41×10^{18} cu. km, 1,410,000,000,000,000,000 km^3 **39.** $A = 55.86$ sq. cm; $P = 37.42$ cm **41.** $A = 76.27$ sq. ft; $P = 34.85$ ft **43.** $A = 30.90$ sq. m; $P = 85.70$ m **45.** 223.8 sq. cm **47.** 530.14 sq. ft **49.** 5 **51.** 77 **53.** −384 **55.** 63 **57.** 66 **59.** 77 **61.** $7\frac{5}{9}$ **63.** $3\frac{1}{3}$ **65.** $7\frac{3}{4}$ **67.** $5\frac{2}{3}$ **69.** 10

Homework 5.3A

1. Like terms have identical variable factors. $3x$ and $5x$ are like terms; $3x$ and $5y$ are not. **3.** False **5.** True **7–9.** Answers will vary **11.** $6y$ **13.** $-4x$ **15.** 0 **17.** $9pq$ **19.** $-79W$ **21.** $-12.8a$ **23.** $-9.4x$ **25.** x **27.** $7ab$ **29.** $2t + 3$ **31.** $12y - 4$ **33.** $-8st + 9s$ **35.** $30d$ **37.** $24h$ **39.** $4a$ **41.** $10m$ **43a.** $4x + 7x = 11x$ is a sum; $4(7x) = 28x$ is a product. **b.** −33; −84 **45.** $10s$ **47.** $16 - 24y$ **49.** $-11h$ **51.** $-4 - 4t$ **53.** $32a - 5$ **55.** 288 **57.** −276 **59.** −32 **61.** −48 **63.** 20 **65.** 12 **67.** 50 **69.** 25 **71.** 32

Homework 5.3B

1. $-3w^3$ **3.** $-3bc$ **5.** $-2pq^2$ **7.** $a^2 - a$ **9.** $2y^2z + 3yz^2$ **11.** cbs **13.** $4m^3 + 3m + 3$ **15.** cbs **17.** m^3 **19.** k^6 **21.** $-30x^5$ **23.** $-12pq^2$ **25.** s^4t^4 **27.** $18x^5z^3$ **29a.** $2x$ **b.** x^2 **31a.** cbs **b.** x^3 **33a.** cbs **b.** x^5 **35a.** cbs **b.** x^2y **37a.** cbs **b.** $6xy$ **39.** $a^2 + a^2 = 2a^2$ **a.** 18 **b.** 18 **c.** 81 **41.** $2v(3v) = 6v^2$ **a.** 150 **b.** 30 **c.** 150 **43.** $5w^2(-2w^2) = -10w^4$ **a.** −160 **b.** 12 **c.** −160 **45.** $12x$ **47.** $60b^2$ **49.** $A = 63v^2$; $P = 32v$ **51.** $\frac{4}{15}$ **53.** $\frac{1}{2}$ **55.** $\frac{4}{15}$ **57.** $\frac{1}{8}$ **59.** $\frac{1}{4}$
Multiply the numerators together. Multiply the denominators together. Reduce if possible. For example $\frac{1}{3}\left(\frac{4}{5}\right) = \frac{1 \cdot 4}{3 \cdot 5} = \frac{4}{15}$.

Homework 5.4A

1a. 6, −6 **b.** 7, −7 **c.** 1, −1 **3a.** 8, −8 **b.** 9, −9 **c.** 12, −12 **5.** $9^2 = 9 \cdot 9 = 81$, but $\sqrt{9} = 3$. To square a number means to multiply it by itself. When we multiply the square root by itself, we get the orignal number. **7.** $\sqrt{x}$ means "the positive square root of x." **9.** No; $\sqrt{25} = 5$ because the symbol $\sqrt{\ }$ means "positive square root." **11a.** 7 **b.** 13 **13a.** −11 **b.** Und. **15a.** −1 **b.** Und. **17a.** $\frac{1}{4}$ **b.** $\frac{1}{5}$ **19a.** $-\frac{14}{5}$ **b.** $\frac{2}{17}$ **21.** 28 **23.** 48 **25.** −3 **27.** −7 **29.** 6 **31a.** $72 + 6h$ **b.** $72 + 6h = 96$, 12 noon **33a.** $50 + \frac{P}{12}$ **b.** $50 + \frac{P}{12} = 68$, \$216 **35a.** $\frac{S - 60}{15}$ **b.** $\frac{S - 60}{15} = 23$, \$405 **37a.** $20 + 0.02B$ **b.** $20 + 0.02B = 340$, \$16,000 **39a.** −15 **b.** cbs **41a.** 20 **b.** 10 **43a.** 21 **b.** 56 **45a.** 2 **b.** 6 **47a.** −26 **b.** 84 **49.** 6 **51.** 9 **53.** 8 **55.** 8

57. 24
Multiply the first fraction by the reciprocal of the second fraction.
For example, $4 \div \frac{1}{3} = 4 \cdot \frac{3}{1} = 12$.

Homework 5.4B

1.

k	0	1	4
$8\sqrt{k} + 12$	12	20	28

3.

h	7	10	42
$3\sqrt{h-6}$	3	6	18

5.

b	0	9	400
$2b - 10\sqrt{b}$	0	−12	600

7.

z	121	144	256
$(12 - \sqrt{z})^2$	1	0	16

9.

x	36	81	225
$-3x + 2x\sqrt{x}$	324	1215	6075

11. ±5 **13.** ±14 **15.** ±4 **17.** ±3 **19.** ±1
21. ±13 **23.** ±4 **25.** ±7 **27.** ±8 **29.** 8
31. −21 **33.** −1 **35.** 3 **37.** 144 **39.** 9 **41.** 81
43. 225
45a.

a	b	$a + b$	a^2
3	4	7	9
5	12	17	25
2	6	8	4

b^2	$a^2 + b^2$	$\sqrt{a^2 + b^2}$
16	25	5
144	169	13
36	40	6.3

b. No

47a.

a	b	$a + b$	$(a + b)^2$
2	3	5	25
3	5	8	64
4	7	11	121

a^2	b^2	$a^2 + b^2$
4	9	13
9	25	34
16	49	65

b. No
49. 4 **51.** 3 **53.** 16 **55.** 9 **57.** 6

Activity

4.8841; 4.9284; 4.9729; 5.0176; 5.0625; 5.1076; 5.1529; 5.1984; 5.2441; 4.977361; 4.981824; 4.986289; 4.990756;4.995225; 4.999696; 5.004169; 5.008644; 5.013121

Homework 5.5A

1.

1	2	3	4	5	6	7	8	9	10
1	1.414	1.732	2	2.236	2.449	2.646	2.828	3	3.162

3a. 6.971 **b.** 5.413 **5a.** 1.183 **b.** 1.342
7a. 20.469 **b.** 28.914 **9.** 85 **11.** 168 **13.** 4082
15. 14.971 **17.** 0.268 **19.** 4.564 **21.** ±2.236
23. ±2.646 **25.** ±1.225 **27.** ±1.960
29.

m	−3	2	9
$9\sqrt{m+5}$	12.728	23.812	33.675

31.

q	3	4	−6
$\sqrt{q^2 - 4}$	2.236	3.464	5.657

33.

a	2	5	12
$\sqrt{a} - \sqrt{a+3}$	−0.822	−0.592	−0.409

35. 193.8 mi **37.** 9.6 sec **39.** 4.5 ft **41.** 17
43. 453 **45.** 39 **47.** 20 **49.** −31 **51.** 0
53. The original number **55.** $\frac{3}{4}$ **57.** $5\frac{1}{3}$ **59.** $1\frac{1}{6}$
61. $2\frac{3}{8}$ **63.** $1\frac{1}{4}$

Homework 5.5B

1. 15 in. **3.** 24 cm **5.** 14.422 mi **7.** 42.650 in. **9.** 3.808 ft **11.** 675.407 mm **13.** 26 ft **15.** 13.856 ft **17.** 4 ft **19.** 7.071 ft **21.** Not a right triangle **23.** a, b, and c are not squared. **25.** $c = 12$, not x. **27.** x is not squared. **29a.** $20x$ **b.** $20x^2$ **c.** $9x$ **d.** cbs **31a.** x^5 **b.** cbs **c.** $2x^3$ **d.** x^6 **33.** In $2x^3$, only x is cubed. **35.** Square of x is $x \cdot x = x^2$; square root of x is a number whose square is x. **37.** $-1^2 = -1$; $(-1)^2 = 1$ **39.** $\sqrt{9+16} = \sqrt{25} = 5$; $\sqrt{9} + \sqrt{16} = 3 + 4 = 7$ **41.** 18 **43.** $\frac{2}{3}$ **45.** 9 **47.** 10 **49.** $\frac{3}{2}$

Challenge Problems

1. About 689 in. or 57 ft

Chapter 5 Review

1. A base raised to an exponent; 6^3 **5.** square inches **7.** base; exponent **9.** $-4^2 = -4 \cdot 4 = -16$; $(-4)^2 = (-4)(-4) = 16$ **11.** True **13.** False **15.** True **17.** False **19.** $7 \cdot 7 \cdot 7$ $aaaaa$ **21.** Move the decimal seven places to the right. **23.** 8.6×10^{12} **25.** $5.6^3 = 5.6(5.6)(5.6) = 175.616$; $5.6 \times 10^3 = 5600$ **27a.** 75 **b.** -22 **29a.** -36 **b.** 49 **31a.** 48 **b.** 144 **33.** -5.1 **35.** $2x^2 - 3x$ **37.** $5x^3 + 2x^3 = 7x^3$ **39.** $2c^5 - 8c^2$; cbs **41.** $p^3(p^5) = p^8$ **43.** a^2b^3; cbs **45a.** 267,035 mi **b.** 3.2×10^{14} cu. mi **47a.** 2.7×10^{16} sq. mi **b.** 583,707,915 mi **c.** 583,708 hours **d.** 3.36×10^{24} cu. mi **49.** Area = 18 sq. ft; Perim = 20.49 ft **51.** Area = 10.93 sq. ft; Perim = 19.20 ft **53.** b **55.** $\sqrt{-16}$ is undefined; $-\sqrt{16} = -4$ **57a.** 1.5 **b.** 5.0625 **59.** -2 **61.** -0.18 **63.** ± 9 **65.** ± 1.58 **67.** $\frac{10}{7}$ sec $\approx$ 1.429 sec **69a.** No **b.** No **c.** Yes **d.** No **71.** $a^2 + b^2 = c^2$ **73.** $\sqrt{72}$ in. **75.** $\frac{2}{3}$ **77.** $\frac{7}{25}$ **79.** 7 **81.** $\frac{4}{3}$ **83.** 36 **85.** 8

Homework 6.1

1a. \$1000 **b.** March, July, November **c.** June, \$1500 **d.** July **3a.** 1,200,000 **b.** 1973 **c.** 5,900,000 **d.** 800,000

5a.

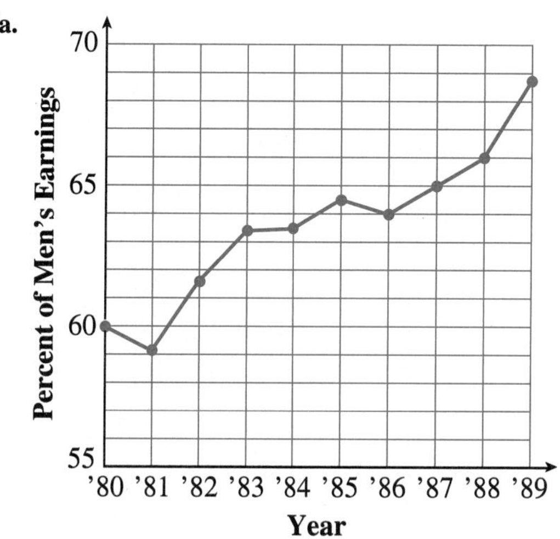

5b. 1982, 1989 **c.** 1981, 1986 **d.** 8.5%

7a.

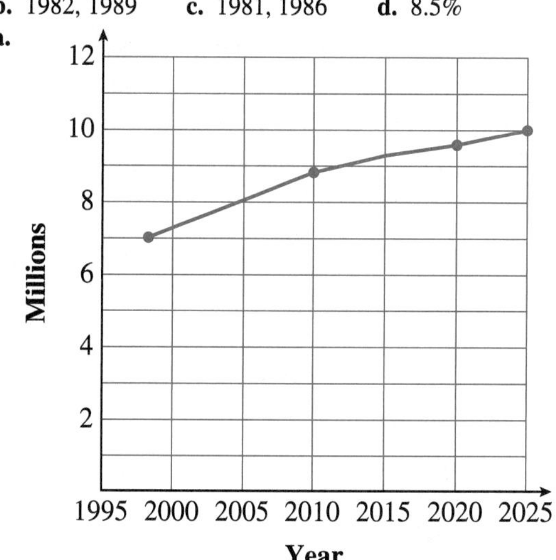

b. The time intervals between bars are not equal.

9a.

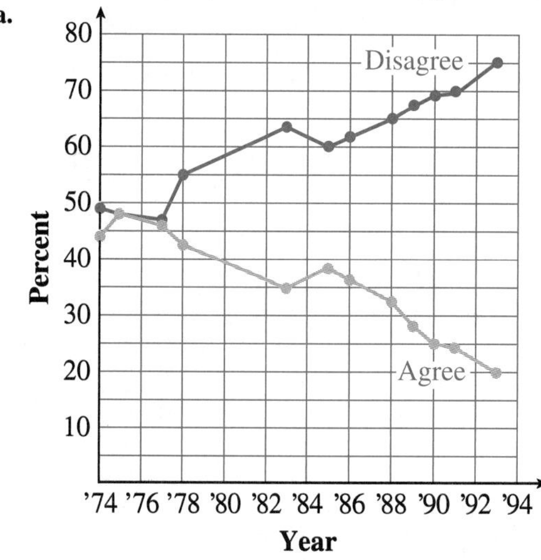

b. The number of people who agree with the statement is decreasing. **c.** Some people were undecided.

11a.

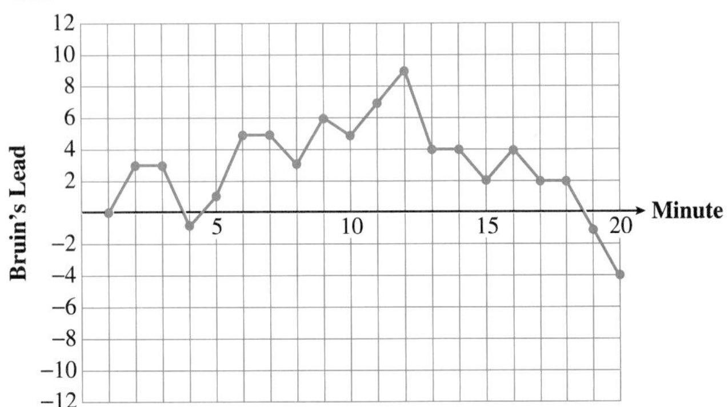

b. Minute 12 **c.** 13 points **d.** Cardinals by 4 **13a.** −1 **b.** 4 **15a.** 6 **b.** 6 **17a.** 9 **b.** 3 **19a.** 9 **b.** 54 **21.** 6 **23.** 6 **25.** 9 **27.** 6 **29.** 15 Divide the old denominator into the new denominator, then multiply the old numerator by the result. For example, $\frac{2}{3} \cdot \frac{2}{2} = \frac{4}{6}$.

Homework 6.2A

1. (a) Extremes: 5, 18; range: 13 (c) Mean: 12; medium: 12; mode: 12

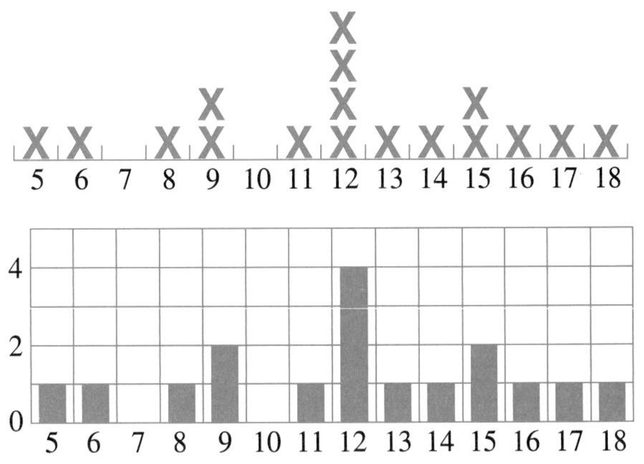

3. (a) Extremes: 6, 40; range: 34 (c) Mean: 20; median: 18; mode: 12

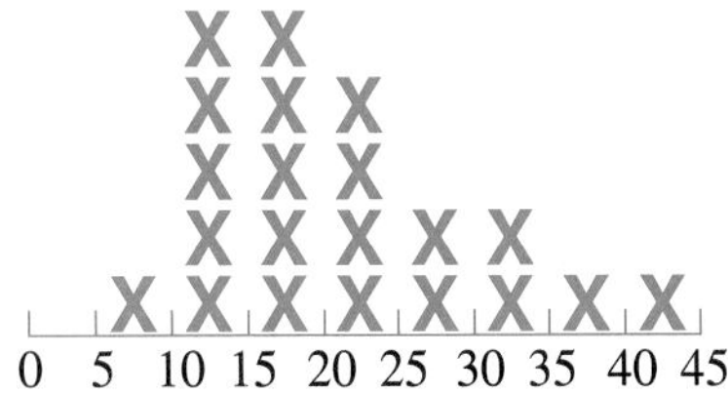

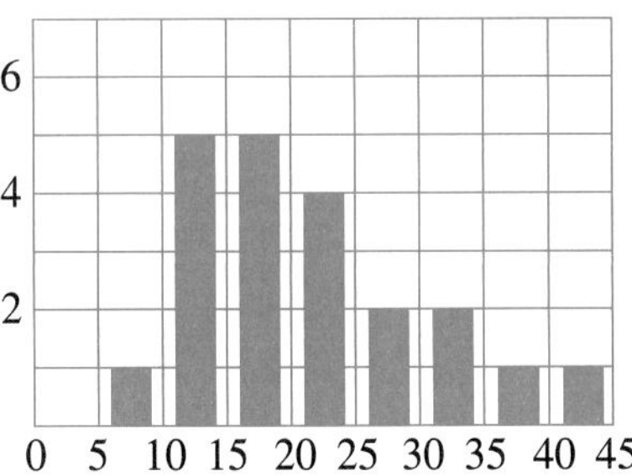

5. (a) Extremes: \$22,684; \$39,482; range: \$16,798 (c) Mean: \$29,998; median: \$29,211; mode: none

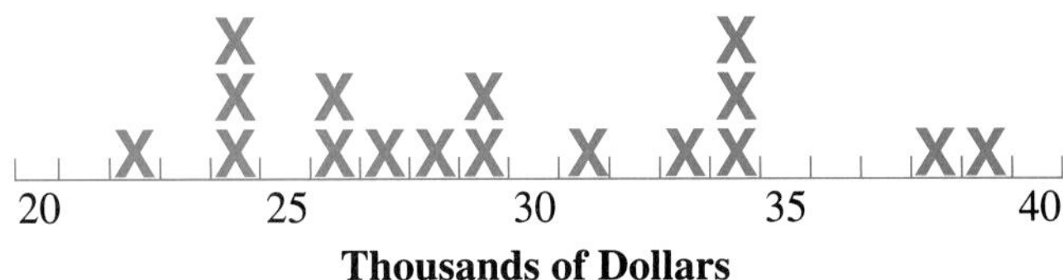

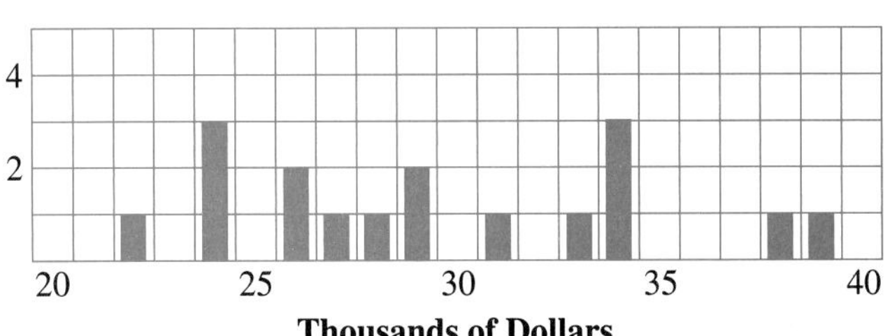

7a. Mean: \$470,000; median: \$120,000; mode: \$100,000 **b.** The median is probably the best indicator of a typical house price. The one extreme value, \$2,500,000, causes the mean to be larger than a typical price. The mode is actually the lowest price in the data. **c.** The median, because it is closest to a typical house price.

9a.

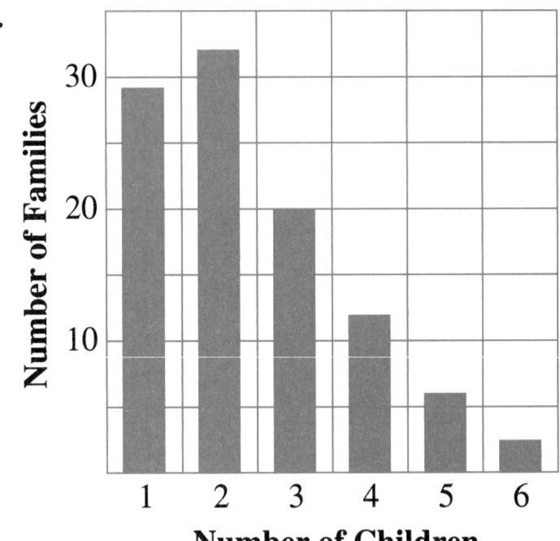

b. 2.42 **11a.** Extremes: 4, 19; range: 15 **b.** 11 **c.** 18 **d.** 50 **e.** 11 **f.** 10% **g.** 10.54 **13.** Undefined, can't divide by 0. **15.** −216 **17.** $3a^2b$ **19.** $6m^3n^2$ **21.** 11 **23.** 4; 9 **25.** 15; 8 **27.** 9; 10 **29.** 4; 9 **31.** 15; 10

Homework 6.2B

1. (a) Median: 149.5; LQ: 143; UQ: 160.5 (b) IQR: 17.5

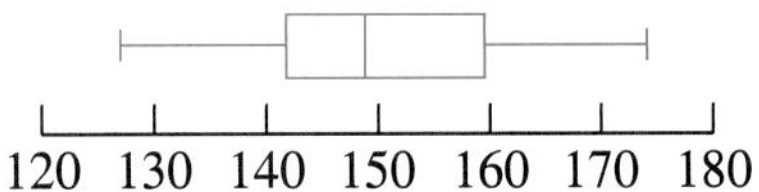

3. (a) Median: 67.5; LQ: 64; UQ: 75 (b) IQR: 11

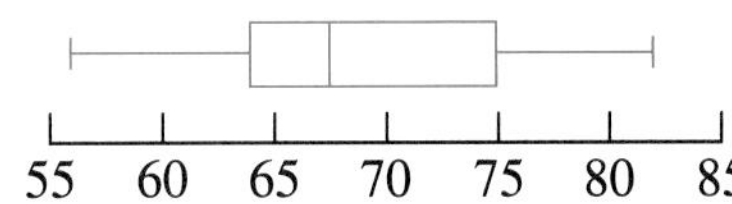

5. (a) Median: 15.3; LQ: 14.7; UQ: 16.7 (b) IQR: 2.0

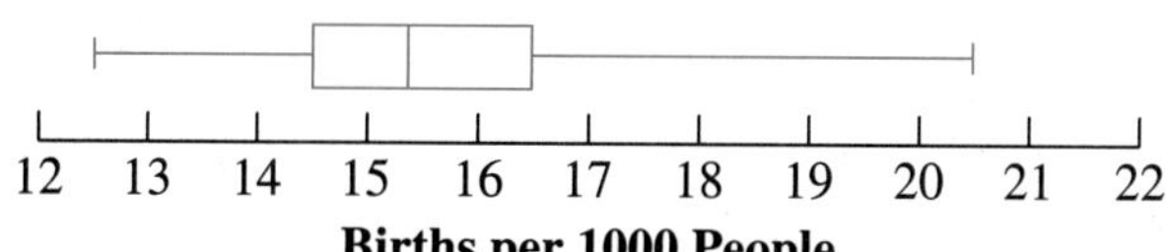

7a. A **b.** B **c.** B **d.** A **e.** B. One-fourth of the children in neighborhood B are between 3 and 5 years old, near Stefanie's age. In neighborhood A, half the children are over 8, and half are between 1 and 8. It seems likely that fewer than one-fourth of the children are between 3 and 5. **9.** $y = -4x$

11. $q = p + 3$ **13.** $r = m - 2$ **15.** $b = \frac{-a}{2}$

17. $p = 3n - 1$ **19.** 2 **21.** 0 **23.** $\sqrt{7}$ **25.** 0

27. Undefined, can't divide by 0. **29.** $\frac{11}{12}$ **31.** $\frac{31}{10}$

33. $\frac{17}{6}$ **35.** $\frac{23}{20}$ **37.** $\frac{16}{15}$

Homework 6.3A

1.

c	d
200	1000
400	800
600	600
800	400
1000	200
1200	0

a. 650
b. 450

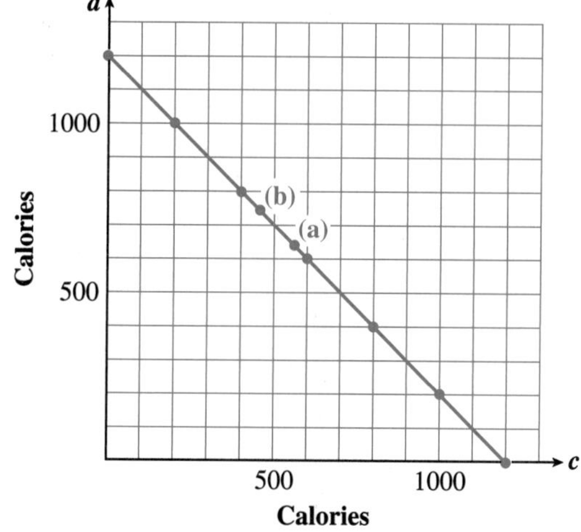

3.

r	C
2	2
3	4.5
4	8
5	12.5
6	18
7	24.5

a. \$4.50
b. 6-in. radius

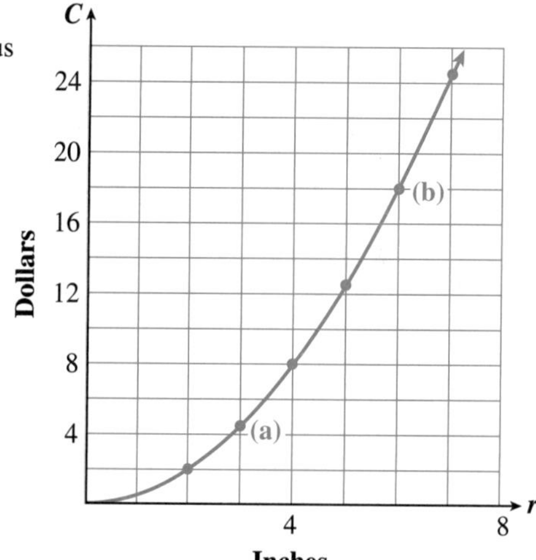

5.

v	t
30	20
50	12
60	10
100	6
120	5
150	4
200	3
300	2

a. 12 hr
b. 200 mph

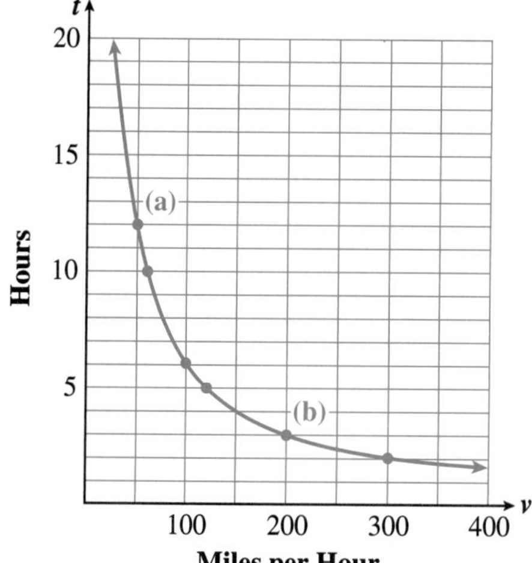

7.

h	d
5,000	86.3
10,000	122
15,000	149.4
20,000	172.5
25,000	192.9
30,000	211.3

a. 122 mi
b. 6719 ft

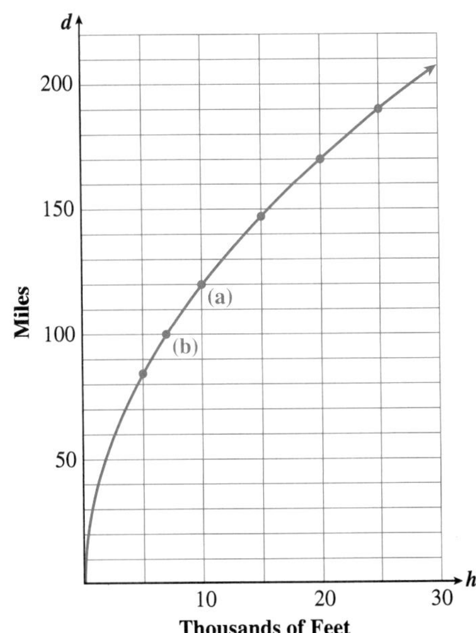

9.

Month	*Sales (thousands)*
April ($m = 0$)	40
May ($m = 1$)	20
June ($m = 2$)	10
July ($m = 3$)	5
August ($m = 4$)	2.5
September ($m = 5$)	1.25
October ($m = 6$)	0.625

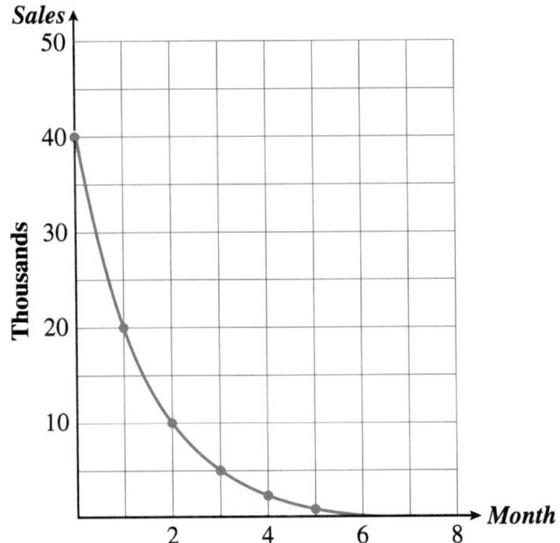

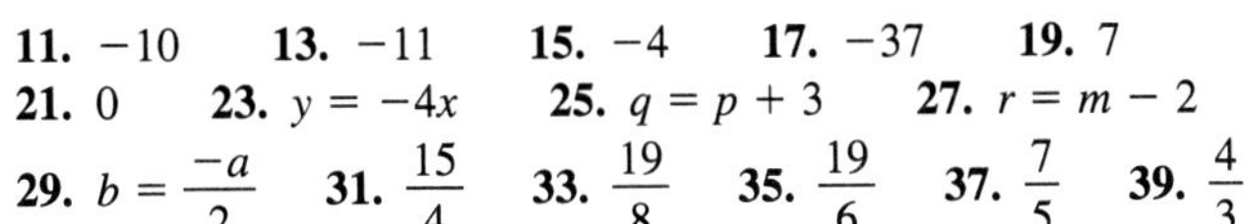
11. -10 **13.** -11 **15.** -4 **17.** -37 **19.** 7
21. 0 **23.** $y = -4x$ **25.** $q = p + 3$ **27.** $r = m - 2$
29. $b = \frac{-a}{2}$ **31.** $\frac{15}{4}$ **33.** $\frac{19}{8}$ **35.** $\frac{19}{6}$ **37.** $\frac{7}{5}$ **39.** $\frac{4}{3}$

Homework 6.3B

3a. x and y are negative **b.** x is negative and y is positive.
5. A horizontal line
7.

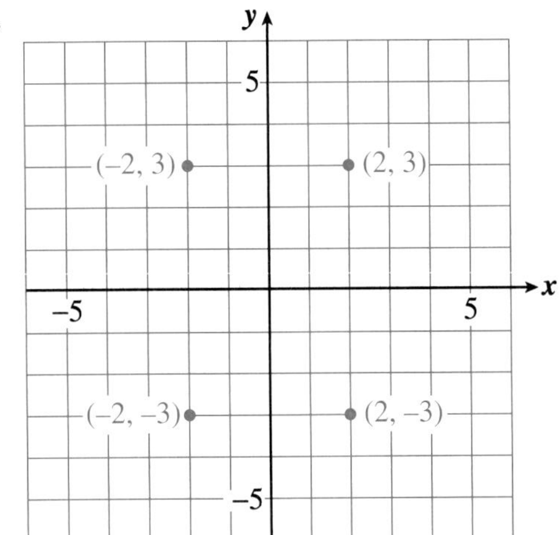

9.

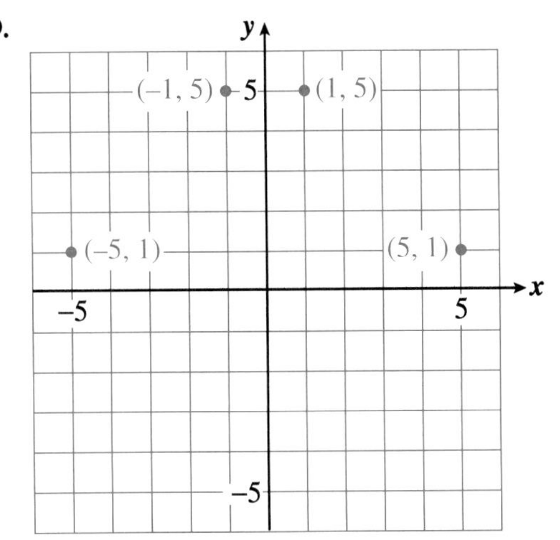

11.

x	-2	-1	$\frac{-1}{2}$	0	$\frac{1}{2}$	1	2
y	-7	0	$\frac{7}{8}$	1	$\frac{9}{8}$	2	9

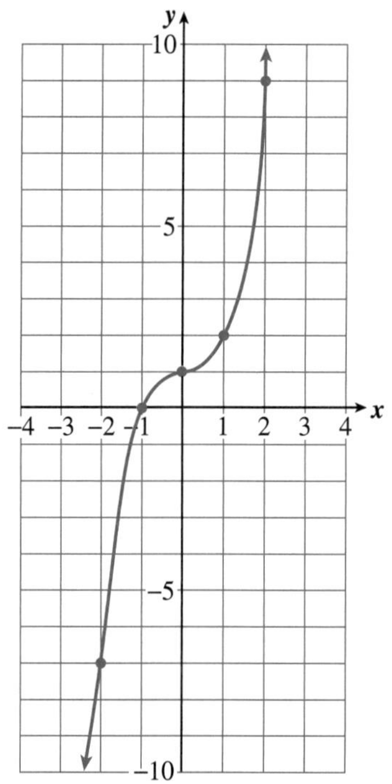

13.

x	-4	-3	-2	-1	$\frac{-1}{2}$	$\frac{-1}{4}$
y	$\frac{-1}{4}$	$\frac{-1}{3}$	$\frac{-1}{2}$	-1	-2	-4

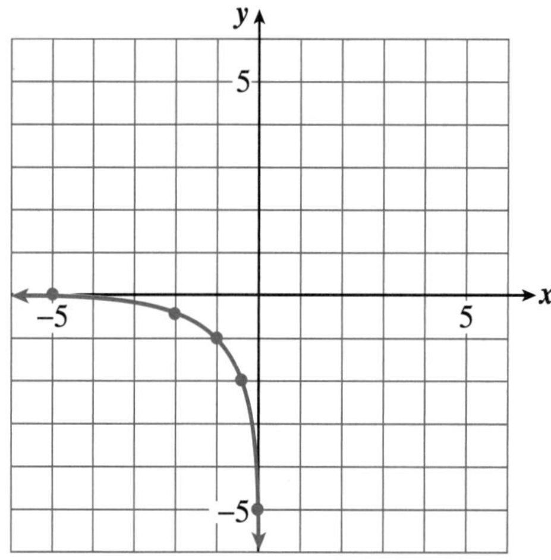

15.

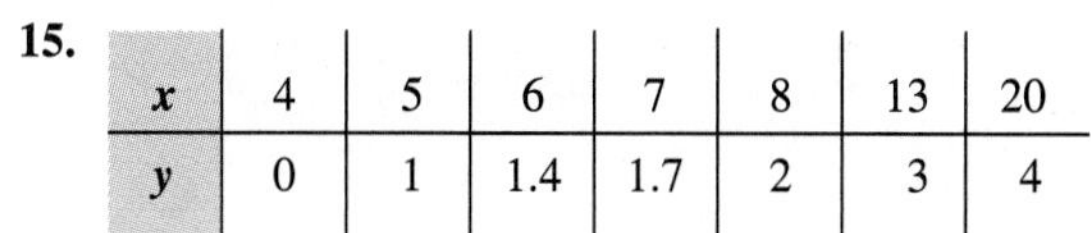

x	4	5	6	7	8	13	20
y	0	1	1.4	1.7	2	3	4

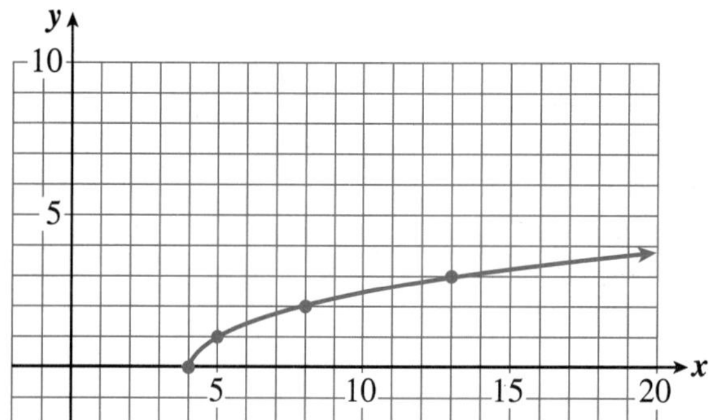

17.

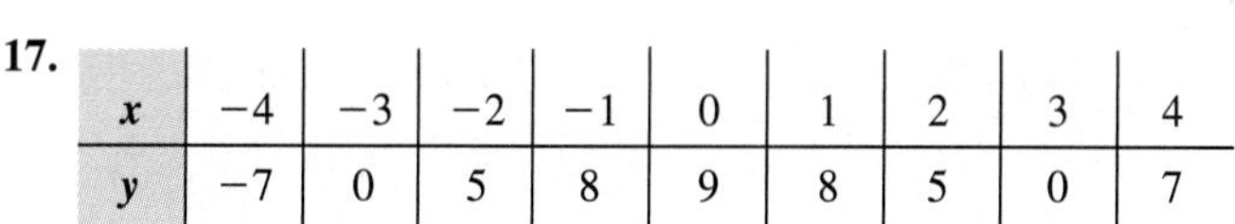

x	-4	-3	-2	-1	0	1	2	3	4
y	-7	0	5	8	9	8	5	0	7

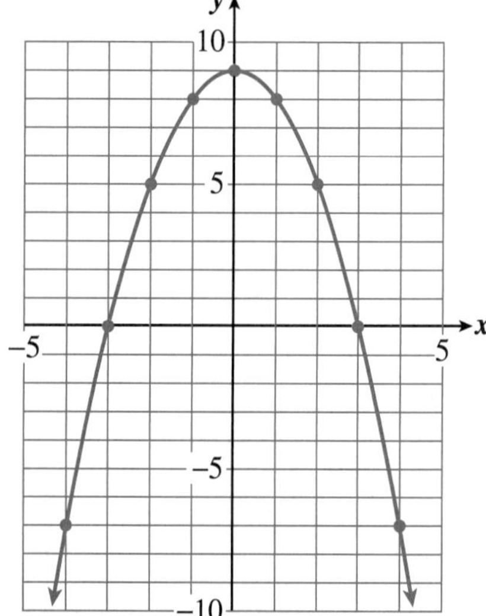

19.

x	-4	-3	-2	-1	0	1	2
y	8	3	0	-1	0	3	8

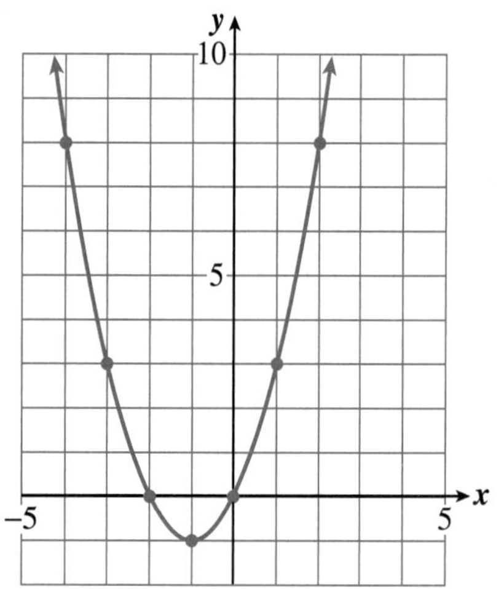

21.

x	y
-2	-1
-1	-2
$\frac{1}{2}$	4
1	2

$y = \frac{2}{x}$

23.

x	y
0	0
1	2
4	4
9	6

$y = 2\sqrt{x}$

25.

x	y
5	3
3	1
0	-2
-1	-3

$y = x - 2$

27. $\frac{x + 25}{5} = 75, x = 350$ **29.** $\frac{x}{5} + 25 = 75, x = \$250$

31. $\frac{7}{6}$ **33.** $\frac{17}{16}$ **35.** $\frac{1}{4}$ **37.** $\frac{2}{5}$ **39.** $\frac{4}{9}$

Homework 6.4

1a. $y = 2x + 3$ **b.** $y = x^2 + 3$ **3.** The x intercept is the point where the graph crosses the x axis. The y intercept is the point where the graph crosses the y axis.

5.

x	-3	3	6
y	-3	3	6

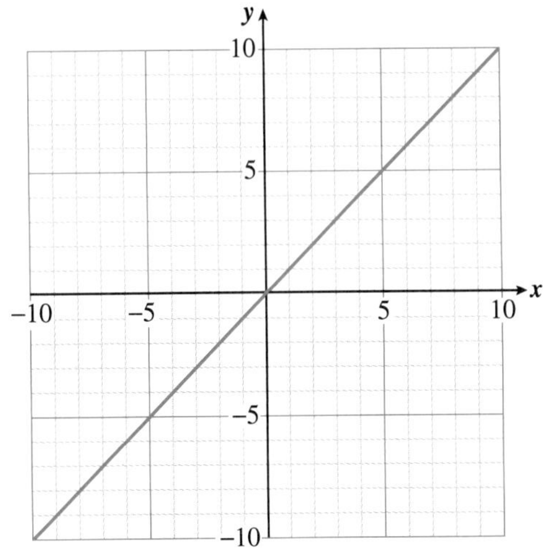

x int.: (0, 0); y int.: (0, 0)

7.

x	-5	0	4
y	-6	-1	3

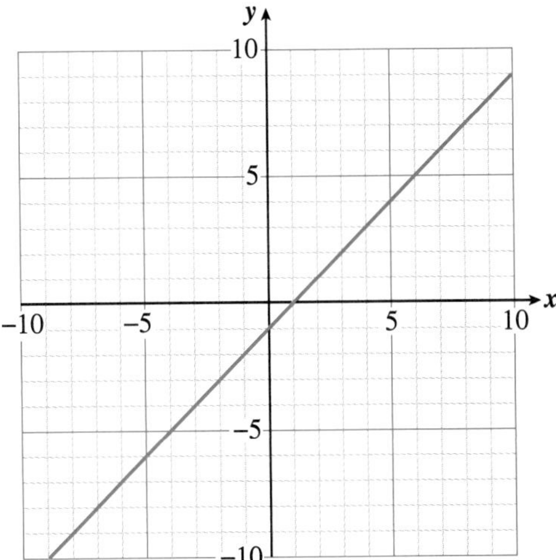

x int: (1, 0); y int: (0, −1)

9.

x	-4	0	4
y	8	0	-8

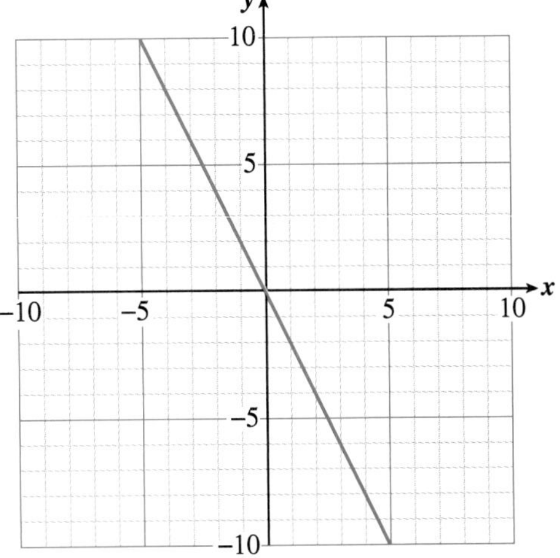

x int: (0, 0); y int: (0, 0)

11.

x	-5	0	5
y	0	-5	-10

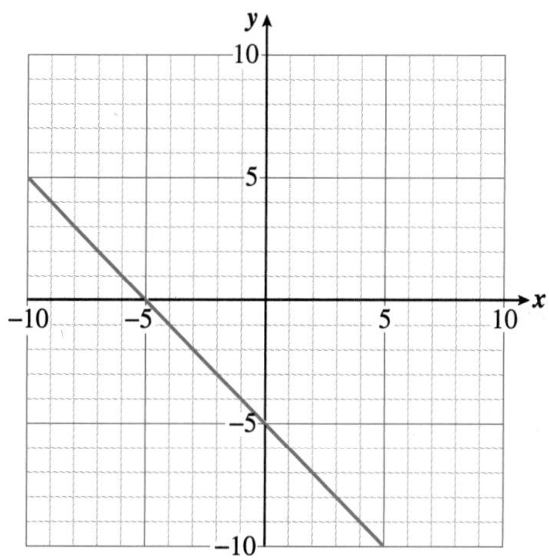

x int: $(-5, 0)$; y int: $(0, -5)$

13.

x	-4	0	2
y	-5	3	7

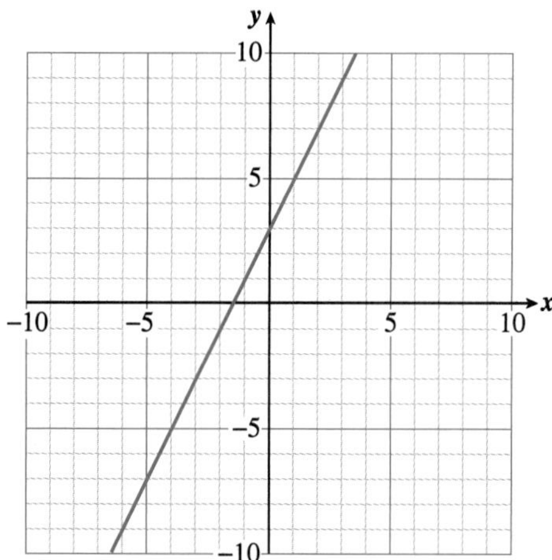

x int: $\left(\frac{-3}{2}, 0\right)$; y int: $(0, 3)$

15.

x	-6	0	6
y	-7	-4	-1

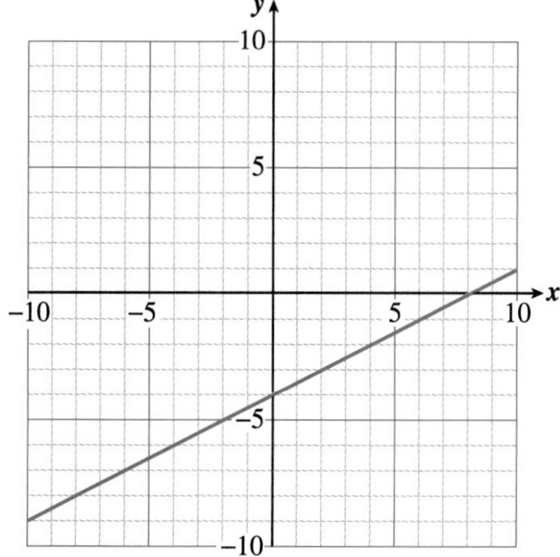

x int: $(8, 0)$; y $(0, -4)$

17a. $y = 20x$

b.

x	y
0	0
10	200
20	400
40	800

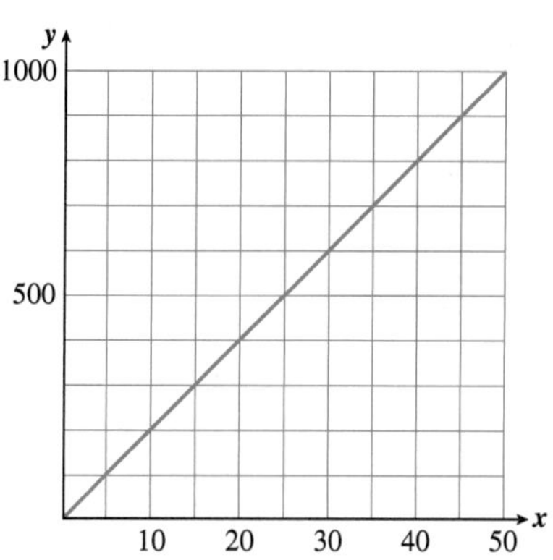

19a. $y = -800 + 5x$ **b.**

x	y
0	-800
50	-550
100	-300
200	200

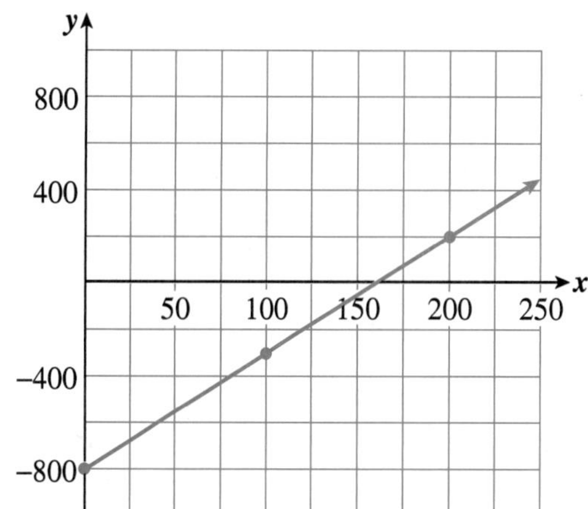

c. $0 = -800 + 5x, x = 160$ **21.** -2 **23.** ± 2
25. $\pm\sqrt{5}$ **27.** $\pm\sqrt{6}$ **29.** $\frac{7}{18}$ **31.** $\frac{9}{20}$ **33.** $\frac{23}{30}$
35. $\frac{5}{24}$ **37.** $\frac{33}{40}$

Homework 6.5

1a. Multiplied by -5, added 2. **b.** Subtract 2, divide by -5.
3a. Divided by 3, added 4. **b.** Subtract 4, multiply by 3.
5. 2 **7.** 11 **9.** $\frac{-8}{7}$ **11.** 18 **13.** 10 **15.** 4
17. 6 **19.** (a) time to reach -150 feet: m
(b) $-45 - 15m = -150$, $m = 7$ (c) It will take 7 minutes
21a. time to wait: t (b) $350 - 11t = 75$, $t = 25$
(c) Alida must wait 25 minutes **23.** (a) point-value of each hand: p (b) $60 - 12p = 96$, $p = 13$
(c) Each hand is worth 13 points **25a.** $y = 750 - 120x$

b.

x	y
0	750
2	510
5	150
7	−90

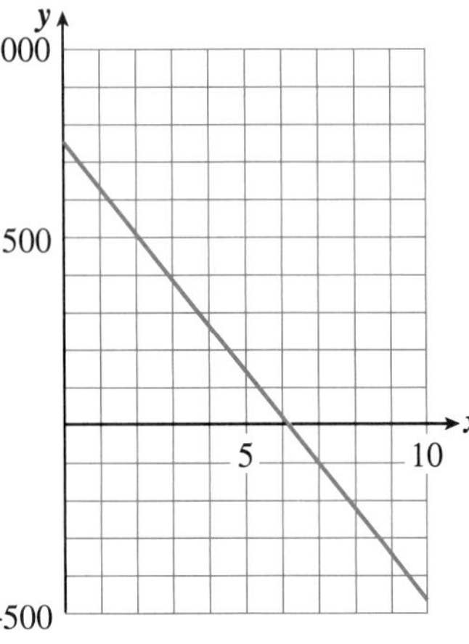

c. $750 - 120x = -210$, $x = 8$ h

27.

C	F
−20	−4
−10	14
0	32
10	50
20	68

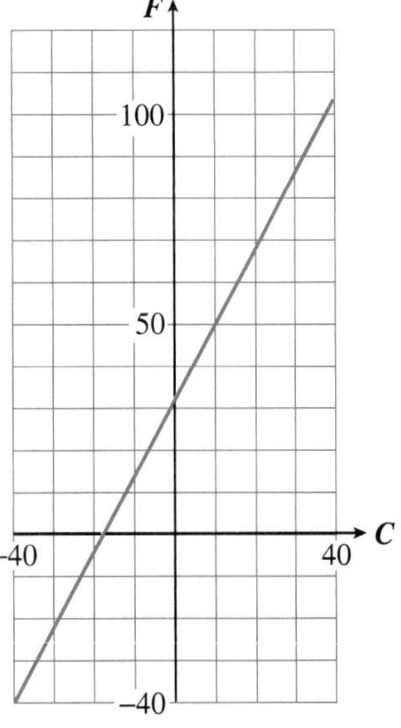

29. 8.2 **31.** 2 **33.** 19 **35.** Undefined **37.** −5
39. 36 **41.** 30 **43.** 60 **45.** 60 **47.** 16

Chapter 6 Review

1. Put a dot at the top of each bar and connect the dots.
3. Mean, median, mode **5.** The interquartile range
7. Independent **9.** An ordered pair **11a.** 22 **b.** 1977
c. 49 **d.** 1972, 1974, 1980, 1981, 1986, 1987
13. (a)

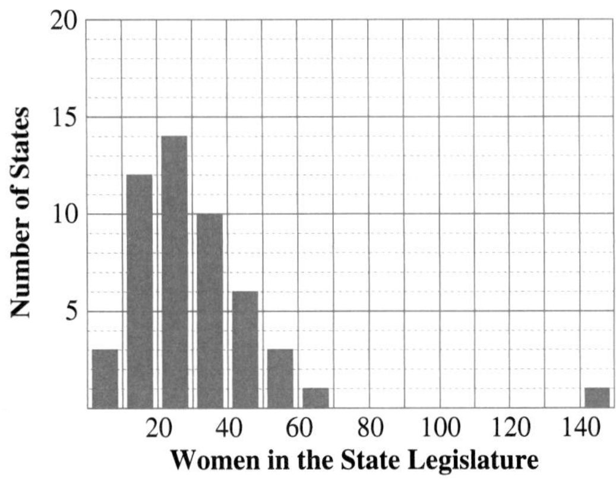

(b) Mean: 30, 48; median: 27.5; mode: 22 (c) LQ: 17; UQ: 37; IQR: 20
(d)

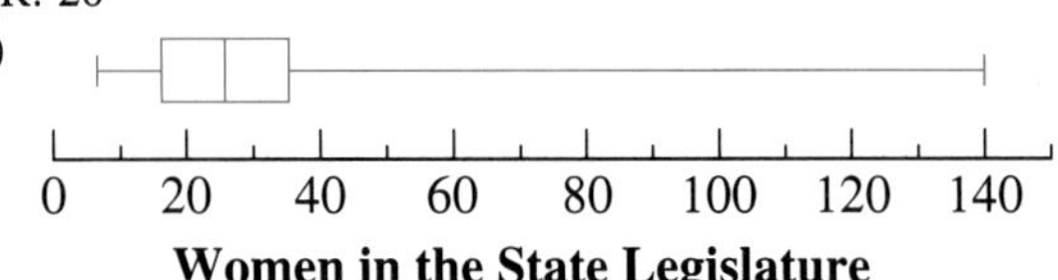

15.

t	*(Computation)*	d
0	$16(0)^2$	0
$\frac{1}{4}$	$16\left(\frac{1}{4}\right)^2$	1
$\frac{1}{2}$	$16\left(\frac{1}{2}\right)^2$	4
$\frac{3}{4}$	$16\left(\frac{3}{4}\right)^2$	9
1	$16(1)^2$	16
$\frac{3}{2}$	$16\left(\frac{3}{2}\right)^2$	36

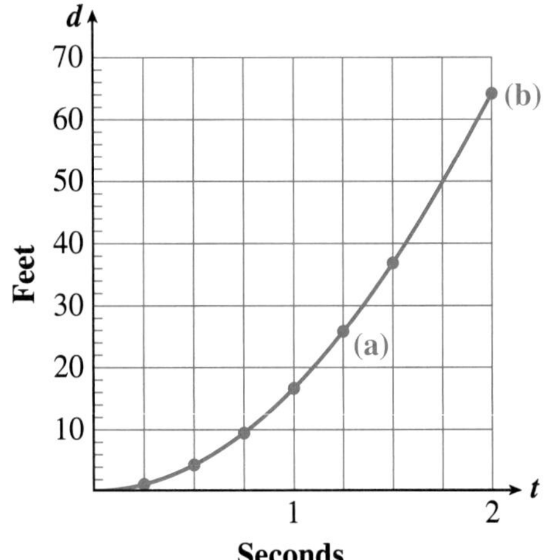

a. 25 ft **b.** 2 sec
17.

x	−4	−3	−2	−1	0	1	2	3	4
y	−11	−4	1	4	5	4	1	−4	−11

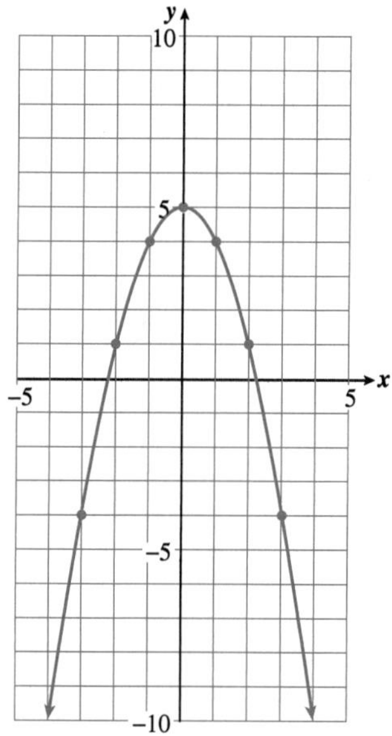

19.

x	y
−8	0
−7	1
−4	2
1	3
8	4

$y = \sqrt{x + 8}$

21. (a)

x	−1	0	3
y	−9	−6	3

(c) (2, 0), (0, −6)

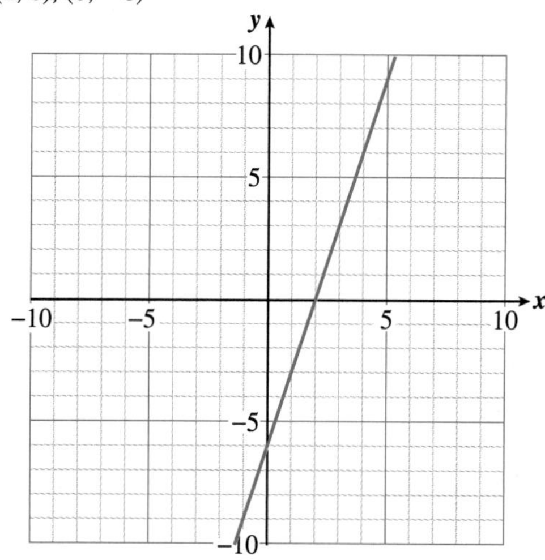

23. (a)

x	−3	0	6
y	−4	−2	2

(c) (3, 0), (0, −2)

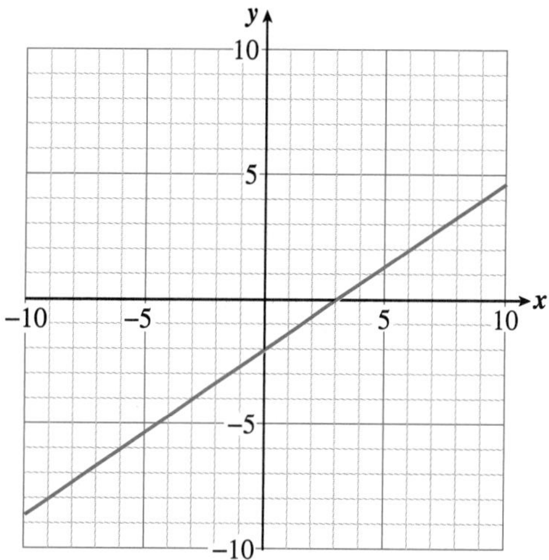

25. −1 **27.** 36 **29.** 2 **31a.** $300 - 6d$
b. $300 - 6d = 126$, $d = 29$

33.

Lisa's age	*Gina's age*
1	5
2	6
5	9
6	10
x	$x + 4$

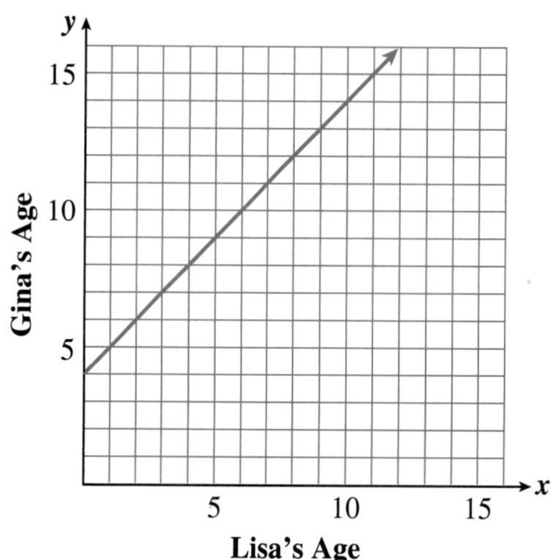

a. $x + 4$ **b.** (0, 4) **c.** Gina was 4 when Lisa was born.

35a. $y = 0.05x$ **b.**

x	y
250	12.5
600	30
800	40

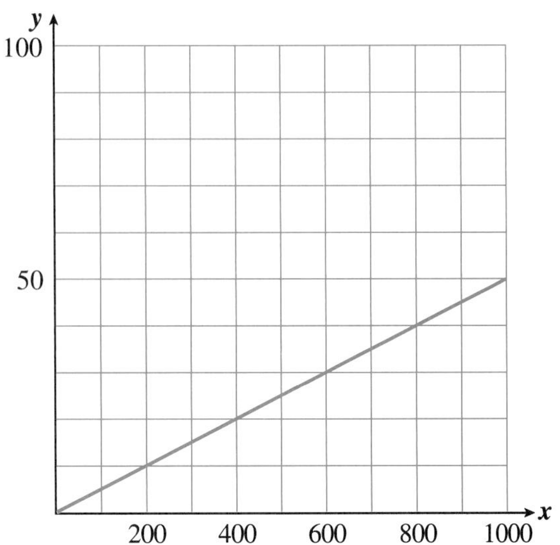

c. \$680,000

37. 6 **39.** 8 **41.** $\frac{22}{5}$ **43.** $\frac{9}{4}$ **45.** $\frac{5}{6}$ **47.** $\frac{1}{15}$

49. $\frac{8}{9}$ **51.** $\frac{1}{18}$ **53.** 36

Homework 7.1A

1. The denominator tells us how many equal pieces to divide the whole into. The numerator tells us how many pieces to take.
3. A rational number is a positive or negative fraction or zero.

5. True. $\frac{4}{5} > \frac{3}{5}$ **7.** True. $-3 = \frac{-3}{1}$ **9.** False. $\frac{-3}{-7} = \frac{3}{7}$

11. $\frac{3}{9} = \frac{1}{3}$ **13.** $\frac{3}{5}$ **15.** $\frac{7}{16}$ **17.** $\frac{2}{8} = \frac{1}{4}$ **19.** $\frac{7}{18}$

21. $\frac{18}{29}$ **23.** $\frac{750}{1200} = \frac{5}{8}$ **25.** $\frac{14}{3} = 4\frac{2}{3}$

27.

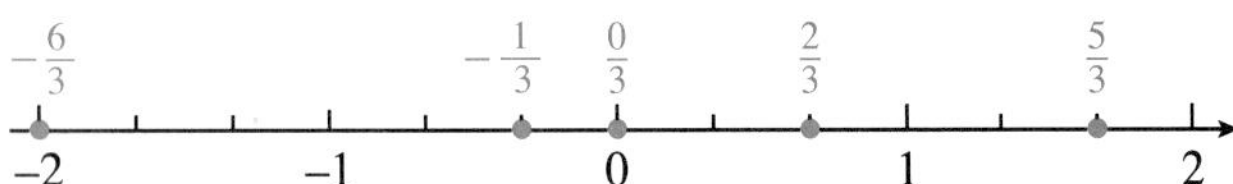

29.

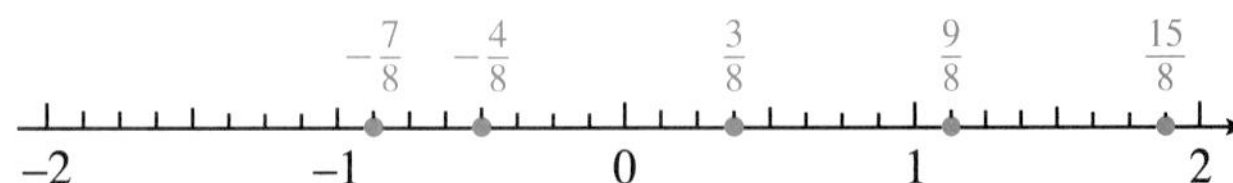

31.

$\frac{3}{12} = \frac{1}{4}$ $\frac{6}{12} = \frac{1}{2}$ $\frac{9}{12} = \frac{3}{4}$

−1 0 1

33. x = immunized first-graders; $\frac{x}{84}$

35. x = Digitronics employees; $\frac{138}{x}$

37. x = gallons gas tank holds; $\frac{480}{3x}$ **39.** $\frac{A}{W}$ **41.** -5

43. 3 **45.** -64 **47.** 45 **49.** $\frac{-15}{4}$ **51.** (b)

53. (b) **55.** (c) **57.** (a) **59.** (a)

Homework 7.1B

1. To reduce or build fractions **3.** factor; term **5.** $\frac{5}{4}$

7. $\frac{-8}{9}$ **9.** $\frac{-4}{3}$ **11.** $\frac{-1}{2}$ **13.** $\frac{5}{x}$ **15.** $\frac{3}{w}$

17. $\frac{12b}{7}$ **19.** $\frac{-5}{6}$ **21.** $\frac{1}{17}$ **23.** $\frac{1}{24v}$ **25.** $\frac{-8}{5}$

27. $\frac{a}{9}$ **29.** $\frac{-3y^2}{14}$ **31.** $\frac{2u^3}{3w}$ **33a.** Cannot cancel terms.

b. $\frac{4}{5}$ **35a.** $\frac{m}{2n}$ **b.** Cannot cancel terms. **37a.** Cannot cancel terms. **b.** $\frac{-1}{3}$ **39a.** -1 **b.** Cannot cancel terms.

41a. Cannot cancel terms. **b.** $\frac{3}{2}$ **43.** $\frac{-25}{30}$ **45.** $\frac{27}{9}$

47. $\frac{2}{4z}$ **49.** $\frac{88}{11m}$ **51.** $\frac{-4b}{19b}$ **53.** $\frac{p^2}{7pq}$ **55.** $\frac{-18d^2}{2dw}$

57. $\frac{x^2}{18x}$ **59.** $\frac{-9g^2}{12g}$ **61.** $\frac{-b}{ab^2}$ **63.** $\frac{10n^2}{2n}$

65. $\frac{-36w^4}{24rw^3}$ **67a.** 10,000 **b.** 1965 **c.** 1955 **d.** 1970

69.

t	v
0.75	16
1	12
1.2	10
1.8	$6.\overline{6}$
2	6
2.5	4.8
3	4
4	3

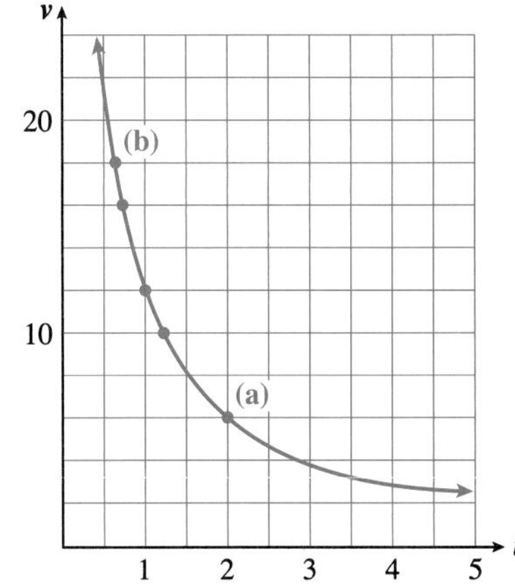

a. 6 mph **b.** $\frac{2}{3}$ h **71.** 0 **73.** $-3y$ **75.** $-H$

77. $-3ab + 7a - 6$ **79.** (b) **81.** (b) **83.** (b)

85. (c) **87.** (a)

Homework 7.2A

1. Divide **3.** $2\frac{1}{4}$ acres **5.** $\frac{2}{27}$ **7.** $\frac{1}{3}$ **9.** $\frac{-c}{d}$

11. $\frac{-4u}{3v}$ **13.** $\frac{3x}{2}$ **15.** $\frac{7w}{8z}$ **17.** $\frac{-17}{3}$ **19.** $\frac{-3d}{2a^2}$

21. $\frac{-3c}{4}$ **23.** $\frac{5}{3m}$ **25.** $\frac{8}{5}$ **27.** $\frac{-k}{4}$ **29.** $9u^2$

31. $\frac{84}{5}$ **33.** $\frac{4}{9z^2}$ **35.** $\frac{n^2}{49}$ **37.** $\frac{25c^2}{4d^2}$ **39.** $\frac{-h^3}{27k^3}$

41.

x	-4	0	4
y	8	4	0

(4, 0); (0, 4)

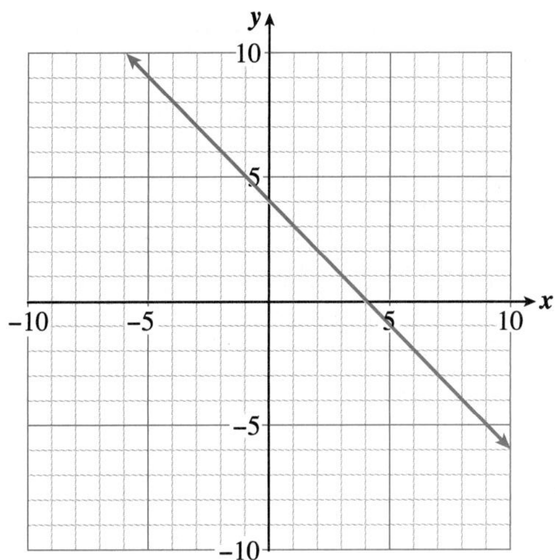

43.

x	-6	0	6
y	-4	-1	2

(2, 0); (0, −1)

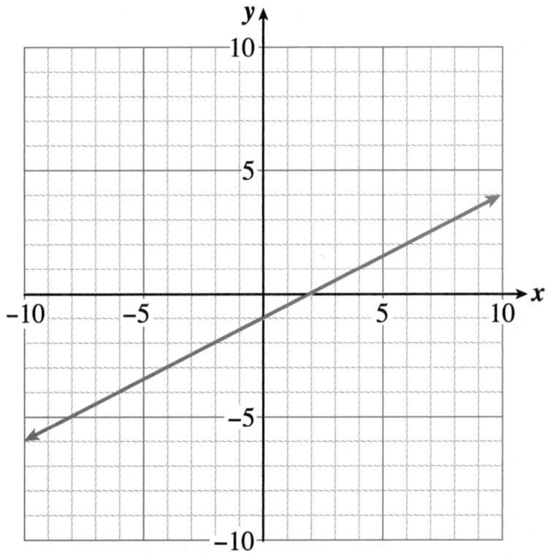

45. (b) **47.** (a) **49.** (a) **51.** (b) **53.** (a)

Homework 7.2B

1. Interchange numerator and denominator. (The denominator of a whole number is 1.) **3.** 15 **5.** $\frac{-4}{3}$ **7.** 10 **9.** $\frac{-1}{6}$
11. $\frac{1}{9}$ **13.** $\frac{-3}{h}$ **15.** $\frac{1}{8p^2}$ **17.** $-2z$ **19.** $\frac{9}{x}$
21. $\frac{-27}{4c^2}$ **23.** $\frac{ad}{bc}$ **25.** $\frac{a}{b}$ **27.** $\frac{AB^2}{3}$ **29.** $\frac{1}{12K^3T}$
31. $\frac{-3m^2}{2}$ **33.** $\frac{p^3}{2}$ **35.** $w = p^2 + 1$ **37.** $K = \frac{-2}{3}m$
39. $2a^3$ **41.** $-9p^2 - 11p$ **43.** $-12b^3$ **45.** $2W^3$
47a. $8m$ **b.** $16m^4$ **49.** (b) **51.** (b) **53.** (c)
55. (c) **57.** (c)

Homework 7.3A

1. Fractions that have the same denominator **3.** In like fractions we add the numerators and keep the same denominator; in like terms we add the coefficients and keep the variable part.
5. $\frac{7}{9}$ **7.** $\frac{1}{2}$ **9.** $\frac{7}{c}$ **11.** $\frac{4}{3q}$ **13.** $\frac{2a}{3}$ **15.** $\frac{-2s}{k}$
17. 0 **19.** $\frac{6p-2}{5}$ **21.** $\frac{10-4h}{3v}$ **23.** $\frac{-8v^3}{9}$
25. $\frac{-1}{ab}$ **27.** $\frac{-2}{x^2}$ **29.** $\frac{7}{b+3}$ **31.** $\frac{-5c}{n-2}$
33. $\frac{-4m}{2m-3}$ **35.** $\frac{4w-3}{w+3}$ **37.** $\frac{4x+3}{x}$ **39.** $\frac{-u+6}{3u}$
41a. $\frac{4}{5}$ **b.** $\frac{-2}{5}$ **c.** $\frac{3}{25}$ **d.** $\frac{1}{3}$ **43a.** $\frac{a+b}{3}$
b. $\frac{a-b}{3}$ **c.** $\frac{ab}{9}$ **d.** $\frac{a}{b}$

47.

x	-3	0	2
y	8	5	3

(5, 0); (0, 5)

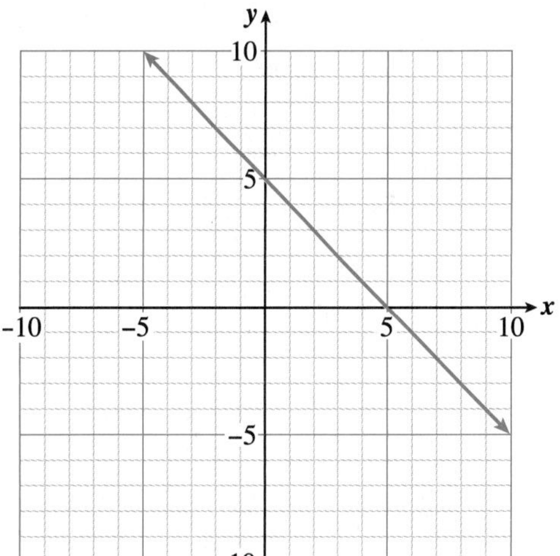

49.

x	-8	0	6
y	-3	1	4

(−2, 0); (0, 1)

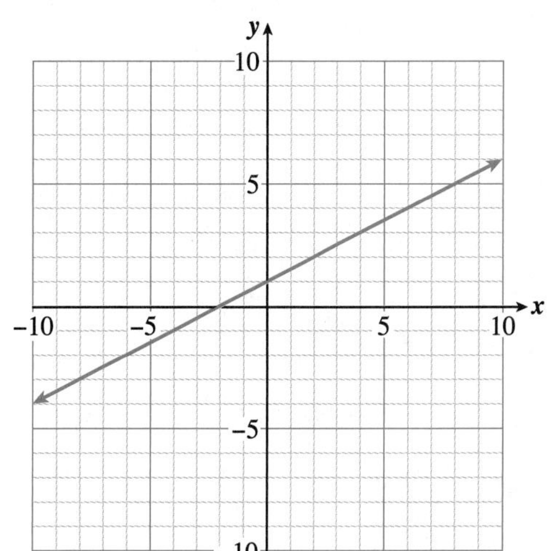

51. 84.33 **53.** 318.67 **55.** 240.64 **57.** 5.5 **59.** 497.36

Homework 7.3B

1. smallest; divide into evenly **3.** 40; 24, 32, 40, 48, 56, 64; 30, 40, 50, 60, 70, 80 **5.** Yes **7.** No **9.** $2 \cdot 2 \cdot 3 \cdot 3 \cdot 5$ **11.** $2 \cdot 2 \cdot 2 \cdot 3 \cdot 3 \cdot 5$ **13.** 45 **15.** 70 **17.** 60 **19.** 840 **21.** $3x$ **23.** $6g$ **25.** $72t$ **27.** $4n^2$ **29.** $15ab$ **31.** w^3 **33.** $12xy^2$ **35.** $24u^2v^2$ **37.** $125p^3$ **39.** rsq **41a.** $\frac{3+b}{5a}$ **b.** $\frac{3b}{25a^2}$ **43a.** $\frac{1}{6p}$ **b.** $\frac{27r^2}{8p^3}$ **45.** 10 **47.** 45 **49.** -32 **51.** $\frac{15}{4}$ **53.** 3 **55.** (a) **57.** (c) **59.** (a) **61.** (b) **63.** (c)

Homework 7.4

1. They are not made up of pieces of the same size. **3.** $\frac{7}{6}$ **5.** $\frac{1}{12}$ **7.** $\frac{19}{18}$ **9.** $\frac{-1}{10}$ **11.** $\frac{5a+3b}{15}$ **13.** $\frac{5m-18}{20}$ **15.** $\frac{-3w-10z}{8}$ **17.** $\frac{-4+15v}{18}$ **19.** $\frac{5b^2-4b}{20}$ **21.** $\frac{9+y}{9}$ **23.** $\frac{3c}{4}$ **25.** $\frac{-x}{40}$ **27.** $\frac{6+pq}{3p}$ **29.** $\frac{b-2a}{ab}$ **31.** $\frac{5-3x}{xy}$ **33.** $\frac{1+s^2}{st}$ **35.** $\frac{-3b-2a^2}{ab}$ **37.** $\frac{2v+1}{v}$ **39.** $\frac{3}{2z}$ **41.** $\frac{-8+5q}{20n}$ **43.** $\frac{-6s+3}{s^2}$ **45.** $\frac{3z^2-2x}{12x^2z}$ **47a.** $\frac{2+x^2}{2x}$ **b.** $\frac{2}{x^2}$ **c.** $\frac{1}{2}$ **d.** $\frac{2-x^2}{2x}$ **49a.** $\frac{3a-4}{2a^2}$ **b.** $\frac{3}{a^3}$ **c.** $\frac{3a+4}{2a^2}$ **d.** $\frac{3a}{4}$ **51.** (b) **53.** (c) **55.** (b) **57.** (a) **59.** (b)

Homework 7.5

1. its reciprocal **3.** 50 **5.** 18 **7.** -20 **9.** $\frac{-14}{3}$ **11.** $\frac{-27}{10}$ **13.** $\frac{2}{7}$ **15.** $\frac{5}{28}$ **17.** $\frac{-21}{10}$ **19.** $\frac{-8}{3}$ **21.** $\frac{3}{2}$ **23.** $\frac{-3}{20}$ **25.** 3 **27.** 10 **29.** $\frac{2}{9}x = 130$, $x = \$585$ **31.** $\frac{2}{3}x = 48$, $x = 72$ **33.** $\frac{3}{20}x = 24$, $x = 160$ **35a.** $\frac{4b-1}{2c}$ **b.** $4b$ **37a.** $\frac{2q}{p}$ **b.** $\frac{v^2+2pq}{pv}$ **39a.** $\frac{2}{3a}$ **b.** $\frac{2a+3a^2}{18b}$ **41.** -12 **43.** 1 **45.** -10 **47.** 1 **49.** 4 **51.** 7 **53.** 32 **55.** 90 **57.** 3000 **59.** 160

Chapter 7 Review

1. False **3.** False **5.** $\frac{-2}{3}$ is negative; $\frac{-2}{-3}$ is positive.

7. The opposite of x is $-x$; the reciprocal of x is $\frac{1}{x}$.

9. $\frac{a}{b} + \frac{c}{d} = \frac{ad+bc}{bd}$ **11.** $\frac{5}{x}$ **13.** $\frac{x}{3}$

15.

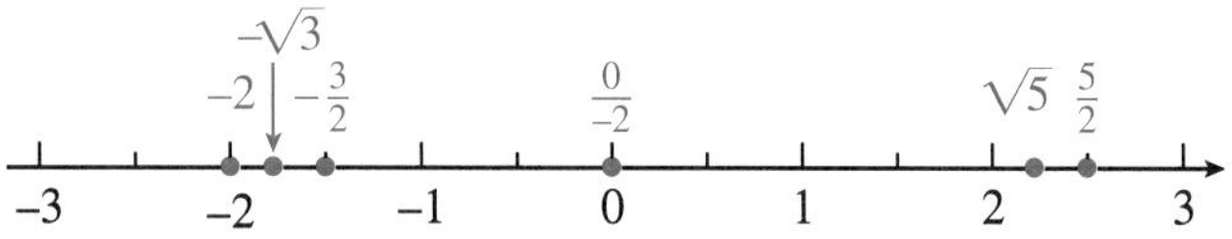

17a. -1 **b.** 4 **19a.** 2 **b.** 2 **21.** (a) and (c) **23.** (b) and (d) **25.** $\frac{-5}{6}$ **27.** $\frac{6}{5}$ **29.** 1 **31.** $\frac{1}{3}$ **33.** $\frac{-9}{10x}$ **35.** $\frac{-9m^3}{5n^2}$ **37.** $\frac{1}{r}$ **39.** $\frac{v}{w^2}$ **41.** $\frac{1}{2}$ **43.** $\frac{a-1}{2a}$ **45.** $\frac{4x-3}{2x}$ **47.** $\frac{6b+4a}{ab^2}$ **49.** $-81b^4$ **51.** 14 **53.** $\frac{-8}{15}$ **55.** $\frac{-9}{10}$ **57.** -6 **59.** -2 **61.** 29 **63.** $\frac{2}{3}x = 4$, $x = 6$ **65.** $\frac{3}{4}x - 150 = 825$, $x = \$1300$ **67.** $\frac{8}{9}x = 1000$, $x = 1125$ **69.** (a) **71.** (a) **73.** (c) **75.** (c) **77.** (c) **79.** (b) **81.** (a)

Homework 8.1A

1. \$22.40 **3.** $83.\overline{3}\%$ **5.** 60 **7a.** 2% **b.** 56% **c.** 94%

d.

Under 20	0.06 × 50,000	=	3,000
21 to 25	0.29 × 50,000	=	14,500
26 to 30	0.27 × 50,000	=	13,500
31 to 35	0.17 × 50,000	=	8,500
36 to 40	0.11 × 50,000	=	5,500
41 to 50	0.08 × 50,000	=	4,000
Over 50	0.02 × 50,000	=	1,000
Total			50,000

9a. 12.3% **b.** Europe **c.** Asia

d.

Europe	0.712 × 71,154	=	50,662
Latin America	0.123 × 71,154	=	8,752
Asia	0.059 × 71,154	=	4,198
Oceania	0.031 × 71,154	=	2,206
Others	0.075 × 71,154	=	5,337
Total			71,155

11. $2t$ **13.** $-3c$ **15.** 3 **17.** $\frac{12w}{7z}$ **19.** $\frac{-3kt}{4m^2}$ **21.** No **23.** Yes **25.** Yes **27.** Yes **29.** No

Homework 8.1B

1. \$594,720 **3a.** \$22,000 **b.** \$24,640 **5a.** 16% **b.** 18% **c.** Edith **7.** School **9.** Mathematics

11a. \$1377 **b.** \$1350 **c.** Catalog **13a.** 9600; 12,960 **b.** 12,400 **c.** No **15a.** Fall **b.** −1.08% **c.** −2.41% **d.** Coca-Cola

17a.

Year	*Population at start of year*	*Increase in population*	*Population at end of year*
1990	20,000	1000	21,000
1991	21,000	1050	22,050
1992	22,050	1103	23,153
1993	23,153	1158	24,311
1994	24,311	1216	25,527
1995	25,527	1276	26,803
1996	26,803	1340	28,143
1997	28,143	1407	29,550
1998	29,550	1478	31,028
1999	31,028	1551	32,579
2000	32,579	1629	34,208

b.

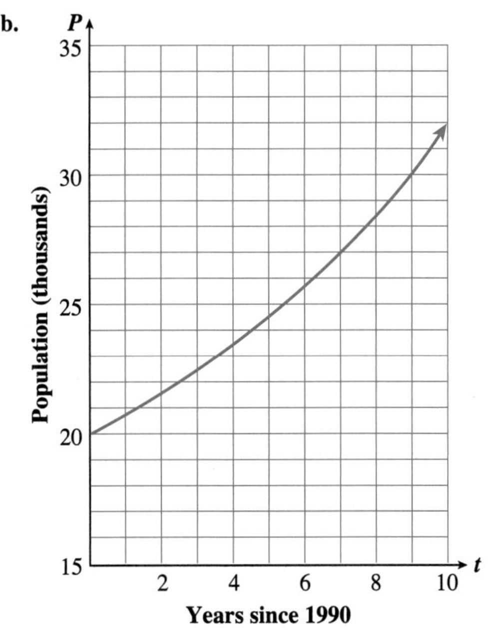

19. $7u^2v^2$ **21.** $6cd^2$ **23.** $70uv$ **25.** $-18p^2q$ **27.** $18x^2y$ **29.** 393.7 **31.** \$44.90 **33.** 94.6 **35.** 454 **37.** 1609.3

Homework 8.2

1a. $\frac{3}{4}$ **b.** $\frac{4}{3}$ **c.** $\frac{3}{7}; \frac{4}{7}$ **d.** 48 **e.** 45

3. 10 and 16, 15 and 24, 20 and 32 **5.** It is greater than 1.

7. $\frac{4}{3}$ **9a.** $\frac{7}{2}$ **b.** $\frac{1}{4}$ **11.** $\frac{3}{2}$ **13a.** $\frac{3}{250}$ **b.** $\frac{247}{3}$

c. 1.2% **15.** 0.262 **17.** 0.625 **19.** 1.6 **21.** 0.6

23. 36 mpg **25.** 8 min/mi; $\frac{1}{8}$ mi/min **27.** 72 sq. ft/h

29a. 6.5 cents/dollar **b.** 6.5% **31.** 1 **33.** $\frac{b}{5}$ **35.** $\frac{p}{2q}$

37. $\frac{-8r^3}{s^3}$ **39.** 23.7 lb **41.** 1248 in. **43.** 4.9 m

45. 0.946 L **47.** 9.38 sec

Homework 8.3A

1. It is constant. **3.** Yes **5.** No **7.** No **9.** No **11.** No **13.** Yes **15.** No **17.** Yes **19.** Yes **21.** No **23.** (a) $d = 15t$

t	*d*
3	45
5	75
10	150

(c) 20 h

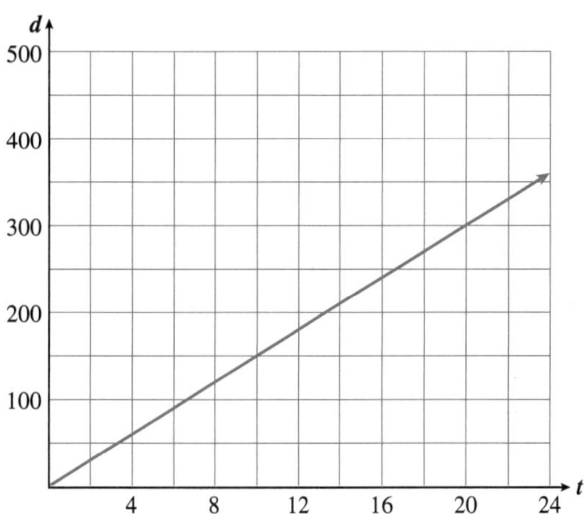

25. (a) $C = 0.80w$

w	*C*
10	8
15	12
20	16

(c) \$17.60

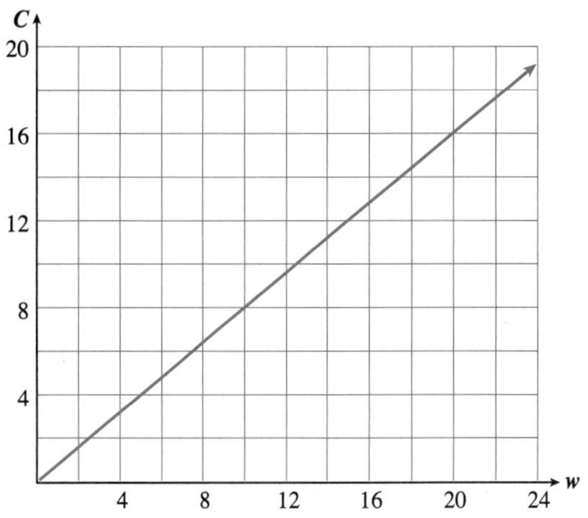

27a. $C = 0.04S$ **b.** \$8000

29a. $P = 4s$ **b.**

s	2	5	8	11
P	8	20	32	44

c.

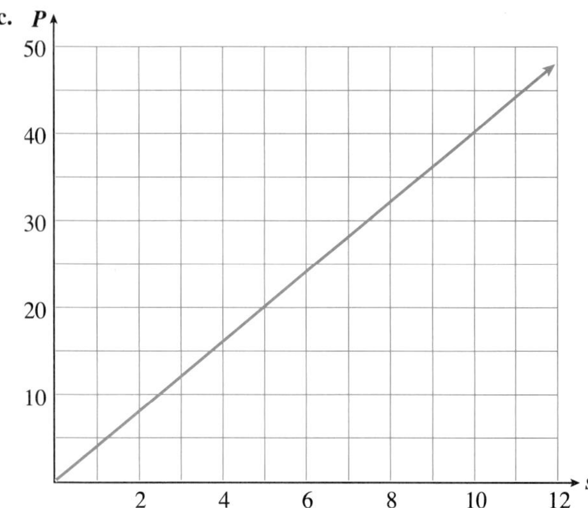

d. Yes

31a. $A = s^2$ **b.**

s	2	5	6	8
A	4	25	36	64

c.

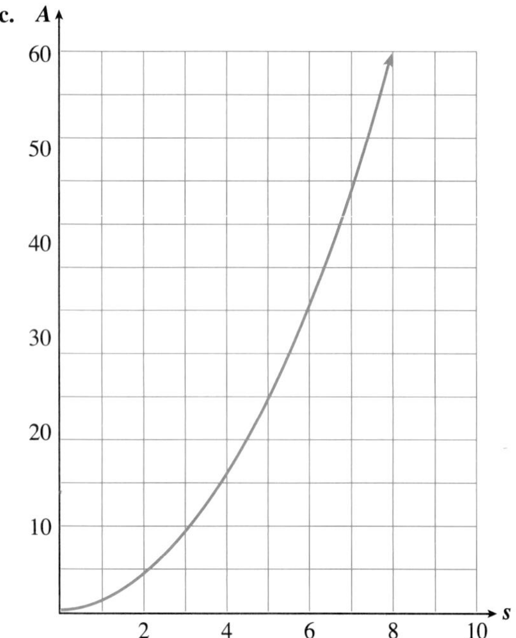

d. No **33.** $\frac{-1}{25w^2}$ **35.** $\frac{a^5}{28h^5}$ **37.** $\frac{adf}{bce}$ **39.** $\frac{ade}{bcf}$
41. 9 **43.** 3 **45.** 15 **47.** 9 **49.** 20

Homework 8.3B

1. An equation that states two ratios are equal. **3.** Yes **5.** No **7.** No **9.** Yes **11.** No **13.** Yes **15.** 49 **17.** 12.6 **19.** 4.5 **21.** $\frac{7}{2}$ **23.** 2 ft 8 in. **25.** 6.5 mi **27.** 22.8 ft **29.** 333 **31.** $A = 112$ sq. m; $P = 44$ m **33.** $A = 18.71$ sq. yd; $P = 21.48$ yd **35.** $A = 56.55$ sq. ft; $P = 30.85$ ft **37.** $A = 21$ sq. cm; $P = 22$ cm **39.** $A = 123$ sq. ft; $P = 56$ ft **41.** \$6 **43.** \$4.80 **45.** (b) **47.** (b) **49.** (b)

Activity 1

a. 9 sq. in. **b.** 36 sq. in.; $\frac{6}{3} = 2$; 2; $\frac{36}{9} = 4$ **c.** 81 sq. in.; $\frac{9}{3} = 3$; $\frac{81}{9} = 9$ **d.** 25; 15 in.; 225 sq. in.; $\frac{225}{9} = 25$; k^2

Activity 2

a. 6 sq. cm **b.** 24 sq. cm; 2; $\frac{24}{6} = 4$ **c.** 54 sq. cm; 3; $\frac{54}{6} = 9$ **d.** 36; 24 cm; 18 cm; 216 sq. cm; $\frac{216}{6} = 36$; k^2

Activity 3

a. 4π sq. in. **b.** 16π sq. in.; 2; $\frac{16\pi}{4\pi} = 4$

c. 36π sq. in.; 3; $\frac{36\pi}{4\pi} = 9$ **d.** 100; 20 in.; 400π sq. in.; $\frac{400\pi}{4\pi} = 100$; k^2

Activity 4

a. 30 cu. in. **b.** 6 in.; 4 in.; 10 in.; 240 cu. in.; 2; 2; $\frac{240}{30} = 8$;

c. 9 in.; 6 in.; 15 in.; 810 cu. in.; 3; $\frac{810}{30} = 27$ **d.** $\frac{1}{8}$; 1.5 in.; 1 in.; 2.5 in.; 3.75 cu in.; $\frac{3.75}{30} = 0.125 = \frac{1}{8}$; k^3

Activity 5

a. 4188.79 cu. ft **b.** 20 ft; 33,510.32 cu. ft; 2; 2; $\frac{33510.32}{4188.79} = 8$

c. 30 ft; 113,097.34 cu. ft; 3; $\frac{113{,}097.34}{4188.79} = 27$ **d.** 3.375; 15 ft; 14,137.17 cu. ft; $\frac{14137.17}{4188.79} = 3.375$; k^3

Homework 8.4

1. (b) and (e) **3.** (a) and (c) **5.** (a) and (e) **7.** 2.25; 33.75 sq. in. **9.** 0.75 sq. ft **11.** 1 to 125,000,000 **13.** One 16-in. pizza **15.** 180,000 **17.** −66 **19.** −18 **21.** −0.385 **23.** 2871.644 **25.** 2 **27.** (a) **29.** (b) **31.** (b) **33.** (b) **35.** (b) **37.** 2 **39.** 2

Homework 8.5

1a. \$50 **b.** (2,300), (6,500); $m = 50$ **c.** dollars/week; the rate at which Lynette's savings grow **3a.** 5 lb; 10 lb **b.** 2.5 lb/mo

c.

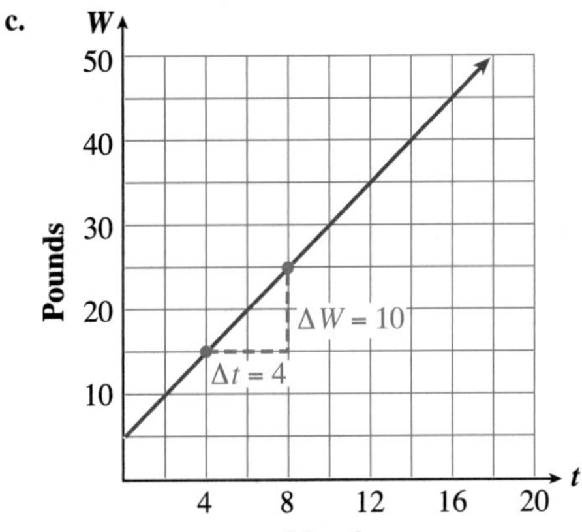

5a. The cost per pound **b.** \$1.50/lb **c.** \$30 **7a.** $\frac{2}{5}$ **b.** steeper **9a.** -75 mph **b.** The speed of the train **c.** 800; Roy started out 800 mi from home. **d.** $10\frac{2}{3}$; Roy's journey took $10\frac{2}{3}$ h. **11a.** -2°/min **b.** How fast the temperature is lowered **c.** 20; the compound started at 20°C **d.** 10; after 10 min, the compound's temperature is 0°C.

13. $\frac{1}{2}$

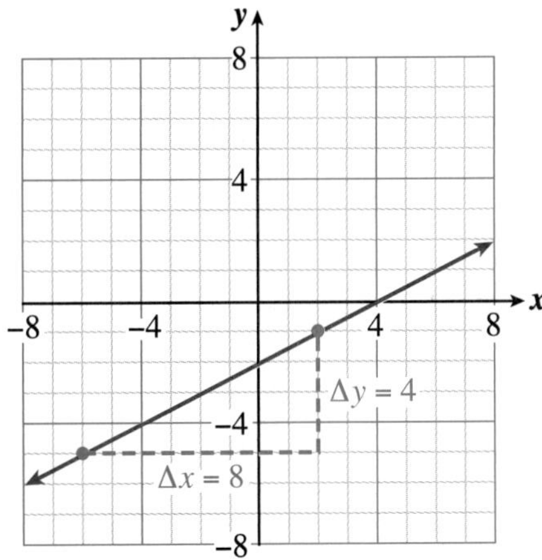

15. $-\frac{2}{3}$

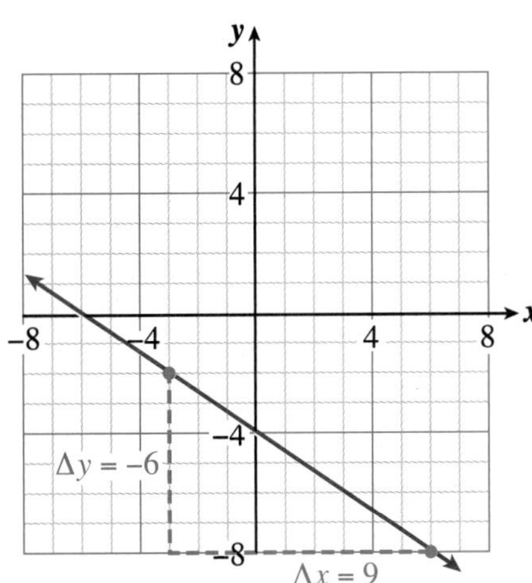

17.

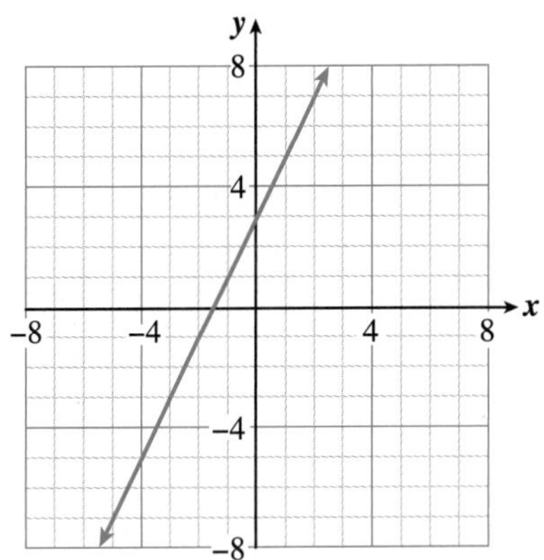

(a)

x	-2	0	2
y	-1	3	7

(b) 2 (c) (0, 3)

19.

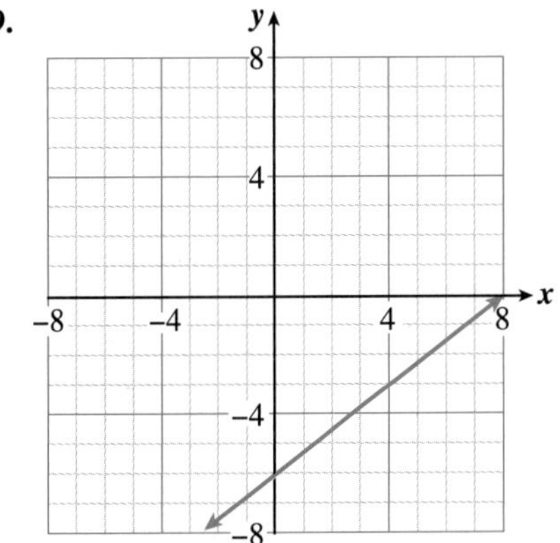

(a)

x	y
-4	-9
0	-6
4	-3

(b) $\frac{3}{4}$ (c) $(0, -6)$

21.

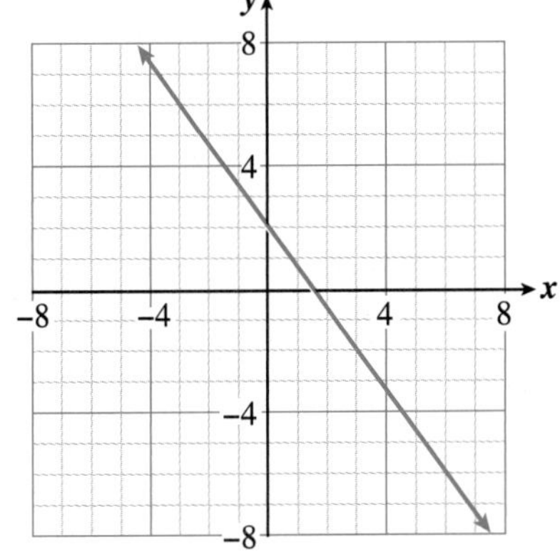

(a)

x	y
-3	6
0	2
3	-2

(b) $\frac{-4}{3}$ (c) (0, 2)

23. $\dfrac{12q + 3p}{8p^2q^2}$ **25.** $\dfrac{g^3 - 9h^2}{3h^2g}$ **27.** $\dfrac{b^2 - 2}{b}$ **29.** $\dfrac{x}{x + 1}$
31. (b) **33.** (c) **35.** (c) **37.** (c) **39.** (b)

Chapter 8 Review

1. 100 **3.** *P*: part; *r*: percentage rate; *W*: whole
5. increase; original amount **7.** One variable is a constant multiple of the other. **9.** Applying the property of proportions; only to a proportion

	Fraction	*Decimal*	*Percent*
11.	$\frac{1}{8}$	0.125	12.5%
13.	$\frac{2}{25}$	0.08	8%
15.	$\frac{2}{20}$	0.35	35%

17. \$1.95; \$11.05 **19.** \$4990; \$5414.15 **21a.** 17.6%
b. Plastics **c.** Paper
d.

Material	*Waste (%)*	*Waste (millions of tons)*
Paper	40	71.84
Food waste	7.3	13.1
Yard waste	17.6	31.6
Metal	8.5	15.3
Glass	7	12.6
Plastic	8	14.4
Other	11.6	20.8

23. \$27,000; \$29,970 **25.** 65 or older **27a.** \$43.19
b. −1.3% **29.** $\frac{13}{60} = 0.21\overline{6}$ **31.** $\frac{13}{2} = 6.5$ costumes/day
33. Yes **35.** No **37.** No **39.** $c = 8p$
c. 80 calories

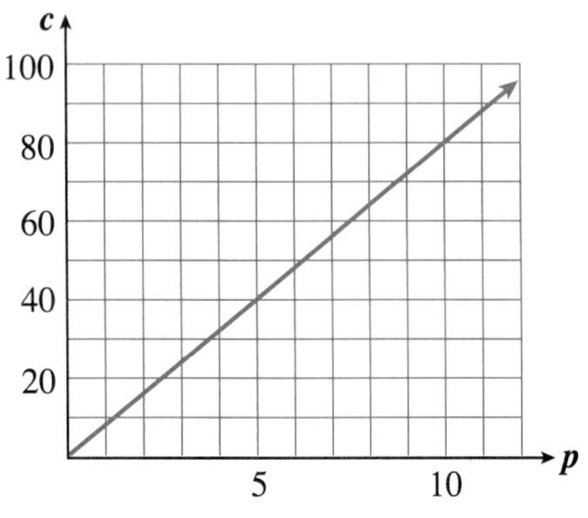

p	*c*
4	32
5	40
10	80

41. $n = 21$ **43.** $x = \pm 6$ **45a.** $\frac{3}{2}$ **b.** 12 **c.** 24
47a. 0.15 **b.** 700 m **c.** The second hill is steeper
49. 14.4 g **51.** I and V are similar. **a.** 2 **b.** 4

53a. 3 m **b.** 0.1 m **c.** $\frac{2}{3}$ sq. m **d.** 0.001 cu. m
55a.

n	0	5	10
A	5	11	17

b. Rani and Larry deposited \$5000 when Colby was born.
c. \$1200/yr **d.** The fund increases by \$1200/yr.
57.

x	0	4	−4
y	−2	3	−7

$\frac{5}{4}$

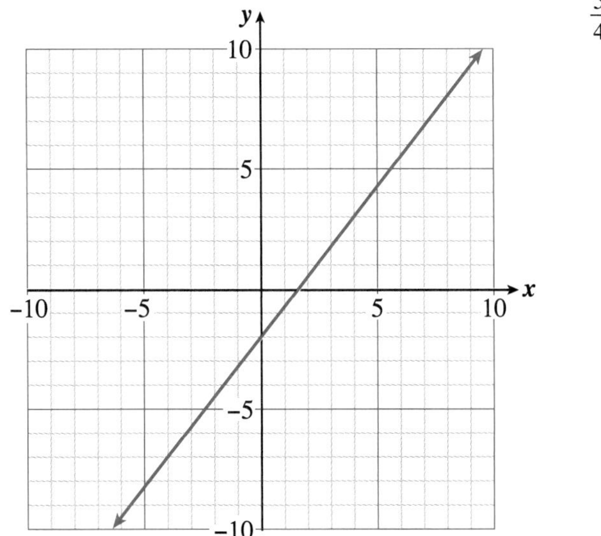

59. Bearclaw Trail **61.** No **63.** 1.6 **65.** 2500 g
67. 60 **69.** \$120 **71.** \$200 **73.** (b)

Homework 9.1A

1. 4 **3.** 11 **5.** −1 **7.** −0.5 **9.** −1.5 **11.** 14
13. 0 **15.** $\frac{-10}{7}$ **17.** 0 **19.** 8 **21.** (a) $3(6y) = 72$
(b) \$4 **23.** (a) $3y + 6 = 72$ (b) 22 **25.** 2100 **27.** 700
29. 3600 **31.** 1200 **33.** 1400

Homework 9.1B

1. −3 **3.** 9 **5.** −7 **7.** −4 **9.** −1 **11.** (a) Ounces of catsup: *c*; (b) $8c = 5c + 12$; (c) He uses 4 oz.
13. (a)Length of post: *m*; (b) $3m + 33 = m + 131$; (c) A post is 49 in. long. **15.** (a) Price of one can: *c*; (b) $12c + 1.80 = 8c + 5$; (c) One can costs \$0.80.
17. (a) Milligrams of calcium in a glass of milk: *m*; (b) $3m - 70 = 2m + 220$; (c) One glass of milk contains 290 mg of calcium. **19.** 54 **21.** 73 **23.** 93 **25.** 75
27. 63

Homework 9.2A

1. $3x - 12$ **3.** $10y - 15$ **5.** $-8x - 16$ **7.** $-20 + 25a$
9. $-5b + 3$ **11.** $36 - 12t$ **13.** Yes **15.** Yes
17. Yes **19a.** \$2.20; \$3.30; \$5.50 **b.** 10 lb; \$5.50

21a. $6x + 15x = 21x$ **b.** $3(7x) = 21x$ **23.** $-4x - 6$
25. $31x - 18$ **27.** $23 + x$ **29.** $3y + 21$ **31.** $-9t + 5$
33a. $5b$ means 5 times b; $5 + b$ means $5 + b$.
b. To simplify $2(5 + b)$, we use the distributive law; for $2(5b)$, we do not. **c.** $2(5b) = 10b$; $2(5 + b) = 10 + 2b$
35a. True **b.** True **c.** True **d.** False **37.** Both
39. Neither **41.** Both **43.** By 2 **45.** Both

Homework 9.2B

1a. $x - 5 = 3$; $x = 8$ **b.** $2x - 10 = 6$; $2x = 16$; $x = 8$
3. 4 **5.** -10 **7.** -1 **9.** $\frac{5}{2}$ **11.** 6 **13.** 5
15. 0 **17.** 12 **19.** 70 **21.** $72 + 6h$ **23.** $300 - 6d$
25. $50 + \frac{P}{12}$ **27.** $\frac{S - 60}{15}$ **29.** $20 + 0.02B$
31. Neither **33.** By 3 **35.** Both **37.** Neither
39. Both

Homework 9.3A

1a. x **b.** $x + 12$ **c.** $2x + 12$ **3a.** p **b.** $p - 20$
c. $2p - 20$ **5a.** d **b.** $2d$ **c.** $3d - 5$ **7a.** c
b. $24 - c$ **c.** $24 - 2c$ **9a.** w **b.** $10 - w$ **c.** $0.12w$
d. $0.55(10 - w)$ **e.** $5.5 - 0.43w$ **11.** -4 **13.** $\frac{-6}{5}$
15. $\frac{1}{3}$ **17.** 0 **19.** 6 **21.** Both **23.** By 5
25. Both **27.** Neither **29.** By 5

Homework 9.3B

1. $x + (x + 57) = 349$; $x = 146$; $x + 57 = 203$
3. $x + 2x = 54$; $x = 18$ **5.** $x + (x - 6) = 64$; $x = 35$
7. $4x + 2(32 - x) = 82$; $x = 9$ **9.** $\frac{1}{2}(4)(x + x + 6) = 28$; $x = 4$; $x + 6 = 10$ **11a.** x; $140 - x$ **b.** $150x$; $230(140 - x)$
c. $150x + 230(140 - x) = 22{,}600$ **d.** $x = 120$; $140 - x = 20$
13. $4y - 20$ **15.** $-14a + 18$ **17.** -17 **19.** 800
21. 36 **23.** 67 **25.** 89 **27.** 49 **29.** 59

Homework 9.4

1. the LCD **3.** -23 **5.** -2 **7.** $\frac{-3}{2}$ **9.** $\frac{-3}{7}$
11. 12 **13.** $\frac{11}{8}x = \frac{22}{3}$, $x = 5\frac{1}{3}$ yd
15. $\frac{2}{5}x + \frac{1}{4}x + 308 = x$, $x = 880$ **17a.** b **b.** $14 - b$
19a. $150b$ **b.** $100(14 - b)$ **c.** $1400 + 50b$ **21a.** k
b. $3k$ **23a.** $20k$ **b.** $9(3k)$ **c.** $47k$ **25a.** a
b. $840 - a$ *c* $0.40a$ *d* $0.10(840 - a)$ **e.** $84 + 0.30a$
27. 120 **29.** 76 **31.** 53 **33.** 27 **35.** 89 **37.** 38

Homework 9.5

1a. $0.05x$ **b.** \$2800 **3.** 12.5% **5.** \$630
7a. x; $25{,}000 - x$ **b.** $0.05x$; $0.084(25{,}000 - x)$
c. $2100 - 0.034x$

9a.

Amount invested in savings ($)	Amount invested in T-bill ($)	Interest from savings ($)	Interest from T-bill ($)	Total interest ($)
100	1100	5	82.50	87.50
200	1000	10	75	85
300	900	15	67.50	82.50
400	800	20	60	80
500	700	25	52.50	77.50
600	600	30	45	75
700	500	35	37.50	72.50
800	400	40	30	70
900	300	45	22.50	67.50
1000	200	50	15	65
1100	100	55	7.50	62.50

b. $0.05x + 0.075(1200 - x)$ **11a.** $5000 - x$
b. $0.12x$; $0.07(5000 - x)$; $350 + 0.05x$ **c.** \$1800; \$3200
13. \$4500 **15.** 36 **17.** 47 **19.** 39 **21.** 19
23. 49

Homework 9.5B

1a. 2.4 gal **b.** $0.3x$ gal
3.

	Number of M & M's	Percent red	Number of red M & M's
First bag	50	0.30	15
Second bag	30	0.70	21
Mixture	80	0.45	36

a. 15; 21 **b.** 80; 36 **c.** 45% **d.** $0.45(80) = 36$
5a. 1 qt; 3 qt **b.** 4 qt; 15 qt **c.** $26.\overline{6}\%$
d.

	Quarts of solution	Strength (% acid)	Quarts of acid
20% solution	5	0.20	1
30% solution	10	0.30	3
Mixture	15	$0.2\overline{6}$	4

7a. $15 - x$
b.

x	0.2	$0.2x$
$15 - x$	0.3	$0.3(15 - x)$
15	0.28	4.2

9.

	Number produced	*Percent meeting standards*	*Number meeting standards*
Cars	50,000	0.3	15,000
Trucks	x	0.8	$0.8x$
All vehicles	$50{,}000 + x$	0.6	$0.6(50{,}000 + x)$

a. Number of trucks **d.** 15,000; $0.8x$
e. $15{,}000 + 0.8x = 0.6(50{,}000 + x)$ **f.** 75,000 trucks
11. 69 **13.** 50 **15.** 70 **17.** 90 **19.** 110

Homework 9.5C

1a. 210 mi
b.

Carl ← 2(45) — MM — 2(60) → Wendy

3a. 40 min **b.** 30 min **5.** 37.5 mph **7a.** $w + 2$
b. $4w$; $4(w + 2)$
c.

Brooke — $4(w+2)$ — X — $4w$ — Claire

9a. $h - 3$ **b.** $45h$; $70(h - 3)$
c.

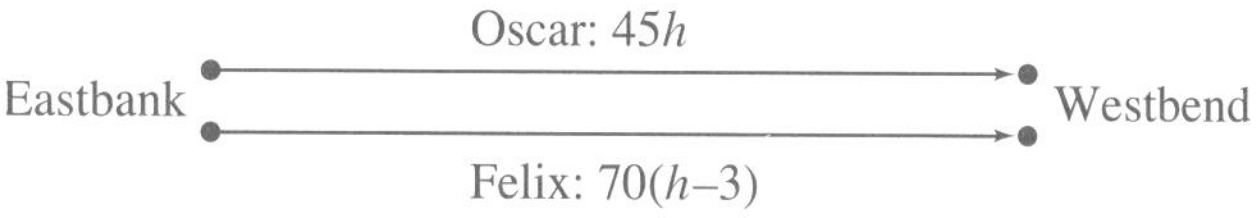

Oscar: $45h$
Eastbank → Westbend
Felix: $70(h-3)$

11a.

	Rate	*Time*	*Distance*
Delbert	s	6	$6s$
Francine	$s - 5$	6	$6(s - 5)$

b. $s - 5$ **c.** $6s$; $6(s - 5)$
d.

Francine ← $6(s-5)$ — Fresno — $6s$ → Delbert

e. $6s + 6(s - 5) = 570$ **f.** 50 mph; 45 mph
13a. Speed of the wind: w
b.

	Rate	*Time*	*Distance*
Initial Trip	$224 - w$	$6\frac{2}{3}$	$\frac{20}{3}(224 - w)$
Return Trip	$224 + w$	5	$5(224 + w)$

c.

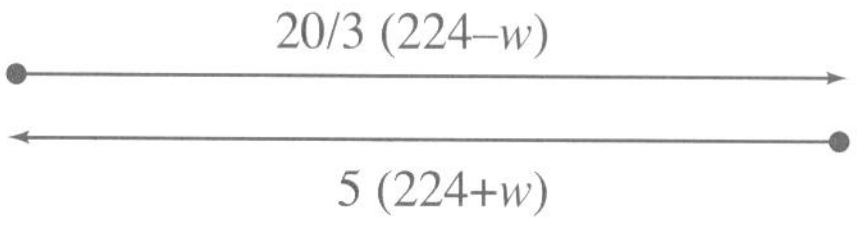

$20/3\ (224-w)$ →
← $5\ (224+w)$

d. $\frac{20}{3}(224 - w) = 5(224 + w)$ **e.** 32 mph **15.** 686
17. 7920 **19.** 5050 **21.** 9090 **23.** 7236

Chapter 9 Review

1. False **3.** True **5.** False **7.** False **9.** True
11. $5x$ **13.** $10 - 16s$ **15.** $-42 + 30w$ **17.** $5 - 8c$
19. $18a + 16$ **21.** $30N - 26$ **23.** 1 **25.** 44
27. $\frac{12}{5}$ **29.** 17 **31.** $x + 7x = 56, x = 7$
33. $2w + 2(38) = 120, w = 22$ m **35.** 4 **37.** 2
39. $\frac{1}{2}$ **41a.** $0.08(20{,}000 - d)$ **b.** \$1328 **c.** \$5000
43a. $\frac{2}{3}(R - 45{,}000)$ dollars **b.** \$50,000 **c.** \$157,500
45a.

Investment (\$) in savings	*Investment (\$) in money market*	*Earnings (\$) from savings*	*Earnings (\$) from money market*	*Total (\$) earnings*
100	1900	4	114	118
500	1500	20	90	110
1000	1000	40	60	100
1200	800	48	48	96
x	$2000 - x$	$0.04x$	$0.06(2000 - x)$	$0.04x + 0.06(2000 - x)$

b. \$960 in savings; \$1040 in money market **47a.** Women; women **b.** 42; 48 **c.** 180; 50% **d.** 24; 120
e. 240; 60%; higher
49a.

	Rate (mph)	*Time (h)*	*Distance (mi)*
Missile	600	t	$600t$
Bat-jet	720	$t - \frac{1}{4}$	$720\left(t - \frac{1}{4}\right)$

b. 1.5 h **c.** Yes **51.** 3100 **53.** 75 **55.** 29
57. 80 **59.** 1794 **61.** By 2, 3

Exercise A.1

1. $\frac{3}{5}$ **3.** $\frac{4}{25}$ **5.** $\frac{1}{3}$ **7.** $\frac{3}{4}$ **9.** $\frac{3}{5}$ **11.** 12 **13.** $\frac{6}{9}$
15. $\frac{12}{8}$ **17.** $\frac{40}{100}$ **19.** $\frac{24}{72}$ **21.** $\frac{90}{144}$ **23.** $\frac{0}{6}$

Exercise A.2

1. $\frac{10}{21}$ **3.** $\frac{4}{5}$ **5.** $\frac{1}{2}$ **7.** $\frac{1}{3}$ **9.** $\frac{1}{2}$ **11.** $\frac{3}{5}$
13. $\frac{6}{5}$ **15.** $\frac{5}{12}$ **17.** $\frac{2}{15}$ **19.** 18 **21.** $\frac{22}{3}$ **23.** 5

Exercise A.3

1. $\frac{4}{11}$ **3.** $\frac{7}{15}$ **5.** $\frac{2}{3}$ **7.** $\frac{1}{2}$ **9.** 1 **11.** $\frac{2}{5}$
13. $\frac{1}{4}$ **15.** $\frac{3}{5}$ **17.** 6 **19.** 12 **21.** 21 **23.** 60
25. 30 **27.** 90 **29.** 36 **31.** 72 **33.** $\frac{5}{6}$ **35.** $\frac{1}{12}$

37. $\frac{17}{24}$ **39.** $\frac{7}{30}$ **41.** $\frac{37}{30}$ **43.** $\frac{73}{120}$ **45.** $\frac{41}{100}$
47. $\frac{479}{840}$

Exercise A.4

1. $\frac{11}{3}$ **3.** $\frac{25}{2}$ **5.** $\frac{104}{5}$ **7.** $\frac{221}{50}$ **9.** $\frac{39}{8}$ **11.** $\frac{149}{12}$
13. $\frac{55}{8}$ **15.** $\frac{139}{24}$ **17.** $\frac{33}{2}$ **19.** $\frac{50}{7}$ **21.** $\frac{5}{12}$
23. 2 **25.** $3\frac{2}{3}$ **27.** $5\frac{3}{8}$ **29.** $6\frac{11}{16}$ **31.** $9\frac{29}{32}$

Exercise A.5

1. $\frac{17}{100}$ **3.** $\frac{7}{100}$ **5.** $\frac{23}{1000}$ **7.** $\frac{3}{5}$ **9.** $\frac{13}{50}$ **11.** $\frac{3}{8}$
13. $\frac{9}{4}$ **15.** $\frac{18}{5}$ **17.** 0.84 **19.** 0.46 **21.** 0.31
23. 0.031 **25.** 0.3125 **27.** 0.375 **29.** $0.8\overline{3}$
31. $0.\overline{27}$

Exercise A.6

1. (a) 10 (b) 14.8 (c) 14.77 (d) 14.774 **3.** (a) 80 (b) 76.3 (c) 76.28 (d) 76.283 **5.** (a) 170 (b) 169.9 (c) 169.90 (d) 169.899 **7.** (a) 5550 (b) 5545.9 (c) 5545.91 (d) 5545.910 **9.** (a) 700 (b) 701.0 (c) 700.96 (d) 700.960 **11.** (a) 20 (b) 20.0 (c) 19.95 (d) 19.951 **13.** (a) 1.9 (b) 1.91 (c) 1.907 **15.** (a) 0.9 (b) 0.92 (c) 0.920 **17.** (a) 0.1 (b) 0.10 (c) 0.099 **19.** (a) 6.2 (b) 6.17 (c) 6.170

Exercise A.7

1. 4.73 **3.** 4 **5.** 1.66 **7.** 13.26 **9.** 6.227
11. 34.45 **13.** 3.47 **15.** 5.74 **17.** 5.07 **19.** 9.74
21. 362.15 **23.** 12.9994 **25.** 3.71 **27.** 542.046
29. 577.08

Exercise A.8

1. 4.08 **3.** 0.128 **5.** 0.0408 **7.** 0.12036 **9.** 3.25
11. 10.368 **13.** 1.7304 **15.** 0.00384 **17.** 4.51
19. 4.21875 **21.** 13.7 **23.** 15.3 **25.** 605 **27.** 230.4
29. $4.\overline{54}$ **31.** 800

Exercise A.9

1. 0.15 **3.** 0.004 **5.** 0.068 **7.** 1.19 **9.** 0.0325
11. 0.004 **13.** 33% **15.** 50.4% **17.** 78.7%
19. 2.01% **21.** 0.8% **23.** 550% **25.** $\frac{7}{20}$ **27.** $\frac{5}{4}$
29. $\frac{3}{5}$ **31.** $\frac{9}{1000}$ **33.** $\frac{3}{8}$ **35.** $\frac{1}{3}$ **37.** 75%
39. 37.5% **41.** 100% **43.** 225% **45.** 240%
47. 0.4%

Exercise A.10

1. 7 **3.** 3 **5.** 147 **7.** 2; 5 **9.** 5 **11.** 50 + 28
13. 30 + 40 **15.** $10 \cdot 7$ **17.** $100 \cdot 13$ **19.** $20 \cdot 30$

Index